CW00801336

Fossil Vertebrates of Arabia

His Highness Sheikh Zayed bin Sultan Al Nahyan, President of the United Arab Emirates, Ruler of the Emirate of Abu Dhabi.

Fossil Vertebrates of Arabia

With Emphasis on the Late Miocene Faunas, Geology, and Palaeoenvironments of the Emirate of Abu Dhabi, United Arab Emirates

Edited by Peter J. Whybrow and Andrew Hill

In collaboration with The Abu Dhabi Company for Onshore Oil Operations

The Ministry for Higher Education and Scientific Research, United Arab Emirates

Yale University Press New Haven and London

Published with generous assistance from The Abu Dhabi Company for Onshore Oil Operations

Designed by In House Production Company, Minneapolis, Minnesota, and set in Galliard type by The Clarinda Company, Clarinda, Iowa. Printed in the United States of America by Edwards Brothers, Ann Arbor, Michigan.

The paper in this book meets the guidelines for permanence and durability of the Committee on Production Guidelines for Book Longevity of the Council on Library Resources.

Library of Congress Cataloging-in-Publication Data

Fossil vertebrates of Arabia : late Miocene faunas, geology, and palaeoenvironments of the Emirate of Abu Dhabi, United Arab Emirates/edited by Peter J. Whybrow and Andrew Hill, in collaboration with the Abu Dhabi Company for Onshore Oil Operations, the Ministry for Higher Education and Scientific Research, United Arab Emirates.
p. cm.
Includes bibliographical references and index.
ISBN 0-300-07183-3 (cl. : alk. paper)
1. Vertebrates, Fossil--Abū Ẓaby (United Arab Emirates : Emirate) 2. Animals, Fossil--Abū Ẓaby (United Arab Emirates : Emirate)
I. Whybrow, Peter J. II. Hill, Andrew P.
III. Sharikat Abū Ẓaby lil-ʻ Amalīyāt al-Batrūlīyah al-Barrīyah. IV. United Arab Emirates. Ministry for Higher Education and Scientific Research.
QE841.F653 1999
566'.095357--dc21 97-41809
CIP

A catalogue record for this book is available from the British Library.

10 9 8 7 6 5 4 3 2 1

To Jonathan, Alex, and Valerie Whybrow;
to May Hill, and to the memory of Rowland Hill;
and also to the memory of Roger Hamilton
and Colin Patterson, FRS

Contents

Foreword

His Excellency Sheikh Nahayan bin Mubarak Al Nahayan, Minister for Higher Education and Scientific Research, United Arab Emirates

For centuries, we who have lived in the Arabian Gulf have drawn strength and inspiration from our natural environment. Inhospitable in some ways, very supportive of human existence in other ways, our natural environment has always been our main natural resource. And, indeed, the distinctive environmental and natural conditions of the region have shaped our history and will continue to shape our future.

The President of the United Arab Emirates, His Highness Sheikh Zayed bin Sultan Al Nahyan, continually expresses our strong national commitment to the study of our past and the importance of studying our human and natural history. It was with his support and encouragement that the first ever conference on the "Fossil Vertebrates of Arabia" was held at Jebel Al-Dhanna, United Arab Emirates, in March 1995. The conference provided an opportunity for prominent scientists from around the world to present and discuss the results of their work on the Miocene fossils from Abu Dhabi's Baynunah Formation. The findings of several of the participating distinguished scholars are quite significant. The fossils themselves, we are pleased to note, have proved to be the most important source of terrestrial vertebrates to be found anywhere in Arabia and they are of international significance.

Inauguration by His Excellency Sheikh Nahayan bin Mubarak Al Nahayan, Minister for Higher Education and Scientific Research, United Arab Emirates (fifth from left) of the First International Conference on the Fossil Vertebrates of Arabia held at the Dhafra Beach Hotel, Western Region, Emirate of Abu Dhabi, United Arab Emirates in March 1995. Back row, from the left, are Hans de Bruijn, Jes de Bruijn, guest, Herbert Thomas, Peter Friend, Véra Eisenmann, Peter Forey, Norman MacLeod, Pascal Tassy, Andrew Hill, John Kingston, Peter Ditchfield, Walid Yasin, Fred Rögl, and Peter Andrews. Front row, from the left, guest, Saif Rashed al Swedi, Finance and Administration Manager, Ministry of Higher Education and Scientific Research, United Arab Emirates, Peter Whybrow, His Excellency Yousef Omair Bin Yousef, Chairman Abu Dhabi National Oil Company, His Excellency Sheikh Nahayan bin Mubarak al Nahayan, Minister for Higher Education and Scientific Research, United Arab Emirates, Kevin Dunne, General Manager, Abu Dhabi Company for Onshore Oil Operations, Sally McBrearty, and other guests. Photograph courtesy of Sah el Baz, Emirates News.

The fossil record of the Arabian Gulf region provides a very long history of important environmental changes. Concerted and collaborative efforts of local and international institutions and scholars, as embodied in this volume, are bringing about a proper understanding of Arabia's past natural history. In this fossil-rich area, we have always known of plants and animals from eras long before humans walked the earth. We are also learning from the diverse fossil record that the area has been a palaeontological bridge between East and West, and we are beginning to understand the processes by which life in Arabia, and in particular human life, evolved in its changing environment. From this understanding, an appreciation emerges of the unique contribution of our region to the natural history of the world.

I am pleased to introduce *Fossil Vertebrates of Arabia*. The chapters contained herein, written by many of the world's top scholars in the field, constitute an important contribution to palaeontology and geology, as well as to the scientific studies of the natural history of the United Arab Emirates. I am confident that these proceedings will stand as a significant work of reference and as a stimulus to further research for many years to come.

PREFACE

KEN W. GLENNIE

One does not normally associate a land covered with sand dunes and salt-covered sabkhas with a wide variety of fossil vertebrates, especially when many of those described between the covers of this book—crocodiles and hippopotamuses, for example—obviously needed water on a scale that is not found in Arabia today. *Fossil Vertebrates of Arabia* represents an important compilation of palaeontological data covering an association of vertebrates, both large and small, aquatic and terrestrial. Most of these fossils were found in one rock unit of Miocene age, the Baynunah Formation of the Western Region, Emirate of Abu Dhabi, United Arab Emirates.

Although multiauthored, the volume is not the usual compendium of isolated articles, but represents the result of carefully planned co-operation between scientists from Arabia, Europe (mostly the United Kingdom), and the United States of America. The driving force in both authorship and in ensuring that the work was undertaken in an efficient manner was the combination of Peter J. Whybrow of The Natural History Museum, London, and Andrew Hill of Yale University, USA; between them, they also authored or coauthored almost a third of the 36 chapters.

Both Whybrow and Hill had visited the original discovery site at Jebel Barakah separately without knowing of the other's work in the area. Once they realised their common interest, co-operation between them was automatic, and further expeditions were undertaken to find more fossils and to study them adequately. They also ensured that the sedimentology of the exposed Miocene host rocks was properly described and evaluated to provide a sound palaeogeographic framework for their fossils. Despite the emphasis on vertebrate fossils, they also had the collaboration of palaeontologists working on associated nonvertebrate fossils of the Baynunah Formation, and of palaeobotanists working in other parts of Arabia where this contributed to the overall palaeoecology.

The importance of the Abu Dhabi vertebrates in further unravelling the migrational pattern of different animal types between Africa, Europe, and Asia during the Tertiary cannot be overemphasised. In this process, Arabia played a pivotal part during the Miocene. First, a partial marine barrier to migration between Africa and Arabia was created by the opening of the Red Sea early in the Miocene; and, second, also during the early Miocene, a land barrier to the migration of Tethyan marine faunas between what is now the Indian Ocean and the Mediterranean Sea was created by the collision of Arabia and Asia, thereby permitting the interchange of terrestrial vertebrate faunas between those two areas for the first time. The newly formed migrational route to Asia was no doubt broadened by a major fall in global sea level during the later Miocene, probably because of the rapidly increasing ice cover of Antarctica. The key role taken by the Miocene vertebrate faunas from the Western Region of the Emirate of Abu Dhabi in understanding the migrations between especially Africa and Asia, will probably not be fully realised until the contents of this volume have been thoroughly digested and compared with studies elsewhere in the Middle East.

It is perhaps unusual for a nonpalaeontologist to be invited to write the preface to a major book on fossil vertebrates.[1] It seems, however, that by mentioning in a 1968 publication the discovery of a proboscidean tooth in gravels at Jebel Barakah, my colleague Brian Evamy and I led Peter Whybrow to visit the site in 1979 and to find evidence of other fossil vertebrates; and as Hill and Whybrow record in Chapter 3, in 1982, Whybrow and I were junior authors in a reappraisal of that tooth by Madden et al. (1982); the rest is history (see Chapter 3).

When Evamy and I wrote our short 1968 paper entitled "Dikaka", we had no idea that it included the first published identification of a Miocene vertebrate fossil in Arabia, and would eventually be followed by the present treatise. And earlier, in 1965, during my first field trip to southeastern Arabia, I certainly had no idea that for the next 30 years or more I would be involved intermittently in trying to unravel some of its geological secrets.

As a Shell research geologist in 1965, my task was to study modern deserts to better understand the Permian (Rotliegend) gas-bearing reservoirs of the Dutch Groningen gasfield, whose great size had only recently been realised; this knowledge was also applied to exploration in the southern North Sea, which then was beginning. That 1965 field trip had already taken Evamy and me through much of interior Oman and the Trucial States (now the United Arab Emirates).

Our direct objective in the western part of the Emirate of Abu Dhabi was to gain a better understanding of Sabkha Matti, an area of widespread saltflats, for comparison with the coastal sabkhas that were being studied by other geologists of Imperial College London, and Shell (see, for example, Purser, 1973). Jebel Barakah acted like a beacon, drawing us to make a brief geological diversion after obtaining a much-needed shower and fresh supplies of water and fuel from the Iraq Petroleum Company (predecessor to the modern Abu Dhabi Company for Onshore Oil Operations—ADCO) base at Jebel Dhanna. What immediately caught our eye was the lightly cemented reddened dune sand riddled with rhizoconcretions (dikaka, the main topic of our 1968 paper) in the coastal cliff, evidence of the close proximity of the water table in an otherwise arid environment. Climbing to the gravels at the top of the cliff resulted in the discovery (by Evamy, if I recall correctly) of the proboscidean tooth referred to above.

The single event that really brought me back into active Middle East geology, however, was an invitation in 1990 from Dr Terry Adams (well known to Whybrow and Hill), then General Manager of ADCO, to give a talk on the geology of the Oman Mountains to the Society of Explorationists in the Emirates. This led to field studies in both the United Arab Emirates and in the Sultanate of Oman (although I could not return to Jebel Barakah until it was vacated by an artillery battery after the Gulf War), to the supervision of Ph.D. students studying desert sediments in the Emirates and Permian glaciogenic rocks in Oman, and to the co-convenorship (and leader of two field trips) of an international conference on "Quaternary Deserts and Climatic Change" in Al Ain, Emirate of Abu Dhabi, in December 1995.

Here, I seem to have come full circle, for the rocks of the Baynunah Formation at Jebel Barakah not only indicate a much wetter climate in the later Miocene than the area experiences today but, with a probable time gap of some 9 million years, is underlain by dune sands and sabkha sediments of the early Miocene Shuwaihat Formation, which are more akin to the product of today's climate. It is perhaps pertinent that the late Quaternary climate in the Emirates fluctuated between hyperaridity at the peaks of high-latitude glaciations and one that is more humid than today's during interglacials (for example, the so-called Climatic Optimum of about 10 000 to 5000 years ago) In this respect, the Shuwaihat dune sands apparently migrated southwards under the influence of a Miocene northern (Shamal) wind, much like the prevailing sand-transporting winds of today.

Apart from their importance in terms of vertebrate evolution and migration in the area, the contributions to this volume provide information of immense value to geologists like me, with an interest in the Neogene history of Arabia in general, and the Emirate of Abu Dhabi in particular. And by including a study of artifacts from the area associated with the most destructive of all vertebrates, human beings, the book directly impinges on my own special interest in the late Quaternary history of the area.

It is good to see that the United Arab Emirates is in the forefront of several aspects of geological research in Arabia. This is in no small measure because of support by the Government of the United Arab Emirates (see the Foreword, by His Excellency Sheikh Nahayan bin Mubarak Al Nahayan) and by local industry. My own recent work and that of Whybrow and Hill would have been impossible without support from the management of ADCO.

The contributions to this book are evidence of today's strong collaboration between scientists from many scientific disciplines. Such interdisciplinary research is now a prerequisite for unravelling the history of the evolving biosphere and lithosphere in many parts of the world—especially Arabia, which is now becoming an important region for studies of past and present climate change.

NOTE

1. Ken Glennie was educated at the University of Edinburgh (D.Sc., 1984) and spent over 32 years working as an exploration geologist for Shell in New Zealand, Canada, Nepal, India, the Middle East, London, and The Hague. His main research interests comprise desert geology (present and past), geology of the Oman Mountains, and geology of the North Sea. Since his "retirement" in 1987, he has continued to be active in these areas. He is an Honorary Lecturer at the University of Aberdeen, Department of Geology and Petroleum Geology.

REFERENCES

Glennie, K. W., and Evamy, B. D. 1968. Dikaka: plants and plant-root structures associated with aeolian sand. *Palaeogeography, Palaeoclimatology, Palaeoecology* 4: 77–87.

Madden, C. T., Glennie, K. W., Dehm, R., Whitmore, F. C., Schmidt, R. J., Ferfoglia, R. J., and Whybrow, P. J. 1982. *Stegotetrabelodon (Proboscidea, Gomphotheriidae) from the Miocene of Abu Dhabi.* United States Geological Survey, Jiddah.

Purser, B. H. 1973. *The Persian Gulf: Holocene Carbonate Sedimentation and Diagenesis in a Shallow Epicontinental Sea.* Springer-Verlag, Berlin.

ACKNOWLEDGEMENTS

The editors and other specialists who have carried out fieldwork in the Emirate of Abu Dhabi have received immense and welcome support from numerous organisations in the Emirate. Rarely during the history of discovery of Miocene terrestrial faunas and floras from the Old World has such support and interest been forthcoming.

We offer our most sincere thanks to The President of the United Arab Emirates, His Highness Sheikh Zayed bin Sultan Al Nahyan, for his enlightened support and continued interest in our work that can now be added to local information concerning the ancient river systems of eastern Arabia.

We are also most grateful to the United Arab Emirates Minister for Higher Education and Scientific Research, His Excellency Sheikh Nahayan bin Mubarak Al Nahayan, for agreeing to be Patron of the First International Conference on the Fossil Vertebrates of Arabia held in the Emirate of Abu Dhabi during March 1995, and to his ministry, especially Saif Rashed al Swedi, for the organisation of the conference in collaboration with the Abu Dhabi Company for Onshore Oil Operations (ADCO).

From ADCO itself, our work could not have been carried out without the encouragement of successive General Managers, Terry Adams, David Woodward, and Kevin Dunne, and their approval of ADCO's grant to The Natural History Museum to support the project. In addition we have received enormous help from ADCO's Public Affairs Department, Nabil Zakhour; General Relations, Hassan M. Al Saigal; Geodectics, El Badri Khalafalla; and Government Relations, Nasser M. Al Shamsi. Nasr M. Salameen, ADCO's Senior Translator, kindly prepared the Arabic section of the volume. We also thank ATA Translations, London, for formatting the Arabic text.

The early work for this project received great help from the Department of Antiquities and Tourism, Al Ain, and we thank the Secretary, His Excellency Saif Ali Dhab'a al Darmaki, for the hospitality and kindness shown by his department at that time, especially from Dr Walid Yasin. We are also grateful to the staff of the Dhafra Beach Hotel, Jebel Dhanna, and its General Manager, Mr Sashi Panikkar, for their logistic help over many years and for their efforts in making the First International Conference on Arabian Fossil Vertebrates, March, 1995, such a great success.

A book such as this could not have been produced without the assistance of many people. We thank Valerie Whybrow, formerly of The Natural History Museum, London, for her initial work on the electronic formatting of manuscripts. To Diana Clements (NHM) we are especially indebted for her diligent and sustained work on texts, figures, and, especially, references. Other colleagues from The Natural History Museum who have assisted are Norman MacLeod, Jeremy Young, Alan Gentry, Mike Howarth, Peter Forey, Phil Crabb, Harry Taylor, and Paul Lund, the last three of the NHM Photographic Unit.

Lastly, we thank the Yale University Press team of Jean Thomson Black, Science Editor, Mary Pasti, Senior Manuscript Editor, and Joyce Ippolito, Production Editor, for their advice concerning the timely production of this book. To our copyeditor, Sarah Bunney, we are especially grateful. Her longstanding experience of books about palaeontological research and her first-hand knowledge of Arabia has greatly improved the content of the volume. We also thank Jean Macqueen for her hard work preparing the index.

Contributors

[†]C. Geoffrey Adams, OBE
The Natural History Museum,
Department of Palaeontology,
Cromwell Road, London SW7 5BD, U.K.

Salim Al-Busaidi
Ministry of Petroleum and Minerals,
Directorate General of Minerals,
P.O. Box 551, Muscat, Sultanate of Oman

Zaher Al-Sulaimani
Ministry of Petroleum and Minerals,
Directorate General of Minerals,
P.O. Box 551, Muscat, Sultanate of Oman

Peter Andrews
The Natural History Museum,
Department of Palaeontology,
Cromwell Road, London SW7 5BD, U.K.

Mustafa Latif As-Saruri
Ministry of Petroleum and Mineral Resources,
Mineral Exploration Board, Aden Branch,
P.O. Box 5252, Ma'alla, Aden,
Republic of Yemen

John C. Barry
Department of Anthropology,
Harvard University, Peabody Museum,
Cambridge, Massachusetts 02138, USA

Deryck D. Bayliss
10, The Fairway, Northwood,
Middlesex, HA6 3DY, U.K.

Laura Bishop
Department of Human Anatomy and Cell Biology,
University of Liverpool,
P.O. Box 147, Liverpool L69 3BX, U.K.

Charlie S. Bristow
Research School of Geological and Geophysical
Sciences, Birkbeck College London,
Gower Street, London WC1 6BT, U.K.

France de Lapparent de Broin
Muséum National d'Histoire Naturelle,
Laboratoire de Paléontologie,
8 rue Buffon, 75005 Paris Cédex 05, France

Hans de Bruijn
Institute of Earth Sciences, Utrecht University,
P.O. Box 80021, 3508 TA Utrecht,
The Netherlands

Diana Clements
The Natural History Museum,
Department of Palaeontology,
Cromwell Road, London SW7 5BD, U.K.

Margaret E. Collinson
Department of Geology,
Royal Holloway College,
University of London, Egham,
Surrey TW20 0EX, U.K.

Peter Paul van Dijk
Department of Zoology,
University College Galway,
Galway, Ireland

Peter W. Ditchfield
Department of Geology,
University of Bristol,
Wills Memorial Building,
Queens Road, Bristol BS8 1RJ

William R. Downs
Bilby Research Center,
Northern Arizona University,
Flagstaff, Arizona 86011, USA

Véra Eisenmann
Muséum National d'Histoire Naturelle,
Laboratoire de Paléontologie,
8 rue Buffon, 75005 Paris Cédex 05, France

Hamed A. El-Nakhal
Department of Environmental Earth Science,
The Islamic University of Gaza, Gaza

[†]Deceased

Lawrence J. Flynn
Department of Anthropology,
Harvard University, Peabody Museum,
Cambridge, Massachussets 02138, USA

Peter L. Forey
The Natural History Museum,
Department of Palaeontology,
Cromwell Road, London SW7 5BD, U.K.

Eberhard "Dino" Frey
Staatliches Museum für Naturkunde Karlsruhe,
Geowissenschaftliche Abteiling,
Erbprinzenstrasse 13, D-76133 Karlsruhe, Germany

Peter F. Friend
Department of Earth Sciences,
University of Cambridge,
Downing Street, Cambridge CB2 3EQ, U.K.

Alan W. Gentry
The Natural History Museum,
Department of Palaeontology,
Cromwell Road, London SW7 5BD, U.K.

Emmanuel Gheerbrant
Laboratoire de Paléontologie des Vertébrés
(URA CNRS 1433),
Université Pierre et Marie Curie,
4 place Jussieu, 75252 Paris Cédex 05, France

Ken W. Glennie
4, Morven Way, Battater, Aberdeenshire
AB35 5SF, Scotland, and the Department of
Geology and Petroleum Geology, King's College,
University of Aberdeen,
Aberdeen AB9 2UE, Scotland

Tom Gundling
Department of Anthropology,
Yale University, P.O. Box 208277,
New Haven, Connecticut 06520, USA

Ernie A. Hailwood
Department of Oceanography,
Palaeomagnetism Laboratory,
University of Southampton,
Southampton SO9 5NH, U.K., and
Core Magnetics, The Green, Sedbergh,
Cumbria LA10 5JS, U.K.

Andrew Hill
Department of Anthropology,
Yale University, P.O. Box 208277,
New Haven, Connecticut 06520, USA

Louis L. Jacobs
Department of Geological Sciences and
Shuler Museum of Paleontology,
Southern Methodist University,
Dallas, Texas 75275, USA

Paul A. Jeffery
The Natural History Museum,
Department of Palaeontology,
Cromwell Road, London SW7 5BD, U.K.

John D. Kingston
Departments of Anthropology and
Geology and Geophysics, Yale University,
P.O. Box 208277,
New Haven, Connecticut 06520, USA

Norman MacLeod
The Natural History Museum,
Department of Palaeontology,
Cromwell Road, London SW7 5BD, U.K.

Sally McBrearty
Department of Anthropology,
University of Connecticut, U-176,
Storrs, Connecticut 06269, USA

Peter B. Mordan
The Natural History Museum,
Department of Zoology,
Cromwell Road, London SW7 5BD, U.K.

Phillip A. Murry
Department of Physical Sciences,
Tarleton State University,
Stephenville, Texas 76402, USA

Ross G. Peebles
Department of Geological Sciences,
University of Durham, Durham DH1 3LE, U.K.,
and Halliburton Energy Services,
800 Halliburton Center,
5151 San Felipe Boulevard,
Houston, Texas 77056, USA

Daniel S. Pemberton
Staatliches Museum für Naturkunde Karlsruhe,
Geowissenschaftliche Abteiling,
Erbprinzenstrasse 13, D-76133 Karlsruhe, Germany

Martin Pickford
Laboratoire de Paléoanthropologie et Préhistoire
(URA CNRS 49), Collège de France,
11 place Marcelin-Berthelot,
75231 Paris Cédex 05, France

Michael Rauhe
Staatliches Museum für Naturkunde Karlsruhe,
Geowissenschaftliche Abteiling,
Erbprinzenstrasse 13,
D-76133 Karlsruhe, Germany

Jack Roger
Bureau de Recherches Géologiques et Minières,
Orléans-la-Source, BP 6009,
45060 Orléans Cédex 2, France

Fred Rögl
Naturhistorisches Museum Wien,
Geologisch-Paläontologische Abteilung,
Burgring 7, A-1014 Wien, Postfach 417,
Austria

Torsten Rossman
Sandstrasse 98, D-64319 Pfungstadt, Germany

Sevket Sen
Laboratoire de Paléontologie des Vertébrés
(URA CNRS 1433),
Université Pierre et Marie Curie,
4 place Jussieu, 75252 Paris Cédex 05, France

Pascal Tassy
Laboratoire de Paléontologie des Vertébrés
(URA CNRS 12),
Université Pierre et Marie Curie,
4 place Jussieu, 75252 Paris Cédex 05, France

Herbert Thomas
Laboratoire de Paléoanthropologie et Préhistoire
(URA CNRS 49), Collège de France,
11 place Marcelin-Berthelot,
75231 Paris Cédex 05, France,

John E. Whittaker
The Natural History Museum,
Department of Palaeontology,
Cromwell Road, London SW7 5BD, U.K.

Peter J. Whybrow
The Natural History Museum,
Department of Palaeontology,
Cromwell Road, London SW7 5BD, U.K.

Walid Yasin
Department of Antiquities and Tourism,
P.O. Box 15715, Al Ain Museum, Al Ain,
Emirate of Abu Dhabi, United Arab Emirates

Sally V. T. Young
The Natural History Museum,
Department of Palaeontology,
Cromwell Road, London SW7 5BD, U.K.

Abbreviations

AABW	Antarctic Bottom Water
ADCO	Abu Dhabi Company for Onshore Oil Operations
ADNOC	Abu Dhabi National Oil Company
AF	alternating field
aff.	affinities with (a species)
A-horizon	a mineral horizon with an accumulation of organic matter formed or forming at or near the surface of a soil profile; typically has lost iron, aluminium, or clays and is enriched in resistant sand- or silt-sized minerals
AMNH	American Museum of Natural History, New York
apw	apparent polar-wander curve
asl	above sea level
AUH	Emirate of Abu Dhabi (in fossil catalogue)
B/L	breadth/length
BMNH	The Natural History Museum, London; formerly the British Museum (Natural History) (in fossil catalogue)
b.p.	before present
BRGM	Bureau de Recherche Géologiques et Minières
c.	circa
C	canine tooth
C_3	generally, a photosynthetic pathway that reflects wooded/forested habitats
C_4	generally, a photosynthetic pathway that reflects tropical grassland habitats
CAM	crassulacean acid metabolism
CENOP	Cenozoic Palaeoceanography Project
cf.	referrable to (a species)
ChRM	characteristic remanent magnetism
chron	unit of time; magnetic time scale
CL	cathodoluminescence
dM	deciduous molar tooth
dP	deciduous premolar tooth
DSDP	Deep Sea Drilling Project
gen. et sp. indet.	genus and species indeterminate
GPS	Global Positioning System
HMC	Holocene Marine Carbonate
I	incisor tooth
IAS	Isotopic Analytical Services Ltd
IGCP	International Geological Correlation Project
KNM	National Museums of Kenya
ka	thousand years
lt	left
M	permanent molar tooth
Ma	millions of years
mA/m	milliamperes per metre
MCZ	Museum of Comparative Zoology, Harvard University
MHNT	Muséum d'Histoire Naturelle de Toulouse

MN Mammals Neogene zones: a series of fossil mammal assemblages from single localities placed in a chronological sequence on the basis of evolutionary stage, entries by migration and exits by extinction of specific taxa. MN zonation is often used for rough correlations in Europe (European land mammal zones), western Asia, and northern Africa; for example, MN 9–13 for the late Miocene

MNHN Muséum National d'Histoire Naturelle, Paris

mT millitesla

n number

N N zones: biostratigraphic zones; for example, N 16–18 for the late Miocene based on the chronological range of certain planktonic foraminifera

NADW North Atlantic Deep Water

NBS National Bureau of Standards, USA

NHM The Natural History Museum, London

NHMW Naturhistorisches Museum, Wien

NN NN zones: biostratigraphic zones based on the chronological range of calcareous nannoplankton; for example, NN 10–11 for the late Miocene

nov. new; as in subgen. nov.—a new subgenus

NRM natural remanent magnetism

n.s. not significant

NSF National Science Foundation, USA

NTIS National Technical Information Service, USA

P P zones; based on planktonic foraminifera

P permanent premolar tooth

P probability

PDB Peedee Formation belemnite standard

PEPC phosphenol pyruvate carboxylase—the enzyme that catalyses the reaction between bicarbonate and phosphenol pyruvate (PEP) to form malate or asparate (4-carbon compounds) in C_4 plants

PIUM Paläontologisches Institut der Universität Mainz

ppm parts per million

rt right

RuBPC ribulose biphosphate carboxylase—the enzyme that catalyses carboxylation in the Calvin cycle during which CO_2 and water combine with ribulose biphosphate to produce a 3-carbon compound (3-phosphoglyceric acid). This enzyme is functional in all photosynthetic organisms and comprises the most abundant protein on earth

s.d. standard deviation

SEM scanning electron microscope

sin. sinistral—from the left

s.l. *sensu lato;* in a wide sense

SMF Senckenbergische Naturforschende Gesellschaft, Naturhistorisches Museum, Frankfurt

SMNK Staatliches Museum für Naturkunde Karlsruhe

SMOW Standard Mean Ocean Water

sp.	species
sp. indet.	species indeterminate
s.s.	*sensu stricto;* in a strict sense.
T	tesla
UAE	United Arab Emirates
UGR	Ungarische Geologische Reichsanstalt
USGS	United States Geological Survey
XRD	X-ray diffraction

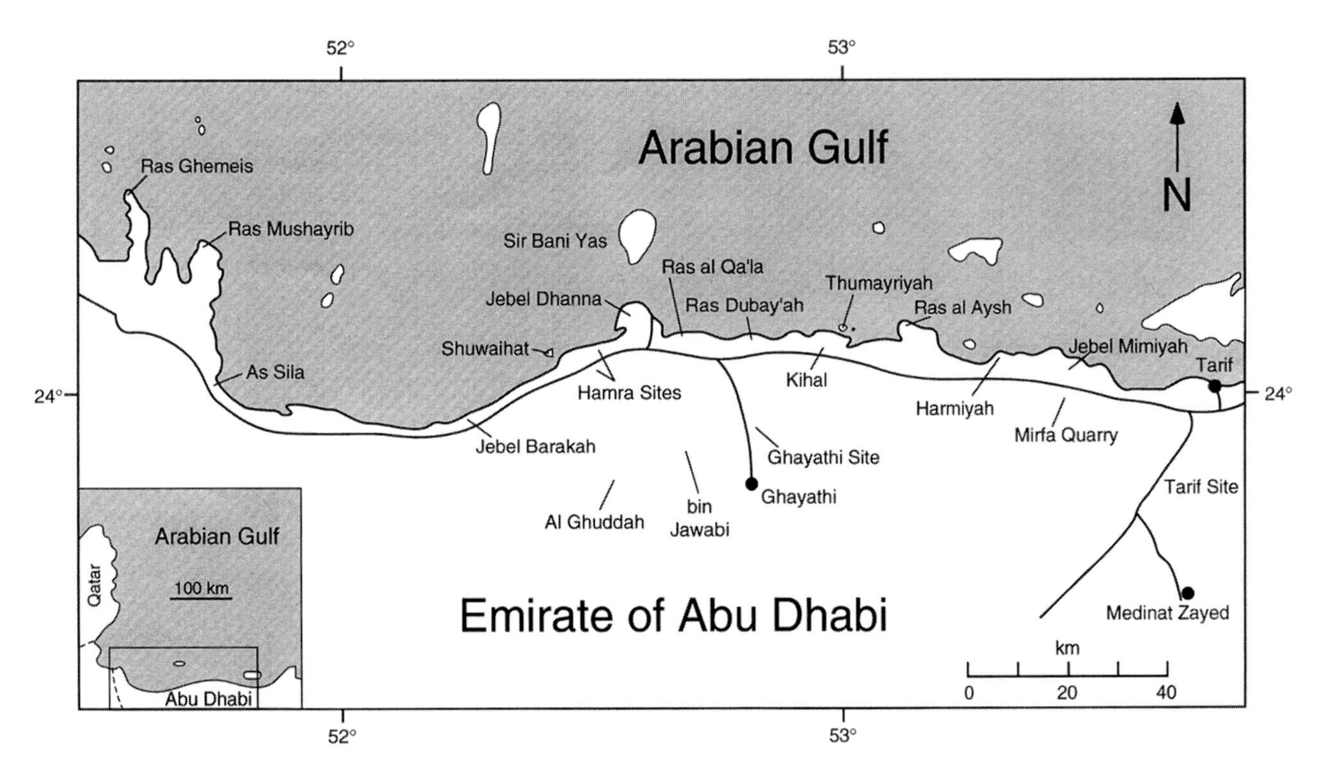

Map of the Western Region, Emirate of Abu Dhabi, United Arab Emirates, indicating the positions of all locations from which both fossils and geological samples have been collected.

Andrew Hill, with a proboscidean scapula, at the Hamra locality, Emirate of Abu Dhabi.

Introduction, Summary, Overview, and History of Palaeontological Research in the Emirate of Abu Dhabi, United Arab Emirates

PART I

Introduction to Fossil Vertebrates of Arabia

1

PETER J. WHYBROW AND ANDREW HILL

Climate appears to limit the range of many animals, though there is some reason to believe that in many cases it is not the climate itself so much as the change of vegetation consequent on climate which produces the effect... Where barriers have existed from a remote epoch, they will at first have kept back certain animals from coming in contact with each other; but when the assemblage of organisms on the two sides of the barrier have, after many ages, come to form a balanced organic whole, the destruction of the barrier may lead to a very partial intermingling of the peculiar forms of the two regions.

—*Wallace (1876)*

In 1876, when Alfred Russel Wallace published his thoughts about the global interrelationship of the changing geosphere with the biosphere, the geography, fauna, and geology of Arabia's interior were unknown. Early publications (Thomas, 1894, 1900; Yerbury and Thomas, 1895) alluded to the dispersal into Arabia of some present-day African and Asian mammal faunas, but nothing at that time could be said about the dispersal of Tertiary continental faunas into or out of Arabia because no fossil faunas were discovered until 1974 (Anon., 1975). In the 1930s the search for Arabian oil provided data for both topographical and geological maps, and explorations by natural historians gave insight into the fauna and flora living in a region of diverse, arid geography. Palaeontological work in the Himalayas and in East Africa dates from the discoveries of Tertiary mammals in those regions in the nineteenth century and the 1920s, respectively. But until recently, Arabia was a palaeobiogeographical gap in our knowledge of low-latitude continental Tertiary vertebrate faunas in the Old World.

This extensively illustrated volume brings together for the first time researches on the fossil vertebrates of the Arabian continent discovered in the United Arab Emirates, the Sultanate of Oman, and the Republic of Yemen.[1] The book provides up-to-date information not only about Arabian faunas and floras but also about Arabian palaeoenvironments, Arabian Miocene palaeomagnetic correlations, stable isotope analyses, and some of Arabia's earliest stone tools. At an interregional level, the Tertiary vertebrate faunas of Arabia are discussed in the context of a North African–Arabian–Southwest Asian faunal belt and new fossil records from eastern Arabia are here presented for the first time. In addition, the timing of the closure of the Tethys seaway in the Middle East is thoroughly discussed; this disconnection formed Afro-Arabia's first land connection with Asia. Although the volume focuses mainly on the late Miocene biota from the Emirate of Abu Dhabi, where leading palaeontologists have carried out work on the Miocene continental sequence since 1979, we have linked this research to associated geological studies (figs 1.1–1.3) on the Abu Dhabi Miocene and neighbouring regions by geochemists, stratigraphers, and magnetochronologists.

This volume should interest specialists whose studies on the dispersal of African and Asian Tertiary faunas have hitherto been incomplete because of the lack of details from the Arabian Peninsula, here available in one publication. Students of Middle East geology and oil-exploration geologists will find it a useful compendium because we have attempted to create the difficult synergy between Tertiary terrestrial and marine events.

The book falls naturally into six parts, each one introduced with a summary of the chapters. The introductions to each part also appear in Arabic. Part I comprises a summary and overview, with a

 ISBN 0-300-07183-3

Figure 1.1.
Miocene jebels,
Emirate of Abu
Dhabi.

Figure 1.2.
The islet of Zab-
but, Arabian Gulf
coast, Emirate of
Abu Dhabi.

Figure 1.3. Miocene siliceous cap-rock, Shuwaihat, Emirate of Abu Dhabi.

history of the Abu Dhabi Project, and Part II presents the geological context from which the vertebrate fossils were collected and the local stratigraphy. The chapters in Part III describe systematic studies on late Miocene invertebrates, reptiles, and mammals (several of them new species) from the Emirate of Abu Dhabi and discusses the palaeogeographical relationships of the fossils. The four chapters in Part IV, deal with topics associated with the fossil biota—taphonomy, carbon isotopes, and Arabian palaeoenvironments—and we include a study on lithic artifacts and the earliest evidence of the genus *Homo* in the region. Part V then links the Abu Dhabi Project to other research in Asia and Africa and includes studies of older faunas from the Sultanate of Oman and the Republic of Yemen. Finally, in Part VI, a broader picture of Arabia in an Old World context is presented. The timing of the disconnection of Tethys is examined as are events in the Mediterranean and Paratethys that relate to the dispersal of mammals. The last chapter provides an up-to-date review of the palaeo-oceanographic factors necessary for an understanding of climatic change during the Miocene.

Arabia does not readily yield evidence of its past terrestrial animal and plant life to the palaeontologist. Apart from the rigours of exploration, where serendipity rules, almost all its rocks are marine.[2] Where rare continental sediments are now exposed, there are no rivers and little rain to erode fossils from the rocks. When erosion exposes a fossil vertebrate, its bones can be fractured and fragmented by extremes of temperature and subsequently scattered and abraded by sand-laden winds. The rarity of Arabian fossil vertebrates is shown by the fact that despite the size of the Arabian geological plate—which is slightly larger than the Indian subcontinent—there are only 11 locations from which significant vertebrate faunas of any geological age have so far been collected.

As we previously mentioned, studies of these faunas and floras originate from the discovery in 1974 by palaeontologists from The Natural History Museum, London—then called the British Museum

(Natural History)—of the first Miocene terrestrial vertebrate fauna from the peninsula, in the eastern part of the Kingdom of Saudi Arabia. Since the 1970s work has been carried out principally by two European teams collaborating with Arab organisations—the oil companies of the countries where the research has been undertaken. The first team, led by Peter Whybrow of The Natural History Museum, London,[3] and Andrew Hill of Yale University (in collaboration with the Abu Dhabi Company for Onshore Oil Operations), has focused on the late Miocene of the Emirate of Abu Dhabi; the second team, led by Herbert Thomas of the Laboratoire de Paléoanthropologie et Préhistoire, Collège de France, Paris (in collaboration with the Ministry of Petroleum and Minerals, Directorate General of Minerals, Sultanate of Oman), has concentrated on Oligocene and Miocene rocks in the Sultanate of Oman. In addition, work by various research groups in the Republic of Yemen (collaborating with the Ministry of Oil and Mineral Resources and the University of Sana'a) has produced the first hints of a Tertiary terrestrial fauna and flora, and the late Jurassic sequence has produced what is believed to be the second record of a dinosaur from Arabia.

Palaeontology has changed dramatically since the first vertebrates were discovered in Arabia. As readers of this book will recognise, vertebrate palaeontology is no longer an independent science. Today, to understand—as far as we ever can—the habitats of extinct biotas, palaeontologists collaborate with other geological specialists; a project such as our research in the Emirate of Abu Dhabi becomes the work of a multidisciplinary team. While the systematic identification of fossils will always be the backbone of vertebrate palaeontology, other specialist studies now provide a multitude of scientific themes that range from evolutionary biology, through palaeobiogeography and the movement of continental plates, to the diagenesis and isotopes within the rocks and fossils themselves.

Explorations for Arabia's palaeontological heritage continue, thanks to the enlightened support of government organisations within the countries of the peninsula. This book is not only the first step in publicising the results of Arabian palaeontological and geological researches to the benefit of the Arabian peoples, but it also provides an in-depth testimonial for the emerging academic role that Arabia now provides for linking studies of Old World Tertiary faunas and environments.

Notes

1. The impetus for this book came from the First International Conference on the Fossil Vertebrates of Arabia held in the Emirate of Abu Dhabi, United Arab Emirates, March 1995.
2. "Serendipity" was coined by Horace Walpole, fourth Earl of Oxford (1717–97), from the Persian fairy tale *The Three Princes of Serendip*, in which the heroes possess this gift.
3. The Abu Dhabi Project, continuing until the year 2000, forms part of The Natural History Museum's Global Change and the Biosphere research programme.

References

Anon. 1975. Mammalian remains from Saudi Arabia. In *Report on the British Museum (Natural History), 1972–1974.* Trustees of the British Museum (Natural History), London.

Thomas, O. 1894. On some specimens of mammals from Oman. *Proceedings of the Zoological Society of London* 1894: 448–55.

———. 1900. On the mammals obtained in Southwestern Arabia by Messrs Percival and Dodson. *Proceedings of the Zoological Society of London* 1900: 95–104.

Wallace, A. R. 1876. *The Geographical Distribution of Animals with a Study of the Relations of Living and Extinct Faunas as Elucidating the Past Changes of the Earth's Surface.* Macmillan, London.

Yerbury, J. W., and Thomas, O. 1895. On the mammals of Aden. *Proceedings of the Zoological Society of London* 1895: 542–55.

Summary and Overview of the Baynunah Fauna, Emirate of Abu Dhabi, and Its Context

2

ANDREW HILL AND PETER J. WHYBROW

How impossible must it be for us to guess, in most cases, at the exact nature of the forces that limit the range of some species and cause others to be rare or to become extinct! All that we can in general hope to do is, to trace out, more or less hypothetically, some of the larger changes in physical geography that have occurred during the ages immediately preceding our own, and to estimate the effect they will probably have produced on animal distribution. We may then, by the aid of such knowledge as to past organic mutations as the geological record supplies us with, be able to determine the probable birthplace and subsequent migrations of the more important genera and families

—*Wallace (1876)*

One of the most interesting biological features of the Arabian Peninsula is the part it has played in Old World biogeography. Arabia lies at the junction of the classic Old World biogeographic divisions—the Ethiopian, Palaearctic, and Oriental regions (Wallace, 1876). The Arabian Peninsula is also a large global area.[1] Situated between 12° and 30°N, and between 35° and 60°E, it occupies an area of just over 3 million square kilometres—almost all of the Arabian continental plate. Arabia is therefore about as large as the Indian subcontinent. The peninsula today provides a diverse set of habitats that range from the mountainous regions of the southwest with their high plateaus and where some peaks reach nearly 3800 metres, to the low-lying sand deserts that occupy most of the eastern area. In the north, bordering the Arabian Gulf, are salt flats, some of which are below sea level.

The region also has a variable climate, within the arid to hyperarid range (Takahashi and Arakawa, 1981). For example, in the high southwest there are 500 mm of rain a year, with a low-temperature range and snow on some high mountains in the cold season. In the central and eastern areas, which are occupied by some of the hottest and most arid deserts on earth, daily mean temperatures reach close to 40 °C and rainfall is slight.

Descriptions of the arduous journeys undertaken by such early western explorers of Arabia as Charles Doughty, St John Philby, William Palgrave, and Wilfred Thesiger provide an apparent authentication of today's common belief in the paucity of Arabian terrestrial animals. Later research has shown, however, that Arabia has a diverse mammalian biota (Harrison and Bates, 1991). Harrison and Bates consider not just the Arabian Peninsula but countries westwards to the Mediterranean and north through Iraq, so in terms of the peninsula the following figures are slightly inflated. But over this whole region eight mammalian orders are represented, comprising 29 families and including 82 genera. Even without bats and other small mammals (insectivores, lagomorphs, and rodents) there remain 14 families and 29 genera. It is true, though, that Arabia lacks species diversity, having only about 40% the number of species present in the smaller area of eastern Africa, for example; but higher taxonomic levels are represented well, even in the peninsular region. Among other nonmarine vertebrates, amphibians and freshwater fish frequent perennial streams and pools in the mountainous areas.

 ISBN 0-300-07183-3

Arabian Biogeography

Serious discussion about the biogeography of Arabia—and the peninsula's possible connections with Asia—began in the early 1930s, coincident with oil exploration. Data on subsurface marine rocks of the Gulf area came from burgeoning oil wells. On the basis of microfossils, Davies (1934) thought that a landbridge between Arabia and southwestern Asia might not be easy to recognise. Savage (1967) concluded that Neogene mammal migrations were strongest between Europe and Asia, weakest between Africa and Asia. This was a development of Wallace's belief (1876) that the ancestors of modern Old World mammalian faunas dispersed from north to south. In this work, Wallace did not refer to Arabia at all, and oddly splits the area between his Ethiopian and Mediterranean (Palaearctic) regions. These regional distinctions were further developed by Bernor (1983) for Miocene faunas.

Until the 1970s comments on Arabian biogeography largely consisted of inferences from marine sequences and from terrestrial faunal change documented in the better-known palaeontological successions on neighbouring landmasses. Neogene terrestrial vertebrate faunas were not known from Arabia itself. But in 1974 middle Miocene mammals were found in the Kingdom of Saudi Arabia (Anon., 1975; Andrews et al., 1978; Hamilton et al., 1978; Whybrow, 1987).[2] Partly based on this new information Whybrow et al. (1982) subsequently developed the notions of Davies and of Savage, and commented on the probability of an early Miocene land connection between eastern Arabia and southwestern Asia. This comment was more fully examined by Adams et al. (1983) from the marine perspective, and expanded by Whybrow (1984) in a terrestrial context. These preliminary data were also used by Bernor (1983: fig. 3). The marine evidence suggests that this land connection occurred at the latest between 19 and 16 million years (Ma) ago.

Such speculations, coupled with some initial field surveys before the early 1980s, generated the impetus for the further collection of Tertiary geological and palaeontological information from Arabia. But despite the great interest of the region, and its important geographic position with respect to other continents, little is known about the fossil fauna and palaeoenvironmental history of the peninsula. In Chapter 33, Whybrow and Clements (1999a) note the paucity of known fossil vertebrate occurrences in Arabia. It is against this background that the importance of the recent discoveries in the Emirate of Abu Dhabi, especially the Baynunah fauna, can best be appreciated. They provide important information about past environments and biotas.

Fossil Vertebrates of Arabia

This volume provides valuable data on the Oligocene and Miocene faunas of the Sultanate of Oman (Thomas et al., 1999—Chapter 30), the first record of an Eocene fauna and flora from the Republic of Yemen (As-Saruri et al., 1999—Chapter 31), and the second record of an Arabian dinosaur (Jacobs, et al., 1999—Chapter 32). Regional contextual information is given in Chapters 34 (Adams et al., 1999), 36 (Rögl, 1999), and 35 (MacLeod, 1999); Chapter 26 discusses the earliest intimations of humans in the Emirates (McBrearty, 1999). But the main theme of the book is the geology and palaeontology of the continental late Miocene that in Arabia has so far been found only in the Emirate of Abu Dhabi.

Ancillary discoveries include the first recognition of the Dam Formation in the Emirates (Whybrow et al., 1999—Chapter 4), previously known only in Saudi Arabia and Qatar. It is a lower to middle Miocene (Burdigalian) marine unit, 19–16 Ma old.

The stratigraphy originally described by Whybrow (1989) has been revised. His Baynunah Formation is now divided into the predominantly aeolian Shuwaihat Formation below, with the largely fluviatile and fossiliferous Baynunah Formation above. The newly described Shuwaihat Formation (Whybrow et al., 1999; Bristow, 1999—Chapters 4 and 6) is characterised by aeolian cross-stratification, representing dunes, separated by mudstones and fine sand interpreted as a sabkha environment. It is likely to be lower to middle Miocene in age, and about 15 ± 3 Ma has been suggested on the basis of palaeomagnetism (Hailwood and Whybrow, 1999—Chapter 8). As Kingston and Hill (1999) point out in Chapter 27, this formation is important for providing tangible evidence of arid conditions in northern latitudes during the mid-Miocene. Also, if similar conditions extended latitudinally across Africa, it provides an explanation for the nature of the postulated barrier leading to the differentiation of north African and sub-Saharan Miocene faunas (Thomas, 1979; Thomas et al., 1982; Hill, 1999—Chapter 29). Homotaxically and in terms of age the Shuwaihat Formation can be broadly linked with the continental clastic Hofuf Formation in Saudi Arabia.

The Baynunah Formation, as redefined (Whybrow et al., 1999—Chapter 4), consists of a sequence of predominantly fluviatile sediments that suggest a low-gradient river made up of numerous small channels separated by low sand banks. The channels were probably no more than 3 metres deep, but the entire braided river network was tens to hundreds of metres wide (Friend, 1999; Ditchfield, 1999—Chapters 5 and 7). This river system drained an area in the interior of the Arabian Peninsula to the northwest of modern Abu Dhabi, and it may have been part of a larger system that includes the modern Tigris and Euphrates rivers. At that time sea level was substantially lower than today, and the marine coastline is thought to have been about 300 km to the east of its present position.

It is clear that the climate changed markedly between Shuwaihat and Baynunah times. The earlier hyperarid conditions ameliorated, and the Baynunah fluvial system developed and provided a habitat for freshwater molluscs (Jeffery, 1999—Chapter 10) and terrestrial gastropods (Mordan, 1999—Chapter 11), fish, aquatic reptiles, birds, and mammals. A permanent flow of water in this river is clear from the presence of large freshwater turtles (Lapparent de Broin and Dijk, 1999—Chapter 13) and crocodiles, including the gharial (Rauhe et al., 1999—Chapter 14), but the presence of catfish suggests that flow was sluggish or intermittent in some of the channels (Forey and Young, 1999—Chapter 12). Occasional flow of a higher velocity is indicated by coarser conglomerates in some of the channels and by the disarticulated and fragmented state of some of the fossil bones (Friend, 1999—Chapter 5). Temperatures were warm during Baynunah times, and calcretes preserved in the sediments indicate that the climate was semiarid, with an annual rainfall of no more than 75 mm (Ditchfield, 1999—Chapter 7). The vegetation consisted of a mixture of grass, shrubs, and trees, including *Acacia*. Trees and shrubs were probably concentrated near the river banks, while a more open grassy vegetation grew farther away from the river itself (Kingston, 1999—Chapter 25).

This habitat supported a rich and diverse group of animals, including ancient forms of elephant, hippopotamus, horse, antelope, wolverine, hyaena, and sabre-tooth cat. In Chapter 23, Whybrow and Clements (1999b) list the elements of the Baynunah fauna. More than 900 specimens have been collected by The Natural History Museum, London/Yale University Project. In total there are 43 vertebrate species belonging to at least 26 families. They include three new species and one new genus: a bagriid fish (Forey and Young, 1999—Chapter 12), a gerbil, *Abudhabia baynunensis* (de Bruijn and Whybrow, 1994), and an hipparionine equid (Eisenmann and Whybrow, 1999—Chapter 19). The other nonmammalian taxa identified are three species of fish, three species of turtles (representing both terrestrial and aquatic forms), three species of crocodiles (including a gharial), and two species of birds. Among mammals there is a total of 31 species documented from 17 or 18 families.

The river system itself and the contained fossil fauna indicate a markedly different environment and climate in Baynunah times than at present, or during preceding Shuwaihat times. Several global factors could be implicated in the shift from Shuwaihat to Baynunah environments. Some of these are discussed by Kingston and Hill (1999) in Chapter 27.

PALAEOBIOGEOGRAPHY AND THE NATURE OF FAUNAL CHANGE

The mammalian fauna is essentially African in character (Hill, 1999—Chapter 29) and particularly northern African (Gentry, 1999a,b—Chapters 21 and 22), but, as might be expected, it also includes Asian elements, as shown by some rodents (de Bruijn and Whybrow, 1994), pigs (Bishop and Hill, 1999—Chapter 20), and bovids (Gentry, 1999b—Chapter 22). Some genera also occur in Europe but there are no definitive links at the species level with late Miocene European faunas, such as those from Greece and those known eastwards through Turkey to northwestern Iran. There are no deer, for example, in the Baynunah collection, and indeed no cervids are known from any of the Miocene and Pleistocene sites in peninsular Arabia, although the Mesopotamian fallow deer and the roe deer are recorded until recent times from northwestern Iran, southern Turkey, and Palestine.

Instead, the Baynunah fauna appears to be part of a late Miocene faunal belt trending west and east between roughly 15°N and 31°N, and including sites in North Africa, Arabia, Pakistan, India, and perhaps Afghanistan. This is a different zoogeographic configuration from the North African and Sub-Paratethyan provinces proposed by Bernor (1983). It suggests that during Baynunah times animals could migrate more easily in an east–west direction, but that north–south movement may have been restricted by barriers presented by ancient deserts, mountains, or river systems.

A systematic revision and comparison of taxa from the relevant regions, particularly northern Africa, will permit the refinement or refutation of recent ideas on the influence of climate on speciation and dispersal, such as those postulated by Vrba (1995). There are numerous problems associated with such comparative work, and with notions regarding the nature and degree of impact of physical extrinsic factors on faunas, their speciation and migration (Hill, 1987, 1995; White, 1995). The primary problem is the imperfection of the fossil record in both space and time. For Arabia, Whybrow and Clements (1999a—Chapter 33) illustrate how sparse is the fossil vertebrate record, a record that has become discovered only during the past 20 years or so. In sub-Saharan Africa as well, there are few late Miocene sites (Hill, 1999—Chapter 29). Not all time is sampled or sampled well, and the sites are predominantly clustered in a small area on the eastern side of Africa. We have no information at all concerning late Miocene events in about 99% of the African continent. Similar calculations may be performed for other regions; for example, areal coverage in Southwest Asia is not much greater.

These limitations of the fossil record constrain our ability to define precisely the first and last appearances of taxa—fundamental data for appreciating the pattern of faunal change and its correlation to extrinsic factors. These estimates translate into time datum lines for fossil appearances in a region, and, additionally, the place of a taxon's first appearance is often assumed to approximate its centre of origin.

From these somewhat insignificant samples, major narratives have been constructed about Old World palaeozoogeographic provinces and dispersal events, and about the pattern and character of faunal change through time. Vrba (1993), for example, predicted that turnover pulses should occur within the fossil record. She postulated that "most lineage turnover has occurred in pulses, *nearly synchronous* across diverse groups of organisms, and in predictable synchrony with changes in the physical environment" (our italics). Even though Vrba later revised her theory (1995), she nevertheless advocated direct and strong environmental forcing of species change. There are problems with demonstrating turnover pulses, given the nature of the fossil record (Hill, 1987, 1995). This kind of change is not easy to see in Africa, Arabia, or Asia, and certainly in Africa some lineages, such as pigs, appear not to conform to Vrba's ideas (Bishop, 1993, 1994; White 1995). Orographic events over a long period of time, such as the uplift of the Himalayan–Tibetan plateau, which perhaps mediate local climate, are also likely to be influencial (Kingston and Hill, 1999—Chapter 27).

The faunal information obtained so far from Arabia, however, does have several implications. One is the strong and novel suggestion that in the late Miocene a belt of faunal affinity stretched across northern Africa, Arabia, and into parts of Asia. This fauna differs from those nearby but north of the Zagros in Iran, and from sub-Saharan Africa. Refining such suggestions will require more fossil material, but the palaeontological exploration of Arabia is just beginning. The Emirate of Abu Dhabi and the Arabian Peninsula as a whole hold vast promise for further research.

NOTES

1. Peninsular Arabia is here considered to include the following countries: the State of Kuwait, the State of Qatar, the Kingdom of Saudi Arabia, the United Arab Emirates, the Sultanate of Oman, and the Republic of Yemen.
2. Further exploration of the Miocene vertebrate sites in the Eastern Province of the Kingdom of Saudi Arabia by European and American palaeontologists has not been possible owing to policy constraints on such work implemented by Saudi authorities since 1974. To our knowledge, no detailed work on these important sites has been carried out by anyone since the mid-1980s. The only studies, published by British and French teams (see Chapter 33), remain incomplete.

REFERENCES

Adams, C. G., Bayliss, D. D., and Whittaker, J. E. 1999. The terminal Tethyan event: A critical review of the conflicting age determinations for the disconnection of the Mediterranean from the Indian Ocean. Chap. 34 in *Fossil Vertebrates of Arabia,* pp. 477–84 (ed. P. J. Whybrow and A. Hill). Yale University Press, New Haven.

Adams, C. G., Gentry, A. W., and Whybrow, P. J. 1983. Dating the terminal Tethyan events. In *Reconstruction of Marine Environments* (ed. J. Meulenkamp). *Utrecht Micropaleontological Bulletins* 30: 273–98.

Andrews, P. J., Hamilton, W. R., and Whybrow, P. J. 1978. Dryopithecines from the Miocene of Saudi Arabia. *Nature* 274: 249–51.

Anon. 1975. Mammalian remains from Saudi Arabia. In *Report on the British Museum (Natural History), 1972–1974.* Trustees of the British Museum (Natural History), London.

As-Saruri, M. L., Whybrow, P. J., and Collinson, M. E. 1999. Geology, fruits, seeds, and vertebrates (?Sirenia) from the Kaninah Formation (middle Eocene), Republic of Yemen. Chap. 31 in *Fossil Vertebrates of Arabia,* pp. 443–53 (ed. P. J. Whybrow and A. Hill). Yale University Press, New Haven.

Bernor, R. L. 1983. Geochronology and zoogeographic relationships of Miocene Hominoidea. In *New Interpretations of Ape and Human Ancestry,* pp. 21–64 (ed. R. L. Ciochon and R. S. Corruccini). Plenum Press, New York.

Bishop, L. C. 1993. Hominids of the East African Rift Valley in a macroevolutionary context. *American Journal of Physical Anthropology,* suppl. 16: 57.

———. 1994. Pigs and the ancestors: hominids, suids and environments during the Plio-Pleistocene of East Africa. Ph.D. thesis, Yale University, New Haven.

Bishop, L., and Hill, A. 1999. Fossil Suidae from the Baynunah Formation, Emirate of Abu Dhabi, United Arab Emirates. Chap. 20 in *Fossil Vertebrates of Arabia,* pp. 254–70 (ed. P. J. Whybrow and A. Hill). Yale University Press, New Haven.

Bristow, C. S. 1999. Aeolian and sabkha sediments in the Miocene Shuwaihat Formation, Emirate of Abu Dhabi, United Arab Emirates. Chap. 6 in *Fossil Vertebrates of Arabia,* pp. 50–60 (ed. P. J. Whybrow and A. Hill). Yale University Press, New Haven.

Bruijn, H. de, and Whybrow, P. J. 1994. A Late Miocene rodent fauna from the Baynunah Formation, Emirate of Abu Dhabi, United Arab Emirates. *Proceedings Koninklijke Nederlandse Akademie van Wetenschappen* 97: 407–22.

Davies, A. M. 1934. *Tertiary Faunas. Vol. 2: The Sequence of Tertiary Faunas.* Thomas Murby, London.

Ditchfield, P. W. 1999. Diagenesis of the Baynunah, Shuwaihat and Upper Dam Formation sediments exposed in the Western Region, Emirate of Abu Dhabi, United Arab Emirates. Chap. 7 in *Fossil Verte-*

brates of Arabia, pp. 61–74 (ed. P. J. Whybrow and A. Hill). Yale University Press, New Haven.

Eisenmann, V., and Whybrow, P. J. 1999. Hipparions from the late Miocene Baynunah Formation, Emirate of Abu Dhabi, United Arab Emirates. Chap. 19 in *Fossil Vertebrates of Arabia,* pp. 234–53 (ed. P. J. Whybrow and A. Hill). Yale University Press, New Haven.

Forey, P. L., and Young, S. V. T. 1999. Late Miocene fishes of the Emirate of Abu Dhabi, United Arab Emirates. Chap. 12 in *Fossil Vertebrates of Arabia,* pp. 120–35 (ed. P. J. Whybrow and A. Hill). Yale University Press, London and New Haven.

Friend, P. F. 1999. Rivers of the Lower Baynunah Formation, Emirate of Abu Dhabi, United Arab Emirates. Chap. 5 in *Fossil Vertebrates of Arabia,* pp. 38–49 (ed. P. J. Whybrow and A. Hill). Yale University Press, New Haven.

Gentry, A. W. 1999a. A fossil hippopotamus from the Emirate of Abu Dhabi, United Arab Emirates. Chap. 21 in *Fossil Vertebrates of Arabia,* pp. 271–89 (ed. P. J. Whybrow and A. Hill). Yale University Press, New Haven.

———. 1999b. Fossil pecorans from the Baynunah Formation, Emirate of Abu Dhabi, United Arab Emirates. Chap. 22 in *Fossil Vertebrates of Arabia,* pp. 290–316 (ed. P. J. Whybrow and A. Hill). Yale University Press, New Haven.

Hailwood, E. A., and Whybrow, P. J. 1999. Palaeomagnetic correlation and dating of the Baynunah and Shuwaihat Formations, Emirate of Abu Dhabi, United Arab Emirates. Chap. 8 in *Fossil Vertebrates of Arabia,* pp. 75–87 (ed. P. J. Whybrow and A. Hill). Yale University Press, New Haven.

Hamilton, W. R., Whybrow, P. J., and McClure, H. A. 1978. Fauna of fossil mammals from the Miocene of Saudi Arabia. *Nature* 274: 248–49.

Harrison, D. L., and Bates, P. J. J. 1991. *The Mammals of Arabia.* Harrison Zoological Museum, Sevenoaks.

Hill, A. 1987. Causes of perceived faunal change in the later Neogene of East Africa. *Journal of Human Evolution* 16: 583–96.

———. 1995. Faunal and environmental change in the Neogene of East Africa: Evidence from the Tugen Hills Sequence, Baringo District, Kenya. In *Paleoclimate and Evolution, with Emphasis on Human Origins,* pp. 178–93 (ed. E. S. Vrba, G. H. Denton, T. C. Partridge, and L. H. Burkle). Yale University Press, New Haven.

———. 1999. Late Miocene sub-Saharan African vertebrates, and their relation to the Baynunah fauna, Emirate of Abu Dhabi, United Arab Emirates. Chap. 29 in *Fossil Vertebrates of Arabia,* pp. 420–29 (ed. P. J. Whybrow and A. Hill). Yale University Press, New Haven.

Jacobs, L. L., Murry, P. A., Downs, W. R., and El-Nakhal, H. A. 1999. A dinosaur from the Republic of Yemen. Chap. 32 in *Fossil Vertebrates of Arabia,* pp. 454–59 (ed. P. J. Whybrow and A. Hill). Yale University Press, New Haven.

Jeffery, P. A. 1999. Late Miocene swan mussels from the Baynunah Formation, Emirate of Abu Dhabi, United Arab Emirates. Chap. 10 in *Fossil Vertebrates of Arabia,* pp. 111–15 (ed. P. J. Whybrow and A. Hill). Yale University Press, London and New Haven.

Kingston, J. D. 1999. Isotopes and environments of the Baynunah Formation, Emirate of Abu Dhabi, United Arab Emirates. Chap. 25 in *Fossil Vertebrates of Arabia,* pp. 354–72 (ed. P. J. Whybrow and A. Hill). Yale University Press, New Haven.

Kingston, J. D., and Hill, A. 1999. Late Miocene palaeoenvironments in Arabia: A synthesis. Chap. 27 in *Fossil Vertebrates of Arabia,* pp. 389–407 (ed. P. J. Whybrow and A. Hill). Yale University Press, New Haven.

Lapparent de Broin, F. de, and Dijk, P. P. van. 1999. Chelonia from the late Miocene Baynunah Formation, Emirate of Abu Dhabi, United Arab Emirates: Palaeogeographic implications. Chap. 13 in *Fossil Ver-*

tebrates of Arabia, pp. 136–62 (ed. P. J. Whybrow and A. Hill). Yale University Press, New Haven.

MacLeod, N. 1999. Oligocene and Miocene palaeoceanography—a review. Chap. 36 in *Fossil Vertebrates of Arabia,* pp. 501–507 (ed. P. J. Whybrow and A. Hill). Yale University Press, New Haven.

McBrearty, S. 1999. Earliest stone tools from the Emirate of Abu Dhabi, United Arab Emirates. Chap. 26 in *Fossil Vertebrates of Arabia,* pp. 373–88 (ed. P. J. Whybrow and A. Hill). Yale University Press, New Haven.

Mordan, P. B. 1999. A terrestrial pulmonate gastropod from the late Miocene Baynunah Formation, Emirate of Abu Dhabi, United Arab Emirates. Chap. 11 in *Fossil Vertebrates of Arabia,* pp. 116–19 (ed. P. J. Whybrow and A. Hill). Yale University Press, New Haven.

Rauhe, M., Frey, E., Pemberton, D. S., and Rossmann, T. 1999. Fossil crocodilians from the late Miocene Baynunah Formation of the Emirate of Abu Dhabi, United Arab Emirates: Osteology and palaeoecology. Chap. 14 in *Fossil Vertebrates of Arabia,* pp. 163–85 (ed. P. J. Whybrow and A. Hill). Yale University Press, New Haven.

Rögl, F. 1999. Oligocene and Miocene palaeogeography and stratigraphy of the circum-Mediterranean region. Chap. 35 in *Fossil Vertebrates of Arabia,* pp. 485–500 (ed. P. J. Whybrow and A. Hill). Yale University Press, New Haven.

Savage, R. J. G. 1967. Early Miocene mammal faunas of the Tethyan region. In *Aspects of Tethyan Biogeography,* vol. 7, pp.247–82 (ed. C. G. Adams and D. V. Ager). Systematics Association, London.

Takahasi, K., and Arakawa, H. 1981. *Climates of Southern and Western Asia. World Survey of Climatology.* Elsevier, Amsterdam.

Thomas, H. 1979. Le rôle de barrière écologique de la ceinture saharo-arabique au Miocène: Arguments paléontologiques. *Bulletin du Muséum National d'Histoire Naturelle, Paris* 1: 127–35.

Thomas, H., Bernor, R., and Jaeger, J.-J. 1982. Origines du peuplement mammalien en Afrique du Nord durant le Miocène terminal. *Geobios* 15: 283–97.

Thomas, H., Roger, J., Sen, S., Pickford, M., Gheerbrant, E., Al-Sulaimani, Z., and Al-Busaidi, S. 1999. Oligocene and Miocene terrestrial vertebrates in the southern Arabian peninsula (Sultanate of Oman) and their geodynamic and palaeogeographic settings. Chap. 30 in *Fossil Vertebrates of Arabia,* pp. 430–42 (ed. P. J. Whybrow and A. Hill). Yale University Press, New Haven.

Vrba, E. S. 1993. Turnover-pulses, the Red Queen, and related topics. *American Journal of Science* 293-a: 418–52.

———. 1995. On the connections between paleoclimate and evolution. In *Paleoclimate and Evolution, with Emphasis on Human Origins,* pp. 178–93 (ed. E. S. Vrba, G. H. Denton, T. C. Partridge, and L. H. Burkle). Yale University Press, New Haven.

Wallace, A. R. 1876. *The Geographical Distribution of Animals with a Study of the Relations of Living and Extinct Faunas as Elucidating the Past Changes of the Earth's Surface.* Macmillan, London.

White, T. D. 1995. African omnivores: Global climatic change and Plio-Pleistocene Hominids and Suids. In *Paleoclimate and Evolution, with Emphasis on Human Origins,* pp. 178–93 (ed. E. S. Vrba, G. H. Denton, T. C. Partridge, and L. H. Burkle). Yale University Press, New Haven.

Whybrow, P. J. 1984. Geological and faunal evidence from Arabia for mammal "migrations" between Asia and Africa during the Miocene. *Courier Forschungsinstitut Senckenberg* 69: 189–98.

———. (ed.) 1987. Miocene geology and palaeontology of Ad Dabtiyah, Saudi Arabia. *Bulletin of the British Museum (Natural History),* Geology 41: 367–457.

———. 1989. New stratotype; the Baynunah Formation (Late Miocene), United Arab Emirates: Lithology and palaeontology. *Newsletters on Stratigraphy* 21: 1–9.

Whybrow, P. J., and Clements, D. 1999a. Arabian Tertiary fauna, flora, and localities. Chap. 33 in *Fossil Vertebrates of Arabia,* pp. 460–72 (ed. P. J. Whybrow and A. Hill). Yale University Press, New Haven.

———. 1999b. Late Miocene Baynunah Formation, Emirate of Abu Dhabi, United Arab Emirates: Fauna, flora, and localities. Chap. 23 in *Fossil Vertebrates of Arabia,* pp. 317–33 (ed. P. J. Whybrow and A. Hill). Yale University Press, New Haven.

Whybrow, P. J., Collinson, M. E., Daams, R., Gentry, A. W., and McClure, H. A. 1982. Geology, fauna (Bovidae, Rodentia) and flora from the Early Miocene of eastern Saudi Arabia. *Tertiary Research* 4: 105–20.

Whybrow, P. J., Friend, P. F., Ditchfield, P. W., and Bristow, C. S. 1999. Local stratigraphy of the Neogene outcrops of the coastal area: Western Region, Emirate of Abu Dhabi, United Arab Emirates. Chap. 4 in *Fossil Vertebrates of Arabia,* pp. 28–37 (ed. P. J. Whybrow and A. Hill). Yale University Press, New Haven.

History of Palaeontological Research in the Western Region of the Emirate of Abu Dhabi, United Arab Emirates

3

Andrew Hill, Peter J. Whybrow, and Walid Yasin

Palaeontological observations first began in what is now the Western Region of the Emirate of Abu Dhabi with the explorations of petroleum geologists. Some references to sediments and to fossils appear in unpublished geological reports of the relevant oil companies. For example, D. A. Holm and R. Layne in an unpublished survey of 1949, reported "probable horse teeth and bones" at Jebel Dhanna. In that survey they also visited Jebel Barakah and produced a geological section. R. A. Bramkamp and L. F. Ramirez, geologists working with the Arabian–American Oil Company (ARAMCO) in the 1940s and 1950s, examined rocks in the Western Region and thought they were middle Miocene in age, about 14 million years (Ma) old. They equated them with formations previously described in eastern Saudi Arabia, the marine Dam Formation, dated between 19 and 16 Ma, and the overlying continental Hofuf Formation, estimated at about 14 Ma. Glennie and Evamy (1968) reported on their visit to Jebel Barakah in the early 1960s while working with Royal Dutch Shell. They had found the tooth of a fossil proboscidean there, and believed the enclosing sediments to be dune sands and wadi conglomerates.[1]

In 1979 Peter Whybrow (The Natural History Museum, London), while working in Qatar, made a one-day visit to Jebel Barakah, where he found crocodilian vertebrae and a proximal ulna of a bovid. He was also impressed by numerous root casts that he interpreted as those of mangroves. He dated the assemblages at about 11 Ma, based upon the earlier correlations provided by geological maps (Whybrow and McClure, 1981). On a revisit to Barakah in 1981, Whybrow discovered two equid teeth, confirming the hints of Holm and Layne. The teeth belonged to the genus *Hipparion,* which is unknown in the Old World until about 11 Ma ago. This demonstrated that the sediments were younger than had previously been thought. He also found a *Hexaprotodon* hippopotamus mandible, the first from Arabia. This phase of his work was carried out in collaboration with the United Arab Emirates (UAE) University, Department of Geology, at Al Ain, and with Professor M. A. Bassiouni (Ain Shams University, Egypt and then at the University of Qatar, Scientific and Applied Research Centre, Doha, Qatar).

In 1982 the proboscidean tooth collected by Glennie and Evamy in 1961 at Jebel Barakah was referred to as *Stegotetrabelodon grandincisivum* by Madden et al. (1982).

In 1983 Walid Yasin of the Department of Antiquities and Tourism, Al Ain, Emirate of Abu Dhabi, was part of an archaeological survey of the Western Region organised by the department and a group of German archaeologists (Vogt et al., 1989). They found interesting later archaeology, but at the same time discovered fossils at Jebel Barakah and at several other sites along the coast, including Jebel Dhanna and Shuwaihat. They believed these to be Miocene in age. By chance, Andrew Hill (Yale University, USA) was informed of this by Hans Peter Uerpmann (University of Tübingen, Germany), who knew both Yasin and the German group. In 1984 Hill was invited by His Excellency Saif Ali Dhab'a al Darmaki, the Under Secretary of the

 ISBN 0-300-07183-3

Department, to help evaluate the fossils, and visited Abu Dhabi for a few days. The collections housed in Al Ain Museum proved to be extremely interesting. There were specimens identifiable as belonging to siluriform fish, crocodiles, turtles, ostrich, *Hipparion, Hexaprotodon,* at least two species of bovid, and a gomphothere proboscidean. There was also fossil wood. Like the results of Whybrow's earlier work, this was an immediate indication that the environment of Abu Dhabi in the past was much different from that of today, and, on the basis of rough comparison with African fossil faunas, Hill estimated the age of the specimens to be somewhere between 8 and 6 Ma. Because of their interest Yasin immediately organised a short visit to the sites. In the limited time available he and Hill visited Jebel Barakah, Hamra, Jebel Dhanna, and a few other localities eastwards along the coast. Another sizeable collection of fossils was recovered.

At that time Hill mistakenly thought that these fossils were the first from the United Arab Emirates. He was aware of Whybrow's work in Arabia, but imagined it to be confined to Saudi Arabia and Qatar. On discussing the finds with Whybrow, however, he learnt of the previous work and discovered that he had missed seeing Whybrow and Peter Andrews (NHM) on Jebel Barakah by only two weeks. In the spring of 1984 Whybrow and Andrews had carried out more detailed investigations on the geology and fossils of Barakah, again in association with the UAE University at Al Ain.

Hill and Whybrow then collaborated, and in 1986 together they submitted a joint report to the Department of Antiquities, with the recommendation that further work be carried out on all of these sites. Some preliminary observations on the Abu Dhabi occurrences were also published by Whybrow and Bassiouni (1986) in a more comprehensive review of the Arabian Miocene.

MIOCENE FAUNAS AND FLORAS OF THE EMIRATE OF ABU DHABI

In 1988 Whybrow and Hill received an invitation and financial assistance from the Department of Antiquities to organise a further expedition. On 1 January 1989 the current project, "Miocene faunas and floras of the Emirate of Abu Dhabi", began.

The aims of the expedition are:

- To locate new fossiliferous sites.
- To recover additional fossils from previously known and new sites so as to find new taxa and better material of taxa already known.
- To document the sedimentary succession and discover the lateral extent of the fossiliferous exposures.
- To interpret the sediments from the point of view of palaeoenvironments.
- To date the fossils, possibly by palaeomagnetic work.

In spring 1989 Hill and Yasin spent the first week of the field season discovering a number of additional sites, and some good fossils, particularly from Shuwaihat, which they had not visited in 1984. The finds included part of the cranium of a crocodile (fig. 3.1). They also explored jebels to the south, away from the coast. They were then joined by Whybrow and Phil Crabb, a photographer from The Natural History Museum, and spent more time in exploration and collecting. Among new fauna found by Yasin, and described in this volume, was a deinothere tooth fragment (AUH 21: Tassy, 1999—Chapter 18) and a mandible of a mustelid (AUH 45: Barry, 1999—Chapter 17). Both of these are the first representatives of their families to be found in Arabia. Whybrow found a cercopithecid canine tooth, one of only two Miocene primate specimens from the peninsula (AUH 35: Hill and Gundling, 1999—Chapter 16). A short report of the work up to this date was published in *Nature* (Gee, 1989). Based upon his previous work and further observations in the spring 1989 season, Whybrow that year published a formal description of the rock unit, naming it the Baynunah Formation after the region in which the outcrops occur (Whybrow, 1989). The type section is at Jebel Barakah.

Whybrow and Hill joined Yasin in Abu Dhabi in December 1989 and remained until mid-February 1990. For parts of that season they were joined by Ernest Hailwood (University of Southampton, U.K.) and by Sally McBrearty (University of Con-

Figure 3.1. Peter Whybrow (left), Andrew Hill, and Walid Yasin examining a crocodile skull found at Shuwaihat in 1989.

necticut, Storrs, USA). Hailwood began work on the palaeomagnetic stratigraphy, collecting specimens at Jebel Barakah, Hamra, and Jebel Dhanna (fig. 3.2). McBrearty began a survey for stone artifacts that might indicate the presence of humans in the Western Region at an early time. These were discovered on Jebel Barakah, Hamra, and Shuwaihat (McBrearty, 1993, 1999—Chapter 26). The group explored additional regions, extending investigations to Ras al Aysh, Thumayriyah, and Kihal in the east, and as far as Ras Mushayrib and Ras Ghemeis near the border of Saudi Arabia to the west. Among the new fossils was the first sabre-tooth cat from Arabia (AUH 202, 241: Barry, 1999—Chapter 17) and some proboscideans (Tassy, 1999—Chapter 18). The finds included cranial parts, and a proboscidean skeleton on Shuwaihat that occupied more time in later seasons (Andrews, 1999—Chapter 24).

Yasin discovered a fine proboscidean tooth (AUH 456) on Ras Dubay'ah (see Tassy, 1999—Chapter 18), which coincided with a visit to Jebel Dhanna by The President His Highness Sheik Zayed bin Sultan Al Nahyan (fig. 3.3). Hill, Whybrow, and Yasin were most honoured to be granted an audience with The President His Highness Sheikh Zayed bin Sultan Al Nahyan at that time, when he was shown the proboscidean tooth and the preliminary results of the project's work were explained. They were greatly encouraged to find His Highness so interested in the research, and the discussion centred on the ancient river systems of Arabia. Also in 1990 Hill presented a report on the research at the annual meetings of the American Association of Physical Anthropologists in Miami (Hill et al., 1990), and a general account incorporating the preliminary results of Hailwood's palaeomagnetic work appeared in the *Journal of Human Evolution* (Whybrow et al., 1990).

Partially owing to events in Kuwait, only Whybrow visited the area in the spring of 1991. This resulted in additional collections, including a partial mandible of *Hipparion,* the type of a new species

Figure 3.2. Ernie Hailwood drilling the Shuwaihat sandstones to obtain samples for palaeomagnetic analyses.

(AUH 270: Eisenmann and Whybrow, 1999—Chapter 19) and a search of new areas, such as an examination of the small island of Zabbut situated just off Shuwaihat. In 1991 a report was published in *Tribulus* (Whybrow et al., 1991). In December of that year an exhibit on the Abu Dhabi fossils opened at The Natural History Museum, London, which gave prominence to the research, and to the importance of the Emirate of Abu Dhabi in understanding the faunas and environments of the Old World.[2]

The next field season began shortly after this, and lasted from December 1991 to January 1992. Hill and Whybrow were joined for some of the time by Yasin, and by Véra Eisenmann (Muséum National d'Histoire Naturelle, Paris), who came particularly to search for horse fossils. In January Robin Cocks, Head of the Palaeontology Department at the NHM visited, along with Gillian Comerford (NHM) and Peter Friend (University of Cambridge, U.K.). Friend began an investigation of the sediments and their palaeoenvironmental implications, particularly at Jebel Barakah, Shuwaihat, and Jebel Mimiyah near Al Mirfa.

Another project initiated that season was the excavation of the proboscidean skeleton that had been discovered on Shuwaihat Island in 1990. This was begun by Comerford and Hill. Among other significant discoveries was the mandible of a juve-

Figure 3.3. Walid Yasin excavating a proboscidean tooth at Ras Dubay'ah (AUH 456).

nile *Hexaprotodon* hippopotamus in good condition (AUH 481: Gentry, 1999a—Chapter 21). Hailwood, on a second visit, sampled other successions for palaeomagnetic purposes (Hailwood and Whybrow, 1999—Chapter 8). Among new potential sites, Whybrow was able to visit Sir Bani Yas very briefly, and to confirm the presence of Baynunah Formation rocks on the island.

In April 1992 the excavation of the proboscidean skeleton continued on Shuwaihat for a short season. The members of this group were Whybrow, Peter and Libby Andrews, Miranda Amour Chelou, Phil Crabb, and Campbell Smith (NHM).

The work of the expedition formed the focus of a film made in 1992 and produced by the Abu Dhabi Company for Onshore Oil Operations (ADCO) entitled *Abu Dhabi—The Missing Link*. This has been shown several times on Abu Dhabi television, and more widely circulated. The work of the project was the theme of ADCO's Annual Report for 1992.

During December 1992 Whybrow, Hill, and McBrearty revisited several of the sites and recovered better fossil material. Some new sites were found. One particularly interesting specimen was located by students working with Ken Glennie, and retrieved by Don Hadley, then a member of the United States Geological Survey/National Drilling Company Groundwater Project. This is a spectacular cranium of *Tragoportax* with both horn cores preserved (AUH 442: Gentry, 1999b—Chapter 22). It came from near Tarif and is the most eastern site so far, documenting fossiliferous exposures now for 140 km along the coast. In January 1993 Peter Friend and Peter Ditchfield (University of Cambridge, U.K.) undertook more detailed studies of the sediments (Friend, 1999; Ditchfield, 1999—Chapters 5 and 7). John Kingston (Yale University, USA) visited to locate soil horizons in the Baynunah Formation, the analysis of which would aid in the interpretation of palaeoenvironments (Kingston, 1999—Chapter 25). Towards the end of the season Hans de Bruijn (University of Utrecht,

The Netherlands) joined the expedition to collect samples for micromammals, which resulted in the description of a new fossil genus and species, *Abudhabia baynunensis* (de Bruijn and Whybrow, 1994). McBrearty's continued investigation of lithic artifacts resulted in the recovery of about 200 additional specimens from several new sites, and later in 1993 she published a preliminary report on her earlier collections (McBrearty, 1993, 1999—Chapter 26).

Another short season took place in March and April 1994 to deal with what remained of the Shuwaihat proboscidean. Members of the group were Peter Whybrow, Peter Andrews, Alan and Anthea Gentry, Phil Crabb, Diana Clements (NHM), Peter Ditchfield, and David Cameron (Australian National University). This resulted in the completion of the detailed excavation of the specimen and its transport to London for preservation and study. In September, Charlie Bristow (Birkbeck College, University of London) studied the fossil aeolian and dune sediments at Jebel Barakah and at Shuwaihat (Bristow, 1999—Chapter 6).

Whybrow, Hill, and Valerie Whybrow spent a few weeks in Abu Dhabi in the spring of 1995 and were able to collect additional fossils and explore jebels to the south of the coastal area. In January, Whybrow was an invited speaker at the International Conference on the Biotic and Climatic Effects of the Messinian Event on the Circum Mediterranean, Benghazi, Libya, where he presented a paper on "Late Miocene palaeofaunas from the United Arab Emirates".

In March 1995 the First International Conference on the Fossil Vertebrates of Arabia was inaugurated by His Excellency Sheikh Nahayan bin Mubarak Al Nahayan, the United Arab Emirates Minister for Higher Education and Scientific Research. The conference was held at the centre for Miocene field operations in the Emirate of Abu Dhabi, the Dhafra Beach Hotel, Jebel Dhanna. Thirty delegates from the United Arab Emirates, Europe, and the United States of America participated and presented papers on topics associated with The Natural History Museum/Yale University project theme—"Miocene faunas and floras of the Emirate of Abu Dhabi". Delegates were shown some of the prime fossiliferous localities (plate 3.1, p. 23).

The joint project will continue for 5 years from 1996 by a continuation of the ADCO 1991 grant to The Natural History Museum, London. Further fossil collections and detailed geological studies will be undertaken from a region occupying a central position between the better-known Miocene fossiliferous sites of southwestern Asia and of Africa. This work will produce data enabling more meaningful observations about barriers to Old World dispersal of Miocene vertebrate faunas.

Acknowledgements

Whybrow's early work was financed by grants from The Natural History Museum—then called the British Museum (Natural History)—and the Qatar Petroleum Producing Company (Onshore). Hill's first visit was sponsored and funded by the Department of Antiquities and Tourism, Al Ain, and he thanks the Under Secretary, His Excellency Saif Ali Dhab'a al Darmaki for the hospitality and kindness shown by his department at that time, especially from Dr Walid Yasin. He would also like to thank Dr Hans Peter Uerpmann (University of Tübingen) for facilitating that visit. The first joint work was also funded by the Department of Antiquities. Whybrow raised additional funds, principally from British Petroleum Exploration, with also a grant from Grindlays Bank to cover our other expenses, and Hill received a grant from the L. S. B. Leakey Foundation, USA. From 1991 onwards the project has been supported by a generous grant to Whybrow from the Abu Dhabi National Oil Company (ADNOC) and the Abu Dhabi Company for Onshore Oil Operations (ADCO). We thank Sohail al Mazrui, former Secretary General, and Rashid Saif al Suwaidi, former Director of Exploration and Production of ADNOC, for their assistance, and Terry Adams, David Woodward, and Kevin Dunne, successive General Managers of ADCO, for their sustained support. During our work we have received the interest and help of many people, both representatives of the organisations already mentioned and others. These include Nabil Zakhour, Head of Public Affairs, ADCO, and Nasser M. Al Shamsi, Government Relations Superintendent, ADCO. Among others are members of the Emirates Natural History Group, especially the late Bish Brown and Peter Hellyer.

We also offer our most sincere thanks to His Excellency Sheik Nahayan bin Mubarak Al Nahayan, the Minister for Higher Education and Scientific Research, Chancellor of the United Arab Emirates University, for agreeing to be the Patron of the First International Conference on the Fossil Vertebrates of Arabia, held in the Emirate of Abu Dhabi during March 1995, and to his ministry and ADCO for their sponsorship of the meeting and for their continuing research support. We are also grateful to the excellent support and sponsorship received from the Abu Dhabi National Hotels Company (especially the Dhafra Beach Hotel, Jebel Dhanna, and its General Manager, Mr Sashi Panikkar).

NOTES

1. The collections of Miocene material from the Western Region, Emirate of Abu Dhabi, and their current locations can be divided into four distinct time periods. The proboscidean tooth collected by Glennie and Evamy in 1961 is currently stored in the Bayerisches Staatsammlung, Munich. A small collection made by Whybrow and others between 1979 and 1984 is incorporated into the palaeontological collections of The Natural History Museum, London, and the collection made by Al Ain Museum in collaboration with German archaeologists is stored in Al Ain Museum, Emirate of Abu Dhabi. Similarly, Al Ain Museum houses the collection made by Hill and Yasin in 1984. The collections made by The Natural History Museum/Yale University team since 1989 (prefix AUH) are the property of the Emirate of Abu Dhabi. It is hoped that type specimens will remain at The Natural History Museum in London; negotiations with the Abu Dhabi authorities are still (1999) in progress.
2. The exhibit included a short video, sponsored by the United Arab Emirates University and by ADCO, which was presented by Sir David Attenborough. The exhibit was officially opened by Dr Abdul Wahab Al Muhaideb, Assistant Under Secretary, Ministry of Health, United Arab Emirates. The Parliamentary Undersecretary at the Foreign and Commonwealth Office, Sir Mark Lennox-Boyd, M.P., gave a short address. The occasion was attended by London-based representatives of the shareholders of ADCO (British Petroleum, Shell, Total, Exxon, Mobil, Partex), Sir Richard Beaumont KCMG, Chairman of Arab British Commerce, Dr Robin Cocks, Head of the Palaeontology Department at The Natural History Museum, and Dr Neil Chalmers, Director of The Natural History Museum, as well as by representatives of British business in Abu Dhabi.

REFERENCES

Andrews, P. J. 1999. Taphonomy of the Shuwaihat proboscidean, late Miocene, Emirate of Abu Dhabi, United Arab Emirates. Chap. 24 in *Fossil Vertebrates of Arabia*, pp. 338–53 (ed. P. J. Whybrow and A. Hill). Yale University Press, New Haven.

Barry, J. C. 1999. Late Miocene Carnivora from the Emirate of Abu Dhabi, United Arab Emirates. Chap. 17 in *Fossil Vertebrates of Arabia*, pp. 204–208 (ed. P. J. Whybrow and A. Hill). Yale University Press, New Haven.

Bristow, C. S. 1999. Aeolian and sabkha sediments in the Miocene Shuwaihat Formation, Emirate of Abu Dhabi, United Arab Emirates. Chap. 6 in *Fossil Vertebrates of Arabia*, pp. 50–60 (ed. P. J. Whybrow and A. Hill). Yale University Press, New Haven.

Bruijn, H. de, and Whybrow, P. J. 1994. Late Miocene rodent fauna from the Baynunah Formation, Emirate of Abu Dhabi, United Arab Emirates. *Proceedings Koninklijke Nederlandse Akademie van Wetenschappen* 97: 407–22.

Ditchfield, P. W. 1999. Diagenesis of the Baynunah, Shuwaihat, and Upper Dam Formation sediments exposed in the Western Region, Emirate of Abu Dhabi, United Arab Emirates. Chap. 7 in *Fossil Vertebrates of Arabia,* pp. 61–74 (ed. P. J. Whybrow and A. Hill). Yale University Press, New Haven.

Eisenmann, V., and Whybrow, P. J. 1999. Hipparions from the Late Miocene Baynunah Formation, Emirate of Abu Dhabi, United Arab Emirates. Chap. 19 in *Fossil Vertebrates of Arabia,* pp. 234–53 (ed. P. J. Whybrow and A. Hill). Yale University Press, New Haven.

Friend, P. F. 1999. Rivers of the Lower Baynunah Formation, Emirate of Abu Dhabi, United Arab Emirates. Chap. 5 in *Fossil Vertebrates of Arabia,* pp. 38–49 (ed. P. J. Whybrow and A. Hill). Yale University Press, New Haven.

Gee, H. 1989. Fossils from the Miocene of Abu Dhabi. *Nature* 338: 704.

Gentry, A. W. 1999a. A fossil hippopotamus from the Emirate of Abu Dhabi, United Arab Emirates. Chap. 21 in *Fossil Vertebrates of Arabia,* pp. 271–89 (ed. P. J. Whybrow and A. Hill). Yale University Press, New Haven.

———. 1999b. Fossil pecorans from the Baynunah Formation, Emirate of Abu Dhabi, United Arab Emirates. Chap. 22 in *Fossil Vertebrates of Arabia,* pp. 290–316 (ed. P. J. Whybrow and A. Hill). Yale University Press, New Haven.

Glennie, K. W., and Evamy, B. D. 1968. Dikaka: Plants and plant-root structures associated with aeolian sand. *Palaeogeography, Palaeoclimatology, Palaeoecology* 4: 77–87.

Hailwood, E. A., and Whybrow, P. J. 1999. Palaeomagnetic correlation and dating of the Baynunah and Shuwaihat Formations, Emirate of Abu Dhabi, United Arab Emirates. Chap. 8 in *Fossil Vertebrates of Arabia,* pp. 75–87 (ed. P. J. Whybrow and A. Hill). Yale University Press, New Haven.

Hill, A., and Gundling, T. 1999. A monkey (Primates; Cercopithecidae) from the late Miocene of Abu Dhabi, United Arab Emirates. Chap. 16 in *Fossil Vertebrates of Arabia,* pp. 198–203 (ed. P. J. Whybrow and A. Hill). Yale University Press, New Haven.

Hill, A., Whybrow, P. J., and Yasin al-Tikriti, W. 1990. Late Miocene primate fauna from the Arabian Peninsula: Abu Dhabi, United Arab Emirates. *American Journal of Physical Anthropology* 81: 240–41.

Kingston, J. D. 1999. Isotopes and environments of the Baynunah Formation, Emirate of Abu Dhabi, United Arab Emirates. Chap. 25 in *Fossil Vertebrates of Arabia,* pp. 354–72 (ed. P. J. Whybrow and A. Hill). Yale University Press, New Haven.

Madden, C. T., Glennie, K. W., Dehm, R., Whitmore, F. C., Schmidt, R. J., Ferfoglia, R. J., and Whybrow, P. J. 1982. *Stegotetrabelodon (Proboscidea, Gomphotheriidae) from the Miocene of Abu Dhabi.* United States Geological Survey, Jiddah.

McBrearty, S. 1993. Lithic artifacts from Abu Dhabi's Western Region. *Tribulus: Bulletin of the Emirates Natural History Group* 3: 13–14.

———. 1999. Earliest stone tools from the Emirate of Abu Dhabi, United Arab Emirates. Chap. 26 in *Fossil Vertebrates of Arabia,* pp. 373–88 (ed. P. J. Whybrow and A. Hill). Yale University Press, New Haven.

Tassy, P. 1999. Miocene elephantids (Mammalia) from the Emirate of Abu Dhabi, United Arab Emirates: Palaeobiogeographic implications. Chap. 18 in *Fossil Vertebrates of Arabia,* pp. 209–33 (ed. P. J. Whybrow and A. Hill). Yale University Press, New Haven.

Vogt, B., Gockel, W., Hofbauer, H., and Al-Haj, A. A. 1989. The coastal survey in the Western Province of Abu Dhabi. *Archaeology in the United Arab Emirates* V: 49–60.

Whybrow, P. J. 1989. New stratotype; the Baynunah Formation (Late Miocene), United Arab Emirates:

Lithology and palaeontology. *Newsletters on Stratigraphy* 21: 1–9.

Whybrow, P. J., and Bassiouni, M. A. 1986. The Arabian Miocene; rocks, fossils, primates and problems. In *Primate Evolution*, pp. 85–91 (ed. J. G. Else and P. C. Lee). Cambridge University Press, Cambridge.

Whybrow, P. J., and McClure, H. A. 1981. Fossil mangrove roots and palaeoenvironments of the Miocene of the eastern Arabian peninsula. *Palaeogeography, Palaeoclimatology, Palaeoecology* 32: 213–25.

Whybrow, P. J., Hill, A., and Yasin al-Tikriti, W. 1991. Miocene fossils from Abu Dhabi. *Tribulus: Bulletin of the Emirates Natural History Group* 1: 4–9.

Whybrow, P. J., Hill, A., Yasin al-Tikriti, W., and Hailwood, E. A. 1990. Late Miocene primate fauna, flora and initial palaeomagnetic data from the Emirate of Abu Dhabi, United Arab Emirates. *Journal of Human Evolution* 19: 583–88.

Plate 3.1. Delegates to the First International Conference on the Fossil Vertebrates of Arabia enjoying a field excursion to Miocene localities in the Western Region, Emirate of Abu Dhabi.

Peter Whybrow gazing at a sea-cliff exposure of the Baynunah Formation, Jebel Barakah, Western Region, Emirate of Abu Dhabi.

Miocene Geology of the Western Region, Emirate of Abu Dhabi, United Arab Emirates

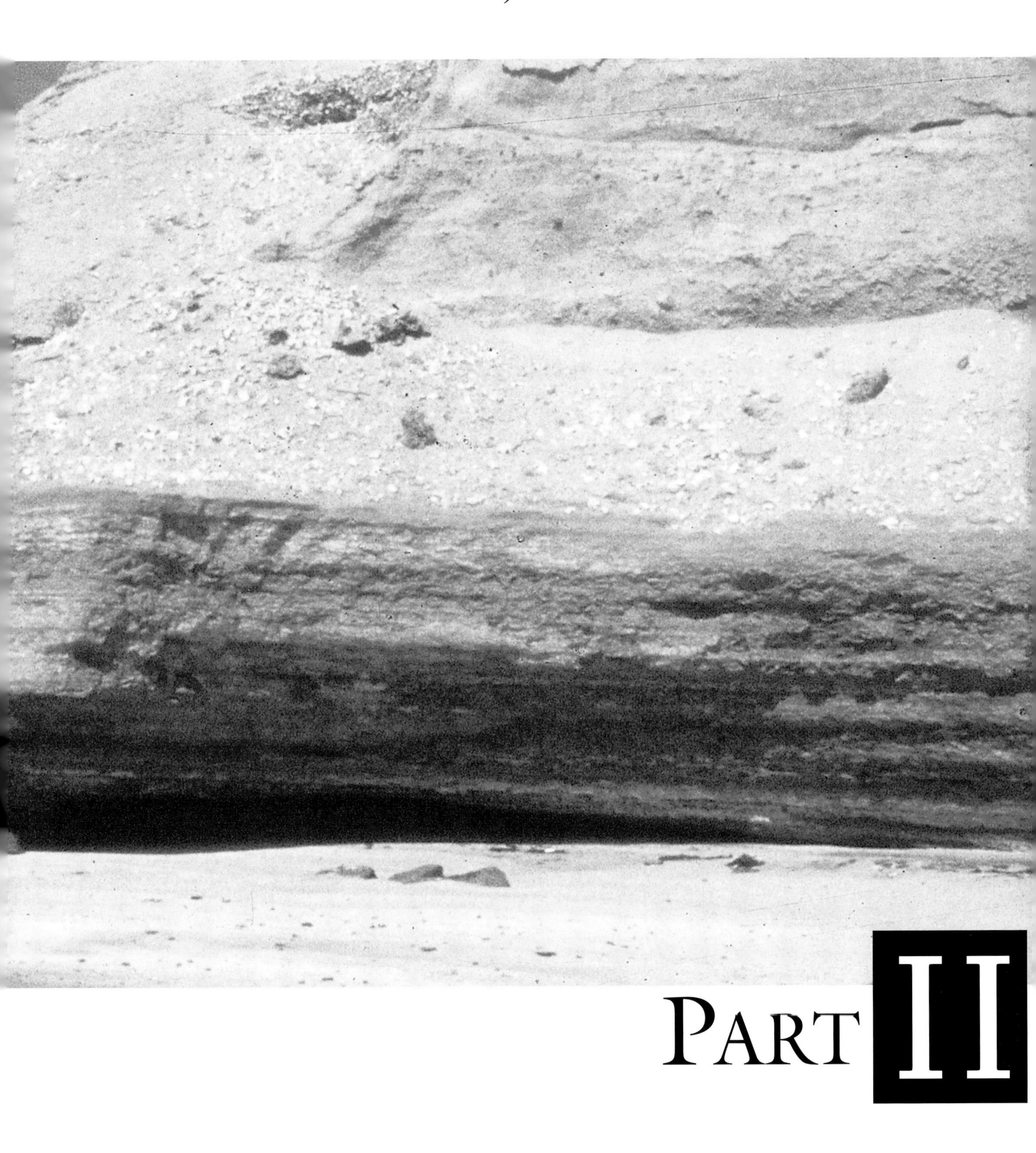

PART II

When Whybrow started his first geological studies of the Miocene exposures in the Western Region in 1979–84 the only publications available were large-scale geological maps of the area and a brief report by petroleum geologists. Much of his work was confined to the coastal exposures where the complex lithologies of the rocks were, in general, more easily seen. Miocene exposures farther inland from the coast, both then and now, are mostly covered with wind-blown sand and eroded Miocene sediments. Since the 1980s, more detailed studies by The Natural History Museum, London/Yale University team have obtained the basic data for any future Miocene researches in the area. Mention must be made of the fact that, although surface exposures reveal only about 60 metres of Miocene sediments, the succession in the Western Region falls into a period of Miocene time from about 19 to 6 million years (Ma) ago. The age of the vertebrate-bearing sediments of the lower part of the Baynunah Formation is believed to be between 8 and 6 Ma.

The first step in any geological study of a previously unknown area is to define the succession and provide names for stratigraphical units if they cannot be easily linked with other regional units. Peter J. Whybrow, Peter F. Friend, Peter W. Ditchfield, and Charlie S. Bristow (Chapter 4) present geological information supporting a new stratigraphic name, Shuwaihat Formation, for sediments predominantly of aeolian origin that overlie the marine Dam Formation, previously described from the Kingdom of Saudi Arabia. The previously defined name, Baynunah Formation, is now restricted to the overlying, predominantly fluvial, unit that contains the important nonmarine fossil fauna described in Part III. Sedimentary logs measured at 14 localities are presented.

Peter F. Friend (Chapter 5) describes the sediments deposited by the river system that once flowed through the area known today as the Baynunah. The lower Baynunah Formation consists mainly of fine-grained sands, but also contains numerous gravel beds and some distinct mud units. Its important fossil vertebrate fauna, along with some molluscs, abundant root marks and soils, confirms that the unit was formed predominantly by river deposition. There is no positive evidence of marine influence.

Detailed examination of unusually clean exposures at Jebel Dhanna Dalma Ferry Terminal and at Jebel Mimiyah (Al Mirfa) provides information about the rivers that deposited some of the sediment. These appear to have been variable in flow, and probably some 3–10 metres deep at time of flood. They were braided in the sense that they contained many sediment bars, and channels of varied size and form. Flow directions appear to have been generally towards the ESE. The river system may have been ancestral to the present-day Tigris–Euphrates system, but this is a speculative suggestion.

Sediments of the Shuwaihat Formation that lie beneath the river sediments of the Baynunah Formation have been studied by Charlie S. Bristow (Chapter 6). Aeolian and sabkha sediments dominate the Shuwaihat Formation, formerly part of the Baynunah Formation. Aeolian dune bedforms are well exposed and indicate that dune morphology was transverse and barchanoid. The dune sands are often interbedded with, and truncated by, sabkha sediments that can be correlated for several kilometres along strike. Two types of sabkha are recognised: small interdune sabkhas (10–100 metres wide), and more extensive playa lake or coastal sabkhas (more than 1000 metres wide). These sediments are occasionally interbedded with and overlain erosionally by fluvial sediments. Palaeocurrent measurements from the aeolian sandstones indicate that the dominant palaeowind was from north to south. Palaeocurrent directions from fluvial sediments indicate a much wider current dispersion with rivers flowing from west to east and south to north, across or opposed to the aeolian transport direction. Palaeogeographic reconstructions indicate that the Shuwaihat Formation was deposited in an intracratonic basin with a semi-enclosed drainage system where rivers debouched into an inland sabkha or playa lake. Changes in lake level that strongly affected the dune systems are attributed to climatic fluctuations during the late Miocene. The presence of reworked foraminifera in some of the aeolian sands suggests that marine conditions were present within the basin, so the overall setting may have been a low-relief coastal plain although no evidence of marine sediments in the Shuwaihat Formation has been found at outcrop.

To understand the palaeoenvironment in which sediments were deposited, it is necessary to undertake detailed analyses of the rocks formed from such sediments. Peter W. Ditchfield's collection of samples (Chapter 7) from the exposed Miocene formations from the Western Region have been analysed petrographically and geochemically to determine the degree of diagenetic alteration and to try to gain information about the depositional environments of these sediments. Petrographic results show that the Baynunah Formation has undergone only minor burial and only localised cementation. Carbonate cements that are present were formed in a meteoric environment, in some cases above any permanent water table. The underlying Shuwaihat Formation also shows little evidence of significant burial or compaction although it containes significant amounts of gypsum cements within some facies. The Dam Formation shows a more complex diagenetic history with the carbonate sediments now pervasively dolomitised. Dolomites from the Dam Formation are geochemically distinct from the rare dolomites found in the overlying units and indicate an evaporitic origin for the dolomitising fluids. Isotopic analyses of well-preserved, calcitic, ratite eggshell samples from the Baynunah Formation indicate that mixed C_3 and C_4 plant communities (wooded grassland and grassy woodland) were present during deposition of the Baynunah Formation.

Palaeomagnetic and magnetic fabric investigations have been carried out, by Ernie A. Hailwood and Peter J. Whybrow (Chapter 8), on five sections that together span most of the exposed part of the Miocene sedimentary succession in the Arabian Gulf coastal region of the Emirate of Abu Dhabi. The objectives of the study were twofold: (1) to define changes in polarity of the geomagnetic field during deposition and so contribute to stratigraphic correlations between sections; and (2) to derive a palaeomagnetic pole position for these formations to help constrain their geological ages.

The Miocene sedimentary sequence in the Baynunah region of Abu Dhabi has been divided into two separate formations, the aeolian Shuwaihat Formation below and the dominantly fluvial Baynunah Formation above. These formations are separated by a disconformity that is clearly visible in the Jebel Barakah section but whose position is less certain in the section on Shuwaihat Island and elsewhere in the region.

Magnetostratigraphic investigations reported in Chapter 8 indicate that the disconformity between the Shuwaihat and Baynunah Formations lies within a long-reverse polarity magnetozone in the Jebel Barakah section. The magnetozone is not well defined in the Shuwaihat Island section, but its inferred position is bracketed by overlying and underlying normal polarity intervals. This magnetostratigraphic correlation implies that the boundary between the Shuwaihat and Baynunah Formations must lie somewhere in the interval 6.75–16 metres above sea level on Shuwaihat Island.

A comparison of the palaeomagnetic pole positions derived from the Shuwaihat and Baynunah Formations with the Cenozoic apparent polar-wander curve for Arabia suggests a probable age of about 15 ± 3 Ma (that is, middle Miocene) for the Shuwaihat formation and about 6 ± 3 Ma (that is, late Miocene) for the Baynunah Formation. This difference in palaeomagnetically determined ages for the two units supports their proposed separation into different geological formations.

Finally in Part II (Chapter 9), Ross Peebles reports that the Miocene of Abu Dhabi can be found as scattered outcrops, mesas, buttes, and in the shallow subsurface. In the west, Upper and Middle Miocene siliciclastics of the Baynunah and Shuwaihat Formations form the mesas and jebels seen along the coast, while in the east Lower to Middle Miocene carbonates and evaporites of the Gachsaran Formation form inland buttes and small isolated outcrops and lie just metres below the surface of the sabkha. The Miocene is also exposed on many of Abu Dhabi's islands, brought to the surface by salt diapirism. Samples from selected locations across the Emirate were analysed for the stable isotope ratios of $^{87}Sr/^{86}Sr$, $\delta^{13}C$, $\delta^{18}O$, and $\delta^{34}S$ to identify an integrated depositional isotopic signature for the Miocene formations of Abu Dhabi. This signature was compared to global isotope curves to determine the chronostratigraphic position of these formations. Based on its isotopic signature, the Gachsaran Formation can be assigned to the early–middle Miocene (Burdigalian–Langhian). The isotopic signatures of the Baynunah and Shuwaihat Formations, found to be the result of diagenetic alteration, are not reflective of depositional conditions; thus they could not be used to determine chronostratigraphic position.

4 Local Stratigraphy of the Neogene Outcrops of the Coastal Area: Western Region, Emirate of Abu Dhabi, United Arab Emirates

PETER J. WHYBROW, PETER F. FRIEND, PETER W. DITCHFIELD, AND CHARLIE S. BRISTOW

The coastline of the Emirate of Abu Dhabi from As Sila in the west to Abu Dhabi city occupies part of the Western Region of the Emirate of Abu Dhabi and extends for some 300 km forming the southernmost edge of the inner Arabian Gulf (fig. 4.1). In this area, isolated, flat-topped hills (jebels) project a few tens of metres above the generally flat coastal plains. Although these hills are low (between 40 and 60 metres high), they contain enough outcrop, especially in hills bordering the present-day sea, to provide collections of Miocene faunas unique to Arabia and sedimentary successions that allow the environments of deposition to be interpreted. This chapter reviews local stratigraphic information that has been gathered recently from these outcrops.

STRATIGRAPHIC BACKGROUND

Steineke et al. (1958) formally named three formations for the Miocene of eastern Saudi Arabia. At base, the early Miocene Hadrukh Formation disconformably overlies Eocene rocks of Ypresian or Lutetian age. The Hadrukh is overlain by the middle Miocene (Burdigalian), marine, Dam Formation. The Dam is disconformably succeeded by clastics of the Hofuf Formation. Powers et al. (1966) state: "The Hofuf Formation is chiefly unfossiliferous although occasional nondiagnostic fresh-water fossils including *Lymnaea* and *Chara* occur. Since the Hofuf represents the closing unit of the Arabian Tertiary deposits, it may be either late Miocene or Pliocene." Thomas et al. (1978; see also Thomas, 1983) provided the first dating of a part of the Hofuf Formation in eastern Saudi Arabia. The carnivores, proboscideans, rhinoceros, suids, giraffids, and especially the bovids that were found at Al Jadidah could be linked to the faunas found at Fort Ternan, East Africa, and thus the age of the Hofuf fossils locality was about 14 million years (Ma). Thomas et al. (1978: 70) note that about 70 metres of Hofuf sandstones overlie the vertebrate-bearing unit, and that sediments at Barj er Rukban (type locality of the Hofuf Formation), about 20 km to the east of the Al Jadidah locality, had important lateral facies variations. Powers et al. (1966: D93) state that thickening of the continental beds "in the south is more drastic, reaching as much as 300 metres before passing under the sands of the Rub' al Khali".

In general, Hofuf continental deposition in eastern Arabia is known from observation of outcrops, in the absence of subsurface information. It commences with extraformational conglomerate overlying the marine Dam Formation, is succeeded by about 30 metres of sandy limestones and clays with minor conglomerates, and underlies 40 metres of unfossiliferous, homogeneous sandstones, whose origin may be either fluvial or aeolian. The top of the Hofuf is the upper limit of the exposures in the Hofuf (Saudi Arabia) area, "commonly an old surface showing strong calcium carbonate enrichment of duricrust-caliche type" (Steineke et al., 1958).

The remaining Miocene/Pliocene unit in eastern Arabia is the Kharj Formation. No type locality exists (Powers et al., 1966: D92) for this unit. Kharj deposits are found around Wadi Sahba, Saudi Arabia, which has been considered to follow the

 ISBN 0-300-07183-3

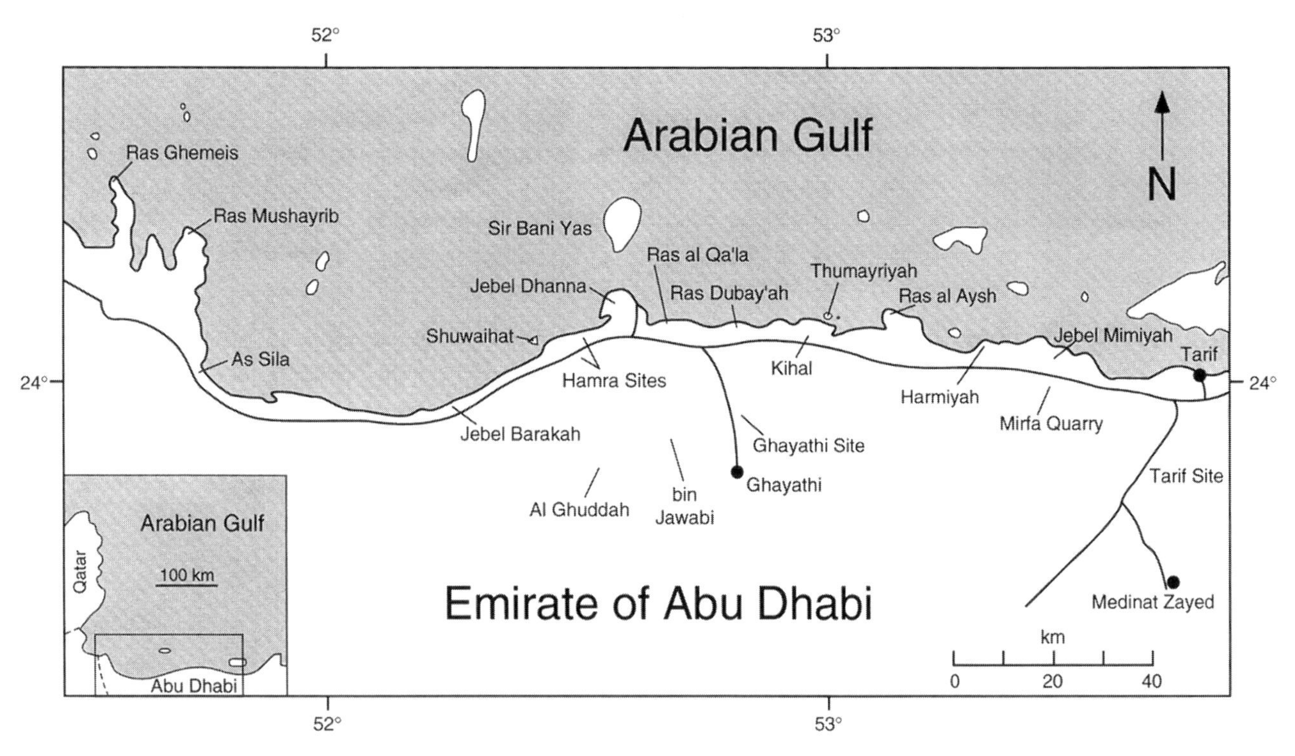

Figure 4.1. Locality map of the coastal region of the western part of the Emirate of Abu Dhabi.

line of a structural trough and represent the path of a ?Miocene river. Philby (1933: 79) states that

> The Sahba is one of the great and long-dead rivers of ancient Arabia, having a total length of more than 500 miles from its head . . . in the central highlands of Najd . . . to its mouth in the Persian Gulf.

Philby's map (1933) shows Wadi Sahba entering the Arabian Gulf at As Sila. Powers et al. (1966: D98) note that some patches of "late Tertiary" sediments are associated with the Kharj Formation near Wadi Sahba and

> these probable links strongly suggest that the channel gravels are remnants of a group of Tertiary rivers which brought down the great gravel flood incorporated in the lower part of the Hofuf Formation. The relationship of the channel gravels to freshwater deposits of the Kharj Formation is unknown even though they are in contact at one locality . . . It is tempting to suggest that the channel gravels may be the result of streams that discharged overflow from lakes in which freshwater sediments were being deposited, but this suggestion is pure speculation.

Stratigraphy of the Western Region

Detailed logs have been measured to represent the bed-by-bed successions in all adequately exposed localities so far discovered. These localities are arranged in order from westernmost to easternmost in table 4.1. Figure 4.1 shows the positions of the localities, and the logs themselves are drawn up in detail in figures 4.2–4.4.

Dam Formation

In the North Sila outcrop (Sila-N, fig. 4.2) north of the main coastal road, we have discovered an

Table 4.1. Localities in the Emirate of Abu Dhabi at which logs have been measured, arranged west to east; code letters and approximate longitude and latitude are given for each locality

Locality	Log no.	Coordinates
As Sila (three logs)	SIL-A	51° 46′ 06.5″ E, 24° 00′ 32.8″ N
	SIL-N	51° 46′ 06.5″ E, 24° 03′ 26.6″ N
	SIL-D	51° 45′ 31.1″ E, 24° 00′ 00.0″ N
Jebel Barakah	JBA	52° 19′ 30.4″ E, 24° 00′ 32.8″ N
Shuwaihat (three logs)	SHU-W	52° 25′ 25.1″ E, 24° 06′ 30.9″ N
	SHU-E	52° 26′ 18.3″ E, 24° 05′ 58.3″ N
	SHU-Q	52° 26′ 00.6″ E, 24° 05′ 59.0″ N
Hamra (two logs)	HAM-N	52° 31′ 37.5″ E, 24° 05′ 42.1″ N
	HAM-S	52° 31′ 55.3″ E, 24° 04′ 36.9″ N
Jebel Dhanna	JDH	52° 37′ 32.2″ E, 24° 10′ 55.5″ N
Giyathi	GIY	52° 47′ 17.4″ E, 23° 58′ 46.7″ N
Ras al Qa'la	RAQ	52° 55′ 31.3″ E, 24° 08′ 57.5″ N
Kihal	KIH	52° 58′ 57.9″ E, 24° 06′ 31.0″ N
Thumariyah	THU	53° 00′ 53.2″ E, 24° 06′ 31.0″ N
Jebel Mimiyah (Al Mirfa)	JMI	53° 29′ 33.4″ E, 24° 04′ 36.9″ N

outcrop of the Dam Formation, previously known from Saudi Arabia (Powers at al., 1966; Powers, 1968) and Qatar (Cavelier, 1975). In these areas the Dam Formation is regarded as middle Miocene in age (Adams et al., 1983). The As Sila outcrop consists of white dolomitic claystones containing bivalves, with well-displayed bioturbation, bored hardgrounds, and gypsum nodules. Similar outcrops have been found further south in Sabkha Matti by Bristow (1999—Chapter 6).

Shuwaihat Formation

This stratigraphic term is herein used and defined for the first time to distinguish a body of fine-grained sandstones with well-defined, relatively thin, mudstones. These sediments form the lowest 7.5 metres of the succession at Jebel Barakah (fig. 4.2), the lowest 6.7 metres exposed in the western part of Shuwaihat (Shuwaihat-W, fig. 4.2), and the lowest 6.1 metres exposed in the southeastern part of Shuwaihat (Shuwaihat-E, fig. 4.2). All the outcrops measured at As Sila (fig. 4.2) are regarded as part of the Shuwaihat Formation, with the exception of the part of the outcrop mentioned above that we regard as Dam Formation.

We propose the name Shuwaihat Formation for these deposits because they are best exposed on the island of Shuwaihat, Jazirat Shuwaihat (and variously spelt on road signs and maps as Shouwihat, Shouwehat, and Shuwayhat), where we regard the particularly clean outcrops as the type succession for this new unit; see Chapter 6 (Bristow, 1999) for details.

The distinctive feature of these Shuwaihat Formation sediments is the presence of aeolian cross-stratification sets, along with other sets with diffuse interbedding of fine sand and muddy laminae that have been deformed into wavy and cuspate folds, and are regarded as the deposits of interdune sabkhas (plates 4.1 and 4.2, pp. 33 and 34). A more detailed discussion of the environments of deposition of these deposits is provided by Bristow (1999). Another characteristic feature of this formation is the presence of distinct grey clay beds,

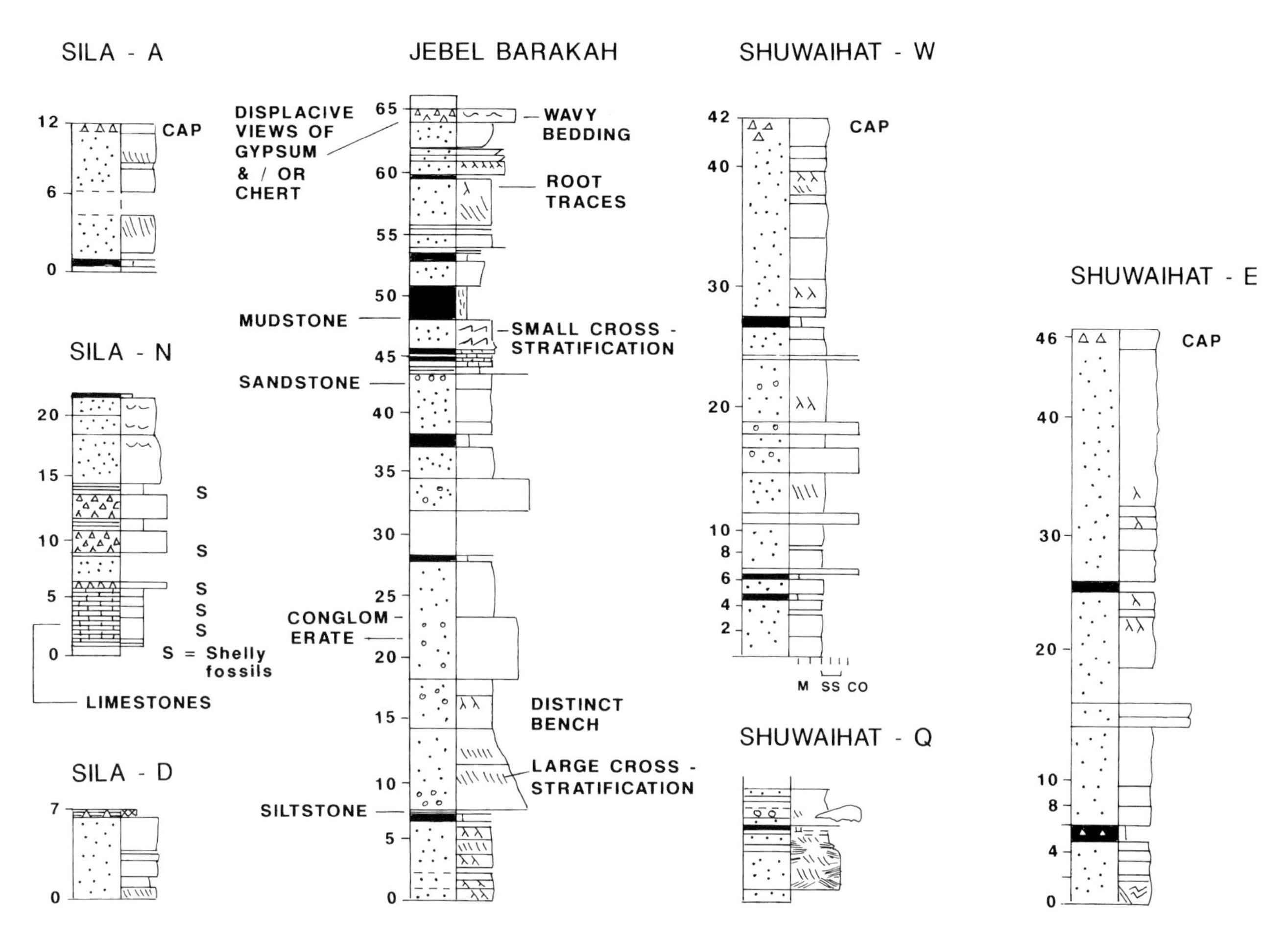

Figure 4.2. Sedimentological logs for the As Sila area, the Jebel Barakah area, Shuwaihat West (W), Shuwaihat Quarry (Q), and Shuwaihat East (E). Figure 4.1 and table 4.1 provide location information.

between 0.35 and 1.1 metres thick, and traceable for several hundreds of metres laterally.

Fluvial strata with climbing ripples, and regular clay beds on major cross-strata are present in East Shuwaihat (fig. 4.2). Rather similar fluvial climbing ripples occur at the base of the outcrops at Jebel Mimiyah (fig. 4.4), but we regard these sediments as part of the fluvial Baynunah Formation (fig. 4.5 and plate 4.3, p. 35).

Whybrow (1989) presented the first detailed outcrop log for the Miocene outcrops of Abu Dhabi. He measured this at Jebel Barakah, through the thickest part of the succession (more than 60 metres) known to outcrop in the region (plate 4.3). Our remeasurement at the same locality is presented on figure 4.2. Whybrow suggested that the lowest 7.6 metres might be separated from the overlying part of the succession by a disconformity, basing this belief largely on the locally greater degree of cementation of the underlying sediments, relative to the overlying sediments. We present a sketch by Bristow of part of the Jebel Barakah coastal outcrop (fig. 4.5) that shows Whybrow's disconformity as an erosional surface with a relief of at least 6 metres that was subsequently covered by the overlying sediment (plate 4.3). The great extent

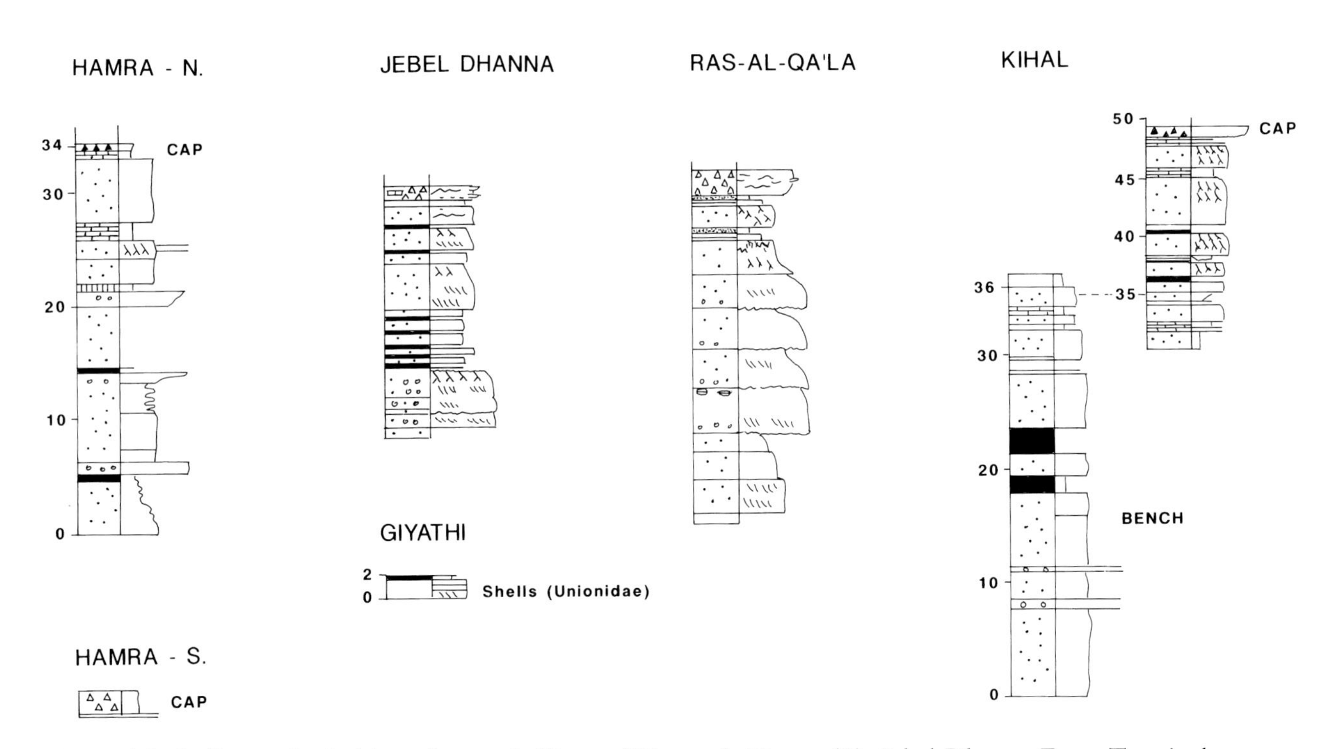

Figure 4.3. Sedimentological logs for north Hamra (N), south Hamra (S), Jebel Dhanna Ferry Terminal, Giyathi, Ras al Qa'la, and Kihal. Figure 4.1 and table 4.1 provide location information.

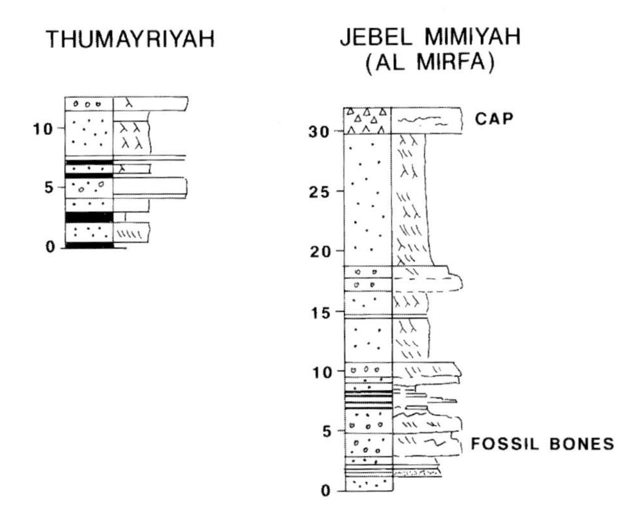

Figure 4.4. Sedimentological log for Thumayriyah and Jebel Mimiyah (Al Mirfa). Figure 4.1 and table 4.1 provide location information.

of this erosional surface, and the distinctive features of the sediments below, support our decision to assign the underlying sediments to our new unit, the Shuwaihat Formation.

Part of the Shuwaihat Formation may be coeval with Miocene clastic deposits—the Hofuf Formation, described from Saudi Arabia by Powers et al. (1966)—that also overlie the Dam Formation. Because no body fossils have yet been recorded from the Shuwaihat Formation, however, we have no means of verifying this conjecture.

Baynunah Formation

This name was first applied by Whybrow (1989) to the outcrops that had yielded late Miocene terrestrial and freshwater fossils in the coastal region of the United Arab Emirates. The Jebel Barakah suc-

Plate 4.1. Lowest part of the Shuwaihat Formation exposed at Shuwaihat Island.

cession was defined as the stratotype (fig. 4.5 and plate 4.3). In the previous paragraphs, we have presented the reasons why we now regard the lowest 7.5 metres of that succession to be sufficiently distinct from the overlying strata to be given a separate formational name, Shuwaihat.

The Baynunah Formation, as now defined, forms most of the successions logged at Jebel Barakah (fig. 4.2) and Shuwaihat (fig. 4.2), and all the successions to the east of these localities (table 4.1, figs 4.3 and 4.4). The Baynunah Formation consists of gravel, fine-grained sandstones, some mudstones, and some thin limestones. The Baynunah Formation differs from the Shuwaihat Formation in that there is no certain evidence for aeolian deposition, although a few isolated beds provide textural and structural suggestions that they are aeolian deposits. We find it useful to distinguish informally between the lower Baynunah Formation, which is dominated by fluvial gravels and fine-grained sandstones, and the upper Baynunah Formation where gravels do not occur, and mudstones and thin limestones are increasingly predominant. In practical terms, we refer to the lower informal division all Baynunah sediments up to, and including, the highest gravel-bearing bed. More detailed discussion of aspects of the environments of deposition of these units is provided in Chapters 5 (Friend, 1999) and 7 (Ditchfield, 1999). We stress, however, that this is an informal distinction. Given the patchy and isolated nature of many of the outcrops, it is impossible to correlate using markers of one of the small number of recurrent sediment types.

Much detailed work has been carried out on the fossils from the Baynunah Formation, and it is continuing. There is general agreement with the earlier conclusion of Whybrow (1989) that the faunas indicate a late Miocene (Vallesian/Tortonian) age between 11 and 5 Ma but probably closer to the latter, between 8 and 6 Ma.

Plate 4.2. Aeolian cross-stratification of the Shuwaihat Formation, Shuwaihat Island.

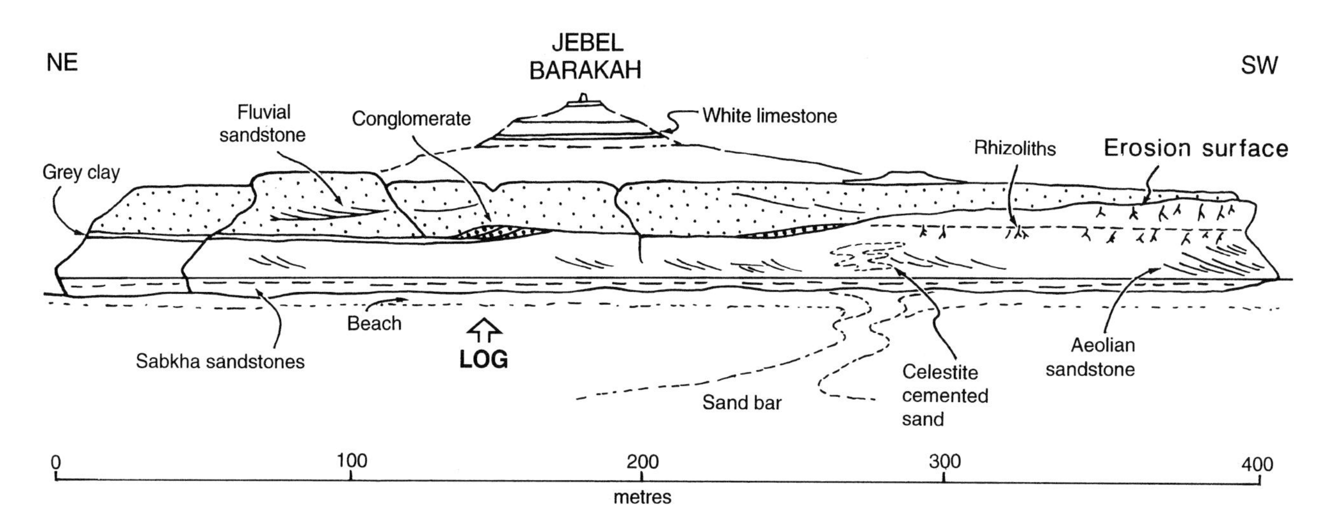

Figure 4.5. Panoramic sketch of cliff section at Jebel Barakah; see also plate 4.3.

Plate 4.3. Sea cliffs at the eastern part of Jebel Barakah; see also figure 4.5. The Shuwaihat Formation, overlain by the Baynunah Formation, lies below the top of the sandstone unit in the centre of the exposure. The highest part of the jebel, a rounded peak about 60 metres above sea level, can just be seen at right centre.

DISCUSSION

Detailed discussion of the environments of deposition represented in this stratigraphy are presented in Chapters 5–7 (Bristow, 1999; Ditchfield, 1999; Friend, 1999). We limit our discussion here, therefore, to some regional aspects of the stratigraphy, especially based on comparisons of the successions along our W–E coastal region profile (fig. 4.6).

In our study area, the Dam Formation has been discovered only at As Sila, close to the United Arab Emirates international border and to the Qatar arch that brings older stratigraphy to the surface (Cavelier, 1975).

Our new Shuwaihat Formation appears to follow a similar distribution in that it outcrops widely round As Sila in the west. In the outcrops further east, the Shuwaihat Formation is distinguished from the overlying Baynunah Formation on lithological grounds, especially the presence of aeolian stratification. In the absence of biostratigraphic evidence, and therefore in strictly lithostratigraphic terms, the Shuwaihat Formation extends eastwards as far as the succession at Ras al Qa'la (fig. 4.3), and not further east. It is possible that this distribution reflects some contemporaneous lateral facies change, but we have no means of testing this.

Many of the outcrops we have discussed are capped by a distinctive resistant unit, 1–2 metres thick, of fine sand and carbonate, full of gypsum and chert veins. This cap-rock seems similar to that described in Saudi Arabia by Steineke et al. (1958), quoted briefly above. Ditchfield (1999—Chapter 7) briefly discusses the origin of this cap-rock.

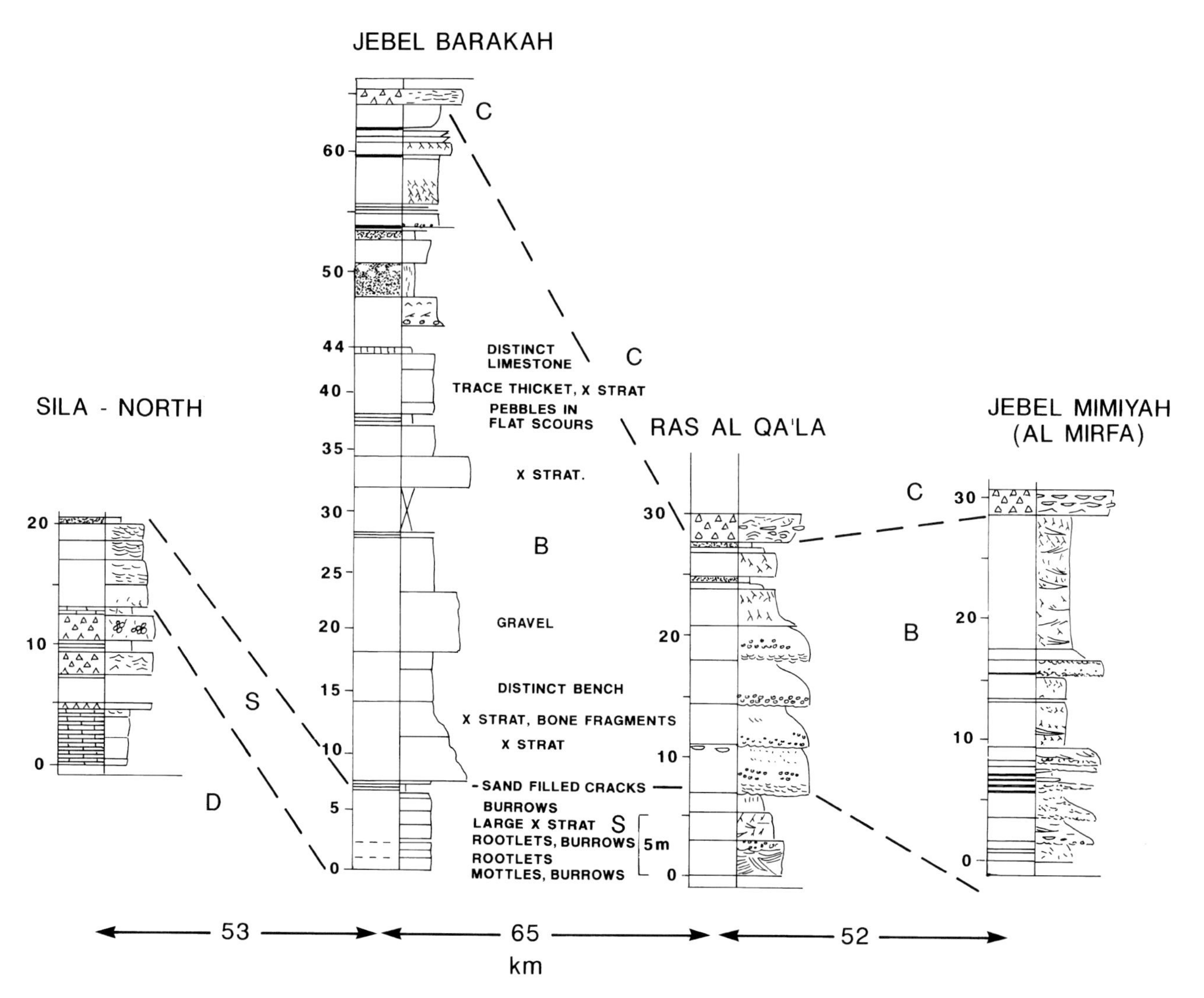

Figure 4.6. Fence diagram suggesting correlation between four of the coastal region logs, at As Sila, Jebel Barakah, Ras al Qa'la, and Jebel Mimiyah. D = Dam Formation; S = Shuwaihat Formation; B = Baynunah Formation; C = cap-rock.

Acknowledgements

We thank the Abu Dhabi Company for Onshore Oil Operations (ADCO) for their continuing grant to The Natural History Museum, London, and for their generous support for Miocene studies in the United Arab Emirates since 1991. We also thank ADCO and the Ministry of Higher Education and Scientific Research, United Arab Emirates, for their invitation to participate in the First International Conference on the Fossil Vertebrates of Arabia held in the Emirate of Abu Dhabi, March 1995. This contribution forms part of The Natural History Museum's Global Change and the Biosphere research programme.

REFERENCES

Adams, C. G., Gentry, A. W., and Whybrow, P. J. 1983. Dating the terminal Tethyan events. In *Reconstruction of Marine Environments* (ed. J. Meulenkamp). *Utrecht Micropaleontological Bulletins* 30: 273–98.

Bristow, C. S. 1999. Aeolian and sabkha sediments in the Miocene Shuwaihat Formation, Emirate of Abu Dhabi, United Arab Emirates. Chap. 6 in *Fossil Vertebrates of Arabia,* pp. 50–60 (ed. P. J. Whybrow and A. Hill). Yale University Press, New Haven.

Cavelier, C. 1975. Le Tertiare du Qatar en affleurement. In *Lexique Stratigraphique Internationale,* pp. 1–120. Centre National de la Recherche Scientifique, Paris.

Ditchfield, P. W. 1999. Diagenesis of the Baynunah, Shuwaihat, and Upper Dam Formation sediments exposed in the Western Region, Emirate of Abu Dhabi, United Arab Emirates. Chap. 7 in *Fossil Vertebrates of Arabia,* pp. 61–74 (ed. P. J. Whybrow and A. Hill). Yale University Press, New Haven.

Friend, P. F. 1999. Rivers of the Lower Baynunah Formation, Emirate of Abu Dhabi, United Arab Emirates. Chap. 5 in *Fossil Vertebrates of Arabia,* pp. 38–49 (ed. P. J. Whybrow and A. Hill). Yale University Press, New Haven.

Philby, H. St J. B. 1933. *The Empty Quarter.* Constable, London.

Powers, R. W. 1968. Arabie Saoudite. In *Lexique Stratigraphique Internationale,* p. 177. Centre National de la Recherche Scientifique, Paris.

Powers, R. W., Ramirez, L. F., Redmond, D., and Berg, E. L. 1966. Sedimentary geology of Saudi Arabia. *United States Geological Survey Professional Paper* 560D: 1–146.

Steineke, M., Hariss, T. F., Parsons, K. R., and Berg, E. L. 1958. Geology of the Western Persian Gulf Quadrangle, Kingdom of Saudi Arabia. *Miscellaneous Geologic Investigations Maps,* I-208A. United States Geological Survey, Washington D.C.

Thomas, H. 1983. Les Bovidae (Artiodactyla, Mammalia) du Miocène moyen de la formation Hofuf (Province du Hasa, Arabie Saoudite). *Palaeovertebrata* 13: 157–206.

Thomas, H., Taquet, P., Ligabue, G., and Del'Agnola, C. 1978. Découverte d'un gisement de vertébrés dans les dépôts continentaux du Miocène moyen du Has (Arabie Saoudite). *Compte Rendu Sommaire des Séances de la Société Géologique de France, Paris* 1978: 69–72.

Whybrow, P. J. 1989. New stratotype; the Baynunah Formation (Late Miocene), United Arab Emirates: Lithology and palaeontology. *Newsletters on Stratigraphy* 21: 1–9.

Plate 5.1. A typical example of rhizoconcretions—a thicket of root traces—in the Baynunah Formation.

Rivers of the Lower Baynunah Formation, Emirate of Abu Dhabi, United Arab Emirates

5

PETER F. FRIEND

In Chapter 4, Whybrow et al. (1999) overview the sediments that outcrop in various low, flat-topped hills in the coastal, Western Region of the Emirate of Abu Dhabi. The object of this chapter is to discuss evidence for the nature of the rivers that deposited the lower part of the Baynunah Formation.

The Baynunah Formation of the Emirate of Abu Dhabi was first defined by Whybrow (1989). This definition has now been modified (Whybrow et al., 1999) to exclude a lower unit of sediments, predominantly of aeolian origin, that is now called the Shuwaihat Formation (see Bristow, 1999—Chapter 6—for a discussion of the aeolian processes of deposition). Whybrow et al. also describe their informal division of the Baynunah Formation into a "lower" part, dominated by gravels and cross-stratified sands, and an "upper" part, with no gravels and thin carbonate beds. This upper part is further discussed in Chapter 7 (Ditchfield, 1999).

There is clear evidence for the fluvial origin of the gravels and cross-stratified sands of the lower Baynunah Formation. Some detailed examples of local evidence from the outcrops will be described later in this chapter, but here, at the outset, three major lines of evidence for a fluvial origin will be outlined: (1) the extensive faunal list (Whybrow and Clements, 1999—Chapter 23) strongly points to a generally terrestrial and freshwater environment: there is a total lack of marine body-fossils. (2) The gravels, which locally contain fossil bones, must have been transported by aqueous flows with speeds of tens of centimetres per second, and their intimate association with cross-stratified pebbly sands and sands, in lenticular, scour-filling units, are highly characteristic of deposition in river channels. (These features—grain size of the gravels and the structure—contrast clearly with those of the aeolian deposits of the underlying Shuwaihat Formation.) (3) Pedogenic horizons, with and without dense "thickets" of root traces, are clear indicators of periodic terrestrial exposure.

Grain Size of the Deposits

Detailed sedimentary logs measured at all the main outcrops of the Baynunah Formation are reproduced in our review of local stratigraphy (Whybrow et al., 1999—Chapter 4). The 14 logs represent a total of about 400 metres of stratigraphic succession, of which some 120 metres are assigned to the gravel-bearing, lower Baynunah Formation. The proportion of gravel beds varies between about 10 and 60% in the intervals that we have assigned to the lower Baynunah; mudrock intervals make up no more than 10% at any locality; the rest (between about 45 and 85%) is composed of a rather uniform fine sand. Indeed, because the fine sand dominates the matrix of the gravels as well as forming most of the sand beds, it makes up an overwhelming proportion of the lower Baynunah. Fine sand is also predominant in the upper Baynunah Formation at most localities.

Local Sedimentary Structures

Study in present-day natural environments and in the laboratory of active systems in which sand is being transported and deposited has done much

 ISBN 0-300-07183-3

recently to advance understanding of the structures formed in sand (Collinson and Thompson, 1982; Tucker, 1982). During transport and deposition, the form of the sediment surface is rarely flat and often develops regular geometrical features, called bedforms. Ripples are one class of bedforms. As transport and deposition continue, the bedforms evolve and migrate, leaving patterns of stratification within the sediment that may be very characteristic. Figure 5.1 illustrates some common simple and

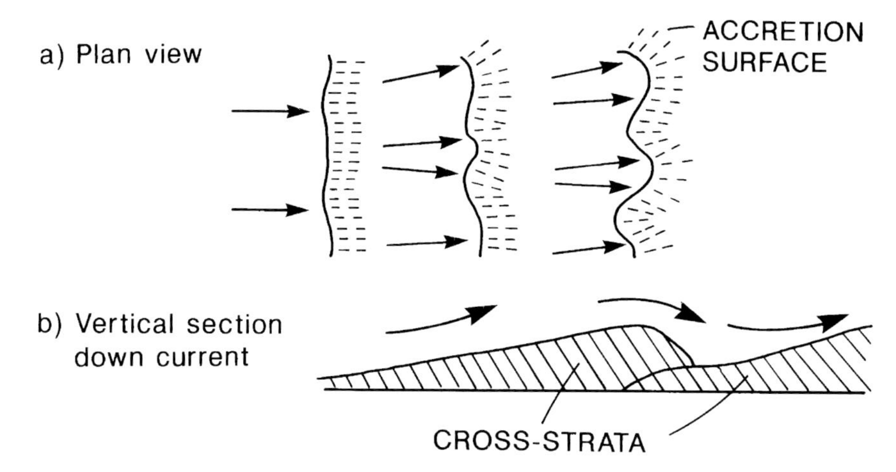

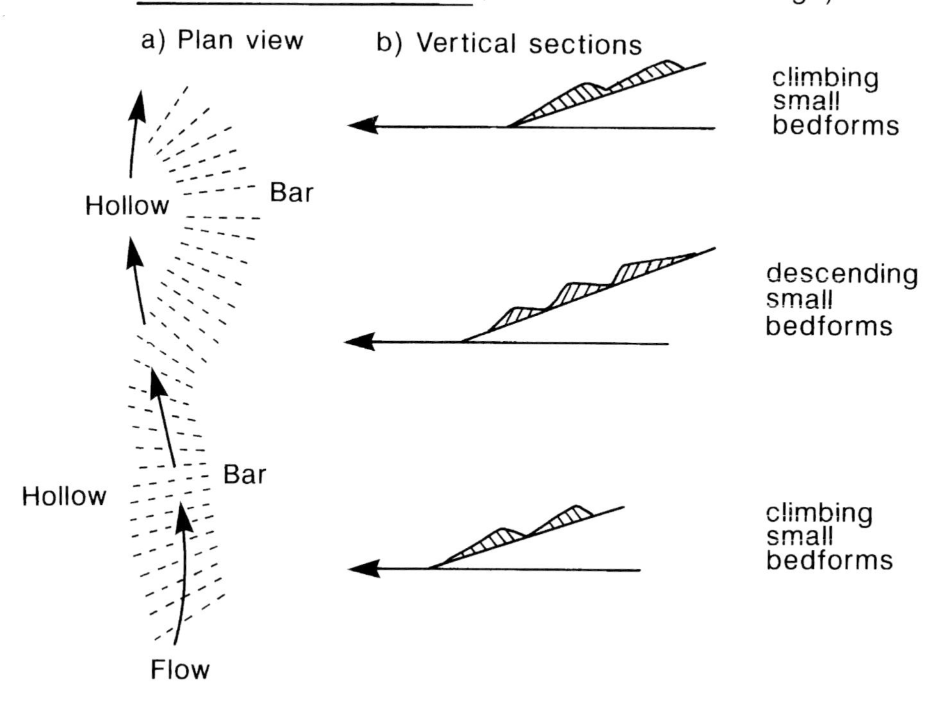

Figure 5.1. Plan views and vertical sections to illustrate some geometrical relationships between flow and bedform geometries and cross-stratification, for simple and compound bedforms.

compound situations, and demonstrates the complexity of the three-dimensional geometries that may be involved. Jackson (1975) proposed a classification of bedforms into microforms (for example, small ripples), mesoforms (for example, megaripples), and macroforms ("geomorphological" scale bedforms, such as channel bars). This classification is particularly useful in fieldwork on the Baynunah. Sometimes the three-dimensional relationships of bedforms of different scales can be complex, as hinted in figure 5.1. Some idea of the complexities involved can be gained from the fact that computer simulation has been used to model the stratification produced by different migrations of various surface geometries (Rubin, 1987).

Unusually clear, man-made, exposures of fluvial structures in the Lower Baynunah Formation have been studied at two localities, and these are discussed next.

Jebel Dhanna Ferry Terminal

Figure 5.2 illustrates exposures of fine sand, pebbly sand, and gravel at the Jebel Dhanna Ferry Terminal locality. At the base of the exposure, a unit of pebbly sand and gravels (2 metres thick) consists mainly of one set of cross-strata that prograded uniformly to the northeast over a distance of at least 80 metres. The upper part of this sheet has been scoured to produce numerous discrete hollows, which have then been filled by cross-stratified pebbly sand that also involved transport in the same northeasterly direction. Within some of the cross-strata in the main set, smaller-scale cross-stratification indicates minor episodes of southeasterly migration of mesoforms over the main northeast dipping surfaces. This situation indicates that the sloping surfaces of a sand bar prograded generally to the northeast in river water that must have been

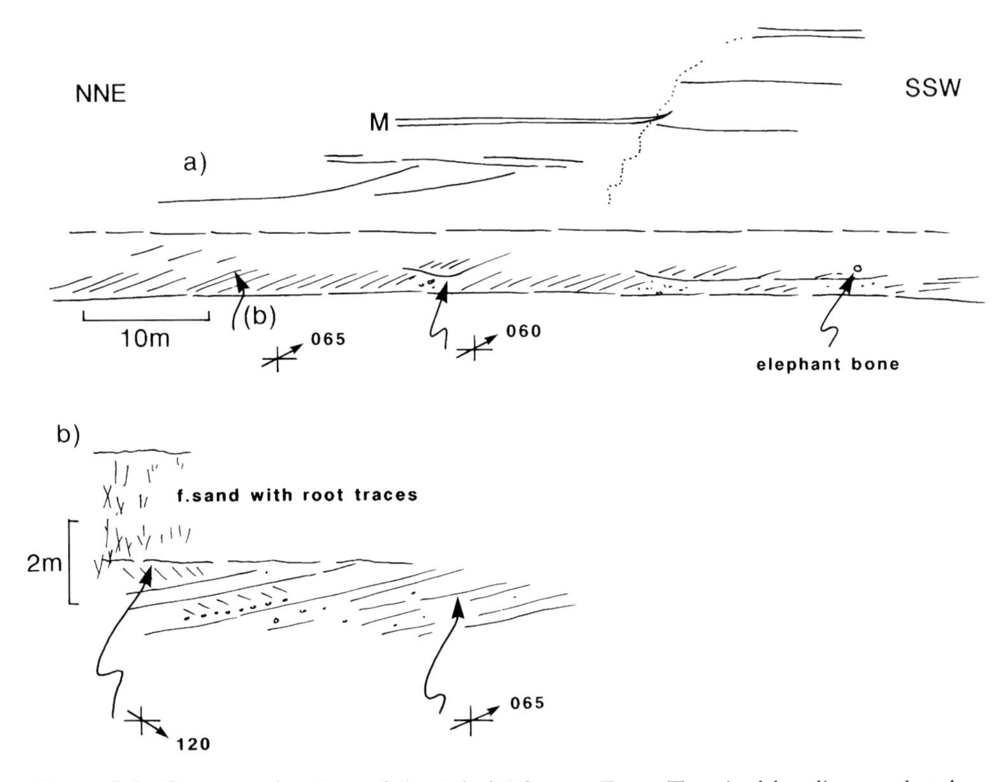

Figure 5.2. Outcrop sketches of the Jebel Dhanna Ferry Terminal locality: *a*, sketch of section; *b*, detail of section as indicated in *a*. M = mudlayer.

more than 2 metres deep, and could have been as much as 8–10 metres deep. At times local sand and small-bedform transport was to the southeast, along the "strike" of the major bar (macroform) surfaces.

This bar deposit was covered by a more uniform sheet of fine sand that shows little clear internal stratification, perhaps because much of it has been penetrated by numerous root traces. The latter provides evidence that the sand sheet was thoroughly colonised by plants during a relatively dry episode.

Above this sheet, the outcrop shows several extensive sloping surfaces dividing the largely fine-sand sediment into broadly lenticular units, 4–6 metres thick. These units appear to represent episodes of channel cutting and river-bar sedimentation, and their thickness provides an indication of the probable depths and heights of the channels and macroform bars, respectively. One distinct mud layer is visible, representing an episode of still water when fine-grained suspended sediment settled out across the channel floor. This layer was subsequently truncated by erosion before further sand was deposited across the truncation.

Jebel Mimiyah

Figure 5.3 contains some outcrop sketches made from the remarkably clean exposures created by excavation by earth-moving machinery, round the bases of the two peaks. These beautiful exposures contain exceptionally clear cross-stratification featuring the full hierarchy of macro-, meso- and microform scales.

Extensive surfaces, sometimes nearly flat but in other cases dipping at about 15° to the east, have been picked out by mud layers. These mud layers, which are 10 cm thick, must represent episodes of slack water, during which suspended mud particles were able to settle out. At one locality (fig. 5.3a), the mud layer was deposited on a bed of fine sand on which wave ripples had developed, probably due to wind shear over the otherwise still water. The layers provide evidence for the occurrence of isolated ponds in hollows between the river bars, presumably in low-stage periods between the main, bar-moving, flood episodes.

The mud layers are distinct features of a wedge, largely of fine sand, that occupies the lower part of one of the exposures (fig. 5.3c). The wedge represents the build-up of a sediment bar by a series of episodes, some involving progradation of bedforms 50 cm high to the southeast, others involving progradation of microforms 2–3 cm high to the northeast (fig. 5.3b). Each produced a distinct layer a few tens of centimetres thick, with a gradual increase in the 2–3 metre-high sloping surface of the bar, which was eventually draped by two of the mud layers already described.

Above and below this fine-sand wedge are units of more pebbly sand and gravel, containing cross-stratification sets ranging from 1 to 2.5 metres thick, uniformly representing progradation towards the southeast (fig. 5.3c). The upper sediment here contains transported bone fragments.

The pebbly sand and gravel material appears to have been deposited in river channels, probably with flood depths of between 2 and 10 metres, and with flow directions towards the southeast. Episodes of channel cutting and sediment bar deposition have left sloping surfaces that indicate relief of several metres for the channels and bars. All the sedimentation has been highly episodic and variable in the size and migration direction of the smaller bedforms. Some episodes involved ponding of standing water in periods when the local channel form was shielded from through-flow. Under these conditions, mud layers accumulated, and under even less-active conditions vegetation colonised some of the sediment surfaces.

Figure 5.4 represents another of the artificially cleaned surfaces at the base of the eastern peak at Jebel Mimiyah. A lower, fine sand interval provides evidence of episodic sedimentation when small bedforms migrated to the northwest, an exceptional direction at this locality. Vegetation developed periodically between some of these episodes. Approximately 3 metres of this earlier material was then scoured out to produce a clear cut-bank geometry, with blocks of the bank material incorporated in the overlying sediment. This sediment accumulated against the bank as a macroform, at least 4 metres high, and deposited largely by climbing ripple

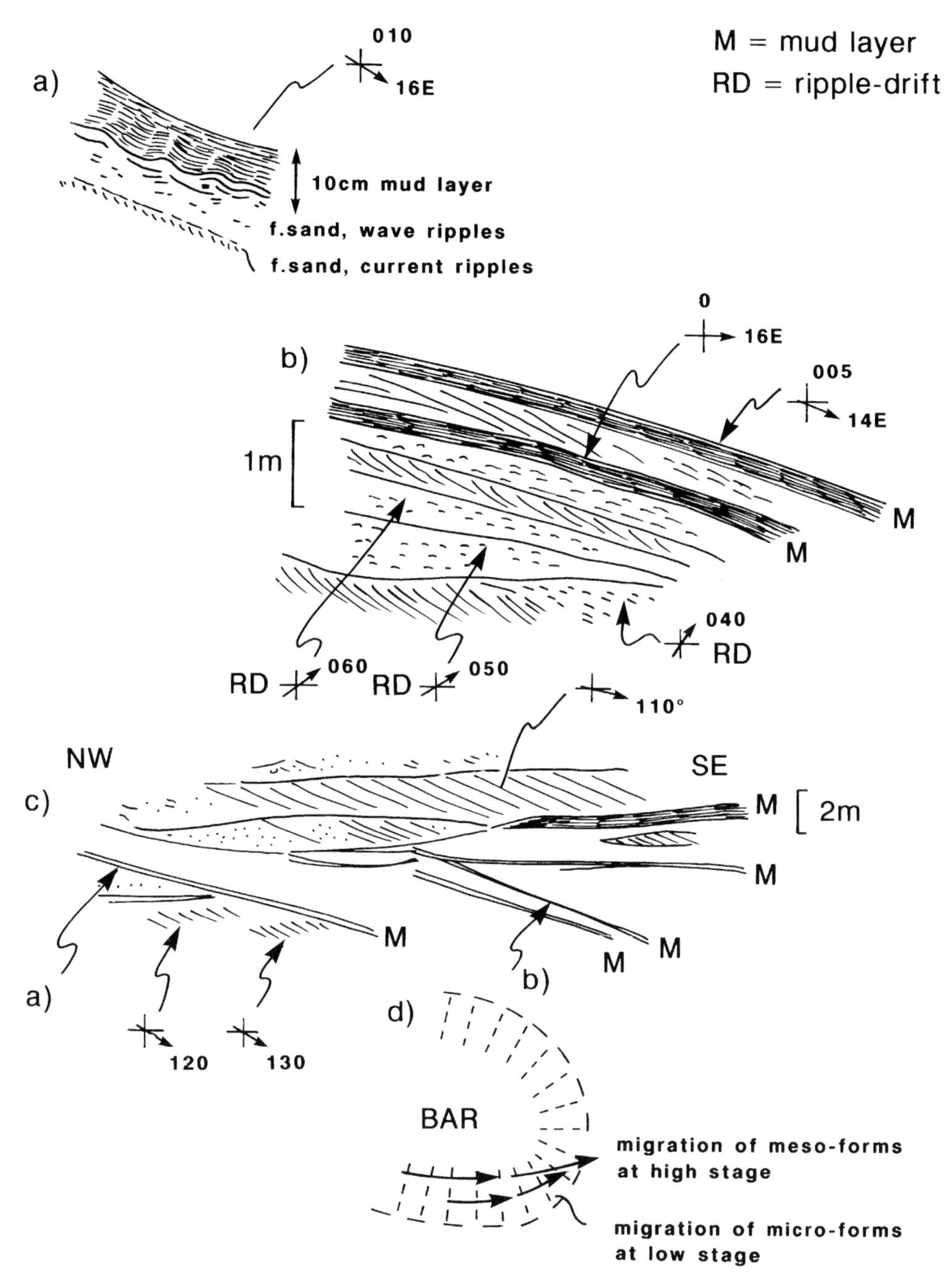

Figure 5.3. Outcrop sketches of the southern side of the base of the eastern peak at Jebel Mimiyah (Al Mirfa, "Twin Peaks"): *a, b,* detail of northwestern section as indicated in *c; c,* sketch of section; *d,* interpretation of *a, b,* and *c.*

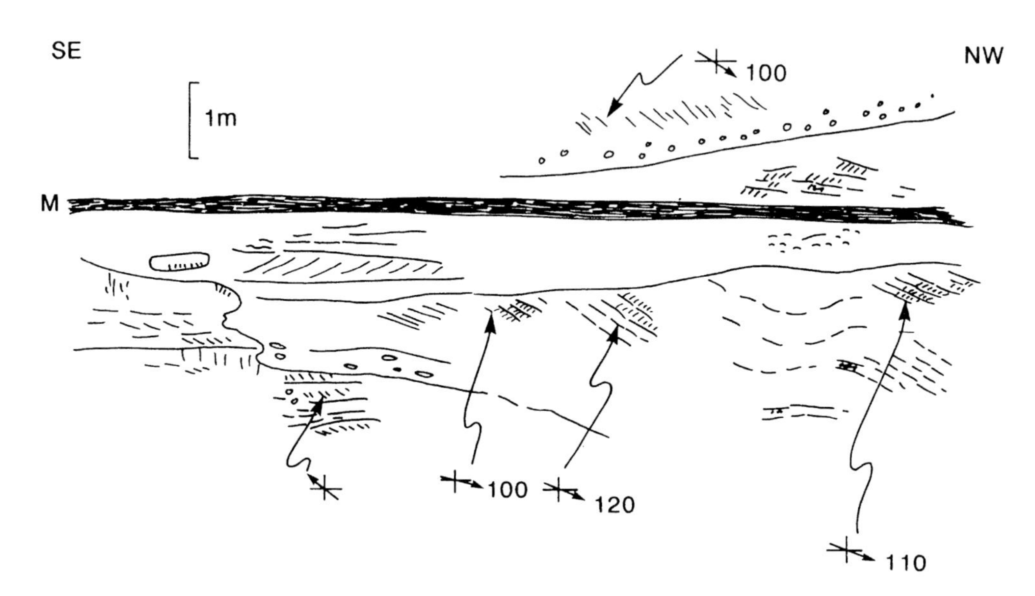

Figure 5.4. Outcrop sketches of the northeastern side of the base of the eastern peak at Jebel Mimiyah (Al Mirfa, "Twin Peaks"). M = mudlayer.

microforms that were generally migrating towards the east.

Subsequent sedimentation involved further episodes of sand accumulation, an episode of slack water, in which a mud layer was deposited, and further climbing ripple sedimentation, before an erosional episode preceded the deposition of pebbly sand and gravel, similar to the material that forms the top of the exposure of figure 5.3 (fig. 5.4).

Biological Evidence

The faunas discovered in the Baynunah Formation come overwhelmingly from the gravels and sands of the lower part of the formation. The terrestrial and freshwater nature of the organisms has already been used in this chapter as evidence that especially the gravels were deposited by rivers, rather than marine, high-energy systems. The object of this section of the chapter is to review evidence from the fossil material that may have implications for the more detailed interpretation of the rivers.

Three genera of fossil fish have been identified (Forey and Young, 1999—Chapter 12), and all appear to be characteristic of freshwater habitats. In the case of one of the genera, *Clarias,* Forey and Young make comparisons with present-day habitats where the genus occurs. Although the habitats vary, the fish appear to have a preference for the periphery of lakes, slow-moving rivers, and swamp environments. They are very hardy, several species being known to be able to withstand limited drought periods (P. L. Forey, personal communication). Two species of unionoid bivalves have been studied by Jeffery (1999—Chapter 10), who states, "The Baynunah unionoids are unornamented, fairly narrow, and more-or-less attenuated, which suggests that they were adapted to life in moderately fast-moving streams or rivers rather than lakes or slow-moving rivers."

In Chapter 24, Andrews (1999) reports the results of the careful taphonomic work that accompanied the excavation of the remarkable, though disarticulated and incomplete, elephantid skeleton at Shuwaihat. All the bones were found on a pebbly gravel layer, and some of the pebbles were banked up against the sides of some of the larger bones. The distribution of the bones, and their relationship to the gravel, seem to imply limited dispersal of the bones during the same episode of river flooding that caused the final transport of the gravel. But it may be that this flood episode was itself the last of several, during which the incomplete carcass of the elephant was transported even though some of the major components of the carcass, such as the head or hind limbs or the two parts together, may not have been transported far in each episode. Multiple cycles of local transport and burial appear to provide a powerful mechanism for producing the rather selective, and well-preserved associations from different components that have been found.

Thickets of root traces, or rhizoconcretions, occur at many levels in the lower Baynunah (plate 5.1, p. 38). They have been described by Glennie and Evamy (1968) from the Quaternary of eastern Arabia and from the Jebel Barakah Baynunah Formation, and were called "dikaka" where they occur in aeolian sandstones. These traces provide clear evidence of the episodic growth of vegetation that must have covered surfaces of the recently deposited sediment, and imply that the sediment was relatively well drained at the time of the growth.

Overbank Sedimentation and Pedogenesis

Elsewhere, many stratigraphic formations of fluvial origin consist of distinct alternations of metres-thick sheets of sandstone, deposited in the channel belts of rivers, and metres-thick sheets of mudstone. These mudstones are generally thought to have been deposited at times of river flood, in the overbank areas that exist on many river floodplains, some distance from the active channel belts.

Most Baynunah localities contain very small amounts of mudstone. In many cases, the only mudstones present are thin sheets, some 10–30 cm thick. In the detailed study of localities described above, at Jebel Dhanna (fig. 5.2) and Jebel Mimiyah (figs 5.3 and 5.4), these thin sheets appear to have formed as minor episodes of still-water sedimentation in the general evolution of the channel belt. In no sense, therefore, can they be regarded as overbank sediments.

At, however, a few localities—Jebel Barakah, Shuwaihat, Hamra, and Kihal (Whybrow et al., 1999—Chapter 4)—intervals of mudstone, 1–5 metres thick, may represent overbank deposition. Kingston (1999—Chapter 25) has identified and analysed palaeosols in some of these mud intervals, and reports carbon stable isotopes from the soil carbonates that suggest the vegetation cover varied from wooded grassland to grassy woodland.

Most of the pebble material in the Baynunah gravels is carbonate-cemented intraformational sediment that must have been derived from the fluvial erosion of earlier deposits in which pedogenic carbonate had grown. This suggests that overbank sediment must have been deposited more widely than indicated by its present occurrence, and was then generally reworked by the channel processes to concentrate the gravel material.

Local Flow Directions

Local flow directions have been determined by measurement of the migrations of bedforms using their stratification geometries. Figure 5.5 presents the measurements (83) made in the lower Baynunah Formation by Peter Ditchfield and Peter Friend. Most of these measurements come from mesoform structures (Jackson, 1975), and they therefore record episodes of flow that must have lasted for tens of minutes or more. At Jebel Mimiyah (fig. 5.5, locality 5), the pattern is rather different because some 24 of the 30 directions measured come from meso- and macroform structures—the sloping margins of scour hollows or channel bars, which are features of rather longer duration. At Jebel Mimiyah microform migration representing just a few minutes of migration is also present, and shows much greater variability. Because of the variety of different structures measured, there is some doubt about the meaning of the measurements, especially the relationship between the structures measured at Jebel Mimiyah, and those further to the west.

In spite of these doubts, figure 5.5 clearly suggests that local flow directions in the lower Bay-

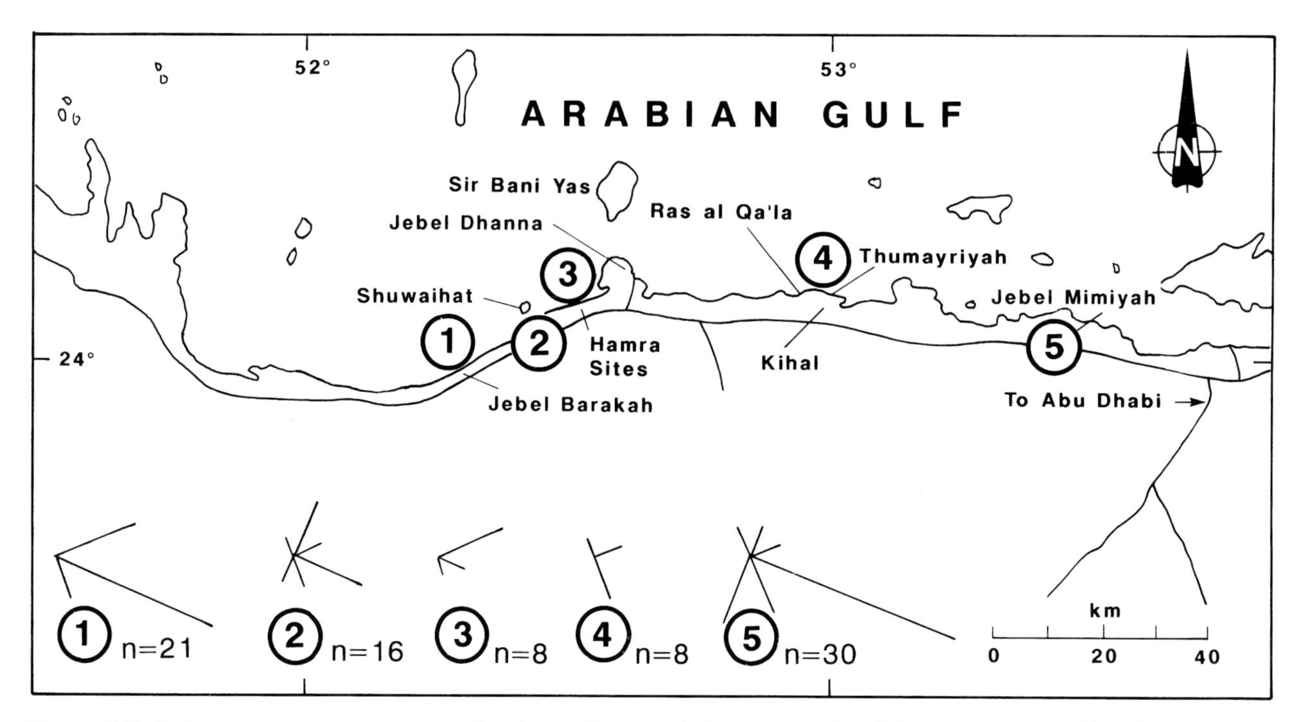

Figure 5.5. Palaeocurrent roses measured at lower Baynunah Formation localities, or groups of localities (as listed by Whybrow et al., 1999—Chapter 4): *1,* Jebel Barakah; *2,* Shuwaihat; *3,* Hamra and Jebel Dhanna; *4,* Kihal, Ras al Qa'la, and Thumayriyah; *5,* Jebel Mimiyah.

nunah deposits tend to average towards the ESE; WNW directions are notably absent. By contrast, aeolian directions from the Shuwaihat Formation are generally northerly.

Regional Patterns

Can wider considerations of a much larger region tell us anything about the Baynunah rivers? Comparison with outcrops outside the Baynunah area requires age determination and time correlation, so that palaeogeographies can be established and compared through time.

The best indicators of the age of the Baynunah Formation are the faunas, especially the vertebrate fossils. Reviews of the fossil material in this volume (Whybrow and Hill, 1999) confirm earlier estimates that the faunas probably lived and died between 8 and 6 million years (Ma) ago, during the Vallesian and/or Turolian, or during the late Tortonian to early Messinian of the late Miocene. Palaeomagnetic work by Hailwood and Whybrow (1999—Chapter 8) indicates that the sedimentation extended over a long-enough period to capture both normal and reverse polarity episodes, but there is no precise correlation with the global magnetic reversal sequence.

Recent general stratigraphic reviews (Stoneley, 1990; Beydoun et al., 1992; Hooper et al., 1994) confirm earlier syntheses (Koop and Stoneley, 1982; Shah and Quennell, 1980) in representing the Western Region late Miocene as the southern edge of a major Zagros foreland basin that trended NW–SE, with the thickest sedimentation in Miocene times, in a zone roughly coincident with the southwest edge of the present Zagros Mountains. In this basin, there was a general change during the middle to late Miocene from marine carbonates and evaporites (the Dam, Lower and Middle Fars, Asmari, and Gachsaran units) to the clastic nonmarine Hofuf unit (Whybrow et al., 1999—Chapter 4), with which part of the Baynunah might be correlated. A late phase of compressional deformation of the Zagros fold and thrust belt then created an unconformity in much of the basin, above which the nonmarine Bakhtiari conglomerates of Pliocene and Pleistocene age were formed.

From the point of view of specifying the Baynunah rivers, it would be good to be able to localise the position of the contemporaneous coastline. There is, however, insufficient age and environmental information to trace the change from marine to nonmarine sedimentation across the region. It is also impossible to establish the location and orientation of the coastline during this change.

The clastic sediments of the Baynunah Formation are siliceous, in contrast with much of the coastal sediment of the present-day Arabian Gulf, which is dominantly carbonate. No detailed petrological work is available on the provenance of the siliciclastic material. It may have been derived by rivers from southwestern Arabia or it may have been derived from source areas in Iraq, Syria, Turkey, and Iran, by rivers that were ancestral to the present Tigris–Euphrates system. This last possibility is presented as a working hypothesis, with the caution that it is entirely speculative (fig. 5.6).

Figure 5.6. Outline map of the present coastline of the Arabian continent, and a speculative location for the lower Baynunah river system.

OVERVIEW OF THE LOWER BAYNUNAH RIVERS

1. The *deposits* of the rivers were predominantly fine-grained sand, with smaller amounts of gravel (of pedogenic carbonate clasts, and bones) and mud. The load of the rivers was probably largely fine sand also, but the presence of the pedogenic clasts implies that larger volumes of overbank muds were temporarily deposited and then eroded and transferred downstream out of the Baynunah area.
2. River *morphology* consisted of channels 2–10 metres deep, with bars separated with scour hollows, generally 2–5 metres in relief. Channels were probably tens of metres wide, up to 100 metres, but probably occurred in belts separated by bars and vegetated islands. The channel belts may locally have been wide enough to compare with the "channel country" of Central Australia.
3. The *flood pattern* of the rivers was very variable. Beds vary in internal structure and direction of transport, suggesting that floods were not only variable but relatively short-lived. Erosional evidence is common. Scour hollows and reaches of channels became isolated and water was sometimes ponded in them.
4. The *regional river system* is presumed to have been a low-gradient one, like the present topography of the area. The flood energy responsible for the transport and erosion of the sediment must have been caused by episodes of high water discharge. It is speculated that the Baynunah formed in part of an ancestral Tigris–Euphrates river system.

ACKNOWLEDGEMENTS

I thank the Abu Dhabi Company for Onshore Oil Operations (ADCO) and the Ministry of Higher Education and Scientific Research, United Arab Emirates, for their invitation to participate in the First International Conference on the Fossil Vertebrates of Arabia held in the Emirate of Abu Dhabi, March 1995.

REFERENCES

Andrews, P. 1999. Taphonomy of the Shuwaihat proboscidean, late Miocene, Emirate of Abu Dhabi, United Arab Emirates. Chap. 24 in *Fossil Vertebrates of Arabia,* pp. 338–53 (ed. P. J. Whybrow and A. Hill). Yale University Press, New Haven.

Beydoun, Z., Hughes-Clarke, M. W., and Stoneley, R. 1992. Petroleum in the Zagros basin: A late Tertiary foreland basin overprinted onto the outer edge of a vast hydrocarbon-rich Paleozoic–Mesozoic passive margin shelf. *American Association of Petroleum Geologists Memoir* 55: 309–39.

Bristow, C. S. 1999. Aeolian and sabkha sediments in the Miocene Shuwaihat Formation, Emirate of Abu Dhabi, United Arab Emirates. Chap. 6 in *Fossil Vertebrates of Arabia,* pp. 50–60 (ed. P. J. Whybrow and A. Hill). Yale University Press, New Haven.

Collinson, J. D., and Thompson, D. B. 1982. *Sedimentary Structures.* Allen & Unwin, London.

Ditchfield, P. W. 1999. Diagenesis of the Baynunah, Shuwaihat, and Upper Dam Formation sediments exposed in the Western Region, Emirate of Abu Dhabi, United Arab Emirates. Chap. 7 in *Fossil Vertebrates of Arabia,* pp. 61–74 (ed. P. J. Whybrow and A. Hill). Yale University Press, New Haven.

Forey, P. L. and Young, S. V. T. 1999. Late Miocene fishes of the Emirate of Abu Dhabi, United Arab Emirates. Chap. 12 in *Fossil Vertebrates of Arabia,* pp. 120–35 (ed. P. J. Whybrow and A. Hill). Yale University Press, New Haven.

Glennie, K. W., and Evamy, B. D. 1968. Dikaka: Plants and plant-root structures associated with aeolian sand. *Palaeogeography, Palaeoclimatology, Palaeoecology* 4: 77–87.

Hailwood, E. A., and Whybrow, P. J. 1999. Palaeomagnetic correlation and dating of the Baynunah and Shuwaihat Formations, Emirate of Abu Dhabi, United Arab Emirates. Chap. 8 in *Fossil Vertebrates of Arabia,* pp. 75–87 (ed. P. J. Whybrow and A. Hill). Yale University Press, New Haven.

Hooper, R. J., Baron, I. R., Agah, S., and Hatcher, R. D. Jr. 1994. The Cenomanian to Recent development of the Southern Tethyan margin in Iran. In *The Middle East Petroleum Geosciences,* pp. 505–15 (ed. M. I. Al-Husseini). Selected Middle East Papers from Geo '94, the Middle East Geoscience Conference, April 1994, Gulf Petrolink, Bahrain.

Jackson, R. G. 1975. Hierarchial attributes and a unifying model of bed forms composed of cohesionless material and produced by shearing flow. *Bulletin of the Geological Society of America* 86: 1523–33.

Jeffery, P. A. 1999. Late Miocene swan mussels from the Baynunah Formation, Emirate of Abu Dhabi, United Arab Emirates. Chap. 10 in *Fossil Vertebrates of Arabia,* pp. 111–15 (ed. P. J. Whybrow and A. Hill). Yale University Press, New Haven.

Kingston, J. D. 1999. Isotopes and environments of the Baynunah Formation, Emirate of Abu Dhabi, United Arab Emirates. Chap. 25 in *Fossil Vertebrates of Arabia,* pp. 354–72 (ed. P. J. Whybrow and A. Hill). Yale University Press, New Haven.

Koop, W. J., and Stoneley, R. 1982. Subsidence history of the Middle East Zagros Basin, Permian to Recent. *Philosophical Transactions of the Royal Society of London* 305: 149–68.

Rubin, D. M. 1987. Cross-bedding, bedforms and paleocurrents. *Concepts in Sedimentology and Paleontology* 1: 1–187. Society of Economic Paleontologists and Mineralogists, Tulsa.

Shah, S. M. I., and Quennell, A. M. 1980. *Stratigraphical Correlation of Turkey, Iran and Pakistan.* Overseas Development Administration, London.

Stoneley, R. 1990. The Middle East Basin: A summary overview. In *Classic Petroleum Provinces,* pp. 293–98 (ed. J. Brooks). Geological Society of London, Special Publication no. 50.

Tucker, M. E. 1982. *The Field Description of Sedimentary Rocks.* Open University Press, Milton Keynes.

Whybrow, P. J. 1989. New stratotype; the Baynunah Formation (Late Miocene), United Arab Emirates: Lithology and palaeontology. *Newsletters on Stratigraphy* 21: 1–9.

Whybrow, P. J., and Clements, D. 1999. Late Miocene Baynunah Formation, Emirate of Abu Dhabi, United Arab Emirates: Fauna, flora, and localities. Chap. 23 in *Fossil Vertebrates of Arabia,* pp. 317–33 (ed. P. J. Whybrow and A. Hill). Yale University Press, New Haven.

Whybrow, P. J., Friend, P. F., Ditchfield, P. W., and Bristow, C. S. 1999. Local stratigraphy of the Neogene outcrops of the coastal area: Western Region, Emirate of Abu Dhabi, United Arab Emirates. Chap. 4 in *Fossil Vertebrates of Arabia,* pp. 28–37 (ed. P. J. Whybrow and A. Hill). Yale University Press, New Haven.

Whybrow, P. J., and Hill, A. eds. 1999. *Fossil Vertebrates of Arabia.* Yale University Press, New Haven.

Aeolian and Sabkha Sediments in the Miocene Shuwaihat Formation, Emirate of Abu Dhabi, United Arab Emirates

CHARLIE S. BRISTOW

The Shuwaihat Formation outcrops in the Western Region of the United Arab Emirates in a series of low hills or jebels, and along the coast in low cliffs and wave-cut platforms. Talus covers many slopes, however, and weathering has often obscured primary sedimentary structures. Furthermore, the outcrops are separated by sabkhas up to 50 km wide, which has made correlation between sections difficult. The sections described here were formerly included in the Baynunah Formation (Whybrow, 1989) but are now redefined as part of the Shuwaihat Formation by Whybrow et al. (1999—Chapter 4). From palaeomagnetic studies by Hailwood (Hailwood and Whybrow, 1999—Chapter 8) the interpolated age of the Shuwaihat Formation is about 15 million years (Ma).

This chapter describes part of the Shuwaihat Formation named after the island of Jazirat Shuwaihat (locally spelt Shouwihat, Shouwehat, and Shuwayhat) where the section is well exposed (Whybrow et al., 1999—Chapter 4). Outcrops at Shuwaihat, As Sila, and Jebel Barakah have been logged (fig. 6.1). The formation consists of continental clastic sediments deposited in fluvial, aeolian, and sabkha environments. The sedimentology and stratigraphy in these localities are described and interpreted. Problems associated with correlating continental facies across areas of no exposure are discussed and illustrated with logs showing lateral variations in facies between well-correlated sections. The potential controls on sedimentation are also discussed and a palaeogeographic reconstruction for the area is presented.

As Sila

The outcrops at As Sila are the most westerly within the United Arab Emirates and possibly the oldest within the Shuwaihat Formation. They rest with probable unconformity on white dolomitic claystones of the middle Miocene (Burdigalian) Dam Formation. The contact is not exposed but outcrops of white dolomite with well-displayed bioturbation, bored hardgrounds, and gypsum nodules occur within a few metres of cross-stratified aeolian sandstones to the east of As Sila. Similar sediments occur as inliers within Sabkha Matti. The dolomites have produced bivalves (*Diplodonta* sp.) (Noel Morris, personal communication, 1995). The aeolian sandstones have an erosive base, overlying striped, pink and grey, partially oxidised siltstones. The sands are fine grained and well sorted and contain medium to large sets of trough cross-stratification with a palaeowind direction towards the south (160–201°) (plate 6.1, p. 57).

To the south of As Sila, near the road to a local refuse dump, there is a good outcrop, about 200 metres long and 6 metres high of pale-pink sandy dolomitic limestones and orange sandstone. Measured sections through the sandstones are shown in figure 6.2. The base of the section is covered by recent talus and gypsum, and the lowest rocks in situ are cross-stratified sandstones with foresets dipping towards the south; these are interpreted as fluvial channel sands. They are overlain by laminated, very fine-grained dolomitic sandy limestones that contain small nodules of gypsum. The dominant

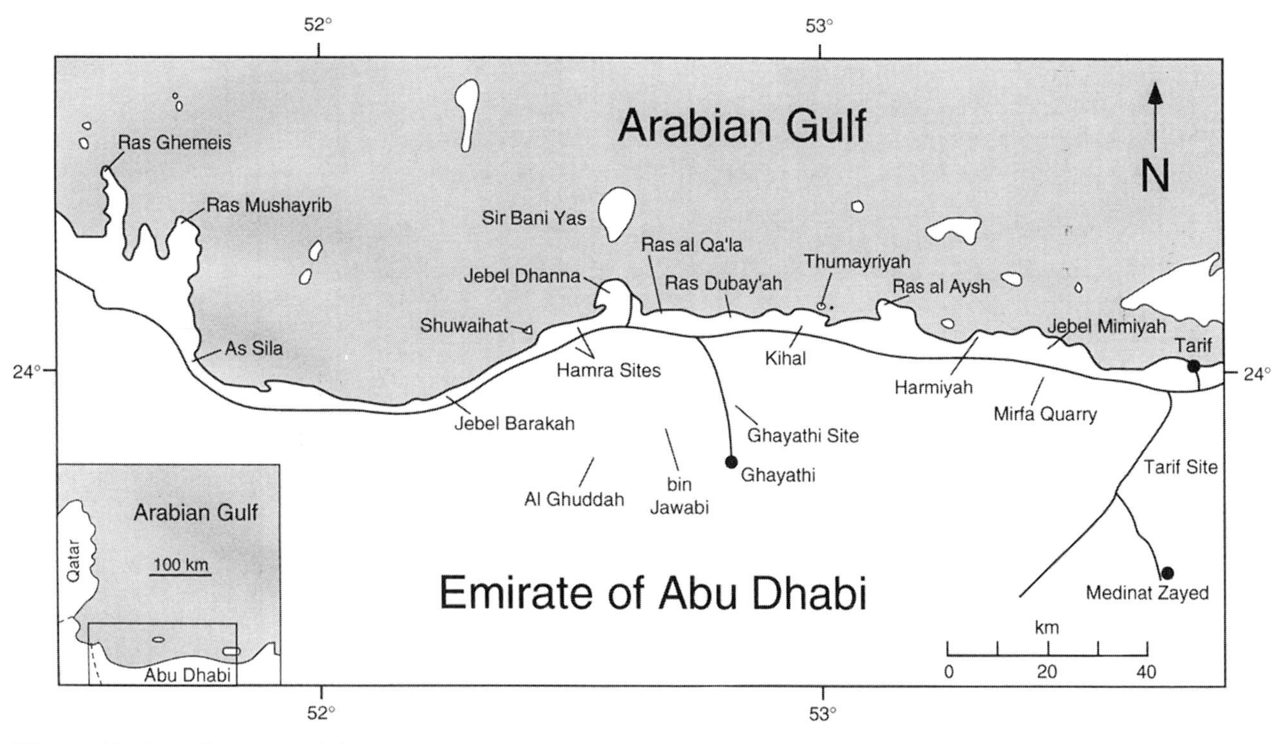

Figure 6.1. Locality map of the Miocene outcrops including Jebel Barakah, Jebel Dhanna, Shuwaihat, and As Sila.

clasts are dolomitised micritic peloids (40%), which could have been washed in from a shallow marine environment but are more likely to be reworked clasts of the underlying Dam Formation. The sediments also contain 30% fine-grained quartz, 10% feldspar (including fresh plagioclase and oligoclase), and minor amounts of epidote. There is one foraminifer with an isopachous fringing cement. The cement indicates a marine diagenetic environment. Both the foraminifer and the cement are partially abraded, which suggests reworking from a marine sediment. The overlying orange sands contain more quartz and feldspar and less dolomitised clasts. They have an erosive base and contain climbing ripple lamination, dewatering structures, contorted laminae, and small-scale sedimentary breccia. These fine-grained sands are interpreted as continental sabkha deposits.

The outcrop is capped by large crystals of selenitic gypsum that may or may not be in situ. Above the gypsum a thin unit of trough cross-stratified sands with a southeasterly palaeoflow was identified. The sands are medium grained, well sorted, and well rounded, with 60% quartz and 15% reworked dolomite clasts and 15% feldspar with minor amounts of epidote and amphiboles. The sands have a poikilotopic gypsum cement and are interpreted as fluvial deposits, but the high roundness and sphericity may indicate some aeolian transport. The overlying sediments, which outcrop towards the Saudi border, are poorly exposed and are believed to be fluvial deposits.

JEBEL BARAKAH

Jebel Barakah is about 64 km east of As Sila, and the outcrops form a 16 metre-high coastal cliff at the base of a 63 metre-high jebel (Whybrow, 1989). The base of the formation is not exposed,

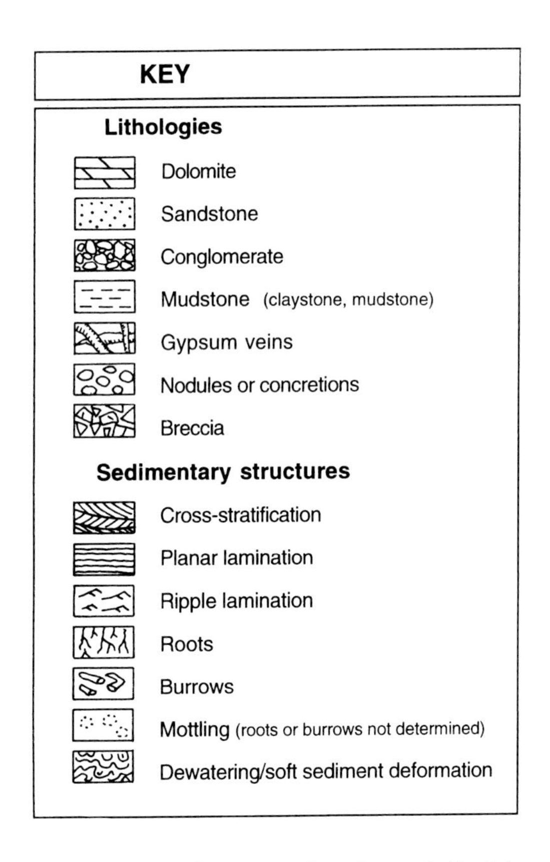

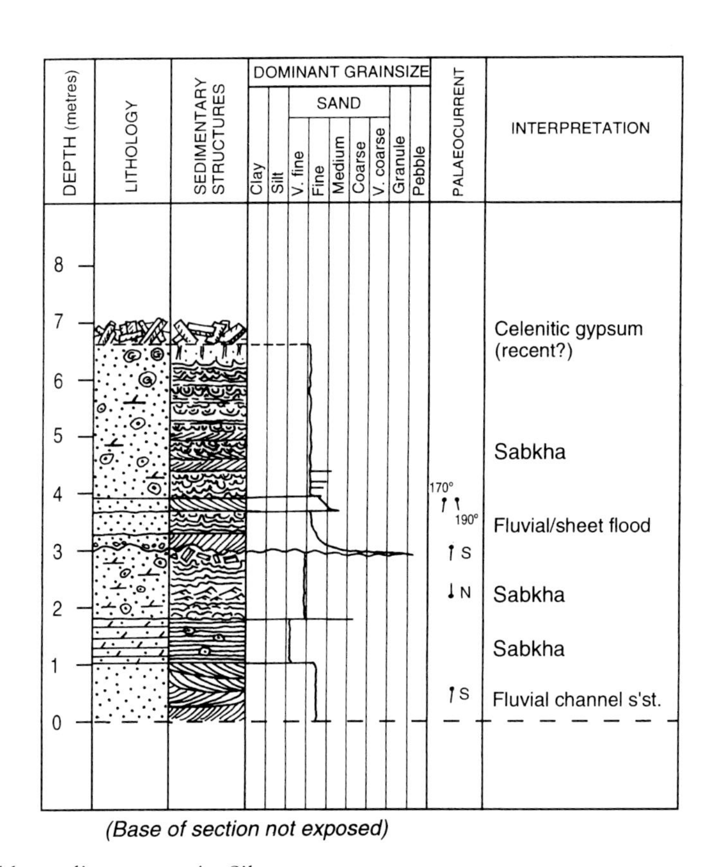

Figure 6.2. Sedimentary log through fluvial and sabkha sediments at As Sila.

and the lowest outcrops at beach level contain very fine dolomitic sandstones with dolomite peloids/lithoclasts similar to those at As Sila and a pervasive fine-grained dolomite cement. These sandstones are overlain by very fine-grained silty sandstones with folded and disrupted lamination interpreted as sabkha deposits. The sabkha deposits are 2–2.5 metres thick and can be traced for more than 400 metres along the strike. They are overlain by fine-grained cross-stratified aeolian sandstones that vary laterally along the strike. At the southwestern end of the outcrop (log 1) the aeolian sands are 11 metres thick although the top 6 metres are heavily overprinted by rhizoliths (fig. 6.3). Two hundred metres further east (log 2, fig. 6.3) the aeolian sandstones are only 5 metres thick. In each case the top of the aeolian sediments is marked by a distinct erosion surface overlain by fluvial conglomerates (the Baynunah Formation; Whybrow et al., 1999—Chapter 4), which can be picked out in the cliff face cutting down from the southwest towards the northeast (fig. 6.4). This erosion surface, which cuts out 6 metres of section, marks significant entrenchment of fluvial sediments into the aeolian sands. The cause of this incision will be discussed later. The conglomerate at the base of the fluvial sands is unusual; the clasts are carbonate-cemented sandstone with irregular, knobly shapes. They are probably reworked concretions. At the northwestern end of the outcrop the aeolian sands are overlain by a grey clay that has yielded remains of freshwater fish (Siluriformes). This clay is interpreted as the deposits of an interdune pond.

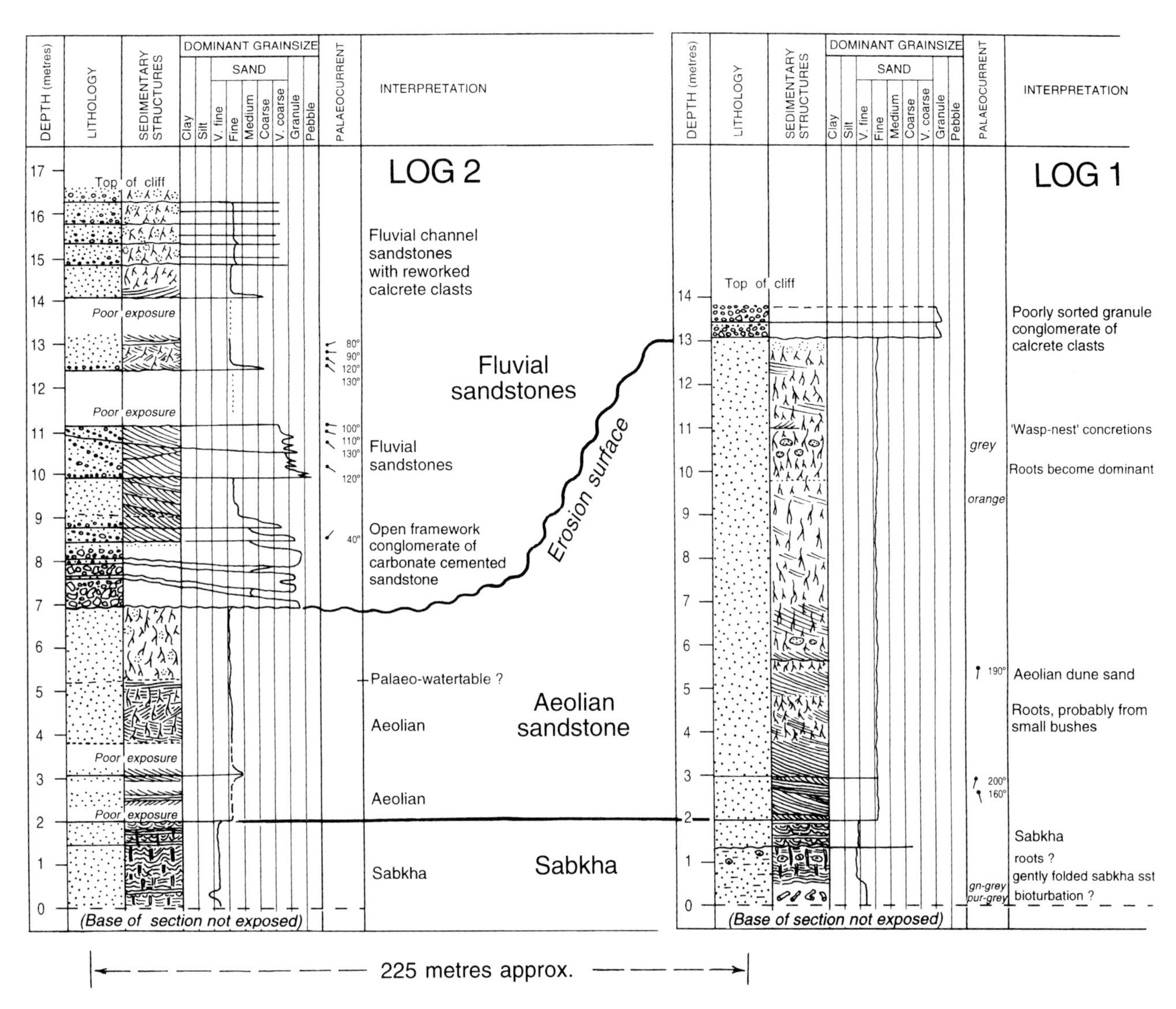

Figure 6.3. Sedimentary logs of the coastal cliff at Jebel Barakah showing basal sabkha facies overlain by aeolian sandstones, which are truncated by fluvial sediments. The fluvial incision has cut out 6 metres of aeolian sands in log 2. For key to lithologies and sedimentary structures, see figure 6.2.

Shuwaihat (Jazirat Shuwaihat)

Shuwaihat is about 10 km east of Jebel Barakah and provides the best exposures of aeolian and sabkha sediments in the Shuwaihat Formation. This area is proposed as the stratotype for the Shuwaihat Formation. The exposures are described from two locations, on the west and east sides of the island, that illustrate the typical facies and lateral variations seen in the Shuwaihat Formation (figs 6.5 and 6.6). On the west coast there is continuous exposure for more than 1 km in the low 5 metre-high cliff, which provides an excellent exposure of aeolian cross-stratification (see fig. 6.7). On the east coast

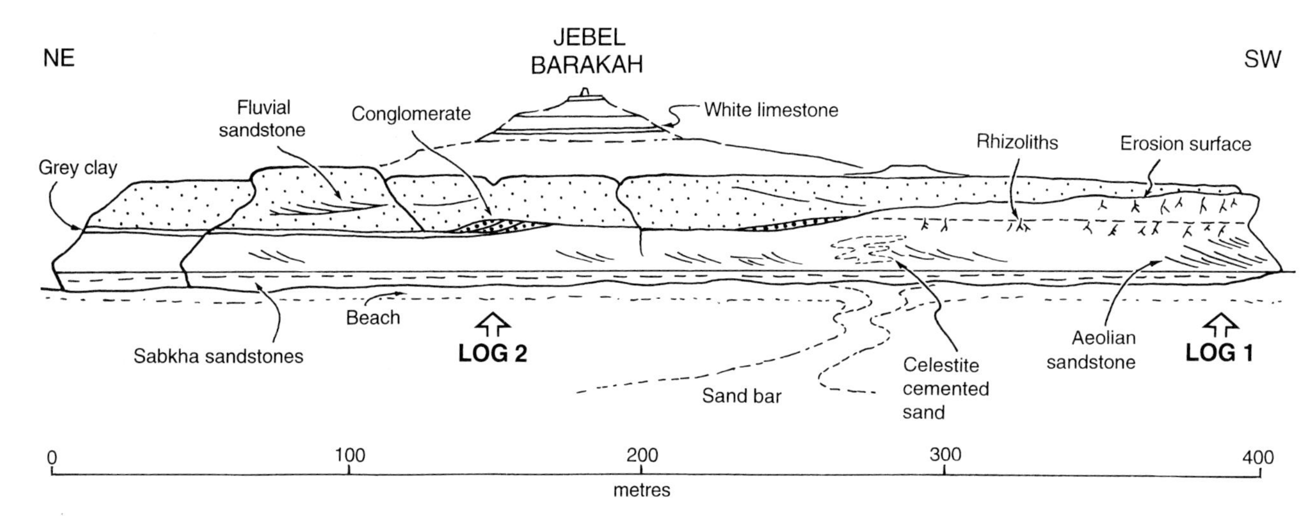

Figure 6.4. Outcrop sketch of Jebel Barakah showing the location of logs 1 and 2 (fig. 6.3) and the fluvial erosion surface that cuts into aeolian sands.

the equivalent section shows more variation with less aeolian cross-stratification, more sabkha sediments, and more fluvial deposits. Four sedimentary logs of the eastern outcrop are shown in figure 6.6; the sections are correlated at the base of a grey claystone with gypsum veins that can be traced along the outcrop.

Western Outcrops of Shuwaihat

On the western side of Shuwaihat dune sandstones are exposed at very low tide; these are overlain by sabkha deposits exposed at beach level (fig. 6.5). The overlying aeolian dune sandstones are exceptionally well exposed in low cliffs and platforms along the coast (Bristow and Hill, 1998). The sandstones are quartz dominated (60%), with fine, subrounded grains of moderate sphericity. They also contain 5% feldspar and 30% dolomite clasts, some of which appear to contain a relict fabric, indicating that they are reworked from the underlying Dam Formation. The sandstones also contain rare reworked bivalve fragments and foraminifera that have been reworked and blown in. The foraminifera are tentatively identified as rotaliids (cf. *Ammonia*), peneroplids, and elphidiids. These foraminifera live in very shallow water associated with algal mats and such like. *Ammonia* and *Peneroplis* spp. are known from the Miocene to the present day. Unfortunately the foraminifera do not help to refine the stratigraphy or the environment of deposition because they are clearly reworked. Comparative studies of modern aeolian dunes in Abu Dhabi indicate that foraminifera occur in dune sands more than 80 km inland (Jonathan Pugh, personal communication, 1995).

Sedimentary structures exposed in planform and cross-sectional outcrops indicate that the dunes were transverse to barchanoid in form, with a dominant palaeowind from north to south (Bristow and Hill, 1998) (plate 6.2, p. 57). Within the aeolian sandstones, sabkha facies of only a few tens of metres in lateral extent have been identified. In one outcrop, sabkha deposits onlap dune morphology, indicating that the sabkha rose around the dune. This contrasts with most of the other sabkha deposits that are much more laterally extensive (about 1000 metres plus) and truncate dune morphology. The sabkha facies that overly the aeolian sandstones on the western coast of Shuwaihat are a typical example. Here the aeolian sandstones are truncated by a sandy sabkha facies overlain by a grey claystone with large selenitic gypsum crystals that cut across the bedding. The clay and gypsum bed forms a good marker horizon along the western coast of Shuwaihat, which can be correlated with the eastern coastal section. The later-

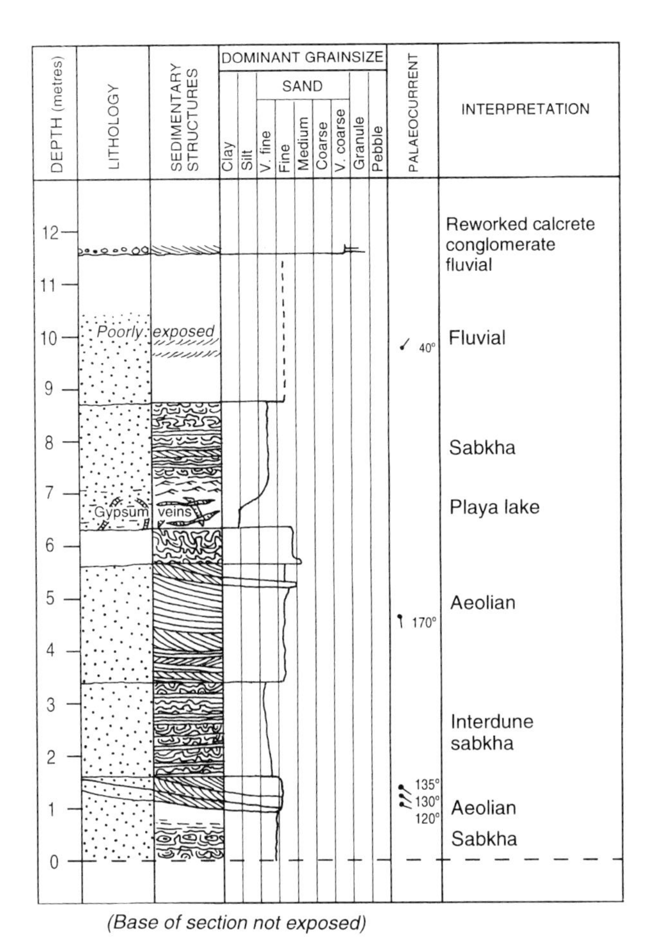

Figure 6.5. Sedimentary log of the western side of Shuwaihat. For key to lithologies and sedimentary structures, see figure 6.2.

ally restricted sabkhas that onlap dune morphology are interpreted as interdune sabkhas whereas the more extensive sabkha facies with gypsum and claystone beds are interpreted as playa lake deposits.

Eastern Outcrops of Shuwaihat

On the eastern coast the outcrops are more variable (fig. 6.6). At the southern end of the outcrop the base of the section consists of fine-grained fluvial sediments with lateral accretion surfaces dipping towards the south and well-displayed current ripple lamination that indicates palaeoflow towards the east. The ripple-laminated sandstones are erosively overlain by cross-stratified sandstones with curled mudstone clasts, dewatering structures, a local breccia of reworked sands, and wave ripples, which are interpreted as ephemeral stream deposits. These are overlain by sabkha facies with a variety of well-exposed, soft-sediment deformation structures. Towards the top of the section a grey claystone with selenitic gypsum veins is exposed. This horizon is used to correlate the measured sections on the eastern side of the island and may correlate with a similar horizon on the western side. In log 1 (fig. 6.6) the claystone is overlain by further sandy sabkha deposits that cap the cliff in this area. Further north (log 2, fig. 6.6), the base of the section consists of grey clay with gypsum veins that may have been deposited in an abandoned fluvial channel. This is overlain by sandy sabkha sediments and a second grey clay with gypsum veins. In section 2 (log 2, fig. 6.6), the clay is erosively overlain by a thin bed of cross-stratified fluvial sandstones with a northerly palaeoflow, and aeolian sandstones with a southerly palaeoflow. This interdigitation with reversing palaeoflows appears to be repeated further up the section, but exposure is poor. Section 3 (log 3, fig. 6.6) has a thin sabkha facies at the base overlain by a set of aeolian sandstone 1.5 metres thick, with cross-stratification dipping towards the south. This is erosively overlain by a gypsum-rich fluvial channel sand. The base of the fluvial channel contains reworked blocks of aeolian sand and gypsum crystals and passes up to inclined heterolithic beds and a grey clay with gypsum veins. In this case the clay was almost certainly deposited as an abandoned channel fill. The top of the fluvial channel fill is overlain by sandy sabkha sediments and the laterally persistent grey clay with gypsum veins. This is erosively overlain by a fluvial channel that cuts out the claystone further south. Unfortunately the outcrop is locally covered by recent talus in this area and it is difficult to distinguish the facies relationship. Log 4 at the northern end of the outcrop has wet interdune sabkha facies at the base overlain by load-cast aeolian sandstones and wind ripple-laminated aeolian sandstones. These are overlain by a thin bed of sabkha sands and a grey clay with gypsum veins. The grey clay is at an equivalent stratigraphic level

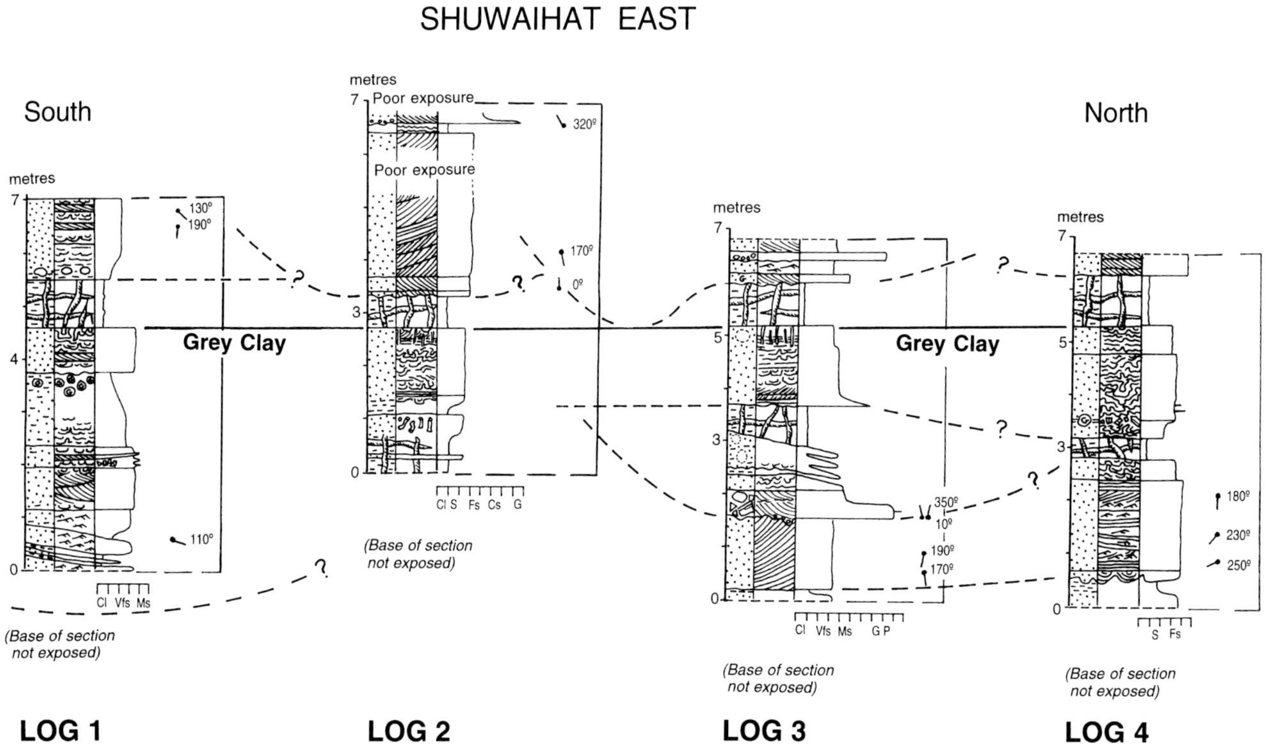

SHUWAIHAT EAST

Figure 6.6. Sedimentary logs of the eastern side of Shuwaihat. For key to lithologies and sedimentary structures, see figure 6.2.

to the abandoned channel fill in log 3 (fig. 6.6) and may be correlated. This is overlain by more sabkha sands with deformation structures, local brecciation, and reworked gypsum clasts. This unit may be ephemeral stream deposits that have been overprinted by sabkha processes. The top of the section is marked by a grey clay with gypsum veins that appears to be correlatable around the island.

Discussion

Correlations

Correlation within continental sequences is often complicated due to a lack of biostratigraphic data. As a result, lithostratigraphic schemes have to be adopted although these are often flawed because of lateral changes in facies. The Shuwaihat Formation is no exception. Locally, laterally continuous outcrops at Shuwaihat have enabled good correlations to be made, notably where a conspicuous sabkha horizon can be correlated around the island while the intervening fluvial and aeolian facies change laterally (fig. 6.6). Facies changes of this local scale make correlations between outcrops over distances of up to 50 km very speculative. Schematic summary lithofacies profiles for four sections throughout the Shuwaihat Formation are shown in figure 6.7. Four potential lithofacies correlation schemes are discussed: (1) correlation by altitude, (2) correlation of fluvial facies, (3) correlation of sabkha facies, and (4) correlation of aeolian facies. Making a simple correlation based on outcrop altitude suggests extensive lateral facies changes, which are possible but unlikely. This model requires no post-depositional folding or faulting (which is improbable), and therefore some form of lithostratigraphic scheme is required. Correlating the aeolian facies is

Plate 6.1. Outcrop of aeolian sandstones at As Sila, with cross-stratification dipping from right to left, towards the south.

Plate 6.2. Outcrop of aeolian cross-stratification at Shuwaihat exposed in planform indicates barchanoid and transverse dune morphology.

a possibility; these are readily identifiable in the field but not always well exposed. The outcrops at Shuwaihat and Jebel Barakah show that aeolian facies can change laterally and are locally eroded by fluvial channels.

At Shuwaihat it has been shown that playa lake-type sabkha facies can be correlated around the island for over 1 km and that aeolian and fluvial facies change laterally over the same distance. It can be argued, therefore, that sabkha facies may be the best units to use for lithostratigraphic correlations

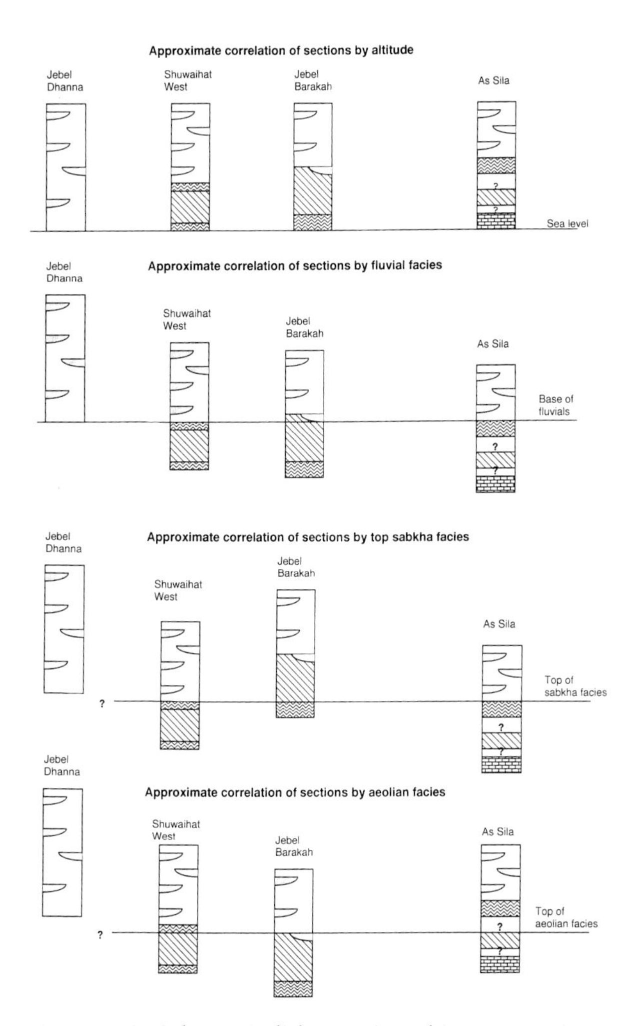

Figure 6.7. Schematic lithostratigraphic sections in the Shuwaihat Formation and possible correlation scenarios.

(fig. 6.7). A correlation based on the top of the highest sabkha facies, however, appears to offset the aeolian units, which is possible, but difficult, to prove. Correlating the base of the fluvial facies could provide the basis for a workable lithostratigraphic scheme of a regional scale, although, like the aeolian correlation, this is flawed at the outcrop scale as shown by the lateral facies changes observed at Shuwaihat (plate 6.2, p. 57). There does appear to be a general tendency, however, for fluvial facies to increase through each section (fig. 6.7). I suggest here that the Shuwaihat and Baynunah Formations should be distinguished by the dominance of fluvial facies in the Baynunah Formation. The contact varies laterally: at Jebel Barakah the base of the fluvial sediments is clearly erosive, at As Sila and Shuwaihat there is a transition with fluvial, aeolian, and sabkha facies interbedded. Defining the top of the Shuwaihat Formation at As Sila and Shuwaihat is difficult. I suggest that in the field the contact should be taken at the top of the highest sabkha facies within the Shuwaihat Formation. The sabkha facies may not be perfectly correlated, but it provides an easily identifiable horizon that can be mapped on a local scale and correlated at the outcrop in the absence of a clearly defined erosion surface. The overlying Baynunah Formation is dominated by fluvial channels and floodplain sediments (with vertebrate fossils) with larger river channels that appear to have been deposited in a less-arid environment (Whybrow et al., 1990). The transition from the arid to semiarid Shuwaihat Formation to the less-arid Baynunah Formation represents a significant change in palaeoclimate.

Palaeogeography

The palaeogeographic reconstruction of the Shuwaihat Formation shows fluvial systems draining into a semi-enclosed basin with barchanoid and transverse dunes in the foreground and draa along the far side of the lake (fig. 6.8). The dunes are migrating towards the south as indicated by palaeocurrent measurements from the Shuwaihat Formation. The river systems are shown flowing from south to north and west to east and locally trending N–S, which is also in agreement with measured

Figure 6.8. Palaeogeographic reconstruction for the Shuwaihat Formation with rivers flowing into an evaporitic playa lake surrounded by transverse and barchanoid dunes.

palaeocurrent directions. The lake was probably ephemeral and its extent is believed to have been controlled by climatic fluctuations. A possible modern analogue could be the inland delta of the River Niger near Timbuktu or Chott el Djerid in Tunisia, which are both inland sedimentary basins where evaporation exceeds precipitation, and fluvial channels, aeolian dunes, and evaporitic lakes or swamps may occur together.

Controls on Sedimentation

The lithostratigraphic correlations described above can be interpreted in a sequence stratigraphic framework (Van Wagoner et al., 1988). The sabkha sediments can be interpreted as transgressive systems tracts with the aeolian sediments accumulating during lowstands. The erosive base of the fluvial channel of the Baynunah Formation at Jebel Barakah cuts out at least 6 metres of section. This requires a lowering of base level of similar magnitude or relative uplift of the basin. The monomict nature of the clasts in the conglomerates, and their unusual form, suggests a lack of transport and a local intrabasinal source that could be due to intrabasinal tectonic uplift. Alternatively, the conglomerates could be derived from intrabasinal erosion due to fluvial incision caused by a fall in base level. In the Shuwaihat Formation this fall in base level could have resulted from desiccation of a lake or a fall in sea level. It is possible that the fluvial incision was produced during a sea-level lowstand and that accumulation occurred during the subsequent transgression. This model cannot be disproved, and although there is no clear evidence of marine sediments in the outcrops the presence of reworked foraminifera indicate that marine sediments were not totally absent from the basin. In this respect base-level changes could be invoked to account for some of the observed facies changes and stratigraphy of the Shuwaihat Formation.

The sabkha facies are interpreted here as inland sabkhas, with interdune ponds and playa lakes. The interdigitation of aeolian and sabkha sediments in the Shuwaihat Formation and the onlap of an aeolian dune by interdune sabkha facies suggest that water levels within the evaporitic playa lakes often varied so that base-level changes can be explained without invoking sea-level change. The interdigitation of aeolian, lacustrine, and fluvial facies is most likely to have been climatically driven with some autocyclic variations due to lateral migration of depositional systems. An intrabasinal tectonic control on base-level change cannot be ruled out, but it is clear that the climate fluctuated during the Miocene and it is possible that the observed stratigraphic changes could be due entirely to climate change.

Conclusions

The Shuwaihat Formation is an unconformity-bounded sequence of fluvial, aeolian, and continental sabkha sediments. The base of the unit is not properly exposed but is believed to be unconformable on presumed middle Miocene carbonates of the Dam Formation. The top of the member is an erosion surface overlain by fluvial sediments of the Baynunah Formation, best exposed at Jebel Barakah. Where there is no clear exposure of an erosive base to the overlying Baynunah Formation the contact is taken at the top of the highest sabkha facies. The fluvial deposits in the Shuwaihat Formation show a dominant palaeoflow from west to east with a wide variance. Aeolian cross-stratification shows much more consistent palaeowind directions towards the south. The aeolian sandstones are usually interbedded with continental sabkha deposits. These include interdune sabkhas that onlap dune topography and more-widespread playa lake sabkhas that truncate dune topography. The transition from an aeolian-dominated Shuwaihat Formation to a fluvial-dominated Baynunah Formation is best explained by a change in climate.

Acknowledgements

This chapter forms part of the Birkbeck/University College London–Natural History Museum Global Change in the Biosphere research theme. I thank Peter Whybrow, Department of Palaeontology at The Natural History Museum, for inviting me to join this project; Nick Hill, Tim Goodall, and Jonathan Pugh for their help in the field; and the Abu Dhabi Company for Onshore Oil Operations (ADCO) for providing vehicles and financial support through their grant to Peter Whybrow, without which this work could not have been undertaken. I also thank John Whittaker (NHM) for identifying the foraminifera.

References

Bristow, C. S., and Hill, N. 1998. Dune morphology and palaeowinds from aeolian sandstones in the Miocene Shuwaihat Formation, Abu Dhabi, United Arab Emirates. In *Quaternary Deserts and Climatic Change,* pp. 553–64 (ed. A. S. Alsharhan, K. W. Glennie, G. L. Whittle, and C. G. St. C. Kendall). A. A. Balkema, Rotterdam.

Hailwood, E. A., and Whybrow, P. J. 1999. Palaeomagnetic correlation and dating of the Baynunah and Shuwaihat Formations, Emirate of Abu Dhabi, United Arab Emirates. Chap. 8 in *Fossil Vertebrates of Arabia,* pp. 75–87 (ed. P. J. Whybrow and A. Hill). Yale University Press, New Haven.

Van Wagoner, J. C., Posamentier, H. W., Mitchum, R. M., Vail, P. R., Sarg, J. F., Loutit, T. S., and Hardenbol, J. 1988. An overview of sequence stratigraphy and key definitions. In *Sea-level Changes: An Integrated Approach* (ed. C. W. Wilgus). *Society of Economic Paleontologists and Mineralogists Special Publication* 42: 39–45.

Whybrow, P. J. 1989. New stratotype; the Baynunah Formation (late Miocene), United Arab Emirates: Lithology and palaeontology. *Newsletters on Stratigraphy* 21: 1–9.

Whybrow, P. J., Friend, P. F., Ditchfield, P. W., and Bristow, C. S. 1999. Local stratigraphy of the Neogene outcrops of the coastal area: Western Region, Emirate of Abu Dhabi, United Arab Emirates. Chap. 4 in *Fossil Vertebrates of Arabia,* pp. 28–37 (ed. P. J. Whybrow and A. Hill). Yale University Press, New Haven.

Whybrow, P. J., Hill, A., Yasin al-Tikriti, W. Y., and Hailwood, E. A. 1990. Late Miocene primate fauna, flora and initial palaeomagnetic data from the Emirate of Abu Dhabi, United Arab Emirates. *Journal of Human Evolution* 19: 583–88.

Diagenesis of the Baynunah, Shuwaihat, and Upper Dam Formation Sediments Exposed in the Western Region, Emirate of Abu Dhabi, United Arab Emirates

PETER W. DITCHFIELD

In this study samples from the various lithological units of the Upper Miocene Baynunah Formation were collected and analysed for their petrographic and geochemical variation to (1) help determine the depositional environments represented by these sediments and (2) determine the degree and type of postdepositional alteration that has affected them. In addition, samples from the underlying Miocene Shuwaihat and Dam Formations were also collected for comparison with those from the Baynunah Formation.

The samples were collected from various outcrops between Al Mirfa in the east and As Sila in the west during the 1993 and 1994 field seasons (fig. 7.1). All of the lithological units of the Baynunah Formation were sampled (see fig. 7.2 for a schematic log through the Baynunah lithofacies).

METHODS

The samples were analysed petrographically by standard thin-section techniques, which included cathodoluminescence (CL) and staining for various carbonate phases (Dickson, 1966). Some samples were also analysed with a scanning electron microscope (SEM). Mineralogies were confirmed by X-ray diffraction (XRD). Trace element analyses of carbonate samples were obtained by inductively coupled plasma atomic emission spectrometry (ICP-AES) and were calibrated against in-house multi-element standards. Carbon and oxygen stable isotopic analyses were also carried out on carbonate samples. Samples for isotopic analysis were extracted using a tungsten carbide micro-drill, then routinely roasted at 400 °C under vacuum to remove any organic contaminants. Carbon dioxide (CO_2) gas for analysis was obtained by reaction with 100% phosphoric acid (H_3PO_4) at 25 °C. Samples were analysed on a VG Isogas mass spectrometer. The results were corrected by the method of Anderson and Arthur (1983) and calibrated to PDB (Peedee Formation belemnite standard) by repeated analysis of a carbonate standard (NBS 19). Analytical precision is better than 0.1 per mil (‰).

RESULTS

Petrography

Much of the sandstone within the Baynunah Formation is composed of fine- to very fine-grained subquartzite with less than 5% feldspar grains (plate 7.1a, p. 64). This lithology is combined with varying amounts of probably soil-derived carbonate clasts, ranging in size from granules to cobbles, to give

 ISBN 0-300-07183-3

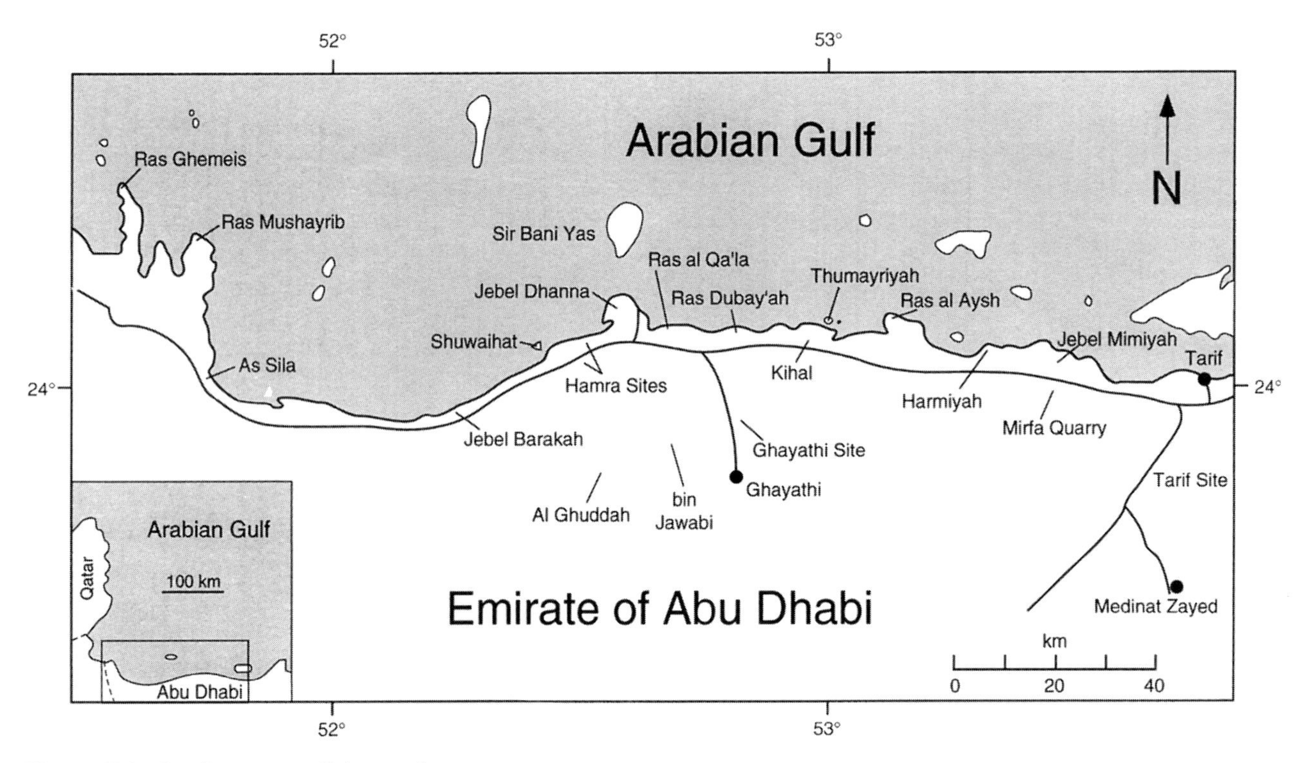

Figure 7.1. Outline map of the study area.

the characteristic conglomerate lithology of the Baynunah Formation (plate 7.1b). More exotic extraformational clasts are seen only in a patchily developed, well-cemented conglomerate facies at the base of the Baynunah Formation (plate 7.1c).

Preservation of porosity within the remaining lithologies of the Baynunah Formation is good. Grain contacts are restricted to points only. The rare biotite grains within the sediment are generally well preserved with no petrographic signs of alteration; these grains are also undeformed by compaction. These features suggest that the sediments of the Baynunah Formation have undergone only a minor amount of burial.

Carbonate cements are observed in samples only where there are locally abundant carbonate (originally aragonitic) bioclasts, in particular unionoid bivalves or in some cases gastropods. These are largely preserved only as mouldic porosity that occasionally contains pore-lining, equant, sparry, calcite cements (plate 7.2a, p. 65). Sparse circumgranular, equant, pore-lining or pore-filling, nonferroan calcite cements are present in the immediate area of the bioclasts, which show a concentrically zoned dull to moderate orange luminescence under CL (plate 7.2b). These cements occasionally show pendant and meniscate morphologies, the latter being especially common within the poorly cemented root casts in the cross-bedded sandstone facies of the upper part of the Baynunah Formation (plate 7.2c). These cement morphologies suggest that dissolution of bioclasts and precipitation of carbonate cements took place soon after deposition and above the permanent water table within the vadose zone.

Similarly, the micritic limestones in the upper part of the Baynunah Formation show significant remobilisation of carbonate only in one location (Al Ghuddah), where originally aragonitic ?cerithid gastropods were locally abundant; these are preserved only as mouldic porosity. At Al Ghuddah, fenestral cavities within the limestone show some

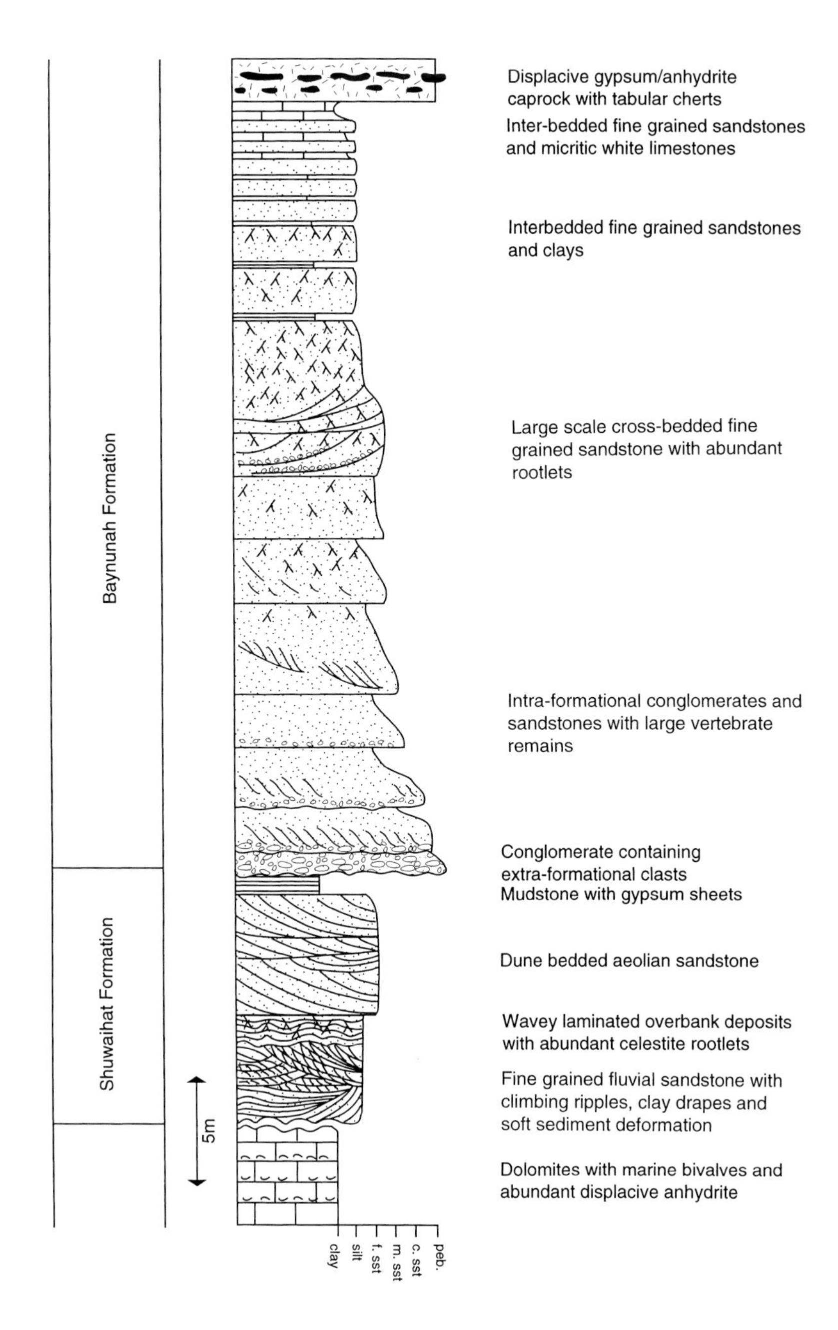

Figure 7.2. Schematic log through the Baynunah, Shuwaihat, and upper part of the Dam Formations showing the characteristic lithofacies in each formation.

pore-filling, nonferroan, dully luminescent cements. These micritic limestones from the upper part of the Baynunah Formation are preserved as nonferroan, low-magnesium calcite and contrast strongly with the micritic limestones of the underlying Dam Formation (as seen at As Sila) that have been pervasively dolomitised.

Sulphate cements within the Baynunah Formation are largely restricted to a patchily developed basal conglomerate and to the "cap-rock" facies.

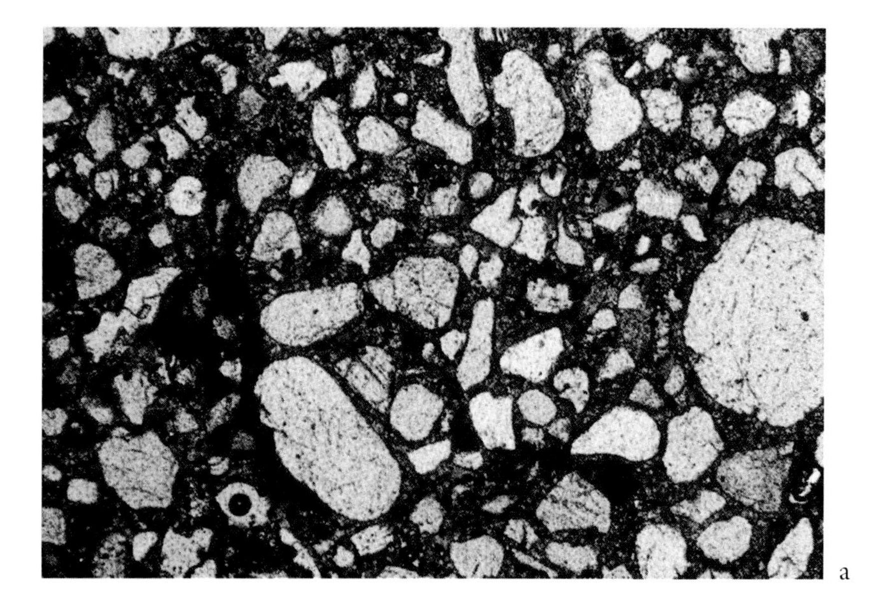

a

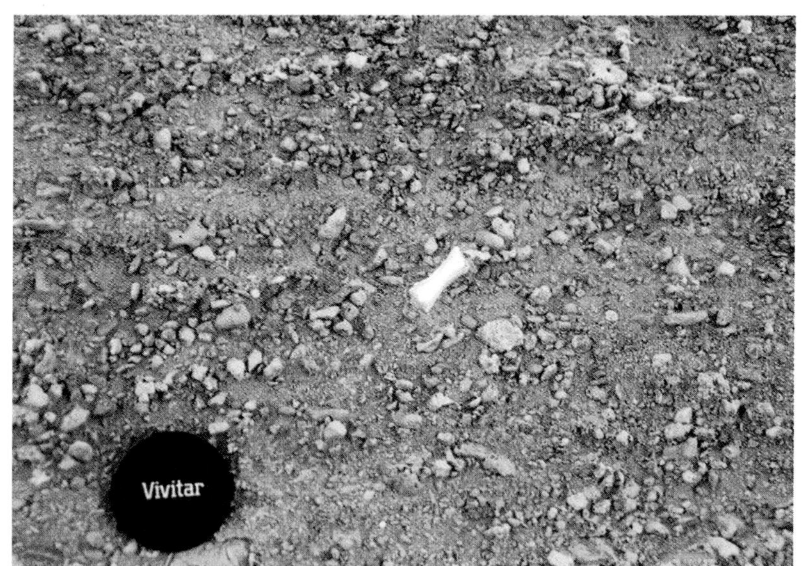

b

c

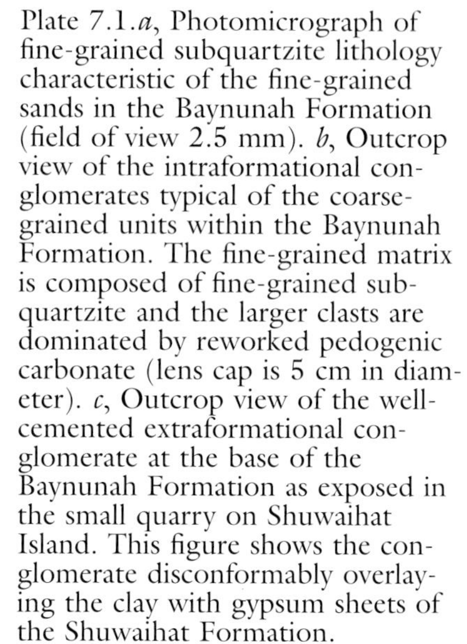

Plate 7.1.*a*, Photomicrograph of fine-grained subquartzite lithology characteristic of the fine-grained sands in the Baynunah Formation (field of view 2.5 mm). *b*, Outcrop view of the intraformational conglomerates typical of the coarse-grained units within the Baynunah Formation. The fine-grained matrix is composed of fine-grained subquartzite and the larger clasts are dominated by reworked pedogenic carbonate (lens cap is 5 cm in diameter). *c*, Outcrop view of the well-cemented extraformational conglomerate at the base of the Baynunah Formation as exposed in the small quarry on Shuwaihat Island. This figure shows the conglomerate disconformably overlaying the clay with gypsum sheets of the Shuwaihat Formation.

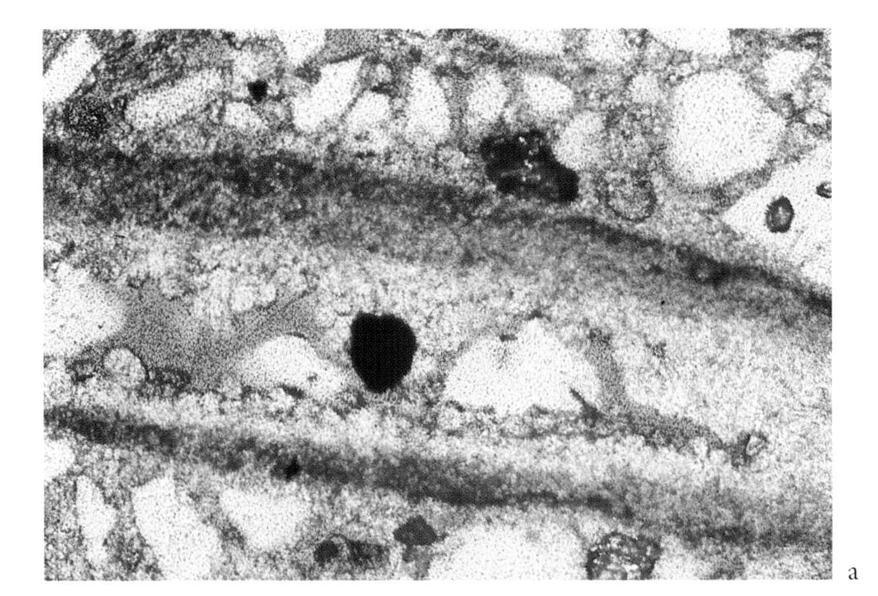

a

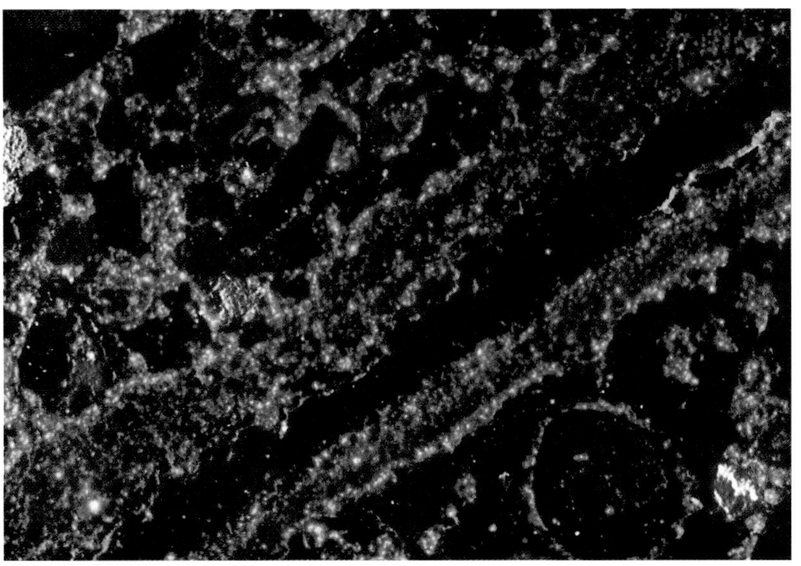

b

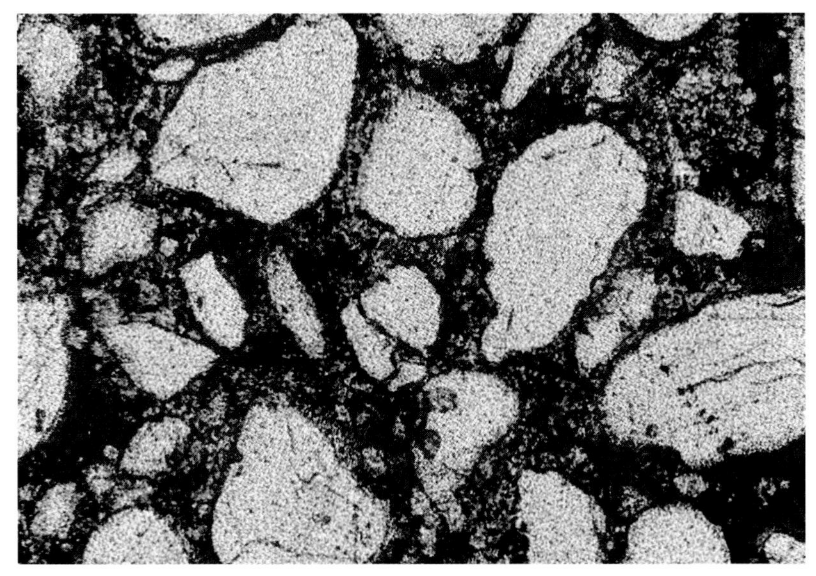

c

Plate 7.2.*a*, Photomicrograph of biomouldic porosity after dissolution of a unionoid bivalve, now partially filled by sparry clacite cement, Baynunah Formation (field of view 5 mm). *b*, Cathodoluminescence photomicrograph of concentrically zoned equant pore-lining cement from the Baynunah Formation (field of view 5 mm). *c*, Photomicrograph of intergranular meniscate calcite cement from the upper Baynunah Formation (field of view 2.5 mm).

Within the basal conglomerate (as seen in the small quarry on Shuwaihat Island) coarse-grained, often poikilotopic gypsum cements are common (plate 7.3a, p. 67). These closely resemble the gypsum cements in the aeolian sandstones of the underlying Shuwaihat Formation. The Baynunah Formation, however, does not contain the abundant celestite cements (replacing rootlets) that occur in the overbank sediments of the Shuwaihat Formation.

The gypsiferous "cap-rock" facies at the top of the succession contains abundant coarsely crystalline, displacive gypsum cements with minor amounts of anhydrite. The classic chicken-wire texture associated with displacive gypsum growth in sabkha environments, however, is only rarely seen. Gypsum crystals are usually arranged as parallel to subparallel equant laths up to 10 mm long. Large-scale displacive features such as "Tee Pee" structures are also only occasionally seen in outcrop (plate 7.3b). The origin of this lithology is unclear, but the lack of characteristic petrographic textures and the absence of dolomite from either the cap-rock or the underlying micritic limestones suggest that this lithofacies does not represent deposition in a classic coastal sabkha-type environment. A more likely possibility is that it represents recent gypsum growth from the evaporation of ground water on, or at, an erosion surface, the stratigraphic position of this surface being controlled by the underlying tabular chert beds.

Gypsum cements are volumetrically important in the dune-bedded aeolian sandstone facies of the underlying Shuwaihat Formation, where large, irregular, single gypsum crystals (up to 10 cm long) form a coarse-grained poikilotopic cement.

Silica cements within the micritic limestones at the top of the section are in the form of replacive cryptocrystalline hydrated cherts (plate 7.4a, p. 68), which often preserve the fine-scale algal laminations of the original limestone. Occasionally, however, as at the Hamra sites, silica cementation replaces large subhorizontal branching (thalassinoides-type) burrow networks (plate 7.4b). These cemented burrows often retain a circular cross-section, which contrasts with the slightly flattened adjacent noncemented burrows and indicates silica cementation early in the diagenetic history prior to the small amount of compaction that has taken place.

Within some of the larger bone fragments silica cements are also developed. These include a fine-grained fibrous chalcedonic quartz phase and a later coarse-grained drusy quartz cement (plate 7.4c). The chalcedonic quartz cements are length-fast. No length-slow chalcedonic quartz—a possible indicator of quartz precipitation in evaporative environments (Folk and Pittman, 1971)—was observed.

Geochemistry of Carbonate Phases

The geochemistry and mineralogy of the analysed carbonate phases are shown in table 7.1. It is clear from these data that the dolomites of the Dam Formation are significantly different from the dolomites of the overlying Shuwaihat and Baynunah Formations. In particular, the dolomites from the Dam Formation are close to stoichiometric dolomites whereas those from the overlying units are enriched in calcium. The Dam Formation dolomites also contain considerably less iron than those from the overlying formations (fig. 7.3). The most striking difference between these two sets of dolomites is in their stable isotopic compositions (fig. 7.4); the samples from the Dam Formation are strongly enriched in $\delta^{18}O$ (3.91–5.63‰), indicating a probable evaporitic origin for the dolomitising fluids. In the dolomite samples from the overlying units, however, the $\delta^{18}O$ values range from −2.71‰ to −5.54‰, which may indicate dolomitisation in a marine/meteoric mixing zone. Samples from the Shuwaihat Formation are more depleted in $\delta^{13}C$ (−2.04‰ to −3.80‰) relative to PDB than the samples from the underlying Dam Formation (−1.08‰ to −2.54‰).

Although the oxygen isotopic data from the Dam Formation dolomites point strongly to an evaporitic origin for the dolomitising fluids, such dolomites are often calcium-rich and poorly ordered. The dolomites of the Dam Formation, however, are stoichiometric and well ordered. This may be the result of recrystallisation during a period of deeper burial before uplift and the forma-

a

b

Plate 7.3.*a*, Photomicrograph of gypsum-cemented fine-grained subquartzite lithology from the Baynunah Formation (field of view 5 mm). *b*, Outcrop view of minor "teepee" structure within the gypsiferous cap-rock of the Baynunah Formation (hammer handle is 30 cm long).

a

b

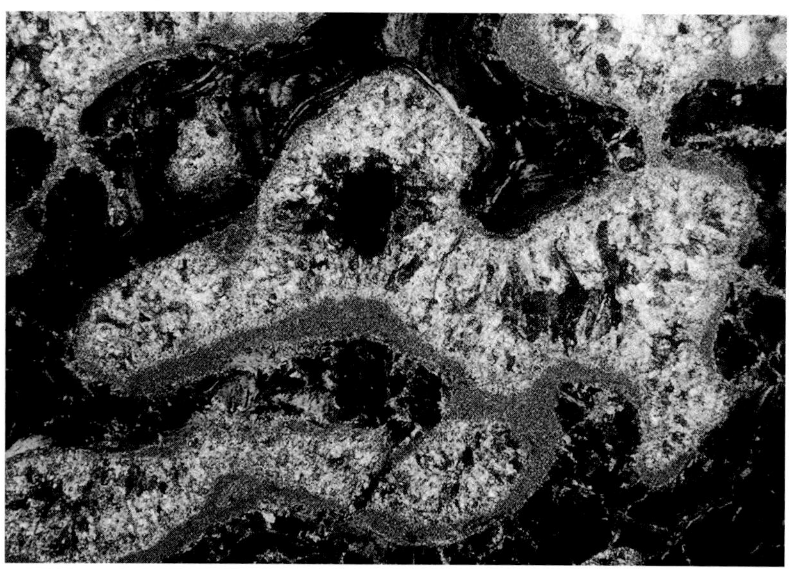
c

Plate 7.4.*a*, Hand specimen of replacive laminated chert after laminated micritic (possibly algal) limestone, from the upper part of the Baynunah Formation. *b*, Outcrop view of replacive chert infilling a thallasinoides-type burrow network in the upper part of the Baynunah Formation. *c*, Photomicrograph of fine-grained chalcedonic quartz cements and later coarser-grained drusy quartz cements within the porosity of a bone fragment from the elephant excavation site in the Baynunah Formation from Shuwaihat Island (field of view 5 mm).

tion of the unconformity separating the Dam Formation from the overlying units. The timing of such uplift is uncertain but could be related to the uplift of the Qatar arch (Alsharhan, 1989).

Fenestral Micrites

Results of stable isotopic and trace element analyses are shown in table 7.1. The isotopic values for the fenestral micrites from the top of the sequence are shown in figure 7.5. The $\delta^{18}O$ values range from −3.99‰ to −5.29‰ with a mean value of −4.40‰ (n = 4). The $\delta^{13}C$ values range from −1.64‰ to −4.68‰, with a mean value of −3.22‰ (n = 4). The sparse faunal evidence suggests that these poorly consolidated micritic limestones were deposited in a marine to slightly brackish-water environment (N. J. Morris, personal communication). The oxygen isotopic data are not incompatible with such an environment, but precipitation of the carbonate would have had to have taken place at temperatures in excess of 35°C, assuming an isotopic composition for the water of 0‰ (relative to Standard Mean Ocean Water—SMOW). Modification of the isotopic composition of the water by evaporation would reduce this temperature estimate. The oxygen isotopic data would also be compatible with diagenetic recrystallisation within a meteoric realm similar to that envisaged for the precipitation of the meteoric cements described below, but there is little petrographic evidence for wholesale recrystallisation of these limestone units.

Calcite Cements

Isotopic results for the rare, sparry calcite cements within the sandstone units of the Baynunah Formation are also shown in figure 7.5. The $\delta^{18}O$ values range from −1.51‰ to −9.44‰, with a mean value of −4.99‰ (n = 7); $\delta^{13}C$ values range from −1.66‰ to −10.40‰, with a mean value of −4.25‰ (n = 7). There is good petrographic evidence to suggest that at least some of these cements precipitated from meteoric water percolating down through the vadose zone (see "Petrography" section, above).

Modern-day precipitation in this area ranges in composition from 0‰ to + 2‰ (SMOW) (Yurtsever, 1975). The oxygen isotopic data are compatible with precipitation of the above cements from waters of such a composition. The moderately negative $\delta^{13}C$ values suggest the incorporation of some carbonate carbon from an organic source as well as from the dissolution of carbonate bioclasts.

Stable Isotopic Analysis of Baynunah Formation Bioclasts

A variety of bioclasts from the Baynunah Formation were analysed petrographically (by SEM) for any textural evidence of recrystallisation. Of these, only the ratite eggshell samples that were originally calcitic preserve their primary texture and therefore possibly their original chemical composition. The carbon stable isotopic signature of ratite eggshell from the Neogene Siwalik sediments of southern Asia has been used to reconstruct a record of environmental (vegetational) change (Stern et al., 1994). In this study the stable isotopic composition of 16 samples of ratite eggshell was analysed to try to deduce the isotopic composition of the dietary carbon ingested by the Baynunah avian fauna. In two of the samples (AUH 15, locality Hamra 3, and AB X2) there was evidence of significant amounts of recrystallisation of the original, main prismatic calcite shell layer. The remaining samples all showed well-preserved prismatic and layered textures characteristic of thick ratite shells (Silyn-Roberts and Sharp, 1986). The preservation of these primary textures was used by Stern et al. (1994) to suggest that no postdepositional recrystallisation or significant isotopic exchange has taken place.

Results

The stable isotopic results for the eggshell analyses are shown in table 7.1 and figure 7.5. The samples with no evidence of recrystallisation have $\delta^{13}C$ values ranging from −3.05‰ to −11.25‰, with a mean value of −7.48‰ (n = 14). The oxygen isotopic results, however, do not show as even a spread as the carbon isotopic data; the $\delta^{18}O$ values range from

Table 7.1. Analysis of mineralogical samples collected from the Emirate of Abu Dhabi

Sample	Type	Mineralogy	$\delta^{18}O$ PDB	$\delta^{13}C$ PDB	Total oxide	Fe (ppm)	Mg (ppm)	Mn (ppm)	Sr (ppm)
Baynunah Formation									
AB 25/1	Fenestral micrite	Calcite	−4.33	−1.64	54.54	178	2 614	334	380
AB 31/1	Fenestral micrite	Calcite	−4.01	−4.45	55.00	224	1 754	374	411
AB 94.8	Fenestral micrite	Calcite	−5.29	−4.68	54.44	302	2 314	544	2 353
AB 15	Fenestral micrite	Calcite	−3.99	−2.10	56.01	256	1 642	294	507
AB 29	Concretion	Dolomite	−3.45	−2.87	50.85	3 997	116 006	3 431	205
AB 94.1	Calcite-cemented sandstone	Calcite	−1.51	−4.20	54.21	120	356	100	452
AB 94.2	Calcite-cemented sandstone	Calcite	−2.03	−10.40	53.68	80	520	98	201
AB 14	Calcite-cemented sandstone	Calcite	−6.09	−3.54	52.27	107	427	50	328
AB 12	Calcite-cemented sandstone	Calcite	−3.47	−3.66	53.67	90	395	64	554
AUH 335/2	Calcite-cemented sandstone	Calcite	−9.44	−1.66	54.88	146	561	66	352
AUH 335/1	Bivalve	Calcite	−7.92	−2.04	54.65	125	567	112	420
AB X1	Bivalve	Calcite	−4.47	−4.25	54.69	134	2 499	515	739
AB X2	Eggshell	Calcite	−3.70	−10.80	54.85	39	505	55	416
AB X3/1	Eggshell	Calcite	8.28	−3.33	53.47	100	657	36	571
AB X3/2	Eggshell	Calcite	8.04	−3.05	53.10	150	453	48	631
AB X4/1	Eggshell	Calcite	7.82	−3.40					
AUH 639 M3	Eggshell	Calcite	1.09	−11.25					
AUH 667 R3	Eggshell	Calcite	−0.08	−9.30					
AUH 472 B2	Eggshell	Calcite	2.81	−7.76					
AUH 315 K1	Eggshell	Calcite	2.13	−10.82					
AUH 487 S4	Eggshell	Calcite	−0.19	−9.83					
AUH 605 JD3	Eggshell	Calcite	2.99	−9.34					
AUH 552 S6	Eggshell	Calcite	6.91	−7.37					
AUH 621 Q1	Eggshell	Calcite	1.64	−8.43					
AUH 446 H6	Eggshell	Calcite	2.33	−5.73					
AUH 15 H3	Eggshell	Calcite	−4.98	3.15					
AUH 612 GU2	Eggshell	Calcite	7.28	−6.62					
AUH 320 TH1	Eggshell	Calcite	8.58	−8.40					

Table 7.1. *(continued)*

Sample	Type	Mineralogy	δ18O PDB	δ13C PDB	Total oxide	Fe (ppm)	Mg (ppm)	Mn (ppm)	Sr (ppm)
Shuwaihat Formation									
AB 7/1	Carbonate crust	Dolomite	−3.51	−2.04	49.32	2 401	116 006	112	120
AB 7/2	Carbonate crust	Dolomite	−2.71	−2.27	50.10	2 215	118 402	98	102
AB 8/1	Carbonate crust	Dolomite	−5.54	−3.19	48.79	2 362	115 836	64	118
AB 10/2	Carbonate crust	Dolomite	−5.04	−3.80	49.22	2 734	114 891	88	126
AB 10/2	Carbonate crust	Dolomite	−4.82	−2.46	50.96	2 540	115 308	102	104
Dam Formation									
AB 94.12/1	Micrite	Dolomite	4.02	−1.64	51.52	1 202	131 392	164	104
AB 94.12/2	Micrite	Dolomite	4.52	−1.98	52.21	1 474	126 552	120	227
AB 94.13/1	Micrite	Dolomite	3.91	−2.54	50.92	1 186	130 293	119	123
AB 94.13/2	Micrite	Dolomite	5.07	−1.08	50.74	1 394	127 409	184	206
AB 94.13/3	Micrite	Dolomite	5.63	1.45	51.63	1 335	128 431	142	154

Note: The AUH refers to a fossil and the suffix following the number refers to the collection locality: B, Jebel Barakah; S, Shuwaihat; GU, Al Ghuddah; H, Hamra; JD, Jebel Dhanna; Q, Ras al Qa'la; R, Ras Dubay'ah; K, Kihal; TH, Thumayriyah; M, Jebel Mimiyah (Al Mirfa). The prefix AB refers to rock samples collected for this analysis.

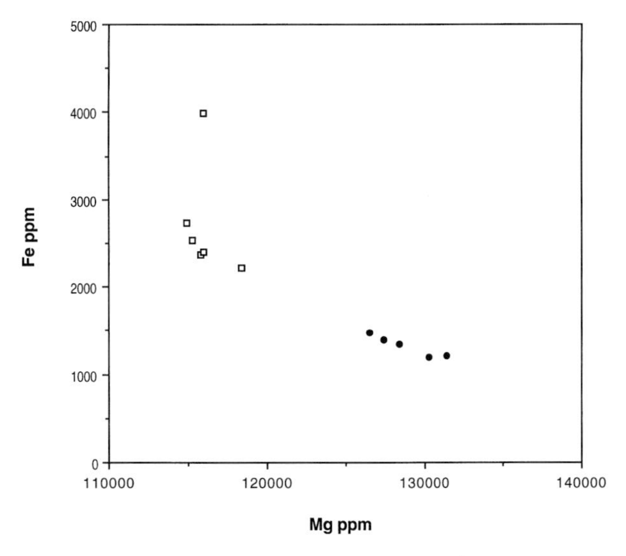

Figure 7.3. Miocene dolomites from the Western Region of Abu Dhabi. Plot showing the iron and magnesium composition of the dolomite samples from the Dam Formation (filled circles) and of the dolomite samples from the Baynunah and Shuwaihat Formations (open squares).

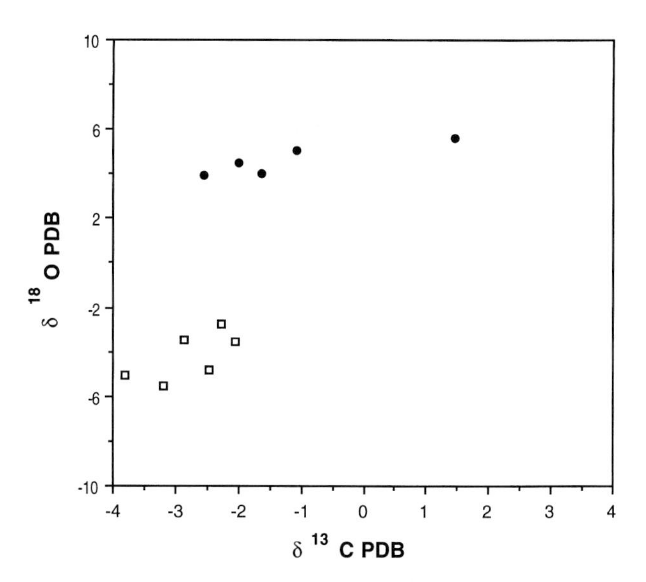

Figure 7.4. Miocene dolomites from the Western Region of Abu Dhabi. Plot showing the oxygen and carbon stable isotopic compositions of the dolomite samples from the Dam Formation (filled circles) and of the dolomite samples from the Baynunah and Shuwaihat Formations (open squares).

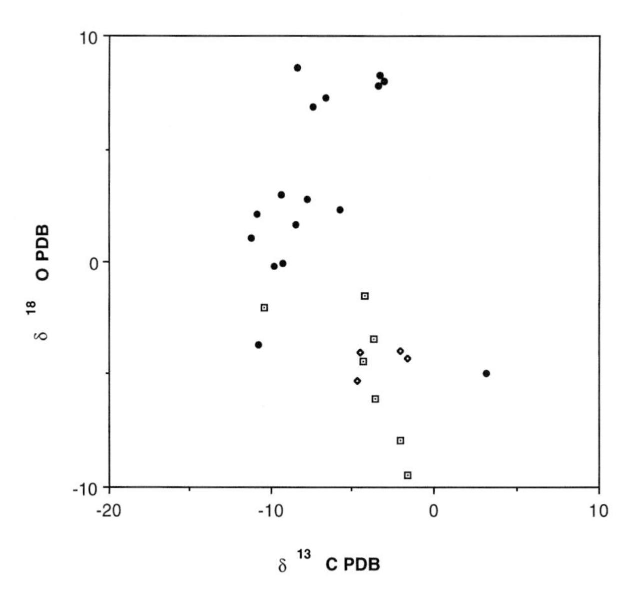

Figure 7.5. Plot of the oxygen and carbon stable isotopic compositions of calcitic samples from the Baynunah Formation. Fenestral micrites (open diamonds), ratite eggshell (open circles), and calcite cements (squares).

−0.19‰ to +8.58‰ and appear to divide the data into two groups: one centred around +7.82‰ ($n=6$) and the other around +1.59‰ ($n=8$).

Interpretation

The metabolic fractionation (vital effect) of carbon stable isotopes between eggshell calcite and dietary organic matter for modern *Struthio camelus australis* is approximately +16‰ (Von Schirnding et al., 1982). Thus, according to Stern et al. (1994), a diet dominated by C_3 plants—with a pre-industrial $\delta^{13}C$ signal of around −27‰ (Friedli et al., 1986)—in a modern ostrich would result in an eggshell $\delta^{13}C$ composition of −11‰, whereas a diet dominated by C_4 plants—with a pre-industrial $\delta^{13}C$ signal of approximately −13‰ (Friedli et al., 1986)—would result in an eggshell $\delta^{13}C$ composition of +3‰. The results from this study clearly fall between these two end members and suggest that both C_3 and C_4 plant materials were available to the avian fauna of the Baynunah Formation. Such a conclusion is in agreement with the carbon isotopic analyses performed on pedogenic carbonates from the Baynunah Formation (Kingston, 1999—Chapter 25). Unfortunately, with the exception of one sample (AUH 552), found during the excavation of the Shuwaihat elephant site, S6 (see Andrews, 1999—Chapter 24), however, all of the eggshell samples were from surface collections from the Baynunah Formation and therefore represent a time averaging of environmental information. Geochemical data on the evolution of environments in the Baynunah Formation must rely, therefore, on other in situ samples, such as palaeosol carbonates or analyses of tooth enamel (Kingston, 1999).

CONCLUSIONS

In general terms the Baynunah Formation is relatively free from cementation as compared with the more tightly cemented underlying sediments of the Shuwaihat and Dam Formations. Much of the conglomerate and sandstone facies of the Baynunah Formation is at best only poorly consolidated, with patchily developed amorphous clay cements. There is no evidence of significant compaction.

Carbonate cements are largely absent or restricted to environments where there were sufficient carbonate (aragonitic) bioclasts to have provided a source of carbonate for remobilisation. Sulphate cements are important only within a locally developed basal conglomerate of the Baynunah Formation and within the much more widespread caprock facies. (Gypsum crusts occur on many exposed surfaces but probably represent the products of recent weathering and groundwater evaporation.) Silica cements are also present but are restricted to replacive bedded cherts within the micritic limestones at the top of the section or to the pore systems of larger bioclasts in some large vertebrate bones.

ACKNOWLEDGEMENTS

I thank Peter Whybrow, through his grant from the Abu Dhabi Company for Onshore Oil Operations (ADCO) to The Natural History Museum, London, for my participation in the Abu Dhabi Miocene Project, and would also like to thank Peter Friend (University of Cambridge, U.K.) for initially involving me in the project.

REFERENCES

Alsharhan, A. S. 1989. Petroleum geology of the United Arab Emirates. *Journal of Petroleum Geology* 12: 253–88.

Anderson, T. F., and Arthur, M. A. 1983. Stable isotopes of oxygen and carbon and their application to sedimentologic and paleoenvironmental problems. In *Stable Isotopes in Sedimentary Geology. SEPM Short Course* no. 10, pp. 1–151 (ed. M. A. Arthur, T. F. Anderson, I. R. Kaplan, J. Veizer, and L. S. Land). Society of Economic Paleontologists and Mineralogists, Tulsa.

Andrews, P. 1999. Taphonomy of the Shuwaihat proboscidean, late Miocene, Emirate of Abu Dhabi, United Arab Emirates. Chap. 24 in *Fossil Vertebrates of Arabia*, pp. 338–53 (ed. P. J. Whybrow and A. Hill). Yale University Press, New Haven.

Dickson, J. A. D. 1966. Carbonate identification and genesis as revealed by staining. *Journal of Sedimentary Petrology* 36: 491–505.

Folk, R. S., and Pittman, J. S. 1971. Length-slow chalcedony: a new testament for vanished evaporites. *Journal of Sedimentary Petrology* 27: 3–26.

Friedli, H., Lötscher, H., Oeschger, H., Siegenthaler, U., and Stauffer, B. 1986. Ice core record of the $^{13}C/^{12}C$ ratio of atmospheric CO_2 in the past two centuries. *Nature* 324: 237–38.

Kingston, J. D. 1999. Isotopes and environments of the Baynunah Formation, Emirate of Abu Dhabi, United Arab Emirates. Chap. 25 in *Fossil Vertebrates of Arabia,* pp. 354–72 (ed. P. J. Whybrow and A. Hill). Yale University Press, New Haven.

Silyn-Roberts, H., and Sharp, R. M. 1986. Preferred orientation of calcite in the *Aepyornis* eggshell. *Journal of Zoology,* London 208: 475–78.

Stern, L. A., Johnson, G. D., and Chamberlain, C. P. 1994. Carbon isotopic signature of environmental change found in fossil ratite eggshells from south Asian Neogene sequence. *Geology* 22: 419–22.

Von Schirnding, Y., van der Merwe, N. J., and Vogel, J. C. 1982. Influence of diet and age on carbon isotope ratios in ostrich eggshell. *Archaeometry* 24: 3–20.

Yurtsever, Y. 1975. *Worldwide Survey of Stable Isotopes in Precipitation. Report of the Section for Isotope Hydrology.* International Atomic Energy Authority, Vienna.

Palaeomagnetic Correlation and Dating of the Baynunah and Shuwaihat Formations, Emirate of Abu Dhabi, United Arab Emirates

ERNIE A. HAILWOOD AND PETER J. WHYBROW

This chapter presents the results of an investigation into the palaeomagnetism and magnetic fabric of the Miocene Baynunah and Shuwaihat Formations of the Emirate of Abu Dhabi. These sediments outcrop in low hills and coastal exposures along the southern shore of the Arabian Gulf, in the Western Region of the Emirate between As Sila and Al Mirfa (fig. 8.1). Their stratigraphy was first documented by Whybrow (1989), who introduced the term Baynunah Formation for the distinctive sequence of entirely nonmarine sediments that outcrop in a 60-metre section at Jebel Barakah. This term subsequently became applied to all of the continental Miocene sediments in this region on the basis of their lithologies. More recently, a major nonsequence has been identified in the Jebel Barakah succession and it has been proposed that the sediments beneath this level may be significantly older than those above (see Whybrow et al., 1999—Chapter 4). The term Shuwaihat Formation has been introduced for the dominantly aeolian sediments beneath this nonsequence; the Baynunah Formation is now restricted to the largely fluvial sediments above it (Friend, 1999; Bristow, 1999—Chapters 5 and 6). A late Miocene age is given to the continental vertebrate fossils from the Western Region (Whybrow and Hill, 1999—this volume) but the general paucity of age-diagnostic fauna in these sediments has hampered attempts at elucidating the relative ages of the two formations, especially the Shuwaihat. Furthermore, rapid lateral facies changes make it difficult to correlate sections across the region on the basis of lithostratigraphy.

The investigation described in this chapter was initiated in 1989 to find out if palaeomagnetic studies would help constrain the age of the vertebrate-bearing sediments. Its aim was to contribute to correlating and dating the exposures of Miocene sediments in the Baynunah region by two means. First, by determining the polarity of the characteristic remanent magnetism in each section, magnetostratigraphy can be used to constrain correlations between sections. Second, a comparison of palaeomagnetic pole positions from the Shuwaihat and Baynunah Formations with the palaeomagnetic apparent polar-wander curve for Arabia has the potential to provide information on the ages of these formations.

SAMPLING

The samples for this palaeomagnetic study were taken during three separate field trips to the area in 1989, 1990, and 1993, respectively. During the first expedition, one large oriented hand sample was collected from an exposure of the Shuwaihat Formation on the northwestern coast of Shuwaihat Island. The sample was taken from a 15 cm bed of indurated fine-grained sandstone, located about 7 metres above the base of the section (Bristow, 1999—Chapter 6). Detailed palaeomagnetic investigations of 12 specimens cut from the sample, using incremental thermal demagnetisation, successfully isolated a stable characteristic magnetisation with a reverse polarity (Whybrow et al., 1990).

 ISBN 0-300-07183-3

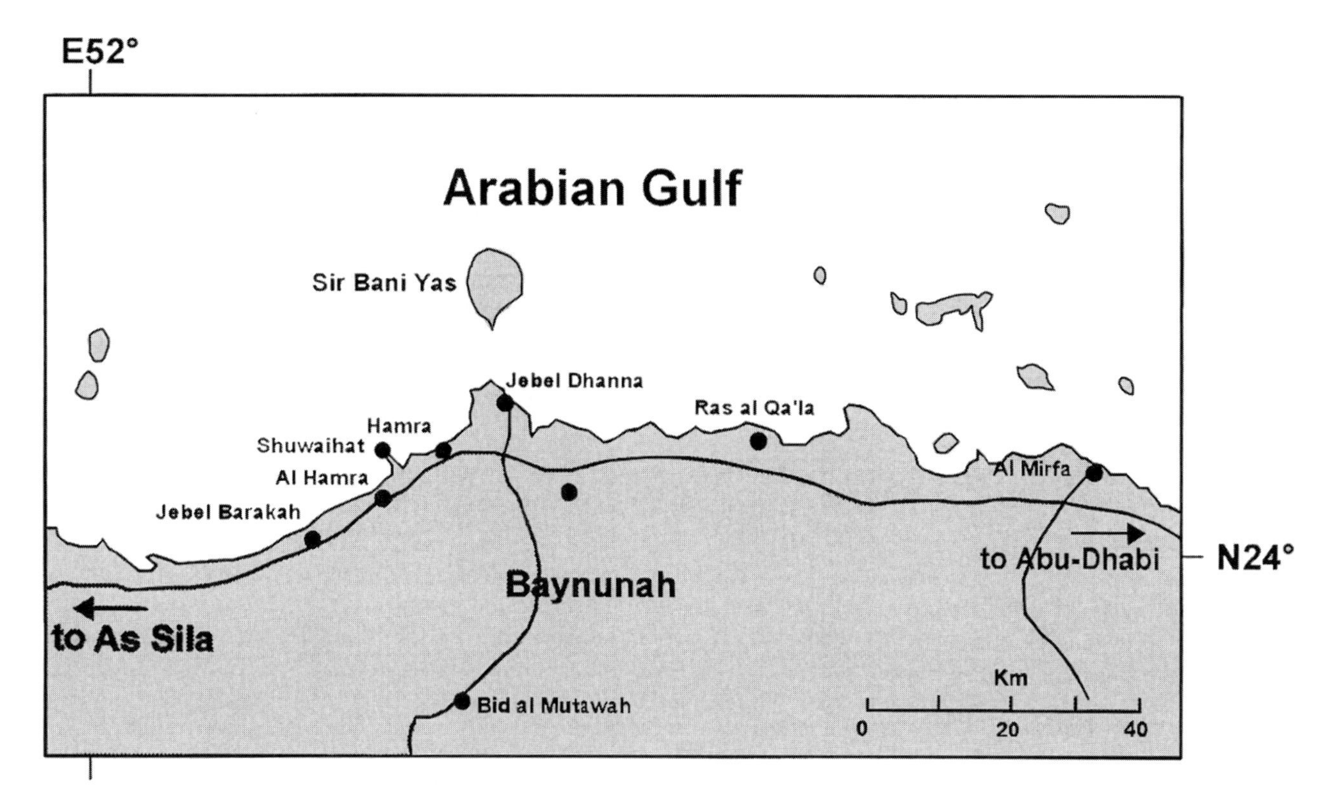

Figure 8.1. Locations of sections sampled for palaeomagnetic investigations in this study.

In the light of this encouraging result, a more extensive collection of samples was made in 1990. This included detailed sampling of the 60-metre type section of the Baynunah Formation at Jebel Barakah (the lower part of which is now assigned to the Shuwaihat Formation) and shorter sections at the Hamra 3 and Jebel Dhanna 3 localities. Oriented mini-core samples (each 25 mm) were taken from the more lithified units in the lower part of the Jebel Barakah section, using a portable petrol-powered rock drill. The softer lithologies higher in the section were sampled by cutting and carefully orienting cube-shaped samples each 8 cm^3 and enclosing these in protective plastic sample boxes. Finally, in 1993, a 13-metre section through the Shuwaihat formation on Shuwaihat Island and a 5-metre section through part of the Baynunah Formation at the Hamra 6 locality were sampled in detail. The latter section outcrops on the southern flank of a low hill and exposes a series of alternating sand and clay layers, each about 10–30 cm thick, possibly formed in a playa lake environment. A few trial samples were taken also at the As Sila and Jebel Mimiyah (Al Mirfa) localities. All of the samples collected on this field trip were oriented hand samples, with volumes of about 1000 cm^3.

Laboratory Palaeomagnetic Analyses

The samples collected during the first two field trips were analysed in the Southampton University palaeomagnetism laboratory, using a two-axis cryogenic magnetometer system manufactured by Cryogenic Consultants Ltd. All of the samples were subjected to incremental demagnetisation analysis, to remove less-stable magnetic overprints and so isolate their characteristic remanent magnetism (ChRM). Alternating field (AF) demagnetisation was used in the first instance, with the applied field being increased in increments of 2.5 or 5 millitesla (mT) up to a maximum of 40 mT. The larger oriented hand samples taken on the 1993 field trip were cut into rectangular blocks, with maximum dimensions of 7 cm, so that they could be analysed on a 2-G "whole-core" cryogenic magnetometer with an integral AF demagnetiser. The larger volumes of these samples provided a stronger magnetic moment, whose direction could be measured more precisely than that of the smaller 8 cm^3 samples taken on the 1990 field trip. Any samples that did not respond to AF treatment were subsequently analysed by the thermal demagnetisation method.

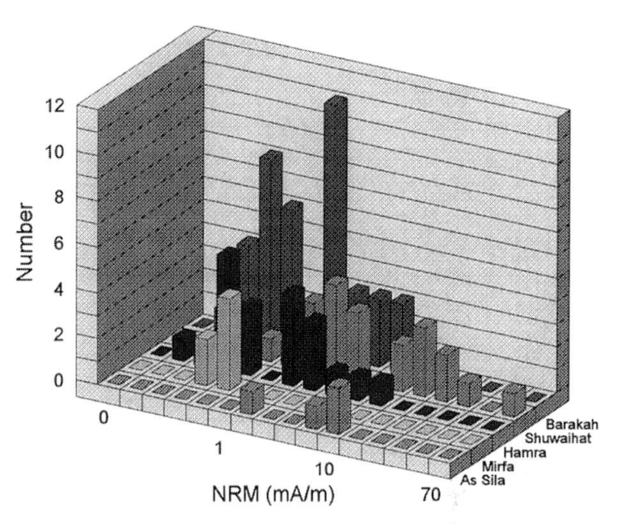

Figure 8.2. Distributions of natural remanent magnetisation (NRM) intensity values in different sections.

Palaeomagnetic Results

Magnetic Intensity

Histograms of the intensity of natural remanent magnetism (NRM) for each of the five principal localities are shown in figure 8.2. The magnetic intensities are quite variable, reflecting differing concentrations (and possibly compositions) of magnetic minerals at different localities. These differences are probably related to differences in the source area of the sediment and the depositional environment. The aeolian sediments at Shuwaihat had the highest values, more than 10 milliamperes per metre (mA/m), whereas more than 50% of the values from the fluvial sediments in the Jebel Barakah section are weaker than 1 mA/m.

Magnetic Stability

Examples of the responses of typical samples to alternating field and thermal demagnetisation are shown in figures 8.3 and 8.4. Most samples respond favourably to AF treatment. For example, sample SH3-1 from Shuwaihat carries a strong magnetic overprint with a westerly declination, defined by the first few demagnetisation steps (fig. 8.3A). Because the direction of this component is totally different from the predicted Miocene or younger geomagnetic field direction in Arabia, it was probably acquired during storage of the sample in a random orientation in the laboratory, prior to measurement. This spurious component is removed progressively during demagnetisation in applied fields up to 20 mT, after which a more stable component with a northerly declination and moderately-steep downwards inclination (corresponding to a *normal polarity*) is isolated. No further significant changes in the direction of magnetisation occur at higher applied fields and the last few points on the vector endpoint diagram lie on a line directed through the origin (right-hand plot in fig. 8.3A). We believe that the ChRM of this sample has been successfully isolated.

Sample Shuwaihat SH4-1, collected from a bed 0.7 metres above Shuwaihat SH3-1, also responds well to AF treatment (fig. 8.3B), but this time the ChRM has a southerly declination and upwards inclination, representative of a *reverse polarity.*

Some samples, such as SL1-2 from As Sila (fig. 8.3C), do not respond well to AF demagnetisation. Thus, although small systematic changes in direction occur during AF treatment up to 40 mT, the

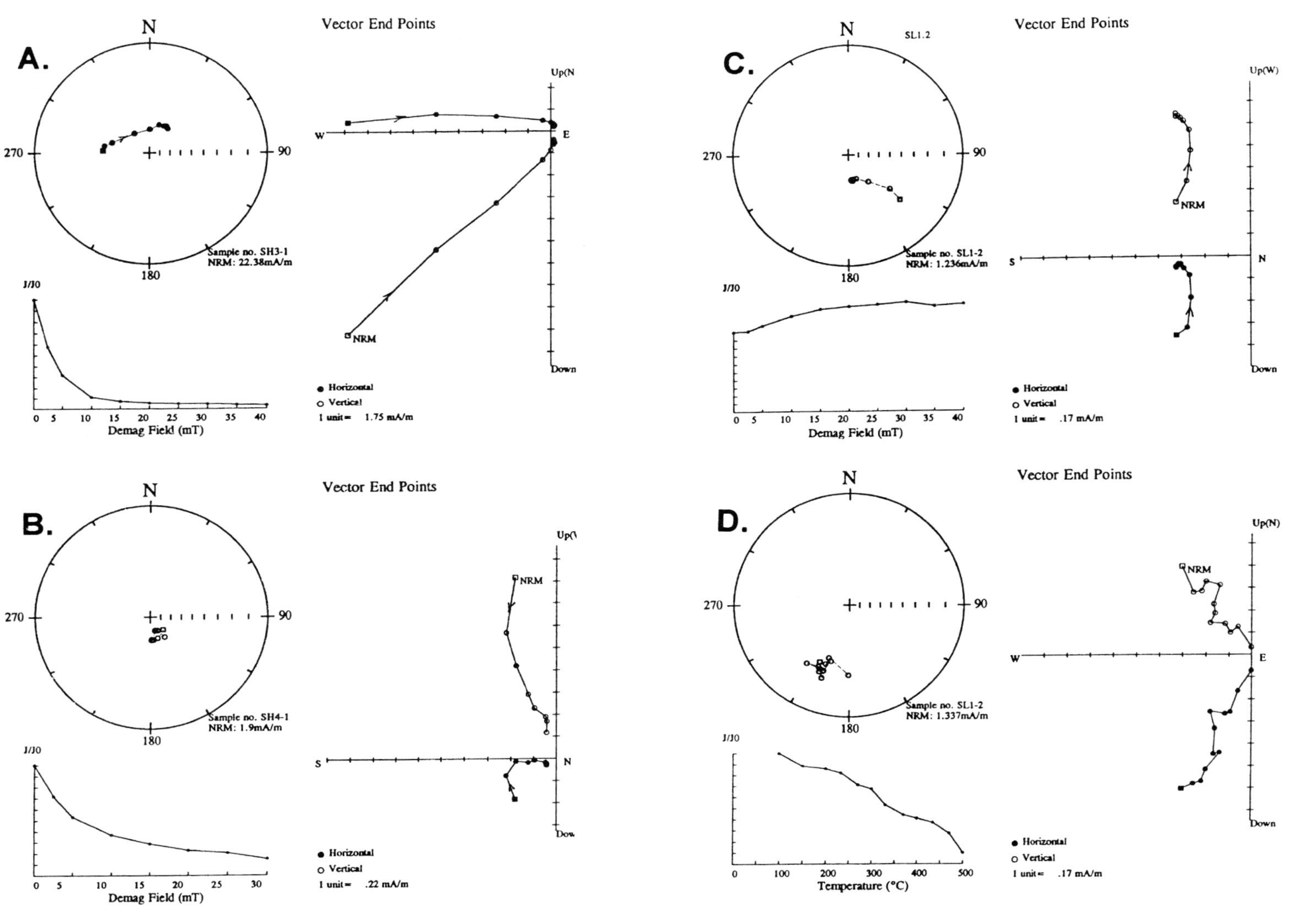

Figure 8.3. Examples of responses of typical samples from Shuwaihat (*A* and *B*) and As Sila (*C* and *D*) to incremental alternating field and/or thermal demagnetisation. For each sample, the directions of magnetisation after each applied field or heating step are shown on a stereographic projection. A graph of the magnetic intensity (J) after each step, normalised by the initial value (J_0) is shown below. On the stereographic projection, closed circles represent positive inclinations and open circles negative inclinations. The square is the NRM direction. The data are shown also on an orthogonal projection (vector endpoint diagram), at the right side of each figure in which the solid and open circles represent projections of the vector onto a horizontal and vertical plane, respectively.

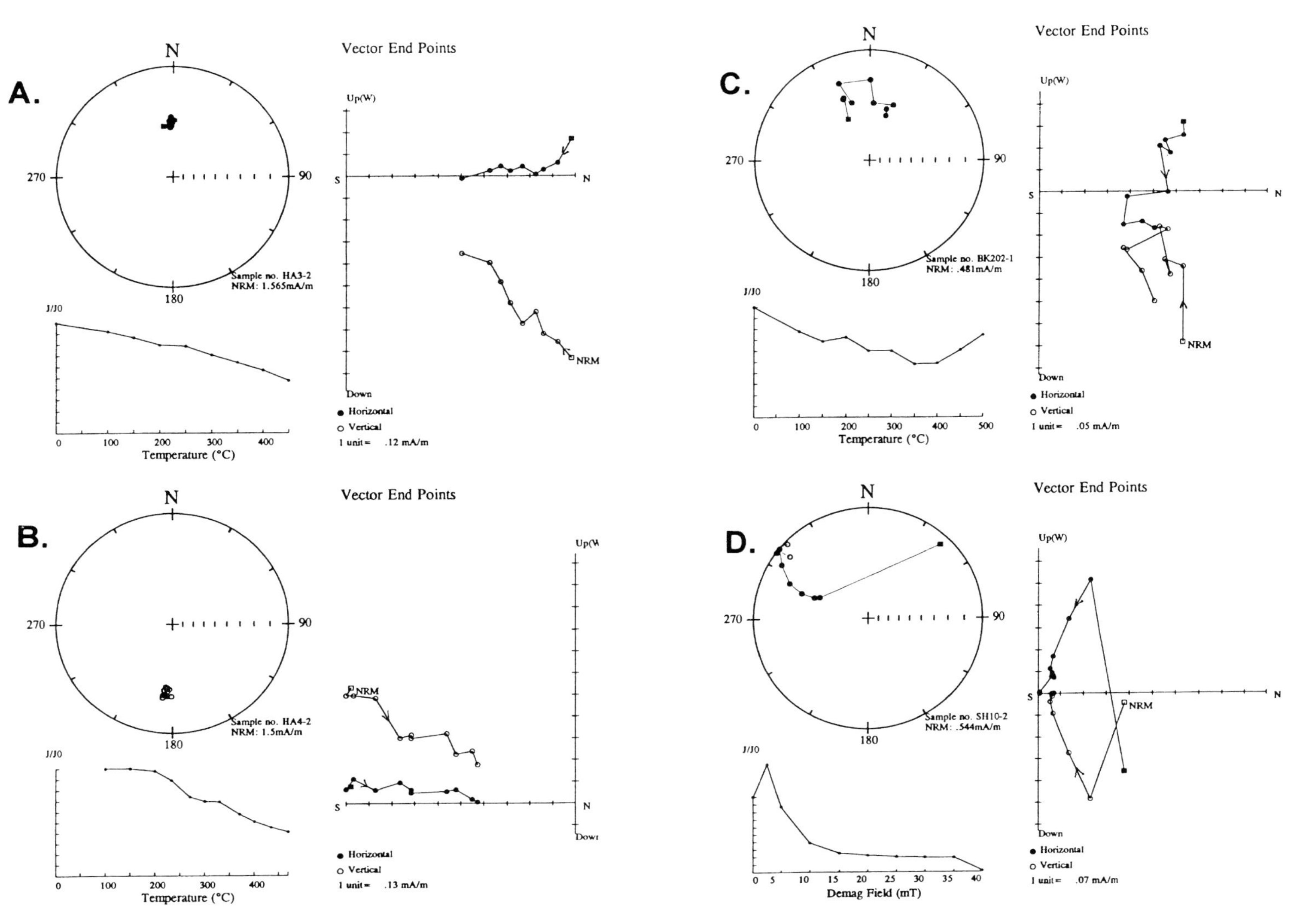

Figure 8.4. Examples of responses of typical samples from Hamra 6 (*A* and *B*), Jebel Barakah (*C*), and Shuwaihat (*D*) to incremental thermal or alternating field demagnetisation. For explanation of plots, see legend to figure 8.3.

magnetic intensity of this sample rises progressively and the points on the vector endpoint diagram are directed *away* from the origin. When this sample was subsequently subjected to thermal demagnetisation, the magnetic intensity was successfully reduced to less than 10% of the NRM value at 500 °C and the ChRM was reasonably well defined by a linear segment through the origin of the vector plot (fig. 8.3D).

Particularly high-quality palaeomagnetic data were obtained from the clay layers sampled at Hamra 6 (fig. 8.4A, B). Thermal demagnetisation successfully isolated a well-defined normal polarity ChRM in sample Hamra HA3-2 (fig. 8.4A) and a reverse polarity ChRM in Hamra HA4-2 (fig. 8.4B).

In general, the palaeomagnetic vectors from the Jebel Barakah section are less well defined than those from the other localities. This is mostly because the Jebel Barakah samples have weaker magnetic intensities and smaller volumes (and hence magnetic moments). An example of the response to thermal demagnetisation of a typical sample from Jebel Barakah (BK202-1) is shown in figure 8.4C. It can be seen that the directional behaviour is more erratic than for the samples from other localities discussed above. The direction does, however, maintain a moderately steep downwards inclination and northerly declination, consistent with a normal polarity. Thus, although palaeomagnetic data from samples such as Barakah BK202-1 are unsuitable for defining the precise palaeomagnetic direction, the magnetic polarity of their ChRM can usually be deduced. Consequently, these samples yield usable magnetostratigraphic information, but cannot contribute to determining the palaeomagnetic pole position for the formation.

Although most samples show quite well-defined normal or reverse magnetic polarities, some exhibit *intermediate* directions. This is the case, for example, with sample Shuwaihat SH10-2 (fig. 8.4D), whose ChRM is defined by quite good linear segments directed through the origin of the vector endpoint plot, but with an anomalous NW declination and very shallow inclination. This particular sample lies between clearly defined normal polarity and reverse polarity zones in the Shuwaihat section (table 8.1). Thus, the intermediate ChRM direction in this sample might represent a record of the transitional geomagnetic field during a polarity reversal. It is equally likely, however, that this direction results from disturbance of the grain structure (and hence remanent magnetism) during sampling. Sev-

Table 8.1. Palaeomagnetic data for the Shuwaihat section

Site no.	Height above sea level (metres)	Mean ChRM direction[a] Dec	Inc	Polarity
SH3	0.7	359.3	59.3	N
SH4	1.40	179.6	–55.5	R
SH5	2.10	352.6	28.3	N
SH6	2.65	356.8	56.7	N
SH8	6.25	18.2	36.0	N
SHA	7.20	175.0	–37.1	
SHB	7.60	186.3	–37.3	R
SH10	10.8	Not defined		No data
SH12	16.2	355.7	49.4	N
SH13	16.7	352.3	39.5	N
SH18	18.2	7.5	56.4	N

[a]Overall mean ChRM direction: Dec = 0.3°, Inc = 45.8°, n = 10, Alpha 95 = 7.6°; pole position: lat = 86.8°, long = 57.0°.

eral samples of soft sand from the Jebel Barakah section showed similar (though rather more erratic) behaviour, which is most likely due to physical disturbance of this type.

MAGNETOSTRATIGRAPHY

Magnetostratigraphy uses records of changes in polarity of the geomagnetic field as a basis for correlation between geological sections of similar age (see Hailwood, 1989). If the polarity reversal sequence in a given formation is particularly well defined and "external" age constraints are available—for example, from biostratigraphy—then it is sometimes possible to match the observed polarity sequence with the appropriate part of the radiometrically calibrated geomagnetic polarity reversal time scale (for example, Cande and Kent, 1995). Numerical ages may be deduced in this way for each of the polarity reversals observed within the sequence.

During the middle to late Miocene, the geomagnetic field frequently reversed. This makes it difficult to match individual reversals with their counterparts on the geomagnetic polarity time scale, thus limiting the use of magnetostratigraphy for dating applications. This high reversal frequency does mean, however, that there is a greater probability of sets of geomagnetic polarity reversals being recorded in Miocene sedimentary sequences, even if the time interval represented by these sequences is quite short. Thus, although magnetostratigraphy is less suitable for dating applications in the Miocene, it is well suited for geological correlations.

The magnetic polarity records defined for all seven sections are shown in figure 8.5. The longest polarity records are those from Jebel Barakah and Shuwaihat Island but quite large gaps, between 1 and 10 metres thick, occur at various levels in these records. These are due either to a lack of suitable, physically coherent, lithologies to sample at these levels, or else to poor magnetic stability and unreliable palaeomagnetic data from samples within the intervals. The reference datum used for stratigraphic height measurements at the Jebel Barakah, Shuwaihat, and Jebel Dhanna 3 sections is the high-water mark on the foreshore. The other localities lie some distance from the coast and their reference datums cannot be tied accurately to sea level.

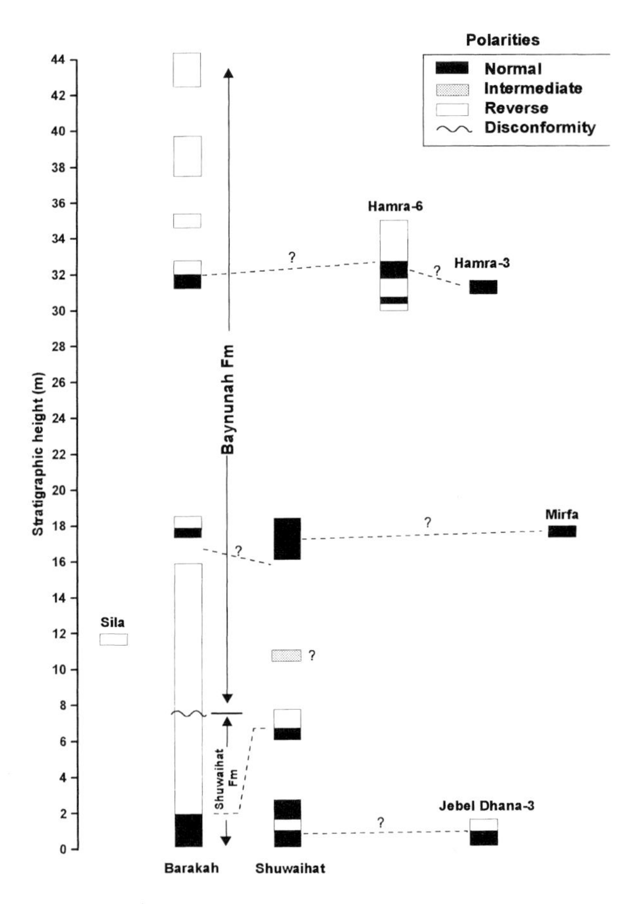

Figure 8.5. Magnetic polarity-reversal sequences at the different localities in the Baynunah and Shuwaihat Formations and tentative magnetostratigraphic correlations.

Correlation Between the Jebel Barakah and Shuwaihat Sections

In the Jebel Barakah section, a short but well-defined normal polarity zone with a maximum thickness of about 2 metres occurs in the lowest part of the exposed Shuwaihat Formation (fig. 8.5). The rest of the Shuwaihat Formation and most of the overlying Baynunah Formation at this locality carry a reverse polarity remanent magnetism, but two further short normal polarity zones occur within the Baynunah Formation, at about 18 metres and 32 metres, respectively, above sea level. The thickness of the lower of these is constrained by adjacent reverse polarity samples to about 1–2 metres. Because of the large gap in the record between 19 and 31 metres above datum, however, the thickness of the uppermost normal polarity zone is poorly defined (fig. 8.5).

The total thickness of the section sampled for magnetostratigraphy on Shuwaihat Island was 19 metres. Only one sample was taken in the interval 8–16 metres above datum, however, and this sample shows an intermediate polarity (fig. 8.4D). Consequently, the magnetic polarity of this interval remains largely unresolved. The data gap is bounded by a clearly defined reverse polarity magnetozone, at least 0.4 metres thick, at the bottom and a normal polarity zone, at least 2 metres thick, at the top (fig. 8.5 and table 8.1).

Two different magnetostratigraphic correlations are possible between the lower parts of the Jebel Barakah and Shuwaihat sections. The simplest of these is illustrated in figure 8.5. This assumes that the intermediate polarity observed in the sample at 10.8 metres above datum in the Shuwaihat section is due to sampling disturbance and that the whole of the (unsampled) interval between 8 and 16 metres actually carries a reverse polarity ChRM, as at Jebel Barakah. The normal polarity interval centred at 17 metres above sea level at Shuwaihat may then be correlated with the thin normal polarity magnetozone at about 17.5 metres in the Jebel Barakah section, and the upwards transition from normal to reverse polarity at about 6.75 metres at Shuwaihat with the top of the normal polarity zone near sea level at Jebel Barakah.

The disconformity that marks the boundary between the Shuwaihat and Baynunah Formations occurs at about 7.5 metres above datum at Jebel Barakah. Thus it lies in the middle of a long reverse polarity interval (fig. 8.5). The position of this formation boundary is less certain at Shuwaihat, but assuming that it is not strongly diachronous, the boundary should lie within a reverse polarity zone, as at Jebel Barakah. If the magneto-stratigraphic correlation proposed in the previous paragraph is correct, then the formation boundary should lie somewhere in the interval 6.75–16 metres at Shuwaihat.

An alternative possible magnetostratigraphic correlation assumes that sediments in the "data gap" between 8 and 16 metres in the Shuwaihat section carry a normal polarity ChRM. In that case, the top of the normal polarity magnetozone near sea level at Jebel Barakah must correlate with a level above 18 metres at Shuwaihat. This would place the boundary between the Shuwaihat and Baynunah Formations some distance above 18 metres at Shuwaihat. On balance, the former magnetostratigraphic correlation is regarded as the most likely one.

Correlation with Other Sections

Although the magnetostratigraphic sections at other localities are much shorter than those at Jebel Barakah and Shuwaihat (fig. 8.5), the observed magnetic polarities can be used to place some constraints on correlations between these sections. In the 5-metre section sampled at Hamra 6, for example, there are two thin but quite distinct normal polarity zones within an otherwise reverse polarity succession. These normal zones have a typical thickness of less than 1 metre but the palaeomagnetic vectors are particularly well defined. Because the shortest known geomagnetic polarity intervals are around 10 000 years, this implies either a slow rate of sediment accumulation, or else intermittent sedimentation at this locality. It is possible that the upper of these normal polarity intervals may correlate with the short normal interval observed at about 31 metres in the Barakah section. The thin normal polarity zone observed near the top of the small roadside jebel at Hamra 3 may also correlate with this interval. These links, however, are poorly constrained.

A 2-metre section sampled on the foreshore about 200 metres west of the Dalma Ferry Terminal (the "Jebel Dhanna 3" locality—now buried by reclamation work) records a clear upwards transition from normal to reverse polarity. This may correlate with the base of the thin reverse magnetozone near sea level at Shuwaihat (fig. 8.5). A well-defined reverse polarity was observed at two closely spaced sites near As Sila—as expected, if this section correlates with the Shuwaihat Formation at Jebel Barakah. Finally, two thin clay layers sampled from the lower part of the isolated pinnacle of Jebel Mimiyah, near Al Mirfa, carry a normal polarity ChRM, which may correlate with the normal polarity units at about 17 metres above sea level at Shuwaihat and Jebel Barakah.

PALAEOMAGNETIC POLE POSITIONS AND DATING APPLICATIONS

Pole Positions for the Shuwaihat and Baynunah Formations

The quality of the palaeomagnetic data from the Barakah section is adequate to establish the general magnetic polarity sequence at this locality, but the directions of ChRM are not defined with sufficient precision for determinations of palaeomagnetic pole positions. The precision of the palaeomagnetic data from the larger samples taken at Shuwaihat and Hamra 6 is much higher and well-defined pole positions have been derived for the Shuwaihat Formation at the former locality and the Baynunah Formation at the latter.

The ChRM directions for all samples at these two localities are plotted in figure 8.6, and the mean directions for each stratigraphic level (site) are listed in tables 8.1 and 8.2. The overall mean direction for each locality is also listed, together with the corresponding palaeomagnetic pole position and 95% confidence angle (Alpha 95). The palaeomagnetic vectors for the two localities fall into two distinct antipodal groups, corresponding with normal and reverse magnetic polarities (fig. 8.6). After inverting the vectors for each reverse polarity site, the overall mean direction for the Shuwaihat Formation at Shuwaihat has a slightly more northerly declination (0.3°) and steeper inclination (45°) than the Baynunah Formation at Hamra 6 (351.6° and 34.3°, respectively). To calculate a meaningful palaeomagnetic pole position for each of these formations, based on the axial geocentric dipole field model, the overall time interval represented by the palaeomagnetic data at each section must exceed 10 000 years (for example, Van der Voo, 1993). The existence of a record of several geomagnetic polarity reversals at each locality confirms that this is the case in our study.

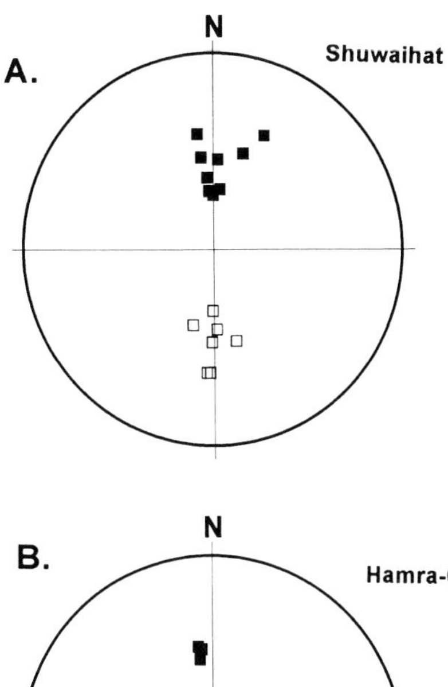

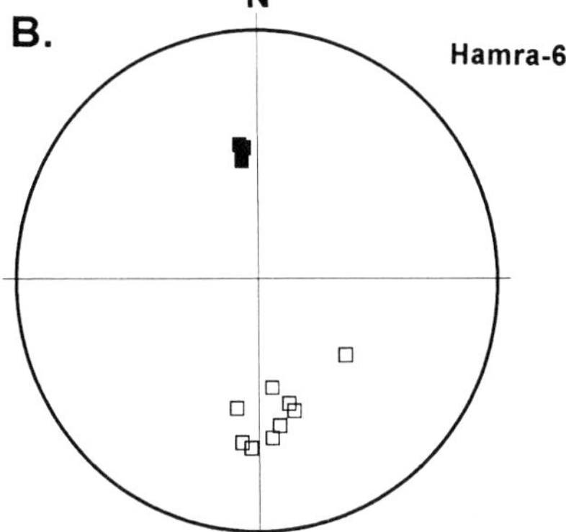

Figure 8.6. Directions of magnetisation in individual samples from Shuwaihat (*A*) and Hamra 6 (*B*) sections. Stereographic projection: solid symbols = lower hemisphere (positive inclination) and open symbols = upper hemisphere (negative inclination).

The Arabian Cenozoic Apparent Polar-Wander Curve

The Cenozoic apparent polar-wander (apw) curve for Arabia, based largely on the compilation of palaeomagnetic data from this plate by Hussain and Bakor (1989), is shown in figure 8.7. The individual poles plotted in this diagram are listed in table 8.3. With the exception of pole 8, derived from the Aden volcanics, these poles fall in a consistent sequence

Table 8.2. Palaeomagnetic data for Hamra 6 section

Site no.	Height above datum (metres)	Mean ChRM direction[a] Dec	Inc	Polarity
HA1	5.0	171.3	–28.2	R
HA2	3.2	171.9	–42.5	R
HA3	2.2	357.0	36.6	N
HA4	1.2	160.5	–34.4	R
HA5	0.8	182.4	–27.0	R
HA6	0.5	353.0	36.0	N
HA7	0	164.5	–34.2	R

[a]Overall mean ChRM direction: Dec = 351.6°, Inc = 34.3°, n = 7, Alpha 95 = 6.0°; pole position: lat = 80.6°, long = 290.3°.

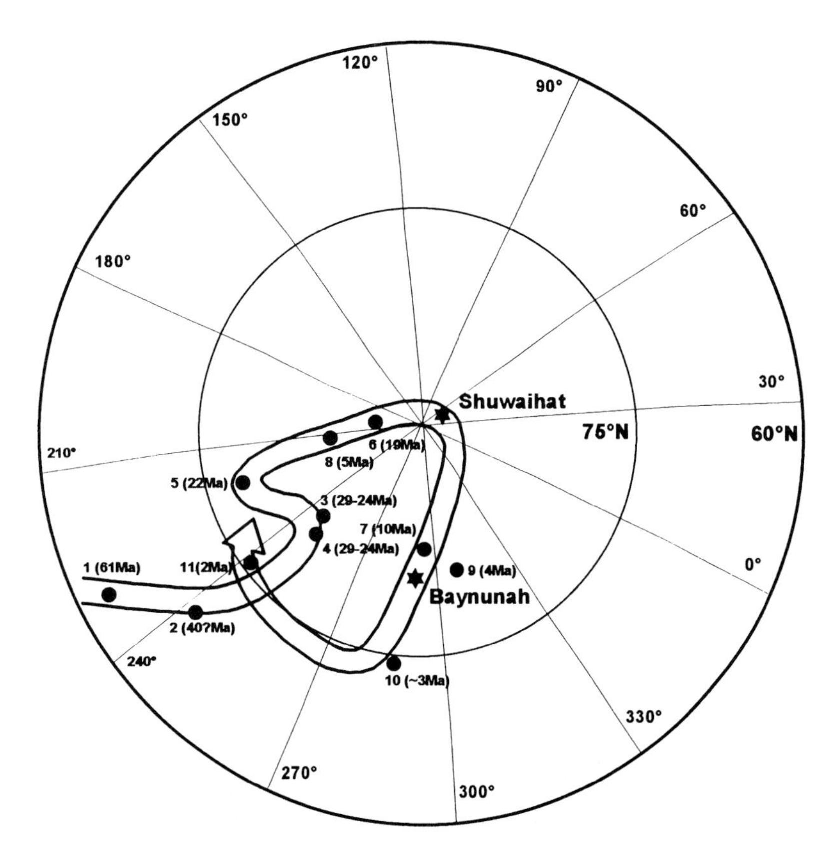

Figure. 8.7. Cenozoic apparent polar-wander curve for Arabia. Pole reference numbers refer to table 8.3; ages are shown in millions of years. The pole positions determined for the Shuwaihat and Baynunah Formations in the present study are shown by stars.

Table 8.3. Cenozoic palaeomagnetic pole positions for the Arabian Plate

Ref. code (fig. 8.7)	Age (Ma)	Pole position		Formation
		Lat (°N)	Long (°E)	
1	60.7	62.9	231.2	Usfan basalts[1]
2	40?	68.0	242.8	Usfan basalts[1]
3	29–24	80.0	247.9	Usfan basalts[1]
4	29–24	78.8	248.8	Sarrat trachytes[2]
5	21.8	75.7	222.0	Sarrat trachytes[2]
6	19	86.3	199.5	Usfan basalts[1]
7	10	81	297	Jebel Khariz volcanics[3]
8	5	83	210	Aden volcanics[3]
9	3.4–3.8	79.3	309.2	Madina Road basalt, Saudi Arabia[4]
10	~3	74.4	285.5	Madina Road basalt, Saudi Arabia[4]
11	1.8	73.8	241.3	Jordan basalts[5]

References: [1]Yousif and Beckmann, 1981; [2]Kellog and Reynolds, 1980; [3]Irving and Tarling, 1961; [4]Hussain and Bakor, 1989; [5]Sallomy and Krs, 1980.

and the apw curve drawn through them describes a clockwise loop between about 40 and 2 million years (Ma) ago. The apw path for Africa shows a similar broad clockwise loop for this period, but displaced about 10° to the east. Reconstruction of the former positions of the Arabian and Nubian plates, by clockwise rotation of Arabia through an angle of 9° about a rotation pole at 29.3° N, 27.1° E to close the Red Sea (Hall, 1979), moves the Arabian and African apw paths into closer conformity. A further shift of the Arabian Plate by 120 km in a SSE direction, parallel to the Red Sea axial trough, closes the Gulf of Aden and improves the fit of the two apw curves (Hussain and Bakor, 1989). The apparent inconsistency of the Aden volcanics pole with other palaeomagnetic data for Arabia may reflect the fact that this pole is based on an early study by Irving and Tarling (1961) before modern methods for palaeomagnetic data analysis were available. Alternatively, it could reflect the effects of local undetected tectonic rotations in the Aden area, or inaccuracies in the age of these volcanics (an age of less than 2 Ma would conform better with other palaeomagnetic data for Arabia).

This overall consistency between the Arabian and African apw curves confirms the validity of the clockwise loop observed in both curves for the interval 40–2 Ma ago. Whether this loop reflects actual motion of the Afro-Arabian Plate during this time, or includes an element of "true" polar wander (that is, departure of the geomagnetic dipole axis from the geographic axis for a brief period), is uncertain. The existence of this characteristic feature in the Arabian apw curve, however, greatly increases the potential value of this curve for palaeomagnetic dating applications.

Palaeomagnetic Dating of the Shuwaihat and Baynunah Formations

Constraints can be placed on the ages of the Shuwaihat and Baynunah Formations of Abu Dhabi by comparing the positions of the palaeomagnetic poles derived from these formations with the age-calibrated apw curve for Arabia (fig. 8.7). The palaeomagnetic pole position for the Baynunah Formation, based on the mean direction of ChRM in the Hamra 6 section, lies just south of poles 7 and 9 (which have ages of about 10 Ma and 4 Ma, respectively) and north of pole 10 (with an age of about 3 Ma). Pole 7, from the Jebel Khariz volcanics (like pole 8 from the Aden volcanics), is based on an early palaeomagnetic study, of uncertain reliability. However, if the published age and position of this pole are accepted at their face value, then the similarity of pole 7 with pole 9, from the Madina Road basalts of Saudi Arabia (dated at about 4 Ma), implies little movement of the palaeomagnetic pole relative to Arabia between 10 Ma and 4 Ma ago. Alternatively, it may reflect uncertainties in the ages of these two geological formations; in which case, the overall mean age of about 7 Ma provides the best estimate of the true age of these two poles.

If the position of the Baynunah Formation pole is interpolated between the mean of poles 7 and 9 (with an assumed age of about 7 Ma) and pole 10 (about 3 Ma), then the resulting age estimate for the Baynunah Formation is about 6 Ma. If the uncertainties in the ages of the adjacent poles are taken into account, however, the associated uncertainty in this age estimate is likely to be about ± 3 Ma.

The pole position from the Shuwaihat Formation is located some 10° north of that from the Baynunah Formation. The Shuwaihat pole falls on the Arabian apw curve, between pole 6 (19 Ma) and poles 7 and 9 (mean age 7 Ma). Assuming a uniform rate of polar motion along the smooth curve through these two poles, the interpolated age of the Shuwaihat Formation is about 15 Ma. As with the Baynunah Formation, the uncertainty in this age estimate is considered to be about ± 3 Ma.

Conclusions

This study of the palaeomagnetism of the Miocene deposits of the Abu Dhabi region has identified the presence of a well-defined characteristic remanent magnetisation, believed to be of primary origin, in most of the samples analysed from the Shuwaihat Formation on Shuwaihat Island and the Baynunah Formation at the Hamra 6 locality. The directions of magnetisation in the more weakly magnetic sediments at Jebel Barakah are less well defined and generally are unsuitable for determining palaeomagnetic pole positions. Usable magnetostratigraphic information was obtained, however, from about 50% of the Jebel Barakah samples.

A tentative magnetostratigraphic correlation, based on a synthesis of the magnetic polarity information from the seven localities studied, is shown in figure 8.5. This interpretation links the normal polarity magnetozone of about 2 metres at the base of the exposed Shuwaihat Formation at Jebel Barakah with the normal polarity magnetozone of about 7 metres in the lowest part of the section exposed on the north shore of Shuwaihat Island. A thin normal polarity magnetozone at about 17.5 metres above datum at Jebel Barakah is tentatively linked with a somewhat thicker normal polarity interval at a similar height in the section at Shuwaihat. Because of a combination of unsuitable lithologies and poor magnetic stability, no usable palaeomagnetic data could be obtained for the interval between 8 and 16 metres above datum at Shuwaihat, but correlations suggest that most of this interval carries a reverse polarity ChRM. This reverse polarity interval, which then spans the boundary between the Shuwaihat and Baynunah Formations, has a maximum thickness of about 9.5 metres in the Shuwaihat section compared with 15 metres at Jebel Barakah. The difference may be explained by the existence of a larger hiatus at the disconformity between the Shuwaihat and Baynunah Formations in the Shuwaihat section than in the Jebel Barakah section.

Possible magnetostratigraphic correlations between the Jebel Barakah and Shuwaihat sections and the shorter sections at the other localities are indicated in figure 8.5.

The geological ages of the Shuwaihat and Baynunah Formations can be constrained by comparing the palaeomagnetic pole positions derived for these formations with the Cenozoic apw curve for Arabia. The results suggest probable ages of about 15 ± 3 Ma (that is, middle Miocene) for the Shuwaihat Formation and about 6 ± 3 Ma (that is, late Miocene) for the Baynunah Formation. This difference in palaeomagnetically determined ages for the two units supports their proposed separation into different geological formations. It is likely, therefore, that part of the Shuwaihat Formation is coeval with the Hofuf Formation of the Kingdom of Saudi Arabia (see Whybrow et al., 1999—Chapter 4).

ACKNOWLEDGEMENTS

We thank the Abu Dhabi Company for Onshore Oil Operations (ADCO) for their continuing grant to The Natural History Museum, London, that has made our work possible and to ADCO and the Ministry of Higher Education and Scientific Research, United Arab Emirates, for their invitation to participate in the First International Conference on the Fossil Vertebrates of Arabia held in the Emirate of Abu Dhabi, March 1995.

REFERENCES

Bristow, C. S. 1999. Aeolian and sabkha sediments in the Miocene Shuwaihat Formation, Emirate of Abu Dhabi, United Arab Emirates. Chap. 6 in *Fossil Vertebrates of Arabia,* pp. 50–60 (ed. P. J. Whybrow and A. Hill). Yale University Press, New Haven.

Cande, S. C., and Kent, D. V. 1995. Revised calibration of the geomagnetic polarity time scale for the late Cretaceous and Cenozoic. *Journal of Geophysical Research* 100, B4: 6093–95.

Friend, P. F. 1999. Rivers of the Lower Baynunah Formation, Emirate of Abu Dhabi, United Arab Emirates. Chap. 5 in *Fossil Vertebrates of Arabia,* pp. 38–49 (ed. P. J. Whybrow and A. Hill). Yale University Press, New Haven.

Hailwood, E. A. 1989. The role of magnetostratigraphy in the development of geological time scales. *Paleoceanography* 4: 1–18.

Hall, S. A. 1979. A total intensity magnetic anomaly map of the Red Sea and its interpretation. *United States Department of the Interior Geological Survey,* Saudi Arabian Mission IR: 1–275.

Hussain, A. G., and Bakor, A. R. 1989. Petrography and palaeomagnetism of the basalts, southwest Harrat Rahat, Saudi Arabia. *Geophysical Journal International* 99: 687–98.

Irving, E., and Tarling, D. H. 1961. The palaeomagnetism of the Aden Volcanics. *Journal of Geophysical Research* 66: 549–55.

Kellog, K. S., and Reynolds, R. C. 1980. Paleomagnetic study of the As Sarat Volcanic Field, Southwestern Saudi Arabia. *United States Department of the Interior Geological Survey,* Saudi Arabian Mission 14: 1–360.

Sallomy, J. T., and Krs, M. 1980. A palaeomagnetic study of some igneous rocks from Jordan. *Proceedings. Evolution Mineralization of the Arabian–Nubian Shield, FES Bulletin* 3: 155–64.

Van der Voo, R. 1993. *Paleomagnetism in the Atlantic, Tethys and Iapetus Oceans.* Cambridge University Press, Cambridge.

Whybrow, P. J. 1989. New stratotype; the Baynunah Formation (Late Miocene), United Arab Emirates: Lithology and palaeontology. *Newsletters on Stratigraphy* 21: 1–9.

Whybrow, P. J. and Hill, A. eds. 1999. *Fossil Vertebrates of Arabia.* Yale University Press, New Haven.

Whybrow, P. J., Friend, P. F., Ditchfield, P. W., and Bristow, C. S. 1999. Local stratigraphy of the Neogene outcrops of the coastal area: Western Region, Emirate of Abu Dhabi, United Arab Emirates. Chap. 4 in *Fossil Vertebrates of Arabia,* pp. 28–37 (ed. P. J. Whybrow and A. Hill). Yale University Press, New Haven.

Whybrow, P. J., Hill, A., Yasin al-Tikriti, W., and Hailwood, E. A. 1990. Late Miocene primate fauna, flora and initial palaeomagnetic data from the Emirate of Abu Dhabi, United Arab Emirates. *Journal of Human Evolution* 19: 583–88.

Yousif, I. A., and Beckmann, G. E. J. 1981. A palaeomagnetic study of some Tertiary and Cretaceous rocks in Western Saudi Arabia: Evidence for the movement of the Arabian Plate. *Bulletin of the Faculty of Earth Sciences, King Abdulaziz University* 4: 89–106.

Stable Isotope Analyses and Dating of the Miocene of the Emirate of Abu Dhabi, United Arab Emirates

Ross G. Peebles

The Miocene of Abu Dhabi can be divided into three formational units (from oldest to youngest): the Gachsaran Formation, consisting chiefly of carbonates and evaporites with minor basal siliciclastics; the Shuwaihat Formation, consisting of aeolian, sabkha, and fluvial siliciclastics; and the Baynunah Formation, consisting of fluvial siliciclastics capped by thin carbonates (Alsharhan, 1989; Whybrow, 1989; Whybrow et al., 1999—Chapter 4). In far western Abu Dhabi, near the border with Saudi Arabia, rare outcrops of carbonates in the Dam Formation (Gachsaran Formation equivalent) have been observed by myself and others (Whybrow et al., 1999; Bristow, 1999; Ditchfield, 1999—Chapters 4, 6, and 7). In Abu Dhabi, the Gachsaran Formation is generally believed to have been deposited in the Early Miocene, based on regional correlation to the Gachsaran Formation of southwestern Iran (Jones and Racey, 1994). A paucity of stratigraphically significant fossils, however, has hindered confident chronostratigraphic assignment of this unit. Recently, Peebles et al. (1995) and Galal (1995) have used strontium isotope stratigraphy to date the Gachsaran Formation from eastern and offshore Abu Dhabi. Peebles et al. (1995) assigned a Burdigalian age to a Gachsaran Formation dolomite–evaporite succession beneath the Abu Dhabi sabkha, and Galal (1995) reported an Early–Middle Miocene age for a Lower Fars (Gachsaran Formation synonym) carbonate–siliciclastic succession outcropping on Dalma Island, offshore Abu Dhabi. Data from these two studies are included in this chapter. The Dam Formation in eastern Saudi Arabia and Qatar is considered to be Early Miocene (Burdigalian, 19–16 million years) in age based on its marine and terrestrial vertebrate fossil assemblage (Adams et al., 1983; Whybrow et al., 1987). Bristow (1999—Chapter 6) reports the presence of the bivalve *Diplodonta* sp. from a Dam Formation outcrop near the town of As Sila, which suggests a Burdigalian age.

The Shuwaihat Formation has only recently been defined as a stratigraphic unit (Whybrow et al., 1999; Bristow, 1999—Chapters 4 and 6). It unconformably underlies the Baynunah Formation at the type locality (western cliffs of Shuwaihat Island) and unconformably overlies the Dam Formation in far western Abu Dhabi (on the western edge of Sabkhat Matti near As Sila). The rich vertebrate fauna present in the overlying Baynunah Formation is not present in the Shuwaihat Formation. Indeed, no fossil evidence whatsoever has been reported. Based on its stratigraphic position above the Early Miocene (Burdigalian) Dam Formation and below the Late Miocene Baynunah Formation, a Middle–Late Miocene age can be proposed for the Shuwaihat Formation.

The siliciclastic Baynunah Formation, which outcrops in western Abu Dhabi, contains an abundance of Late Miocene vertebrate fossils (Whybrow and McClure, 1981; Whybrow, 1989; Whybrow et al., 1990). Whybrow (1989) suggested an age range of 11.2–5 million years (Ma) for the Baynunah Formation based on the presence of the equid *Hipparion,* while Whybrow et al. (1990)

 ISBN 0-300-07183-3

refined that estimate to a date of around 8 Ma based on the total fossil vertebrate assemblage.

In an effort to further define and refine the stratigraphic position of the Miocene of Abu Dhabi, a stable isotope study was performed to identify isotopic signatures for the Gachsaran, Shuwaihat, and Baynunah Formations. If these signatures are representative of depositional conditions, they can then be correlated to global isotope chronostratigraphic curves to produce an age determination for each formation. Such a determination would provide an independent assessment of the stratigraphic age of these formations that could then be compared with palaeontological (biostratigraphic) and palaeomagnetic (magnetostratigraphic) data.

Analysis Programme

Dolomite and marine evaporite (sulphate) samples of the Gachsaran Formation were collected from a series of shallow boreholes (about 30 metres deep) drilled into the Abu Dhabi sabkha near the city of Abu Dhabi, from outcrops in eastern and western Abu Dhabi, and from outcrops on two offshore islands (fig. 9.1). Sulphate samples from the Shuwaihat and Baynunah Formations were also collected from outcrops at Shuwaihat Island and near As Sila in western Abu Dhabi (fig. 9.1). These samples were measured for one or more stable isotope ratios (table 9.1). Analytical results are listed in tables 9.2 and 9.3.

Methodology

Strontium isotope analyses were performed by Isotopic Analytical Services, Ltd (IAS) using a VG Sector 54 thermal ionisation mass spectrometer. Samples were run on tantalum filaments loaded in H_3PO_4. IAS reports a measured value of 0.709 190 for the Holocene Marine Carbonate (HMC) standard and a mean measured value of 0.710 266 for the NTIS (NBS) 987 standard (see table 9.3 for a detailed listing). Carbon, oxygen, and sulphur isotope analyses were performed by the Russian Academy of Sciences.

Samples for carbon and oxygen isotope analysis were first powdered and then reacted with 100% orthophosphoric acid under a vacuum at 75 °C (McCrea, 1950). The CO_2 gas that evolved was simultaneously collected via freezing using liquid nitrogen, purified, and analysed using a mass spectrometer. Carbon and oxygen isotope ratios are expressed relative to the Peedee Formation belemnite (PDB) standard. Oxygen isotope ratios can also be expressed relative to the Standard Marine Ocean Water (SMOW) standard; however, in this study all $\delta^{18}O$ values are reported as per mil (‰) PDB. The $\delta^{18}O$ value for dolomite samples was corrected for temperature effects that occur during reaction (Hoefs, 1980) by subtracting 9.5‰ (SMOW) from the initial measured value to derive the "true" $\delta^{18}O$ value for dolomite (Swart et al., 1991). The standard deviation determined for replicate analyses was

Table 9.1. Number of samples collected from Miocene formations in the Emirate of Abu Dhabi

Isotope analysis	Gachsaran Formation*	Shuwaihat Formation	Baynunah Formation
$^{87}Sr/^{86}Sr$ strontium isotope ratio	18	2	1
$^{18}O/^{16}O$ oxygen isotope ratio ($\delta^{18}O$)	89	2	1
$^{13}C/^{12}C$ carbon isotope ratio ($\delta^{13}C$)	89	2	1
$^{34}S/^{32}S$ sulphur isotope ratio ($\delta^{34}S$)	33	2	1

*This includes a Dam Formation sample that underwent strontium isotope analysis. Analytical results are listed in tables 9.2 and 9.3.

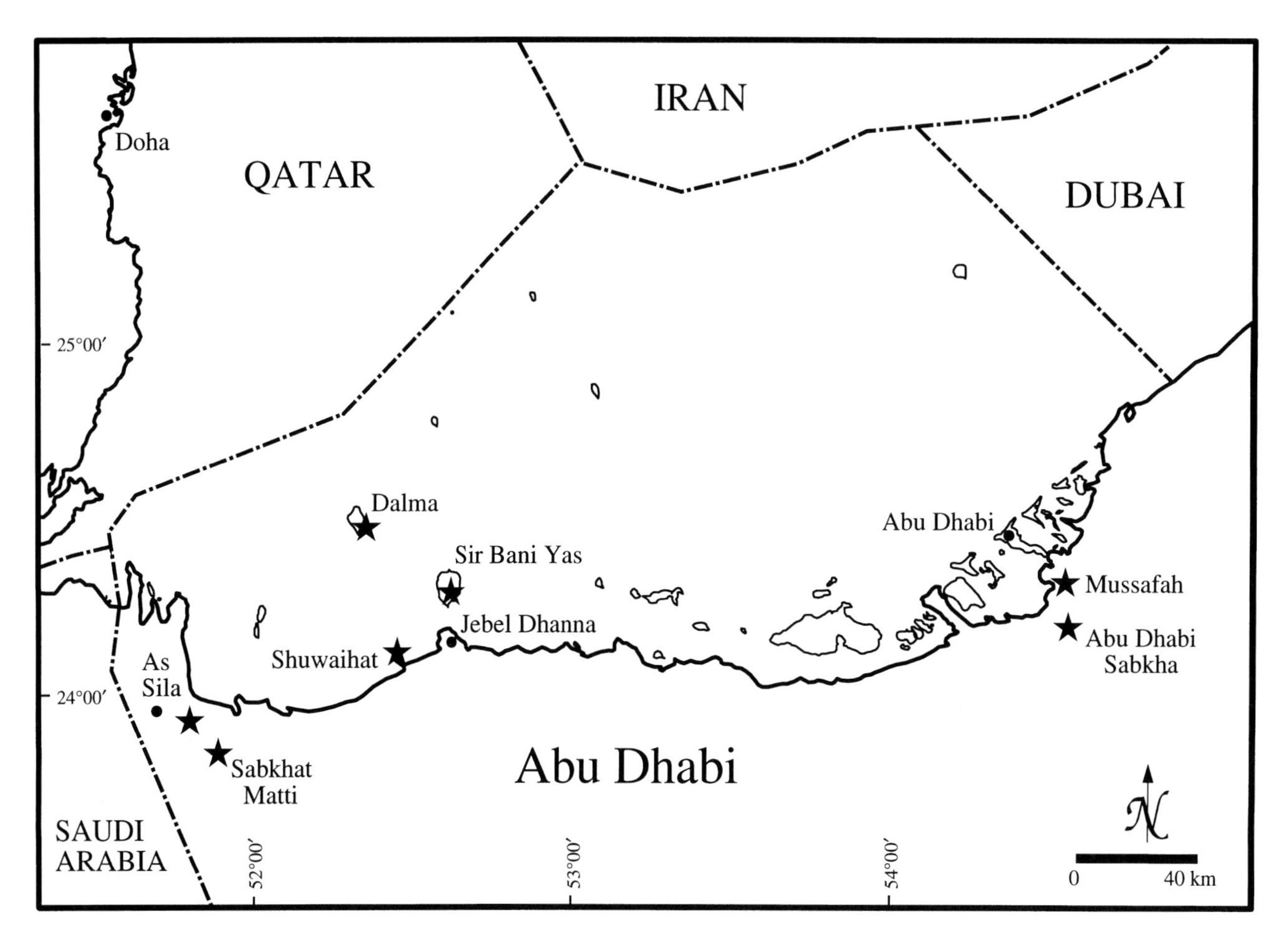

Figure 9.1. Map of the study area. Sample locations are starred. All samples were collected from outcrop except Gachsaran Formation core samples, which were collected from shallow boreholes at Mussafah (near the city of Abu Dhabi).

± 0.2‰ with an analytical precision of measurement always less than ± 0.06‰. The sulphur isotope ratio was measured by mixing the powdered sample with iron and charcoal and then heating at 75 °C to reduce the sulphate. The mixture was then treated with HCl to release H_2S followed by precipitation of CdS. Dried CdS was burned under a vacuum in the presence of CuO at 800 °C to produce SO_2. The ion current of masses 64 and 66 were measured by mass spectrometer. The sulphur isotope ratio measurements are reported as $\delta^{34}S$ values (‰) relative to the meteorite standard. The standard deviation determined for replicate analyses was usually less than ± 0.3‰ with an analytical precision of measurement estimated to be ± 0.02‰.

ISOTOPIC SIGNATURES

Gachsaran and Dam Formations

A total of 51 dolomite and 38 sulphate (35 gypsum, 3 anhydrite) samples from the Gachsaran Formation were taken from the shallow subsurface below the Abu Dhabi sabkha and analysed for $\delta^{13}C/\delta^{18}O$. All revealed strongly positive $\delta^{18}O$ values and positive to strongly positive $\delta^{13}C$ values (table 9.2, fig. 9.2). The dolomites form a relatively closely spaced grouping with $\delta^{18}O$ values ranging from 3.8 to 8.6‰ (average 6.6‰) with $\delta^{13}C$ values ranging from 1.5 to 7.0‰ (average 4.9‰). The sulphates—mainly gypsum with a few anhydrite and celestite samples— are less well clustered, but still reflect positive to strongly positive $\delta^{13}C/\delta^{18}O$ values. Oxygen isotope values for the sulphates range from 1.1 to 7.7‰ (average 5.8‰) while the $\delta^{13}C$ values are more variable, ranging from −2.1 to 7.5‰ (average 3.4‰). Measurement of the $\delta^{34}S$ of 33 gypsum samples produced a coherent cluster of values ranging from 18.7 to 23.5‰ (average 21.2‰). Strontium isotope ratios of 4 dolomite and 14 sulphate (8 gypsum, 4 anhydrite, 2 celestite) samples from outcrops, islands, and the subsurface form a consistent and tightly grouped dataset (table 9.3, figs 9.3 and 9.4), with values ranging from 0.708 438 to 0.708 846 (average 0.708 630).

A crossplot of the sulphur, carbon, and oxygen isotope values presents a generally coherent cluster (grouping), suggesting that all samples were precipitated or dolomitised under common hydrological

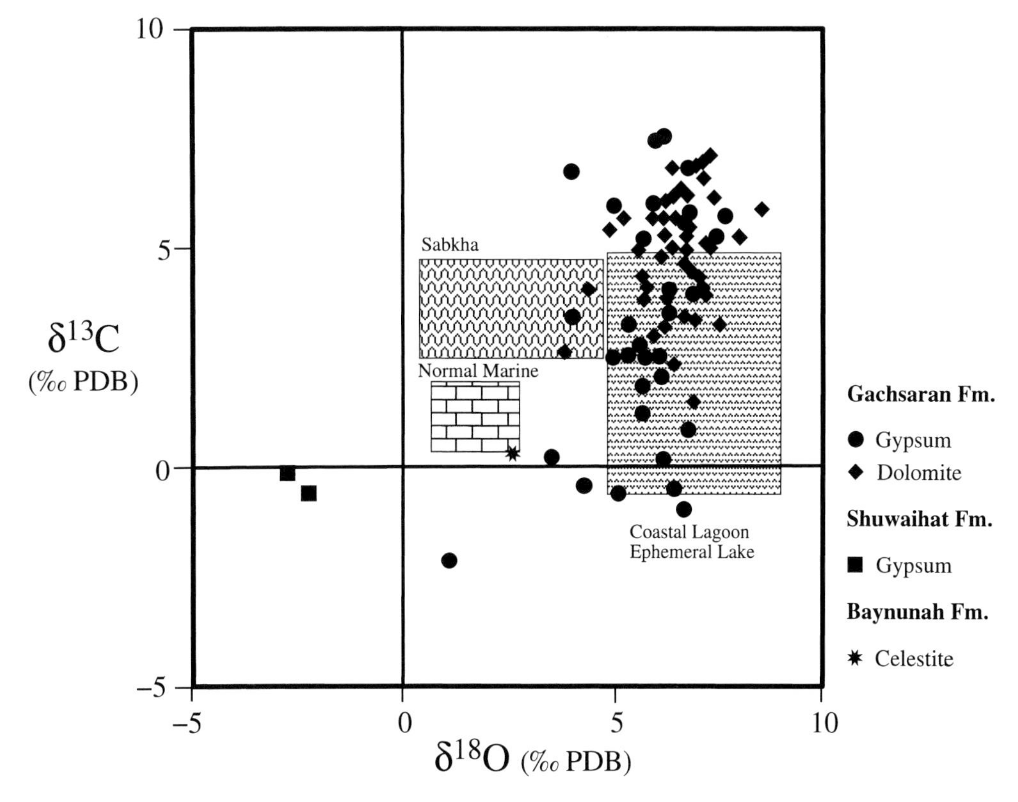

Figure 9.2. Carbon/oxygen crossplot. Carbon and oxygen isotope values for the Gachsaran, Shuwaihat, and Baynunah Formations are plotted. The ranges of $\delta^{13}C/\delta^{18}O$ values for limestones formed in normal marine conditions and dolomites formed in sabkha and coastal lagoon and lake environments are denoted by shaded boxes. These isotopic ranges were derived from Clayton et al. (1968), McKenzie (1981), Allan and Wiggins (1993), and Hodell and Woodruff (1994). Note that interstitial, microcrystalline dolomite is the source of carbon (from which $\delta^{13}C$ is measured) in the evaporite samples.

Table 9.2. **Stable isotope database: Mocene, Emirate of Abu Dhabi**

Location	Depth (metres)	$\delta^{13}C$ (‰ PDB)	$\delta^{18}O$ (‰ PDB)	$\delta^{34}S$ (‰)	Lithology
Outcrop: Shuwaihat and Baynunah Formations					
Shuwaihat Island	Outcrop	−0.6	−2.2	13.8	Gypsum
	Outcrop	0.3	2.6	16.0	Celestite
As Sila	Outcrop	−0.1	−2.7	14.1	Gypsum
Subsurface: Gachsaran Formation					
Borehole M-2	9.0	3.9	7.2	—	Dolomite
	10.0	4.4	6.8	—	Dolomite
	11.0	3.3	6.9	—	Dolomite
	12.0	4.9	7.1	—	Dolomite
	13.0	6.1	6.5	—	Dolomite
	14.0	5.6	6.7	—	Dolomite
	15.0	5.3	6.7	—	Dolomite
	16.0	5.0	7.2	21.3	Gypsum
	17.0	5.2	5.8	20.1	Gypsum
	18.0	4.9	5.7	—	Dolomite
	19.0	2.4	5.8	20.8	Gypsum
	20.0	1.2	5.7	21.3	Gypsum
	21.0	3.4	4.0	20.2	Gypsum
	22.0	5.6	6.5	—	Dolomite
	23.0	6.1	6.7	—	Dolomite
	24.0	7.4	6.1	19.1	Gypsum
	25.0	4.9	6.4	20.3	Gypsum
	26.0	5.0	7.3	—	Dolomite
	27.0	5.6	6.3	—	Dolomite
	28.0	6.5	7.1	—	Dolomite
	29.0	6.0	6.3	—	Dolomite
	29.7	—	—	—	Dolomite
Borehole M-3	9.0	2.2	6.4	20.4	Gypsum
	10.0	6.0	6.0	20.8	Gypsum
	11.0	3.4	6.3	—	Gypsum
	12.0	3.1	6.3	—	Dolomite
	13.0	5.5	6.7	20.2	Gypsum
	14.0	5.5	6.0	—	Dolomite
	15.0	5.4	4.9	—	Dolomite
	16.0	6.7	6.9	—	Dolomite
	17.0	5.2	7.4	21.6	Gypsum
	18.0	5.6	6.1	20.5	Gypsum
	19.0	5.2	8.0	—	Dolomite
	20.0	3.9	6.9	—	Dolomite
	21.0	2.7	5.7	22.3	Gypsum
	22.0	4.8	6.5	21.4	Gypsum
	23.0	6.2	6.6	—	Dolomite
	24.0	4.2	7.0	—	Dolomite
	25.2	3.8	6.3	—	Dolomite
	26.0	5.9	5.0	21.3	Gypsum
	27.0	5.5	6.1	20.4	Gypsum
	28.0	4.9	6.4	—	Dolomite
	29.0	5.7	7.7	21.3	Gypsum
	29.9	6.8	7.2	—	Dolomite

Table 9.2 *(continued)*

Location	Depth (metres)	$\delta^{13}C$ (‰ PDB)	$\delta^{18}O$ (‰ PDB)	$\delta^{34}S$ (‰)	Lithology
Borehole M-4	8.0	3.2	7.5	—	Dolomite
	9.0	5.6	5.2	—	Dolomite
	10.0	2.9	6.0	—	Dolomite
	11.0	3.6	6.3	—	Dolomite
	12.0	6.8	7.0	—	Dolomite
	13.0	6.7	6.5	—	Dolomite
	14.0	5.8	8.6	—	Dolomite
	15.0	4.3	5.7	—	Dolomite
	16.0	4.6	6.6	—	Dolomite
	17.0	6.7	4.0	—	Anhydrite
	18.0	2.5	5.3	—	Anhydrite
	19.0	7.5	6.2	—	Anhydrite
	20.0	4.0	6.3	—	Gypsum
	21.0	3.4	6.7	—	Dolomite
	22.0	5.8	6.8	—	Gypsum
	23.0	2.6	3.8	—	Dolomite
Borehole M-5	8.0	0.8	6.7	18.7	Gypsum
	9.0	–0.5	6.4	22.0	Gypsum
	10.0	–0.6	5.1	20.4	Gypsum
	11.0	–2.1	1.1	21.8	Gypsum
	12.0	–1.0	6.6	22.5	Gypsum
	13.0	0.2	3.5	22.0	Gypsum
	14.0	1.8	5.7	23.4	Gypsum
	15.0	2.5	6.2	20.5	Gypsum
	16.0	4.7	6.2	—	Dolomite
	17.0	3.8	6.7	—	Dolomite
	18.0	3.2	5.4	22.0	Gypsum
	19.0	6.8	6.7	21.2	Gypsum
	20.0	5.6	6.0	—	Dolomite
	21.0	5.2	6.3	—	Dolomite
	22.0	4.8	6.7	—	Dolomite
	23.0	4.0	5.9	—	Dolomite
	24.0	2.0	6.2	21.4	Gypsum
	24.9	2.5	5.0	23.5	Gypsum
Borehole M-6	8.0	—	—	19.6	Gypsum
	9.0	–0.4	4.3	22.3	Gypsum
	10.0	0.1	6.2	22.0	Gypsum
	11.0	1.5	6.9	—	Dolomite
	12.0	2.3	6.3	—	Dolomite
	13.0	2.3	6.4	—	Dolomite
	14.0	3.9	6.9	21.9	Gypsum
	15.0	6.1	7.4	—	Dolomite
	16.0	6.9	6.9	—	Dolomite
	17.0	5.7	6.7	—	Dolomite
	18.0	7.0	7.3	—	Dolomite
	19.0	4.0	4.4	—	Dolomite
	19.5	3.7	5.8	—	Dolomite

Note: Carbon, oxygen, and sulphur isotope analyses were performed by the Russian Academy of Science, Moscow.

conditions (fig. 9.5). As the $\delta^{13}C/\delta^{18}O$ values are much higher than those associated with Miocene normal marine precipitates (fig. 9.2) (Williams, 1988; Hodell and Woodruff, 1994), their strongly positive $\delta^{18}O$ values can be attributed to evaporitic depositional conditions that generate isotopically heavy waters, enriched in ^{18}O, with strongly positive $\delta^{18}O$ values. The positive $\delta^{13}C$ values are the result of microbial or bacterial methane production that generates isotopically heavy CO_2, creating pore waters with positive $\delta^{13}C$ values. Extensive inter-supratidal or lacustrine microbial communities, or algal "mats," can generate significant methane production. The $\delta^{13}C/\delta^{18}O$ values observed in the Gachsaran Formation are typical of sediments produced in arid coastal environments (fig. 9.2) (Morse and Mackenzie, 1990; Allan and Wiggins, 1993).

Based on detailed sedimentological and geochemical analyses, Peebles et al. (1995) interpreted the Gachsaran Formation in Abu Dhabi to be composed of multiple shallowing-upward cycles (parasequences) of dolomitised (shallow subtidal to supratidal) carbonates and evaporites (gypsum and anhydrite) formed in an arid

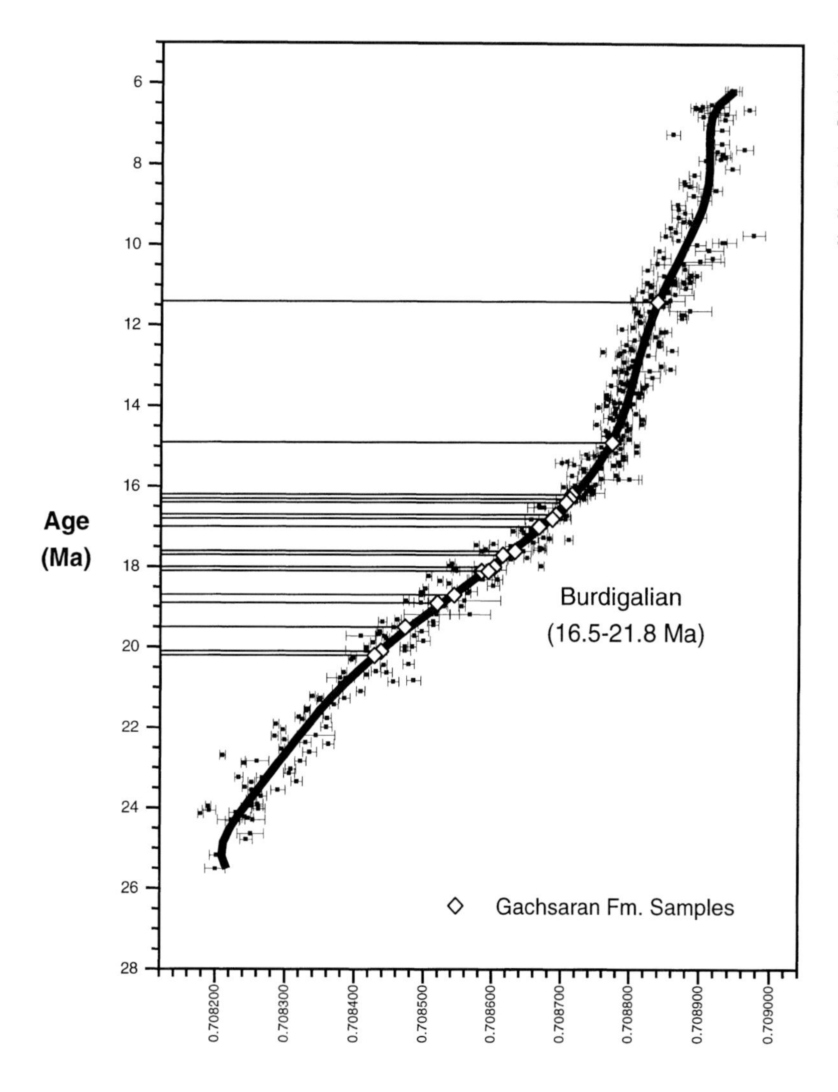

Figure 9.3. Strontium isotope stratigraphy of the Miocene, with data plotted for the Gachsaran and Dam Formations. The curve is after Hodell et al. (1991), Hodell and Woodruff (1994), and Oslick et al. (1994). The time scale is after Berggren et al. (1985) and Cande and Kent (1992).

Table 9.3. Strontium isotope stratigraphy database: Miocene, Emirate of Abu Dhabi

Location	Depth (metres)	Sample description	$^{87}Sr/^{86}Sr$ ratio	Error (2 s.d)	Age (Ma)[1]	Stage	Epoch	Formation
Outcrop								
Shuwaihat Island	Outcrop–base	Gypsum vein	0.707 788[a]	0.000 011	NA	NA	NA	Shuwaihat
	Outcrop–middle	Celestite-cemented rhizoconcretion	0.708 054[a]	0.000 013	NA	NA	NA	Baynunah
As Sila	Outcrop	Gypsum vein	0.708 017[a]	0.000 013	NA	NA	NA	Shuwaihat
Sabkhat Matti	Outcrop	Dolomitised bivalve grainstone (coquina)	0.708 528[b]	0.000 014	18.9	Burdigalian	Early Miocene	Dam
Dalma Island	Outcrop	Gypsum	0.708 720[c]	0.000 017	16.3	Langhian	Middle Miocene	Gachsaran
Sir Bani Yas Island	Outcrop	Gypsum	0.708 700[d]	0.000 011	16.7	Burdigalian	Early Miocene	Gachsaran
Abu Dhabi Sabkha	Outcrop–top	Dolomitised bivalve grainstone (coquina)	0.708 611[b]	0.000 014	18.9	Burdigalian	Early Miocene	Gachsaran
	Outcrop–middle	Massive gypsum	0.708 715[e]	0.000 014	16.4	Langhian	Middle Miocene	Gachsaran
	Outcrop–base	Massive-nodular anhydrite	0.708 447[e]	0.000 014	20.1	Burdigalian	Early Miocene	Gachsaran
Subsurface (Mussafah)								
Borehole M-3	14.9	Dolomitised bivalve grainstone (coquina)	0.708 723[f]	0.000 010	16.2	Langhian	Middle Miocene	Gachsaran
	26.0	Celestite vein	0.708 603[f]	0.000 011	18.1	Burdigalian	Early Miocene	Gachsaran
	26.2	Massive gypsum	0.708 641[f]	0.000 011	17.6	Burdigalian	Early Miocene	Gachsaran
Borehole M-4	12.7	Massive-nodular anhydrite	0.708 846[f]	0.000 011	11.4	Serravalian	Middle Miocene	Gachsaran
	18.4	Massive anhydrite	0.708 782[f]	0.000 013	14.9	Langhian	Middle Miocene	Gachsaran
	19.8	Massive gypsum	0.708 593[f]	0.000 013	18.1	Burdigalian	Early Miocene	Gachsaran
Borehole M-5	16.2	Dolomitised mudstone	0.708 695[f]	0.000 013	16.8	Burdigalian	Early Miocene	Gachsaran
Borehole M-6	9.9	Massive anhydrite	0.708 676[f]	0.000 013	17.0	Burdigalian	Early Miocene	Gachsaran
	10.6	Massive gypsum	0.708 482[f]	0.000 013	19.5	Burdigalian	Early Miocene	Gachsaran
	10.6	Celestite vein	0.708 438[f]	0.000 014	20.2	Burdigalian	Early Miocene	Gachsaran
	10.8	Massive gypsum	0.708 533[f]	0.000 010	18.7	Burdigalian	Early Miocene	Gachsaran
	15.7	Anhydrite/gypsum nodule	0.708 624[f]	0.000 013	17.7	Burdigalian	Early Miocene	Gachsaran

Note: Strontium isotope analysis was performed by Isotopic Analytical Services Ltd, Aberdeen, U.K. Standard: mean Holocene marine carbonate = 0.709 186.
[1]Time scale after Cande and Kent (1992).
[a]NTIS 987 = 0.710 245; [b]NTIS 987 = 0.710 333; [c]NTIS 987 = 0.710 257; [d]NTIS 987 = 0.710 252; [e]NTIS 987 = 0.710 256; [f]NTIS 987 = 0.710 255.
NA = not applicable. The sample is composed of diagenetic material whose isotope ratio is not representative of the ratio present during deposition.

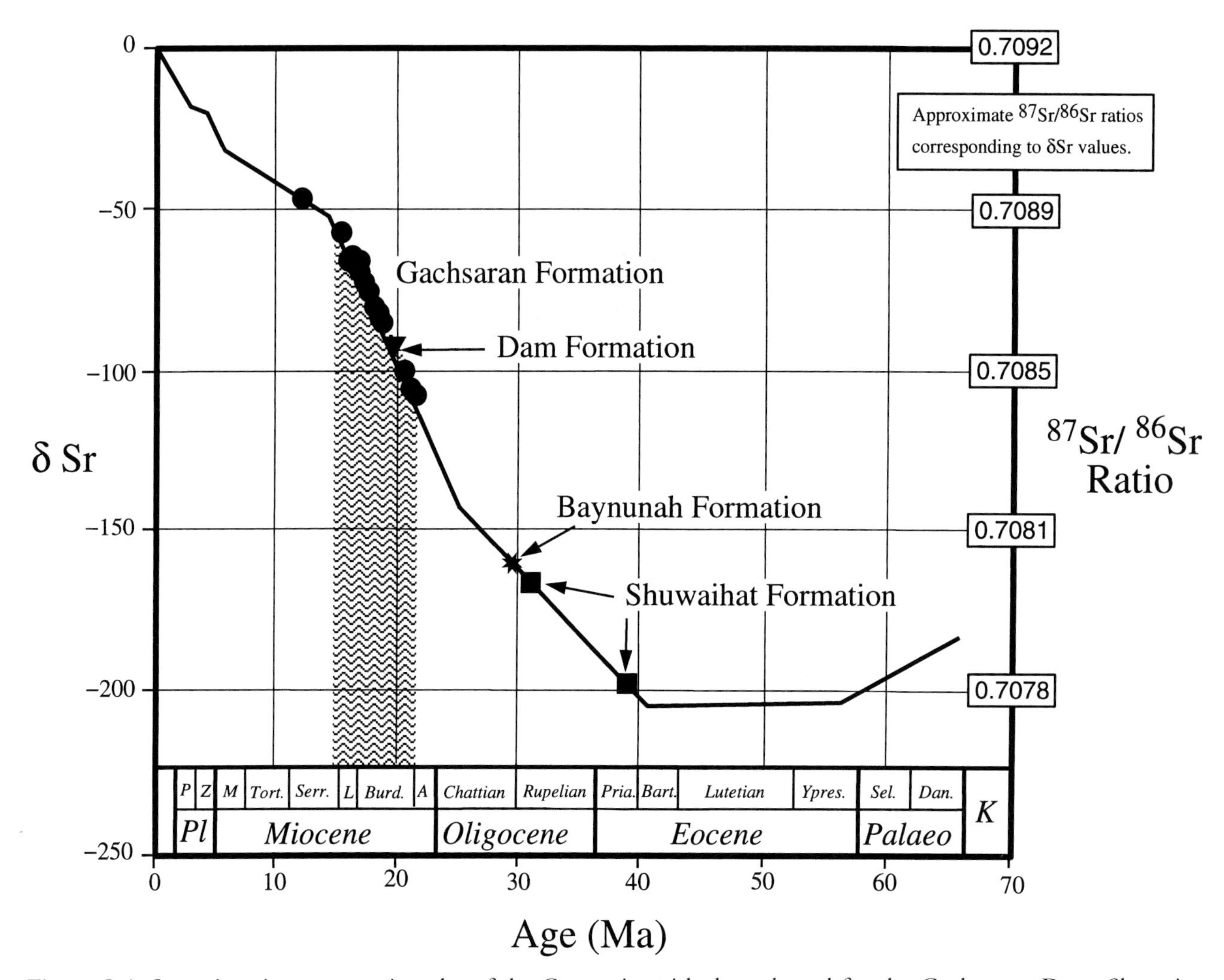

Figure 9.4. Strontium isotope stratigraphy of the Cenozoic, with data plotted for the Gachsaran, Dam, Shuwaihat, and Baynunah Formations. The stratigraphic interval identified by the Gachsaran Formation results is shaded. The curve was developed from the data of DePaolo (1986), Hess et al. (1986), and Hodell et al. (1989, 1990, and 1991) and was provided by Dr Euan Mearns of Isotopic Analytical Services Ltd. The time scale is after Berggren et al. (1985).

coastal setting. While the Gachsaran Formation dolomites have been diagenetically altered from their original limestone lithology, the strongly positive $\delta^{13}C/\delta^{18}O$ values indicate low temperatures of formation (nonburial temperatures) (Allan and Wiggins, 1993). The dolomite samples also show an isotopic signature very similar to that of dolomites associated with modern evaporitic lagoons and lakes in South Australia (fig. 9.2) (Clayton et al., 1968; Rosen et al., 1988). This suggests that dolomitisation was penecontemporaneous with deposition. The depositional model for the Gachsaran Formation proposed by Peebles et al. (1995) involved an arid coastal setting dominated by large, hypersaline, coastal lagoons and ephemeral lakes. Thus, although the $\delta^{13}C/\delta^{18}O$

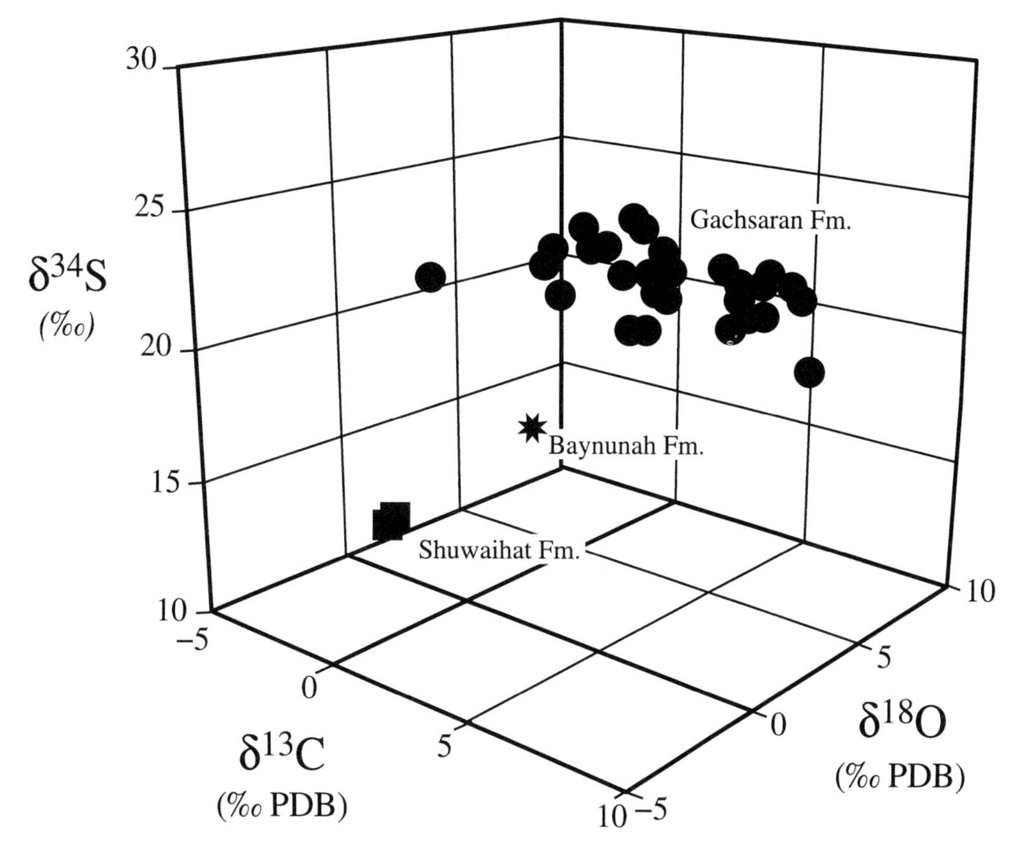

Figure 9.5. Crossplot of carbon, oxygen, and sulphur isotopes for the Gachsaran, Shuwaihat, and Baynunah Formations. All samples are sulphates (gypsum, anhydrite, or celestite).

values for the Gachsaran Formation are atypical of open marine values, they are representative of values produced by the depositional conditions that created the Gachsaran Formation.

Sulphur and strontium isotope values support the interpretation that the isotopic signature of the Gachsaran Formation reflects depositional conditions. The $\delta^{34}S$ values of 18.7 to 23.5‰ and the $^{87}Sr/^{86}Sr$ ratios of 0.708 438 to 0.708 846 are characteristic of the Early–Middle Miocene when the Gachsaran Formation is proposed to have been deposited.

Geochemical analyses of shallow marine dolomites from the Dam Formation, collected from outcrops near As Sila (Ditchfield, 1998—Chapter 7), reveal strongly positive $\delta^{18}O$ values ranging from 3.9 to 5.6‰ (average 4.6‰) with slightly negative $\delta^{13}C$ values ranging from 1.5 to −2.5‰ (average −1.2‰). These values suggest shallow burial or penecontemporaneous dolomitisation from evaporative pore waters. Such dolomitisation is compatible with the depositional and diagenetic framework developed for the Gachsaran Formation. Further, strontium isotope analysis of a dolomitised bivalve from the Dam Formation produced a $^{87}Sr/^{86}Sr$ ratio of 0.708 528, characteristic of the Early Miocene (Burdigalian) when the Dam Formation is proposed to have been deposited.

Shuwaihat and Baynunah Formations

A crossplot of the sulphur, carbon, and oxygen isotope values shows the Shuwaihat and Baynunah Formations to have distinct isotopic signatures that are clearly different from the Gachsaran Formation (fig. 9.5). Isotopic analyses of two gypsum samples from Shuwaihat Formation outcrop localities over 70 km apart (Shuwaihat Island and As Sila) produced nearly identical $\delta^{13}C$, $\delta^{18}O$, and $\delta^{34}S$ values, suggesting they formed under similar conditions. These samples have negative $\delta^{18}O$ values (−2.2‰, −2.7‰), slightly negative $\delta^{13}C$ values (−0.6‰, −0.1‰), and nearly identical $\delta^{34}S$ values (13.8‰, 14.1‰) (table 9.2). Isotopic analyses of a celestite sample from a Baynunah Formation outcrop on Shuwaihat Island revealed a unique isotopic signature with a mildly positive $\delta^{18}O$ value (2.6‰), a slightly positive $\delta^{13}C$ value (0.3‰), and a $\delta^{34}S$ value of 16.0‰ (table 9.2). Strontium isotope analysis of all three samples produced $^{87}Sr/^{86}Sr$ ratios ranging from 0.707 788 to 0.708 054 (average 0.707 953), with no clear distinction between the Shuwaihat and Baynunah Formation samples (table 9.3).

The negative $\delta^{18}O$ values associated with the gypsum samples from the Shuwaihat Formation are much lower than $\delta^{18}O$ values associated with Miocene normal marine precipitates (fig. 9.2) and suggest freshwater (meteoric) influence—rainwater is enriched in ^{16}O, which produces a negative $\delta^{18}O$ value. The slightly negative $\delta^{13}C$ values are also atypical of Miocene normal marine $\delta^{13}C$ values (fig. 9.2) and imply near-surface oxidising conditions. Together, the $\delta^{13}C/\delta^{18}O$ isotope values suggest that the gypsum samples were precipitated in the shallow subsurface from meteoric or mixed (meteoric–marine) groundwaters. A likely scenario would involve the leaching of existing (?depositional) gypsum by meteoric (or mixed) groundwaters undersaturated in $CaSO_4$ with reprecipitation as diagenetic gypsum. This interpretation is supported by outcrop observation of vertical (and subordinate horizontal) gypsum veins in distinct stratigraphic intervals within the Shuwaihat Formation. Further, Ditchfield (1999—Chapter 7) reports strongly negative $\delta^{18}O$ values (−2.7 to −5.5‰, average −4.3‰) and negative $\delta^{13}C$ values (−2.0 to −3.8, average −2.8‰) from his analysis of dolomitic crusts in the Shuwaihat Formation. These values indicate dolomitisation in a meteoric–marine mixing zone, supporting the diagenetic evolution proposed for the gypsum samples.

The celestite sample from the Baynunah Formation has strikingly different $\delta^{13}C/\delta^{18}O$ values, implying markedly different conditions of precipitation. Positive $\delta^{18}O$ values and near-zero $\delta^{13}C$ values suggest formation in marine porewaters. The celestite's $\delta^{18}O$ value of 2.6‰ is consistent with Late Miocene and Plio-Pleistocene normal marine values (Williams, 1988; Hodell and Woodruff, 1994). Outcrop occurrence of celestite as small veins and as cement in rhizoconcretions (Glennie and Evamy, 1968; Friend, 1999—Chapter 5) characterises it as a diagenetic precipitate. The fact that its $\delta^{13}C/\delta^{18}O$ signature is distinctly different from the Shuwaihat Formation gypsum samples but similar to Late Miocene or Plio-Pleistocene marine precipitates suggests that the celestite was precipitated in the shallow subsurface from marine groundwaters, perhaps during a transgression in the Pliocene or at the beginning of the Pleistocene.

None of the Shuwaihat or Baynunah Formation samples has a $\delta^{34}S$ value consistent with Miocene (or even Tertiary) deposition, which provides further evidence of the diagenetic nature of these sulphates (fig. 9.6). As with the $\delta^{13}C/\delta^{18}O$ data, the $\delta^{34}S$ values of the two Shuwaihat Formation gypsum samples are similar (14.1 and 13.8‰) and the Baynunah Formation celestite sample is distinctly different (16.0‰). This reinforces the belief that the gypsum and celestite precipitated from different pore fluids during different hydrological conditions. The $^{34}S/^{32}S$ isotope ratio is controlled by the input of two types of sulphur: oxidised sulphate (that concentrates ^{34}S) and reduced sulphide (that concentrates ^{32}S). Because the $\delta^{34}S$ values for all three samples are lower than the "normal" value associated with Miocene marine sulphate, it can be assumed that they precipitated from a pore fluid

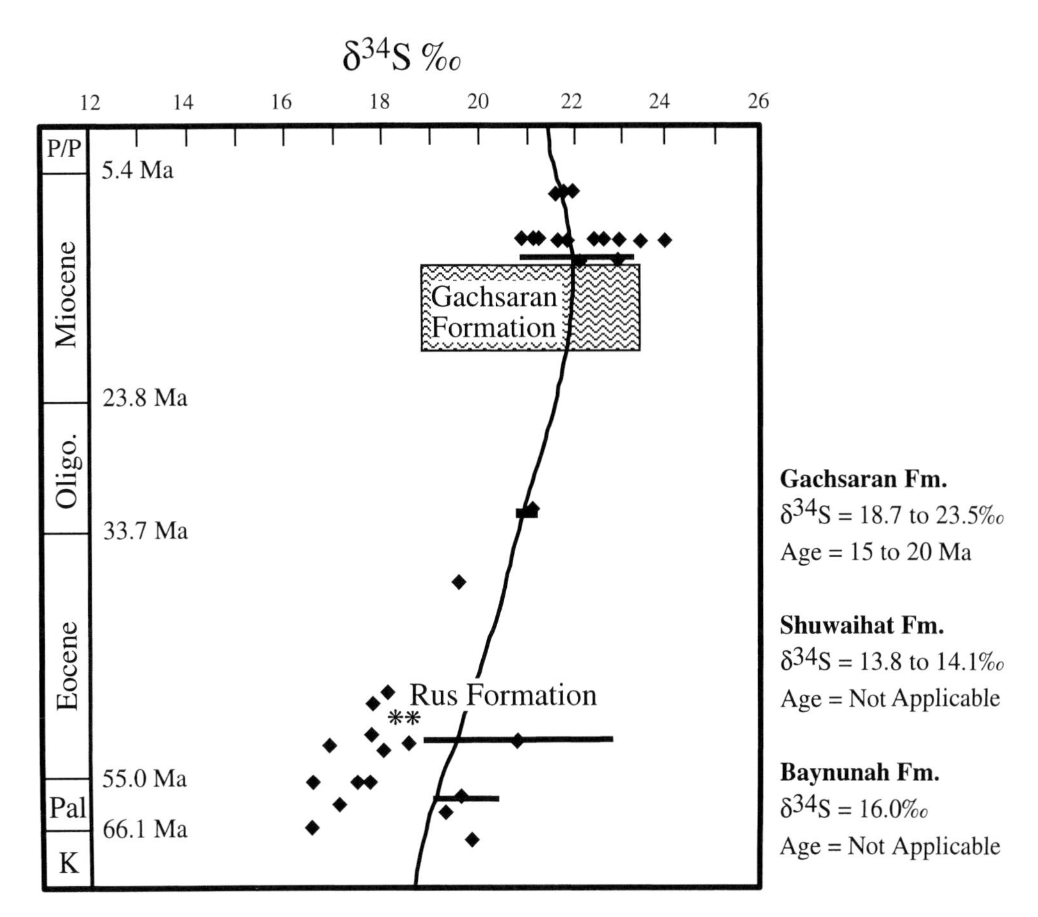

Figure 9.6. Sulphur isotope stratigraphy of the Cenozoic. The range of values measured from 33 Gachsaran Formation gypsum samples is shown by the shaded box. All samples were collected from core material taken from shallow boreholes at Mussafah (near the city of Abu Dhabi). The curve and diamond-denoted data points are from Claypool et al. (1980). The solid lines represent published data ranges and are also from Claypool et al. (1980). The asterisk-denoted data points are from the Early Eocene Rus Formation of Kuwait (Sakai, 1972). The time scale is after Cande and Kent (1992).

with an elevated concentration of sulphides, typical of shallow groundwater conditions.

The strontium isotope ratios determined from the diagenetic sulphates (gypsum and celestite) from the Shuwaihat and Baynunah Formations range from 0.707 788 to 0.708 054 and are shown in figure 9.4 plotted against a global Cenozoic seawater strontium curve and with Gachsaran and Dam Formation data. Note that none of the three samples falls within the Miocene portion of the curve and that the sample range brackets a Late Eocene to Middle Oligocene chronostratigraphic position. As the samples are believed to have formed during shallow burial diagenesis, the $^{87}Sr/^{86}Sr$ value reflects the ratio in the shallow subsurface groundwater from which these samples precipitated, rather than the strontium isotope ratio typical of each formation's depositional conditions. The original marine seawater $^{87}Sr/^{86}Sr$ isotope ratio can be altered by mixing with groundwaters that carry the strontium isotope signature of other (older) strata and by percolation through certain lithologies. Siliciclastic material (aluminosilicates such as potassium feldspars and illite) can be enriched in ^{87}Sr, which produces a higher $^{87}Sr/^{86}Sr$ isotope ratio, while volcaniclastic or basaltic rocks are enriched in ^{86}Sr, which results in a lower $^{87}Sr/^{86}Sr$ isotope ratio. Although the Shuwaihat and Baynunah Formations are siliciclastic with the potential to bias the ratios higher, the measured ratios are much lower than expected for the Miocene, which suggests that the samples are the result of mixing with strontium from older, underlying strata.

The Lower Miocene Gachsaran Formation and the Eocene Rus Formation are two carbonate-evaporite units that underlie the Shuwaihat and Baynunah Formations. Strontium isotope ratios from the Gachsaran Formation and the global Eocene strontium ratio that would be typical of the Rus Formation are shown in figure 9.4. Note that the strontium ratios for the Shuwaihat and Baynunah Formations fall between the ratios associated with the Rus (Eocene) and Gachsaran (Early Miocene) Formations. It can be postulated that dissolution and mixing of the evaporites from these two units produced a pore fluid with a strontium ratio similar to that measured from evaporites of the Shuwaihat and Baynunah Formations.

CHRONOSTRATIGRAPHY

The stable isotope ratios of carbon ($^{13}C/^{12}C$), oxygen ($^{18}O/^{16}O$), sulphur ($^{34}S/^{32}S$), and strontium ($^{87}Sr/^{86}Sr$) have varied over geological time. Curves that display these isotopic values versus time have been published and may be used to determine the chronostratigraphic position or age of a sample based on its isotopic signature.

Gachsaran Formation

Eighteen strontium isotope ratios measured from Gachsaran Formation samples collected across the Emirate of Abu Dhabi range from 0.708 438 to 0.708 846 (average 0.708 630) and are characteristic of an Early to Middle Miocene age (fig. 9.3). As the samples represent four different lithologies, a qualitative examination of the dataset was carried out to test for lithological bias. Celestite samples from veins of obvious diagenetic origin are excluded from the chronostratigraphic evaluation. Of the four anhydrite samples, two samples had values clearly higher than the dataset cluster (fig. 9.7). For this reason, the strontium ratios of all of the anhydrite samples are considered suspect (probably contaminated by shallow burial diagenetic fluids) and were excluded from statistical and chronostratigraphic evaluation. The $^{87}Sr/^{86}Sr$ ratios of the dolomite and gypsum samples form a coherent data cluster or trend, ranging from 0.708 482 to 0.708 723 (average 0.708 629). From these data the Gachsaran Formation can be assigned an Early–Middle Miocene (Burdigalian to Early Langhian) date with an age range of 19.5–16.2 Ma (table 9.3 and fig. 9.3).

The global chronostratigraphic curve of $\delta^{34}S$ produced by Claypool et al. (1980) was based on values obtained from marine sulphate samples. Correlation of the $\delta^{34}S$ values from marine and marginal marine sulphate (gypsum) samples of the Gachsaran Formation with this curve is shown in figure 9.6. The data cluster defined by these samples ranges from 18.7 to 23.5‰ (average 21.2‰) and also indicates an Early–Middle Miocene chronostratigraphic position for the Gachsaran Formation with an age range of about 20–12 Ma (fig. 9.6). Note that the $\delta^{34}S$ curve has a much lower resolution than the strontium isotope curve with stratigraphic resolution based on 1–2‰ variations in the $\delta^{34}S$ value (William Holser, personal communication, 1994).

Global chronostratigraphic curves of $\delta^{13}C$ and $\delta^{18}O$ values are based on carbonates (chiefly foraminifer tests) formed in normal or open marine conditions. As the dolomites and evaporites of the Gachsaran Formation were formed under different conditions (restricted and evaporative coastal lagoons and lakes), their measured $\delta^{13}C$ and $\delta^{18}O$ values are not correlative to these curves and cannot be used to determine the age of the formation.

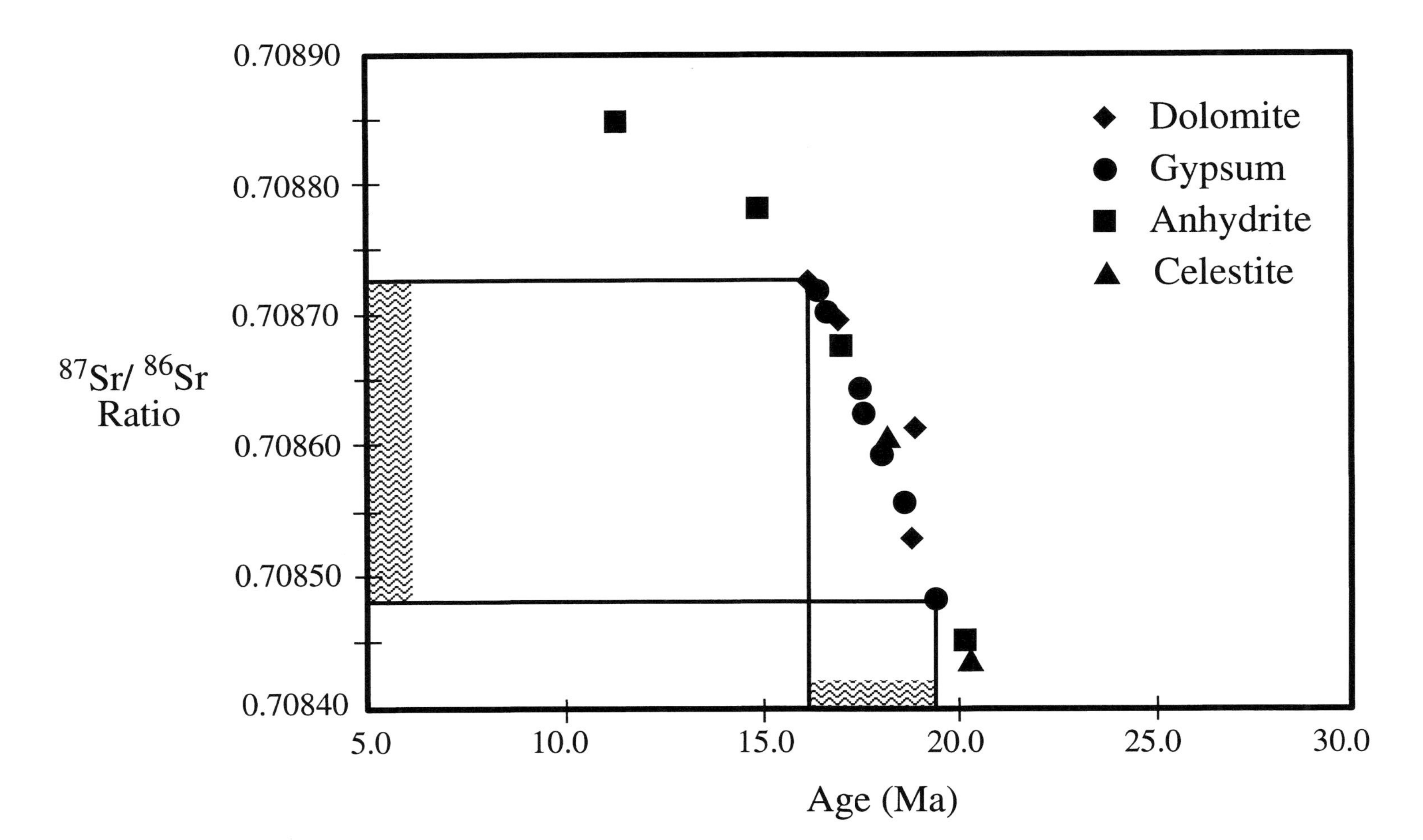

Figure 9.7. Relationship between lithology and strontium isotope ratio for Gachsaran Formation samples. Although the ratios of four different lithologies were measured, a clear stratigraphic trend can be observed in the dataset. The values show a clear correlation between decreasing strontium isotope ratio and increasing age.

Shuwaihat and Baynunah Formations

The isotopic signatures of the Shuwaihat and Baynunah Formation samples were produced during shallow burial diagenesis and are not reflective of the isotopic conditions that were present during deposition. Therefore, the measured isotope values cannot be used to determine ages for the Shuwaihat and Baynunah Formations.

Regional Stratigraphic Correlation

In southwestern Iran, the Gachsaran Formation has been assigned a latest Early–earliest Middle Miocene (Burdigalian–Langhian) age based on the occurrence of the index taxa of the *Austrotrillina howchini–Peneropolis evolutus* and *Neoalveolina* (*Borelis*) *melo curdica* zones (Wynd, 1965; Jones and Racey, 1994). The Gachsaran Formation of southwestern Iran has been correlated to the Dam Formation of eastern Saudi Arabia and Qatar (Adams et al., 1983). The Dam Formation has been assigned a Burdigalian age, dated as 19–16 Ma based on its marine and terrestrial fossil assemblage (Adams et al., 1983; Whybrow et al., 1999—Chapter 4). The Gachsaran Formation of Abu Dhabi, based on its depositional history and chronostratigraphic position, appears to be the western continuation of the coastal carbonate–evaporite depositional system that produced the Gachsaran Formation in southwestern Iran. It also appears to be laterally correlative to the marine carbonates of the Dam Formation in Qatar and the marine carbonates and coastal siliciclastics of the Dam Formation in eastern Saudi Arabia.

Conclusions

1. Isotope-dating has shown that the carbonate–evaporite strata of the Gachsaran Formation that outcrop in Abu Dhabi are Early–Middle Miocene (Burdigalian–Langhian) in age, ranging from 19.5 to 16.2 Ma.
2. Isotope analyses of three sulphate samples from the Shuwaihat and Baynunah Formations reveal these samples to be diagenetic minerals whose isotopic signatures do not reflect their periods of deposition. Therefore, the isotope values derived from these samples cannot be used to constrain the chronostratigraphic positions (the "ages") of these formations. But these data do provide interesting insight into the diagenetic history of the formations.
3. Isotope chronostratigraphic analyses of the Gachsaran Formation in Abu Dhabi has provided evidence of correlation between the Miocene of Abu Dhabi, the Dam Formation of Qatar and eastern Saudi Arabia, and the Gachsaran Formation of southwestern Iran.
4. Stable isotopic analyses of carbonate and evaporite successions can provide valuable chronostratigraphic information of intervals that typically have poor to nonexistent fossil assemblages.

Acknowledgements

I thank the Abu Dhabi Company for Onshore Oil Operations (ADCO) and the Ministry of Higher Education and Scientific Research, United Arab Emirates, for their invitation to participate in the First International Conference on the Fossil Vertebrates of Arabia held in the Emirate of Abu Dhabi, March 1995.

Carbon, oxygen, and sulphur isotope analyses were performed by the Russian Academy of Sciences under the supervision of Dr E. Morgun. Strontium isotope analysis was performed by Isotopic Analytical Services Ltd under the supervision of Dr John McBride. Abu Dhabi Oil Company Ltd (Japan) provided access to the subsurface (borehole) material and funding for isotope analyses. Peter Whybrow of The Natural History Museum, London, provided samples from the Shuwaihat and Baynunah Formations and partial funding for analysis. Muriel Shaner provided technical and editorial assistance. Dr Maurice Tucker and Dr Norman MacLeod reviewed draft versions of this manuscript. Finally, I thank Halliburton Energy Services for providing the time and resources necessary to conduct this study.

REFERENCES

Adams, C. G., Gentry, A. W., and Whybrow, P. J. 1983. Dating the terminal Tethyan events. In *Reconstruction of Marine Environments* (ed. J. Meulenkamp). *Utrecht Micropaleontological Bulletins* 30: 273–98.

Allan, J. R., and Wiggins, W. D. 1993. *Dolomite Reservoirs: Geochemical Techniques for Evaluating Origin and Distribution.* American Association of Petroleum Geologists, Tulsa.

Alsharhan, A. S. 1989. Petroleum geology of the United Arab Emirates. *Journal of Petroleum Geology* 12: 253–88.

Berggren, W. A., Kent, D. V., and Flynn, J. J. 1985. Cenozoic geochronology. *Bulletin of the Geological Society of America* 96: 1407–18.

Bristow, C. S. 1999. Aeolian and sabkha sediments in the Miocene Shuwaihat Formation, Emirate of Abu Dhabi, United Arab Emirates. Chap. 6 in *Fossil Vertebrates of Arabia,* pp. 50–60 (ed. P. J. Whybrow and A. Hill). Yale University Press, New Haven.

Cande, S. C., and Kent, D. V. 1992. A new geomagnetic polarity time scale for the Late Cretaceous and Cenozoic. *Journal of Geophysical Research* 97: 13917–51.

Claypool, G. E., Holser, W. T., Kaplan, I. R., Sakai, H., and Zak, I. 1980. The age curves of sulfur and oxygen in marine sulfate and their mutual interpretation. *Chemical Geology* 28: 199–260.

Clayton, R. N., Skinner, H. C. W., Berner, R. A., and Rubinson, M. 1968. Isotopic compositions of recent South Australia carbonates. *Geochimica et Cosmochimica Acta* 32: 983–88.

DePaolo, D. J. 1986. Detailed record of the Neogene Sr isotopic evolution of seawater from DSDP Site 590B. *Geology. Geological Society of America* 14: 103–6.

Ditchfield, P. W. 1999. Diagenesis of the Baynunah, Shuwaihat, and Upper Dam Formation sediments exposed in the Western Region, Emirate of Abu Dhabi, United Arab Emirates. Chap. 7 in *Fossil Vertebrates of Arabia,* pp. 61–74 (ed. P. J. Whybrow and A. Hill). Yale University Press, New Haven.

Friend, P. F. 1999. Rivers of the Lower Baynunah Formation, Emirate of Abu Dhabi, United Arab Emirates. Chap. 5 in *Fossil Vertebrates of Arabia,* pp. 38–49 (ed. P. J. Whybrow and A. Hill). Yale University Press, New Haven.

Galal, M. S. 1995. Hydrocarbon potential of the diapiric islands in the United Arab Emirates (Case study, The Dalma Island). *Selected Middle East Papers from Geo '94, the Middle East Geoscience Conference, April 1994.* Gulf Petrolink, Bahrain.

Glennie, K. W., and Evamy, B. D. 1968. Dikaka: Plants and plant-root structures associated with aeolian sand. *Palaeogeography, Palaeoclimatology, Palaeoecology* 4: 77–87.

Hess, J., Bender, M. L., and Snelling, J. G. 1986. Seawater $^{87}Sr/^{86}Sr$ evolution from Cretaceous to present—applications to paleoceanography. *Science* 231: 979–84.

Hodell, D. A., and Woodruff, F. 1994. Variations in the strontium isotopic ratio of seawater during the Miocene: Stratigraphic and chemical implications. *Paleoceanography* 9: 405–26.

Hodell, D. A., Mead, G. A., and Mueller, P. A. 1990. Variation in the strontium isotopic composition of seawater (8 Ma to present): Implications for chemical weathering rates and dissolved fluxes to the oceans. *Chemical Geology* 80: 291–307.

Hodell, D. A., Mueller, P. A., and Garrido, J. R., 1991. Variations in the strontium isotopic composition of seawater during the Neogene. *Geology. Geological Society of America* 19: 24–27.

Hodell, D. A., Mueller, P. A., McKenzie, J. A., and Mead, G. A. 1989. Strontium isotope stratigraphy and geochemistry of the late Neogene ocean. *Earth and Planetary Science Letters* 92: 165–78.

Hoefs, J. 1980. *Stable Isotope Geochemistry.* Springer-Verlag, Berlin.

Jones, R. W., and Racey, A. 1994. Cenozoic stratigraphy of the Arabian Peninsula and Gulf. In *Micropalaeontology and Hydrocarbon Exploration in the Middle East,* pp. 273–307 (ed. M. Simmons). Chapman and Hall, London.

McCrea, J. M. 1950. On the isotopic chemistry of carbonates and a paleotemperature scale. *Journal of Physical Chemistry* 18: 849–57.

McKenzie, J. 1981. Holocene dolomitization of calcium carbonate sediments from the coastal sabkhas of Abu Dhabi, U.A.E.: A stable isotope study. *Journal of Geology* 89: 185–98.

Morse, J. W., and Mackenzie, F. T. 1990. *Geochemistry of Sedimentary Carbonates.* Elsevier, Amsterdam.

Oslick, J. S., Miller, K. G., Feigenson, M. D., and Wright, J. D. 1994. Oligocene-Miocene strontium isotopes: Stratigraphic revisions and correlations to an inferred glacioeustatic record. *Paleoceanography* 9: 427–43.

Peebles, R. G., Suzuki, M., and Shaner, M. 1995. The effects of long-term, shallow-burial diagenesis on carbonate–evaporite successions. *Selected Middle East Papers from Geo '94, the Middle East Geoscience Conference, April 1994.* Gulf Petrolink, Bahrain.

Rosen, M. R., Miser, D. E., and Warren, J. E. 1988. Sedimentology, mineralogy and isotopic analysis of Pellet Lake, Coorong region, South Australia. *Sedimentology* 35: 105–22.

Sakai, H. 1972. Oxygen isotopic ratios of some evaporites from PreCambrian to Recent ages. *Earth and Planetary Science Letters* 15: 201–5.

Swart, P. K., Barnes, S. J., and Leder, J. J. 1991. Fractionation of stable isotopes of oxygen and carbon in carbon dioxide during the reaction of calcite with phosphoric acid as a function of temperature and technique. *Chemical Geology (Isotope Geoscience Section)* 86: 89–96.

Whybrow, P. J. 1989. New stratotype; the Baynunah Formation (Late Miocene), United Arab Emirates: Lithology and palaeontology. *Newsletters on Stratigraphy* 21: 1–9.

Whybrow, P. J. and Hill, A. eds. 1999. *Fossil Vertebrates of Arabia.* Yale University Press, New Haven.

Whybrow, P. J., and McClure, H. A. 1981. Fossil mangrove roots and palaeoenvironments of the Miocene of the eastern Arabian peninsula. *Palaeogeography, Palaeoclimatology, Palaeoecology* 32: 213–25.

Whybrow, P. J., Friend, P. F., Ditchfield, P. W., and Bristow, C. S. 1999. Local stratigraphy of the Neogene outcrops of the coastal area: Western Region, Emirate of Abu Dhabi, United Arab Emirates. Chap. 4 in *Fossil Vertebrates of Arabia,* pp. 28–37 (ed. P. J. Whybrow and A. Hill). Yale University Press, New Haven.

Whybrow, P. J., Hill, A., Yasin al-Tikriti, W., and Hailwood, E. A. 1990. Late Miocene primate fauna, flora and initial palaeomagnetic data from the Emirate of Abu Dhabi, United Arab Emirates. *Journal of Human Evolution* 19: 583–88.

Whybrow, P. J., McClure, H. A., and Elliott, G. F. 1987. Miocene stratigraphy, geology, and flora (algae) of eastern Saudi Arabia and the Ad Dabtiyah vertebrate locality. *Bulletin of the British Museum (Natural History),* Geology 41: 371–82.

Williams, D. F. 1988. Evidence for and against sea-level changes from the stable isotopic record of the Cenozoic. In *Sea-level Changes: An Integrated Approach,* pp. 31–36 (ed. C. K. Wilgus, B. S. Hastings, C. G. St. J. C. Kendall, H. W. Posamentier,

C. A. Ross, and J. C. Van Wagoner). SEPM Special Publications no. 42. Society of Economic Paleontologists and Mineralogists, Tulsa.

Wynd, J. G., 1965. Biofacies of the Iranian Oil Consortium Agreement Area, unpublished report cited in Jones and Racey (1994).

Excavation of the Shuwaihat proboscidean, Emirate of Abu Dhabi.

Miocene Fossil Fauna from the Baynunah Formation, Emirate of Abu Dhabi, United Arab Emirates

PART III

Paul A. Jeffery describes (Chapter 10) part of the invertebrate fauna collected from the Baynunah Formation. The bivalves are assigned to two species in two separate unionoid superfamilies. The presence of these and the nature of their enclosing sediments support the inference of a rapidly changing, possibly seasonal, river system in this area during the late Miocene. The fact that these species were previously unknown in this area and are absent from the modern Arabian fauna suggests that the climate and geography of the region during the late Miocene allowed communication with geographical provinces from which the Arabian Peninsula is now effectively isolated.

Peter B. Mordan then describes (Chapter 11) an internal cast of a fossil land-snail shell. This has been provisionally referred to the pulmonate family Buliminidae. Its occurrence does not conflict with present-day geographical distribution patterns at the family level, and suggests wetter, perhaps seasonal, conditions during the late Miocene.

Among the aquatic fauna of the Baynunah river system are three ostariophysan fish taxa described by Peter L. Forey and Sally V. T. Young (Chapter 12). All are based on large collections of fragmentary remains: a new species of *Bagrus* (Siluriformes: Bagridae) and *Clarias* sp. (Siluriformes: Clariidae) are represented chiefly by cranial and pectoral girdle remains, while *Barbus* sp. (Cypriniformes: Cyprinidae) is known only by pharyngeal teeth. It is suggested that the river in which these fishes lived was slow moving with occasional periods of flash flooding. The occurrence of *Clarias* in Neogene deposits of Afro-Arabia at numerous localities casts doubt on traditional theories that these catfishes migrated to Africa from Asia in the Pliocene, while the phylogeny of bagrid catfishes strongly suggests that a pre-Miocene dispersal from Asia to Africa took place.

Part of the reptile fauna inhabiting the Baynunah river and the surrounding area are three taxa of turtles, described in detail by France de Lapparent de Broin and Peter Paul van Dijk (Chapter 13). They appear to be older forms than the extant–Pliocene species and later than the early–middle Miocene forms of the Arabian Plate. The terrestrial *Geochelone* (*Centrochelys*) aff. *sulcata* belongs to a group already established from the early Miocene in the Afro-Arabian Plate and conforms to a species also present in Sahabi, Libya. The two freshwater turtles appear to be, during late Miocene times, new Laurasiatic immigrants in the Arabian Gulf area or recently arrived in Africa, depending on the age of the locality. The preserved fragments of *Trionyx* s.l. do not allow determination to the subgenus or subgenera (*Rafetus* or other undefined genus?; also *Trionyx* s.s.?) and no precise comparison is possible with other localities, although the species (or the two species?) seems to be new. *Mauremys* belongs to a western species group and is not specifically defined but its evolutive state is that of a late Miocene form. Close relations with northern Africa (Tunisia, Libya, Egypt) are possible as well as with the eastern Mediterranean basin (Iraq). No close relationship with the Indian subcontinent and Afghanistan is shown. The freshwater turtles indicate a complete change relative to the early–middle Miocene faunas of the Arabian Plate owing to the lack of any cyclanorbine, carettochelyid, and pleurodire forms, now intertropical.

Michael Rauhe, Eberhard "Dino" Frey, Daniel S. Pemberton, and Torsten Rossmann (Chapter 14) describe the osteology of fossil crocodilians from the formation. The systematic and taxonomic status of these fossils is compared with African, Asian, and South American crocodilians. The fragmentary material is identified as *Crocodylus* sp. indet., *Crocodylus* cf. *niloticus*, *?Ikanogavialis*, and Gavialidae gen. et sp. indet. From these results palaeoecological conclusions can be drawn to clarify the association of the gavialid and crocodylid species in the environment of the Baynunah river system.

After screening and sorting many kilograms of sediments from the Baynunah Formation, Hans de Bruijn (Chapter 15) found three isolated insectivore incisors and 45 isolated rodent cheek teeth. The rodents represent five subfamilies and seven species. This small sample indicates the presence of a diverse assemblage dominated by the gerbil *Abudhabia baynunensis*. The occurrence of *Abudhabia*, which is otherwise known from the Kabul basin (Afghanistan), and of an unidentified genus of the Zapodinae, suggests Asiatic influence; the presence of two species of *Dendromus* suggests African influence in this fauna.

The evolutionary stage of the rodents from the Baynunah Formation does not allow precise biostratigraphical correlation because the Neogene succession of Arabian faunas is poorly known. The small rat *Parapelomys* and the primitive gerbil *Abudhabia* suggest a late Turolian–early Ruscinian Age for the fluvial part of the formation. The sedimentary environment and comparison of the habitat requirements of extant species that are considered to be closely related to the fossil rodents suggest a palaeoenvironment of a large river system with dense vegetation of reeds and shrubs on the banks of the channels in an otherwise open, semiarid plain.

The rarity of primates in any Miocene fauna is well known. In Chapter 16, Andrew Hill and Tom Gundling describe the only fossil primate from the United Arab Emirates—indeed the only fossil monkey from the Arabian Peninsula. The specimen is represented by a tooth of a male cercopithecoid and was found in the Baynunah Formation. It is placed in the context of late Miocene monkey evolution as currently known.

John C. Barry states (Chapter 17) that Miocene carnivores from Abu Dhabi include *Plesiogulo praecocidens,* two species of hyaena, and a large sabre-tooth felid. The assemblage is typical of late Miocene and Pliocene sites in Eurasia and Africa, and could indicate that relatively open habitats were present at that time on the Arabian Peninsula.

The larger herbivores include an elephantid, a deinothere, and a non-stegotetrabelodont (perhaps "*Mastodon*" *grandincisivus*). There is probably only one species of a primitive elephantid allocated to *Stegotetrabelodon*. One specimen, described by Pascal Tassy in Chapter 18, consists of scattered remains from Shuwaihat, site S6, and belongs to a single individual. This discovery offers a rare and invaluable example of associated characters of various anatomical systems: dental, cranial, and postcranial. Characters of the mandible, lower tusks, and molars match those of African species of the genus *Stegotetrabelodon: S. syrticus* and *S. orbus*. No trait matches with those of putative non-African stegotetrabelodonts described by various authors. Especially, the traits noted on the animal from site S6 refute the allocation of the binomen *Mastodon grandincisivus* Schlesinger 1917 to the genus *Stegotetrabelodon*. *Mastodon grandincisivus*—first described in Iran and later in Eastern Europe, Emirate of Abu Dhabi, and the Democratic Republic of Congo (formerly Zaire)—is sometimes alleged to be the ancestor of African stegotetrabelodonts, a hypothesis not supported here. Although the origin of the clade *Stegotetrabelodon* (*Primelephas* and other elephantines) could be outside Africa, its differentiation is considered to have occurred in Africa itself. Understanding the detailed evolution of the genus *Stegotetrabelodon* itself needs more discoveries.

In Chapter 19, Véra Eisenmann and Peter J. Whybrow report that the genus *Hipparion* is represented in the Baynunah Formation by mandibles, teeth, and fragmented limb bones. The sizes of the bones and teeth show that two species are represented; one small or middle-sized and one larger. Two mandibular fragments of the smaller species have been compared with, and are distinct from, similar material from Eurasia and Africa. In size and proportions they differ from other *Hipparion* mandibles and are here referred to as a new species of the genus.

Fossil suids, described by Laura Bishop and Andrew Hill (Chapter 20), provide a unique opportunity to investigate aspects of phylogeny, taxonomic diversity, and biogeography in late Miocene Suidae. The chapter briefly examines the late Neogene evolution and distribution of fossil suid taxa in the Old World and describes specimens from the Baynunah Formation. They belong to a minimum of two taxa. Larger suids are assigned to the tetraconodont species *Nyanzachoerus syrticus* (Leonardi, 1952). This species is known from numerous late Miocene localities in Africa. Smaller specimens are suine and are identified as *Propotamochoerus hysudricus* (Stehlin, 1899), known from the Siwalik sequence on the Indian subcontinent. There are no identifiable European elements in the Abu Dhabi suid fauna. The presence of these two taxa suggests faunal interchange between Asia, Africa, and Arabia during the late Miocene. The two identified taxa are consistent with a terminal Miocene age for the formation.

Alan W. Gentry (Chapter 21) describes the remains of late Miocene hippopotamuses that belong to a species more primitive than *Hexaprotodon harvardi* of Lothagam, Kenya. Compared with more poorly known fossil hippopotamuses, they appear to be closest to *Hexaprotodon sahabiensis* and *H. iravaticus*. He then identifies (Chapter 22) the ruminant species. These are three species of Giraffidae (*?Palaeotragus* sp., *?Bramatherium* sp., and sp. indet.) and six species of Bovidae (*Tragoportax cyrenaicus* and *Pachyportax latidens* from the tribe Boselaphini; Bovidae sp. indet. from a tribe indet.; *Prostrepsiceros* aff. *libycus*, *Prostrepsiceros* aff. *vinayaki*, and *Gazella* aff. *lydekkeri* from the tribe Antilopini). Taken together, they support a terminal Miocene age for the Baynunah Formation, perhaps MN 13 in terms of the European Mein zone scale and around 6 million years old. They show more affinity with faunas from the late Miocene of North Africa, Siwaliks, and Piram Island than with those of the Graeco-Iranian Turolian or of sub-Saharan Africa. They could be part of a latitudinally differentiated fauna lying across the vast tract of land to the south of that inhabited by the Graeco-Iranian faunas. The presence of giraffines, boselaphines, spiral-horned antilopines, and a gazelle in Abu Dhabi could indicate rather park-like country. There might have been a southern or hot-country climate, but not one with any substantial development of aridity.

Finally in Chapter 23, Peter J. Whybrow and Diana Clements list the fauna found at each Baynunah Formation locality. This list will be useful both for future studies and as a historical record. The pace of development in the Emirate of Abu Dhabi is rapid and some sites are not now available for additional collecting. Geographical descriptions of the localities are given with the GPS co-ordinates of specific sites within them.

Late Miocene Swan Mussels from the Baynunah Formation, Emirate of Abu Dhabi, United Arab Emirates

Paul A. Jeffery

Collecting by The Natural History Museum/Yale University team from the late Miocene Baynunah Formation (Whybrow et al., 1998—Chapter 4) of the coastal region of the Emirate of Abu Dhabi in the period 1979–94 yielded a rich terrestrial and fluvial flora and fauna, amongst which are unionoid bivalves previously recorded as ?*Anodonta*. Closer examination of specimens collected from the stratotype section at Jebel Barakah, and also at Jebel Dhanna, Hamra, Shuwaihat, and Ghayathi, revealed two species, which can be placed in two separate unionoid superfamilies.

Material and the Rock Matrix

Most of the fossil mussels are preserved as recrystallised shells in a more or less poorly lithified arenaceous matrix; specimens from Shuwaihat are preserved as internal casts and external moulds, in some cases partially infilled with microcrystalline calcite. They are preserved in a weakly indurated sandstone comprising subangular to well-rounded and polished sand-grade clasts in a calcareous matrix. The clasts are principally transparent colourless quartz with occasional translucent pink ?quartz, mid-green ?olivine, reddish-brown ?garnets, and small grains of dark, probably ferromagnesian minerals. In some cases subrounded intraformational clasts of pebble grade are incorporated; some of these are closely similar in composition to the main fabric of the matrix, whereas others are relatively sand-free carbonate pellets up to 1 cm across.

The nature of the rock suggests a subaqueous depositional environment at varying energy levels (Friend, 1999; Forey and Young, 1999—Chapters 5 and 12), with carbonate precipitation during periods of nondeposition and evaporation, through periods of balanced deposition, to episodes of active erosion and reworking of existing sediments. The degree of roundness of the clasts indicates a fluvial environment of deposition, which supports conclusions based on the flora and fauna. The small-scale variability of the sediments may reflect seasonality or longer-term climatic fluctuations.

Shell Preservation

A lamellar shell structure is apparent in all specimens with shell preserved. Organic laminae are lost, while the inorganic (originally probably aragonitic) laminae appear to have undergone dissolution and replacement by crystalline calcite. The remaining structural relics suggest an originally nacreous shell that has been dissolved by weathering and partly replaced by secondary calcite infill. The interior of some valves is visible, showing weakly developed hinge teeth and isomyarian muscle scars. These facts, together with observations on the nature of the enclosing matrix, the accompanying fauna, and shell size and shape, indicate placement of these bivalves in the Unionoida.

In many cases the specimens are preserved with both valves intact and together, suggesting rapid burial—possibly in situ while the individuals were still alive. In the case of material from Jebel

 ISBN 0-300-07183-3

Barakah the shells, especially of the larger species, are sometimes damaged and preserved with the valves imbricated upon one another. This accords with other evidence that material from this site has suffered a degree of postdepositional reworking, probably shortly after deposition, as the result of flooding in the river system.

Environment and Palaeoclimate

Unionoids are an extant group of exclusively non-marine bivalves. Modern representatives are present in a variety of freshwater habitats, from fast-flowing rivers and streams to lakes and stagnant, almost anoxic ponds; typically they live partly embedded in the sediment. They pass through a larval stage or glochidium that for some of its existence is parasitic on fish, which explains the widespread distribution of some forms. As might be expected shell shape and ornament reflect the habitat; inflated and highly ornamented forms develop in slow-moving rivers and lakes, whereas smoother, narrower, and more attenuated forms are more typical of faster-moving rivers and streams. The Baynunah unionoids are unornamented, fairly narrow, and more-or-less attenuated, which suggests that they were adapted to life in moderately fast-moving streams or rivers rather than lakes or slow-moving rivers.

The late Miocene was a time of global climatic change as increases in polar ice mass were accompanied by a corresponding drop in sea level, climatic cooling, and increased aridity in lower latitudes. The fossiliferous part of the Baynunah Formation appears to just pre-date the terminal Miocene cooling phase—the unionoids could not have flourished without a steady water supply. The top 5.2 metres of the type section at Jebel Barakah, however, consists of evaporitic deposits (Whybrow, 1989) that may coincide with the commencement of the terminal Miocene event. Furthermore, Whybrow (1989) dates the Baynunah Formation at between 11 and 5 million years (Ma), probably closer to 5 Ma. According to van Zinderen Bakker and Mercer (1986), oceanic data from off northwest Africa show that a relatively warm interval about 5.8 Ma ago was followed by two substantial cooling events. The timing of the warm interval seems to coincide with events in the Baynunah Formation; better dating may establish whether they are coeval.

Systematics

Unionoida
Mutelacea
Mutelidae
Mutela (?subgen. nov. aff. *Chelidonopsis*) sp.
(fig. 10.1a, 1b)

Material

Shuwaihat. Site S3, N 24° 06′ 03.5″, E 52° 26′ 19.0″: AUH 275. Jebel Barakah. Site B1, N 24° 00′ 24.9″, E 52° 19′ 48.6″: AUH 454. Site B2, N 24° 00′ 13.6″, E 52° 19′ 35.5″: BMNH LL41637, from 1979 collection.

Samples from Shuwaihat and Jebel Barakah include a rather elongate species of unionoid. This is up to 75 mm long, 35 mm high, and the inflation across both valves when united is about 25 mm. The shell wall is less than 0.5 mm thick so this was perhaps a thin-shelled species with a thick organic periostracum. The valves are elongate, subtrapezoidally ovate in outline, and anterodorsally produced, with low umbos and ornamented with growth lines only. The hinge structure of this species is not conclusively demonstrated in any of the available specimens, but it is apparently edentulous. This and its transversely elongate shape suggest placement in the family Mutelidae. Its shape is reminiscent of the African genus *Mutela,* especially the subgenus *M.* (*Chelidonopsis*) from the Kwango River (a tributary of the Congo), but it seems to lack the distal tube of that subgenus. This is apparently the first record of mutelid unionoids from the Arabian Peninsula.

The characteristic narrow elongate shape of mutelids is probably an adaptation to life on a fairly soft bottom in rivers and streams. The narrow attenuated shape would be easy to bury fairly deeply while in the exposed portion of the shell it would provide reduced resistance to water flow.

Unionoida
Unionacea
Unionidae
Leguminaia (*Leguminaia*) sp. (fig. 10.2a, 2b)

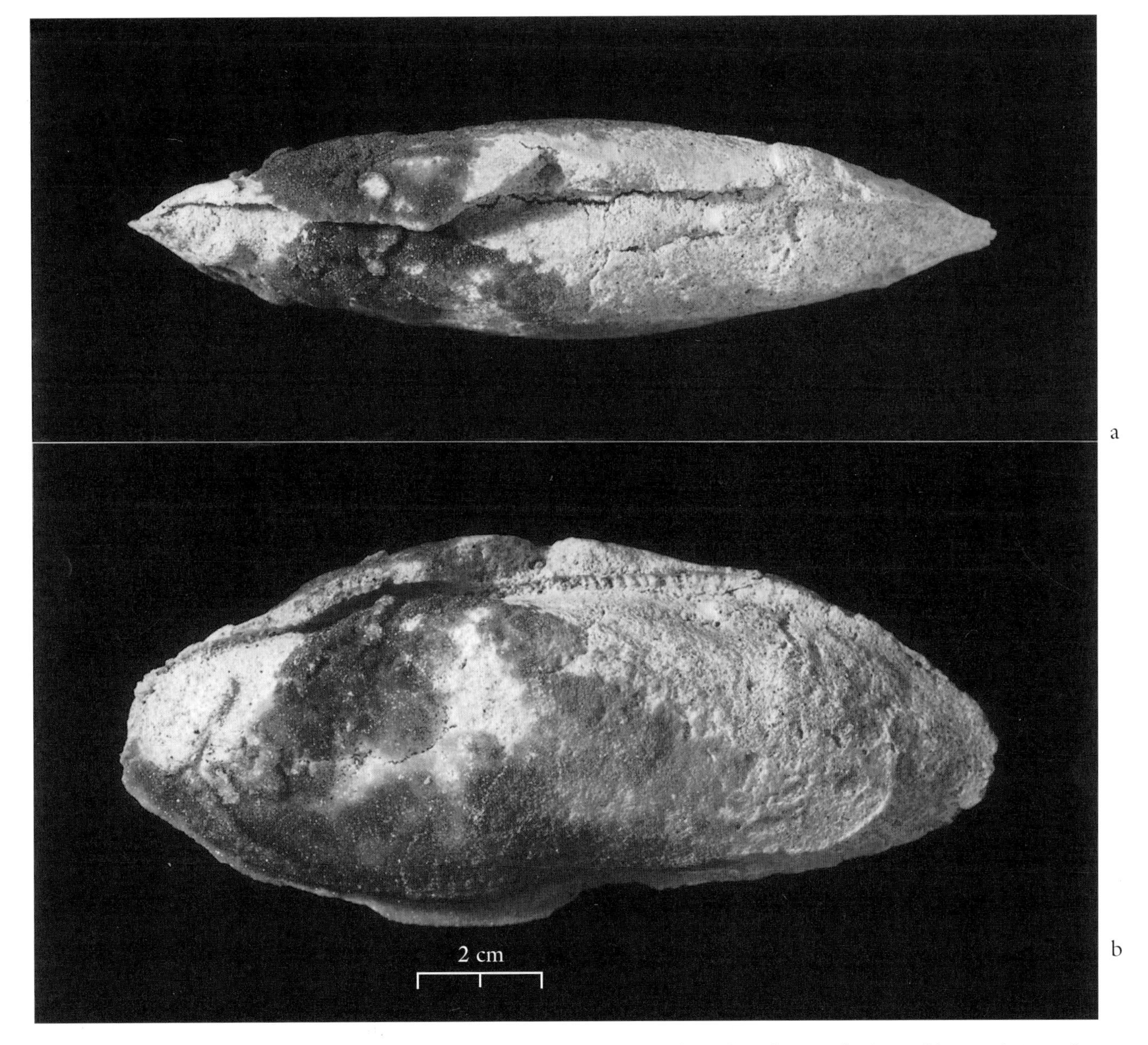

Figure 10.1. *Mutela* (?subgen. nov. aff. *Chelidonopsis*) sp., AUH 275: *a*, dorsal view; *b*, view of internal cast of left valve.

Material

Jebel Dhanna. Site JD3, N 24° 10′ 31.1″, E 52° 34′ 21.0″: AUH 290, AUH 323. Site JD5, N 24° 10′ 22.9″, E 52° 34′ 38.5″: AUH 137.

Jebel Barakah. Site B1, N 24° 00′ 24.9″, E 52° 19′ 48.6″: AUH 454a. Site B2, N 24° 00′ 13.6″, E 52° 19′ 35.5″: AUH 306, BMNH LL41637, from 1979 collection.

Ghayathi. Site G1, N 23° 54′ 30.7″, E 52° 48′ 15.9″: AUH 335.

Hamra. Site H5, N 23° 06′ 04.7″, E 52° 31′ 38.5″: AUH 418.

Figure 10.2. *Leguminaia (Leguminaia)* sp., AUH 323: *a,* internal views of right and left valves; *b,* external views of right and left valves.

The samples from Jebel Barakah, Jebel Dhanna, Hamra, and Ghayathi contain a second species belonging in a separate genus. It is smaller, more regularly ovate, and less elongate than the preceding species, with a thicker shell and a well-developed umbo. The shell is ornamented externally with growth lines only; any trace of umbonal sculpture has been lost. The largest example, a single left valve from Jebel Dhanna, is 43 mm long, 26 mm high with an inflation of 10 mm, and the thickness of the shell is up to 3 mm in some individuals. Several individuals from Jebel Dhanna show the interior of the shell: the muscle attachments are subisomyarian with a separate anterior pedal retractor scar behind the anterior adductor scar. The hinge dentition is relatively strong with well-developed cardinals and posterior laterals in both valves. These features are consistent with the family Unionidae, of which the subgenus *Leguminaia (Leguminaia)* is most similar. This subgenus constitutes an important record as it is otherwise known only from the Recent of Iraq (where it is found in the River Tigris); the Baynunah specimens thus increase both its age range and geographical distribution significantly.

The shorter and somewhat more tumid shape of *Leguminaia* suggests it favoured less fast-flowing water courses than *Mutela,* and, in support of this, the two species do not often occur together. Although samples from Jebel Barakah contain both species in association, they are in a matrix rich in intraformational clasts, suggesting that one or both species may have been derived into the assemblage by erosion of pre-existing sediments.

CONCLUSIONS

The presence of unionoid bivalves in the sediments of the Baynunah Formation supports the inference that they were deposited under a fluvial regime. The bivalves are assigned to genera of the Mutelidae and the Unionidae not previously known from the Arabian Peninsula. The mutelids are known from the Cretaceous to Recent of Africa (and possibly Europe) while *Leguminaia* (s.s.) is known only from the Recent of Iraq. The presence in the Arabian fossil fauna of two species of unionoids whose nearest relatives are known only from geographically separate provinces is partly explained by the parasitic lifestyle of the larvae, which ensure wide distribution by attaching themselves to fish. Several species of fish are known from the contemporary fauna, and the unionoids' vector may have been among them. The unionoids were dispersed, however, so it is clear that environmental conditions and distributional pathways existed in this area during the Miocene, which due to subsequent climatic changes are no longer available today.

ACKNOWLEDGEMENTS

I thank John Cooper and Noel Morris (The Natural History Museum, London) for critically reading the manuscript of this chapter and for their helpful suggestions for its improvement.

REFERENCES

Forey, P. L., and Young, S. V. T. 1999. Late Miocene fishes of the Emirate of Abu Dhabi, United Arab Emirates. Chap. 12 in *Fossil Vertebrates of Arabia,* pp. 120–35 (ed. P. J. Whybrow and A. Hill). Yale University Press, New Haven.

Friend, P. F. 1999. Rivers of the Lower Baynunah Formation, Emirate of Abu Dhabi, United Arab Emirates. Chap. 5 in *Fossil Vertebrates of Arabia,* pp. 38–49 (ed. P. J. Whybrow and A. Hill). Yale University Press, New Haven.

van Zinderen Bakker, E. M., and Mercer, J. H. 1986. Major Late Cainozoic climatic events and palaeoenvironmental changes in Africa viewed in a world wide context. *Palaeogeography, Palaeoclimatology, Palaeoecology* 56: 217–35.

Whybrow, P. J. 1989. New stratotype; the Baynunah Formation (Late Miocene), United Arab Emirates: Lithology and palaeontology. *Newsletters on Stratigraphy* 21: 1–9.

Whybrow, P. J., Friend, P. F., Ditchfield, P. W., and Bristow, C. S. 1999. Local stratigraphy of the Neogene outcrops of the coastal area: Western Region, Emirate of Abu Dhabi, United Arab Emirates. Chap. 4 in *Fossil Vertebrates of Arabia,* pp. 28–37 (ed. P. J. Whybrow and A. Hill). Yale University Press, New Haven.

A Terrestrial Pulmonate Gastropod from the Late Miocene Baynunah Formation, Emirate of Abu Dhabi, United Arab Emirates

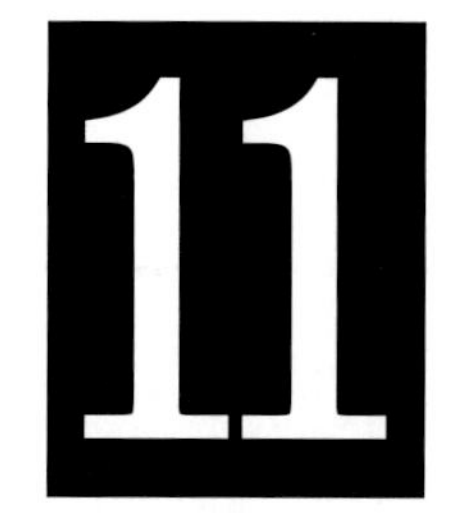

PETER B. MORDAN

A single specimen of an internal cast of a land-snail shell was collected by The Natural History Museum/Yale University team from the Baynunah Formation, Western Region of the Emirate of Abu Dhabi, United Arab Emirates (Shuwaihat, site S1, N 24° 06′ 38.1″, E 52° 26′ 09.6″; specimen number AUH 272). The sample is dated as late Miocene, about 8–6 million years (Ma). Figures 11.1 and 11.2 show photographs and line drawings, respectively, of the cast, which measures 8.2 mm maximum height by 4.2 mm maximum diameter. Five whorls are present, but the apex is broken. There is no evidence of any apertural dentition from the cast, although the aperture region is also rather eroded.

I have been unable to trace other published records of Neogene fossil land Mollusca from the Arabian Gulf region. Certainly the best-documented account of Miocene land molluscs in the broad geographical vicinity is that of Verdcourt (1963), who described a large assemblage of nonmarine molluscs from the early Miocene of Kenya. The most striking thing about this Kenyan material is how similar most of it is to the Holocene fauna of that area. This is true both of the entire family-level composition of the assemblage, as well as of the individual taxa, all of which were referable to extant genera, and in some cases to extant species. Even those species described as new were extremely similar to living forms. Moreover, the material referred to by Verdcourt is considerably older than the Abu Dhabi specimen, with a potassium/argon date in the Burdigalian of some 18 Ma (Mordan, 1992).

From Verdcourt's findings it would not be unreasonable to expect to be able to relate the fossil from Abu Dhabi to an extant genus or even a species with some degree of confidence. As Verdcourt (1963: 2) has pointed out, however, there is a major difficulty with the identification of fossil land Mollusca in that for many cases accurate determination of the recent taxa can be based only on an examination of the soft anatomy, especially where shell characters are rather few. Second, there is the obvious problem of interpreting a cast, where all the details of shell surface sculpture are lost. Together, these two factors make a precise determination impossible in this case.

Taxonomic Position

There are several recent accounts of the land molluscs of the Arabian Gulf region: Biggs (1937, 1962, 1971) for Iran; Biggs (1959) for Iraq; Mordan (1980a, 1986) for the Sultanate of Oman; and Mordan (1980b) for the Kingdom of Saudi Arabia. Together these give a comprehensive picture of the local fauna, and the few museum records of land molluscs from the United Arab Emirates include only species already listed in these published accounts.

From the shape of the fossil cast from the Baynunah Formation it is possible to exclude most of the land-snail families occurring in the region. Only the pulmonate families Subulinidae and Buliminidae (= Enidae) show broadly the right shell proportions. Several genera of buliminid occur in the Ara-

 ISBN 0-300-07183-3

Figure 11.1. Fossil internal cast of shell AUH 272 from Shuwaihat site S1: *A*, apertural view; *B*, adapertural view. Height 8.2 mm, diameter 4.2 mm.

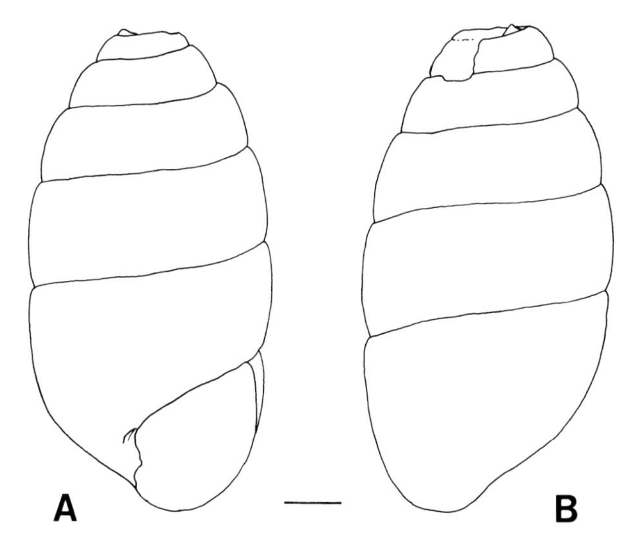

Figure 11.2. Drawing of fossil internal cast shown in figure 11.1: *A*, apertural view; *B*, adapertural view. Scale bar = 1 mm.

bian Gulf area, but only one of a similar appearance, *Imparietula jousseaumei* (Smith), occurs locally, apparently being restricted to northern Oman (Mordan, 1986). Similarly, only one of the subulinid species, *Zootecus insularis* (Ehrenberg), is close in shape to the fossil, and is a widely distributed coastal species found around the Arabian Gulf region, including Bahrain. I have taken internal casts of shells of both *Imparietula* and *Zootecus* (fig. 11.3), but both are clearly different from the fossil.

Internal casts were also taken from several Iranian buliminids listed by Biggs (1937, 1962, 1971), as well as some not mentioned by him. The most similar of these was *Subzebrinus oxianus* (Martens) from near Isfahan, Iran (fig. 11.4). This species has been synonymised with *Pseudonapaeus albiplicatus* (Martens) by Shileyko (1984), but the whole complex of buliminid species and genera from the Middle East must be considered unresolved taxonomically. Most are known from shells alone, but as Solem (1979: 24) has warned with respect to the inclusion of various buliminid shells from Afghanistan in the genus *Subzebrinus:* "Inclusion . . . is a matter of convenience. Until they can be dissected, no meaningful generic reference is possible." If this is the situation with recent shells, clearly one must be considerably more circumspect with a Miocene fossil cast. It is possible to state with some confidence that it does not belong to

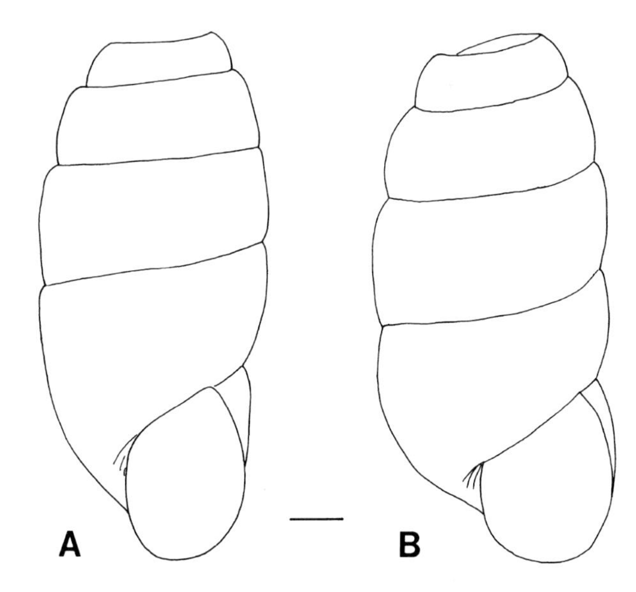

Figure 11.3. Drawings of internal casts of Holocene shells. *A*, *Imparietula jousseaumei* from Wadi Qatam, Saiq, Oman. Collected by M. D. Gallagher, 17.x.1984. *B*, *Zootecus insularis* from Bahrain. Collected by S. Green, 1994. Scale bar = 1 mm.

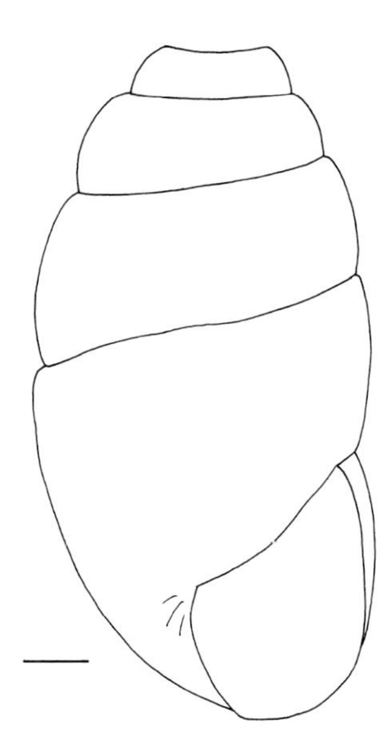

Figure 11.4. Drawing of internal cast of Holocene *Subzebrinus oxianus* from Isfahan, Iran. Collected by E. S. Brown, 22.xi.1958. Scale bar = 1 mm.

species or even genera currently occurring in the immediate vicinity of the Emirate of Abu Dhabi. It almost certainly belongs to the family Buliminidae Kobelt, 1880, but the genus remains uncertain; *Subzebrinus* Westerlund, 1887 and *Pseudonapaeus* Westerlund, 1887, both of which belong to the subfamily Pseudonapaeinae Schileyko, 1978 (which, incidentally, also includes *Imparietula* Lindholm, 1925), appear to be the most likely genera.

BIOGEOGRAPHY AND ECOLOGY

The genera within the Pseudonapaeinae, and certainly those three mentioned above, are all characteristic of rather xeric habitats, but tend to occur in situations where there is regular seasonal humidity, either in the form of precipitation at rather high elevations, or lower down close to water at the base of wadis. *Imparietula jousseaumei* is a case in point, being restricted to sites above about 500 metres in the northern mountains of the Sultanate of Oman; it is found at lower elevations only in synanthropic situations (Mordan, 1986). It does not occur in the United Arab Emirates today. These snails spend the dry season in aestivation attached to rocks, trees, and shrubs well above ground level, where they can avoid the hottest conditions, and only become active during wet periods when feeding and reproduction take place. The presence of a pseudonapaeine during the late Miocene suggests wetter, perhaps seasonal, conditions at that time. This notion compares well with other evidence of habitats near to the fluvial system of the Baynunah Formation (Friend, 1999; Jeffery, 1999—Chapters 5 and 10).

The Pseudonapaeinae, as a subfamily of the Buliminidae, are a group with Palaearctic and Oriental affinities. This is in contrast to the Cerastidae, a "southern" family that occurs in Africa, Madagascar, the Seychelles, India, and Australia, and was previously included in the Buliminidae but is now thought to be distinct (Mordan, 1984, 1992). The Holocene geographical distributions of these two families hardly overlap and there is a sharp line of demarcation in Oman, with the Buliminidae occurring in the northeast and the Cerastidae occurring in the mountains of Dhofar in the southwest. The presence in the Emirate of Abu Dhabi during the Miocene of a putative species of Pseudonapaeinae does not conflict with present-day distribution patterns at the family level.

REFERENCES

Biggs, H. E. J. 1937. Mollusca of the Iranian plateau. *Journal of Conchology* 20: 342–50.

———. 1959. Some land Mollusca from northern Iraq. *Journal of Conchology* 24: 342–47.

———. 1962. Mollusca of the Iranian plateau—II. *Journal of Conchology* 25: 64–72.

———. 1971. Mollusca of the Iranian plateau—III. *Journal of Conchology* 27: 211–20.

Friend, P. F. 1999. Rivers of the Lower Baynunah Formation, Emirate of Abu Dhabi, United Arab Emirates. Chap. 5 in *Fossil Vertebrates of Arabia,* pp. 38–49 (ed. P. J. Whybrow and A. Hill). Yale University Press, New Haven.

Jeffery, P. A. 1999. Late Miocene swan mussels from the Baynunah Formation, Emirate of Abu Dhabi, United Arab Emirates. Chap. 10 in *Fossil Vertebrates of Arabia,* pp. 111–15 (ed. P. J. Whybrow and A. Hill). Yale University Press, New Haven.

Mordan, P. B. 1980a. Land Mollusca of Dhofar. In *Journal of Oman Studies. Special Report no. 2: The Scientific Results of the Oman Flora and Fauna Survey, 1977,* pp. 103–11. Ministry of Information and Culture, Sultanate of Oman.

———. 1980b. Molluscs of Saudi Arabia—land molluscs. *Fauna of Saudi Arabia* 2: 359–67.

———. 1984. Taxonomy and biogeography of the southern Arabian Enidae *sensu lato,* (Pulmonata: Pupillacea). In *World-wide Snails: Biogeographical Studies on Non-marine Mollusca,* pp. 124–33 (ed. A. Solem and A. C. van Bruggen). Brill, Leiden.

———. 1986. A taxonomic revision of southern Arabian Enidae *sensu lato,* (Mollusca: Pulmonata). *Bulletin of the British Museum (Natural History),* Zoology 50: 207–71.

———. 1992. The morphology and phylogeny of the Cerastinae (Pulmonata: Pupilloidea). *Bulletin of the British Museum (Natural History),* Zoology 58: 1–20.

Shileyko, A. A. 1984. Mollusca: Terrestrial molluscs of the suborder Pupillina of the fauna of the USSR (Gastropoda, Pulmonata, Geophila). *Fauna SSSR,* n.s. 130: 1–399.

Solem, A. 1979. Some mollusks from Afghanistan. *Fieldiana, Zoology* no. 1301: 1–89.

Verdcourt, B. 1963. The non-marine Mollusca of Rusinga Island, Lake Victoria and other localities in Kenya. *Palaeontographica* 121A: 1–37.

Late Miocene Fishes of the Emirate of Abu Dhabi, United Arab Emirates

12

Peter L. Forey and Sally V. T. Young

Late Miocene fishes were reported from Abu Dhabi by Whybrow et al. (1990) and Whybrow (1989). These reports were based on several collecting trips and we use these results, together with new collections made by The Natural History Museum/Yale University team from 1989, to present an up-to-date summary of the Miocene ichthyofauna of Abu Dhabi and its significance. Specifically we address three areas of interest: (1) local palaeoecology of the fish-bearing sediments; (2) the significance of the finds in understanding the development of the African freshwater fish fauna; and (3) the part, if any, that these Miocene fishes play in explaining the current fish fauna of the Arabian Peninsula.

Material and Preservation

The fish remains were recovered from various localities within the fluvial part of the Baynunah Formation (Whybrow, 1989; Whybrow et al., 1999—Chapter 4), dated from the contained mammal fauna as Turolian (European Mammal Stage MN 13–14). Deposition thus took place about 7–6 million years (Ma) ago—see Parts I and II of this volume (Whybrow and Hill, 1999). The geology of the sites is described in Chapters 4 and 5 (Whybrow et al., 1999; Friend, 1999), and the localities are listed in Chapter 23 (Whybrow and Clements, 1999).

The fish remains are all fragmentary and are found on scree slopes beneath weathered sandstones and conglomerates within the Baynunah Formation. It is impossible, therefore, to relate fish occurrences to beds. At each of the sites the fish collections contain weathered, rounded bones as well as sharply fractured bones. The rounded bones are usually white whereas the sharply angled bones are pale brown. The rounded white bones are surely the result of recent erosion and weathering. Together, the size ranges and shapes of the bones are very variable and this implies that the original depositional environment was one of high energy, quick burial, and little water sorting.

The fishes found in the Baynunah Formation are here referred to three taxa: two catfishes, *Clarias* sp. and a new species of *Bagrus*, and a cyprinid, *Barbus* sp. We have decided to begin this report with general statements about the fishes: ecological preferences of modern relatives; biogeographic significance of the Baynunah fishes; and some statements about the modern fish fauna of the Arabian Peninsula. This is followed by systematic descriptions.

Modern Distributions and Ecology

The modern distribution of *Clarias* is coextensive with that of the family Clariidae (100 species), which extends throughout Africa, Syria, peninsular India, Southeast Asia, and the Philippines. The family is very rare in northern and Saharan Africa. Of the 43 species, the majority (32) are found in Africa, one in Syria, four in India, and seven in Asia (*C. batrachus* is widespread in both India and Southeast Asia). Where ecological data are available then most species prefer quiet, shallow waters; many prefer swamp conditions (Talwar and Jhingran, 1991) and can withstand intermittent periods of drought.

 ISBN 0-300-07183-3

The Bagridae, as traditionally recognised, are a large group of primitive catfishes with a fossil record back to the Thanetian (White, 1926). This may be a misleading statement, however, because the bagrids may not be a monophyletic group. Mo (1991) recently revised their systematics and split the family into three monophyletic families: a restricted Bagridae, the Clareotidae, and the Austroglanididae. The restricted Bagridae contains *Bagrus,* the only African genus, with 10 species, and 15 Asiatic genera (125 species). Mo (1991) considers that the Asiatic *Aorichthys* (with two species distributed in India and Burma) is the sister group of *Bagrus.* Species of *Bagrus* are widely distributed in a variety of freshwater habitats but most are bottom dwellers in relatively slow-moving waters.

Barbus is a widespread genus with species in Africa, Asia, and Europe. Members inhabit a very wide variety of habitats and although individual species do show habitat preferences, no meaningful generalisations may be made about the ecology of the genus.

Biogeographic Implications from the Fossil Record

The geographic position of Abu Dhabi at or near a possible migration route between Africa and Asia is clearly relevant to the theories of historical biogeography that use dispersal (migration) as the causal explanation. Much has been written about the mammalian faunal exchange between Africa and Asia during the Miocene (for example, Bernor et al., 1987; Steininger et al., 1985; Rögl and Steininger, 1984). The dating and direction of postulated migration depend in large part on the fossil occurrences, palaeoecological interpretations of the putative corridors, and precise dating, which, in turn, may be correlated with theories of earth history derived from lithological comparisons, palaeomagnetism, and absolute dating.

In the absence of the last categories of evidence or where there is considerable dispute between models of earth history—see, for example, earth histories proposed by Smith et al. (1991) and Owen (1983)—palaeontological evidence has to stand on its own. Of course there is some degree of reciprocal illumination in dating geological events from fossils and then postulating dates of migrations. In the case of the proposed connection between Asia and Afro-Arabia, however, there is considerable and varied palaeontological evidence drawn from continental sediments containing mammals (Bernor et al., 1987) and intercalated marine sediments with foraminifera (Adams et al., 1983) and echinoids (Ali, 1983). Opinions as to whether there was a single event leading to the contact between Asia and Afro-Arabia (Adams et al., 1983) or two events separated by a marine interval (Rögl and Steininger, 1984; Thomas, 1985) would seem peripheral to theories of the distribution of the fishes considered here. What is of more interest is the timing of the first occurrences of the fishes in Africa and Asia and what relevance this might have to the direction of migration.

Menon (1950), when discussing the biogeographic history of clariid catfishes, adopted a strictly palaeontological centre-of-origin approach. At that time the earliest fossils of clariid fishes (*Clarias* and *Heterobranchus*) were known from the middle Pliocene of the Siwalik Hills (Lydekker, 1886), followed by middle Pliocene occurrences in Egypt and Pleistocene occurrences in Java. He concluded, therefore, that *Clarias* originated in the Siwalik Hills and migrated west to North Africa with a subsequent range extension to central, western, and southern Africa during the Pleistocene. Menon's paper was written within the paradigm of fixed continents. He explained the absence of *Clarias* (and *Heterobranchus*) from the present Arabian Peninsula by suggesting that this part of the world had submerged after the dispersal event from Asia to Africa sometime in the middle Pliocene. A similar scenario has been suggested for *Barbus* (Menon, 1964; Gayet, 1982), which supposedly migrated from Asia to Africa as late as Plio-Pleistocene times.

The severest test of Menon's methodology came with subsequent finds of *Clarias* in the Miocene of the Arabian Peninsula and North Africa (see table 12.1). Accepting the fossil record at face

Table 12.1. **Neogene localities yielding *Clarias* (●), *Bagrus*, (◆), and *Barbus* (■)**

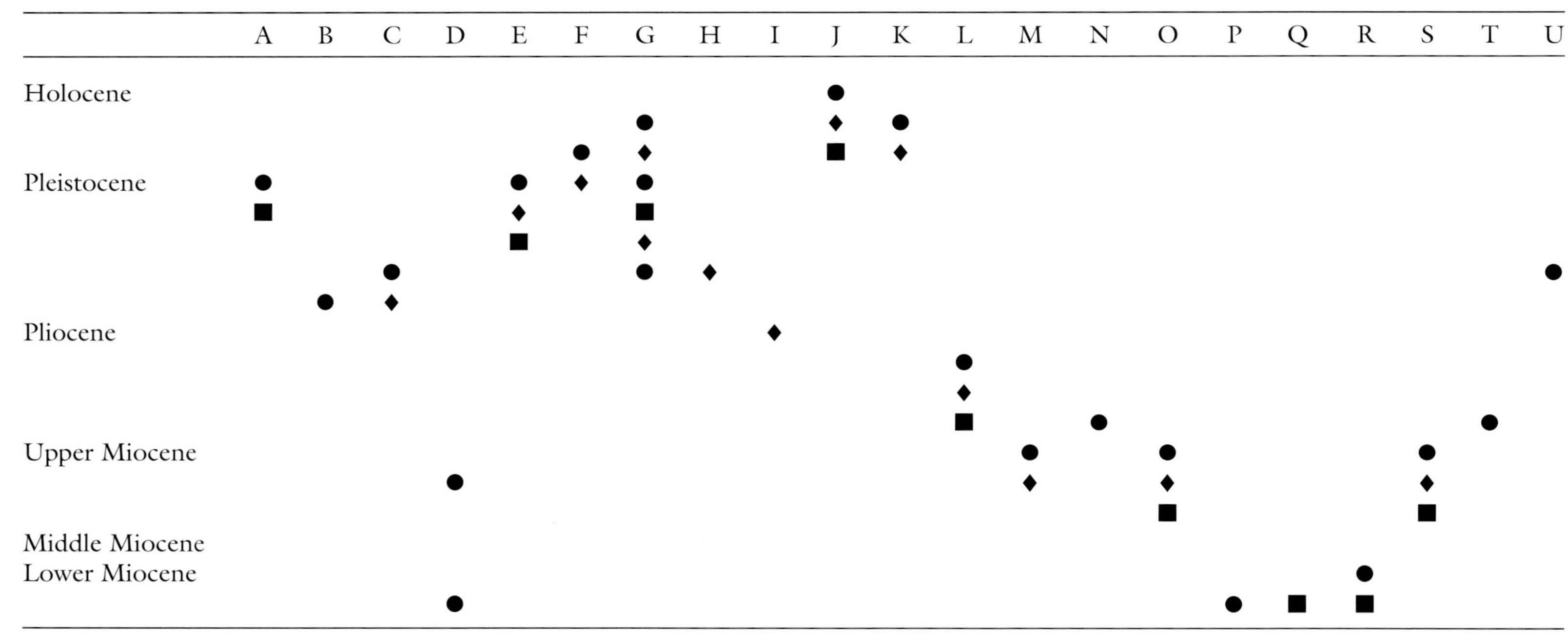

	A	B	C	D	E	F	G	H	I	J	K	L	M	N	O	P	Q	R	S	T	U
Holocene										●											
							●			◆	●										
						●	◆			■	◆										
Pleistocene	●				●	◆	●														
	■				◆		■														
					■		◆														
			●				●	◆													●
		●	◆																		
Pliocene									◆												
												●									
												◆									
												■		●						●	
Upper Miocene													●		●				●		
				●									◆		◆				◆		
															■				■		
Middle Miocene																					
Lower Miocene																		●			
				●												●	■	■			

A, Lake Eyasi, Tanzania, Upper Pleistocene (Greenwood, 1957); B, Olduvai, Tanzania, Lower Pleistocene (Greenwood and Todd, 1970); C, Lake Omo, Kenya, Lower Pleistocene (Arambourg, 1947); D, Lake Baringo, Kenya (Bishop et al., 1971); E, Lake Turkana (Rudolf), Kenya, Upper Pleistocene (Thomson, 1966); F, Chad, Sahara, Upper Pleistocene to Lower Holocene (Joleaud, 1935; Daget, 1958, 1961); G, Lake Albert, Zaire, Upper Pliocene (Greenwood, 1959); H, Lake Albert, Zaire, Lower Pleistocene (Greenwood and Howes, 1975); I, Lake Albert, Uganda, Upper Pliocene (White, 1926); J, Nubian Sudan, Pleistocene or Holocene (Greenwood, 1968); K, Fayum, Egypt, Holocene (Stromer, 1904); L, Wadi Natrun, Egypt, Middle Pliocene (Greenwood, 1972); M, Chalouf, Egypt, Upper Miocene (Priem, 1914); N, Sahabi, Libya, Lower Pliocene (Gaudant, 1987); O, Bled ed Dourah, Tunisia, Upper Miocene (Greenwood, 1973); P, Huqf, Sultanate of Oman, Lower Miocene (Roger et al., 1994); Q, Ad Dabtiyah, Eastern Province, Kingdom of Saudi Arabia, Lower Miocene (Greenwood, 1987); R, As Sarrar, Eastern Province, Kingdom of Saudi Arabia, Lower Miocene (Thomas et al., 1982); S, Emirate of Abu Dhabi, Upper Miocene (Forey and Young, 1998); T, Siwalik Hills, Middle Pliocene (Lydekker, 1886); U, Java, ?Pleistocene (Koumans, 1949).

value would place the centre of origin on the Arabian Plate/North Africa with migration to Asia. Discoveries of *Barbus* in the late Miocene of Tunisia (Greenwood, 1973) and Abu Dhabi also might upset Menon's hypothesis for the history of *Barbus*. Table 12.1 is a plot of the Neogene freshwater fish-bearing localities containing *Clarias, Bagrus,* and *Barbus* of Africa–Asia. It is an attempt to update that provided by Greenwood (1974).

Table 12.1 demonstrates just how incomplete, varied, and sparse is the sampling of Neogene fish history. Absence of a particular fossil takes on considerable relevance in the "fossil record and centre-of-origin approach". But absence is a difficult observation to square with reality. For instance, there are just three fishes currently known from the Baynunah Formation. Considering the relative richness of the remainder of the fauna, as described in this volume (Whybrow and Hill, 1998), the paucity of the fishes may reflect collecting effort (unlikely), taphonomic, or palaeoecological bias, or it may be a true reflection of the faunal content.

More particularly, the determination of the centre of origin using fossils is susceptible to new fossil discoveries, difficulties of dating, or the redating of already known localities. There may be instances where the fossil record of a particular group of organisms can be assumed to be complete enough to tell a "true" story (for example, foraminifera—Adams et al., 1983). The freshwater fish record of the African/Asian Neogene is, however, so patchy (and we do not know how incomplete it may be), that it would be unwise to base theories of origin and migration with any degree of confidence.

The fossil record does show that by late Miocene times *Clarias* was a widespread genus, the range of which extended from Tunisia in the west to the Arabian Peninsula in the east and to the Siwalik Hills on the Asian Plate in the middle Pliocene. *Clarias* was also present in Africa in the early Miocene. The genus *Bagrus* is limited to Africa, where it is nearly always associated with *Clarias* (table 12.1). This might suggest that the biogeographic histories of *Clarias* and *Bagrus* were congruent.

Putting the stratigraphical distribution to one side then, in trying to account for the biogeographic history of the three fishes within the Abu Dhabi fauna we meet three explanations based on three separate methodological premises. The distributional history of *Clarias* is explained (Menon, 1950) by assuming that because the greatest species diversity of the genus is in Africa this is the furthest place from the centre of origin (Asia). For *Barbus* the scenario is the reverse, with Europe/Asia being the assumed centre of origin because it has the highest species diversity. Until we have species-level phylogenies of *Barbus* and *Clarias* it would appear futile to try and choose between these alternative historical explanations. For *Bagrus,* accepting the phylogeny proposed by Mo (1991), the African *Bagrus* is the most derived member of a family that is otherwise distributed in Asia. Furthermore, the species of *Bagrus* described here shows some plesiomorphic similarities with the sister genus *Aorichthys* (see below). This suggests that there was single migration from Asia to Africa and might provide an example of Hennig's (1966) progression rule.

THE MODERN FAUNA

The modern freshwater fish fauna of the Arabian Peninsula is very depauperate compared with neighbouring parts of the world. Banister and Clarke (1977) record nine species for the entire peninsula. These represent eight cyprinids (three species of *Barbus,* three species of *Gara,* two species of *Cyprinion*) and one cyprinodont, *Aphanius dispar* (Rüppell). The cyprinodont is a euryhaline species, tolerant of seawater, and has a widespread distribution from northeast Africa to northwest India. It is therefore of limited significance in understanding the biogeographic history of Arabia. No freshwater catfishes are present on the modern Arabian Peninsula: this may well reflect the absence of slow-moving rivers.

The modern fish fauna shows three different biogeographic affinities according to Banister and Clarke (1977). Two species, now living in the western part of the Arabian Peninsula, have affinities with relatives in the Horn of Africa; one has a distribution spanning the Arabian Gulf; the remaining species (distributed along either side of the Arabian Peninsula) have affinities with species living in the Levant. It should be stressed that "affinities", as used by Banister and Clarke (1977), does not mean cladistic affinities. Most of the resemblances that those authors show between species are phenetic similarities, which may or may not have phylogenetic significance.

Even if we accept this phenetic resemblance as meaningful, the mixture of biogeographic relationships suggests either that the modern distribution patterns arose at completely different times and were layered upon one another or, more likely, that the species were once widespread and that increasing aridity has caused different but more restricted distributions. The absence of *Clarias,* the species of which are extremely hardy, might reinforce this idea. If the Arabian Peninsula were effectively "sterilised" by harsh aridity then there may have been subsequent invasions from three separate directions: from the Ethiopian highlands, from the Levant, and from Asia. Because aridity was a relatively recent event (Kassler, 1973) these invasions must have occurred within historic times. The idea of widespread extinction followed by reinvasion is also suggested for the North African fish fauna. Here, the modern fish fauna consists mostly of southern European fishes such as the cyprinids *Phoxinellus* and *Barbus capito* Pfeffer as well as the cyprinodont *Aphanius fasciatus* (Valenciennes), all of which may well have migrated via the Iberian Peninsula. This complement is totally unlike the fishes from Miocene and Pliocene times. The exception to this scenario are the fishes of the Nile that appear to have had a relatively stable history throughout the Neogene.

Parenthetically, it is worth noting that the modern and diverse herpetofauna of the Arabian Peninsula, which consists of some 135 species (Arnold, 1987), shows some interesting parallels with the fish fauna. Some species of amphibians and reptiles of southwest Arabia have their closest relatives in the Horn of Africa. According to Arnold the detailed phylogeny of semaphore geckos (*Pristurus*) suggests multiple movements between these regions. Some other species of amphibians and reptiles living in the northern Oman Highlands have conspecifics living in Iran across the Arabian Gulf; while others, living in the southwestern highlands, are most closely related to species in the eastern Mediterranean, Iraq, Iran, and even Pakistan and India. Arnold (1987) suggests that the last category of herpetofauna is absent from the central part of the Arabian Peninsula because of ecological conditions. All three of these patterns parallel those among the modern fishes. The reptile fauna has an additional component: the arid-adapted species that inhabit the central deserts of Saudi Arabia and have relatives in North Africa. The relatively moist Nile valley may provide an effective ecological barrier to these species.

SYSTEMATICS OF THE BAYNUNAH FISHES

Superorder Ostariophysi
Series Otophysi
Order Siluriformes
Family Clariidae
Genus *Clarias* Scopoli

Material[1] (figs 12.1–12.5)

Shuwaihat. Site S1, N 24° 06′ 38.1″, E 52° 26′ 09.6″: AUH 114j, cleithrum; AUH 117b, dermethmoid; AUH 112, dermethmoid plus lateral ethmoids; AUH 114c, 117d, 280g, frontals; AUH 117c, 117m, 753, lateral ethmoids; AUH 117f, nasal; AUH 114d, 117e, 280e, 752, supraoccipitals; AUH 117, centrum; AUH 117j, pectoral fin spine. Site S2, N 24° 06′ 41.7″, E 52° 26′ 04.0″: AUH 125d, dermethmoid; AUH 125e, lateral ethmoids. Site S4, N 24° 06′ 44.7″, E 52° 26′ 12.7″: AUH 732, cleithrum; AUH 728, dermethmoids; AUH 731, frontal; AUH 726, lateral ethmoid; AUH 730, post-temporal; AUH 729, quadrate; AUH 727, supraoccipitals; AUH 725, centrum; AUH 733, coracoid. Site S6, N 24° 07′ 06.6″, E 52° 26′ 32.7″: AUH 493, frontal; AUH 493a, 617a, lateral ethmoids; AUH 493b, sphenotic; AUH 617, supraoccipital.

Jebel Dhanna. Site JD3, N 24° 10′ 31.1″, E 52° 34′ 21.0″: AUH 134b, 134c, 134e, 606a, cleithra; AUH 134e, frontal; AUH 134c, 274b, lateral ethmoids; AUH 655, skull roof bones; AUH 134f, pterotic; AUH 134a, 274c, sphenotics; AUH 134d, 274d, supraoccipitals. Site JD5, N 24° 10′ 22.9″, E 52° 34′ 38.5″: AUH 140i, 747, cleithra; AUH 745, dentary; AUH 749, dermethmoid; AUH 659b, dermopterotics; AUH 659c, 748, lateral ethmoids; AUH 140b, lateral ethmoid and pterotic; AUH 751, post-temporal; AUH 750, sphenotic; AUH 140f, 140g, 659d, supraoccipitals; AUH 744, centrum; AUH 746, pectoral fin spine.

Hamra. Site H3, N 23° 04′ 28.7″, E 52° 31′ 37.3″: AUH 276f, dermethmoid.

Kihal. Site K1, N 24° 07′ 23.2″, E 53° 00′ 27.9″: AUH 309, frontal and lateral ethmoid; AUH 309a, supraoccipital.
Jebel Barakah. Site B2, N 24° 00′ 13.6″, E 52° 19′ 35.5″: BMNH P62047, dermopterotic; AUH 469a, BMNH P62048, P62049, frontals; AUH 469c, post-temporal; AUH 469b, BMNH P62046, P62050, supraoccipitals; BMNH P62051, pectoral fin spine.
Jebel Mimiyah (Al Mirfa). Site M1, N 24° 04′ 58.2″, E 53° 26′ 07.6″: AUH 632, cleithrum; AUH 668, sphenotic.
Ras Dubay'ah. Site R2 (no co-ordinates available): AUH 645, broken skull bones; AUH 645b, cleithra; AUH 645c, dermethmoid; AUH 646a, R2, dermethmoid and lateral ethmoid; AUH 646b, lateral ethmoid; AUH 646, vomer.

The collections contain many isolated bones of this large catfish. The commonest recognisable elements are the supraoccipital, sphenotic, dermopterotic, lateral ethmoid, dermethmoid, post-temporal, and orbitosphenoid amongst the head bones in addition to pectoral spines and cleithra. Some of these are illustrated in figures 12.1–12.5, alongside a skull of the Holocene *Clarias anguillaris* (Linnaeus). The head bones are thick, coarsely ornamented with pronounced tubercles that rarely run together and show only faint regular patterning relative to ossification centres.

The skull bones suggest that these remains belong to a species of the genus *Clarias* rather than the closely related *Heterobranchus* for the following reasons: the anterior suture between the supraoccipital and frontals is simple rather than complex and digitate; and the posterior end of the dermethmoid is a simple posteriorly directed V-shape rather than W-shaped.

It is impossible to identify these *Clarias* remains with any living species because so few parts have been found. Equally, it would be unwise to erect a new species on such disarticulated pieces. But it is possible to add some comparative comments. The African Holocene species of *Clarias* have been revised by Tuegels (1986). He did not establish a phylogeny but did recognise several groups of species (his subgenera) based on the combination of shapes of the frontal and occipital foramina and the pattern of denticulation on the pectoral spine. In the Abu Dhabi *Clarias* the frontal fontanelle is long and narrow (for example, AUH 112), the occipital foramen is located anterior to the supraoccipital process and therefore lies in front of the posterior margin of the skull, and the serrations on the pectoral spine are restricted to the outer edge. This combination of features is seen in *C. anguillaris*, *C. ngamensis* Castelnau, and *C. lamottei* Daget and Planquette (Roberts, 1989, questions the taxonomic status of the last-mentioned species). *Clarias ngamensis* is restricted to southern Africa but *C. anguillaris* is one of the more widely distributed species, being found from Senegal, along the southern edges of the Sahara to Ethiopia, and along the Nile to the Mediterranean. This corresponds to the Nilo-Sudan ichthyofaunal province of Roberts (1975). Other near-contemporaneous occurrences of *Clarias* in Africa are in the late Miocene deposits of Bled ed-Douarah, Tunisia (Greenwood, 1973), late Miocene of Chalouf, Egypt (Priem, 1914), and early Pliocene of Sahabi, Libya (see table 12.1). All these have been identified as *Clarias* sp. and they are closely comparable with the *Clarias* remains from Abu Dhabi. Modern species names have been applied only to fossil *Clarias* from the Pleistocene of Africa (Greenwood, 1974; Thomson, 1966).

Clarias is also known as fossils from the middle Pliocene of the Siwalik Hills (Lydekker, 1886) and from the Pleistocene of Java (Koumans, 1949).

Superorder Ostariophysi
Series Otophysi
Order Siluriformes
Family Bagridae (*sensu* Mo, 1991)
Genus *Bagrus* Bleeker, 1858
Bagrus shuwaiensis sp. nov. (figs 12.6–12.16)

Diagnosis

A species of *Bagrus* in which the ornament on the supraocciptal, pterotic, and sphenotic consists of coarse, irregular honeycomb rugosities; the supraoccipital spine is relatively broad (transverse width at base is equal to 18–20% of the length) and equal to about 40% of the total length of the supraoccipital; the lateral profile of the supraoccipital is flat and lies in continuity with the remainder of skull roof.

Holotype

Emirate of Abu Dhabi AUH 678; rear half of a braincase from the posterior level of the orbit to the occiput and showing the entire length of the

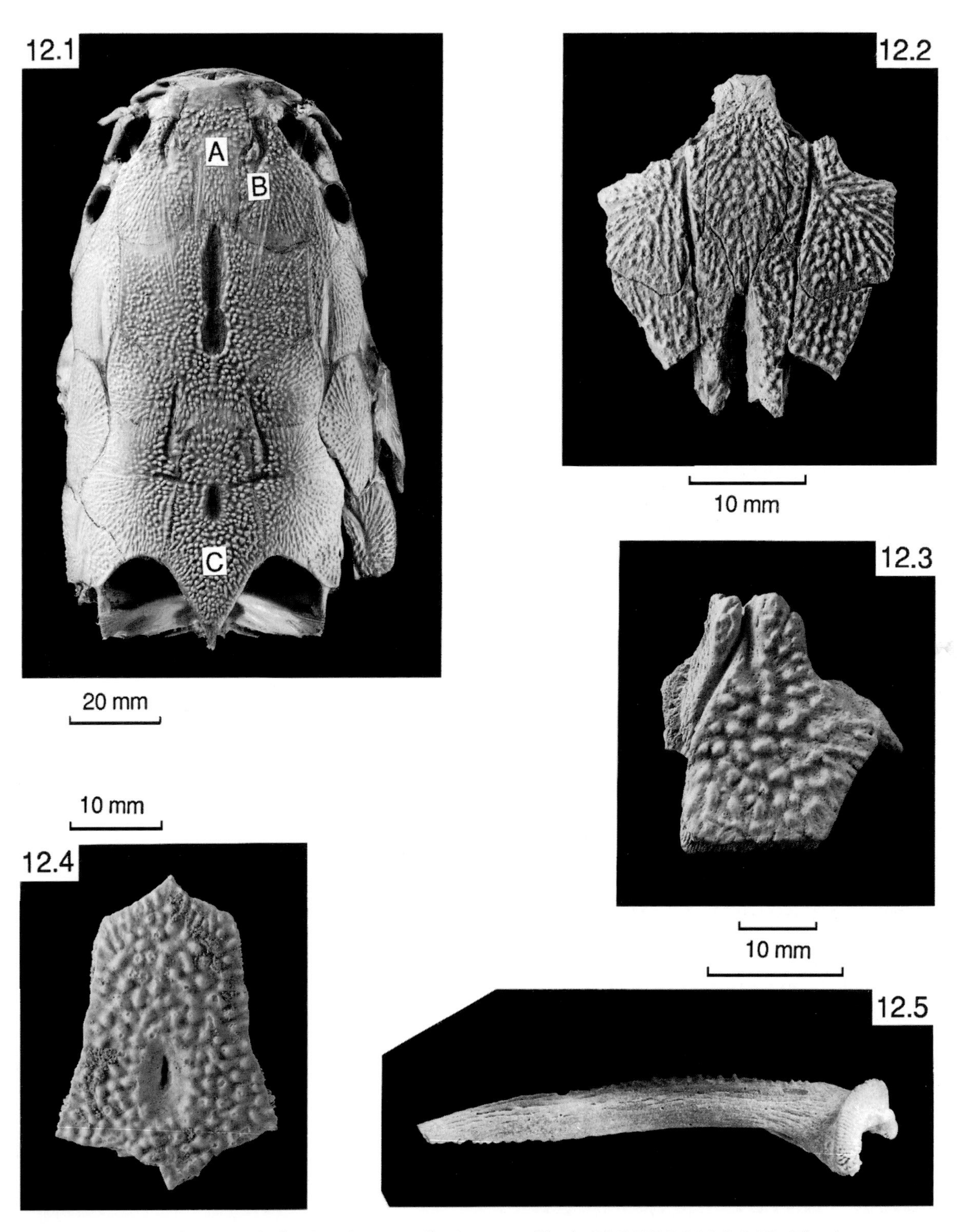

Figures 12.1–12.5. *12.1,* Skull of Holocene *Clarias anguillaris,* BMNH 1866.9.9.70. The letters refer to parts of the skull represented by fossil specimens illustrated in figures 12.2–12.4. *12.2–12.5*: *Clarias* sp., Baynunah Formation; *12.2,* dermethmoid, lateral ethmoids, and anterior portions of frontals, AUH 112 (portion A in fig. 12.1); *12.3,* right lateral ethmoid, AUH 117m (portion B in fig. 12.1); *12.4,* partial supraoccipital, AUH 140 (portion C in fig. 12.1); *12.5,* left pectoral spine in dorsal view (note restriction of the denticles to the outer edge; cf. *Bagrus* fig. 12.6), AUH 117j.

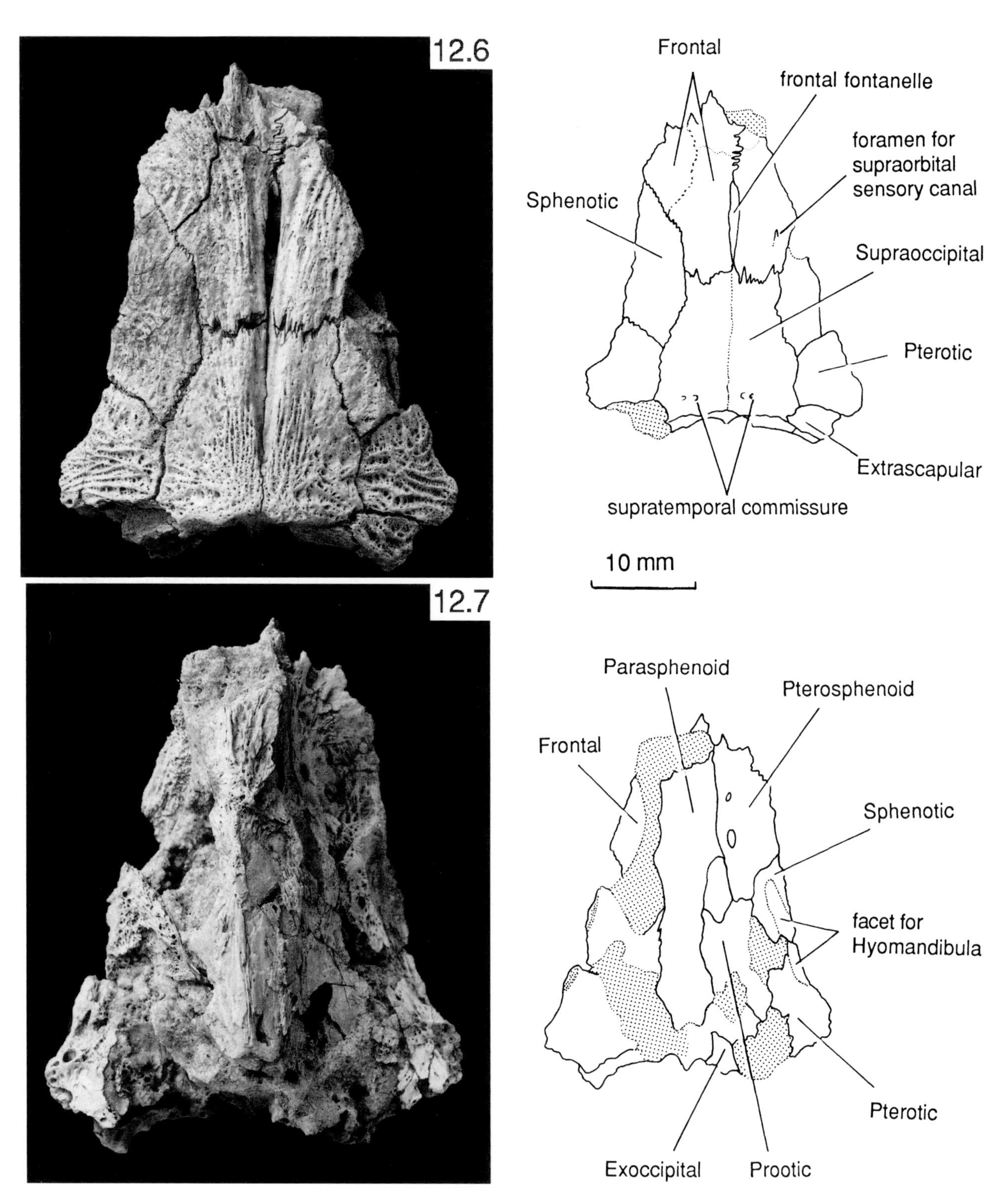

Figures 12.6 and 12.7. *Bagrus shuwaiensis* sp. nov., braincase, holotype AUH 678: *12.6*, dorsal view; *12.7*, ventral view. Scale bar applies only to photographs.

posterior fontanelle but lacking the supraoccipital spine. Late Miocene, Baynunah Formation, west side of Shuwaihat, Abu Dhabi.

Etymology

Named after a contraction of the name of find place of the holotype.

Material

There are many isolated bones representing those of a bagrid catfish. Isolated supraoccipitals, sphenotics, and frontals can be directly compared with the holotype and thus referred to this species. All other isolated elements can be referred only on their association in the same deposits. There is, however, no reason to suspect that more than one species is present. Besides the holotype there is one articulated piece of skull (AUH 108), which consists of part of the frontal, sphenotic, pterotic, and half of the supraoccipital of the left side of a skull considerably larger than the holotype. The many isolated bones are listed below and represent the following elements: supraoccipital, sphentoic, frontal, basioccipital, parasphenoid, anguloarticular, dentary, three fused occipital vertebrae, post-temporal, cleithrum, second dorsal fin spine, pectoral fin spine, and centra. All come from various localities within the late Miocene Baynunah Formation of Abu Dhabi.

Specimens Assigned to Bagrus shuwaiensis *sp. nov.*[1] (figs 12.6–12.16)

Shuwaihat. Site S1, N 24° 06′ 38.1″, E 52° 26′ 09.6″: AUH 114h, basioccipital; AUH 108, braincase; AUH 114a, 117l, 117n, cleithra; AUH 33b, dentaries; AUH 114g, 742, frontals; AUH 117g, supraoccipital; AUH 114b, 114i, centra; AUH 114, fused occipital vertebrae; AUH 33, 117h, 280f, second dorsal fin spines; AUH 114e, median fin spine; AUH 280i, second median dorsal fin spines; AUH 33a, 114f, 117i, 117k, 280h, 416, 743, pectoral fin spines. Site S2, N 24° 06′ 41.7″, E 52° 26′ 04.0″: AUH 678, braincase; AUH 125b, cleithra; AUH 125, frontals; AUH 125c, median fin spine; AUH 125a, pectoral fin spine. Site S4, N 24° 06′ 44.7″, E 52° 26′ 12.7″: AUH 723, cleithrum; AUH 719, dentaries; AUH 724, frontal; AUH 718, hyomandibular; AUH 722, post-temporal; AUH 717, sphenotic; AUH 720, sphenotic and frontal; AUH 754, supraoccipital; AUH 721, centra; AUH 716, second dorsal fin spine; AUH 715, pectoral fin spines. Site S6, N 24° 07′ 06.6″, E 52° 6′ 32.7″: AUH 492c, anguloarticular; AUH 492, frontal; AUH 492d, post-temporal; AUH 488a, 489, 492b, 554a, median fin spines; AUH 488, 492a, 554, pectoral fin spines.

Jebel Dhanna. Site JD3, N 24° 10′ 31.1″, E 52° 34′ 21.0″: AUH 134i, cleithra; AUH 134j, dentary; AUH 606b, frontal; AUH 655a, 655b, supraoccipitals; AUH 606e, anterior fused centra; AUH 134, fused occipital vertebrae; AUH 134h, 606d, median fin spines; AUH 134g, 606c, pectoral fin spines. Site JD5, N 24° 10′ 22.9″, E 52° 34′ 38.5″: AUH 140k, 737, anguloarticular; AUH 741, basioccipitals; AUH 140d, 736, cleithra; AUH 140h, 140j, 659, dentaries; AUH 140, 659a, 735, frontals; AUH 140a, parasphenoid; AUH 140c, 740, sphenotic; AUH 738, supraoccipital; AUH 140e, 739, pectoral fin spines.

Hamra. Site H3, N 23° 04′ 28.7″, E 52° 31′ 37.3″: AUH 276g, 276h, cleithra.

Jebel Barakah. Site B2, N 24° 00′ 13.6″, E 52° 19′ 35.5″: AUH 471, cleithrum; BMNH P62053, dentary; BMNH P62052, supraoccipital; AUH 469, pectoral fin spine.

Jebel Mimiyah (Al Mirfa). Site M1, N 24° 04′ 58.2″, E 53° 26″ 07.6″: AUH 669, dorsal pterygiophore; AUH 634, pectoral fin spines.

Harmiyah. Site Y2, N 24° 04′ 38.4″, E 53° 19′ 29.0″: AUH 347, median dorsal fin spine.

Ras Dubay'ah. Site R2 (no co-ordinates available): AUH 645a, pectoral fin spine.

Remarks

In a revision of the bagrid genera Mo (1991) identified a single synapomorphy for the genus *Bagrus;* namely, a dorsal fin consisting of eight soft fin rays. This cannot be checked in the material described here. Nevertheless there are several other features commonly found in *Bagrus* and the closely related *Aorichthys.* The skull is relatively flat and shallow beneath the orbit. The frontal fontanelle (fig. 12.6) is elongate and continues posteriorly as a groove reaching close to the base of the supraoccipital spine. The groove forms a plane of weakness and many specimens, including the holotype, are broken along this line. The supraoccipital spine ends posteriorly without a notch, implying that the nuchal plate does not overlap the supraoccipital. The post-temporal fossa is relatively deep and the opening is slit-like (fig. 12.9). The frontal (fig. 12.6) is narrow posteriorly where it is flanked by the sphenotic and it broadens above the orbit. The parasphenoid is poorly preserved in the holotype but, on the left side (fig. 12.8), it appears to form part of the border of the trigeminal foramen (this is a feature of bagrid genera other than *Bagrus* or *Aorichthys*). The dentary (fig. 12.11) is shallow throughout and lacks the pronounced coronoid

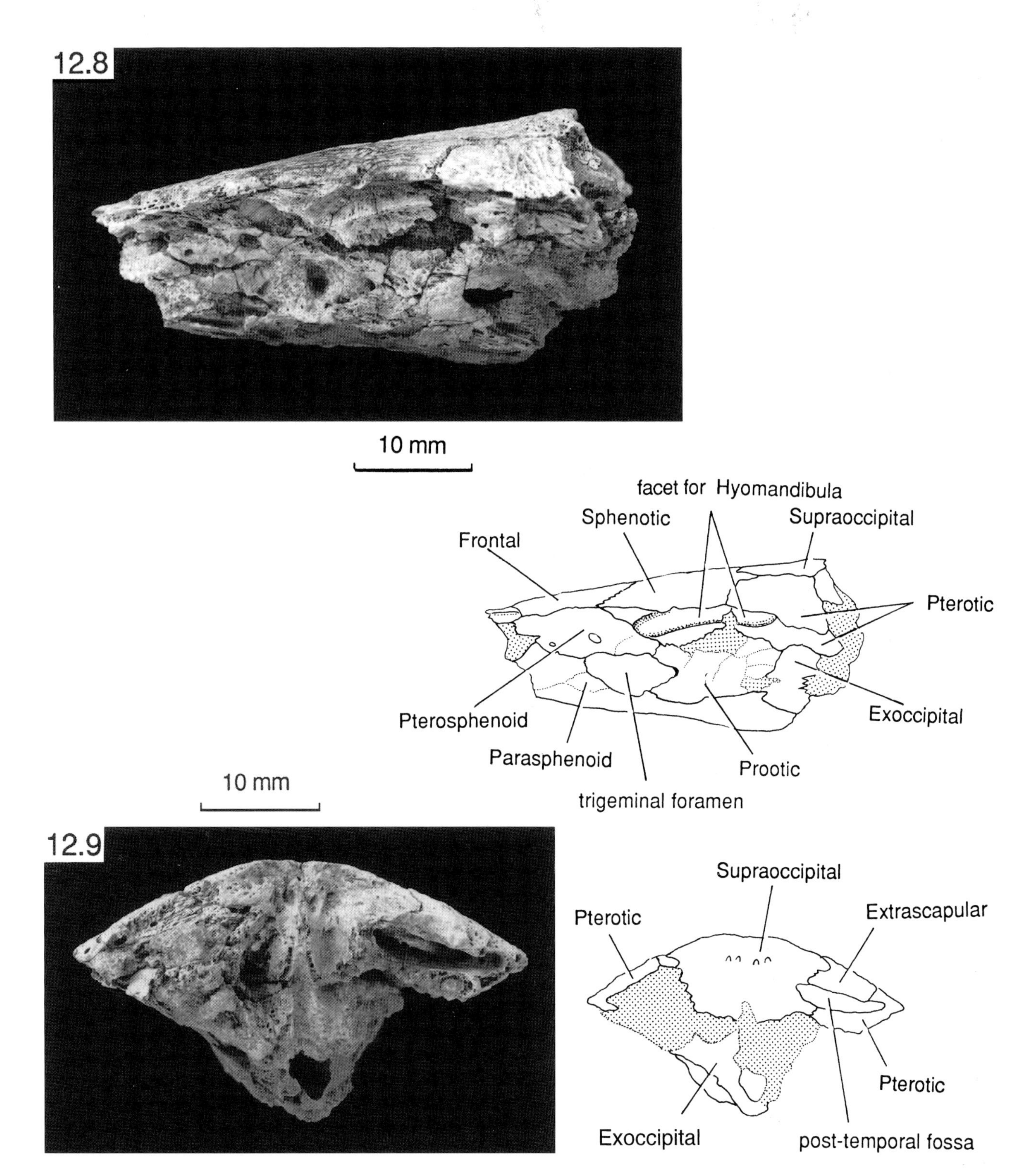

Figures 12.8 and 12.9. *Bagrus shuwaiensis* sp. nov., braincase, holotype AUH 678: *12.8,* left lateral view; *12.9,* posterior view. Scale bars apply to photographs.

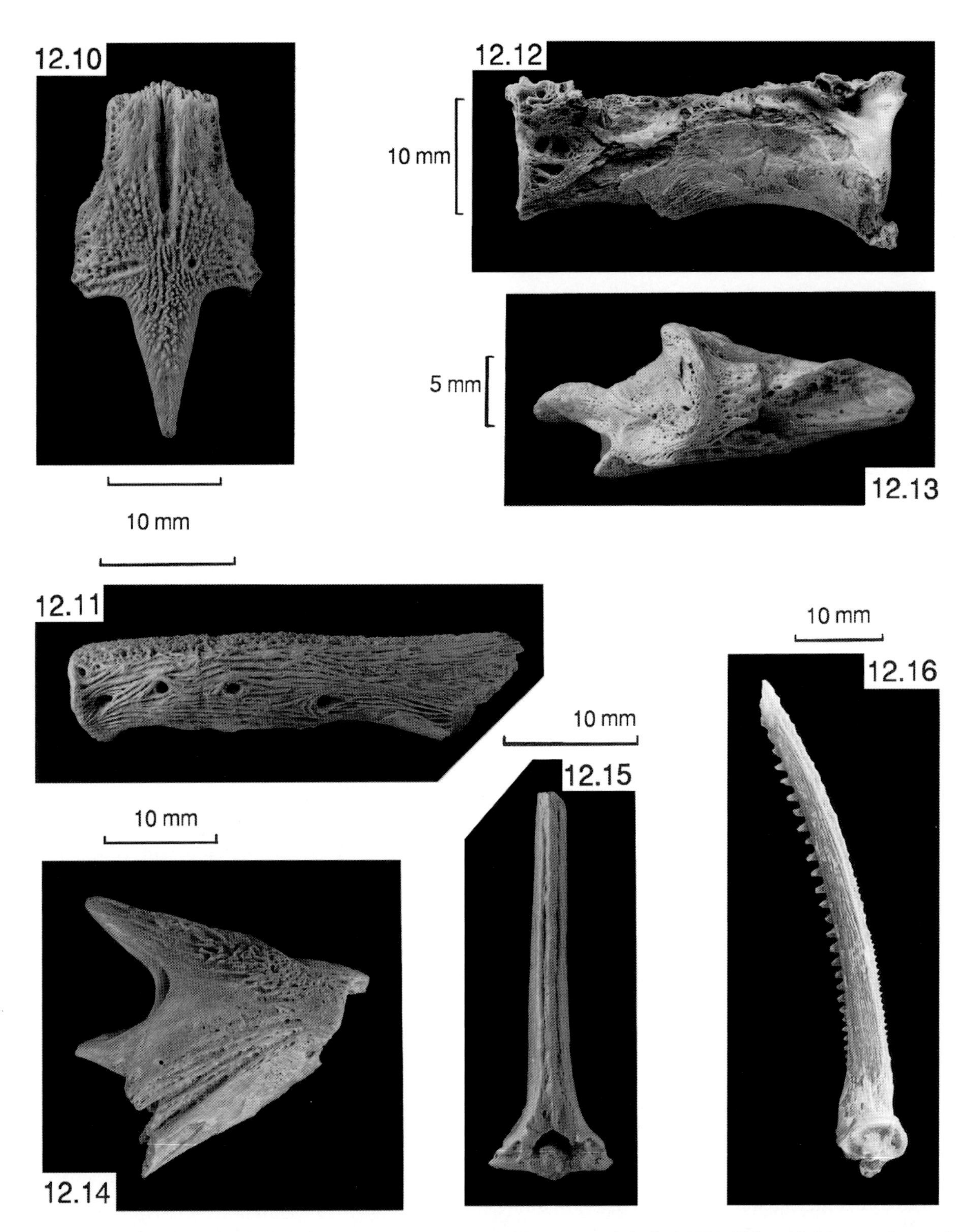

Figures 12.10–12.16. *Bagrus shuwaiensis* sp. nov.: *12.10,* supraoccipital, AUH 655a; *12.11,* anterior end of left dentary, lateral view, AUH 140j; *12.12,* three fused occipital vertebrae in right lateral view, AUH 114; *12.13,* anguloarticular in right lateral view, AUH 140k; *12.14,* right cleithrum, lateral view, AUH 117n; *12.15,* second dorsal fin spine, anterior view, AUH 117h; *12.16,* left pectoral spine in dorsal view (note denticles along mesial edge; cf. *Clarias,* fig. 12.5), AUH 117i.

process present in most other bagrids. The post-temporal is pierced by a foramen that receives the anterodorsal prong of the cleithrum (a feature also seen in *Mystus;* Mo, 1991) rather than a notch, which is more usual in bagrids.

The proportions of the frontal relative to the sphenotic and pterotic are much closer to those of *Bagrus* spp. than to *Aorichthys* spp. There is no notch between the pterotic and the extrascapular, as is seen in *Aorichthys,* and the post-temporal shows no inflated bulla, which in *Aorichthys* receives an anterior limb of the swimbladder. For these reasons we consider the Abu Dhabi bagrid to belong to the genus *Bagrus.* The proportions of the supraoccipital spine, however, and the spine's flat contour are unlike those in other species of *Bagrus* and more like those of *Aorichthys* spp. and other genera such as *Mystus* and *Chrysichthys.* This might suggest that *Bagrus shuwaiensis* is a relatively primitive species of the genus.

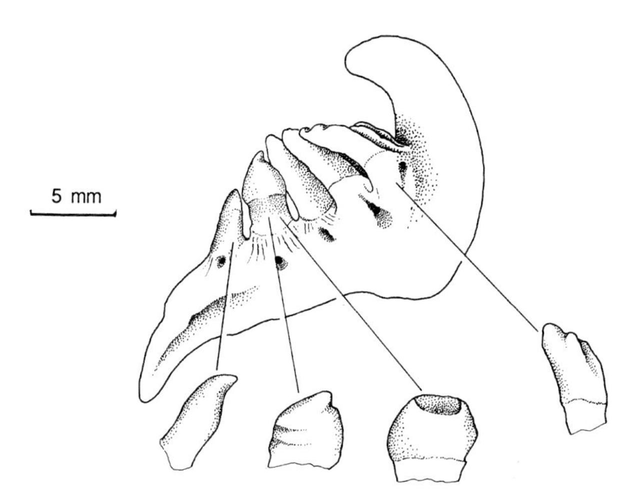

Figure 12.17. *Barbus.* Above is a drawing of the right lower pharyngeal of Holocene *Barbus bynni* (Forskål). Below are some variants of the isolated teeth found in the Baynunah Formation.

Superorder Ostariophysi
Series Otophysi
Order Cypriniformes
Family Cyprinidae
Genus *Barbus*

Material[1] (fig. 12.17)

Shuwaihat. Site S1, N 24° 06′ 38.1″, E 52° 26′ 09.6″: AUH 111, teeth. Site S2, N 24° 06′ 41.7″, E 52° 26′ 04.0″: AUH 123, teeth. Site S4, N 24° 06′ 44.7″, E 52° 26′ 12.7″: AUH 734, teeth.

The collection contains an assortment of isolated pharyngeal teeth of *Barbus.* Some are mammilliform, some molariform, and some elongate with recurved and contoured tips. Together they represent the normal array of teeth expected in a pharyngeal dentition of many of the large *Barbus* species. The size of the teeth suggest that they belong to a species referable to the "large *Barbus* group" of Banister (1987), which are generally over 200 mm standard length and show parallel or converging striae on the scales. There is no implication that these form a monophyletic group.

There are several "large *Barbus*" modern species living today in East Africa and along the Nile as far as the Nile Delta, in which the dentition consists of markedly different teeth, including all the differently shaped teeth found in the Abu Dhabi collection. *Barbus bynni* (Forskål), *B. intermedius* Rüppell, *B. macrolepis* Pfeffer, and *B. mirabilis* Pappenheim and Boulenger all show such dentitions and it is noticeable that where a large growth series has been studied (for example, *B. bynni*—see Banister, 1973: fig. 19) the large teeth are worn into concave crowns, similar to some of the teeth in the Abu Dhabi collection.

Both Greenwood (1987) and Gayet (1982) have referred early Miocene pharyngeal teeth from Saudi Arabia to *B. bynni,* a species that currently lives in the Nile drainage system. This is a possible identification for the Abu Dhabi teeth but, in our opinion, all of the Miocene teeth agree with teeth of all of the species mentioned above. As Banister (1973) pointed out, there is considerable variation in tooth morphology within a species and during the life of an individual. Tooth shapes may also be related to diet. Therefore species-level identification of individual *Barbus* teeth would seem very difficult.

Conclusions

1. Only three kinds of fishes can be recognised in the late Miocene Abu Dhabi fauna; only one has been identified to species level.
2. *Clarias* and *Bagrus* are genera whose modern representatives prefer slow-moving waters.
3. The nature of preservation of the fishes suggests that they may have been deposited at times of flooding.
4. The occurrence of *Clarias* in Neogene deposits of Afro-Arabia at numerous localities casts doubt on traditional theories that these catfishes migrated to Africa from Asia in the Pliocene. The phylogeny of bagrid catfishes strongly suggests that there was a dispersal from Asia to Africa but the fossil record suggests that this took place before Afro-Arabia met Asia.
5. Distributions of the modern fish fauna of the Arabian Peninsula may owe more to reinvasion after "sterilisation" than to continuity of historical events reaching back to the Miocene.

Acknowledgements

Peter Forey thanks Peter Whybrow and Andrew Hill for the opportunity to travel to the Emirate of Abu Dhabi to participate in the First International Conference on the Fossil Vertebrates of Arabia, March 1995, and to examine the Baynunah Formation at first hand. He also thanks the Abu Dhabi Company for Onshore Oil Operations (ADCO) and the Ministry of Higher Education and Scientific Research, United Arab Emirates, for their invitation to attend the conference. We thank William Lindsay, Palaeontological Laboratory, The Natural History Museum, London, for preparing the holotype specimen of *Bagrus shuwaiensis*. The photographs were taken by Phil Crabb (Photo Studio, NHM).

Note

1. In the list of identifiable fish material from the Baynunah Formation, AUH numbers refer to specimens that will be deposited in the Emirate of Abu Dhabi; numbers prefixed by BMNH P are in the palaeontological collections of The Natural History Museum, London.

References

Adams, C. G., Gentry, A. W., and Whybrow, P. J. 1983. Dating the terminal Tethyan events. In *Reconstruction of Marine Environments* (ed. J. Meulenkamp). *Utrecht Micropaleontological Bulletins* 30: 273–98.

Ali, Mohamed S. M. 1983. Tertiary echinoids and the time of collision between Africa and Eurasia. *Neues Jahrbuch für Geologie und Paläontologie, Monatshefte* 1983: 213–27.

Arambourg, C. 1947. Contribution à l'étude géologique et paléontologique du bassin du lac Rodolphe et de la basse Vallée de l'Omo, 2. Paléontologie. In *Mission scientifique de l'Omo, 1932–1933,* vol. 1. *Géologie—Anthropologie,* fasc. 3, pp. 231–562. Muséum National d'Histoire Naturelle, Paris.

Arnold, E. N. 1987. Zoogeography of the reptiles and amphibians of Arabia. In *Proceedings of the Symposium on the Fauna and Zoogeography of the Middle East,* pp. 245–56 (ed. F. Krupp, W. Schneider, and R. Kinzelbach). Dr Ludwig Reichert Verlag, Wiesbaden.

Banister, K. E. 1973. A revision of the large *Barbus* (Pisces, Cyprinidae) of East and Central Africa. Studies of African Cyprinidae Part II. *Bulletin of the British Museum (Natural History),* Zoology 26: 1–148.

———. 1987. The *Barbus perince–Barbus neglectus* problem and a review of certain Nilotic small *Barbus* species (Teleostei, Cypriniformes, Cyprinidae). *Bulletin of the British Museum (Natural History),* Zoology 53: 115–38.

Banister, K. E., and Clarke, M. A. 1977. The freshwater fishes of the Arabian Peninsula. In *The Scientific Results of the Oman Flora and Fauna Survey, 1975,* pp. 111–54. Ministry of Information and Culture, Sultanate of Oman.

Bernor, R. L., Brunet, M., Ginsburg, L., Mein, P., Pickford, M., Rögl, F., Sen, S., Steininger, F., and Thomas, H. 1987. A consideration of some major topics concerning Old World Miocene mammalian chronology, migrations and palaeoecology. *Geobios* 20: 431–39.

Bishop, W. W., Chapman, G. R., Hill, A., and Miller, J. A. 1971. Succession of Cainozoic vertebrate assemblages from the northern Kenya Rift Valley. *Nature* 41: 389–94.

Daget, J. 1958. Sur la présence de *Porcus* cf. *docmac* (poisson Siluriformes) dans le gisement néolithique saharien de Faya. *Bulletin de l'Institut Fondamental d'Afrique Noire, Dakar* 20A: 1379–86.

———. 1961. Restes de poissons du Quaternaires saharien. *Bulletin de l'Institut Fondamental d'Afrique Noire, Dakar* 20A:182–91.

Friend, P. F. 1999. Rivers of the Lower Baynunah Formation, United Arab Emirates. Chap. 5 in *Fossil Vertebrates of Arabia,* pp. 38–49 (ed. P. J. Whybrow and A. Hill). Yale University Press, New Haven and London.

Gaudant, J. 1987. A preliminary report on the osteichthyan fish-fauna from the Upper Neogene of Sahabi. In *Neogene Paleontology and Geology of Sahabi,* pp. 91–99 (ed. N. T. Boaz, A. El-Arnauti, A. W. Gaziry, J. de Heinzelin, and D. D. Boaz). Alan R. Liss, New York.

Gayet, M. 1982. Pisces. *ATLAL, The Journal of Saudi Arabian Archaeology* 5: 116.

Greenwood, P. H. 1957. Fish remains from the Mumba Cave, Lake Eyasi. *Mitteilungen aus dem Geologischen Staatsinstitut in Hamburg* 26: 125–30.

———. 1959. Quaternary fish fossils. *Exploration du Parc Albert* 4: 3–80.

———. 1968. Fish remains. In *The Prehistory of Nubia,* vol. 1, pp. 100–9 (ed. F. Wendorf). Fort Burgwin Research Center and Southern Methodist University Press, Dallas.

———. 1972. New fish fossils from the Pliocene of Wadi Natrun, Egypt. *Journal of Zoology, London* 168: 503–19.

———. 1973. Fish fossils from the Late Miocene of Tunisia. *Extrait des Notes du Service Géologique, Tunis* 37: 41–72.

———. 1974. Review of Cenozoic freshwater fish faunas in Africa. *Annals of the Geological Survey of Egypt, Cairo* 4: 211–32.

———. 1987. Early Miocene fish from eastern Saudi Arabia. *Bulletin of the British Museum (Natural History),* Geology 41: 451–53.

Greenwood, P. H., and Howes, G. J. 1975. Neogene fossil fishes from the Lake Albert-Lake Edward Rift (Zaire). *Bulletin of the British Museum (Natural History),* Geology 26: 72–127.

Greenwood, P. H., and Todd, E. J. 1970. Fish remains from Olduvai. In *Fossil Vertebrates of Africa,* vol. 2, pp. 225–42 (ed. L. S. B. Leakey and R. J. G. Savage). Academic Press, London.

Hennig, W. 1966. *Phylogenetic Systematics.* University of Illinois Press, Urbana.

Joleaud, L. 1935. Gisements de vertébrés quaternaires du Sahara. *Bulletin de la Société d'Histoire Naturelle de l'Afrique du Nord* 26: 23–39.

Kassler, P. 1973. The structural and geomorphic evolution of the Persian Gulf. In *The Persian Gulf,* pp. 11–32 (ed. B. H. Purser). Springer-Verlag, Berlin.

Koumans, F. P. 1949. On some fossil fish remains from Java. *Zoölogische Mededeelingen* 30: 77–88.

Lydekker, R. 1886. Indian Tertiary and post-Tertiary Vertebrata. *Memoirs of the Geological Survey of India. Palaeontologia Indica* 10: 24–258.

Menon, A. G. K. 1950. Distribution of clariid fishes, and its significance in zoogeographical studies. *Proceedings of the National Institute of Science, India* 17: 291–99.

———. 1964. Monograph of the cyprinid fishes of the genus *Garra* Hamilton. *Memoirs of the Indian Museum, Calcutta* 14: 173–260.

Mo, T. 1991. Anatomy, relationships and systematics of the Bagridae (Teleostei: Siluroidei) with a hypothesis of siluroid phylogeny. *Theses Zoologicae* 17: 1–216.

Owen, H. G. 1983. *Atlas of Continental Displacement.* Cambridge University Press, Cambridge.

Priem, R. 1914. Sur les poissons fossiles et en particulier des siluridés de Tertiare supérieur et des couches récentes d'Afrique. *Memoires de la Société Géologique de France* 21: 1–13.

Roberts, T. R. 1975. Geographical distribution of African freshwater fishes. *Zoological Journal of the Linnean Society of London* 57: 249–319.

———. 1989. Book review of "A systematic revision of the African species of the genus *Clarias* (Pisces; Clariidae)", by G. G. Tuegels. *Copeia* 1989: 530–32.

Roger, J., Pickford, M., Thomas, H., de Lapparent de Broin, F., Tassy, P., Van Neer, W., Bourdillon-de-Grissac, C., and Al-Busaidi, S. 1994. Découverte de vertébrés fossiles dans le Miocène de la région du Huqf au Sultanat d'Oman. *Annales de Paléontologie, Paris* 80: 253–73.

Rögl, F., and Steininger, F. F. 1984. Neogene Paratethys, Mediterranean and Indo-Pacific seaways. Implications for the paleobiogeography of marine and terrestrial biotas. In *Fossils and Climate,* pp. 171–200 (ed. P. Brenchly). Wiley, London.

Smith, A. G., Hurley, A. H., and Briden, J. C. 1991. *Phanerozoic Palaeocontinental Maps.* Cambridge University Press, Cambridge.

Steininger, F. F., Rabeder, G., and Rögl, F. 1985. Land mammal distribution in the Mediterranean Neogene: A consequence of geokinematic and climatic events. In *Geological Evolution of the Mediterranean Basin,* pp. 559–71 (ed. D. J. Stanley and F.-C. Wezel). Springer-Verlag, New York.

Stromer, E. 1904. Nematognathi aus dem Fajum und dem Natronthale in Aegypten. *Neues Jahrbuch für Mineralogie, Geologie und Paläontologie, Stuttgart* 1904: 1–7.

Talwar, P. K., and Jhingran, A. 1991. *Inland Fishes of India and Adjacent Countries.* Oxford and IBH Publishing, Delhi.

Thomas, H. 1985. The early and middle Miocene land connection of the Afro-Arabian Plate and Asia: A major event for hominid dispersal? In *Ancestors: The Hard Evidence,* pp. 42–50 (ed. E. Delson). Alan R. Liss, New York.

Thomas, H., Sen, S., Khan, M., Battail, B., and Ligabue, G. 1982. The lower Miocene fauna of Al-Sarrar (Eastern Province, Saudi Arabia). *ATLAL, The Journal of Saudi Arabian Archaeology* 5: 109–136.

Thomson, K. S. 1966. Quaternary fish fossils from the west of Lake Rudolf, Kenya. *Breviora* no. 243: 1–10.

Tuegels, G. G. 1986. A systematic revision of the African species of the genus *Clarias* (Pisces: Clariidae). *Musée Royale de l'Afrique Centrale Tervuren, Belgique, Zoologische Wetenschappen* 247: 1–199.

White, E. I. 1926. Pisces. *Occasional Papers. Geological Survey of Uganda, Entebbe* 2: 45–51.

Whybrow, P. J. 1989. New stratotype; the Baynunah Formation (Late Miocene), United Arab Emirates: Lithology and palaeontology. *Newsletters on Stratigraphy* 21: 1–9.

Whybrow, P. J., and Clements, D. 1999. Late Miocene Baynunah Formation, Emirate of Abu Dhabi,

United Arab Emirates: Fauna, flora, and localities. Chap. 23 in *Fossil Vertebrates of Arabia,* pp. 317–33 (ed. P. J. Whybrow and A. Hill). Yale University Press, New Haven.

Whybrow, P. J., and Hill, A. eds. 1999. *Fossil Vertebrates of Arabia.* Yale University Press, New Haven.

Whybrow, P. J., Friend, P. F., Ditchfield, P. W., and Bristow, C. S. 1999. Local stratigraphy of the Neogene outcrops of the coastal area: Western Region, Emirate of Abu Dhabi, United Arab Emirates. Chap. 4 in *Fossil Vertebrates of Arabia*, pp. 28–37 (ed. P. J. Whybrow and A. Hill). Yale University Press, New Haven.

Whybrow, P. J., Hill, A., Yasin al-Tikriti, W., and Hailwood, E. A. 1990. Late Miocene primate fauna, flora and initial palaeomagnetic data from the Emirate of Abu Dhabi, United Arab Emirates. *Journal of Human Evolution* 19: 583–88.

Chelonia from the Late Miocene Baynunah Formation, Emirate of Abu Dhabi, United Arab Emirates: Palaeogeographic Implications

13

FRANCE DE LAPPARENT DE BROIN AND PETER PAUL VAN DIJK

INTRODUCTION

Chelonian material collected by The Natural History Museum/Yale University team from the late Miocene, Baynunah Formation, Western Region, Emirate of Abu Dhabi, consists of isolated, rather small, fragments of skull (one), lower jaw (one), appendicular skeleton (two), and shell (around 75). Owing to the lack of sufficiently important parts of either skull or shell material, any precise specific and generic determinations are not yet possible; further study will be needed to try to resolve the systematic problems.

Three genera of continental turtles are represented, one terrestrial tortoise, and two freshwater turtles. In our first examination of this material in 1994, we considered that only one species was represented in each genus (Lapparent de Broin and Dijk, unpublished report 1994). This might prove incorrect for *Trionyx* s.l. after further study and collection of more material. In modern turtle populations, several species of the same genus often occur in the same area, although usually in different ecological niches. The discovery in the Baynunah Formation of scarce fragments of several individuals of different sizes at several sites clearly indicates that the specimens had not been fossilised in their habitat.

For the names and distribution of Chelonia see, amongst others:

1. For extant species see Bour (1980, modified 1985a); Iverson (1992); Loveridge and Williams (1957); Meylan (1987); and Wermuth and Mertens (1961, 1977). The names of genera or subgenera given here are listed in the above works with some modifications because they do not include analyses of the fossil forms.
2. For literature dealing with fossil forms see Broin (1977); Hay (1908); Hummel (1932); Kuhn (1964); and Lydekker (1889b).

The familial nomenclature is that of Bour and Dubois (1986) and the European mammal (MN) zonation is from Mein (1990).

The specimens will be housed by the Emirate Abu Dhabi, and are currently stored (1997) in The Natural History Museum, Department of Palaeontology, London. Prefixes to specimen numbers are AUH (Emirate of Abu Dhabi) and BMNH R, the latter prefix to collections made in 1979. The comparative material of extant species is housed in The Natural History Museum (Zoology of Reptiles), BMNH; the Muséum d'Histoire Naturelle, Toulouse, MHNT; the Muséum National d'Histoire Naturelle, Paris (AC, Anatomie Comparée; P, Paléontologie, Zoologie des Reptiles et Amphibiens), MNHN; and the Naturhistorisches Museum, Wien, Zoology, NHMW.

Sites with Turtles in the Baynunah Formation (fig. 13.1)

The sites were first described in Whybrow (1989):

Jebel Barakah. Site B2, N 24° 00′ 13.6″, E 52° 19′ 35.5″: *Trionyx* s.l. sp., *Mauremys* sp.

Hamra. Site H2, N 23° 06′ 06.9″, E 52° 31′ 42.5″: *Mauremys* sp. Site H3, N 23° 04′ 28.7″, E 52° 31′ 37.3″: *Trionyx* s.l. sp., *Mauremys* sp., cf. *Geochelone* aff. *sulcata*. Site H5, N 23° 06′ 04.7″, E 52° 31′ 38.5″: *Mauremys* sp.

Jebel Dhanna. Site JD3, N 24° 10′ 31.1″, E 52° 34′ 21.0″: *Trionyx* s.l. sp., *Mauremys* sp., cf. *Geochelone* aff. *sulcata*. Site JD5, N 24° 10′ 22.9″, E 52° 34′ 38.5″: *Trionyx* s.l. sp., *Mauremys* sp.

Kihal. Site K1, N 24° 07′ 23.2″, E 53° 00′ 27.9″: *Mauremys* sp., *Geochelone* aff. *sulcata*.

 ISBN 0-300-07183-3

Jebel Mimiyah. Site M1, N 24° 04′ 58.2″, E 53° 26′ 07.6″: *Mauremys* sp.
Mirfa Quarry. Site M2, N 24° 02′ 18.6″, E 53° 27′ 45.4″: *Trionyx* s.l. sp.
Ras Dubay'ah. Site R2 (no co-ordinates available), *Trionyx* s.l. sp., *Mauremys* sp.
Shuwaihat. Site S1, N 24° 06′ 38.1″, E 52° 26′ 09.6″: *Trionyx* s.l. sp., *Mauremys* sp., *Geochelone* aff. *sulcata*. Site S4, N 24° 06′ 44.7″, E 52° 26′ 12.7″: *Trionyx* s.l. sp., *Mauremys* sp., cf. *Geochelone* aff. *sulcata*.
Thumariyah. Site TH1, N 24° 09′ 25.1″, E 53° 00′ 32.6″: *Trionyx* s.l. sp.

Age

Late Miocene, between 5 and 11 million years (Ma) ago (Whybrow, 1989; Whybrow et al., 1990); estimated at 8–6 Ma at the First International Conference on the Fossil Vertebrates of Arabia, March 1995, Emirate of Abu Dhabi (Whybrow and Hill, 1999—this volume).

Systematics

Infraorder Cryptodira Cope, 1868

Family Trionychidae Fitzinger, 1826

Subfamily Trionychinae Fitzinger, 1826
Genus *Trionyx* Geoffroy Saint-Hilaire 1809, s.l.
Trionyx s.l. sp.

Material by Sites (fig. 13.1)

There are around 20 isolated specimens, representing at least 12 individuals:

B2: BMNH R10977, two fragments of pleurals.
H3: AUH 14, lateral part of pleural (right 5 or 6?); AUH 409, anterior fragment of right xiphiplastron.
JD3: AUH 606, medio-left posterior part of nuchal of a young specimen.

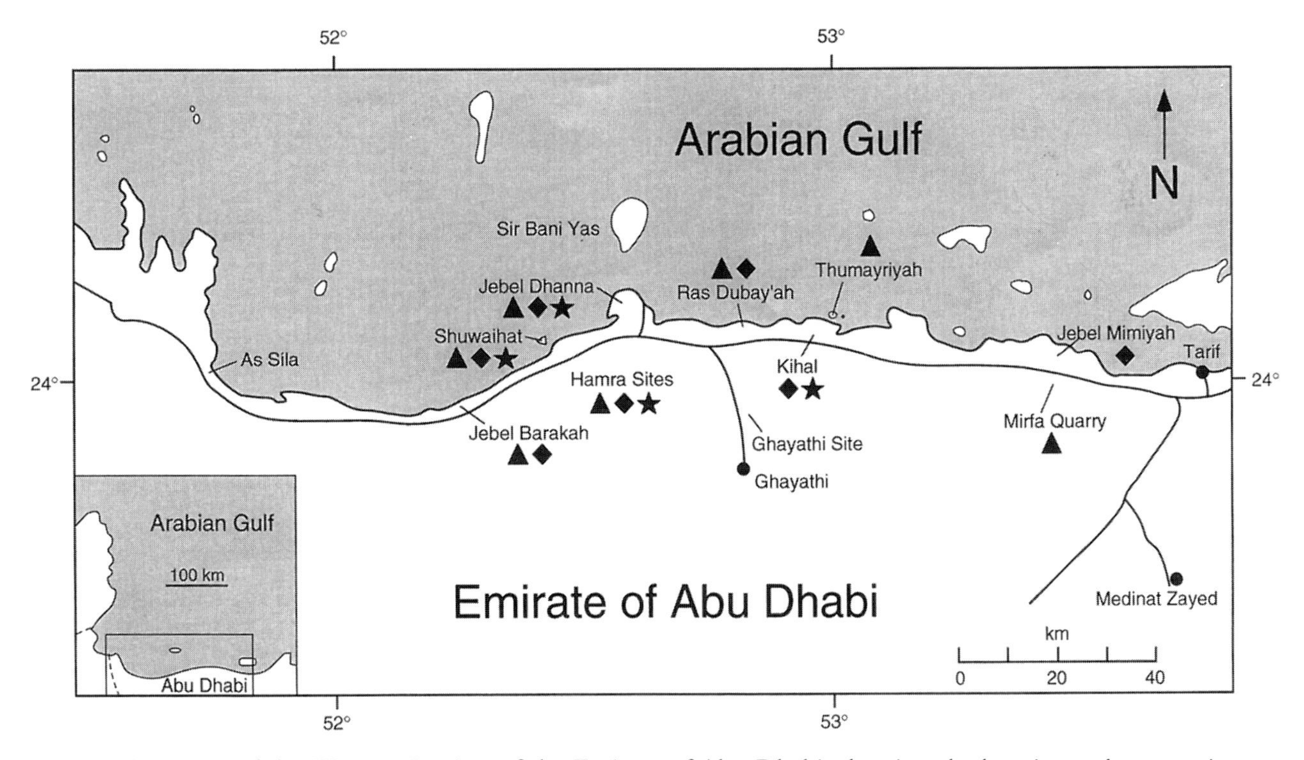

Figure 13.1. Map of the Western Region of the Emirate of Abu Dhabi, showing the locations where specimens of turtles have been found: *Trionyx* s.l. (solid triangles), *Geochelone* aff. *sulcata* (stars), and *Mauremys* sp. (diamonds).

JD5: AUH 145, right scapula (without the precoracoid); AUH 148, right pleurals 6 and 7, three fragments of other pleurals, a fragment of neural; AUH 148e, extremity of left squamosal; AUH 422, fused right hyo–hypoplastron.
M2: AUH 301, pleurals 1 and 2, neurals 1 and 2.
R2: AUH 462, medial part of pleural.
S1: AUH 62a, fragment of left hypoplastron; AUH 93, symphysis and right dentary S4; AUH 680, right hyoplastral fragment; AUH 681, free extremity of a rib; AUH 682, medial part of pleural; AUH 683, distal part of pleural; AUH 684, medial part of pleural.
TH1: AUH 319, right pleural 1; AUH 319a, fragment of pleural.

Description

Dimensions

The adult specimens (at least 12) are estimated to have had a shell (pleural discus) length of between 35 and 40 cm. This size is consistent with that of adults of the genus *Trionyx* s.l. (see Broin, 1977); the extant, very small-shelled *Dogania subplana* is excluded. They are smaller than the maximum adult size of subgenera such as the extant *Trionyx* s.s. (eastern Mediterranean Basin and Africa) and consistent with the largest specimens of the extant *Rafetus euphraticus* (Tigris–Euphrates Basin) (see Siebenrock, 1913).

Decoration

Pleural and plastral decoration is similar in all the preserved fragments (figs 13.2 and 13.3). It is consistent with that of many species of the various subgenera of the trionychine genus *Trionyx* (s.l.). The decoration consists of crests, parallel to the lateral border of the shell and to the medial border of the plastron, which unite to produce polygonal or elongated pits; sometimes the crests are cut into round tubercles. The size of the pits, larger in the dorsal shell than in the plastron, is moderate and the crests are narrow and rounded. In eroded specimens, the crests are sharper than normal. This decoration is especially similar to that of the extant *Rafetus euphraticus* and *Trionyx* s.s., for example, and also to that of the western European fossil *T. stiriacus* (see the middle Miocene form from Artenay, France, MN 4, in Broin, 1977). It seems to be a little finer than in the extant *Trionyx triunguis*. It is unlike (amongst others) the decoration of the extant Indian species *Aspideretes gangeticus,* which has more rounded and wider crests, or, for example, that of the adult south Asiatic *Amyda cartilaginea,* with wider pits, and *Dogania subplana,* with very small pits and very thin crests.

In the *Trionyx triunguis* lineage, which developed from the Paleocene to the Pliocene in western Europe before reaching Africa, the decoration of the plastron is frequently more tubercular than that of the dorsal shell. This is unlike the four preserved parts of plastron from the Baynunah Formation. We do not know the variation of the decoration in the *Rafetus euphraticus–R. swinhoei* lineage, which is northern to middle Asiatic (extant, respectively, in the Tigris–Euphrates Basin and in China). *Rafetus* is not yet well defined as far as the fossil record is concerned. It has been suggested (Broin, 1977) that a possible lineage exists between "*Aspideretes maortuensis* Yeh, 1965", Lower Cretaceous of Inner Mongolia, "*Amyda gregaria* Gilmore, 1934", Oligocene of Mongolia, and the two extant species of *Rafetus.* The polarity of the shared characters is not easy to establish, however; some of them may be plesiomorphic (skull) or pedomorphic (plastron).

In fact, the most important fact is that the Baynunah Formation material exhibits no cyclanorbine decoration, which consists mainly of rounded crests cut in round or elongated tubercles (some polygonal pits or some sharp crests being present in some parts of the two African Cyclanorbinae, *Cycloderma* and *Cyclanorbis*). This is important because the Cyclanorbinae are the first representatives of Trionychidae known in Africa, from the early Miocene (Burdigalian) of East Africa (Koru, around 20 Ma), before the arrival of *Trionyx,* known only from late Miocene times of Sahabi, Libya (Erasmo, 1934; Wood, 1987) and Wadi Natrun, Egypt (Reinach 1903; Dacqué, 1912; Lapparent de Broin and Gmira, 1994) and possibly earlier from the late Miocene (MN 9, Vallesian) of Bled ed-Douarah, Tunisia (Robinson and Black, 1974; see also Geraads, 1989).

Carettochelyidae are not present either in the Abu Dhabi material. This is another trionychoid family with decorated dermal plates, with rounded low sinuous ridges separated by low sinuous sulci and some rounded or elongated granulations in the

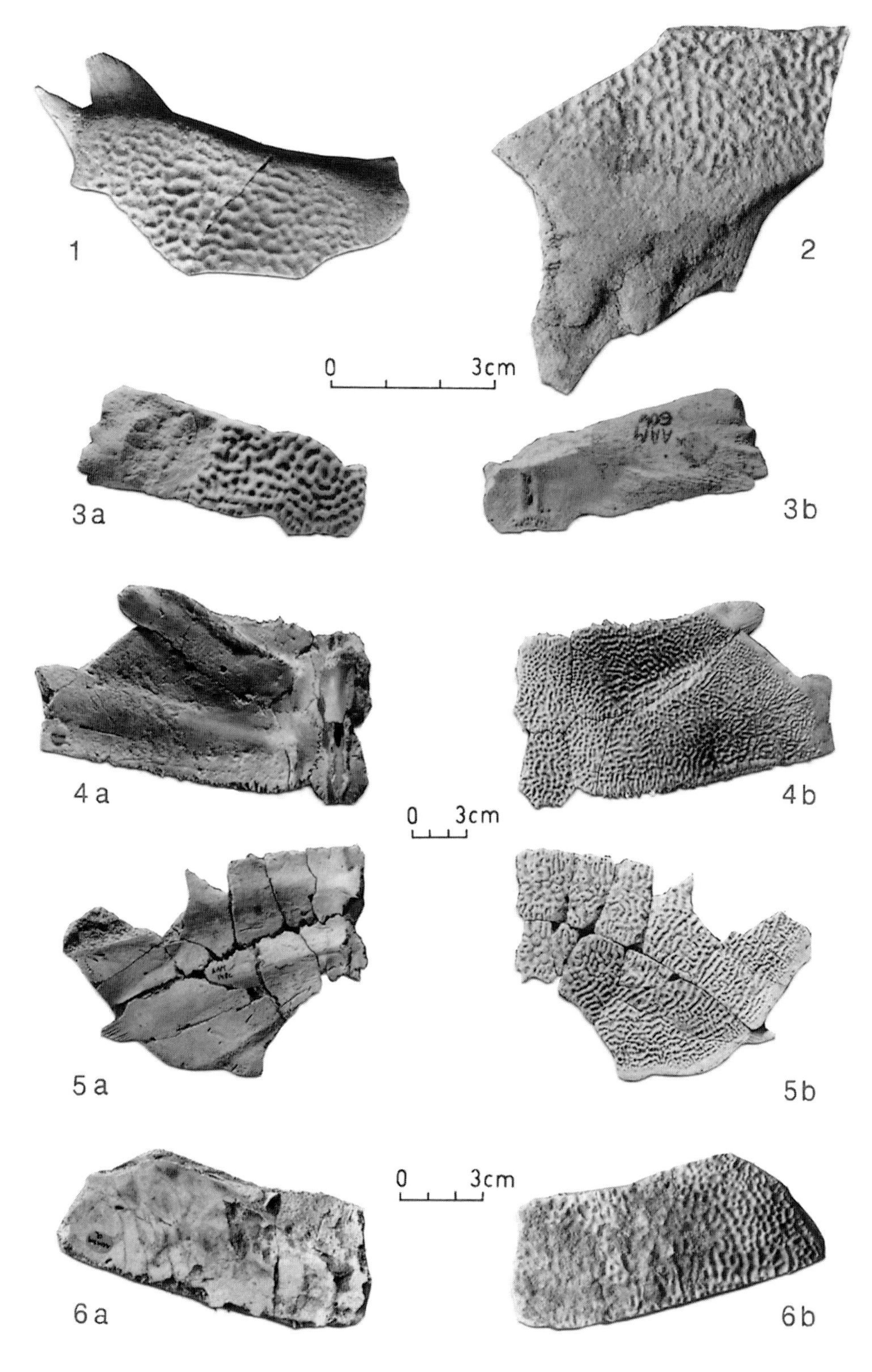

Figure 13.2. *Trionyx* s.l. sp. *1,* AUH 409, anterior part of right xiphiplastron, ventral view. *2,* AUH 62a, medial part of left hypoplastron, ventral view. *3,* AUH 606, left part of a nuchal; a, b, dorsal, ventral views. *4,* AUH 301, neurals 1–2, right pleurals 1–2; a, b, ventral, dorsal views. *5,* AUH 148, right pleurals 6–7; a, b, ventral, dorsal views. *6,* AUH 319, right pleural 1; a, b, ventral, dorsal views.

peripheral elements, found in the early Miocene, Burdigalian, of Egypt at Wadi Faregh, MN 3 (Dacqué, 1912, *Cyclanorbis*?), Saudi Arabia at As Sarrar, MN 5, and in Oman at Ghaba, late early–early middle Miocene (Thomas et al., 1982; Roger et al., 1994).

Dorsal Shell

Material: nuchal, AUH 606; pleurals 1 and 2 with neurals 1 and 2, AUH 301; pleural 1, AUH 319; pleural 5 or 6, AUH 14; pleurals 6 and 7, AUH 148; parts of undefined pleurals, AUH 681–684, 148, 462, 319a, BMNH R10977; undefined neural (2–7), AUH 148.

The nuchal bone is partly preserved in one immature specimen, AUH 606 (fig. 13.2.3). From an estimated original width of 11 cm only 5.4 cm is preserved. It is a wide nuchal, consistent with *Trionyx* s.l. The decorated part is incomplete. It is partly eroded on the left as well as the distal part of the lateral margin and separated into four prongs, consistent with *Rafetus*. Other species have the prongs subdivided into many fine points, as in the fossil and extant *Trionyx* s.s. (see Ammon, 1911; Peters, 1855; Villiers, 1958) but an intraspecific variability is possible here, as in, for example, *Amyda cartilaginea*. The presence of a posterior left nucho-pleural fontanelle, lateral to the neural one, the small size of the nuchal, and the small pits indicate a rather young individual. In *Rafetus* and *Trionyx* s.s. the fontanelles are still present in specimens of this size. Unfortunately, the anterior border is lacking: in *Rafetus*, it is apomorphically long medially and widely rounded (see Heude, 1880; Salih, 1965; comparative material: BMNH, MNHN), the callosity being well developed and prematurely covering nearly all the lateral prongs; but in *Trionyx* s.s., it is anteriorly concave or straight, not covering the lateral points in specimens of this size.

Pleural 1 is preserved in two specimens. In AUH 319 (right pleural 1; fig. 13.2.6), its anterior border is transverse, parallel to the posterior one. In AUH 301 (pleurals 1 and 2, neurals 1 and 2; fig. 13.2.4), its lateral border is strongly narrowed so that the anterior border is transverse and the posterior border is oblique; consequently, the lateral border of pleural 2 is strongly widened, relative to its medial border. The narrowing of the lateral border of pleural 1 is present in some individuals in many lineages of *Trionyx* s.l. This is a parallel tendency within different genera of the subfamily Trionychinae. We have observed this tendency in one fossil specimen of *Trionyx stiriacus* from Artenay, four specimens of five different fossil species from the lineage of *Apalone* in North America (see Hay, 1908), and one specimen of extant *Amyda cartilaginea* and *Rafetus euphraticus*, but we do not know about its frequency in each species. We observe that this lateral reduction of pleural 1, in the Baynunah Formation form, is much more pronounced than in the above-mentioned species, except for *Trionyx leucopotamicus* Cope, 1891 (Cypress Hills, Alberta, Lower Oligocene, which is an *Apalone*). As this feature is present in only one of the two specimens, we cannot attribute a specific importance to it, but we acknowledge the possibility of a specific tendency for pleural 1 to be reduced. The important fact is that this reduction is the opposite of the condition in the Cyclanorbinae, where, specifically or generically, pleural 1 is very often laterally much widened—with a long semicircular nuchal bone in the Indo-Asiatic *Lissemys* and in some species of the African *Cycloderma* and *Cyclanorbis*—and never reduced. The variable enlargement of pleural 1 also occurs in some individuals of many Trionychinae.

Specimen AUH 301 (fig. 13.2.4) indicates that there is no more preneural bone, a derived condition, as in most of the extant subgenera of Trionychidae, but not found in *Aspideretes* (sensu Broin, 1977), nor the Cyclanorbinae or fossil species of several lineages: here it is already lost.

Specimen AUH 148 (right pleurals 6 and 7; fig. 13.2.5) demonstrates that there are eight pairs of pleurals. The eighth pleural (around 5.5 cm long and 6 cm wide) is reduced in width but not in length, being medially longer than pleurals 7 and 6 (2.9 cm and 3.7 cm long, respectively) as in several subgenera—for example, *Trionyx* and *Rafetus* (see Yuen in Heude, 1880). These two subgenera may also have pleurals 8 shorter than pleurals 7, or, in *Rafetus*, pleurals 8 may be both shorter and narrower, meanwhile the other extant Asiatic subgenera have wide or medially long pleurals 8. Here, the ribs of pleurals 7 and 8 are anteriorly removed, ending with the seventh between pleurals 6 and 7 and the eighth outside pleural 7 (instead or between pleurals 7 and 8). Although the position of the ribs, relative to the border of the pleurals, is highly variable in Trionychidae, the present position appears unique in our preliminary investigation. We do not know its importance, whether it is an individual variation or a specific character. This specimen shows that the neural series is long, with eight neurals, the last one being pentagonal, and with pleurals 8 meeting behind it.

Plastron

Material: hyo–hypoplastron, AUH 422; partial hypoplastron, AUH 62a; partial xiphiplastron, AUH 409; AUH 680, partial hyoplastron.

Specimen AUH 422 (fig. 13.3.1), a medial part of the right hyo–hypoplastron, has a typical trionychine character (see Siebenrock, 1902), as well as its trionychine decoration: the posterior lateral prong of the hypoplastron is externally bordered by the anterior lateral point of the xiphiplastron. In the Cyclanorbinae, the posterolateral hypoplastral prong borders the anterolateral xiphiplastral prong externally. The specimen, AUH 422, is rare because of the fusion of the two plates, hyoplastron and hypoplastron, and this fusion is a result of the individual having grown rather old, indicated by the thickness (1.3 cm maximum) and large size (14.5 cm in medial length) of the fossil. This feature is not seen in specimen AUH 62a (fig. 13.2.2), a part of the left hypoplastron, which is smaller, thinner, and with the decorated callosity not so well developed. The fusion is almost the rule in Cyclanorbinae (but not present in Miocene Saudi Arabian and Indian forms; see Thomas et al. 1982); it is exceptional in Trionychinae, as an individual variation. The two specimens of hyo–hypoplastron and hypoplastron indicate a form that retains the primitive (and juvenile) condition of the hypoplastral prongs not united in a continuous mesioposterior border, unlike most of *Trionyx* s.s., especially *Trionyx triunguis*, two southern Asiatic trionychid genera *Pelochelys* and *Chitra*, in extant adult forms (Villiers, 1958; Siebenrock, 1902), and some fossil forms from North America (in Hay, 1908, at least) where the character developed in parallel. The Tertiary (Paleocene–Pliocene) European *Trionyx* adult specimens have (when the hypoplastron is known) the continuous mesioposterior border of prongs, more or less rounded (Ammon, 1911; Broin, 1977; Peters, 1855), the gaps between the anterior and posterior sets of prongs being filled by the decorated callosity. At least one species is an exception (see Broin,

1977). *Trionyx bohemicus* Liebus, 1930 from the early Miocene of Preschen bei Bilin, Bohemia (and see Sandelzhausen, MN 6; Schleich, 1981) has few developed callosities in specimens of this size. This character and the absence of xiphiplastral callosity may be considered as pedomorphic. The Bohemian form also shares with *Rafetus* the absence of xiphiplastral callosity, differing from the Baynunah preserved xiphiplastron and from *Trionyx* s.s. But in young adult specimens of fossil *Trionyx*, with the callosity and the decoration still not completely developed, as in specimen AUH 62a (fig. 13.2.2), the medial and posterior groups of prongs may remain separate, as seen, for example, in specimen ART 284 of *T. stiriacus* from Artenay, France, that is two-thirds of the maximum known width. To resolve the problem of the generic status of the Bohemian form it would be necessary to prepare the ventral side of the skull and examine its morphology.

Specimen AUH 409 (fig. 13.2.1), an anterior fragment of the right xiphiplastron, has the two anterolateral prongs (those encircling the posterolateral hypoplastral prong) unusually developed: the more lateral is less developed than the more medial. We do not know the likelihood of variation in this character, which we have observed only in one fossil and in one extant specimen of *Trionyx*. The anterior medial development of the callosity is consistent with many forms including *Trionyx* s.s. but not *Rafetus euphraticus* (see Siebenrock 1902, 1913) nor *R. swinhoei* (see Heude, 1880; Siebenrock, 1902), nor the above-mentioned fossil form from Preschen. The specimen AUH 680 is a lateral part of a right hyoplastron, very thick (11 mm) for the width of the pits (around 1–2 mm) and of the crests (0.1 mm), which indicates a full adult of small size with a well-developed callosity.

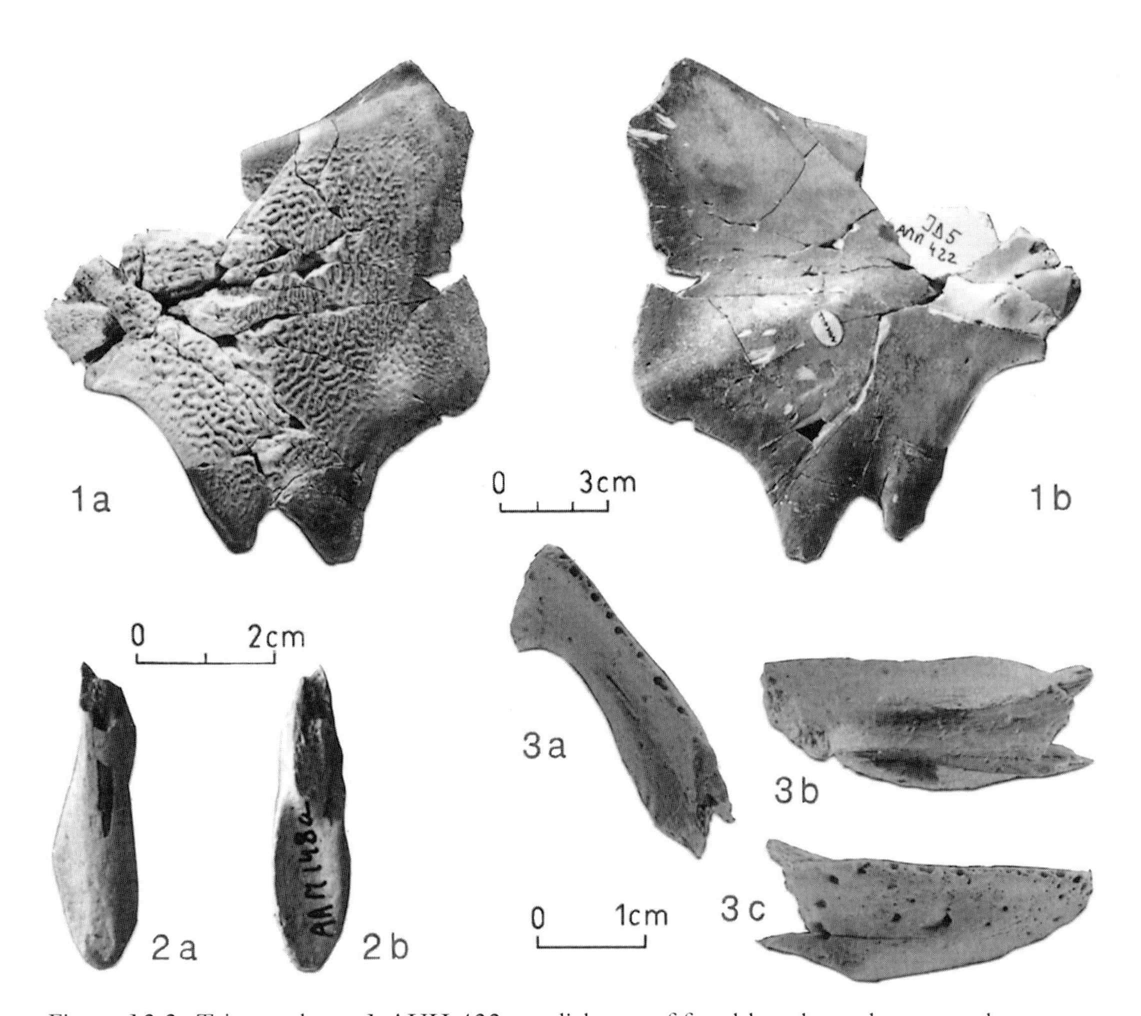

Figure 13.3. *Trionyx* s.l. sp. *1,* AUH 422, medial part of fused hyo–hypoplastron; a, b, ventral, dorsal views. *2,* AUH 148a, extremity of left squamosal; a, b, medial, lateral views. *3,* AUH 93, symphysial and right masticatory part of a lower jaw; a, b, c, dorsal, medial, lateral views.

Skull

AUH 148a is the extremity of a left squamosal (fig. 13.3.2). Its dorsal edge is sharp and posteromedially curved; seen laterally, its posterior extremity is rounded. A preliminary comparison shows us great diversity in extant (Gray, 1873a,b; Loveridge and Williams, 1957) and fossil Trionychidae, with individual variations: a more or less sharp dorsal edge, its extremity curved inwards or outwards and more or less pointed. The free extremity, posterior to the suture with the opisthoticum, is too long for a Cyclanorbinae and consistent only with a part of the Trionychinae that includes *Trionyx* s.s. and *Rafetus.*

Lower Jaw

The specimen AUH 93 (symphysis and right partial dentary, fig. 13.3.3) shows the fusion of the dentaries, as in all the cryptodiran turtles as well as in the pleurodiran Pelomedusoides. The slightly expanded dorsal masticatory surface, with a line of vascular foramina along its external border, is typical of the Trionychidae. The nearly horizontal dorsal surface in the symphyseal area and the crested medial border, medial to the weak dentary masticatory fossa (in each dentary, posterior to the symphysis), are typical of Trionychinae. In Cyclanorbinae, the symphysis area is deeply inclined downwards and backwards. The triangular shape of the external symphysis border and the rounded shape of the internal symphysis border, the very slightly elongated symphysis, and the triangle of slight rugosities at the posterior part of the symphysis taken together characterise the species of the Baynunah Formation and are different from the morphology of all the extant species. A similar condition is seen in *Rafetus,* which also has a rather short symphysis but with a longitudinal symphysial crest—medial in *R. swinhoei* from China and medioposterior in *R. euphraticus*—more or less weak. This symphysis excludes *Trionyx* s.s., which has a long (derived) and smooth symphysis. In *Trionyx* s.s., the long and smooth symphysis fits with a long anterior palate, more or less narrow according to the species; the skull is shorter and wider in *Rafetus.* The present morphology implies a skull with a short triangular palate and a triangular posterior palatine sulcus that absolutely excludes *Trionyx* s.s. (see Atatür and Üçünçü, 1986; Broin, 1977; Eiselt, 1976; Gilmore, 1934; Gray, 1873a,b; Heude, 1880; Villiers, 1958; osteological material from BMNH, MNHN, and NHMW).

Anterior Girdle

Specimen AUH 145, a right scapula, without the precoracoid, is consistent with Trionychinae because of its regular ventrodorsal narrowing. It differs from the extant African Cyclanorbinae, which have an abrupt narrowing of the medial border at mid-length dorsoventrally. The glenoid facet is, as in all the observed Trionychinae, a little longer than in the three genera of Cyclanorbinae.

Discussion

The family Trionychidae is composed of two subfamilies, the Trionychinae and the Cyclanorbinae. All the trionychid material from the Baynunah Formation belongs to the Trionychinae. From the decoration, all the specimens may belong to one species, which appears to be new. Without more complete material, especially the anterior part of the skull and the complete nuchal bone and associated elements of the plastron, it is difficult to define the species and, at present, it is impossible to attribute it to one defined subgenus within *Trionyx* s.l.. Of the characters seen on the preserved fragments, the short symphysis eliminates *Trionyx* s.s. and fits with *Rafetus,* and the tall hypoplastron, with two widely separated sets of prongs, ought also to eliminate *Trionyx* s.s., at least the *T. triunguis* lineage. Additionally the developed plastral callosities ought to eliminate *Rafetus,* but the weak posterior crest or rugose area of the lower jaw symphysis may fit with *Rafetus.* The subgenus might be *Rafetus* by reversion of the neotenic character of reduction of the callosities and by reduction of the symphysial crest.

The other possibility is that two subgenera are present, with the same decoration: the first, *Trionyx* s.s., characterised by the decorated xiphiplastron; the other one, possibly related to *Rafetus,* characterised by the big hypoplastron with two lots of well-separated prongs and the short-symphysed dentary, represented by a species differing from *R. euphraticus.* This hypothesis appears dubious because of the absence of an unquestionable specimen of *Trionyx* s.s. in the collected specimens from Abu Dhabi.

Until now, the only known Trionychidae from the Arabian Plate were: those from the late early Miocene (MN 5) of As Sarrar, Saudi Arabia, Dam Formation (Thomas et al., 1982), and from the late early–early middle Miocene from Ghaba, Huqf, Oman (Roger et al., 1994), consisting of only one genus of Cyclanorbinae, aff. *Cycloderma* with a separated hyo- and hypoplastron; and trionychid fragments housed and observed in London, BMNH R52, from "beds of unknown (?Tertiary) age at Chalon, in the Arabian desert" (Lydekker, 1889b: 21; see also Hummel, 1932), which are also clearly cyclanorbine and not trionychine. From the Lower Oligocene of the Sultanate of Oman fossil turtles have been found at Thaytiniti and Taqah (Thomas et al., 1989; Thomas et al., 1991) but they do not include Trionychidae and comprise only Pelomedusoides and a Testudininei.

The Trionychidae is a Laurasiatic family, first known by the Trionychinae. Members of this family appear in Asia, at least in the Lower Cretaceous of Mongolia (Yeh, 1965), and in Europe in the Paleocene (Hummel, 1932; Kuhn, 1964; Broin, 1977). Trionychidae invaded Africa and Arabia in at least two waves (Broin, 1983; Lapparent de Broin and Gmira, 1994). First, the oldest record of the family (which belongs to the subfamily Cyclanorbinae) is from the early Miocene of Koru, Kenya, around 20 Ma ago (Pickford et al., 1986) with an undefined form, aff. *Cycloderma* sp., housed in the NHM collections. Second, the oldest verified record of the Trionychinae from the late Miocene of Sahabi, Libya (Erasmo, 1934; Wood, 1987) and Wadi Natrun, Egypt) (Reinach 1903; Dacqué, 1912) is represented by the *Trionyx triunguis* lineage, which is particularly well known from the Sahabi form, *Trionyx* cf. *triunguis* in Wood (1987). *Trionyx* is also listed by Robinson and Black (1974) from the late Miocene (around MN 9) from Bled Douarah, Tunisia (see also Geraads, 1989), but its presence has not been verified. If true, it would be the oldest record of *Trionyx* in Africa. Later records (Plio-Pleistocene) are given in Lapparent de Broin and Gmira (1994).

Records of *Trionyx* from the early Miocene of Moghara, Egypt, *T. senckenbergianus* Reinach, 1903, are also dubious. One of them (Reinach, 1903: pl. 17, fig. 6), and probably all of them (Reinach, 1903: pl. 17, figs 2 and 5), are Cyclanorbinae. In our first study (Broin, 1983), we supposed the presence of a carettochelyid among them, but the presence of *Trionyx* is improbable. *Trionyx melitensis,* Lydekker, 1891, from the Miocene of Malta, is not a *Trionyx* of the *Aspideretes gangeticus* lineage and clearly appears to be a cyclanorbine by its size, the shape of the preneural and first neural, and the decoration. It is similar to Mio-Pliocene Ugandan and East African forms, "*Cycloderma*" or *Cyclanorbis* (see Broin, 1979; Lapparent de Broin and Gmira, 1994; Meylan et al., 1990).

All of the Mio-Pliocene circum-Mediterranean records are either of the *Trionyx triunguis* lineage, or they are undefined and possibly of *Rafetus.* An undefined species of *Trionyx* s.l. was found in the late Miocene (late Vallesian, around MN 10) from Injana, Jebel Hamrin, Iraq, Agha Jari Formation, but we could not (Thomas, Sen, and Ligabue, 1980) attribute it either to *Trionyx* s.s. or to *Rafetus.* It consists of an eroded undefined plate with a coarse decoration (?*Trionyx*) and a fragment of right pleural 7 (MNHN, IJN, 209, 211) of *Trionyx* with the same decoration as in the Baynunah specimens. The medial border is around 3.3 cm long and pleural 8 ought to be no more than 1.5 cm long: it is much reduced in relation to the seventh, which is more congruent with *Rafetus* than with *Trionyx* s.s. In the Baynunah Formation, an alternative to the presence of a *Trionyx* of the lineage of *T. triunguis* ought to be the presence of *Rafetus,* which is now found in the Tigris–Euphrates river basin, western Orient, in southeastern Turkey, northeastern Syria, Iraq, and southwestern Iran (see Iverson, 1992). If there are two forms in the Baynunah Formation, the one that is not *Trionyx* s.s. but which does not have exactly the same characters as the extant *Rafetus,* might, in fact, be related to it. We have seen that if no synapomorphic characters can be observed between the fossil and the extant species, there is no inconsistency in their belonging to the same genus.

Family Testudinidae Batsch, 1788

Infrafamily Geoemydininei Theobald, 1868
Mauremys Gray, 1869
Mauremys sp.

Material by Sites (fig. 13.1)

There are 34 catalogued specimens, representing at least 31 different individuals:

B2: BMNH R10973, left pleural 5; BMNH R10974, left hyoplastron; AUH 470, partial right hypoplastron; AUH 470a, partial right hypoplastron; AUH 470b, entoplastron; AUH 470c, partial right hyoplastron.
H2: AUH 8, left peripheral 1.
H3: AUH 276a, left peripheral 2; AUH 276b, neural 4, left part of nuchal and posterior part of right hyoplastron; AUH 276c, medial part of a juvenile right pleural 5 (or 3).
H5: AUH 50, left hypoplastron.
JD3: AUH 274, neural 1; AUH 274a, partial right juvenile hyoplastron.
JD5: AUH 148b, partial left hypoplastron; AUH 148c, plastral inguinal buttress.

K1: AUH 310, fragment of pleural.
M1: AUH 363, proximal part of a right humerus.
R2: AUH 383, left hypoplastron; AUH 412, partial left peripheral 2; AUH 412a, partial left peripheral 9.
S1: AUH 62, right hypoplastron; AUH 82, left peripheral 8; AUH 280, juvenile neural 5, juvenile neural 4 or 6, fragment of juvenile right peripheral 8; AUH 280a, medial part of left pleural 5.
S4: AUH 679, left xiphiplastron; AUH 685, anterior part of juvenile right hyoplastron; AUH 686, upper part of right peripheral 5; AUH 687, fragment of left hypoplastron; AUH 688, right half nuchal; AUH 689, fragment of left hypoplastron.

Description

This sample consists of small specimens with undecorated morphology (no pits, no tubercles, no crests). All the specimens are consistent with a cryptodire Geoemydininei conforming to the North African–European–Middle Eastern species of the genus *Mauremys* (see McDowell, 1964; Iverson, 1992). They have the same general shape, relative proportions (length, width, and thickness of the bones), relationships between sutures and scutes sulci, and individual variations as in the extant *Mauremys leprosa,* the type species of the genus, *M. caspica,* and as in the *Mauremys* from the Mio-Pliocene of Europe (see the figures in Ammon, 1911; Bachmayer and Młynarski, 1983; Bergounioux, 1935; Broin, 1977; Depéret, 1885; Depéret and Donnezan, 1890; Kotsakis, 1980; Peters, 1869; Portis, 1890; Purschke, 1885; Sacco, 1889; Schleich, 1981). *Mauremys pygolopha,* for example, from Artenay, France, MN 4, is preserved with a large number of specimens useful for comparison (see Broin, 1977). None of the elements conforms to pleurodiran Pelomedusoides, the usual inhabitants of Africa, from the Cretaceous to the present time.

Dimensions

Estimated shell lengths: 15–22 cm.

Decoration

The fragments of shell are almost smooth, finely punctuated, some of them pitted by parasitism like the extant western *Mauremys.* The scute sulci are thin, straight, or slightly sinuous.

Dorsal Shell

Material: nuchals, AUH 276b, 688; peripheral 1, AUH 8; peripherals 2, AUH 412, 276a; peripheral 5, AUH 686; peripherals 8, AUH 82, 280; peripheral 9, AUH 412a; pleural 5 or 3, AUH 276c; pleurals 5, AUH 280a, BMNH R10973; undefined pleural, AUH 310; neural 1, AUH 274; neural 4, AUH 276b; neural 4 or 6, AUH 280; neural 5, AUH 280.

Nuchals AUH 276b and 688 (figs 13.4.3 and 13.4.4) allow a reconstruction of a complete adult plate. It is slightly dorsally convex and slightly ventrally concave. Its width is greater than its length and it has a long marginal 1 and cervical covering. On each lateral border, the plate is covered by marginal 1 on about 78% of its length, and the cervical, on its lateral border, covers a third of the length of the plate. Furthermore, the cervical is primitively wide, wider than it is long. This conforms: (1) to the extant species *Mauremys caspica caspica* from the eastern Mediterranean Basin (see Iverson, 1992) (11 out of 14 specimens examined from MNHN have a greater width than length)—this species is now present as far as latitude 25° N in Saudi Arabia (Gasperetti et al., 1993); (2) partly to *M. c. rivulata,* also from the eastern Mediterranean Basin (see Iverson, 1992) (width greater than length in four out of six specimens examined from MNHN); (3) to Pliocene forms of *Mauremys* (sufficiently known for this character) like *Mauremys gaudryi* Depéret, 1885 (and Bergounioux, 1935; Depéret and Donnezan, 1890) from the Pliocene of Perpignan, MN 15; and (4) to *M.* aff. *gaudryi* from the late Miocene, MN 10, of Kohfidisch Höhle (Austria) in Bachmayer and Młynarski (1983).

It differs from the extant western adult form of the Mediterranean Basin, *Mauremys leprosa,* which acquired a narrow cervical scute, longer than its width (19 adults out of 24 specimens examined from the MNHN; cervical wider or of equal width in the five juvenile specimens), a derived character already recognised in specimens from Moghreb (Pleistocene–Holocene), Aïn Boucherit, Algeria (MNHN collection), and Doukkala II in Morocco (see Gmira, 1995). But, like *M. caspica, M. leprosa* has a long lateral covering of the nuchal plate by marginals 1 (69–95% of the lateral border). The Baynunah form differs also from the Miocene forms (when the cervical covering on the nuchal bone is known), *M. pygolopha* from Artenay (48–70%), *M. sarmatica* Purschke, 1885, from Hernals bei Wien, Austria, and *M. sophiae* from Sandelzhausen, Germany, MN 6 (Schleich, 1981), which have a more primitive shorter covering of the smaller cervical (although with width greater than length) and marginals 1 on the nuchal. So, in the Baynunah form, this plate is derived by its long and tall scutes, and is primitive by its wide cervical, as in Pliocene forms and in the extant *Mauremys caspica.*

Consistent with the two nuchal plates, the first peripheral, AUH 8 (fig. 13.4.2) indicates a form with the vertebral 1 much wider than the nuchal bone, contacting the marginal scute 2, as in the late Miocene (from Sandelzhausen, MN 6, of Germany at least), Pliocene and extant forms. In early to middle Miocene forms (including *M. pygolopha*), the vertebral 1 just covers the nuchal and we believe that the enlargement is secondary in the later forms. The other peripherals, AUH 276a (fig. 13.4.1), AUH 412, AUH 412a (fig. 13.4.7), and AUH 82 (fig. 13.4.8) also indicate that the scute covering is important on the plates, as in the fossil and extant forms (longer on the anterior plates in Pliocene to extant forms than in Oligocene to Miocene forms).

AUH 310, a pleural fragment, shows growth annuli. BMNH R10973 (pleural 5, fig. 13.4.9) shows the position of the inguinal buttress of the hypoplastron between pleurals 5 and 6. The ventral part of the plate, which includes the rib, is strongly raised, growing from the medial part of the plate to the lateral extremity (the elevation takes around 66% of the width of the plate); the inguinal process is linked by suture to the lateral part of this elevation, but

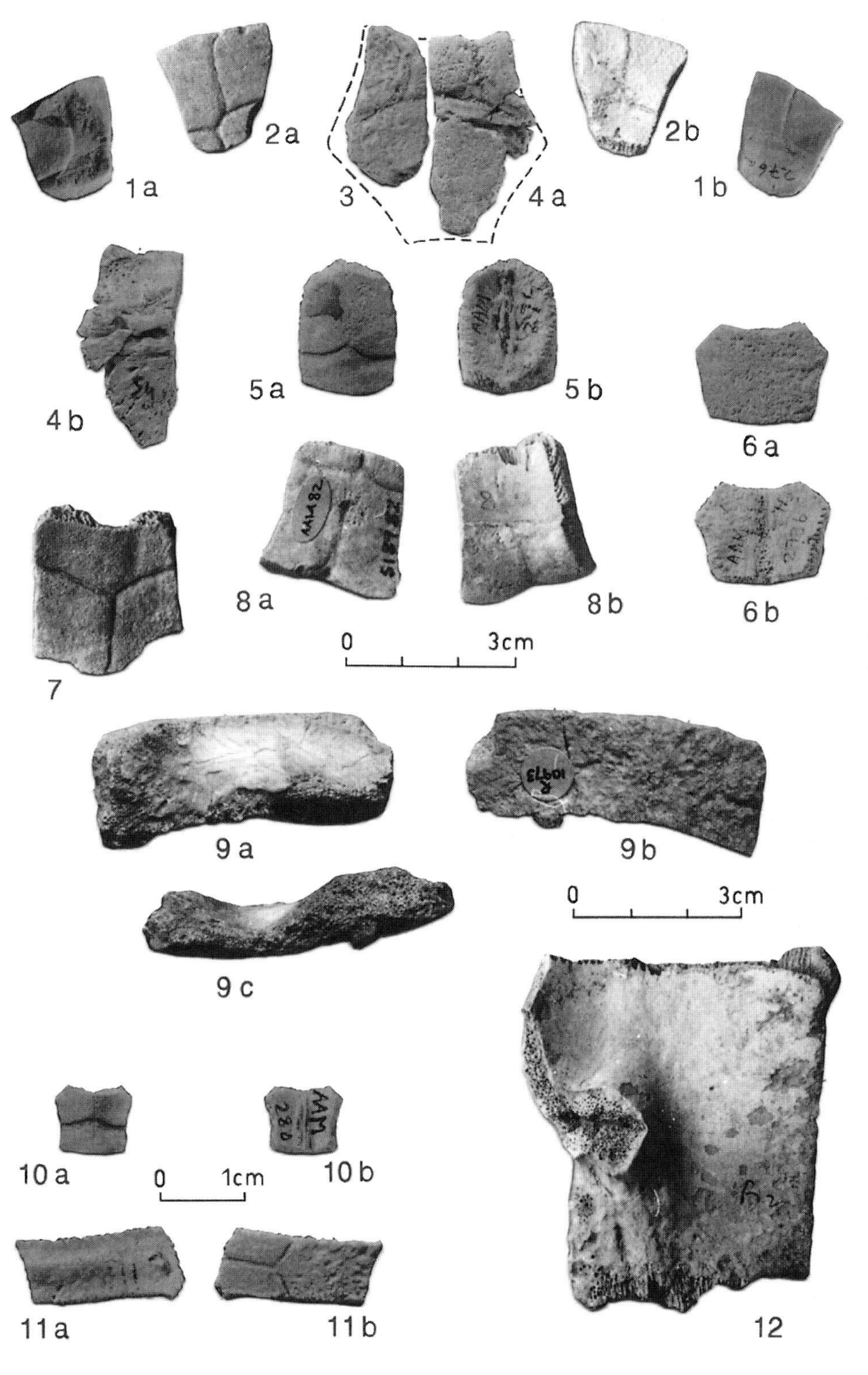

Figure 13.4. *Mauremys* sp. *1,* AUH 276a, second left peripheral; a, b, dorsal, ventral views. *2,* AUH 8, first left peripheral; a, b, dorsal, ventral views. *3,* AUH 276b, left part of nuchal; dorsal view. *4,* AUH 688, right part of nuchal; a, b, dorsal, ventral views. *5,* AUH 274, neural 1; a, b, dorsal, ventral views. *6,* AUH 276b, neural 4; a, b, dorsal, ventral views. *7,* AUH 412a, left peripheral 9; dorsal view. *8,* AUH 82, left peripheral 8; a,b, dorsal, ventral views. *9,* BMNH R10973, pleural 5; a, b, c, ventral, dorsal, posterior views. *10,* AUH 280, juvenile neural 4 or 6; a, b, dorsal, ventral views. *11,* AUH 276c, juvenile right peripheral 5; a, b, ventral, dorsal views. *12,* AUH 383, left hypoplastron; dorsal view.

in fact it is not well developed medially (the suture of the buttress is around 40%, see fig. 13.4.9b, 9c); the lung chambers ought not to be not well developed either. Specimen AUH 148b, a fragment of hypoplastral inguinal buttress, conforms to this relatively short development of the buttress under the highly raised ventral part of the plate. In extant western forms—*M. caspica caspica, M. c. rivulata,* and *M. leprosa*—the inguinal buttress may be linked only under pleural 5 or, as well, between pleurals 5 and 6 as exhibited in the Baynunah specimen and the fossils from Artenay and Sandelzhausen. In extant forms, in the Holocene form from Doukkala II, and in the type of the Pliocene *M. gaudryi,* which is a juvenile specimen—not known in the unprepared adults "*Paralichelys carinata*" and "*Clemmys romani*" (see Bergounioux, 1935), and in the Italian Plio-Pleistocene forms from Toscana (Portis, 1890), and Piemonte (Sacco, 1889)—the elevation of the plate towards the buttress is variably pronounced. The width of this elevation and of the suture of the buttress on the plate is, however, much smaller and much more lateral: the elevation is around 48%, the suture is 8–15% the width of the plate in *M. leprosa.* With its wide elevation and suture, the Baynunah specimen is consistent with the Miocene forms from Artenay (elevation around 55–65%, suture around 40% the width of the plate) and Sandelzhausen (not known in the other Miocene forms). Wide elevation and suture of buttresses are also known in the fossil form *Palaeochelys* s.s. from the Oligocene to early Miocene of Europe, a plesion of *Mauremys,* and in exant eastern Asiatic *Mauremys* forms, *M. iversoni* Pritchard and McCord, 1991, *M. mutica,* and *Annanemys annamensis* (included in *Mauremys* by Iverson and McCord, 1994; studied in the Bourret collection, MHNT) as well as in *Ocadia,* an exclusively Asiatic genus similar in shell shape to *Mauremys* that is a Batagurinae with secondary palate and not a Geoemydinei (see Hirayama, 1985). Many fossil *Mauremys* and *Palaeochelys* from Europe have, until recently, been attributed to this genus (see Schleich, 1981, 1984; "*Ocadia*" *sophiae*).

Neurals 1 (AUH 274) and 4 (AUH 276b) are adult, without any keel, whereas the juvenile neural 4 or 6 and the neural 5 (AUH 280) bear a longitudinal keel, rather smooth and wide, as in the western *Mauremys.* Here, the three neurals 4–6 are hexagonal, short-sided in front. In *Mauremys,* their shape is individually variable, regular hexagonal short-sided in front or behind, or irregular, alternatively seven-sided and five-sided. But in *Ocadia* they are regular, hexagonal, and short-sided in front. As in extant western *Mauremys,* the adult neural 4 (fig. 13.4.6) is rather short and wider than it is long—a difference from *Ocadia,* which has neurals 1–4 longer than, or as long as, the width in the studied material (MNHN; MHNT; BMNH). In western *Mauremys,* these neurals may also be approximately as wide as they are long.

Plastron

Material: entoplastron, AUH 470b; hyoplastrons, AUH 274a, 276b, 470c, 685, BMNH R10974; hypoplastrons, AUH 50, 62, 148b, 383, 470, 470a, 687, 689; inguinal buttress, AUH 148c; xiphiplastron, AUH 679.

The entoplastron, AUH 470b (fig. 13.5.3), has two anterior borders, oriented at 90° to each other, and a rounded posterior border, like *Mauremys* (although there is individual variation in the proportions) but not Pelomedusoides where it is rhomboid. The specimen shows that two meeting gulars are sufficiently long to cover a part of the entoplastron and that the pectorals cover the posterior third of the entoplastron. Consequently, they are longer than the humerals, which are short medially between the gulars and the pectorals. The fragments of hyoplastra (figs 13.5.2 and 13.5.4) confirm that the humero-pectoral sulcus cuts the entoplastron in its posterior third. The specimen BMNH R10974 (fig. 13.5.4) has a wide sinuous humero-pectoral sulcus ending laterally in a weak notch, similarly present in fossil and in exant species of *Mauremys.* In the other, small, younger specimens (fig. 13.5.2a), the notch is variably accentuated. The fragment of hyoplastron AUH 274a (fig. 13.5.2a) shows the medial obliquity behind the pectoro-abdominal sulcus (pectorals longer medially).

The fragmentary hypoplastra AUH 62 (fig. 13.5.5), AUH 383 (fig. 13.5.6), AUH 50 (fig. 13.5.1), and AUH 148b have a medial obliquity behind the abdomino-femoral sulcus (shortening the femorals). The length of the abdominal scute on the hypoplastron is variable in the same proportion as in western fossil and extant species of the genus. The dorsal covering of the hypoplastra by the femoral scute is also consistent with the genus: moderate width of covering, a little variable within the individual, and no strong bone elevation under this covering (figs 13.4.12 and 13.5.1a). The length of the bone relative to its width is also variable, the plastron varying in width within individuals (in general wider and flat in the females); the obliquity of the suture of the hypoplastron with the hyoplastron, variable from back to front, also conforms to *Mauremys.* The suture between hyoplastron and hypoplastron is laterally straight. This indicates that there is no lateral mesoplastron, which occurs in the Tertiary pleurodiran Pelomedusoides and not in the Tertiary cryptodiran turtles.

The xiphiplastron AUH 679 (fig. 13.5.7) exactly conforms to fossil and extant western *Mauremys* with its femoro-anal notch, its deep anal rounded notch (although these notches can be variable in depth), the anal longer than the part of the femoral scute on the plate, and the moderate lateral covering of the dorsal part of the plate by the scutes (fig. 13.5.7).

The inguinal buttress AUH 148c is strong at its base and does not cover much of the pleural width although it covers more than in the extant Mediterranean species *Mauremys caspica* and *M. leprosa.* It is like the Miocene European species *M. pygolopha* specimens from Artenay, France, MN 4 (Broin, 1977), and *M. sophiae* specimens from Sandelzhausen, Germany, MN 6 (Schleich, 1981), as mentioned above.

Humerus

The head of humerus AUH 363 (fig. 13.5.8) is eroded but conforms to *Mauremys* in the sharp angle between the trochanters, the weak inter-trochanteric fossa, the light dorsal obliquity of the head, and the prolongation of the small trochanter along part of the head in a lip.

Discussion

Shells of the pleurodiran turtles, which constitute the northern Gondwana population (see Broin, 1988), are known from at least the early Cretaceous in Africa to the present, south to latitude 20° N, including the extremity of the southwestern Arabian Plate (Gasperetti et al., 1993). Pleurodirans are present in the late early and early middle

Figure 13.5. *Mauremys* sp. *1,* AUH 50, left hypoplastron; a, b, dorsal, ventral views. *2,* AUH 274a, right hyoplastron; a, b, dorsal, ventral views. *3,* AUH 470b, entoplastron; ventral view. *4,* BMNH R10974, partial left hyoplastron; ventral view. *5,* AUH 62, right hypoplastron; ventral view. *6,* AUH 383, left hypoplastron; ventral view. *7,* AUH 679, left xiphiplastron; a, b, dorsal, ventral views. *8,* AUH 363, proximal part of a right humerus; a, b, c, external, dorsal, ventral views.

Miocene of As Sarrar (Saudi Arabia) and Ghaba (Oman) and are principally characterised by the sutured link of the shell with the three parts of the pelvis that is seen on the xiphiplastra and on pleurals 7 and 8. The preserved xiphiplastron (fig. 13.5.7a) shows a pelvis (pubis and ischion) free from this plate. Other pleurodiran characters are also lacking. The absence of lateral mesoplastra between the hyoplastra and hypoplastra, and the shape of the plates (particularly the elements of the plastron and plates 5–6 bearing the inguinal buttress) are not of a pleurodiran turtle and they are consistent with the Geoemydinei *Mauremys*. The entoplastron shape fully agrees with *Mauremys*.

Mauremys is a Laurasiatic Geoemydinei, Batagurinae without a secondary palate, unlike *Ocadia* (see Hirayama, 1985), with a vertebral 1 as wide as the nuchal (as in *Ocadia*) or wider. The extant Asiatic Geoemydinei *Annanemys* (included in *Mauremys* by Iverson and McCord, 1994), *Sacalia, Notochelys,* and the western fossil forms, *Palaeochelys* s.s., Oligocene–early Miocene, that are plesions of *Mauremys,* also have a wide vertebral 1. *Mauremys* has a rather flat nuchal bone, slightly concave on the ventral side. We have seen the evolution of marginals 1 and cervical size on the nuchal. In *Ocadia,* the cervical remains small and short when the lateral covering by marginals 1 is long. Like *Sacalia* and *Notochelys,* neurals of *Mauremys* are sometimes regular but not always (see above). The Baynunah form conforms to these characters.

In *Mauremys,* there is no exact link between each pleural and each peripheral from peripheral 4 and pleural 2 back (a plesiomorphic character present in *Ocadia* and *Palaeochelys*). There is generally contact between costal 3 and marginal 6 as in *Palaeochelys* and many other Batagurinae, or near contact (contact between costal 3 and marginal 7 as in the more derived Geoemydinei, from *Heosemys* to *Geoemyda,* and as in *Ocadia*); the vertebral 5 overlaps the suprapygal 2 and not the pygal, like the other Geoemydinei, *Rhinoclemys* and *Notochelys* excepted: these characters cannot be seen in the Baynunah form.

As mentioned above, the plastron has a slight femoro-anal notch and a deep or rather deep anal notch (not *Palaeochelys, Sacalia,* and *Notochelys*). The entoplastron is as described above, the gulars are as long as they are wide or longer and more or less overlapping the entoplastron. The epiplastrons have a wide and slightly protruding or not protruding gular part: in its preserved fragments; the Baynunah form conforms to these characters.

The preserved fragments are sufficient to attribute the Baynunah form to *Mauremys* without any doubt. *Mauremys* is known in Europe from Oligocene times and it is known to be present in Africa from the late Miocene of Wadi Natrun (Egypt) (Dacqué, 1912: "*Ocadia* nov. sp. ind.", poorly known, figured by the anterior lobe part only) and possibly of Sahabi, Libya (Erasmo, 1934: fig. 15). Furthermore, *Mauremys* is now distributed all around the Mediterranean Basin with *M. leprosa* from France to western Libya in the west, and *Mauremys caspica* from Yugoslavia to northern Arabia, along the Mediterranean coast and inland in other countries in the east (see Iverson, 1992; Gasperetti et al., 1993). The two extant western species of *Mauremys* have diverged, at least from Pleistocene times (*M. leprosa* lineage in North Africa) or possibly before.

Neontologists have given discriminant characters for the extant species but they are based principally on the decoration of the scutes and the colours, which are, of course, not preserved in fossil forms. They are also concerned with related proportions that are based on complete specimens and calculated on the mean averages of populations (Busack and Ernst, 1980; Iverson and McCord, 1994; Pritchard and McCord, 1991). With isolated fragments or few specimens, we cannot use these characters. We can differentiate the specimens only by the bony shell, the relative proportions of the impressions of the horny scutes on the bones (sulci), and the sutures of the plates. But we do not have the full proportions of the dorsal marginal and ventral scutes that may be longer than the bones on the border of the shell. On the bony shell, it appears that, in the type species *M. leprosa,* the cervical scute on the nuchal bone of the adult is longer than its width in most of the specimens (on all the 19 adult specimens examined), and the femorals are medially shorter than the pectorals in most of the specimens (14 out of 16 specimens examined, males or females). This is valid for the living specimens, still with their scutes, as well as for the bony shells. These two characters are apomorphic. In *M. caspica,* as in most of the fossil forms studied, the width of the cervical is still greater than its length and the femorals are longer than the pectorals, and much longer than the pectorals in most of *M. caspica caspica* compared with *M. c. rivulata,* where the femorals can be as long as the pectorals. We cannot consider these characters in the Baynunah Formation form, which is represented only by isolated fragments of the shell. But in studied fossils from western Europe (France, Italy, Germany, Austria) and North Africa, the condition is that of *M. caspica* from the Oligocene to Recent, except in:

1. *M.* aff. *gaudryi* from the "Pontian" of Kohfidisch Höhle (MN 10), Austria (Bachmayer and Młynarski, 1983), known by one juvenile specimen, which has a long cervical, although its width is still greater than its length, and femorals shorter than pectorals.
2. *M. portisi* Sacco, 1889, a specimen with femorals shorter than pectorals (and narrow vertebral 1).
3. *M. gaudryi etrusca* Portis, 1890 (see Kotsakis, 1980), a specimen with femorals equal to pectorals (Plio-Pleistocene).
4. *M. leprosa* from the Rharbian (late Soltanian or Holocene) of Doukkala II, Morocco, with a cervical longer than its width and femorals longer than pectorals; one specimen only (Gmira, 1995), which agrees with some extant *M. leprosa.* In fact, the sample is too poor in most of the fossils to define the species with these two characters. But if we consider the derived long covering of the nuchal by the cervical and marginals 1, the extremely wide vertebral 1, and the primitive wide cervical and strong inguinal buttress, we can then assume that the species is of an evolutive state comparable with a hypothetical late Miocene form, after the Miocene (MN 6) of Sandelzhausen (short covering, strong buttress but already wide vertebral 1), surely before the Pliocene (MN 14–15) of Montpellier-Perpignan (long covering, weaker buttresses, wide vertebral 1), and possibly at the same time or before the "Pontian" (late Vallesian, MN 10) of Kohfidisch (long covering, wide vertebral 1, buttresses unknown). Too many fossil forms are insufficiently prepared or figured to compare them with the Baynunah *Mauremys* sp. and to determine this to species level. Another problem is the reduction in numbers, followed by the extinction of fossil *Mauremys* species in Europe at the end of the Miocene (MN 6–8 to MN 13) (Lapparent de Broin, in press; Schleich, 1984), possibly due to aridity, which impedes comparisons with localities of equivalent age.

In conclusion, although possibly new, the Baynunah Formation *Mauremys* species remains indeterminate. In the late Miocene, the extant species are still not differentiated and we do not know at what period in geological time the differentiation occurred. It is impossible to know if the Abu Dhabi species is the ancestor of the extant population of *M. caspica* in northern Arabia or if the extant population is the result of another invasion much later than the age of the Baynunah Formation, from eastern Europe–western Orient (see Gasperetti et al., 1993). Furthermore, the fossil form might have arrived in the United Arab Emirates from the west, due to the presence of *Mauremys* in Wadi Natrun (Egypt) and possibly in Libya (Sahabi), instead of from the eastern Mediterranean. We do not, however, know the relative age of these localities and the reverse is just as possible. During the end of the Miocene, between MN 6 (in France) to MN 8 (in Germany) and MN 13 (in Spain and Italy) to MN 14–15 (in southern France), western Europe endured climatic change such as increasing aridity, so that *Mauremys* (like *Trionyx* from MN 6–9 and MN 10–13) disappeared or was extremely reduced (verified presence in Kohfidisch only, late MN 10) until MN 13. We find few turtles at these times, only terrestrial (giant and small) and semiterrestrial small forms. Were these two genera still present in southern Europe and/or the eastern Mediterranean, from where they invaded the Arabian Plate and North Africa, and then returned north into Europe?

Infrafamily Testudininei Batsch, 1788
Geochelone Fitzinger, 1835
Subgenus *Centrochelys* Gray, 1872
Geochelone (*Centrochelys*) aff. *sulcata* (Miller, 1779)

Material by Sites (fig. 13.1)

There are six identified specimens and four unidentified fragments, representing at least seven individuals.

H3: AUH 276, fragment of peripheral.
JD3: AUH 295, lateral fragment of pleural 1, 2, or 4.
K1: AUH 388, left epiplastron.
S1: AUH 107, right part of the nuchal bone associated with the peripheral 1 and the medial part of the peripheral 2; AUH 284a, part of a right pleural 8 and AUH 284, part of an unidentified plate.
S4: AUH 768, a juvenile peripheral; AUH 769, 770, 771, three unidentified fragments.

Description

Dimensions

Estimated shell lengths: 52–58 cm long. Some thick undefined specimens in the locality indicate longer shells.

Decoration

Apparently smooth; very finely punctiform, granulous with vascular foramina, and small vascular sulci. Most of the scute sulci are straight with sharp borders, sometimes elevated on a crest (figs 13.6.1b, 2b) but not always (fig. 13.6.2c). This deco ration is common to the terrestrial forms of Testudininei. The raised sulci are sometimes found in the African *Pelusios*, a Pelomedusidae.

Dorsal Shell

Material: part of the nuchal, peripheral 1, and peripheral 2, AUH 107; fragment of peripheral, AUH 276; fragment of pleural 1, 2, or 4, AUH 295; part of pleural 8, AUH 284a; juvenile peripheral, AUH 768; part of an unidentifed plate, AUH 284; three unidentified fragments, AUH 769, 770, 771.

The anterior border of the shell, AUH 107, is a right part of the nuchal bone associated with right peripheral 1 and the medial part of right peripheral 2 (fig. 13.6.1a) The width of the fragment is 12.5 cm. The specimens belong to a tall species of tortoise (terrestrial form) of the genus *Geochelone* due to the size (maximum here around 58 cm long), the slight anterior notch of the shell (deeper in the animals provided with their marginal scutes, see *Centrochelys sulcata* in Villiers, 1958), wide vertebral 1 (see the transverse sulcus between marginal 1 and vertebral 1, fig. 13.6.1b), which covers the nuchal (although in some *Geochelone* the vertebral 1 may also be anteriorly narrower than the nuchal bone and widened posteriorly on pleural 1), and the thickness of the ventral lip of the nuchal and of the peripherals (3 cm) covered by the long ventral border of the marginal scutes (fig. 13.6.1a). If complete, the notched nuchal should be seen without the presence of a cervical scute. The plates are thick, without secondary thinning (the lateral part of the pleural AUH 295 measures 2.1 cm).

Plastron

Material: left epiplastron, AUH 388; unidentified part of a plate, AUH 284.

The epiplastron AUH 388 (length 10.4 cm; thickness at the posterior end of the dorsal lip 3.4 cm; fig. 13.6.2) is similar to that of a female of the extant *Geochelone* (*Centrochelys*) *sulcata* from the pre-Sahelian and Sahelian part of Africa (Senegal to Ethiopia) and southwestern Arabian Plate (see Gasperetti et al., 1993; Iverson, 1992; Roset et al., 1990) because of the shape of the gular area and of the dorsal lip. The dorsal lip is moderately bent back (fig. 13.6.2a, 2b), in this case not covering the entoplastron; the ventral face of the gular area is rounded (fig. 13.6.2c), and the dorsal face of the lip is concave and posteriorly narrow with a rounded posterior border. Comparison is possible with a specimen of extant *Geochelone sulcata* without its scutes figured in Roset et al. (1990: pl. 3, figs 1 and 3). It is not wide, not convex, and lacks a double posterior convexity, differing from *Geochelone* (*Stigmochelys*) from eastern and southern Africa, the other tall tortoise from Africa, from Ethiopia–Somalia to southern Africa (Iverson, 1992). Like many tortoises, including *Geochelone* s.l., the protruding gular lip is notched for a partial subdivision of the gular. The gular covers all the medial part of the epiplastron along the symphysis and extends only slightly over the entoplastron. We have to decide if the long covering of the epiplastron by the gular is a plesiomorphic or reversal character for *Geochelone*. The short gulars are primitive in Testudinidae. The long covering is present in most *Geochelone* (*Stigmochelys*); for instance, the extant *G. pardalis*, where the covering is variable, sometimes extending over the entoplastron, as in *G.* (*S.*) *brachygularis* from Laetoli (Tanzania) (Meylan and Auffenberg, 1987). In the six extant *G. sulcata* specimens studied (MNHN, AC, P), without scutes, the gular is shorter than the epiplastral symphysis length, and does not cover all the epiplastron backwards, as in a Holocene specimen from Niger (Roset et al., 1990). In the Sahabi specimen—an undefined species of "cf. *Geochelone*" (Wood, 1987: fig. 5), clearly a young *Geochelone* of the *sulcata* lineage and exactly the same as the Baynunah form—the gulars just touch the entoplastron. If the long covering is derived, the Baynunah form is more derived than the extant and Holocene specimens. Information about the extant *G. sulcata* is given (amongst others) in Lambert (1993), Loveridge and Williams (1957), and Villiers (1958), but they do not include figures of the bony shell. Further investigation of more than six extant specimens will indicate if we have here, and in Sahabi, an individual variation within the *G. sulcata* species or a characteristic of a late Miocene new species, more apomorphic than the extant *G. sulcata* for this character. Nevertheless, a parallel development of the covering is evident between the subgenera of terrestrial tortoises.

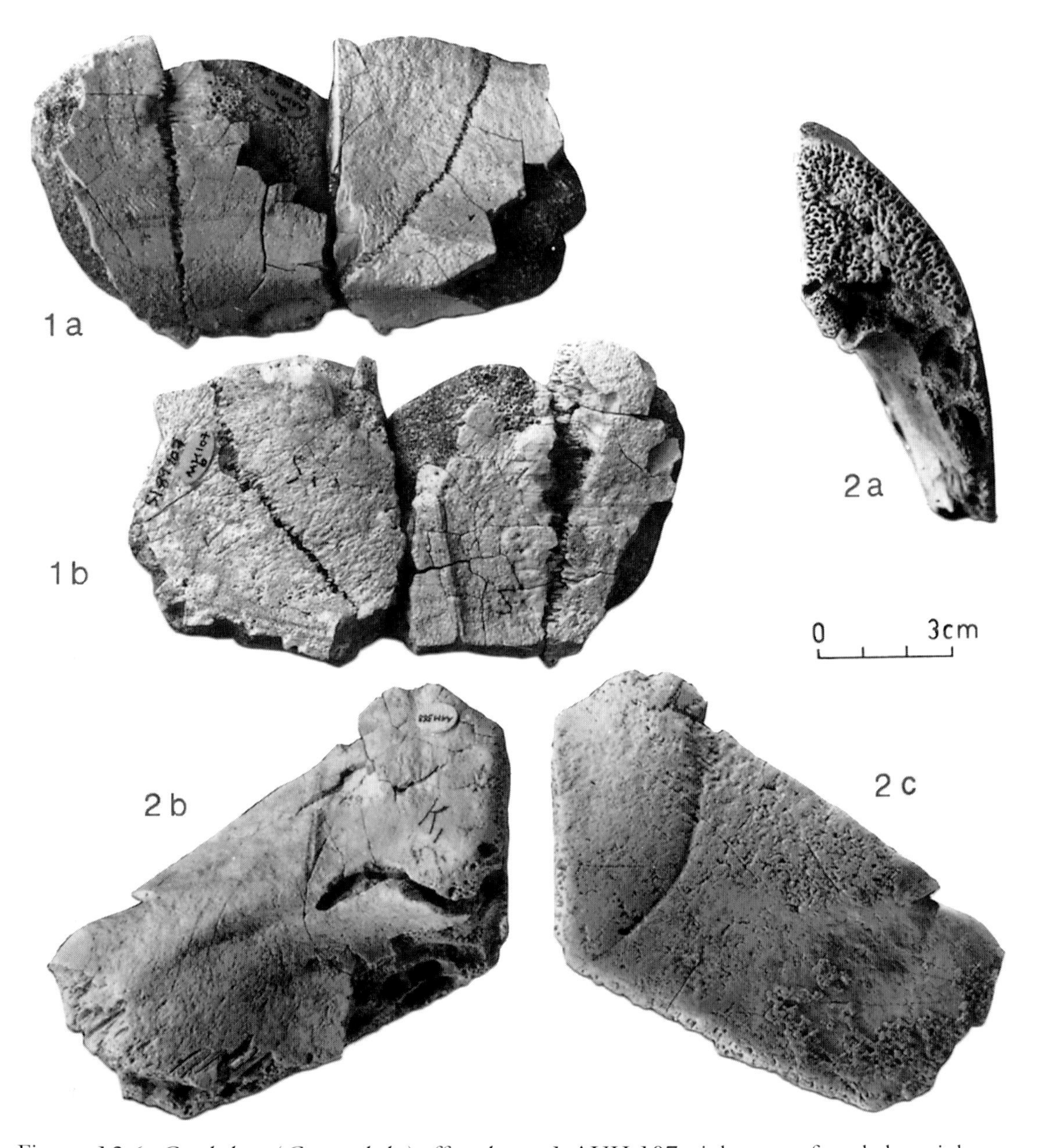

Figure 13.6. *Geochelone* (*Centrochelys*) aff. *sulcata*. *1,* AUH 107, right part of nuchal, peripheral 1 and left part of peripheral 2; a, b, ventral, dorsal views. *2,* AUH 388, left epiplastron; a, b, c, medial (symphysial), dorsal, ventral views.

2579 and BMNH R7729) (Tchernov and Van Couvering, 1978: 858, pl. 99, figs 3 and 4) and Maccagno's concerning *C. checchiai* (Maccagno, 1947) are based on only two specimens. In the case of *C. lloydi* we have also only a few specimens; some are very fragmentary and cannot be reliably assigned to this species. For the same reason it is impossible to make reliable statements about the intraspecific variability of skull and mandible features (and therefore species differences) for these species based on our present state of knowledge.

Although *C. pigotti* sensu Tchernov and Van Couvering, 1978 is documented from several isolated incomplete mandibles—KNM-RU 2580, 2582, and 2583 (Tchernov and Van Couvering, 1978: 858)—this material has not been fully described and remains unfigured. The reference of a mandible to *C. pigotti* (AS 979) by Buffetaut (1984) is presumably based on the brief and taxonomically insufficient description of the mandible of *C. pigotti* (KNM-RU 2580, 2582, and 2583) by Tchernov and Van Couvering (1978: 861). The reason for the determination of the mandibular fragment by Buffetaut (1984) is unclear and the coincidence in stratigraphical age with material from Kenya is insufficient for reliable species identification. Thus, the association of the Saudi Arabian material with *C. pigotti* is questionable.

Osteology and Comparative Anatomy of the Taxonomically Relevant Specimens—*Crocodylus* sp. indet., *Crocodylus* cf. *niloticus*, and Crocodylidae gen. et sp. indet.

Crocodylus articeps and *C. megarhinus* are excluded from the comparison with the Baynunah crocodilians, because both these taxa in Africa extend only into the early Oligocene. The descriptions and figures of *C. checchiai* (Maccagno, 1947) do not allow a reliable taxonomic comparison with the material from Abu Dhabi. According to Kälin (1933) the recent *Crocodylus* species with a massive jaw construction share many similar characters in skull and mandibular dentition, and *C. niloticus* and *C. palustris* are especially similar in these aspects. It is unnecessary to compare the Abu Dhabi material with the skulls of all extant *Crocodylus* species, due to the distribution history of *Crocodylus* in Africa. Until now only two fossil *Crocodylus* species are with certainty known from northern and eastern Africa that reach into modern times—*C. niloticus* and *C. cataphractus* (Tchernov, 1986)—and we therefore have no problem in restricting comparison of the Abu Dhabi material to these two species.

Specimen AUH 285—Edentulous Left Maxilla (table 14.1 and fig. 14.1)

AUH 285 is a fragment 91 mm long consisting of the first eight and part of the ninth alveolus. A vestige of the premaxillary–maxillary suture on the palate (median suture of the premaxillaries; Tchernov, 1986: 19) is preserved. The posterior point of this suture is at the level of the anterior border of the second maxillary alveolus. The dorsolateral process of the premaxillary–maxillary suture extends posteriorly to the level of the anterior border of the second maxillary alveolus. The arrangement, shape, and dimension of the alveoli are as follows:

1. The alveoli increase somewhat irregularly in size from alveoli 1 to 5; the fifth alveolus is by far the largest.
2. Alveolus 6 is about equal to the fourth in size.
3. Alveoli 2, 3, 7, and 8 are nearly equal in size to each other.
4. Posterior to alveoli 3, 4, and 6 the interalveolar grooves are linguolabially elongated and distinctly wider than long. The interalveolar groove between the fourth and fifth alveolus is much larger (mesiodistal diameter 6.3 mm) and deeper than the interalveoloar groove 3 (mesiodistal diameter 4.5 mm).
5. Posterior to alveoli 8 and 9 the interalveolar grooves are circular, rather deep, and almost as large as the neighbouring alveoli.
6. Interalveolar spaces 6 and 7 are the largest; interalveolar space 8 is the next largest. Interalveolar grooves 6 and 7 are distinctly shallower than interalveolar groove 8.
7. Alveoli 2 and 7 are a compressed oval in cross-section; all other alveoli are almost circular.
8. Neither the lingual nor the labial borders of the alveoli are supported by a stout buttress.

Comparison of C. niloticus *(SMF 47244, PIUM S11, Tchernov, 1986: pl. 3, fig. 1; 20, table 1) with AUH 285* (table 14.1 and fig. 14.1)

Similarities

—Interalveolar space 7 is largest. Interalveolar space 6 is the second largest. Interalveolar spaces 3 and 4 are larger than 1 and 2. Interalveolar spaces 3 and 4 are equal in size to each other.
—Alveoli 2 and 3 are of about the same size as 7 and 8.

Figure 14.1. *Crocodylus* sp. indet. Fragment of an edentulous left maxilla, AUH 285; palatinal view. Scale bar = 20 mm.

—The most posterior extension of the palatinal part of the premaxillary–maxillary suture reaches the centre of the second maxillary alveolus (PIUM S11).
—Small, linguolabially elongated interalveolar grooves are situated median between alveoli 3–5, large circular ones between alveoli 7–9.

Differences between C. niloticus *and AUH 285*

The dorsolateral processus of the premaxillary–maxillary suture in *C. niloticus* extends posteriorly to the fourth–fifth maxillary alveolus (Tchernov, 1986: 20, table 1); in AUH 285 it reaches only the second. According to Kälin (1933: 557), who contrary to Tchernov (1986) studied recent individuals of *C. niloticus* from different geographical areas, the length of the premaxillary–maxillary suture is highly variable between the second and fourth tooth; in PIUM S11 the dorsolateral processus of the premaxillary–maxillary suture extends posteriorly to the anterior border of the third tooth.

Comparison of C. pigotti *Tchernov and Van Couvering, 1978 (BMNH R7729, Tchernov, 1986: pl. 1, figs 2, 4; Tchernov and Van Couvering, 1978: 858–64, pl. 99, figs 1–4) with AUH 285* (table 14.1 and fig. 14.1)

Similarities

—Interalveolar space 7 is the largest. Alveoli 5–6 are separated by a small interalveolar space. Interalveolar space 6 is much smaller than 7.
—The most posterior extension of the palatinal part of the premaxillary–maxillary suture reaches the second maxillary alveolus.
—The dorsolateral processus of the premaxillary–maxillary suture extends posteriorly to the second maxillary alveolus.

Differences of C. pigotti *Tchernov and Van Couvering, 1978 from AUH 285*

—In *C. pigotti* alveoli 2–5 (and possibly 1) have a stout buttress labially and lingually, and the teeth therefore appear to emerge from long protruding collars.
—There is a steep wall between the alveoli row and palate (BMNH R7729): the palate is sunk below the level of the alveoli row so that there is a near vertical wall posteriorly (maxillary teeth 10–7) and a steeply sloping wall anteriorly (maxillary teeth 7–1).
—Interalveolar spaces 7 and 8 are nearly equal in size.

Comparison of C. lloydi *(BMNH R5893, R5893, R8333, and R4697, Tchernov, 1986: 28–30, pl. 4, fig. 1; 20, table 1) with AUH 285* (table 14.1 and fig. 14.1)

Similarities

—Alveoli 2 and 8 are nearly equal in size.
—Interalveolar space 7 is the largest; interalveolar spaces 6 and 8 are the second largest and nearly equal in size; interalveolar space 5 is small.
—There are large interalveolar grooves 6 (linguolabially elongated) and 7 (circular).
—Interalveolar spaces 1, 2, and 5 are smaller than the others.
—The most posterior extension of the palatinal part of the premaxillary–maxillary suture reaches the middle of the second maxillary alveolus (BMNH R8333); according to Tchernov (1986: 20, table 1): "1st to the mid 1st".

Differences of C. lloydi (BMNH R8333) from AUH 285

—Alveoli 3 and 4 in *C. lloydi* are much larger than 7 and 8.
—Interalveolar grooves 1–5 are small, trigonal in shape, and located far lingually; in AUH 285 interalveolar pits 3 and 4 are located directly between the alveoli, they are linguolabially elongated, and relatively larger than those in *C. lloydi.*
—Interalveolar groove 8 is small, circular, and located far lingually from alveoli 8 and 9; in AUH 285 interalveolar groove 8 is as large as interalveolar groove 7 and is located median between alveoli 8 and 9.
—There is a small lingual trigonal interalveolar groove 9 located lingually between alveoli 9 and 10 in *C. lloydi.*
—The dorsolateral processus of the premaxillary–maxillary suture extends posteriorly to the fourth maxillary alveolus.

Table 14.1. Measurements (in mm) of the maxilla of *Crocodylus* sp. indet., AUH 285

Alveoli	1	2	3	4	5	6	7	8	9
Linguolabial diameter	4.0	4.0	4.5	5.5	8.0	5.0	4.0	4.0	?
Mesiodistal diameter	5.0	5.5	5.3	6.0	9.0	6.0	6.0	5.5	?
Interalveolar space	2.0	2.0	3.2	3.7	2.0	4.8	7.0	4.5	?
Compression index	1.25	1.37	1.17	1.1	1.12	1.2	1.5	1.37	?

Taxonomic Conclusions on Maxillary Fragment AUH 285 (table 14.1 and fig. 14.1)

Fragment AUH 285 reveals both similarities to and distinct differences from *C. lloydi* in alveoli size, relative size of the interalveolar spaces, shape, relative size, and position of interalveolar grooves, and the extension of the dorsolateral processus of the premaxillary–maxillary suture. The resemblance with *C. pigotti* and *C. niloticus* in the aforementioned characters is much greater. These similarities are, however, shared by many *Crocodylus* species and are therefore not diagnostic at species level (Kälin, 1933; Wermuth, 1953). Otherwise *C. pigotti* shows striking differences from AUH 285 in the long collars of alveoli 2–5 and the steep wall between the tooth row and palatinal roof. Considering all noted differences we can exclude AUH 285 as belonging to *C. pigotti* or *C. lloydi*. In general, however, it is extremely difficult to distinguish fossil and extant *Crocodylus* species on the basis of even complete upper jaws (premaxilla and maxilla) with the exception of *C. cataphractus* (Kälin, 1933: 592) and *C. johnsoni* (Wermuth, 1953). Therefore we identify the fragment as *Crocodylus* sp. indet.

Specimen AUH 616—Fragment of an Edentulous Left Dentary (table 14.2 and fig. 14.2.1a–1d)

The 13 preserved alveoli extend over a distance of 176 mm. The symphysis extends posteriorly to the level of the anterior border of the fifth alveolus. The margo dorsalis of the symphysis is concave, the margo ventralis convex. The anterior and posterior ends of the symphysis are small and rounded, the cranial end being somewhat smaller. The large lobus symphysialis dorsalis extends 11.5 mm caudal to the distinctly smaller lobus symphysialis ventralis. The canalis meckeli runs parallel to the long axis of the jaw. The fossa meckelian forms an angle of about 35° with the ventral border of the symphysis. It is impossible to determine to which alveoli the splenial posteriorly extends. The dentary has a small, shallow, longitudinal canal on its medial surface that slopes dorsally in a convex path from alveoli 10 to at least 13. The posterior outline of the symphysis forms a parabolic bow. The largest foramen dentale is slit-like, 4 mm long, and situated median between alveoli 8 and 9. The measurements of AUH 616 are as follows:

1. The length of the symphysis is 52 mm.
2. The height of the symphysis is 20.5 mm.
3. The maximum width of the symphysis (across alveolus 4) is $26.5 \times 2 = 53$ mm.
4. The distance between alveoli 4 and 10 (measure points: tip of the crown or centre of the alveolus) is 79 mm.
5. The minimum height of the mandible (between alveoli 6 and 7) is about 16 mm.

The ratio of symphysial length to height is 253.6%. The ratio between the minimum height of the mandible and the distance between the fourth and tenth dentary alveolus is 20% (see Berg, 1966: 45). The ratio of symphysial length to width is 0.98 (see Tchernov, 1986).

The upper border of the lower jaw is vertically festooned and there are moderately pronounced elevations at the level of alveoli 1, 4, and 11 with valleys in between. Posterior to alveolus 11 the dorsal border of the dentary slopes somewhat ventrally—an indication that there was a third depression of the jaw posterior to the most anterior two jaw depressions. The three largest dentary alveoli are nearly equal in size to each other: 1, 4, and 11; the smallest alveolus is the eighth. By far the largest interalveolar space is the eighth; the second largest interalveolar spaces 1, 2, and 7 are nearly equal in size to each other. Between alveoli 9 and 14 the interalveolar grooves are situated at the level of the labial border of the alveoli. All interalveolar grooves are circular. At the lateral surface of the jaw there are two shallow vertical grooves between alveoli 5 and 6, and 7 and 8. Interalveolar groove 9 is situated distinctly below the level of the tenth and eleventh interalveolar groove. Interalveolar grooves 9, 10, and 11 are nearly equal in size (mesiodistal diameter 2.6 mm) to each other and are situated at the level of their neighbouring alveoli. The mesiodistal diameter of interalveolar groove 12 is 3.8 mm, and 5 mm for interalveolar groove 13. Alveoli 7, 8, 12, and 13 are a compressed oval in cross-section, all other alveoli are almost circular. The facies dorsalis of the dentary is developed between alveoli 1 and 2 and 3 and 4 into a median interalveolar ridge.

Comparison of C. niloticus *(SMF 41042, 47244, 68150, 28158, 41073, 47732, 28155, 46921, 41072, BMNH R3266, PIUM S335, Tchernov, 1986: pl. 3—fig. 3; 20, table 1; 22–23, table 2) with Mandibular Fragment AUH 616* (table 14.2 and fig. 14.2.1a–1d)

Similarities

—Interalveolar spaces 1, 2, 7, and 8 are larger than the others. Interalveolar space 8 is the largest of these as in most *Crocodylus* species. Interalveolar space 12 is the fifth largest (as in *C. niloticus* SMF 68150 and 47244). Interalveolar space 1 is the third largest and interalveolar space 2 the second largest.
—Alveolus 8 is the smallest, a character shared by many specimens in most *Crocodylus* species (SMF 47244, 68150,41072).
—Alveoli 1, 4, and 11 are nearly equal in size to each other and are the largest alveoli of the mandible. The sockets of these alveoli reach higher compared with the other alveoli.
—Alveolus 3 is distinctly smaller than alveolus 2. This difference is especially obvious in *C. niloticus* (SMF 47244, 68150; Tchernov, 1986: 29).
—The mandibular symphysis extends posteriorly to the anterior margin of alveolus 5. With a symphysial length:width ratio of 0.98 the Baynunah *Crocodylus* falls within the range of the variability of *C. niloticus* (0.7861–1.1115) (Tchernov, 1986). The table published by Tchernov (1986: 22–23, table 2) is based on the measurements of 38 specimens. One of us (M. R.) measured seven specimens of *C. niloticus* and calculated a symphysial length:width ratio ranging from 0.8 to 1.0. According to Tchernov (1986: 23, table 2), the symphysial length:width ratio in *C. palustris* ranges from between 0.2821 and 0.4461. This value is clearly distinct from *C. niloticus* and the Baynunah *Crocodylus.*
—The posterior margin of the mandibular symphysis forms with the medial border of the mandibular rami a parabolic bow (reconstructed for AUH 616). Instead of this some specimens of *C. niloticus* (PIUM S335) sometimes have a trapezoid outlined posterior symphysis.
—The anterior margin of the symphysial area forms an angle of 20–25° with the symphysis suture (see Tchernov, 1986: 35, fig. 13).
—The length–height symphysis index of 253.6% lies within the intraspecific range of seven specimens of *C. niloticus* in Frankfurt (SMF) (230–265%).
—The largest foramen dentale (4 mm long) is almost halfway median between alveoli 8 and 9. This was also the case in most *C. niloticus* specimens examined.
—Interalveolar spaces 9 and 10 each have a small rounded interalveolar groove approximately at the level of the labial margin. Interalveolar spaces 12 and 13 each have a large rounded interalveolar groove (in *C. niloticus* SMF 47244, interalveolar spaces 13 and 14).

Differences of C. niloticus *from AUH 616*

—In *C. niloticus* a longitudinal furrow, 1 mm wide, originates close to the border of the tenth alveolus at the facies medialis of the dentary and extends at least to the posterior border of the thirteenth alveolus.
—The recessus on the dorsal margin of the dentale between alveoli 1 and 4 is missing. The recessus on the dorsal margin of the dentale between alveoli 4 and 10 is less pronounced compared with AUH 616.

Comparison of C. cataphractus *with AUH 616*

Only the differences of *C. cataphractus* from *C. niloticus* and the Abu Dhabi specimens will be mentioned here:
—The symphysis extends posteriorly as far as the seventh tooth.
—The range of tooth size for maxillary teeth 2–5 is smaller.
—The interalveolar spaces in maxillary teeth 6 and 7 are only somewhat larger than the others.
—The interalveolar spaces are relatively larger.

Comparison of C. lloydi *(BMNH R5892, R5893, R8333, R4697, Tchernov, 1986: 20, table 1; 22, table 2; pl. 4, fig. 3; pl. 6, fig. 5) with AUH 616* (table 14.2 and fig. 14.2)

Similarities

—The mandibular alveoli of BMNH R8333 are too badly preserved for a reliable comparison, so we refer in this respect to the results of Tchernov (1986: 26–35).

Table 14.2. Measurements (in mm) of the dentary of *Crocodylus* cf. *niloticus*, AUH 616

Alveoli	1	2	3	4	5	6	7	8	9	10	11	12	13
Linguloabial diameter	9.0	7.0	5.7	8.5	5.5	5.5	5.0	3.5	5.4	7.0	8.0	6.5	5.5
Mesiodistal diameter	9.0	8.0	6.5	9.5	6.5	6.0	6.5	6.0	6.5	6.0	9.5	8.5	8.0
Interalveolar space	7.5	8.0	4.0	5.0	5.0	5.5	7.3	13.7	5.5	3.3	4.9	6.7	?
Compression index	1.0	1.1	1.1	1.1	1.1	1.1	1.3	1.7	1.2	0.9	1.2	1.3	1.45

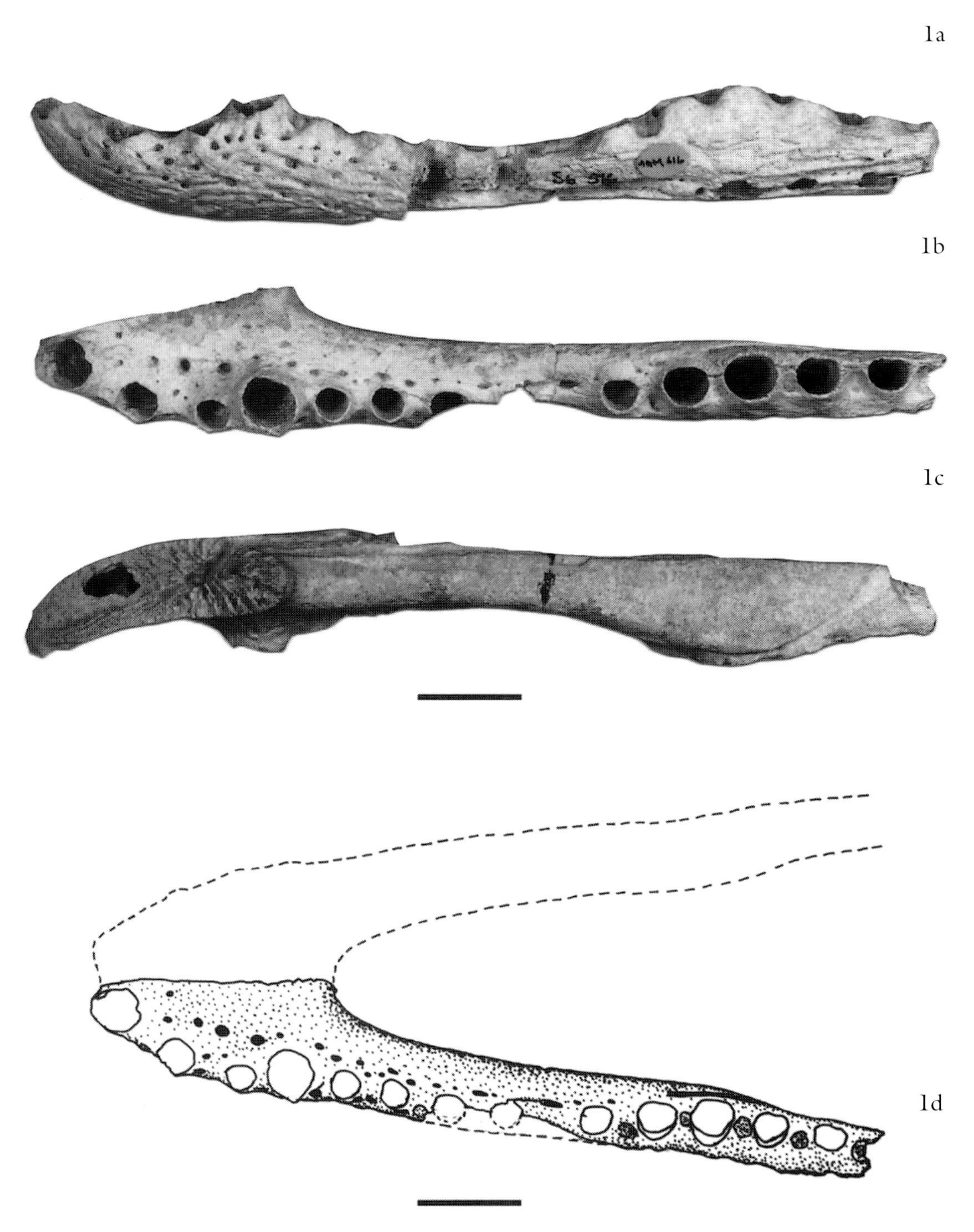

Figure 14.2. *Crocodylus* cf. *niloticus*. Fragment of an edentulous left dentary, AUH 616: *1a,* lateral view; *1b,* dorsal view; *1c,* medial view; *1d,* reconstruction of the dentary, dorsal view. Scale bars = 20 mm.

—The length:height ratio of the symphysis is 235% (BMNH R8333) and lies together with AUH 616 within the range of seven specimens of *C. niloticus* measured in Frankfurt (SMF).
—Interalveolar space 8 is the largest.
—Alveoli 5, 6, and 7 are almost equal in size (see Tchernov, 1986: 29).
—The largest dental foramen (13 mm long) is close to and lingual to the anterior border of alveolus 9 (BMHN R5892).

Differences of C. lloydi *from AUH 616*

—In *C. lloydi* alveoli 3 and 4 are spaced close together; alveoli 2 and 3 are equal in size to each other (Tchernov, 1986: 29); alveoli 10 and 11 are equal in size to each other. In AUH 616 alveolus 11 is much larger than the 10; alveoli 4 is larger than both the tenth and eleventh; alveoli 4 and 11 are nearly equal in size to each other and distinctly larger than the tenth alveolus.
—The mandibular symphysis extends posteriorly as far as alveoli 3 and 4. In BMNH R5892 the symphysis extends posteriorly to the posterior border of the fourth alveolus (anterior border of alveoli 5 in AUH 616). According to Tchernov (1986: 22, table 2) the length:width ratio of the mandibular symphysis in the Miocene form of *C. lloydi* ranges between 0.5647 and 0.6512, in Plio-Pleistocene forms between 0.5804 and 0.8058 compared with the same ratio in the Baynunah *Crocodylus* of 0.98.
—The symphysis is distinctly higher in relation to its length than in AUH 616; in medial view the outline of the symphysis is rectangular with an equal-sized anterior and posterior end (see Aoki, 1992: 83, pl. 3, fig. 6). AUH 616 in medial view has a symphysis with a convex margo ventralis and concave margo dorsalis. Both ends of the symphysis are small and rounded.
—The apparent absence of interalveolar grooves (BMNH R5892). According to Müller (1927: 69–70) this is a highly variable feature in *C. niloticus*.
—The posterior margin of the symphysis forms with the medial border of the mandibular rami a widely rounded arch (see Tchernov, 1986: 28; 35, fig. 13).
—The anterior margin of the symphysial area forms an angle of 45° with the symphysial suture (see Tchernov, 1986: 28; 35, fig. 13).

Comparison of C. pigotti *Tchernov and Van Couvering, 1978, sensu Buffetaut 1984 (BMNH R7729, Buffetaut, 1984: 515; 516, fig. 1A–C) with AUH 616* (table 14.2 and fig. 14.2.1a–1d)

Similarities

—The symphysis is rather slender in longitudinal section with small rounded anterior and posterior ends.
—Interalveolar spaces 2 and 7 are nearly equal in size to each other. Interalveolar space 8 is by far the largest, being almost twice as large as interalveolar space 7.
—The length:width ratio of the mandibular symphysis is 1.0 compared with 0.98 in AUH 616.
—The symphysis extends posteriorly to the anterior border of alveolus 5. In *C. pigotti* Tchernov and Van Couvering (1978: 860, table 1) the symphysis reaches the posterior margin of the fifth mandibular tooth.
—The posterior margin of the mandibular symphysis is short and transverse with respect to the longitudinal axis of the skull. The posterior border of the mandibular symphysis forms a parabolic bow (reconstructed for *C. pigotti* and AUH 616).
—Posterior to each of alveoli, 9, 10, and 11 is a small rounded interalveolar groove at the level of the alveolus's labial margin. A large rounded interalveolar groove lies between alveoli 13 and 14.
—The recessus of the margo dorsalis of the dentale between alveolus 1 and the anterior border of alveolus 4, and from the posterior border of alveolus 4 to the posterior border of alveolus 10, are of similar depth to those of AUH 616.
—The largest dental foramen is situated closely anterior to alveolus 9, but compared with AUH 616 this foramen is slightly larger than the others.

Differences of C. pigotti *sensu Buffetaut (1984) from AUH 616*

—In *C. pigotti* alveoli 3 and 4 are separated by a small septum; alveoli 2 and 3 are of equal size to each other and the smallest of thirteen alveoli; alveoli 11 and 12 are of equal size to each other and alveolus 11 is only slightly larger than 10. In AUH 616 the alveolus 2 is much larger than 3; alveoli 2, 3, 5, and 6 are smaller than 8; alveolus 8 is the smallest of thirteen; alveoli 11 is much larger than 12 and 11 is much larger than 10 (table 14.2).
—Alveoli 5–7 are closely spaced. In AUH 616 these alveoli are distinctly separated.

Taxonomic Conclusions on Mandibular Fragment AUH 616

AUH 616 corresponds with *C. niloticus* in almost every anatomical detail—for example, alveolus size relations; number, position, shape, and relative size of the interalveolar grooves; construction and proportions of the mandibular symphysis; and the position and relative size of the interalveolar spaces. Only the slightly greater degree of vertical festooning of the dorsal margin of the dentale distinguishes it from *C. niloticus.*

The number of differences between AUH 616 and *C. lloydi* in the mandibular anatomy is much greater than those between AUH 616 and *C. pigotti* sensu Buffetaut, 1984.

Of all the fossil crocodilian specimens known from Africa, AUH 616 shows by far the most similarities to *C. niloticus.* The specimen is, however, too fragmentary to assign it with confidence to *C. niloticus* and we therefore determine AUH 616 as *C.* cf. *niloticus* until more complete specimens are found to support this determination.

Specimen AUH 32—an Almost Complete Posterior Part of a Skull with Skull Table and Uncompressed Occiput (fig. 14.3.1a–1c)

The facies articulares of the condylus occipitalis and the condyli mandibulares of the quadrata are slightly weathered. Quadratojugalia and pterygoidea are missing.

The margines laterales of the skull table are straight and diverge posteriorly. The skull table is heavily sculptured, mainly with closely spaced circular, triangular, or polygonal pits up to 3 mm deep and from 1 to 10 mm long. There are also many pits, which are anteroposteriorly and anteromedially extended. The ovoid foramina supratemporales externa are 35 mm long (40% the length of the skull table). The minimum distance between the foramina supratemporales externa is 15 mm. With the exception of the straight margo lateralis all margins of the foramen supratemporalis externum are concave. The foramen supratemporalis externum leads into the steeply anterolaterally running canalis supratemporalis, ending in the ovoid foramen supratemporalis internum, which is about 35 mm long. The long axes of the foramina supratemporales are nearly parallel to the median axis of the skull. The skull table is 85 mm long, 114 mm wide posteriorly, and 98 mm wide anteriorly.

The part of the supraoccipital that participates in forming the skull table is a small triangular bone about one-third the width of the parietal. The open foramina posttemporales are slit-like, each being 23 mm wide (50% the width of the parietal).

The facies posterior of the occiput is approximately 101 mm high and nearly vertical to the transverse plane. Both cristae posttemporales of the supraoccipital are covered by the skull table. The margo posterior of the skull table is biconcave with a median and posteriorly orientated tip. The facies posterior of the occiput above the foramen magnum is deeply concave, about 40 mm high and built up into a short posteriorly orientated process directly roofing the foramen magnum.

The dorsal processes of the exoccipitals are 8 mm wide and reach the condylus occipitalis.

The width of the parietal is approximately one-quarter the width of the skull table's margo posterior.

The margo lateralis of the exoccipital is deeply concave, the margo laterales of the basioccipital slightly convex. There is a well-developed crista basioccipitalis 15 mm high and 6.3 mm wide; it is restricted to the ventral half of the basioccipital. The crista basioccipitalis at its highest point forms a small straight ridge; from this ridge the flanks of the crista slope laterally at a steep angle. The lateral regions of the basioccipital are extended slightly posteriorly.

Of the three large foramina in the facies posterior of the exoccipital the high foramen, 14 mm wide and 8 mm high, is the largest. The foramen caroticum posterius and the smaller foramen hypoglossale are nearly circular. A canal 1 mm in diameter is on the medial wall of the canalis vagi and is about 1 mm anterior to the foramen vagi.

The canalis hypoglossale lies 3.5 mm dorsolateral to the foramen vagi. The foramen caroticum posterius lies 2.5 mm ventromedial to the foramen vagi. Of the three occipital foramina the canalis hypoglossale is closest to the margo lateralis of the foramen magnum—it lies 0.7 mm lateral to it. The greatest distance between the occipital foramina is 19 mm between the foramen hypoglossale and foramen caroticum posterius.

The condylus occipitalis (27 mm wide and 24 mm high) is built up mainly of the basioccipital with only two small dorsal processus of the exoccipitalia contributing. The basisphenoid and basioccipital are nearly vertical in transversal plane. The dorsum sellae of the basisphenoid is perforated medially by two circular and parallel foramina carotica anteriores and dorsolaterally by the foramina abducidens.

The foramen ovale is 10 mm long and 6 mm wide; its long axis is dorsoventrally orientated.

The recessus oticus externus is roofed by a thick squamosal. The squamosal is thickest (14 mm) at its lateral margin above the anterior half of the recessus oticus externus.

The condyli mandibulares of the quadrates are 46 mm wide, 16 mm high medially, and 19 mm high laterally. The margo dorsalis et ventralis of the condylus mandibularis is concave, the margo lateralis et medialis of the condylus mandibularis convex. The facies articularis of the condylus mandibularis is convex laterally and saddle-shaped medially. The facies ventralis belonging to the processus articularis of the quadrate is only well preserved at its posterior end, at the centre of which is a small area of rugosities.

The foramen aerum of the quadrate lies posterior to and 3 mm apart from the concave margo medialis of the quadrate.

Comparisons of Posterior Skull Fragment AUH 32 (see Tchernov, 1986: 22–23, table 2; 31, table 3) (fig. 14.3.1a–1c)

The proportions of the skull fragment AUH 32 lie within the variability of *C. niloticus.* The size, position, and shape of the foramen supratemporalis are coincident with those of *C. niloticus.* The same holds true for the construction of the facies posterior of the occiput with an oblique anteroventrally orientated occipital supraoccipital and the arrangement and size ratios of the occipital foramina. The facies lateralis of the basioccipital of *C. palustris* is much more bulged compared with *C. niloticus.*

Extant specimens of *C. cataphractus* clearly differ from AUH 32 in the shape and the proportions of the skull table and foramina supratemporales. Only the morphology of the margo caudalis of the skull table of AUH 32 and *C. cataphractus* is similar; a slender but prominent processus postoccipitalis is present. In *C. pigotti, C. checchiai,* and *C. lloydi* this processus is much more massive when compared with AUH 32. In *C. niloticus* and most other extant *Crocodylus* species, however, the processus postoccipitalis is very small. Striking similarities to AUH 32 in the construction and proportions of the skull table and relative position, size, and shape of the foramina supratemporales are found in a Plio-Pleistocene specimen referred to *C. cataphractus* from Kenya (Tchernov, 1986: pl. 4, fig. 4).

Comparison of C. lloydi *(BMNH R8333), occipital part of the skull, with AUH 32* (fig. 14.3.1c)

Similarities

The foramen magnum is trigonal in shape. There is nearly the same arrangement, relative size, and shape of the occipital foramina. The foramen vagi is the largest of the occipital foramina; housed in its

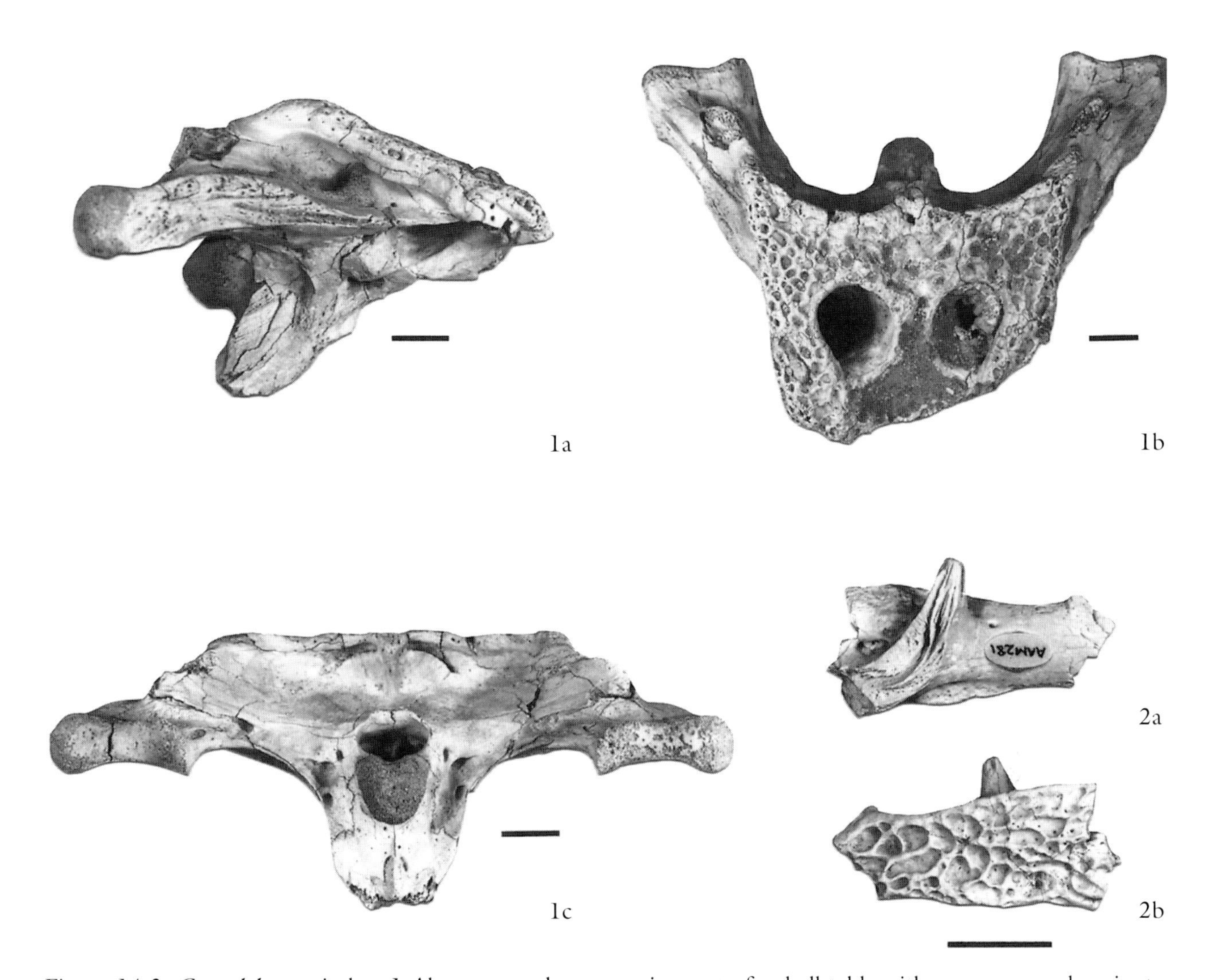

Figure 14.3. *Crocodylus* sp. indet. *1,* Almost complete posterior part of a skull table with uncompressed occiput, AUH 32: *1a,* lateral view; *1b,* dorsal view; *1c,* occipital view. *2,* Fragment of a right jugal, AUH 281: *2a,* medial view; *2b,* lateral view. Scale bars = 20 mm.

medial wall is a small canal. The foramen caroticum posterius is smaller than the foramen hypoglossale; both are circular.

Differences of C. lloydi *(BMNH R8333) from AUH 32* (fig. 14.3.1c)

C. lloydi has a large and long crista basioccipitalis and a cuirass-like shaped basioccipital (AUH 32 has a short crista basioccipitalis restricted only to the ventral part of the basioccipital; the basioccipital is tongue-shaped without any incisurae at its lateral borders).

Comparison of C. niloticus *(SMF 41072, 47244, PIUM S11) with AUH 32* (fig. 14.3.1c)

Similarities

—Nearly the same arrangement, shape, and relative size of the occipital foramina.
—A short crista basioccipitalis restricted to the ventral half of the basioccipital. Tongue-shaped basioccipital without any incisurae at its lateral borders.

—The crista basioccipitalis lies in a slight depression on the posterior surface of the basioccipital. An identical "unusual" kind of sculpturing of the basioccipital has been noticed by one of us (M. R.) in the *C. niloticus* specimen PIUM S11. At the present state of knowledge we cannot say whether this is an individual variation of the basioccipital only of *C. niloticus* or also of other species of *Crocodylus*.

Taxonomic Conclusions on Skull Fragment AUH 32 (fig. 14.3.1a–1c)

AUH 32 differs from *C. niloticus* only in the massiveness of the processus postoccipitalis. Hence an identification of the specimen as *C. niloticus* is impossible, because a fossil specimen of *C. cataphractus* is similar to *C. niloticus* with respect to the proportions of the skull table and foramina supratemporales. The *C. cataphractus* specimen, like AUH 32, has a massive processus postoccipitalis. The processus postoccipitalis in *C. lloydi* and *C. pigotti* is much stronger compared with that seen in AUH 32 and among the extant *Crocodylus* species the processus is small. Because of the lack of dental characters of AUH 32 we can assign the specimen only to *Crocodylus* sp. indet.

Summary of AUH 285, AUH 616, and AUH 32

All three fragments referred to *Crocodylus* share many similarities with *C. niloticus*. The current state of knowledge concerning the ecology of recent crocodilians indicates that one would not expect to find three different constructionally identical crocodilian species in the same type of habitat (Cott, 1961; Gorzula, 1978; Rauhe, 1990, 1995; Webb et al., 1982, 1983). It is much more likely that all three specimens come from a single species, which, in terms of morphology, is strikingly similar to *C. niloticus* but clearly different from *C. pigotti* and *C. lloydi*. If more-complete future finds reveal additional diagnostically relevant characters to confirm the identification, the Baynunah *Crocodylus* would be the oldest-known record of *C. niloticus*.

Specimen AUH 281—right jugal fragment (fig. 14.3.2a–2b)

The specimen is 55.5 mm long. At its base the postorbital bar is 8.2 mm wide, at its dorsal end the bar is 4.4 mm wide. The highest point of the jugal, 10.7 mm anterior to the postorbital bar, is 25 mm; the lowest point, 10.5 mm posterior to the postorbital bar, is 16 mm. In lateral view the dorsal border of the jugal is slightly sinusoidal and the margo dorsalis in the area of the postorbital bar is not set off by a step from the margo dorsalis of the posterior bar. The margo dorsalis of the jugal gets narrower in anterior direction from 3.3 mm (3.5 mm anterior to posterior end of the infratemporal fenestra) to 1.1 mm (17.6 mm anterior to the postorbital bar). The margo dorsalis posterior to the postorbital bar is flat with rounded medial and lateral edges but the anterior 28 mm of the margo dorsalis of the jugal is distinctly convex and ridge-shaped.

On the facies medialis of the jugal there are three apertures: two small circular foramina jugales interna posteriores of diameter 1.0 mm and 0.8 mm. They are 3.3 mm apart from one another. The most anterior one is situated at the level of the base of the postorbital bar; the second one lies 2.2 mm ventral from the former foramen and 4.8 mm posterior to the base of the postorbital bar.

The elliptical foramen jugalis internum anterior is 6 mm long and 3.6 mm high. It lies 8 mm anterior to the base of the postorbital bar and 5.7 mm lateral to the deepest point of the postorbital bar. It is 8.3 mm ventral to the margo dorsalis of the jugal. The thickness of the infratemporal bar of the jugal is maximum 7.4 mm: its facies medialis is convex.

The facies lateralis of the jugal is sculptured mostly with large (between 2.7 mm and 10.8 mm) mainly irregular pits. They are spaced close together separated by strong thick convex ridges 0.7–1.9 mm thick. The pits are maximum 2 mm deep.

Taxonomic Conclusions on the Jugal Fragment AUH 281 (fig. 14.3.2a–2b)

The large foramen jugalis internum anterior is, among other characters, characteristic for the Crocodylidae but not for the Alligatoridae. The position of the postorbital bar at the facies medialis of the jugal excludes *Gavialis*. Thus we determine AUH 281 as *Crocodylus* sp. indet.

Osteology and Comparative Anatomy of the Taxonomically Relevant Specimens ?*Ikanogavialis* and Gavialidae gen. et sp. indet.

Specimen AUH 56—A Right Mandibular Fragment (fig. 14.4.1a–1c and table 14.3)

This specimen is 140 mm long and is from the anterior part of a lower jaw. Nine anterior alveoli are preserved and the symphysis extends posteriorly to the end of the fragment and presumably would have extended beyond alveolus 9 in life.

All teeth are conical and pointed. The surface of the tooth crown is cannulated with apicobasally orientated ridges. Of all the ridges on the tooth crown the lingual and labial ridges are the most prominent. None of the tooth crown ridges is serrated.

Figure 14.4. ?*Ikanogavialis.* Fragment of a right dentary, AUH 56: *1a,* lateral view; *1b,* dorso-lateral view; *1c,* medial view. Scale bar = 20 mm.

The facies symphysalis is vertically orientated and sculptured with anteroposteriorly running ridges.

All the alveoli are laterally directed with the exception of the fourth alveolus. The apertures of nearly all the alveoli lie therefore within the facies lateralis and not the facies dorsalis. Alveolus 4 faces more dorsally being orientated at an angle of about 30° to the transverse plane. Within the dentary the alveoli run linguoposteriorly in a slightly concave curve. Alveolus 1 is procumbent. All apertures of the alveoli are nearly circular. The edges of the alveoli, especially the ventral one, extend outwards as low collars surrounding the aperture. The interalveolar spaces are concave, the degree of concavity increasing posteriorly. The margo lateralis is at its most concave between alveoli 8 and 9 of the dentary fragment.

Interalveolar space 2 is the largest (13.2 mm) and interalveolar space 1 the smallest (about 2.5 mm). The height of the dentary at the second alveolus is 11.5 mm and at the ninth alveolus 12.5 mm.

The anterior tip of the dentary possessing the first three alveoli is laterally expanded. The dentary fragment is narrowest in the region between aveoli 3 and 4 and 4 and 5. Posteriorly the width of the dentary fragment slightly increases to 16 mm between alveoli 8 and 9.

A row of foramina nutricia lies lingual to the alveoli. The remaining lingual part of the facies dorsalis of the dentary is smooth and plane.

The sculpturing of the facies ventralis of the dentary is weak and consists only of low sulci and some irregularly arranged foramina nutricia. The sulci of the facies ventralis are anteroposteriorly orientated. At the anterior end of the facies ventralis of the dentary there are only foramina nutricia.

Measurements for comparison with *Charactosuchus* Kugleri (see Berg, 1969: 734) are as follows:

1. The width of the mandible at the level of alveolus 2 is about 20 mm.
2. The width of the mandible at the level of alveolus 4 is 16.9 mm.
3. The width of the mandible at the level of alveolus 6 is 18 mm.
4. The width of the mandible between alveoli 3 and 4 is 15.6 mm.

Table 14.3. Measurements (in mm) of the dentary of ?*Ikanogavialis*, AUH 56

Alveoli	1	2	3	4	5	6	7	8	9
Linguolabial diameter	8.0	7.1	6.3	7.8	6.1	6.2	6.1	6.1	6.7
Mesiodistal diameter	12.5	6.7	6.6	6.9	5.0	5.5	5.2	5.7	5.8
Interalveolar space	2.5	13.2	12.2	8.0	7.6	8.9	8.5	9.7	?
Compression index	1.55	0.94	1.05	0.89	0.81	0.89	0.85	0.93	0.86

Comparison of Euthecodon *(Tchernov and Van Couvering, 1978: 865, textfig. 2; Aoki, 1992: 85, pl. 4, fig. 9; Molnar, 1982: 678, figs 3A–B) with AUH 56* (table 14.3 and fig. 14.4.1a–1c)

Differences of AUH 56 from Euthecodon

—In AUH 56 the collars of the alveoli are flatter in comparison, consequently the concavity of the interalveolar spaces in general is shallower than in *Euthecodon*. The concavities of the interalveolar spaces increase posteriorly at least until alveolus 8 (whereas in *Euthecodon* they are of equal depth throughout the same section of the mandible).

—The alveoli are oriented not so much laterally as in *Euthecodon*; alveolus 4 is oriented semivertically, in *Euthecodon* nearly vertically.

—Because a full description of the lower jaw and a complete record of *Euthecodon* are missing a detailed comparison with *Euthecodon* is impossible.

Comparison of Ikanogavialis *(Pliocene, Venezuela; Pleistocene, Murua (Woodlark) Island, Salomon Islands (Sill, 1970: 155, fig. 3; Molnar, 1982: 678, figs 3A–B) and AUH 56* (table 14.3 and fig. 14.4.1a–1c)

Similarities

—The concavities of the interalveolar spaces increase posteriorly at least until alveolus 8.

—Alveolus 4 is oriented semivertically.

Differences between AUH 56 and Ikanogavialis

—Interalveolar space 1 is the smallest in AUH 56; in *Ikanogavialis* it is the largest (Sill, 1970: 155), but Sill's own drawing (1970: 156, fig. 3) contradicts his description so that an evaluation of this character by us is impossible.

—Interalveolar spaces 2 and 3 are the largest in AUH 56; in *Ikanogavialis* the smallest.

—The width at the level of alveoli 2 and 6 is nearly identical; in *Ikanogavialis* the width of the lower jaw at the level of alveolus 6 is distinctly wider than that of alveolus 2.

—The depth of the interalveolar spaces between alveoli 5 and 10 is relatively distinctly shallower in AUH 56.

Comparison of Charactosuchus *Langston, 1965 (Miocene, Columbia) (Berg, 1969: 731–35, pl. 1, figs A, B; Langston, 1965: 44, fig. 14) with AUH 56* (table 14.3 and fig. 14.4)

Differences of AUH 56 from Charactosuchus kugleri

—Alveoli 4–9 in AUH 56 are much more laterally orientated.

—The collars of the alveoli are flatter, and hence the concavity of the interalveolar spaces in general is, at least until alveolus 5, much shallower.

—No longitudinal furrow-like depressions are present on the dorsal surface of the jaw parallel to and between alveolus 4 and 9 (Berg, 1969: 733, pl. 1b).

—Interalveolar space 1 is relatively much smaller; interalveolar space 4 is relatively much smaller than those between 2 and 4; in *Charactosuchus kugleri* the conditions are the other way round (Berg, 1969: 732, textfig., pl. 1b).

—Alveoli 2 and 3 are almost equal in size to each other, alveolus 5 is the smallest; in *Charactosuchus kugleri* alveolus 3 is much smaller than alveolus 2, alveoli 3 and 9 are the smallest (Berg, 1969: 735).

—The lower jaw is only slightly smaller between alveoli 3 and 4 than at the level of the sixth; in *Charactosuchus kugleri* the lower jaw is distinctly wider at the level of alveolus 6 than between 3 and 4.

Taxonomic Conclusions on Mandibular Fragment AUH 56 (table 14.3 and fig. 14.4.1a–1c)

The specimen is too fragmentary for clear identification at the generic level. The extreme lateral orientation of the teeth and deep concavity of the interalveolar spaces do not allow the specimen to be assigned to *Gavialis*, *Tomistoma*, or *Crocodylus*. These characters are shared only by *Ikanogavialis* (?Gavialidae: Steel, 1973; Gavialidae: Kraus, 1996), *Charactosuchus* (Tomistominae: Berg, 1969), and *Euthecodon* (Thoracosaurinae: Steel, 1973; Tomistomidae: Tchernov, 1986), but the gradual posteriorly increasing depth of the interalveolar spaces (at least until alveolus 8) is at present known only from *Ikanogavialis.* With respect to interalveolar space 2, however, the width of the lower jaw at the level of alveoli 2 and 6, and the relatively deeper interalveolar spaces 5, 6, 7, 8, and 9 *Ikanogavialis* is quite different from AUH 56. There are no other similarities, as mentioned above, between *Charactosuchus, Euthecodon,* and AUH 56. We therefore feel it reasonable to refer the specimen provisionally to ?*Ikanogavialis* (see also under AUH 334).

Specimen AUH 334—Skull Table and Occiput (figs 14.5.1a–1d)

Specimen AUH 334 is strongly weathered with a nearly vertically orientated facies posterior 85 mm in height. The margines laterales of the basioccipital are strongly bent outwards. These lateral regions of the basioccipital are a maximum 11 mm thick and are heavily sculptured with ridges. The maximum distance between the margines laterales of the concave facies posterior of the basioccipital is 52 mm. The margo ventralis of the basioccipital is sinusoidal in shape with a median concavity. Ventrally the processus basioccipitales are separated by a shallow depression 9.5 mm wide, anterior to which lies the 4 mm-wide and 5.3 mm-long eustachian canal. The foramen hypophysio-basicranialis lies 8 mm anterior to the posterior end of the median point of the facies ventralis basioccipitalis. The foramen hypophysio-basicranialis leads into the 5 mm-wide canalis hypophysio-basicranialis, which runs anterodorsally at a low angle to the transverse plane.

The foramen magnum is a trigonally shaped aperture 24 mm broad and 15 mm high. The condylus occipitalis is semicircular: 27.5 mm wide and 19.5 mm high. The posterior processes of the exoccipitals are 12 mm wide and reach the condylus occipitalis. The three large foramina of each exoccipital lie close together in the vicinity of the foramen magnum. They are located not higher than the mid-height of the foramen magnum. The largest of these foramina is the undivided foramen vagi, which is transversely elongated, 10 mm wide and 4 mm high. The foramen hypoglossale lies 2 mm dorsal to the foramen vagi; 0.4 mm ventral to the foramen vagi lies the foramen caroticum posterius. The canalis hypoglossale is connected with the foramen vagi by a small circular canal located on the medial wall of the canalis hypoglossale. Of the three occipital foramina the foramen vagi is closest to the foramen magnum; it is situated 5.2 mm lateral to it. The greatest distance between the occipital foramina is 11 mm between the canalis hypoglossale and the foramen caroticum posterius. Both foramina are nearly circular and about 3 mm in diameter.

The facies posterior of the supraoccipital does not reach the foramen magnum and is triangular. The cristae posttemporales are not covered by the skull table and are, like the median part of the supraoccipital, expanded posteriorly. Both foramina posttemporales are slit-like openings with a minimum width of 19 mm. Only an 11 mm-wide and 4 mm-long part of the supraoccipital participates in forming the skull table. The facies posterior of the processus paroccipitales of the exoccipitals is only slightly concave.

The minimum distance between the foramina supratemporales externa is 13 mm. Both foramina supratemporales externa are about 40 mm long and lie 12 mm anterior to the margo posterior of the skull table. The smooth facies laterales of the parietal are anteroposteriorly concave, slope laterally, and are not overlain by the facies dorsalis of the parietal.

The nearly circular foramen ovale lies in the lateral wall of the braincase.

The part of the facies posterior dorsal to the foramen magnum is 32 mm high, and the part of the facies posterior ventral to the condylus occipitalis is 21 mm high. The margo lateroventralis of the condylus occipitalis is semicircular; its margo dorsalis is sinusoidal with a median concavity.

Comparisons of Posterior Skull Fragment AUH 334 (figs 14.5.1a–1d)

The skull fragment is in many respects similar to the same portion of the skull in *Gavialis,* the yet-unnamed SMNK Pal 1282 (gen. et. sp. nov.; sp. nov. at present under study by R. Kraus, Mainz) and *Ikanogavialis:* the margines laterales of the basioccipital are laterally strongly convex and protrude laterally and ventroposteriorly as massive processus basioccipitales. The processus basioccipitales at their ventral midline are separated by a depression in front of which lies the eustachian canal. The cristae posttemporales are not overlapped by the skull roof as it is in the case of *Euthecodon, Gavialosuchus americanus* Sellards, 1915, *Tomistoma africana, Tomistoma gavialoides,* and the recent species of the Crocodilia. The foramina supratemporales are extremely large (and form in AUH 334 about 85% of the length of the skull table). According to Kälin (1933: 613, 645) and Hecht and Malone (1972: 282, table 1), the above-listed similarities are characteristic for a gavialid skull.

Similarities between Gavialis *and AUH 334*

—In ventral view the processus basioccipitales form two lateral rhomboids.

—The dorsal, rectangular supraoccipital forms only a very small part of the skull table and it is wedged into the parietal. The median processus postoccipitalis is formed by the dorsal part of the supraoccipital.

—Nearly the same arrangement, relative size, and shape of foramina occipitales as figured by Langston (1965: 22, fig. 6b) for "*Gavialis*" *colombianus* (= *Gryposuchus colombianus* sensu Buffetaut (1982b), Steel (1989)—see Iordansky, 1973: 216, fig. 5D = *Gavialis gangeticus*), juvenile specimen of *Gavialis gangeticus* SMNK 486 (see this chapter).

—The occipital part of the supraoccipital is orientated obliquely anteroventrally.

Differences of AUH 334 from Gavialis gangeticus

—The height of the condylus occipitalis in AUH 334 is only one-third larger than the height of the foramen magnum. In this feature the specimen resembles recent *Crocodylus* species. In *Gavialis gangeticus* the height of the condylus occipitalis is at least twice as large as the height of the foramen magnum.

—The frontoparietal suture in AUH 334 seems not to participate in forming the supratemporal foramina; at best the suture barely contacts the supratemporal foramina—but the same condition is described by Langston (1965: 22, fig. 6a) for "*G.*" *colombianus.* According to Kälin (1933) this feature varies between individuals in extant crocodilians.

—The processus basioccipitales in AUH 334 are more robust and diverge more posterolaterally.

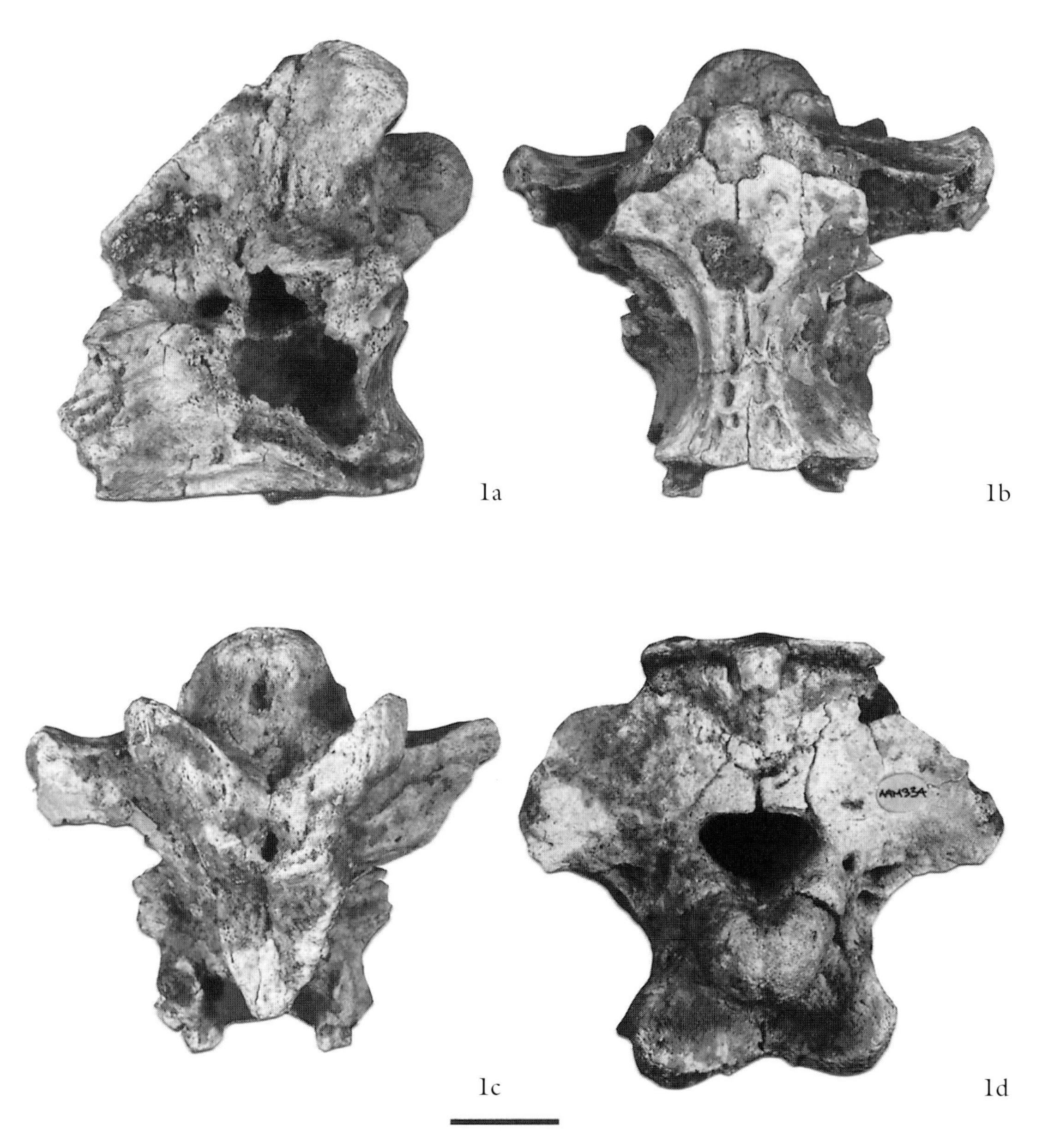

Figure 14.5. Gavialidae gen. et sp. indet. A strongly weathered posterior fragment of a skull table, AUH 334: *1a,* lateral view; *1b,* dorsal view; *1c,* ventral view; *1d,* occipital view. Scale bar = 20 mm.

Differences of Ikanogavialis *from AUH 334 (Sill, 1970: 152, fig. 1; 153, pl. 1, fig. B; 154)*

—In *Ikanogavialis* the supraoccipital is not incorporated into the skull table.
—The processus postoccipitalis is small and acute.
—The condylus occipitalis is twice as large as the height of the foramen magnum.
—The medial margin of the foramina supratemporales is much more concave.

Taxonomic Conclusions on Skull Fragment AUH 334

The presence of a cristae posttemporales (= processus postoccipitales sensu Kälin, 1933), which is not overlapped by the skull table, and the laterally and ventroposteriorally pointing terminal processus of the basioccipital, which reach ventrally the posterior end of the pterygoid, confirm that AUH 334 must be assigned to the Gavialidae. The lack of further diagnostic criteria allows only for identification at the family level: Gavialidae gen. et sp. indet.

Summary of AUH 334 and AUH 56

It seems highly likely that the two fragments AUH 334 and AUH 56 come from a single gavialid species because none of the extant longirostrine crocodilians shares their habitat with another species. The posterior part of the skull of *Ikanogavialis* is insufficiently described and figured, however, so that the comparison must be superficial. The differences in the construction and shape of the margo posterior of the skull table, the relative size of the condylus occipitalis, and the shape of the medial border of the foramina supratemporales between *Ikanogavialis* and AUH 334 indicate against a reference to *Ikanogavialis.*

The similarities with *Gavialis* are striking but we cannot confidentally refer AUH 334 to this genus, principally because the features are probably present in non-*Gavialis* forms of the family Gavialidae. Fossil crocodilian remains referred to *Gavialis* that are both geographically and temporally close to the Abu Dhabi form are from an indeterminable species from the Upper Miocene of Iraq, Bakhtiari Formation (Steel, 1989: 136).

Most of the South American gavialid material is very fragmentary and/or insufficiently described or figured. Thus the South American forms cannot be identified with confidence as belonging to *Gavialis* or even to the Gavialidae. If, however, AUH 334 and AUH 56 belong to a single species (which we believe), then these two fragments cannot be referred to *Gavialis* but to a new gavialid genus.

According to this definition the following fossil crocodilians belong to the family Gavialidae: "*Gavialis*" *colombianus, Ikanogavialis,* AUH 334, and SMNK Pal 1282 gen. et sp. nov. (Kraus, 1996). But, first, more detailed anatomical knowledge of the taxonomically questionable South American and North African Tertiary crocodilian material, referred until now to the Gavialidae or Tomistominae(idae), is needed to supply us with further diagnostic criteria for the family Gavialidae.

We also hope to get an answer to the question of whether the following features are restricted only to the genus *Gavialis:* position of the postorbital bar at the dorsal border of the jugal; smooth facies lateralis of the jugal; presence of bullae pterygoideae nariales; and exclusion of nasals from contact with the premaxillae.

We agree with Gürich (1912: 69) that forms morphologically intermediate between the recent Gavialidae and Crocodylidae (including *Tomistoma* and *C. cataphractus*) must be referred to a separate family and it is likely that *"Tomistoma" africana* and *"Tomistoma" gavialoides* could be members of such a family.

PALAEOECOLOGY OF THE ABU DHABI CROCODILES

During the late Tertiary crocodilians of the genera *Crocodylus* and *Gavialis* were much more widespread than today (Guggisberg, 1972; Neill, 1971; Ross et al., 1989; Steel, 1973, 1989; Tchernov, 1986). In the Baynunah Formation determinable remains of *Crocodylus* sp. are much more abundant than those of *Gavialis* sp. The size that can be reconstructed for *Crocodylus* specimens ranges from 1250 mm to 4000 mm maximum length, whereas for the gavialid crocodilian a size range of only 2500–3000 mm length is reported. The fact that no fossil hatchlings, egg shells, or nest remains occur in the Miocene of Abu Dhabi might be explained by the breeding habit found among similar extant crocodilians where nest sites are some distance from the river banks to avoid flooding.

Crocodylus remains from the Baynunah Formation are about three times as abundant as the remains of the gavialid crocodilian, but from their range in size both species appear to have had about the same potential for fossilisation. This need not only reflect a difference in abundance between *Crocodylus* and the gavialid, however, as it could be the result of different habitats.

The long and slender rostrum and the pointed teeth of the gavialid skull do not allow these animals to seize large prey with well-developed defensive abilities. For this reason even gharials as long as 7 metres can cope only with small prey and therefore almost exclusively feed on fish and sometimes small mammals and birds (Alderton, 1991; Daniel, 1983; Neill, 1971; Penny, 1991; Thorbjarnarson, 1990; Whitaker and Basu, 1983). An important dietary element of *Gavialis gangeticus* is the catfish *Clarias* sp. Compared with other crocodilians the extremely long and slender rostrum produces very little drag during a lateral predation stroke. The pointed head, the crests on the caudal margin of the fore limbs and hind limbs, and the short single crest make gharials excellent swimmers over distances but with a reduced manoeuvrability. The feeding habit and the swimming abilities require large, deep and open water bodies free of obstacles such as submerged plants or rocks. Such water bodies can be either lakes or slow- to fast-moving large rivers (Alderton, 1991; Basu, 1979; Bustard and Singh, 1977; Frey, 1982; Grenard, 1991; Neill, 1971: 711; Penny, 1991; Ross et al., 1989; Singh and Bustard, 1977; Steel, 1989; Trutnau, 1994; Thorbjarnarson, 1990; Whitaker and Basu, 1983). Because of the construction of its bracing system *Gavialis gangeticus* cannot use the high walk as a locomotion mode (Bustard and Singh, 1977; Frey, 1988a,b; Singh and Bustard, 1977). They therefore depend on low-gradient river banks to creep ashore for basking or nesting. Such river banks are flooded extensively even when the water level has not risen much, resulting in wide and sandy banks with only little perennial pioneer vegetation. Female gharials have to creep high upshore for safe nesting sites and therefore the breeding season must not fall within the rainy season (Alderton, 1991; Bustard, 1980; Maskey and Mishra, 1981; Maskey and Schleich, 1992; Penny, 1991; Ross et al., 1989; Singh and Bustard, 1977; Whitaker, 1987).

A similar life-style has to be reconstructed for the constructionally identical Baynunah gavialid. The constructional coincidence (not the taxonomical one!) is demonstrated by the morphology of rostrum, braincase, and osteoderms (Frey, 1982, 1988b). No matter in which environment the fossil gavialid remains were found, the animals were restricted to the deep river sections with low-lying sandy banks. *Clarias* remains are documented among other fish remains (Forey and Young, 1999—Chapter 12), indicating that the gavialid crocodilian had an appropriate food resource.

The *Crocodylus* remains are allocated to crocodilians constructionally identical to the robust forms of present-day *Crocodylus*, such as *Crocodylus niloticus* or *C. palustris*. With massive jaws and strong conical teeth these animals can kill and dismember large prey in addition to the fish and small animals that also make up their diet. The ability for high walking enables this ecotype to inhabit river sections with steep banks. In summary, the massive *Crocodylus* ecotype has more options to utilise habitat than the *Gavialis* ecotype (Alderton, 1991; Basu, 1979; Bustard and Singh, 1977; Cott, 1961; Daniel, 1983; Guggisberg, 1972; Johnson, 1973; Penny, 1991; Ross et al., 1989; Singh and Bustard, 1977; Webb and Manolis, 1989; Webb and Messel, 1978; Webb et al., 1983; Whitaker, 1987). This suggests that the fossil association does not reflect a trophic habitat partitioning.

The co-occurrence of gavialids and crocodilians of the robust *Crocodylus* type in the same biotope is known from modern times: in India, *Crocodylus palustris* and *Gavialis gangeticus* occur sympatrically (Basu, 1979; Daniel, 1983; Trutnau, 1994). Such a co-occurrence is stable, however, only because these two species live in different river systems. Furthermore, they are different in their breeding/mating seasons and feeding habits.

An encounter between a *Gavialis* and a *Crocodylus*-type crocodilian in most cases would be fatal for *Gavialis*. *Gavialis, Tomistoma,* and even less-longirostrine crocodilians such as *Crocodylus cataphractus* and *Crocodylus johnsoni* inhabit river systems where no other crocodile forms live (Cott, 1961; Galdikas and Yaeger, 1984; Jelden, 1985;

Johnson, 1973; Ross et al., 1989; Taylor, 1970; Waitkuwait, 1981; Webb and Manolis, 1989; Webb and Messel, 1978; Webb et al., 1983; Whitaker, 1980; Whitaker and Basu, 1983). Further examples are known where recent crocodile species occur sympatrically: *Caiman crocodilus* and both *Paleosuchus* species (Rio Apaporis, Colombia), *Crocodylus porosus* and *Crocodylus johnsoni* (Adelaide River, Northern Territory, Australia), and *Alligator mississippiensis* and *Crocodylus acutus* (Arch Creek close to the estuary into the Biscayne Bay, southeast coast of Florida).

In all these cases the different crocodile forms live in different ecological habitats: whereas *Crocodylus porosus* is a typical inhabitant of coastal brackish waters, *Crocodylus johnsoni* lives in inland freshwaters; whereas *Alligator mississippiensis* inhabits nearly all freshwaters in Florida, *Crocodylus acutus* lives only in the coastal region of South Florida. In some recent ecosystems it is known that such encounters between two different crocodilian forms can result in the breakdown of the population of the physically weaker form (Cott, 1961; Jelden, 1985; Johnson, 1973; Medem, 1971; Otte, 1978; Webb and Manolis, 1989; Webb and Messel, 1978; Webb et al., 1983; Vanzolini and Gomez, 1979).

Geographically only four extant crocodilian species co-exist: in the Amazonian Basin, the largest crocodilian ecosystem (7 million square kilometres, one-fifth of the world's freshwater sources). It is important to repeat, however, that such a co-existence can be stable only when the species live in different habitats and/or have different breeding seasons or nesting sites (Cott, 1961; Jelden, 1985; Johnson, 1973; Neill, 1971; Webb and Manolis, 1989; Webb and Messel, 1978; Webb et al., 1983; Whitaker, 1980; Vanzolina and Gomez, 1979). The social and, especially, the territorial behaviour may be useful for avoiding fatal intraspecific interactions.

For the Miocene Baynunah river system the coexistence of *Gavialis* and *Crocodylus* ecotypes could be explained (1) by habitat splitting and territorial control or (2) by time-condensing of the fossilised remains by reworking. The co-occurrence of the two crocodilian ecotypes in the Baynunah river system remains a problem. It might be explained (1) by a territorial behaviour not existing in recent crocodilian communities or (2) by time-averaging mixing two distinct assemblages. Taphonomic data suggest little reworking and support the co-existence of the crocodilians. We therefore prefer the former explanation.

ACKNOWLEDGEMENTS

We present our sincere thanks to Peter Whybrow (The Natural History Museum, London) for loaning us the crocodile material from Abu Dhabi, for references to the newest literature on the geology of the Baynunah Formation, and for the improvements to our English manuscript. We also thank Alexandra Anders (Staatliches Museum für Naturkunde, Karlsruhe) for her preparation work and Colin M. Wight for translation of the *Ikanogavialis* paper. The photographs were taken by D. F. and D. P., with developing and printing by Volker Griener (SMNK).

REFERENCES

Alderton, D. 1991. *Crocodiles and Alligators of the World.* Blandford, London.

Andrews, C. W. 1901. Preliminary note on some recently discovered extinct vertebrates from Egypt. *Geological Magazine,* London 8: 436–44.

———. 1905. Notes on some new crocodilia from the Eocene of Egypt. *Geological Magazine, London* 11: 481–84.

———. 1906. *A Descriptive Catalogue of the Tertiary Vertebrates of the Fayum (Egypt).* Trustees of the British Museum (Natural History), London.

Aoki, R. 1992. Fossil crocodilians from the Late Tertiary strata in Sinda Basin, Eastern Zaire. *African Study Monographs,* suppl. 17: 67–85.

Arambourg, C. 1947. Contribution à l'étude géologique et paléontologique du bassin du lac

Rodolphe et de la basse vallée de l'Omo. 2, Paléontologie. In *Mission scientifique de l'Omo, 1932–1933,* vol. 1. *Géologie—Anthropologie,* fasc. 3, pp. 5–562. Muséum National d'Histoire Naturelle, Paris.

Basu, D. 1979. Indien kämpft um den Ganges-Gavial. *Sielmanns Tierwelt* 3: 4–13.

Berg. D.E. 1966. Die Krokodile, insbesondere *Asiatosuchus* und aff. *Sebecus* aus dem Eozän von Messel bei Darmstadt. *Abhandlungen des Hessischen Landes-Amtes für Bodenforschung* 52: 1–105.

———. 1969. *Charactosuchus kugleri,* eine neue Krokodilart aus dem Eozän von Jamaica. *Eclogae Geologicae Helvetiae* 62: 31–735.

Boulenger, G. 1920. Sur le gavial fossile de l'Omo. *Comptes Rendus de l'Académie des Sciences, Paris* 170: 1–914.

Buffetaut, E. 1982a. Crocodylia. In *The Lower Miocene fauna of Al-Sarrar (Eastern Province, Saudi Arabia).* ATLAL, *The Journal of Saudi Arabian Archaeology* 5: 109–36.

——— 1982b. Systématique, origine et évolution des Gavialidae Sud-Américains. *Geobios, Memoirs Spéciale* 6: 127–40.

———. 1984. On the occurrence of *Crocodylus pigotti* in the Miocene of Saudi Arabia, with remarks on the origin of the Nile crocodile. *Neues Jahrbuch für Geologie und Paläontologie, Monatshefte* 1984: 513–20.

Bustard, H. R. 1980. A note on the nesting behaviour in the Indian Gharial, *Gavialis gangeticus. Journal of the Bombay Natural History Society* 77: 514–15.

Bustard, H. R., and Singh, L. A. K. 1977. Studies on the Indian gharial *Gavialis gangeticus* (Gmelin) (Reptilia, Crocodylia): Change in terrestrial locomotory pattern with age. *Journal of the Bombay Natural History Society* 74: 534–35.

Cott, H. B. 1961. Scientific results of an enquiry into the ecology and economic status of the Nile crocodile (*Crocodylus niloticus*) in Uganda and Northern Rhodesia. *Transactions of the Zoological Society of London 29:* 211–356.

Daniel, J. C. 1983. *The Book of Indian Reptiles.* Bombay.

Forey, P. L., and Young, S. V. T. 1999. Late Miocene fishes of the Emirate of Abu Dhabi, United Arab Emirates. Chap. 12 in *Fossil Vertebrates of Arabia,* pp. 120–35 (ed. P. J. Whybrow and A. Hill). Yale University Press, New Haven.

Fourteau, R. 1920. *Contribution à l'étude des vertébrés miocènes de l'Egypte.* Geological Survey of Egypt, Cairo.

Frey, E. 1982. Der Bau des Bewegungsapparates der Krokodile und seine Funktion bei der aquatischen Fortbewegung. Diploma thesis, Fakultät für Biologie der Universität Tübingen.

———. 1988a. Anatomie des Körperstammes von *Alligator mississippiensis* Daudin. *Stuttgarter Beiträge zur Naturkunde, ser. A,* 424: 1–106.

———. 1988b. Das Tragesystem der Krokodile—eine biomechanische und phylogenetische Analyse. *Stuttgarter Beiträge zur Naturkunde, ser. A,* 426: 1–60.

Frey, E., Laemmert, A., and Riess, J. 1987. *Baryphracta deponiae* n.g. n.sp. (Reptilia, Crocodylia), ein neues Krokodil aus der Grube Messel bei Darmstadt (Hessen, Bundesrepublik Deutschland). *Neues Jahrbuch für Geologie und Paläontologie, Monatshefte* 1: 15–26.

Friend, P. F. 1999. Rivers of the Lower Baynunah Formation, Emirate of Abu Dhabi, United Arab Emirates. Chap. 5 in *Fossil Vertebrates of Arabia,* pp. 38–49 (ed. P. J. Whybrow and A. Hill). Yale University Press, New Haven.

Galdikas, B. M. F., and Yaeger, C. P. 1984. Crocodile predation on a crab eating macaque in Borneo. *American Journal of Primatology* 6: 49–51.

Ginsburg, L., and Buffetaut, E. 1978. *Euthecodon arambourgi* n. sp. et l'évolution du genre *Euthecodon,* crocodilien du Néogène d'Afrique. *Géologique Méditerrané* 5: 291–302.

Gorzula, S. G. 1978. An ecological study of *Caiman crocodilus crocodilus* inhabiting Lagoons in the Venezuelan Guayana. *Oecologica* 35: 21–34.

Grenard, S. 1991. *Handbook of Alligators and Crocodiles.* Krieger, Malabar, Fla.

Gürich, G. 1912. *Gryposuchus jessei,* ein neues schmalschnauziges Krokodil aus den jüngeren Ablagerungen des oberen Amazonas-Gebietes. *Jahrbuch der Hamburgischen Wissenschaftlichen Anstalten* 29: 59–71.

Guggisberg, C. A. W. 1972. *Crocodiles: Their Natural History, Folklore and Conservation.* David and Charles, Newton Abbot.

Hecht, T. M. K., and Malone, B. 1972. On the early history of the gavialid crocodilians. *Herpetologica* 28: 281–84.

Iordansky, N. N. 1973. The skull of the Crocodilia. In *Biology of the Reptilia,* vol. 4, pp. 201–62 (ed. C. Gans and T. S. Pearson). Academic Press,London.

Jelden, D. 1985. Brutbiologie und Ökologie von *Crocodylus porosus* und *Crocodylus n. novaeguineae* am mittleren sepik (Papua Neuguinea). *Stuttgarter Beiträge zur Naturkunde,* ser. A, 378: 1–32.

Jesus, de M., Fuentes, E. J., Fincias, B., del Prado, J. M., and Alanso, E. M. 1987. Los crocodylia del Eoceno y Oliogoceno del la cuenca del Duero. Dientes y osteoderms. *Revista Española de Paleontología* 2: 95–108.

Johnson, C. R. 1973. Behaviour of the Australian crocodiles, *Crocodylus johnsoni* and *Crocodylus porosus. Journal of the Linnean Society* 52: 315–36.

Joleaud, L. 1920. Sur la présence d'un Gavialidae du genre *Tomistoma* dans le Pliocène de l'eau douce de l'Ethiopie. *Comptes Rendus de l'Académie des Sciences, Paris* 70: 816–18.

Kälin, J. A. 1933. Beiträge zur vergleichenden Osteologie des Crocodilidenschädels. *Zoologische Jahrbucher (Abteilung Anatomie und Ontogenie)* 57: 535–714.

Kraus, R. 1996. *Piscgavialis jugaliperforatus* n. gen. n. sp. ein neuer Gavialide aus der Pisco-Formation (oberes Miozän, Peru). Diploma thesis, University of Mainz.

Laemmert, A. 1990. Körperpanzerung, die ersten drei Halswirbel und die ersten beiden Halsrippen der Messeler Diplocynodonten (Crocodylia, Eozän, Messel). Diploma thesis, Institut für Geologie und Paläontologie, Tübingen.

Laemmert, A. 1993. Dorsal and ventral armour and various positions of embedding in *Diplocynodon* (Crocodilia). *Kaupia (Darmstädter Beiträge zur Naturgeschichte)* 3: 35–40.

Langston, W. 1965. Fossil crocodilians from Colombia and the Cenozoic history of the Crocodilia in South America. *Publications of the University of California* 52: 1–157.

Maccagno, A. M. 1947. Descrizione di una nuova specie di *Crocodilus* del giacimento di Sahabi (Sirtica). *Atti della Reale Accademia* (*Nazionale*) *dei Lincei. Memorie, Roma* 1: 61–95.

Maskey, T. M., and Mishra, H. R. 1981. Conservation of gharial (*Gavialis gangeticus*) in Nepal. In *Wild is Beautiful,* pp. 185–196 (ed. T. Ch Majupuria). Bangok.

Maskey, T. M., and Schleich, H. H. 1992. Untersuchungen und Schutzmaβnahmen zum Ganges gavial in Südnepal. *Natur und Museum* 122: 258–67.

Medem, F. 1971. Biological isolation of sympatric species of South American Crocodilia. *IUCN Publications, n.s., Suppl. Paper* 32: 152–58.

Molnar, R. E. 1982. A longirostrine crocodilian from Murua (Woodlark), Solomon Sea. *Memoirs of Queensland Museum* 20: 675–85.

Müller, L. 1927. Ergebnisse der Forschungsreisen Prof. E. Stromer in den Wüsten Ägyptens V. Tertiäre Wirbeltiere. 1. Beiträge zur Kenntnis der Krokodilier des ägyptischen Tertiärs. *Abhandlungen der Bayerischen Akademie der Wissenschaften (Naturwissenschaftliche Abteilung)* 31: 1–96.

Neill, W. T. 1971. *The Last Ruling Reptiles, Alligators, Crocodiles and their Kin.* Columbia University Press, New York.

Otte, K. C. 1978. *Untersuchungen zur Biologie des Mohrenkaimans (Melanosuchus niger Spix 1825) aus dem Nationalpark con Manu (Peru).* Munich.

Penny, M. 1991. *Alligators and Crocodiles.* Crescent Books, New York.

Rauhe, M. 1990. Habitat-Habitus-Wechselbeziehung von *Allognathosuchus gaudryi* Stefano 1905 (=*Allognathosuchus haupti* Weitzel 1935). *Geologisches Jahrbuch Hessen* 118: 53–61.

———. 1993. Postkranialskelett und Taxonomie des Alligatoriden *Allognathosuchus haupti* (Mitteleozän von Messel, Darmstadt) unter Berücksichtigung der Anatomie und Altersvariationen von *Allognathosuchus* cf. *haupti.* PhD thesis, University of Mainz.

———. 1995. Die Lebensweise und Ökologie der Geiseltal-Krokodilier-Abschied von traditionellen Lehrmeinungen. *Hallesche Jahrbuch für Geowissenschaften* B17: 65–80.

Rauhe, M., and Rossmann, T. 1995. News about fossil crocodiles of the Middle Eocene from Messel and Geiseltal. *Hallesche Jahrbuch für Geowissenschaften* B17: 81–92.

Ross, C. A., Garnett, S., and Pyrzakowski, T. 1989. *Crocodiles and Alligators.* Merehurst Press, London.

Sill, W. D. 1970. Nota preliminar sobre un nuevo Gavial dal Plioceno de Venezuela y una discusion de los Gaviales Sudamericanos. *Amerghiniana* 7: 151–59.

Singh, L. A. K., and Bustard, H. R. 1977. Locomotory behaviour during basking and spoor formation in the gharial (*Gavialis gangeticus*). *British Journal of Herpetology* 5: 673–76.

Steel, R. 1973. Part 16: Crocodilia. In *Handbuch der Paläoherpetologie* (ed. O. Kuhn). Gustav Fischer, Stuttgart.

——— 1989. *Crocodiles.* Christopher Helm, London.

Taylor, E. H. 1970. The turtles and crocodiles of Thailand and adjacent waters. *Bulletin of the University of Kansas* 49: 87–179.

Tchernov, E. 1986. *Evolution of the Crocodiles in East and North Africa.* Edition du Centre National de la Recherche Scientifique, Paris.

Tchernov, E., and Van Couvering, J. A. H. 1978. Crocodiles from the Early Miocene of Kenya. *Journal of Paleontology* 21: 857–67.

Tchernov, E., Ginsburg, L., Tassy, P., and Goldsmith, N. F. 1987. Miocene mammals of the Negev (Israel). *Journal of Vertebrate Paleontology* 7: 284–310.

Thomas, H., Sen, S., Khan, M., Battail, B., and Ligabue, G. 1982. The Lower Miocene fauna of Al-Sarrar (Eastern Province, Saudi Arabia). *ATLAL, The Journal of Saudi Arabian Archaeology* 5: 109–36.

Thorbjarnarson, J. B. 1990. Notes on feeding behavior of the gharial (*Gavialis gangeticus*) under seminatural conditions. *Journal of Herpetology* 24: 99–100.

Trutnau, L. 1994. *Krokodile—Alligatoren, Kaimane, Echte Krokodile und Gaviale,* Neue Brehm Bücherei 593. Westarp Wissenschaften, Magdeburg.

Vanzolini, P. E., and Gomez, N. 1979. Notes on the ecology and growth of Amazonian caimans. *Papeis Avulsos de Zologie* 32: 205–16.

Waitkuwait, E. 1981. Untersuchungen zur Brutbiologie des Panzerkrokodils (*Crocodylus cataphractus*) im Tai-Nationalpark in der Republik Elfenbeinküste. Diploma thesis, University of Heidelberg.

Webb, G. J., and Manolis, S. C. 1989. *Crocodiles of Australia.* Reed Books, New South Wales, Australia.

Webb, G. J., Manolis, S. C., and Buckworth, R. 1982. *Crocodylus johnstoni* in the McKinlay River area, N. T. I. Variation in the diet and a new method of assessing the relative importance of prey. *Australian Journal of Zoology* 30: 877–90.

———. 1983. *Crocodylus johnstoni* in the McKinlay River area, N. T. II. Dry season habitat selection and an estimate of the total population size. *Australian Journal of Zoology* 10: 373–82.

Webb, G. J., and Messel, H. 1978. Movement and dispersal pattern of *Crocodylus porosus* in some rivers of Arnhem Land, Northern Australia. *Australian Wildlife Research* 5: 263–83.

Wermuth, H. 1953. Systematik der rezenten Krokodile. *Mitteilungen aus dem Zoologischen Museum der Humboldt-Universität in Berlin* 29: 375–514.

Whitaker, R. 1980. *Status and Distribution of Crocodiles in Papua New Guinea.* Wildlife Division, Department of Lands and Environment, Papua New Guinea, FAO/UNDP field document, Port Moresby.

———. 1987. The management of crocodilians in India. In *Wildlife Management—Crocodiles and Alligators.* Chipping Norton, New South Wales, Australia.

Whitaker, R., and Basu, D. 1983. The gharial (*Gavialis gangeticus*): a review. *Journal of the Bombay Natural History Society* 79: 531–48.

Whybrow, P. J. 1989. New stratotype; the Baynunah Formation (Late Miocene), United Arab Emirates: Lithology and palaeontology. *Newsletters on Stratigraphy* 21: 1–9.

Whybrow, P. J., Hill, A., Yasin al-Tikriti, W., and Hailwood, E. A. 1990. Late Miocene primate fauna, flora and initial palaeomagnetic data from the Emirate of Abu Dhabi, United Arab Emirates. *Journal of Human Evolution* 19: 583–88.

A Late Miocene Insectivore and Rodent Fauna from the Baynunah Formation, Emirate of Abu Dhabi, United Arab Emirates

HANS DE BRUIJN

The small collection of rodent remains from the Baynunah Formation (Whybrow, 1989) made in 1992 (de Bruijn and Whybrow, 1994) was, with few exceptions, obtained by dry-screening sediment from the surface of blow-holes. Because of the technique used during the reconnaissance for small mammals the exact provenance of the specimens is unknown and the enamel is often corroded. The opportunity presented to revisit the Shuwaihat localities after participating in the First International Conference on the Fossil Vertebrates of Arabia in Abu Dhabi, March 1995 was therefore welcomed. During the second collecting expedition we restricted ourselves to the S1 and S4 sites at Shuwaihat, which presumably represent about the same lithostratigraphical horizon. After cleaning a section near the northern limit of the S1 outcrop two conglomerate beds separated by an erosional surface could be seen. Both these beds show cross-stratification and seem to have accumulated in river bars. The clasts of the lower bed are small caliche pebbles; those of the overlying bed are on average larger caliche pebbles and clay-balls. Three different lithologies—the lower conglomerate, the upper conglomerate minus the clay-balls, and the clay-balls—were collected separately and subsequently wet-screened after drying. A trench dug at about 30 metres to the south of this small outcrop revealed the presence of only one, much thinner conglomerate bed, which was also sampled. These four samples from site S1—named S1-low, S1-high, S1-balls, and S1-trench—all appeared to contain some rodent remains and fragments of fish bones, crocodile teeth, eggshell, internal casts of gastropods, and oogonia of Characaea.

A trench dug in the small escarpment of site S4 revealed the presence of one narrow fossiliferous conglomerate bed. The clasts of this bed are caliche pebbles and small indurated clay-balls.

The amount of sediment processed from S1-low, S1-high, S1-trench, and S4 is of the order of magnitude of about 200–250 kg per site; that from S1-balls is about 50 kg. Exact data are not available, because the sediment was dry-screened on a 0.5 mm mesh at the locality before wet-screening.

METHODS

The small-mammal material described below comprises the specimens collected in 1992 and 1995. Although the various sites are not strictly contemporaneous, the rodent material is treated as a sample from one community because there is no evidence that the fluvial part of the Baynunah Formation comprises a geologically long time interval.

Measurements of the teeth were made using a Leitz Ortholux measuring microscope with mechanical stage. All measurements are given in 0.1 mm units. The teeth figured are all ×40 and illustrated as if they are from the left side.

SYSTEMATICS

Thryonomyidae Pocock, 1922 (cane rats)
Thryonomyidae gen. et sp. indet. (fig. 15.1)

 ISBN 0-300-07183-3

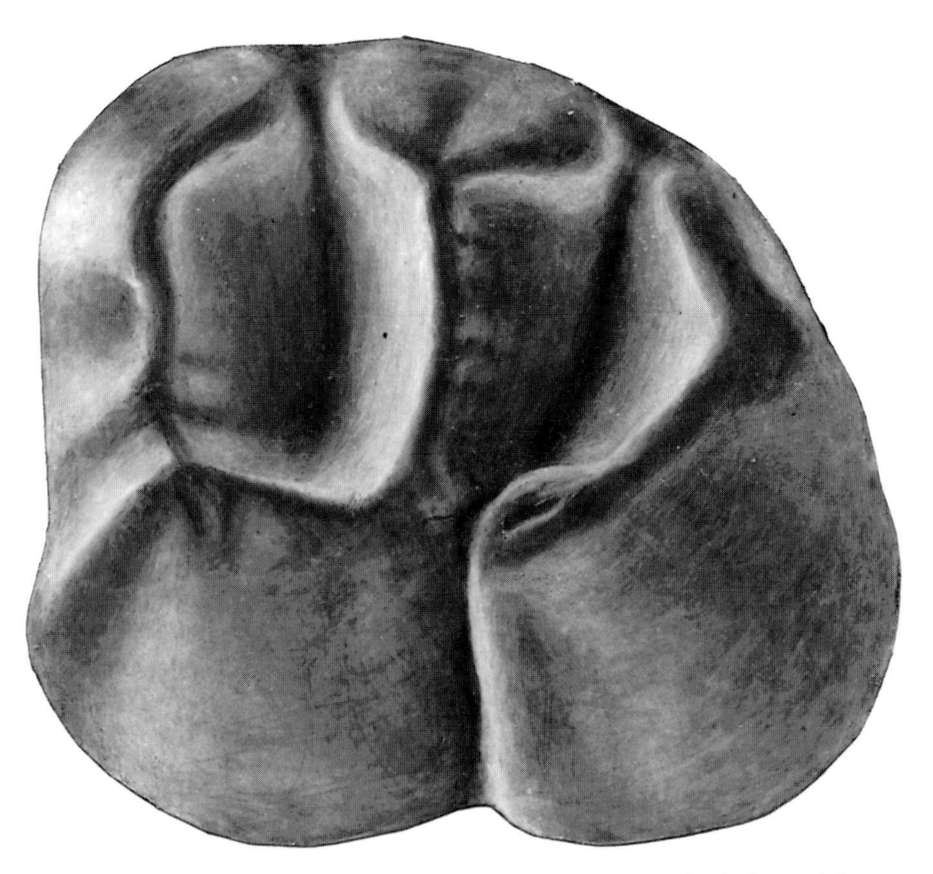

Figure 15.1. Thryonomyidae gen. et sp. indet. M_3 (AUH 571), figured from the side, ×40.

Material and Measurements

One M_3 (30.0 × 28.0), AUH 571 (site S4).

Remarks

The single tooth represents a rather small thryonomyid that shows close resemblance in size and morphology to the M_3 of *Neosciruomys stromeri* (Hopwood, 1929) and would fit well also with the dP_4–M_2 of the holotype of *Neosciuromys africanus* Stromer, 1922, both from the early Miocene of Namibia.

The morphology of the tooth from Abu Dhabi is rather modern relative to Miocene specimens from Kenya (Flynn et al., 1983) and from the Siwaliks of Pakistan (de Bruijn and Hussain, 1985) in having lost all traces of the anteroconid and the posterior arm of the protoconid. The M_3 is in fact very similar to its homologue in extant *Thryonomys,* but much smaller than *T. gregorianus* Thomas, 1894, the lesser cane rat.

Gerbillidae Alston, 1876 (sand rats)

Remarks

Since Jaeger (1977) suggested that the sand rats are derived Myocricetodontinae, this hypothesis has been endorsed by later studies (Tong, 1989; Tong and Jaeger, 1993). Chaline et al. (1977) recognise three subfamilies in the Gerbillidae: the Gerbillinae Alston, 1876; the Taterillinae Chaline et al. 1977;

and the Myocricetodontinae Lavocat, 1961. This arrangement was followed by Tong and Jaeger (1993), but was considered to be premature by Carleton and Musser (1984). Their major objections are the wide variety of dental patterns in the Myocricetodontinae and the absence of fossils with a dental pattern that is intermediate between unquestionable Myocricetodontinae and Gerbillinae. This last objection is no longer valid and it seems probable that the root of the Gerbillinae is in the Myocricetodontinae. I therefore retain the three subfamilies of Chaline et al. (1977) with the observation that the Gerbillinae and Taterillinae cannot be differentiated on the basis of fossils of which the ear region is unknown.

The contents of the subfamilies Myocricetodontinae and Gerbillinae plus Taterillinae differ considerably among authors. Whereas Tong (1989) places all fossil genera (*Mascaramys* excepted) in the Myocricetodontinae, de Bruijn et al. (1970), Sen (1977, 1983), Jaeger (1977), and others allocate the genera *Pseudomeriones* and *Protatera* to the Gerbillinae s.l. Although the classification of fossils that are known only by teeth remains necessarily speculative, I am inclined to include those species that were previously described as myocricetodontines, but that lack (1) the enterocone in the upper molars; (2) have very reduced longitudinal ridges; and (3) have more or less opposing cusps in the cheek teeth, as gerbils (= Gerbillinae and Taterillinae). That decision refers to *"Myocricetodon" magnus* Jaeger, 1977 and *"Myocricetodon" ultimus* Jaeger, 1977.

Gerbillinae s.l.
Abudhabia de Bruijn and Whybrow, 1994

Remarks

The additional material collected in 1995 shows that the original diagnosis and differential diagnosis of *Abudhabia* need slight changes.

Emended Diagnosis

Abudhabia is a medium-sized rodent with dental characters that are intermediate between the Gerbillinae and the Myocricetodontinae. The M_1 always has the posterior cingulum developed as an isolated cusp. The M^2 and M_2 have remnants of the anterior cingulum. Cusp-pairs of the M^1, M^2, and M_2 form transverse ridges. The M_1 has alternating main cusps and an anteroconid with a posterolabially directed crest as in most cricetids. The upper incisor has one longitudinal groove.

Differential Diagnosis

Abudhabia is most similar to *Taterillus* among the extant Gerbillinae. These two genera differ by the presence in *Abudhabia* of a central posterior cusp in the M_1 and of remnants of the anterior cingulum in the M^2 and M_2.

Abudhabia is among fossils most similar to the insufficiently known species *"Myocricetodon" magnus* from Pataniak (Morocco), *"Myocricetodon" ultimus* from Kendek-el-Quaich (Morocco) and *"Protatera"* sp. nov. (Munthe, 1982) from Sahabi, Libya. It is not impossible that some or all of these species will eventually be referable to *Abudhabia* as is the case with *"Protatera" kabulense* Sen, 1983.

Abudhabia baynunensis de Bruijn and Whybrow, 1994 (figs 15.2–15.5)

Material and Measurements (tables 15.1 and 15.2)

M^1, AUH 572, 591 (damaged) (site S1), AUH 573 (site S4); M^2, AUH 574, 576, 592, 593, 594 (damaged) (site S1); M^3, AUH 575 (site M1), AUH 595, 596; ?M^3, AUH 779 (site S1); M_1, AUH 566 holotype, AUH 567, 781 (fragment) (site S4), AUH 597, 598, 599 (damaged) (site S1); M_2, AUH 568, 600, 755, 756, 765 (damaged) (site S1), AUH 782 (site S4); M_3, AUH 569, 570, 766, 767, 775 (site S1), AUH 783 (site S4)

Remarks

With the exception of the M^1 (fig. 15.2) the material of *A. baynunensis* collected in 1995 is similar to the specimens described (de Bruijn and Whybrow, 1994). The stage of wear of the complete M^1 from site S1-low is somewhat more advanced than in the specimens figured on plate 1, figures 7 and 8 of de Bruijn and Whybrow (1994), which may be the reason that it does not show a posterior cingulum and has an incipient connection between the anterocone and the protocone. In both these respects this tooth

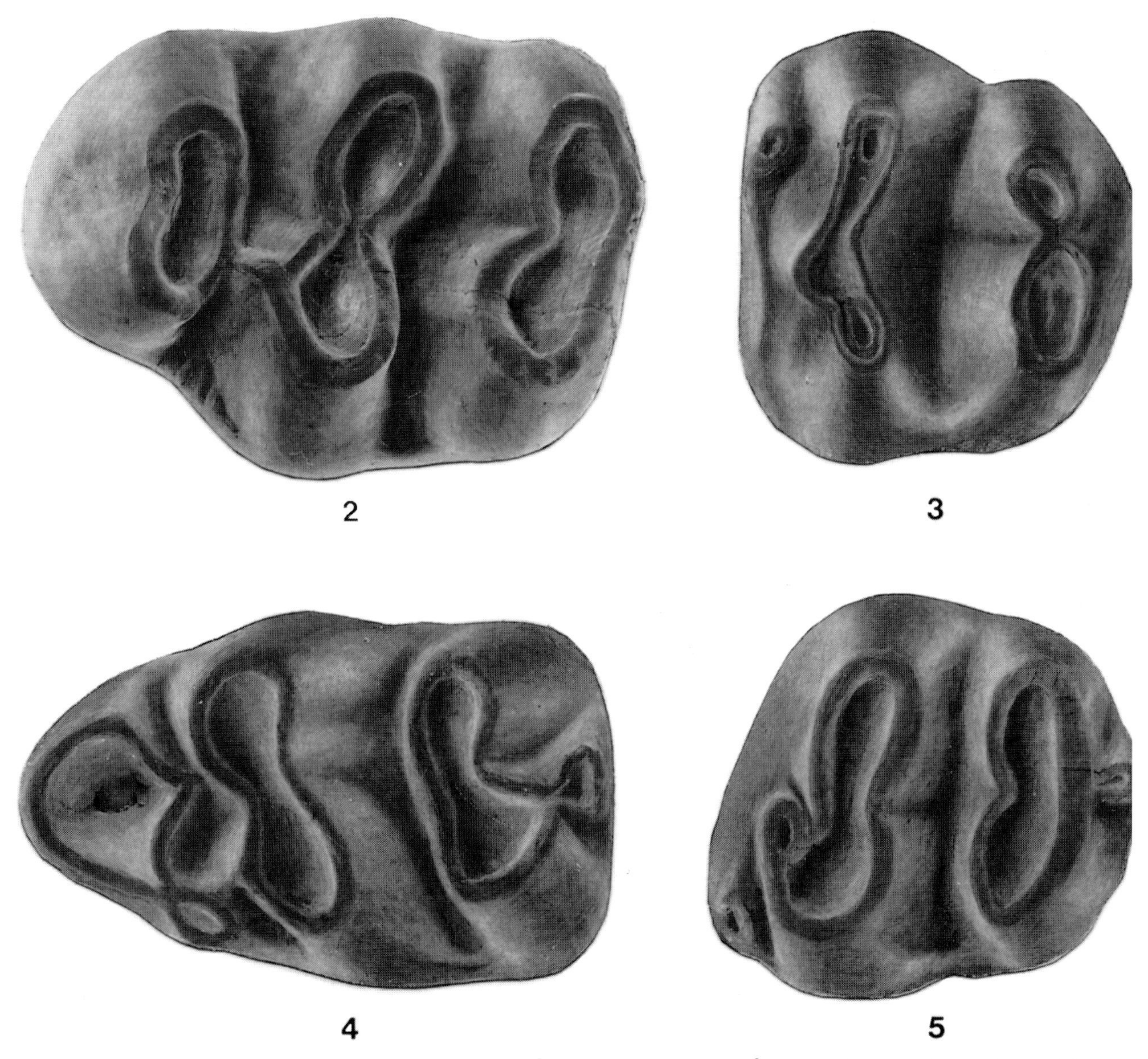

Figures 15.2–15.5. *Abudhabia baynunensis*. *15.2*, M^1, AUH 573; *15.3*, M^2, AUH 593; *15.4*, M_1, AUH 597; *15.5*, M_2, AUH 568. All ×40 and figured as left teeth regardless of which side the tooth really is. Figures 15.4 and 15.5 are from the right side.

Table 15.1. Summary of the small mammals collected from the Baynunah Formation by site and taxon; the numbers refer to almost complete cheek teeth from rodents and to molars and incisors for the Soricidae

Small mammal taxa	S1-undivided	S1-low	S1-high	S1-clay-balls	S1-trench	S4-undivided	S4-trench	M1	K1	Total
Thryonomyidae gen. et sp. indet.	—	—	—	—	—	1	—	—	—	1
Abudhabia baynunensis	5	7	2	1	2	2	2	1	—	22
Parapelomys cf. *charkensis*	2	—	—	—	1	1	—	2	—	6
Dendromus aff. *melanotus*	4	3	—	—	—	—	—	—	—	7
Dendromus sp.	1	1	1	—	—	—	—	—	—	3
Myocricetodon sp.	—	1	1	1	1	—	—	1	—	5
Zapodinae gen. et sp. indet.	—	—	1	—	—	—	—	—	—	1
Soricidae gen. et sp. indet.	—	1	—	—	—	—	—	—	2	3
Total from sites	12	13	5	2	4	4	2	4	2	48

Sites: S1 = Shuwaihat on the western flank of the eastern jebel; S4 = Shuwaihat on the eastern flank of the jebel known as Jebel Mershed (see Whybrow and Clements, 1999—Chapter 23), about 1 metre above site S6, the proboscidean excavation (see Tassy, 1999—Chapters 18 and 24; Andrews, 1999); M1 = Jebel Mimiyah (Al Mirfa); K1 = Kihal.

is different from the specimens collected earlier. The development of longitudinal connections at advanced stages of wear is common among Gerbillinae, but the presence of a posteromedial cusp in the M_1 is unexpected because this feature is lost at an early stage of evolution in the Myocricetodontinae, the alleged ancestral group of the Gerbillinae. The material available, however, is insufficient to be able to decide whether the posteromedial cusp is present in unworn M^1 and gets lost due to abrasion or whether the M^1 figured on plate 1, figure 8 of de Bruijn and Whybrow (1994) has an aberrant morphology.

Table 15.2. Dental measurements of *Abudhabia baynunensis* de Bruijn and Whybrow, 1994 (units = 0.1 mm)

	Length				Width		
	Min.	Max.	Mean	*n*	Mean	Min.	Max.
M^1	—	—	24.7	2	17.5	—	—
M^2	15.5	15.9	15.8	4	16.0	15.4	17.5
M^3	8.2	9.7	8.8	3	10.2	9.5	11.0
M_1	22.2	24.6	23.4	4	15.2	14.5	15.9
M_2	15.0	16.7	15.9	3	15.2	13.7	16.0
M_3	7.6	9.8	8.7	6	10.5	9.5	12.0

n = number of teeth measured.

Abudhabia baynunensis is of particular interest because its dental characteristics are intermediate between the Myocricetodontinae and the Gerbillinae. Structurally *Abudhabia* makes a perfect ancestor for *Taterillus*. It is peculiar that *Abudhabia baynunensis* is in many respects more primitive than the much older *"Myocricetodon" magnus* from the Upper Aragonian of Morocco. This suggests that the evolutionary development of the gerbil dentition from the *Myocricetodon*-type of dentition has taken place several times. This conclusion is endorsed by the *"Protatera"* material that has become known from the Upper Turolian of southern Spain. These specimens, from strata that are roughly of the same age as the Baynunah Formation, are different in having strong longitudinal ridges and prismatic cusps and obviously belong to a different group of Gerbillinae.

The polyphyly of the Gerbillinae s.l. is further enhanced by including the genus *Pseudomeriones,* because the fossil record of this genus strongly suggests that it derives from an Asiatic myocricetodontine, whereas the true gerbils seem to have originated in Africa.

Family Muridae Gray, 1821
(Old World rats and mice)
Murinae Gray, 1821
Parapelomys Jacobs, 1978
Parapelomys cf. *charkensis* Brandy, 1979
(figs 15.6 and 15.7)

Material and Measurements (table 15.1)

One M^1 (23.3 × ?), AUH 776 (damaged) (site S1); 1 M^2 (16.8 × 15.9), AUH 581 (site M1) plus three other specimens: AUH 757 (site S1), AUH 582 (damaged) (site M1), and AUH 583 (damaged) (site S4); 1 M_2 (15.7 × 15.1), AUH 579 (plus AUH 758, too badly damaged to measure) (site S1) and 1 M_3 (15.2 × 13.0), AUH 580 (plus AUH 777, worn) (site S1)

Remarks

The specimens from the Baynunah Formation referred to the Murinae represent one species. Their simple dental pattern, the large difference in size

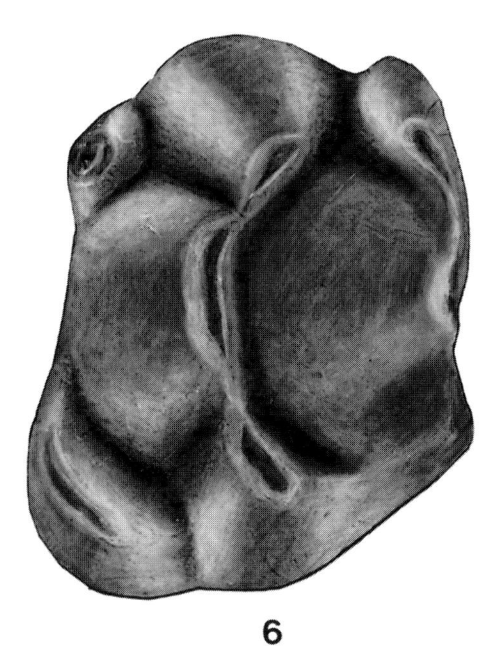

Figures 15.6 and 15.7. *Parapelomys* cf. *charkensis*. *15.6,* M^2, AUH 581; *15.7,* M_2, AUH 579. Both ×40 and figured as left teeth although both are from the right side.

between the t1 and t3 of the M^2, the absence of a posteroloph in the M^2, and the inflated t2, t5, and t8 of the M^1 clearly show that these specimens represent a small rat. Although the material is poor, its similarity in size and dental pattern to *P. charkensis* is striking. I therefore list these specimens as *Parapelomys* cf. *charkensis.*

Dendromurinae Alston, 1876 (African forest mice)
Dendromus Smith, 1829
Dendromus aff. *melanotus* Smith, 1834
(figs 15.8–15.11)

Synonym: *Dendromus* sp. 1
(de Bruijn and Whybrow, 1994)

Material and measurements (tables 15.1 and 15.3)

M^1, AUH 584, 587 (site S1); M^2, AUH 759 (site S1); M_1, AUH 760 (site S1); M_2, AUH 585, 588, 761 (site S1)

Remarks

These teeth, which represent a small species of *Dendromus,* have about the same size and relative dimension as the extant species *D. melanotus.* The fossil cheek teeth show minor differences in details of the dental pattern. The anterocentral cusp in front of the anterocone of the M^1 is smaller and the paracone–metacone connection of the M^1 is absent.

Similarly the endolophid between the metaconid and the entoconid is absent in the M_1 and M_2

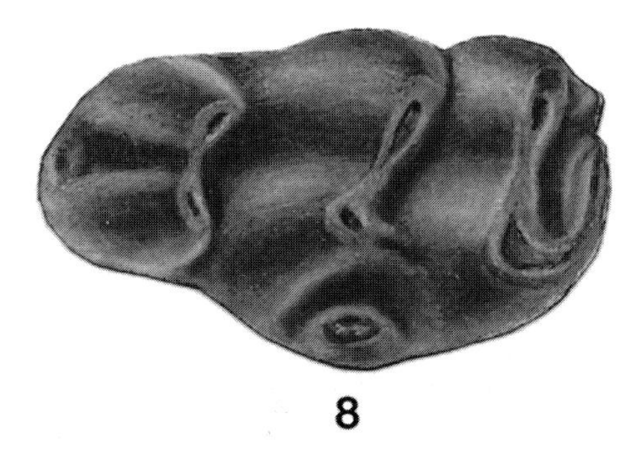

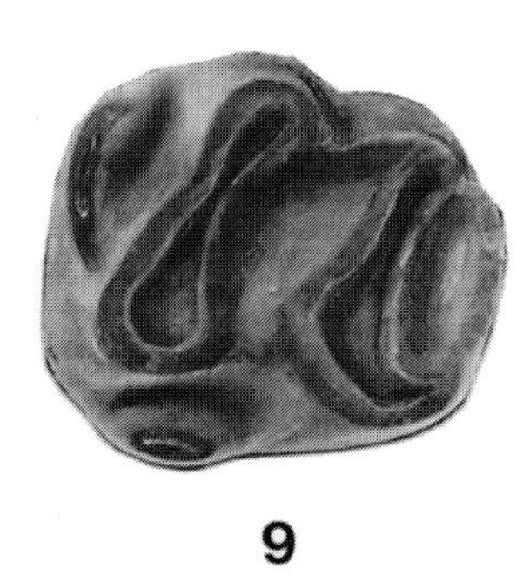

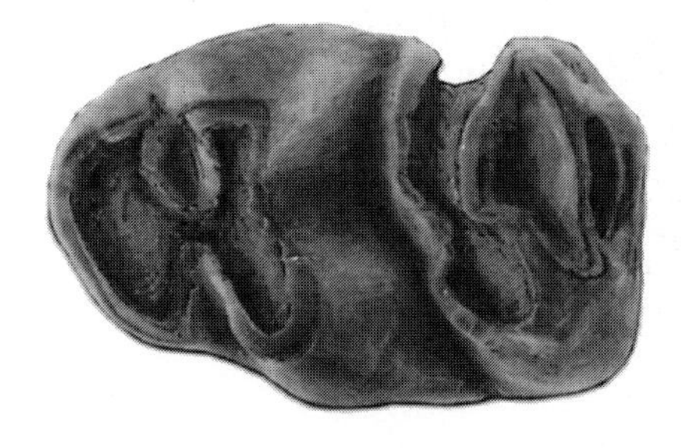

Figures 15.8–15.11. *Dendromus* aff. *melanotus. 15.8,* M^1, AUH 587; *15.9,* M^2, AUH 759; *15.10,* M_1, AUH 760; *15.11,* M_2, AUH 761. All ×40 and figured as left teeth regardless of which side the tooth really is. Figures 15.8 and 15.10 are from the right side.

Table 15.3. Dental measurements of *Dendromus* aff. *melanotus;* synonym *Dendromus* sp. 1 (de Bruijn and Whybrow, 1994) (units = 0.1 mm)

	Length				Width		
	Min.	Max.	Mean	*n*	Mean	Min.	Max.
M^1	15.7	16.8	16.3	2	9.2	9.0	9.3
M^2	—	—	9.4	1	8.3	—	—
M_1	—	—	13.2	1	8.1	—	—
M_2	9.5	10.0	9.8	3	8.4	7.9	9.0

n = number of teeth measured.

and the labial cingulum of the M_1 is much weaker in the material from the Baynunah Formation than in the extant material that I have seen.

The material is very limited, however, so I cannot verify whether or not the differences from the extant species are consistent and of sufficient importance to define a new species. I prefer tentatively to assign the fossils from the Baynunah Formation to the extant species *D. melanotus* because of their surprising similarity. True dendromurids apparently developed much earlier than hitherto thought.

Dendromus sp. (figs 15.12 and 15.13)

Synonym: *Dendromus* sp. 2 (de Bruijn and Whybrow, 1994)

Material and Measurements (table 15.1)

One M^2 (11.4–11.7 × 10.3–10.8), AUH 586 (site S1); two M_2 (12.9 × 11.5), AUH 762, 772 (site S1).

Remarks

The two teeth referred to *Dendromus* sp. differ sufficiently in size and morphology from *D.* aff. *melanotus* to establish the presence of a second species of *Dendromus* in the Baynunah Formation. The central sinus of the M^2 and the sinusid of the M_2 are more transverse, the metaloph of the M^2 is not connected to the metacone, and the posterior cingulum is developed as a cusp. These differences relative to *Dendromus* aff. *melanotus* and their about 25% larger size are far beyond the range of

12

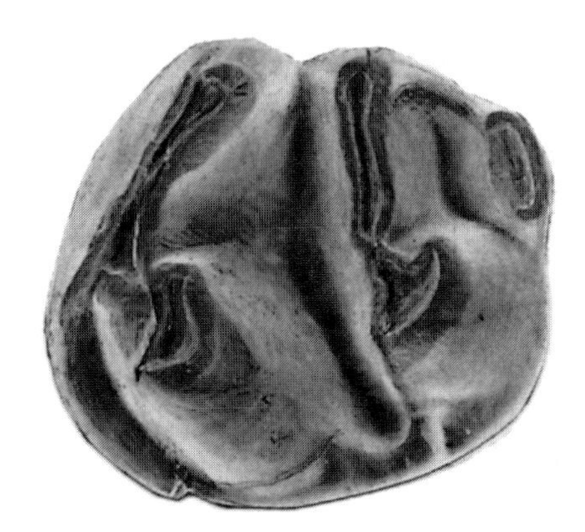

13

Figures 15.12 and 15.13. *Dendromus* sp. *15.12,* M^2, AUH 586; *15.13,* M_2, AUH 772. Both ×40 and figured as left teeth regardless of which side the tooth really is. Figure 5.13 is from the right side.

variation expected in one population even when we allow for some minor age difference between these fossils. The material is insufficient for identification at species level.

Myocricetodontinae Lavocat, 1961
Myocricetodon Lavocat, 1952
Myocricetodon sp. nov. (figs 15.14–15.16)

Material and Measurements (table 15.1)

Two M^1 (14.0 × 8.3, ± 13.8 × 10.1), AUH 773 (site S1), AUH 589 (damaged) (site M1), AUH 778 (too badly damaged to measure) (site S1); two M_2 (9.5 × 9.0, 9.2 × 9.0), AUH 763, 780 (site S1).

Remarks

The three M^1s show that the *Myocricetodon* from the Baynunah Formation had a weakly split anterocone, an enterocone that is either very weak or absent, and a short posteriorly directed protoloph. The Baynunah specimens are small and within the range of *M. parvus* Lavocat, 1961 from Beni Mellal (Morocco) but differ from that species in not having the protocone and paracone aligned and confluent. The two M_2s both have a strong anterolabial cingulum and, for a myocricetodontine, a distinctly bunodont dental pattern. The posterior cingulum is developed as an individual cusp, a configuration that is also seen in *M. trerki* and *M. ouedi* from the Vallesian of Oued Zra (Morocco). The teeth from the Baynunah Formation seem to document a new species of *Myocricetodon*. The available material is, however, too scant to serve as a basis for a new name.

Dipodidae Fischer, 1817 (jerboas and jumping mice)
Zapodinae Coues, 1875
Zapodinae gen. et sp. indet. (fig. 15.17)

Material and Measurements (table 15.1)

One M_1 (11.5 × 6.9), AUH 774 (site S1).

Remarks

This small, damaged tooth from site S1-high is interpreted as a right first lower molar of a zapodine. The peculiar position of the metaconid in front of the protoconid is, in fact, more similar to the configuration seen in Allactaginae than in Zapodinae, which usually have a more symmetrical pattern of the anterior part of the M_1. The ectomesolophid and the hypoconulid of the M_1 from the Baynunah Formation are strong. The tooth has these characteristics in common with many Zapodinae and Allactaginae. The anteroconid of the M_1 is absent. The entoconid is broken but was probably high.

Soricidae Gray, 1821 (shrews)
Soricidae gen. et sp. indet.
(see de Bruijn and Whybrow, 1994)

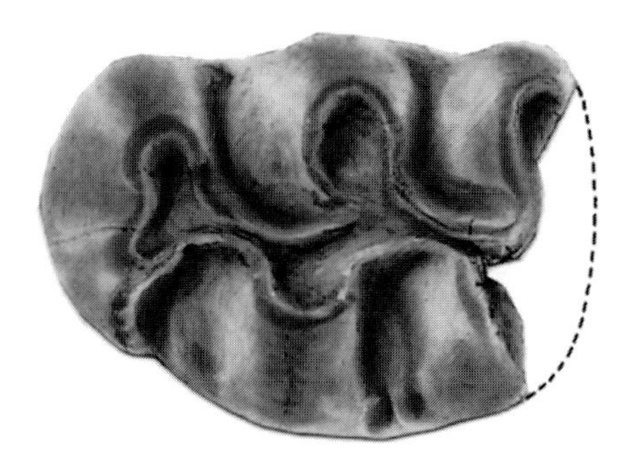

Figures 15.14–15.16. *Myocricetodon* sp. *15.14*, M^1, AUH 773; *15.15*, M_2, AUH 763; *15.16*, M_2, AUH 780. All × 40 and figured as left teeth regardless of which side the tooth really is. Figures 15.15 and 15.16 are from the right side.

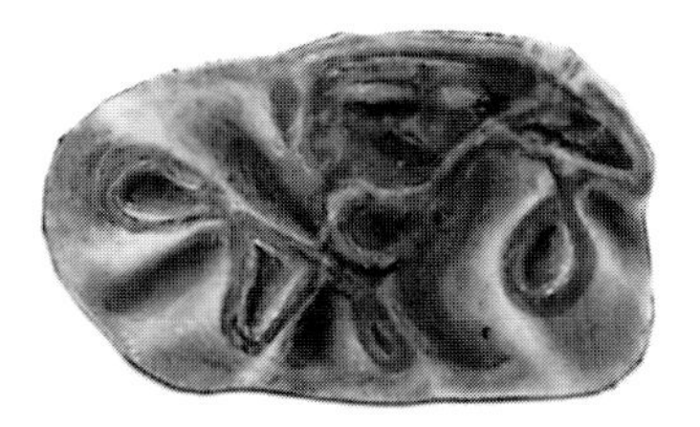

17

Figure 15.17. Zapodinae gen. et sp. indet. M_1, AUH 774, figured as a left tooth ×40 although it is from the right side.

Material (table 15.1)

Three incisors: AUH 577, 578 (site K1), AUH 764 (site S1).

BIOSTRATIGRAPHY

The age of the small association of rodents from the Baynunah Formation is difficult to assess at present because the faunal succession of the Arabian Neogene is virtually unknown (Whybrow et al., 1982). The presence of the small rat *Parapelomys* cf. *charkhensis* suggests that the fauna is not older than late Turolian. A late Turolian age is also suggested by the stage of evolution of *Abudhabia baynunensis* because the dental pattern of this gerbil is close to *A. kabulense* from the Lower Ruscinian of Pul-e Charki, Afghanistan, but more primitive. The thryonomyid M_3 is derived relative to the specimen from the middle Miocene of As Sarrar, Saudi Arabia, figured by Flynn et al. (1983) in lacking the anteroconid.

Dendromus aff. *melanotus* is not essentially different from the extant species giving the association a modern aspect. *Myocricetodon* sp. nov. is a small representative of the genus with archaic characteristics, such as nonalignment of the protocone and paracone in M^1, and a bunodont M_2. This species seems, at first sight, to suggest a Vallesian age for the association, but if the assumption is correct that extant *Calomyscus* with its primitive dental characteristics is a living myocricetodontine, this argument loses appeal.

The zapodine M_1 is among the comparative material available to me most similar to specimens from the early Miocene of Turkey. The zapodine subfamily has many representatives that have conservative dental characteristics, which makes them inappropriate for biostratigraphy.

In conclusion, the presence of a small rat and of two species of *Dendromus* suggests that the Baynunah Formation cannot be older than late Turolian. An early Ruscinian age cannot be excluded.

BIOGEOGRAPHY

The Thryonomyidae, now restricted to Africa south of the Sahara, had a range during the Miocene that included Arabia and the Indian subcontinent. The range of the Gerbillinae during the late Miocene was even more extensive and included major parts of Central Asia. It is striking, however, that the only other species of *Abudhabia* known comes from Afghanistan (but see Flynn and Jacobs, 1999—Chapter 28), where it occurs associated with a small rat that is also found in the Baynunah Formation. The genus *Myocricetodon* is known from Mio-Pliocene deposits of Pakistan, Turkey, North Africa, and East Africa. *Dendromus,* although represented by few specimens in some late Turolian fissure fillings in eastern Spain, is basically restricted to Africa south of the Sahara. The Zapodinae, on the other hand, is typical of the Asiatic fauna.

The association from the Baynunah Formation is evidently composed of a mixture of faunal elements of southwestern Asiatic and African origin. This suggests that the Red Sea did not yet function as a barrier when the fluvial sequence of the formation was deposited.

ECOLOGY

The reconstruction of biotope on the basis of small mammals remains necessarily vague when samples are too small for statistical analysis. In the case of the Baynunah association the situation is worse because most fossils have obviously been transported before burial and can be shown to have been redeposited and are, therefore, not strictly contemporaneous. It is thus not surprising that different faunal elements suggest the presence of very different biotopes adjacent to the river in which the sediment was deposited.

The dominant gerbil *Abudhabia* and the zapodine suggest the presence of open landscapes; the thryonomyid and the dendromurines indicate thick cover of at least tall grasses. Rodents that need either real forests to survive (Petauristinae) or the year-around availability of water bodies (Castoridae) are lacking in the association.

The rodent association therefore indicates a river with a dense cover of reed and possibly shrubs along its banks in an otherwise open landscape. Rainfall may well have been restricted to a particular season.

Acknowledgements

I thank Peter Whybrow, Peter Forey, and my wife Jes for helping to collect. Lodewijk IJlst centrifuged the concentrates. W. den Hartog made the SEM photographs that were subsequently retouched by Jaap Luteijn.

References

Andrews, P. 1999. Taphonomy of the Shuwaihat proboscidean, late Miocene, Emirate of Abu Dhabi, United Arab Emirates. Chap. 24 in *Fossil Vertebrates of Arabia,* pp. 338–53 (ed. P. J. Whybrow and A. Hill). Yale University Press, New Haven.

Bruijn, H. de, and Hussain, S. T. 1985. Thryonomyidae from the Lower Manchar Formation of Sind, Pakistan. *Proceedings Koninklijke Nederlandse Akademie van Wetenschappen* 88: 155–66.

Bruijn, H. de, and Whybrow, P. J. 1994. A Late Miocene and rodent fauna from the Baynunah Formation, Emirate of Abu Dhabi; United Arab Emirates. *Proceedings Koninklijke Nederlandse Akademie van Wetenschappen* 97: 407–22.

Bruijn, H. de, Dawson, M. R., and Mein, P. 1970. Upper Pliocene Rodentia, Lagomorpha and Insectivora (Mammalia) from the isle of Rhodes (Greece), Part II. *Proceedings Koninklijke Nederlandse Akademie van Wetenschappen* 73: 535–95.

Carleton, M. D., and Musser, G. G. 1984. Muroid rodents. In *Orders and Families of Recent Mammals of the World,* pp. 289–379 (ed. S. Anderson and J. K. Jones). Wiley, New York.

Chaline, J., Mein, P., and Petter, F. 1977. Les grandes lignes d'une classification évolutive des Muroidea. *Mammalia, Paris* 41: 245–52.

Flynn, J., and Jacobs, L. L. 1999. Late Miocene small-mammal faunal dynamics: The crossroads of the Arabian Peninsula. Chap. 28 in *Fossil Vertebrates of Arabia,* pp. 412–19 (ed. P. J. Whybrow and A. Hill). Yale University Press, New Haven.

Flynn, J., Jacobs, L. L., and Sen, S. 1983. La diversité de *Paraulacodus* (Thryonomyidae, Rodentia) et des groupes apparentés pendant le Miocène. *Annales de Paléontologie, Paris* 69: 355–66.

Hopwood, A. T. 1929. New and little known mammals from the Miocene of Africa. *American Museum Novitates* no. 344: 1–9.

Jaeger, J. J. 1977. Les Rongeurs du Miocène Moyen et Supérieur du Maghreb. *Palaeovertebrata* 8: 1–166.

Munthe, J. 1982. Small mammal fossils from the Pliocene Sahabi Formation of Libya. *Garyounis Scientific Bulletin* 4: 33–39.

Sen, S. 1977. La faune de rongeurs pliocènes de Calta (Ankara, Turquie). *Bulletin du Muséum National d'Histoire Naturelle, Paris* 61: 89–171.

———. 1983. Rongeurs et lagomorphes du gisement pliocène de Pul-e Charkhi, bassin de Kabul, Afghanistan. *Bulletin du Muséum National d'Histoire Naturelle* 5C: 33–74.

Tassy, P. 1999. Miocene elephantids (Mammalia) from the Emirate of Abu Dhabi, United Arab Emirates: Palaeogeographic implications. Chap. 18 in *Fossil Vertebrates of Arabia,* pp. 209–33 (ed. P. J. Whybrow and A. Hill). Yale University Press, New Haven.

Tong, H. 1989. Origine et évolution des Gerbillidae (Mammalia, Rodentia) en Afrique du Nord. *Mémoires de la Société Géologique de France* 155: 1–120.

Tong, H., and Jaeger, J. J. 1993. Muroid rodents from the Middle Miocene Fort Ternan locality (Kenya) and their contribution to the phylogeny of muroids. *Palaeontographica* 229: 51–73.

Whybrow, P. J. 1989. New stratotype; the Baynunah Formation (Late Miocene), United Arab Emirates: Lithology and palaeontology. *Newsletters on Stratigraphy* 21: 1–9.

Whybrow, P. J., and Clements, D. 1999. Late Miocene Baynunah Formation, Emirate of Abu Dhabi, United Arab Emirates: Fauna, flora, and localities. Chap. 23 in *Fossil Vertebrates of Arabia,* pp. 317–33 (ed. P. J. Whybrow and A. Hill). Yale University Press, New Haven.

Whybrow, P. J., Collinson, M. E., Daams, R., Gentry, A. W., and McClure, H. A. 1982. Geology, fauna (Bovidae, Rodentia) and flora from the Early Miocene of eastern Saudi Arabia. *Tertiary Research* 4: 105–20.

A Monkey (Primates; Cercopithecidae) from the Late Miocene of Abu Dhabi, United Arab Emirates

16

ANDREW HILL AND TOM GUNDLING

This short note reports a specimen of no great intrinsic interest, but which assumes more significance as the only fossil primate found in Abu Dhabi, and the only fossil monkey so far known from the Arabian Peninsula. The fossil, AUH 35, was found by Peter Whybrow (The Natural History Museum, London) in sediments of the Baynunah Formation, in the western region of Abu Dhabi on 10 January 1989. It came from site Jebel Dhanna 3, from a spot that is now obscured by subsequent development. Other common fossils at the site include fish, turtles, crocodilian teeth, and struthionid eggshell. In addition there are hippopotamuses, equids, giraffids, and bovids (Whybrow and Clements, 1999—Chapter 23). The specimen received a short notice in *Nature* (Gee, 1990), and was mentioned in Hill et al. (1990) and Whybrow et al. (1990).

DESCRIPTION

The tooth is a lower left canine measuring 16.36 mm in maximum buccal height, 8.36 mm in mesiodistal length, and 5.10 mm in width (both measured at the cervix). The crown is intact with negligible wear, showing a helical twist along its long axis, and the tip of the root is broken about 12 mm from the cervix. The elliptical cross section, the strong distal heel at the cervix, and the well-developed mesial groove extending along the crown and onto the root show that it is a canine of a cercopithecoid monkey (plate 16.1, p. 201).

MIOCENE CERCOPITHECOIDS

The earliest Old World monkeys, constituting the Victoriapithecidae (Benefit, 1987, 1993), appear during the early Miocene in eastern and northern Africa, and are not known after 12.5 million years (Ma) ago (Hill et al., in press). The time of separation and divergence of the two modern subfamilies of Cercopithecidae, the Cercopithecinae and Colobinae, is poorly constrained, and colobines are recognised in the record before cercopithecines (Gundling and Hill, in press).

The earliest known cercopithecid from Africa is the colobine from the top of the Ngorora Formation, Tugen Hills, Kenya. Material from the site of Ngeringerowa has been described as *Microcolobus tugenensis* by Benefit and Pickford (1986). This sequence is now dated by the Baringo Paleontological Research Project between 10 and 8 Ma (Hill, 1995; A. Deino personal communication). The premolar from slightly older in the Ngorora Formation alluded to by Benefit and Pickford (1986) and by Delson (1994) as possibly colobine, belongs to an unnamed new species of victoriapithecine (Hill et al., in press). The colobine genus *Mesopithecus* is well represented at several European late Miocene sites, especially in Greece (Bonis et al., 1990), and one of these may be as old as the African example (Delson, 1994). Another colobine, *Libypithecus,* comes from Wadi Natrun in Egypt (Stromer, 1913), and possibly from Sahabi, the latest Miocene to early Pliocene site in Libya (Meikle, 1987). The type of the species *Macaca flandrini,* described by Arambourg (1959) from Menacer (formerly Marceau), Algeria (Thomas and Petter, 1986) is in fact a colobine, dated at about 7 Ma. In the near east, *Mesopithecus* is known from Molayan in Afghanistan (Heintz et al., 1981) and from

 ISBN 0-300-07183-3

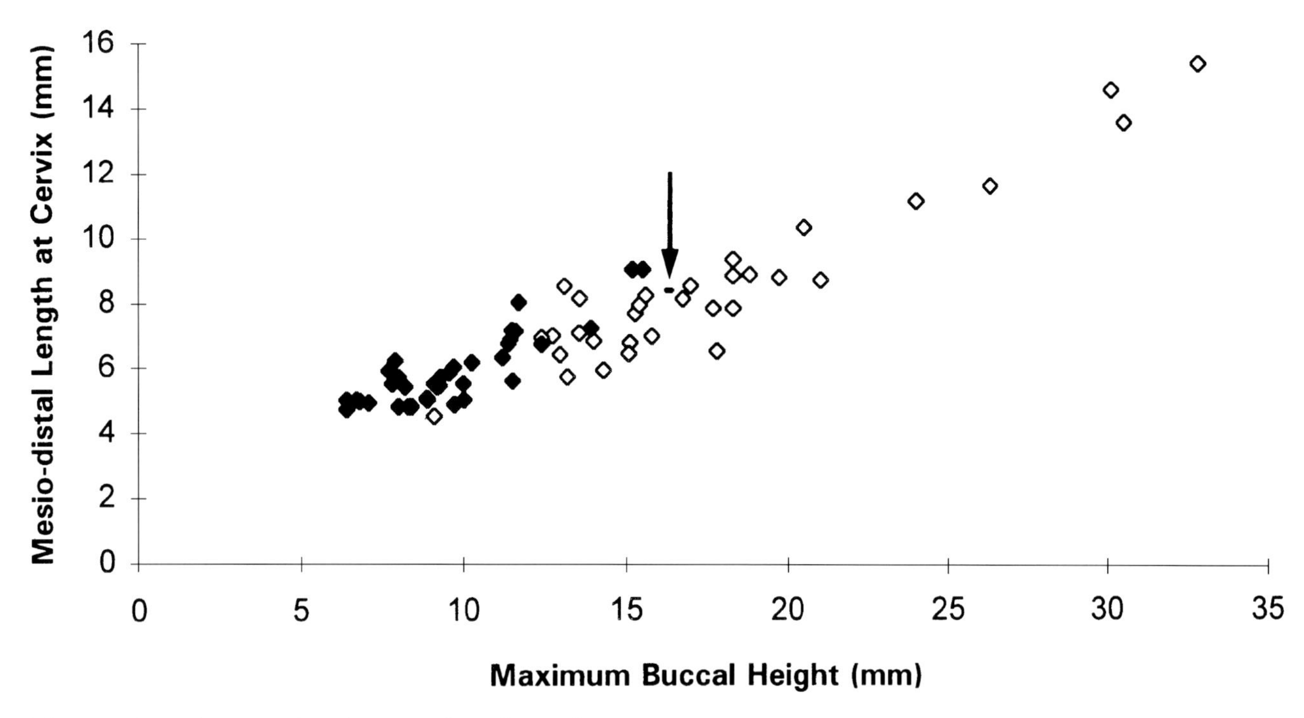

Figure 16.1. Bivariate plot showing the relation between lower canine height and length from a variety of cercopithecid male (open diamonds) and female taxa (solid diamonds). Note the almost complete separation of the two groups. The Abu Dhabi specimen, arrowed, falls within the male cluster.

Maragheh in Iran (Mecquenem, 1924–25). At Maragheh, Bernor (1986) suggests that the specimens are younger than 9 Ma. Elsewhere in Asia, monkeys do not appear in the Siwalik sequence of Pakistan before 8 Ma ago (Barry, 1987), and at present the best estimate is 7.8 Ma (M. Morgan, personal communication).

The earliest known cercopithecines are part of Arambourg's hypodigm of *?Colobus flandrini* from Menacer, Algeria (see above) (Arambourg, 1959), dated at about 7 Ma, and designated by Delson (1980) as *?Macaca* sp. These monkeys, like late Miocene cercopithecines from other regions, do not show any derived traits that would securely ally them with later species, therefore *?Macaca* is used simply to indicate a primitive taxon. Dental remains from presumed cercopithecines are also known from Ongoliba, Zaire (Hooijer, 1963), Wadi Natrun, Egypt (Stromer, 1913), and, probably at a little later in time, Sahabi, Libya (Meikle, 1987).

DISCUSSION

As the estimated age of the Baynunah Formation may be close to the presumed divergence time of the cercopithecid subfamilies, it would be interesting to know to which of these the specimen belongs. This distinction is difficult to make for canine teeth on the basis of overall size or morphology. We hoped that the proportions of canine teeth might decisively discriminate between colobines and cercopithecines. Accordingly we adopted the metric methods developed by Kelley (1995a,b) to solve an analogous problem. He used canine tooth proportions to differentiate the sexes of modern and fossil apes on the basis of canines, and successfully applied the results to distinguish multiple species of fossil hominoids from sexually dimorphic representatives of the same species.

The data given in figure 16.1 document the simple relationship between tooth height and breadth in a variety of modern cercopithecids. The sample consisted of a total of 29 species from 16 genera: *Colobus, Procolobus, Piliocolobus, Presbytis, Simias, Nasalis, Rhinopithecus, Macaca, Theropithecus, Mandrillus, Lophocebus, Cercocebus, Cercopithecus, Allenopithecus, Papio,* and *Erythrocebus.* These data show that by these means sexes of monkeys can be distinguished with a high degree of confidence. There is little overlap in measurements between the males and females. The Baynunah specimen is clearly a male. This ratio also partially discriminates female colobines and female cercopithecines. It is, however, impossible to separate the males of the two subfamilies in this way. As AUH 35 is a male, this makes it impossible to allocate it to subfamily using this ratio.

Conclusions

AUH 35 is a male cercopithecid. Although only a single specimen, it demonstrates that monkeys had occupied the Arabian Peninsula by 8–6 Ma ago. The lack of resolution on the date of the Baynunah Formation precludes knowing whether this occurrence is earlier than the first appearance of monkeys in the Asian Siwalik succession.

Acknowledgements

Many individuals and organisations have assisted with the fieldwork that has led to the retrieval of the Baynunah specimens. In particular, A. H. thanks the Undersecretary of the Department of Antiquities and Tourism, Al Ain, His Excellency Saif Ali Dhab'a al Darmaki, for the help provided by his department, especially from Walid Yasin. Subsequent work has been largely funded by grants to the project from the Abu Dhabi National Oil Company (ADNOC) and the Abu Dhabi Company for Onshore Oil Operations (ADCO). A.H. also received funding from the L. S. B. Leakey Foundation, USA. We thank Jay Kelley for helpful discussions, and also for providing some unpublished data that he generously allowed us to use in this analysis. T. G. thanks Ross McPhee and Bryn Mader at the American Museum of Natural History, New York, for providing access to comparative samples of extant monkeys. Other modern skeletal material came from the Peabody Museum of Natural History, Yale University. We thank Michelle Morgan for information about Siwalik cercopithecoids and Eric Delson for his advice and comments on the manuscript.

References

Arambourg, C. 1959. Vertébrés continentaux du Miocène supérieur de l'Afrique du Nord. *Publications du Service Carte Géologique de l'Algérie,* n.s. Paléontologie, Alger 4: 1–161.

Barry, J. C. 1987. The history and chronology of the Siwalik cercopithecids. *Human Evolution* 2: 47–58.

Benefit, B. 1987. The molar morphology, natural history, and phylogenetic position of the middle Miocene monkey *Victoriapithecus.* Ph.D. thesis, New York University.

———. 1993. The permanent dentition and phylogenetic position of *Victoriapithecus* from Maboko Island, Kenya. *Journal of Human Evolution.* 25: 83–172.

Benefit, B., and Pickford, M. H. L. 1986. Miocene fossil cercopithecoids from Kenya. *American Journal of Physical Anthropology* 69: 441–64.

Bernor, R. L. 1986. Mammalian biostratigraphy, geochronology, and zoogeographic relationships of the Late Miocene Maragheh fauna, Iran. *Journal of Vertebrate Paleontology* 6: 76–95.

Bonis, L. de, Bouvrain, G., Geraads, D., and Koufos, G. 1990. New remains of *Mesopithecus* (Primates, Cercopithecoidea) from the Late Miocene of Macedonia (Greece), with the description of a new species. *Journal of Vertebrate Paleontology* 10: 473–83.

Delson, E. 1980. Fossil macaques, phyletic relationships and a scenario of deployment. In *The Macaques: Studies in Ecology, Behavior and Evolution,* pp. 10–30 (ed. D. G. Lindburg). Van Nostrand Reinhold, New York.

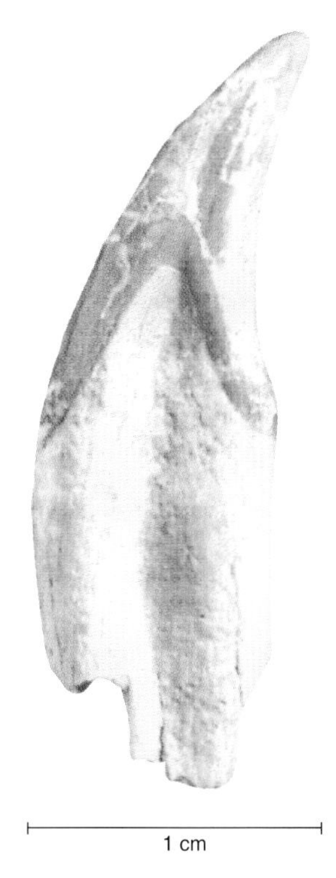

Plate 16.1. Cercopithecidae gen. et sp. indet. canine, AUH 35; mesial view.

———. 1994. Evolutionary history of the colobine monkeys in paleoenvironmental perspective. In *Colobine Monkeys: Their Ecology, Behaviour and Evolution,* pp. 11–43 (ed. A. G. Davies and J. F. Oates). Cambridge University Press, Cambridge.

Gee, H. 1989. Fossils from the Miocene of Abu Dhabi. *Nature* 338: 704.

Gundling, T., and Hill, A. (in press). Geological context of fossil Cercopithecoidea from eastern Africa. In *Old World Monkeys* (ed. P. F. Whitehead and C.J. Jolly). Cambridge University Press, Cambridge.

Heintz, E., Brunet, M., and Battail, B. 1981. A cercopithecid Primate from the Late Miocene of Molayan, Afghanistan, with remarks on *Mesopithecus. International Journal of Primatology* 2: 273–84.

Hill, A. 1995. Faunal and environmental change in the Neogene of East Africa. In *Paleoclimate and Evolution,* pp. 178–93 (ed. E. S. Vrba, G. H. Denton, T. C. Partridge, and L. H. Burkle). Yale University Press, New Haven.

Hill, A., Leakey, M., Kingston, J., and Ward, S. in press. New cercopithecids and a hominoid from 12.5 Ma in the Tugen Hills succession, Kenya. *Journal of Human Evolution.*

Hill, A., Whybrow, P. J., and Yasin al-Tikriti, W. 1990. Late Miocene primate fauna from the Arabian

Peninsula: Abu Dhabi, United Arab Emirates. *American Journal of Physical Anthropology* 81: 240–41.

Hooijer, D. A. 1963. Miocene Mammalia of Congo. *Annales du Musée Royale d'Afrique Centrale, ser. 8,* 46: 1–71.

Kelley, J. 1995a. Sex determination in Miocene catarrhine primates. *American Journal of Physical Anthropology* 96: 391–417.

———. 1995b. Sexual dimorphism in canine shape among extant great apes. *American Journal of Physical Anthropology* 96: 365–89.

Mecquenem, R. de. 1924–25. Contribution à l'étude des fossiles de Maragha. *Annales Paléontologie Vertébrés, Paris* 13/14: 135–60.

Meikle, W. E. 1987. Fossil Cercopithecidae from the Sahabi Formation. In *Neogene Paleontology and Geology of Sahabi,* pp. 119–27 (ed. N. T. Boaz, A. El-Arnauti, A. W. Gaziry, J. de Heinzelin, and D. D. Boaz). Alan R. Liss, New York.

Stromer, E. 1913. Mitteilungen ber die Wirbeltierreste aus dem Mittlepliocän des Natrontales (Ägypten). *Zeitschrift der Deutschen Geologischen Gesellschaft, A, Abhandlungen, Berlin* 65: 349–61.

Thomas, H., and Petter, G. 1986. Revision de la faune de mammifères du Miocène supérieur de Menacer (ex-Marceau), Algérie: Discussion sur l'âge de gisement. *Geobios* 19: 357–73.

Whybrow, P. J., and Clements, D. 1999. Late Miocene Baynunah Formation, Emirate of Abu Dhabi, United Arab Emirates: Fauna, flora, and localities. Chap. 23 in *Fossil Vertebrates of Arabia,* pp. 317–33 (ed. P. J. Whybrow and A. Hill). Yale University Press, New Haven.

Whybrow, P. J., Hill, A., Yasin al-Tikriti, W., and Hailwood, E. A. 1990. Late Miocene primate fauna, flora and initial palaeomagnetic data from the Emirate of Abu Dhabi, United Arab Emirates. *Journal of Human Evolution* 19: 583–88.

Late Miocene Carnivora from the Emirate of Abu Dhabi, United Arab Emirates

John C. Barry

A small collection of fossil carnivores from late Miocene sites in Abu Dhabi has four taxa—*Plesiogulo praecocidens*, a large and a medium-sized hyaena, and an indeterminate machairodontine felid. Although small, the collection adds to our otherwise scant knowledge of the fossil mammals of the Arabian Peninsula. The Abu Dhabi carnivore assemblage is typical of the latest Miocene of Eurasia and Africa.

Systematics

Carnivora
Mustelidae
Plesiogulo Zdansky, 1924
Plesiogulo praecocidens Kurtén, 1970

Material

AUH 45, a left mandible with crowns of the canine and P_2 through M_1; AUH 702, a right mandible fragment with P_3 and broken P_4. Sites: Hamra, H5 (AUH 45) and H6 (AUH 702).

Description

The specimens referred to this taxon are mandibles of a large mustelid, comparable in tooth size to a male wolverine *(Gulo gulo)* with an estimated weight between 25 and 30 kg. The most complete specimen is damaged and weathered, with broken ascending ramus and symphysis and badly corroded tooth crowns entailing loss of some detail of crown morphology (figs 17.1 and 17.2). I have, however, been able to reconstruct the contact and orientation of the canine and the anterior part of the ascending ramus. The carnassial is moderately worn, with widely exposed dentine on the trigonid and talonid. The other teeth show little sign of wear, which suggests that the animal was a young adult when it died.

In contrast to most large mustelids, the mandibular corpus is relatively long. There is a large mental foramen under P_2 and a second, smaller one between P_3 and P_4. The indistinct anterior margin of the deep masseteric fossa lies under M_2.

The premolar crowns are in contact with each other, but are not as crowded as in other large mustelids. As a consequence their anteroposterior axes lie along a single line. P_1 fell out of the jaw before burial, but judging from the size and position of the alveolus must have contacted the back of the canine. P_1 had a single root, whereas the substantially larger P_2 is two-rooted. The premolars form a uniformly graded morphological and size series (table 17.1). All the premolars have an elongated oval basal outline when viewed from above.

The canine is large, with traces of a prominent mesial cingulum connecting anteriorly with a strong mesial longitudinal ridge that extends to the apex of the crown. Although badly corroded, the canine surface preserves traces of the corrugated enamel common to large mustelids. The preserved crown of P_2 has a low, anteriorly situated principal cusp and a lobe-like, laterally expanded talonid. P_3 and P_4 both have more centrally located and taller principal cusps, and more expanded talonids. P_4 has a very faint anterior cingulum. Neither P_4 nor P_3 shows any traces of anterior or posterior accessory cusps, nor other secondary features such as longitudinal crests.

M_1 is a well-developed, robust, shearing carnassial with a large and anteriorly placed paraconid. Although the crown is damaged, the metaconid was clearly absent, its place being taken by a posterolingually sloping crest descending from the apex of the protoconid. The relatively long talonid is heavily worn, but was broad and apparently shallowly basined. The missing M_2 probably had a crown about the same size as the M_1 talonid.

The second specimen is a fragment of a right mandible with a well-preserved P_3 and the anterior part of P_4 (fig. 17.3) The lateral face of the mandible is broken, exposing a very large mandibular canal and part of the canine alveolus under the P_3. The crown of P_3 is complete, with expanded talonid and weak longitudinal crests. The P_4 has a weak anterior cingulum and longitudinal crest.

Comparisons

The ratel *(Mellivora capensis)* has also lost the M_1 metaconid, but differs substantially from the Abu Dhabi specimens in other features. *Mellivora capensis* lacks both P_1 and M_2, has a short and trenchant M_1 talonid, and has premolars that are wide relative to their length. In addition, P_4 has salient anterior and posterior accessory cusps, while P_2 and P_3 have traces of an anterior cingulum. The fossils differ from large extinct Eurasian species of *Eomellivora*

 ISBN 0-300-07183-3

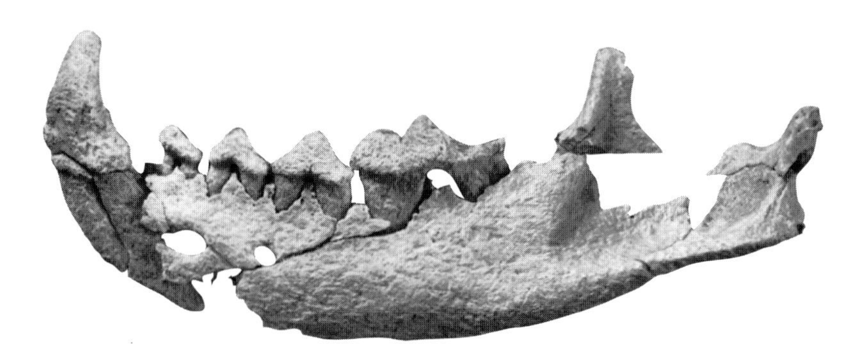

Figure 17.1. *Plesiogulo praecocidens,* left mandible, AUH 45; lateral view.

and *Perunium* and North American *Beckia* and *Ferinestrix* (all of which have strongly sectorial carnassials with reduced metaconids) in some or all of the same features. These include the presence in the fossil taxon of a long, basined, nontrenchant M_1 talonid, a more loosely packed and longer premolar series, premolars with an oval basal outline, and the absence of anterior and posterior accessory cusps on P_3 and especially P_4.

The Abu Dhabi mustelid is most like species of *Plesiogulo* and the closely related extant wolverine, *Gulo gulo*. Like the fossil taxon, *Plesiogulo* and *Gulo* retain P_1 and M_2, and both have simple premolars and a reduced M_1 metaconid. The premolar series of *Gulo gulo* is more crowded, however, with the axes of the teeth being imbricated along the tooth row. In addition, in *Gulo gulo* the premolars are more labially expanded and wider than those of the fossil, giving them a more triangular basal outline. The M_1 of *Gulo gulo* is also wider and more robust, with a much shorter and more trenchant talonid. These same features are similarities shared between *Plesiogulo* and the Abu Dhabi fossils and indicate reference to that taxon.

Kurtén (1970) reviewed the Eurasian species of *Plesiogulo,* and recognised six species, one of which was new. He diagnosed the species primarily on the basis of size, and features of the P^4 and M^1, plus the presence or absence of an M_1 metaconid. Hendey (1978) and Harrison (1981) subsequently reviewed African and North American species, adding two North American species. I have referred the Abu Dhabi species to *Plesiogulo praecocidens* Kurtén, 1970 because of their relatively small size within the genus, and absence of the carnassial metaconid. This species is otherwise known only from the type and one referred specimen, both from Baode (originally spelt "Paote") in the late Miocene of China. Hendey (1978), however, has questioned the use of metaconid development as a taxonomic character, seeing it as being variably present in at least one widespread form, that if treated as a single species would have ranged throughout Africa and all of Eurasia. Resolution of the issue

Figure 17.2. *Plesiogulo praecocidens,* left mandible, AUH 45; occlusal view.

Table 17.1. Dental measurements (in cm) of *Plesiogulo praecocidens*

	P_2	P_3	P_4	M_1[a]
Tooth AUH 45				
Length	0.70	0.83	1.09	2.12
Width	0.47	0.54	0.61	0.81
Tooth AUH 702				
Length		0.92		
Width		0.60		

[a]Depth of mandible below M_1 on lingual face = 2.12 cm.

Figure 17.3. *Plesiogulo praecocidens*, right mandible fragment, AUH 702, lateral view.

obviously depends on acquiring much more material from throughout the Old World.

Hyaenidae
Genus indeterminate
Very large indeterminate species

Material

AUH 294, a left mandible with unerupted canine, intact dP_2, and fragment of dP_3. Site: Jebel Dhanna, JD3.

Description

AUH 294 belongs to a very large species as evidenced by the size of the dP_2 compared to that of *Crocuta crocuta* (table 17.2). The fossil has a well-preserved dP_2 crown and the anterior portion of dP_3, as well as a partially exposed but unerupted crown of the adult canine (plate 17.1). The root of the canine had not begun to form. The mandible is deep below the dP_2, with a single small mental foramen. A short diastema separates the dP_2 from the partially preserved alveolus of the deciduous canine. The crown of the dP_2 is slender and noticeably wider across the talonid than at the front due to a slight lingual expansion of the talonid. The principal cusp is low and has anterior and posterior longitudinal crests. A large anterior accessory cusp is positioned slightly lingually at the front of the tooth, while a small posterior accessory cusp is situated on the rear of the principal cusp. The fragment of the dP_3 indicates it had a large, well-separated anterior accessory cusp as well.

Comparisons

The morphology of dP_2 is not particularly distinctive among hyaenas. The fossil's form compares closely to the dP_2 of *Crocuta crocuta*, except for having a relatively larger and more separated anterior accessory cusp. The size of the fossil, however, precludes reference to *Crocuta crocuta* or, indeed, almost all known hyaenids. Juvenile mandibles of *Pliocrocuta perrieri*, as described and illustrated by Viret (1954), have a generally similar dP_2 morphology, although they appear to be narrower and have less distinct cusps. These mandibles also appear to be significantly smaller, although Viret does not give measurements.

Genus indeterminate
Medium-sized indeterminate species

Table 17.2. Measurements (in cm) of dP_2 in AUH 294 and *Crocuta crocuta*

	Length	Width
AUH 294	1.62	0.90
MCZ 25403	0.86	0.42
MCZ 25404	0.83	0.39

Plate 17.1. Hyaenidae gen. indet., very large indeterminate species, fragment of left mandible, AUH 294; lateral view.

Material

AUH 370, an incomplete right ulna. Site: Jebel Mimiyah (Al Mirfa), M1.

Description

AUH 370 comprises a right ulna, lacking the olecranon process and about one-fifth of the distal end of the shaft. The specimen shows evidence of postmortem abrasion in the rounding of the broken ends and polished surface, and perhaps gnawing by a carnivore. The ulna is gracile, suggesting an animal slightly smaller than *Crocuta crocuta*. The radial tubercle is much reduced, while the coronoid process is mediolaterally compressed and extends anteriorly onto the dorsal surface of the shaft as a supporting buttress. The articular surface for the head of the radius is a shallow arc, indicating that supination of the radius was restricted. The greater sigmoid notch comprises a narrow trochlear articulation and a more extensive capitular articulation. There is a large, deep fossa just anterior to the radial tubercle. Distally the shaft is broken just proximal to the expansion for the styloid process.

Comparisons

The specimen is demonstrably hyaenid in the conformation of the coronoid process and radial tubercle. It compares in size to a subadult *Crocuta crocuta* with fully erupted dentition in the Museum of Comparative Zoology, Harvard collections (MCZ 14558). In the fossil the diameter of the greater sigmoid notch is slightly smaller, but the preserved portion of the shaft is longer and more gracile than the corresponding part of the ulna of MCZ 14558. Although not determinable to genus or species, the specimen clearly indicates the presence of a second hyaenid species in the Abu Dhabi faunas.

Felidae
Machairodontinae
Genus and species indeterminate

Material

AUH 241, a complete right calcaneum, and AUH 202, a last lumbar vertebra. Sites: Kihal, K1 (AUH 241) and Ras Dubay'ah, R2 (AUH 202).

Description

The specimen belongs to a large species, about the size of *Panthera leo* or *Panthera tigris*. It is broad for its length (fig. 17.4). The anterior and posterior parts of the inferior sustentacular facet are continuous, not separated as in many extant felids. Both facets are separated from the superior sustentacular facet by a wide sulcus. The medially directed process supporting the posterior part of the inferior sustentacular facet is reduced in size, as is the area of the inferior susentacular facet. The superior sustentacular facet is lower and flatter than in *Panthera leo*. There is a large, distinct navicular facet, with is set at a 45° angle to the cuboid facet.

AUH 202 comprises the centrum and dorsal arch, with the bases of the spinal and transverse processes. It is wide relative to its length, with the distance between the posterior zygapophyses being greater than that between the anterior zygapophyses. Both features suggest it is a last lumbar vertebra. The construction and orientation of the base of the transverse process indicate that it is a felid, not a large hyaenid. It is of a size to fit the calcaneum.

Comparisons

The generally heavy construction of the calcaneum and the presence of a distinct navicular facet indicate that this large felid is a machairodont. The specimen compares in size to *Smilodon* (Merriam and Stock, 1932), but differs in morphology.

Discussion of the Abu Dhabi Taxa

Four carnivore taxa are identified in the collections from Abu Dhabi. They include *Plesiogulo praecocidens;* a very large, indeterminate species of hyaenid;

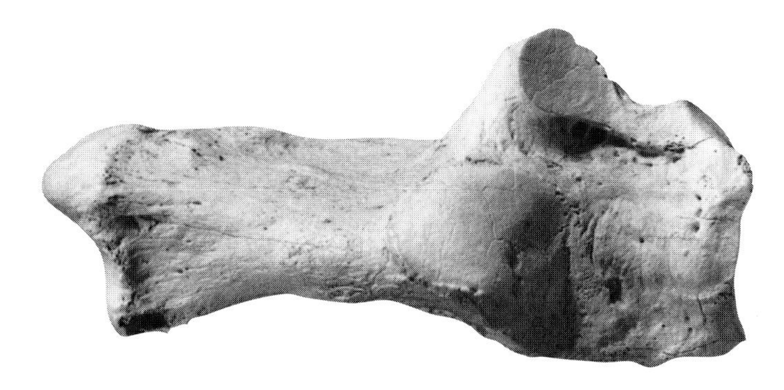

Figure 17.4 Machairodontinae gen. et sp. indet., complete right calcaneum, AUH 241.

a second medium-sized, indeterminate hyaenid; and an indeterminate machairodontine felid. With low population densities and moderately high diversity within communities, individual carnivore species tend to be represented in fossil assemblages by one or at most a few specimens. A fossil assemblage may therefore have a seemingly disproportionate number of carnivore species given the paucity of remains overall. The Abu Dhabi carnivores illustrate this double principle, with four taxa for six specimens. It is also noteworthy that the known taxa are of medium to large size. Presumably many smaller taxa were also present, but were not preserved in the fossil assemblages due to chance and biases against the preservation of small animals.

The documented Abu Dhabi carnivore species are typical of the late Miocene and Pliocene of Eurasia and Africa. Medium-sized hyaenas and sabre-tooth cats are widely distributed by the late Miocene, about 9 or 10 million years (Ma) ago. Large species of both groups, however, more typically are not present until the end of the Miocene or even within the Pliocene—that is, in the time range between 7 and 4 Ma.

The appearance of *Plesiogulo* near 6 Ma ago is one of several events used to define the base of the late Hemphillian Land Mammal Age in North America, where the genus has a restricted time range (Harrison, 1981; Tedford et al., 1987). In the Old World, however, the genus is both younger and older, with an approximate time range from 9 to 4 Ma ago. The early Pliocene species are larger and markedly different from the taxon at Abu Dhabi. They include *Plesiogulo monspessulanus* from Europe and *P. major* from China. The oldest Eurasian record I know for the genus is an unpublished occurrence from Pakistan that is c. 9 Ma, while Solounias (1981) records a possible occurrence at Pikermi in Greece that may be of similar age. The only African record to date is *P. monspessulanus* from South Africa at about 4 Ma (Hendey, 1978).

As noted, *Plesiogulo praecocidens* is based on two specimens, both from Baode in China. Flynn et al. (1995) have suggested on the basis of a faunal analysis that most of the Baode localities are late Miocene in age, and are probably older than 6.5 Ma. The closely similar species *Plesiogulo crassa* also occurs in China in the Baode and younger levels, and is documented in Pakistan as coming from sediments lying in the lower normal zone of chron C3A of the magnetic time scale, that is between 6.6 and 6.3 Ma.

The Abu Dhabi carnivores as a whole do not suggest any particular habitat. *Plesiogulo* is similar to, and possibly an ancestor of *Gulo,* which has usually been considered as indicative of boreal forest or woodland. Harrison (1981), however, has pointed out that wolverines are also animals of open tundra, and that North American species of *Plesiogulo* are most abundant in fossil assemblages with hypsodont and cursorial ungulates. She interprets such assemblages as indicating grassy and open habitats, which might as well characterise the habitats of the Abu Dhabi fossils.

Acknowledgements

I thank Phil Crabb (Photographic Studio, The Natural History Museum, London) and Al Coleman (Peabody Museum, Harvard University) for photographs of the specimens.

References

Flynn, L. J., Qiu, Z., Opdyke, N., and Tedford, R. H. 1995. Ages of key fossil assemblages in the Late Neogene terrestrial record of northern China. In *Geochronology, Time Scales, and Stratigraphic Correlation.* SEPM Special Publication no. 54: 365–73. Society of Economic Paleontologists and Mineralogists, Tulsa.

Harrison, J. A. 1981. A review of the extinct wolverine, Plesiogulo (Carnivora: Mustelidae), from North America. Smithsonian Contributions to Paleobiology 46: 1–27.

Hendey, Q. B. 1978. Late Tertiary Mustelidae (Mammalia, Carnivora) from Langebaanweg, South Africa. Annals of the South African Museum, Cape Town 76: 329–57.

Kurtén, B. 1970. The Neogene wolverine Plesiogulo and the origin of Gulo (Carnivora, Mammalia). Acta Zoologica Fennica 131: 1–22.

Merriam, J. C., and Stock, C. 1932. The Felidae of Rancho La Brea. *Publications of the Carnegie Institution of Washington* 422: 1–231.

Solounias, N. 1981. Mammalian fossils of Samos and Pikermi: Resurrection of a classic Turolian fauna. *Annals of Carnegie Museum* 50: 231–69.

Tedford, R. H., Skinner, M. F., Fields, R. W., Rensberger, J. M., Whistler, D. P., and Galusha, B. E. 1987. Faunal succession and biochronology of the Arikareean through Hemphillian interval (Late Oligocene through earliest Pliocene Epochs) in North America. In *Cenozoic Mammals of North America,* pp. 153–210 (ed. M. O. Woodburne). University of California Press, Berkeley.

Viret. J. 1954. Le loess a bancs durcis de Saint-Vallier (Drôme) et sa faune de mammifères villafranchiens. *Nouvelles Archives du Muséum d'Histoire Naturelle de Lyon* 4: 1–200.

Zdansky, O. 1924. Jungtertiäre Carnivoren Chinas. *Palaeontologia Sinica* C, 2: 1–155.

Miocene Elephantids (Mammalia) from the Emirate of Abu Dhabi, United Arab Emirates: Palaeobiogeographic Implications

18

PASCAL TASSY

The first discovery of an elephantoid in the Neogene of Abu Dhabi was an isolated tooth found at Jebel Barakah by Glennie and Evamy (1968) and identified by Madden et al. (1982) as *Stegotetrabelodon grandincisivum* (see Whybrow, 1989). Later, Peter Whybrow and Andrew Hill visited Abu Dhabi and collected fossils in the Baynunah region, at Jebel Barakah, Shuwaihat, and Jebel Dhanna (Whybrow et al., 1990). In this chapter I describe the elephantoid specimens, all of them found since 1989 in the Baynunah Formation, mainly at Shuwaihat.

Proboscideans from the Shuwaihat locality were found in several sites. Site S6 yielded two individuals allocated to *Stegotetrabelodon*. The first is an incomplete skeleton, the taphonomy of which is given by Andrews (1999—Chapter 24). This skeleton is composed of the skull (cranium and mandible), atlas, 7 thoracic vertebrae, 1 lumbar vertebra, 11 nearly entirely preserved ribs, right scapula, partial right radius, right femur, and right and left tibiae. Size and growth stage of these different elements are in agreement, so that it is very likely that they belong to the same individual (a young adult, with M^3 not erupted, and epiphyses of long bones not fused). An isolated left P^4 (site S6; AUH 560) is thought to belong to another individual although its state of wear is compatible with that of M^1–M^2. The reasons are: (1) the roots are eroded and the tooth was probably transported; (2) on the cranium there is no clearly defined alveolus close to the M^1s (but the bone is damaged), so that the premolars were already lost.

Other localities have yielded elephantoid remains. One edentulous juvenile mandible (AUH 475) can be allocated to *Stegotetrabelodon*, as can isolated teeth (except one fragment, which is a deinothere).

Teeth that match those seen on the skull found at site S6 are a partial indeterminate molar ?M^2 (AUH 234), a right M_3 (AUH 456), and a right P_4 (AUH 342). Isolated postcranial elements are a magnum (AUH 271), a navicular (AUH 240), and a metapodial without proximal extremity (AUH 479). They all show elephantine traits compatible with their allocation to *Stegotetrabelodon*.

Only one partial enamel fragment can be allocated to *Deinotherium* sp. (AUH 21), already listed by Whybrow et al. (1990).

In conclusion, two proboscidean species are present in the collection found by The Natural History Museum/Yale University team, *Stegotetrabelodon* being the dominant taxon.

In addition, the only tooth from Jebel Barakah (1965 I 112, Bayerisches Staatsammlung, Munich) found by Glennie and Evamy (1968) belongs to another species, perhaps *"Mastodon" grandincisivus*, but not a stegotetrabelodont.

I have devoted this chapter to systematics to obtain a better understanding of the biogeographic history of *Stegotetrabelodon*. I focus on cranial and dental remains because these are the anatomical parts previously used to discuss the taxonomy and relationships of the genus, which are controversial.

 ISBN 0-300-07183-3

Stegotetrabelodon syrticus from the Baynunah Formation

Proboscidea Illiger, 1811
Elephantoidea Gray, 1821
Elephantidae Gray, 1821
Stegotetrabelodon Petrocchi, 1941
Stegotetrabelodon syrticus Petrocchi, 1941

Material

Cranium with right and left I^2, M^1–M^2, erupting left M^3 (AUH 502); mandible with right I_2, ?M_2–erupting M_3 (AUH 503); atlas (AUH 555); thoracic vertebrae (AUH 501, 504, 505, 535, 536, 545, 550); lumbar vertebra (AUH 563); ribs (AUH 532, 538, 542, plus eight additional ribs not yet conserved); right scapula (AUH 528); distal epiphysis of right radius (AUH 500); right femur (AUH 506); right tibia (AUH 534); left tibia (AUH 524); left P^4 (AUH 560); right P_4 (AUH 342); right M_3 (AUH 456); portion of edentulous juvenile mandible (AUH 475); navicular (AUH 240); magnum (AUH 271); portions of $M^{?2}$ (AUH 234).

Description

The skull from Shuwaihat (site S6) belongs to a young adult (figs 18.1–18.3). M^1–M^2 are in function, with worn four lophs and postcingulum of M^1, worn first and second lophs of $M^{?2}$ (wear

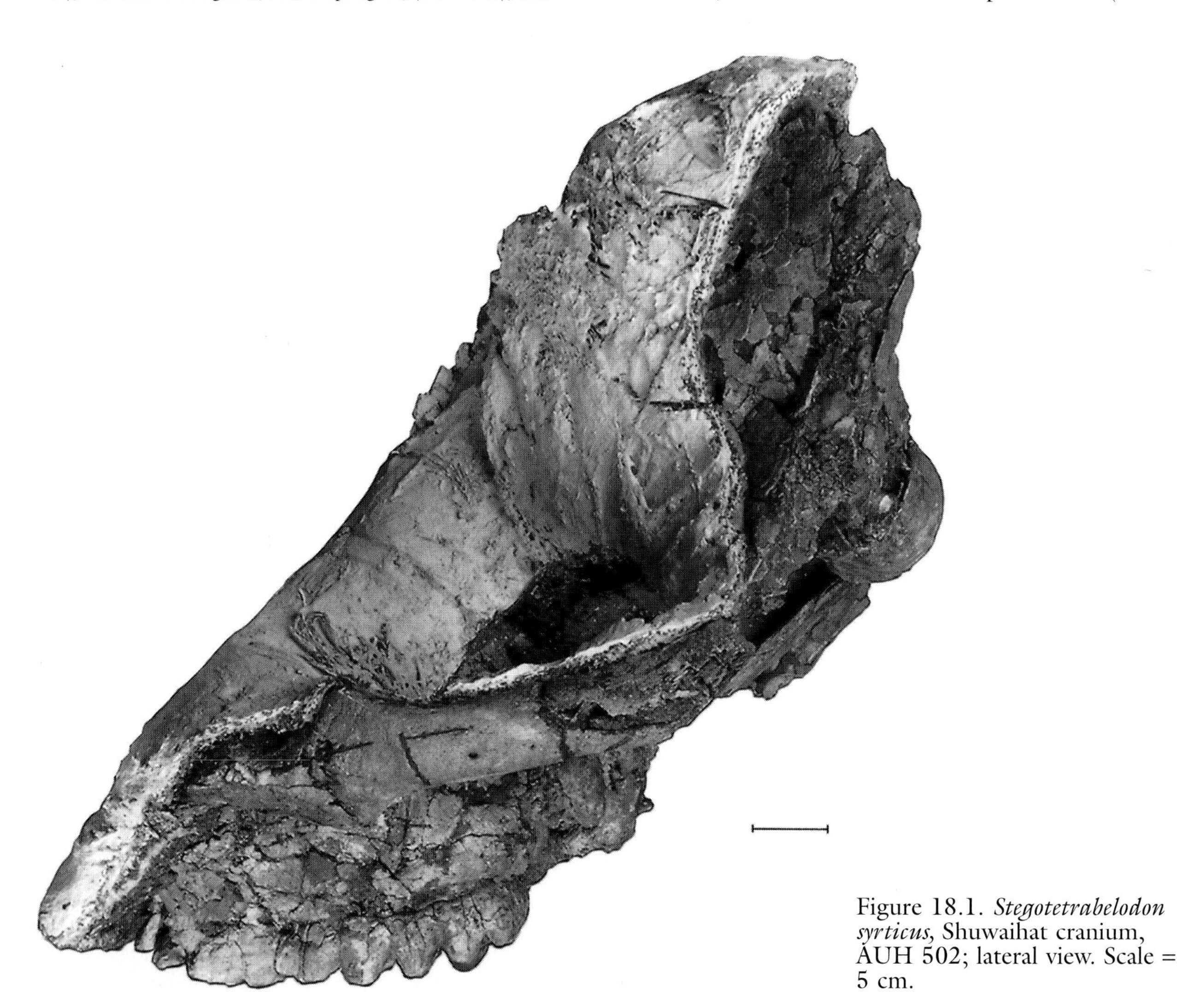

Figure 18.1. *Stegotetrabelodon syrticus*, Shuwaihat cranium, AUH 502; lateral view. Scale = 5 cm.

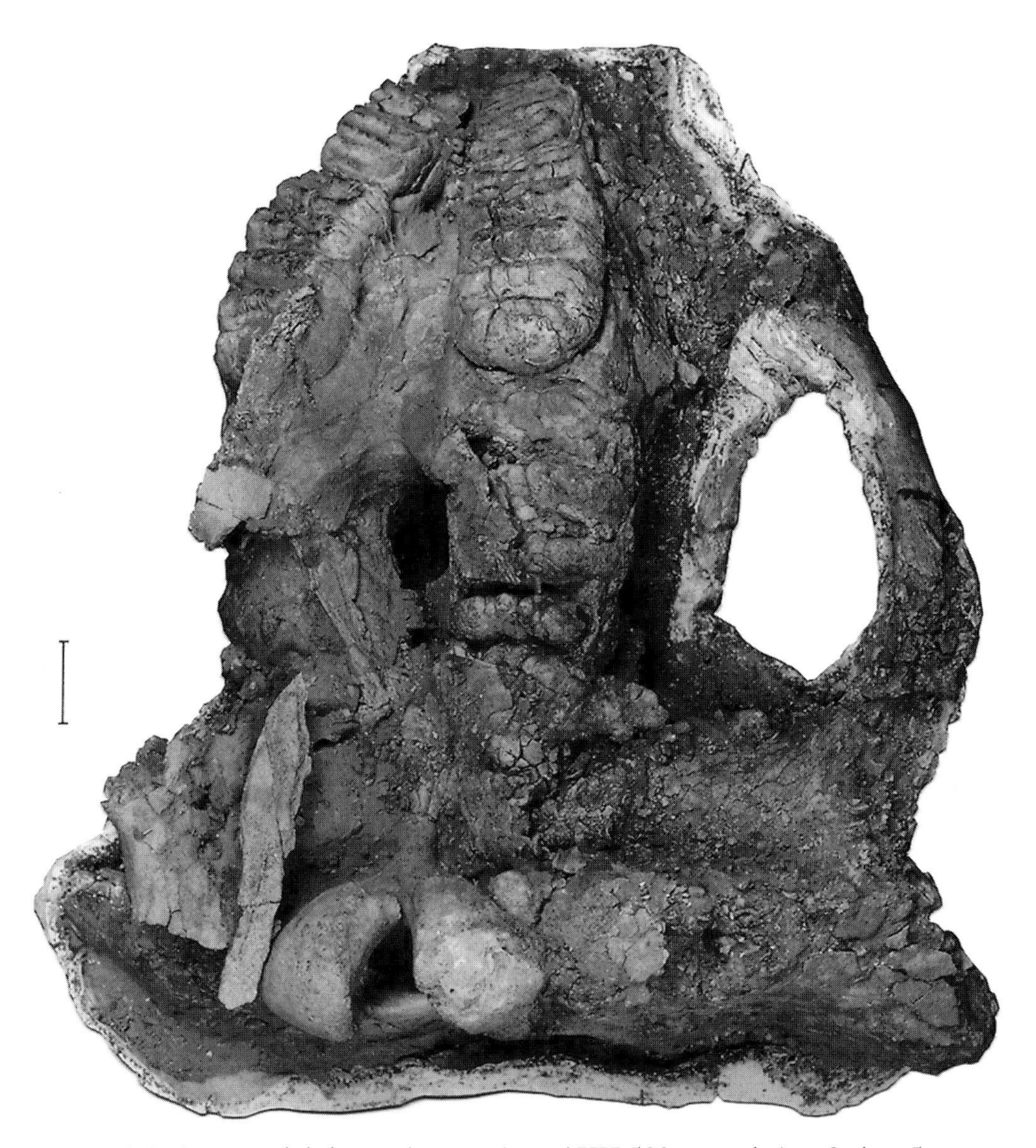

Figure 18.2. *Stegotetrabelodon syrticus,* cranium, AUH 502; ventral view. Scale = 5 cm.

figures), and slightly worn third loph (dentine visible at the top); the alveolus of M^3 has opened with the first loph visible. Consequently, its African-age equivalent is 20–30 years (Saunders, 1977).

The cranium is heavily eroded, particularly the dorsal face. The rostrum is crushed (but the incisors were retained). The left zygomatic arch is preserved, although its anterior root is broken. The right arch is crushed on the posterior part of M^2, the alveolus of M^3, and the occipital condyle. The basicranium is eroded so that no detailed anatomical features can be seen. The paroccipital area overhangs the postglenoid fossa but the

Figure 18.3. *Stegotetrabelodon syrticus,* cranium, M^1–M^2, AUH 502; occlusal view. Scale = 5 cm.

weathering of the bone is such that it is difficult to estimate whether the postglenoid fossa is deep (a derived elephantid feature) or not. The basicranium is not distorted, however, and is elevated (the angle with the palatine plane is about 50–60°).

The measurements of the cranium (in mm) are as follows:

Palatal length from the anterior border of M^1 to the choanae	300
Length of basicranium from the choanae to the foramen magnum	300
Maximum width taken at the zygomatic arches	c. 820
Width of the basicranium taken at the lateral borders of the glenoid fossae	c. 820
Maximum width of the choanae	90
External maximum width of the palate	265
Internal width taken between the first loph of the M^2s	58
Internal width taken between the fourth loph of the M^2s	75
Length of the basicranium from the condyles to the area of the pterygoid process	c. 350

The mandible is crushed ventrally at the symphysis but it is entirely preserved with the right tusk in situ (fig. 18.4). The symphysis is strong, its tip is well preserved but the total length could not be measured with precision because its posterior face is crushed (it is estimated to be 370 mm). The dorsal face of the mandible, especially the mid-portion, is heavily weathered and still embedded in its plaster jacket (including the occlusal surface of the molars), and therefore this face was not observed during this study.

The upper tusks are long, straight, oval in cross-section, and devoid of an enamel band (figs 18.5 and 18.6). The right tusk is the best preserved and its tip is nearly complete. The overall length is 1020 mm. The cross-section is 85.6 × 70.3 mm at 270 mm from the tip. The right lower tusk is preserved. It is long, protruding out of the symphysis. The visible length of the tusk is 530 mm, longer than the symphysis length, which is estimated at

Figure 18.4. *Stegotetrabelodon syrticus,* mandible, AUH 503 (same individual as AUH 502); ventral view. Scale = 20 cm.

c. 370 mm. The cross-section is oval with a longitudinal dorsal sulcus (= subpiriform cross-section) (figs 18.7 and 18.8). The dentine of the lower tusk has a concentric lamellar structure.

The intermediate molars (M^1 and M^2) belong to the tetralophodont grade (fig. 18.9, tables 18.1 and 18.2). M^1 is tetralophodont *sensu stricto.* M^2 is nearly pentalophodont; its postcingulum is strongly inflated, forming a narrow and incipient fifth loph but this loph is much lower than the fourth and is only followed by an enamel bump at the cervix (on the right M^2 only). In conclusion, the fifth loph of M^2 is not complete and is a perfect intermediate between tetralophodont and pentalophodont molars.

Five lophs of the left M^3 are visible on the labial side but the posterior part of the molar lies in the bone so that it is impossible to ascertain the complete ridge formula (fig. 18.10 and tables 18.1 and 18.3). The lophs of the upper molars show a primitive plate-like pattern typical of primitive elephantids—that is, a mosaic of primitive ("gomphotheriid" traits) and derived (elephantid) traits.

Derived elephantid traits are as follows: the cusps forming the lophs are more or less of equal size, including the mesoconelets (conelets that are close to the median sulcus); the cusps are transversely aligned—when marked, the wear of the pretrite and posttrite parts is nearly equal so that it forms a plate-like wear figure, not, as for gomphotheres, a typical trefoiled one (especially clear on the worn M^1s); cement is plentiful.

Primitive "gomphotheriid" traits are as follows: the cusps are rather low; the persistence of a median sulcus; the lophs are made of few cusps—four cusps of M^2 (two on each half-loph), except the fourth of right M^2, which shows five apical digitations (three on the pretrite half-loph), five to six on M^3 (three on the pretrite half and two on the posttrite, except

Figure 18.5. *Stegotetrabelodon syrticus,* right upper tusk of cranium AUH 502; lateral view.

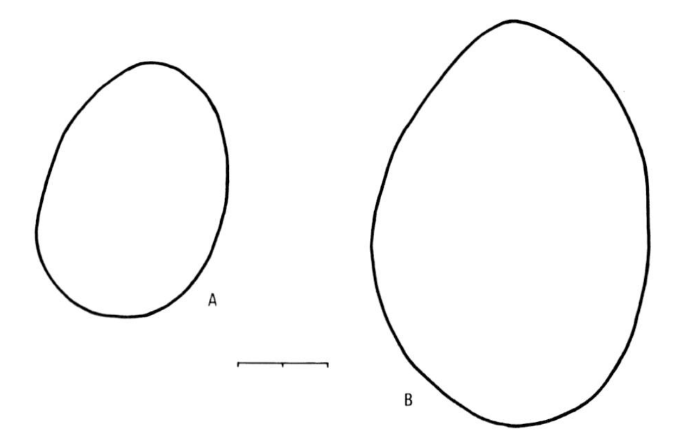

Figure 18.6. *Stegotetrabelodon syrticus*, cranium AUH 502. Cross-sections of the right upper tusk taken at 70 mm from the tip *(A)* and 270 mm from the tip *(B)*; rear views. Scale = 2 cm.

Figure 18.7. *Stegotetrabelodon syrticus*, mandible AUH 503. Cross-section of the right lower tusk taken at 500 mm from the tip.

Plate 18.1. *Stegotetrabelodon syrticus*, right M_3, AUH 456: *top*, occlusal view; *bottom*, labial view (front is right). Scale bar = 20 cm.

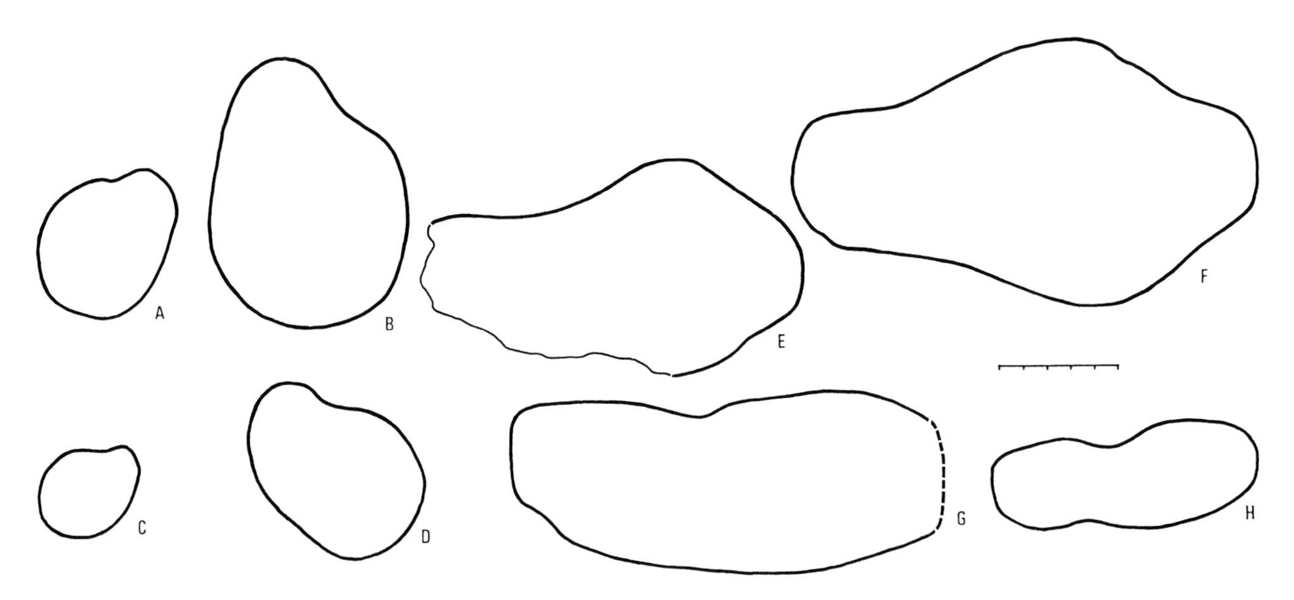

Figure 18.8. *A,* Cross-section of the lower tusk of *Stegotetrabelodon syrticus* from Shuwaihat, AUH 503, compared with *(B)* cf. *Stegotetrabelodon* from Jebel Semene; *C,* Elephantidae indet. (cf. *Stegotetrabelodon* seu *Primelephas*), Lukeino; *D, Tetralophodon longirostris,* Grossweissendorf; *E, "Mastodon" grandincisivus,* Maragheh (holotype); *F,* Pestszentlörincz; *G,* Kertch; *H,* Sahabi. Rear views are *(B)* from Bergounioux and Crouzel (1956), *(D)* from Steininger (1965), *(E)* redrawn from Schlesinger (1917), *(F)* redrawn from Schlesinger (1922), *(G)* redrawn from Pavlow (1903), and *(H)* from Gaziry (1987). Scale = 5 cm.

Table 18.1. Measurements (in mm) of the molars of the Shuwaihat cranium

	P	L	W	H	LF	ET	HI
rt M^1	4	119.0	$73.6^{(3)}$	—	3.9	4.0	—
lt M^1	4	116.7	$72.1^{(4)}$	—	3.8	—	—
rt M^2	4–5	174.4	c.$101^{(4)}$	c.$46^{(4)}$	3.0	—	c.45.5
lt M^2	4–5	176.9	$101.4^{(4)}$	50.2	2.9	—	49.5
lt M^3	5+	—	c.$103^{(4)}$	$59.7^{(4)}$	2.9	—	60.3

Abbreviations: P = number of plates; L = length; W = width; H = height; LF = laminar frequency; ET = enamel thickness; HI = height index. Superscripts on measurements denote the ridge number on which the measurement was taken.

Table 18.2. Comparative measurements (in mm) of M^2

	P	L	W	H	LF	ET	HI
Shuwaihat	4–5	174.4–176.9	101.4	50.2	2.9–3.0	—	49.5
S. syrticus	4–5	158.0–172.0	96.0	49.0	—	—	—
S. orbus	5	145.0–174.0	75.8–93.1	60.9	3.2–4.0	5.0–6.4	70.0

For key to abbreviations see Table 18.1.
Sources of data: S. syriticus from Sahabi (Gaziry, 1987); *S. orbus* from Lothagam and Adu-Asa (Maglio, 1973; Kalb and Mebrate, 1993).

for the fourth loph which has three on each half-loph); the persistence of the pretrite conules, although they are weakly developed, seen on M^2 (on the posterior face of the second loph and, as a bump, on the anterior face of third loph) and on M^3 (one on the posterior face of the first loph and two on the posterior face of the second loph); and the shape of the lophs convex/convex (according to the criteria of Kalb and Mebrate, 1993).

Isolated teeth found in the Baynunah region share the same morphological pattern with those of the skull: an incomplete upper molar, made of six parts (AUH 234), fits with the M^2 seen on the cranium. The cusps are massive, the enamel is thick (6.5–7.3 mm), and the cement is plentiful.

An isolated M_3 (AUH 456) has eight lophids and a narrow postcingulum made of two small cuspules (Plate 18.1, p. 214 and table 18.4). The lophids are convex/convex (except the second which is nearly rectilinear/convex). The apex of the lophids are subdivided into five or six cusps. The mesoconelets are as developed as the main cusps. The pretrite trefoil in the first interloph is complete, with a strong conule on the posterior face of

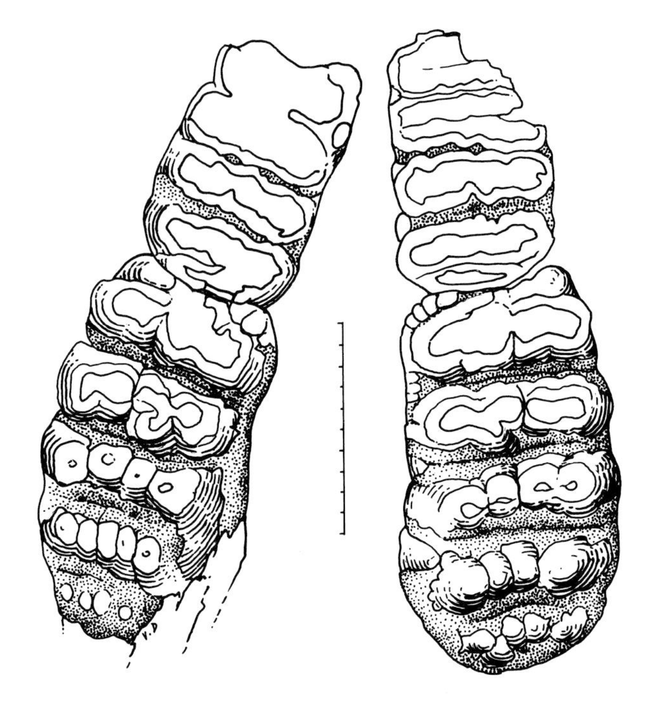

Figure 18.9. *Stegotetrabelodon syrticus*, right and left M^1–M^2 of cranium AUH 502; occlusal view. Scale = 10 cm.

Figure 18.10. *Stegotetrabelodon syrticus*, dissected germ of the left M^3 in cranium AUH 502; lateral view. Scale = 3 cm.

Table 18.3. Comparative measurements (in mm) of M^3

	P	L	W	H	LF	ET	HI
Shuwaihat	5+	—	c.103	59.7	2.9	—	60.3
S. syrticus	6	232.0–242.0	109.8–126.1	73.0–80.1	2.8–3.2	6.0–7.1	65.6–67.2
S. orbus	6	212.0–255.9	93.4–110.5	69.8–75.1	2.4–3.3	5.0–6.6	68.4–74.8

For key to abbreviations see Table 18.1.
Sources of data: S. syrticus from Sahabi and *S. orbus* from Lothagam (Maglio, 1973).

Table 18.4. Comparative measurements (in mm) of M_3

	P	L	W	H	LF	ET	HI
AAM 456	8	277.0	109.1	62.0	3.0	6.6	56.4
S. syrticus	7	280.0–317.4	115.0–123.4	57.0–74.1	2.6–3.1	5.8–6.0	60.1
S. orbus	7	234.3–285.0	81.6–109.3	75.1–81.6	2.8–3.3	5.6–6.4	63.1–83.4
Stegotetrabelodon sp.	6–8	241.6+	94.0–102.0	81.0	3.0	5.4–6.8	79.0

For key to abbreviations see Table 18.1; + = probable underestimate.
Sources of data: S. syrticus from Sahabi (modified from Maglio, 1973); *S. orbus* from Lothagam (Maglio, 1970, 1973); *Stegotetrabelodon* sp. from Uganda (Tassy, 1995).

Table 18.5. Comparative measurements (in mm) of P^4 and P_4

	P	L	W	H	WI
P^4 AUH 560	2	49.7	47.6	28.2	95.8
P^4 *S. syrticus*	2–3	(45.0)–54.0	(41.0)–44.0	—	81.5–(91.1)
P^4 *S. orbus* (KNM-LT 343)	2–3	51.0+	46.5	—	91.2–
P_4 AUH 342	2–3	40.1+	33.0	—	82.3–
P_4 *S. orbus* (KNM-MP 47)	2–3	49.0+	43.0	—	87.7–

For key to abbreviations see Table 18.1. Measurements in parentheses are estimates (usually for a partly damaged tooth); + = probable underestimate; – = probable overestimate (due to the underestimated length).
Sources of data: S. syrticus from Sahabi (Gaziry, 1987); *S. orbus* from Lothagam and Mpesida (Tassy, 1986).

Figure 18.11. *Stegotetrabelodon syrticus,* left P^4, AUH 560: *top,* labial view; *bottom,* occlusal view (front is left).

Figure 18.12. *Stegotetrabelodon syrticus,* right P_4, AUH 342; occlusal view (front is left).

the first lophid. It is made of posterior conules only in the second to fourth interlophids. There is no posttrite conule. The lophids are moderately high. The two first lophids are slightly worn, the wear facets are nearly as pronounced on the pretrite and posttrite halves. The enamel is thick (6.6 mm). The occlusal face is deeply concave, a derived trait.

Both the number of lophids (which exceeds that of the M_3 of tetralophodont gomphotheres) and the plate-like pattern fit with the morphology of upper molars. The concavity of the occlusal face is associated with an open erupting angle of molars, a trait coherent with the elevated basicranium of Shuwaihat.

Two isolated premolars (P^4 and P_4) belong to the tetralophodont grade (figs 18.11 and 18.12 and table 18.5). The left P^4 (AUH 560) comes from the same site as the skull. It looks rather primitive. It is slightly worn and distinctively shows two lophs and a strong postcingulum, which, nevertheless, does not form a true third loph. The cusps are round; the pretrite and posttrite half-lophs are well separated by a strong median sulcus. The second loph is lower than the first, which is reminiscent of a P^3, but the large size of this tooth make it unlikely to be a P^3. The postcingulum is made of two cusps connected to the pretrite cusp of the second loph.

The right P_4 from Hamra (site H1, AUH 342) is damaged. The two anterior lophids show a bunodont pattern, a primitive state compatible with that of P^4 (AUH 560), but also to any taxon of tetralophodont grade. On the second lophid the posttrite half is made by two cusps. The lophids are not truly plate-like. The postcingulum is strong and could form a true third lophid, but the enamel is lacking on the posterior side of the tooth so that it is impossible to see the development of this postcingulum towards a third lophid. The pretrite cusp of this postcingulum/third lophid is almost entirely preserved. It is large and subdivided at the apex. The posttrite half-lophid is broken but appears to be developed. This P_4 is tentatively allocated to *S. syrticus.*

Specimen AUH 475—Edentulous Mandible

A portion of an edentulous mandible (AUH 475) from Jebel Barakah with a broken symphysis belongs to a young individual (fig. 18.13). It shows the roots of a right P_4, the nearly resorbed alveole of a left P_4 (only the anterior root is partly preserved), the roots of a right and left M_1, the partial anterior root of a right M_2, and fragments of an erupting left M_2. The size of the broken roots of the right P_4 fits with that of the P_4 AUH 342 (length = 46.1 mm; posterior width = 30.2 mm). The symphysis is broken but what is preserved shows the ventral curve seen on the mandibles of *Stegotetrabelodon*. It is rather narrow as in *Stegotetrabelodon* and tetralophodont gomphotheres, and not enlarged as in amebelodontids. Premolars are known to be retained in *Stegotetrabelodon* and two are present in the Abu Dhabi collection. A complete plate of the broken erupting left M_2 still in the alveole has been dissected. It shows the nearly elephantine plate-like structure of stegotetrabelodont molars. The cusps are columnar and nearly identical. The plate is low (length = c. 85 mm; height = 44.7 mm). The height index (c. 52.6) is compatible with that of the M_3 AUH 456 (height index = 56.4). Hence, this specimen can be allocated to the same taxon as the Shuwaihat skull. The measurements (in mm) of AUH 475 are as follows:

Posterior symphysial width	177.6
Internal width between the anterior grinding teeth (here P_4)	101.0
Width of the horizontal ramus taken at the anterior root of P_4	57.0 (right) 53.0 (left)
Width of the horizontal ramus taken at the M_1 (= at the posterior mental foramen)	67.2 (right) 62.3 (left)
Maximum width of the horizontal ramus	110
Maximum height of the horizontal ramus	146.6 (right) 154.0(left)

Postcranial Elements

The elephantid postcranial remains found in the Baynunah Formation do not give precise taxonomic information. The limb bones are more gracile than those of gomphotheres but the proximal epiphyses are larger than those of extant elephants, as seen on the femur and tibiae. These traits can be compared with those of the postcranials of *S. syrticus* described at Sahabi (Petrocchi, 1954). This morphology, somehow intermediate between "mastodonts" and elephants, is expected in primitive elephantids such as *Stegotetrabelodon*.

Affinities of the Shuwaihat Material

The skull from Shuwaihat compares best with *Stegotetrabelodon syrticus* Petrocchi, 1941, from Sahabi, Libya. Isolated teeth are all compatible with *S. syrticus*. The Abu Dhabi sample can be considered more primitive than the Sahabi sample as described by Petrocchi (1943, 1954) and Gaziry (1987), but the differences are not profound.

At Sahabi, *S. syrticus* is represented by several individuals, including a cranium and mandible of the same individual described in detail by Petrocchi (1954). Size differences between both skulls from Shuwaihat and Sahabi are mainly due to individual age. The Sahabi skull is older (the M3s are worn, M2s are lost) so that the ever-growing tusks are much larger. Nevertheless, in both localities the tusks are very close in proportion and morphology. The lower tusk from Shuwaihat, with the length of its protruding part longer than the symphysis, displays an apomorphic condition (autapomorphy of the genus). Although protruding lower tusks are known in the late Miocene European species *Tetralophodon longirostris* (skull from Esselborn, Vallesian, Germany; see Tobien, 1978: pl. 13), they are not so protruding—they are shorter than the symphysial length, independently of the individual age. The oval cross-section and concentric lamellar dentine of lower tusks from Sahabi, Lothagam (Kenya), and Shuwaihat are identical, and close to cf. *Stegotetrabelodon* from Tunisia, Elephantidae indet. from Lukeino (Kenya), and *T. longirostris* (fig. 18.14). Upper tusks from Abu Dhabi (figs 18.5 and 18.6) and Sahabi are straight without an enamel band and with an oval to round section. The upper tusks of the late Miocene species *T. longirostris* from Europe and *Paratetralophodon hasnotensis* from the Siwaliks of Pakistan (Tassy, 1983) have the same cross-section (as for trilophodont gomphotheres). They also lack enamel, but they are curved ventrally (plesiomorphic condition).

The isolated M_3, with its plate-like lophids, matches the M_3 of *S. syrticus*, with comparable lamellar frequency and only slightly lower height index ("hypsodonty index"); it is outside the variation seen in *Tetralophodon longirostris*. Some primitive traits are present on the upper molars and premolars found in the Baynunah Formation. The M^1 is tetralophodont as in primitive tetralophodont gomphotheres and *S. syrticus* from Sahabi. The fifth loph of M^2 is low, narrow, and not connected with a distinct postcingulum; it forms the posterior part

Figure 18.13. *Stegotetrabelodon syrticus,* partial edentulous mandible with roots of P_4, M_1, M_2, and partial right M_2, AUH 475; *left,* occlusal view; *right,* lateral view.

of the tooth. True pentalophodont M^2s (that is with a postcingulum separated from the fifth loph) are not present in primitive tetralophodont gomphotheres, and are common at Sahabi. But an isolated M^2 from Sahabi with exactly the same formula as that of the M^2 of the skull from Shuwaihat is described by Gaziry (1987).

Two traits combined on the Shuwaihat molars, probably primitive, are reminiscent of *Stegolophodon,* a genus of tetralophodont grade known in Asia from the early Miocene up to the Pliocene. The molars are relatively low and the lophs are made of few big cusps (four on the M^2s). Intraspecific variation was already suspected as "*Stegolophodon sahabianus* sp. nov." (one M^3 only) was described by Petrocchi (1954) at Sahabi and later considered a synonym of *Stegotetrabelodon syrticus* (Maglio, 1973). The height index is much lower at Abu Dhabi than in *S. orbus* from Lothagam (table 18.2). In any case, the shape and height of the unworn upper M^3 of the skull from Shuwaihat (estimated height index is only slightly lower than in *S. syrticus* from Sahabi), and the morphology of the isolated M_3 are clearly outside the range of *Stegolophodon* of any geological age. Although molars of *S. syrticus* with loph(ids) showing numerous apical digitations (up to six or seven cusps) are known at Sahabi, simpler molars with four cusps on each loph are also found in this locality (three M^2s described by Gaziry, 1987: 194).

The premolars found in the Baynunah Formation seem more primitive than those described at Sahabi by Gaziry (1987) and at Lothagam and Mpesida (Kenya) by Maglio and Ricca (1977) and Tassy (1986). The posterior loph of P^4 is not platelike, and this tooth could be allocated to any other tetralophodont genus, such as *Tetralophodon.* The intraspecific variation of the premolars of trilopho-

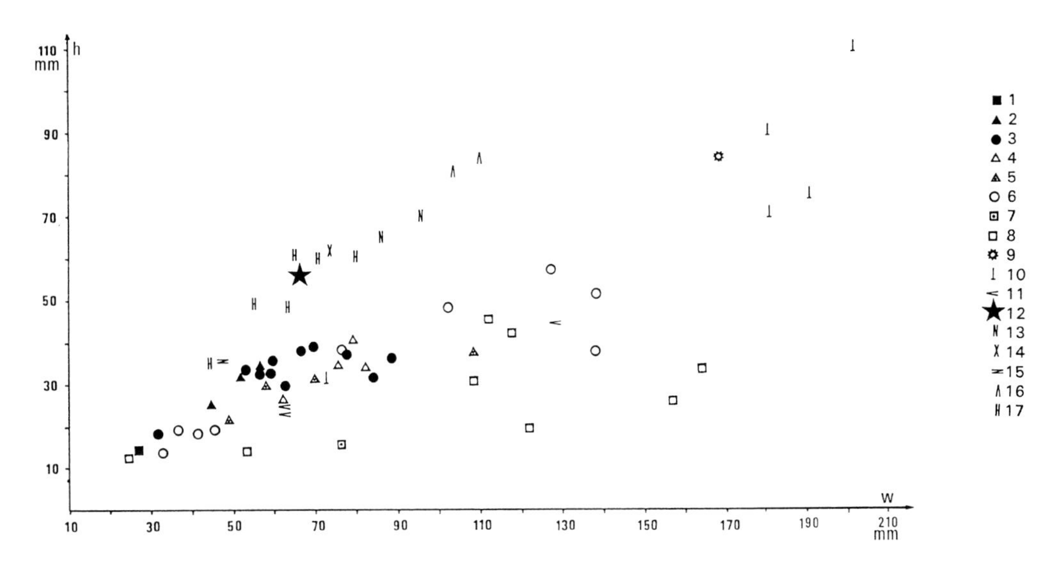

Figure 18.14. Scatter diagram of lower tusk (cross-section = width × height) of *Stegotetrabelodon* (12, 13, 14) compared with Elephantidae indet. (cf. *Stegotetrabelodon* seu *Primelephas*) (15), cf. *Stegotetrabelodon* (16), *Tetralophodon longirostris* (17), and Amebelodontidae (1–11) including "*Mastodon*" *grandincisivus* (10, 11). 1, cf. *Archaeobelodon* (Rusinga-Hiwegi; Kenya); 2, *Archaeobelodon* aff. *filholi* (Buluk, Mwiti; Kenya); 3, *Archaeobelodon filholi* (Western Europe); 4, *Protanancus macinnesi* (Maboko, Alengerr; Kenya); 5, *Protanancus chinjiensis* (Chinji, Pakistan) and *Protanancus* sp. (Yürükali, Turkey); 6, *Amebelodon* (North America); 7, *Platybelodon* sp. (Loperot, Kenya); 8, *Platybelodon* and *Torynobelodon* (China; Georgia; Turkey; North America); 9, "*Mastodon*" *grandincisivus*, holotype (Maragheh, Iran); 10, "*M.*" *grandincisivus* (Kertch, Ukraine; Orjachovo, Bulgaria; Pestszenlörincz, Hungary; Dhok Pathan, Pakistan); 11, "*Amebelodon cyrenaicus*" (Sahabi, Libya); 12, *Stegotetrabelodon syrticus* (Shuwaihat, Emirate of Abu Dhabi); 13, *S. syrticus* (Sahabi, Libya); 14, *S. orbus* (Lothagam, Kenya); 15, Elephantidae indet. = cf. *Stegotetrabelodon* seu *Primelephas* (Lukeino, Kenya); 16, cf. *Stegotetrabelodon* (Jebel Semene, Tunisia); 17, *Tetralophodon longirostris* (Germany; Austria).

dont gomphotheres is known to be important (Tassy, 1985), and the same can be expected for tetralophodont gomphotheres and primitive elephantids. It would be premature to conclude that the evolutionary stage of the premolars from Abu Dhabi indicates the presence of a species more primitive than *Stegotetrabelodon syrticus*. Yet, this trend observed on premolars (especially P^4) and molars (especially M^2) indicate that *S. syrticus* is possibly more primitive in the Baynunah Formation (and, consequently, perhaps older) than at Sahabi.

Gaziry (1987) considers that *S. syrticus* and *S. orbus* are synonymous. Indeed, diagnostic features of *S. orbus* (according to Maglio, 1973) are very tenuous and some do not apply—such as the higher number of lophids of M_2 (which is based on KNM-LT 342, a *Primelephas* tooth not a *Stegotetrabelodon*: Tassy, 1986; Gaziry, 1987; Kalb and Mebrate, 1993), and the central conules only present on the two anterior lophs of M_3 (they are on five lophids of one M_3, KNM-LT 352). Other features, such as the absolute size, are very variable. In the sample from Lothagam, however, there are no molars (especially M^2) as simple as those from Abu Dhabi and as several of those from Sahabi. The height index of M^3 and M_3 of *S. orbus* is higher than at Sahabi and Abu Dhabi (tables 18.3 and 18.4), which is a derived trend. I therefore provisionally conclude that *S. orbus* can be retained as a small, more-evolved member of *Stegotetrabelodon*.

Elephantoidea indet. (?"*Mastodon*" *grandincisivus* Schlesinger, 1917) from Jebel Barakah

The M_3 found by Glennie and Evamy (1968) at Jebel Barakah and described by Madden et al. (1982) under the binomen *Stegotetrabelodon grandincisivus* remains the only tooth from the Baynunah Formation that does not belong to a stegotetrabelodont elephantid. This molar is different from those of the Shuwaihat skull and from the partial M_2 of the mandible AUH 475, also found at Jebel Barakah.

This large, anteriorly damaged tooth (fig. 18.15) shows five preserved lophids and a small postcingulum (preserved length = 195.6 mm; width taken at the second lophid = 88.2 mm). The cusps are cone-like, bulbous, and do not form plate-like lophids. Unequal wear divides the lophids into distinct pretrite and posttrite halves. Gomphotheroid pretrite trefoils are present, associated in the two anterior lophids with complete posttrite trefoils. Cement is present, especially in the posterior interlophids. This molar does not exhibit elephantid features, seen in the Shuwaihat sample on the M_3, AUH 456, and in stegotetrabelodont molars in general. Its allocation to the species "*Mastodon*" *grandincisivus* is possible though not warranted. In particular, the Jebel Barakah molar does not show pseudo-anancoid contacts between successive pretrite and posttrite half-lophids, a feature present, though variable, in the molars of "*Mastodon*" *grandincisivus* (see below for status of *Mastodon* (*Bunolophodon*) *grandincisivus* Schlesinger, 1917).

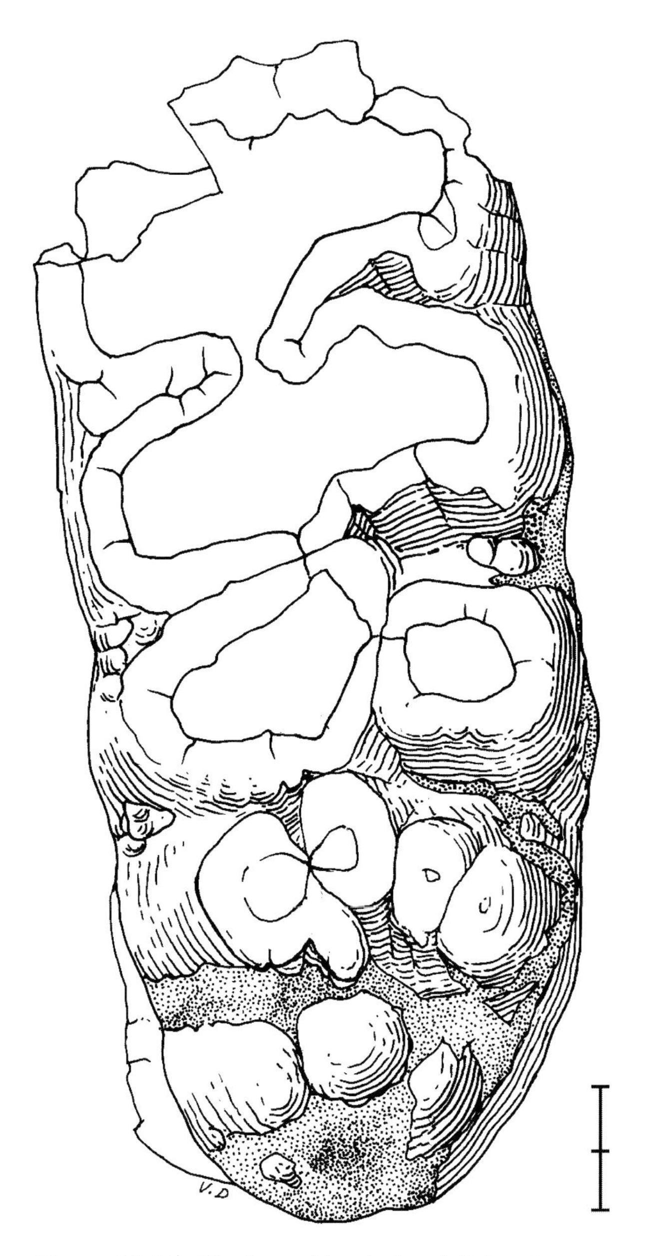

Figure 18.15. Elephantoidea indet. (cf. Amebelodontidae, ?"*Mastodon*" *grandincisivus*), from Jebel Barakah, left M_3 (1965, I 112, Bayerisches Staatssammlung, Munich). Scale = 2 cm.

Systematics and Palaeobiogeography

Nomenclatural Precision on the Genus *Stegotetrabelodon* Petrocchi, 1941 and the Type Species *S. syrticus* Petrocchi, 1941

The diagnostic features and extent of the genus *Stegotetrabelodon* are controversial. In discussing the taxonomy of *Stegotetrabelodon* we must bear in mind the fact that the definition and extent of the genus should be based first on the traits—derived traits—of the type species. The allocation of any specimen to the genus *Stegotetrabelodon* means that such a specimen shows a derived trait found in *Stegotetrabelodon* and not in other elephantoid taxa. These traits must be checked with reference to the type species. Even if, as usual, the diagnosis of the genus is emended in relation to successive discoveries, at least one character of *Stegotetrabelodon* not found elsewhere should make the diagnosis of the genus. This diagnostic trait should, however, be present in the type species.

In respect to African species (the type species *S. syrticus* Petrocchi, 1941, and *Stegotetrabelodon orbus* Maglio, 1970), the genus *Stegotetrabelodon* differs from other elephantoids of tetralophodont grade in having the following:

1. Strongly elongated lower tusks (oval in cross-section), the protruding part of the tusks being longer than the symphysis.
2. Straight upper tusks, not curved ventrally.
3. Pentalophodont upper and lower M2s.
4. Pattern of the loph(id)s of molars outlining the elephantine pattern; that is, more or less equal development of the tubercles, forming linear ridges ("plate-like pattern").

The fourth character is a probable elephantine synapomorphy. The third character is also an elephantine synapomorphy (but a possible homoplasy found in Stegodontidae). The second character is a probable elephantid apomorphy. The first character is an autapomorphy of *Stegotetrabelodon,* but the oval (that is, also suboval, piriform, round) cross-section is a plesiomorphy inside the group made up of gomphotheres and elephants (including stegodontids).

Inside the elephant–tetralophodont elephantoid group, *Stegotetrabelodon* displays the following plesiomorphies:

1. Upper tusks not twisted upwards, without an enamel band.
2. Mandibular symphysis directed downwards.
3. Upper and lower dP4s and upper and lower M1s with four lophs.
4. Persistence of upper and lower P4s (perhaps P3s, see Gaziry, 1987).
5. Persistence of some basic gomphothere traits in the molars with central conules and median sulcus still distinct.

These plesiomorphic traits are seen in *S. syrticus* from Sahabi. Gaziry (1987) described at Sahabi tetralophont dP_4 and M_1. Several upper and lower P4s that clearly display a strong postcingulum typical of the tetralophodont grade are known from Sahabi (Gaziry, 1987) and from Mpesida and Lothagam (Maglio and Ricca, 1977; Tassy, 1986).

In conclusion, from a cladistic viewpoint, the elongation of the lower tusks is the only autapomorphy of *Stegotetrabelodon* (Tassy, 1986; Kalb and Mebrate, 1993).

In respect to the type species of *Stegotetrabelodon,* a linguistic error occurred in Maglio (1973) and was frequently repeated. I recall here the correction I have already given (Tassy, 1986) because the error appears in later papers (that is, Gaziry, 1987; Tobien et al., 1988). According to Petrocchi (1941: 110) the type species of *Stegotetrabelodon* is *S. syrticus.* Maglio (1973) claimed that another name appeared in the same paper on a previous page and that the binomen of the type species should be *S. lybicus* Petrocchi, 1941. In fact no *S. lybicus* appears in this publication (*S. lybicus* appears in Petrocchi, 1943). Only the name "*M. angustidens* var. *lybica*" appears in Petrocchi (1941: 110) for a totally different taxon (a gomphothere of trilophodont grade). The valid binomen is *Stegotetrabelodon syrticus.*

Palaeobiogeography and Taxonomic Difficulties

A palaeobiogeographic understanding of the genus *Stegotetrabelodon,* and its impact on the understanding of the faunal relationships between Africa and Eurasia in the late Miocene/early Pliocene epoch, depends heavily on basic taxonomic knowledge. In the particular case of *Stegotetrabelodon,* however, there is no consensus on the definition and extant of the genus.

To make things simple, two "schools" of diagnostic thought exist: one recognises *Stegotetrabelodon* as an exclusive African genus (Maglio, 1970, 1973; Tassy, 1985, 1986; Kalb and Mebrate, 1993); the other recognises *Stegotetrabelodon* as a diverse genus in the Old World—that is, not only Africa but also Europe and Asia (Sarwar, 1977; Tobien, 1978, 1980; Zhou and Zhang, 1983; Gaziry, 1987; Tobien et al., 1988). These two schools give the genus different attributes (that is, different definitions or diagnoses) and different extents (that is, allocate different sets of species to the genus). Hence, different palaeobiogeographic scenarios are proposed by the two groups of authors.

Stegotetrabelodon as an African Genus

Two species of *Stegotetrabelodon* from Africa are universally recognised—the type species *S. syrticus* Petrocchi, 1941 from Sahabi, Libya, and the East African species *S. orbus* Maglio, 1970 from Lothagam 1, Kenya. *Stegotetrabelodon orbus* has been recognised by Kalb and Mebrate (1993) in Ethiopia (Middle Awash, lower Adu-Asa Formation, and Kuseralee Member). *Stegotetrabelodon* sp. (most probably *S. orbus*) is also present in Kenya in the Mpesida Beds (Tassy, 1986) and a somewhat primitive *Stegotetrabelodon* unidentified at the species level has also been found in the Kakara and Oluka Formations of Uganda (Tassy, 1995). Bergounioux and Crouzel (1956; see Tobien, 1978) described a partial mandibular symphysis from the late Miocene of Jebel Semene (Tunisia), with its tusks sectioned (first erroneously interpreted as premaxillae with upper tusks). The cross-section of the tusks, in shape and size, is comparable to that of *Stegotetrabelodon*. The known chronological range of *Stegotetrabelodon* in Africa is restricted to the late Miocene, and perhaps early Pliocene in Ethiopia (a partial uncollected specimen from Kuseralee Member described by Kalb and Mebrate, 1993: 43). Sahabi is dated late Miocene/early Pliocene—that is, about 5 million years (Ma). According to Geraads (1982, 1989) it is late Miocene, whereas, according to Boaz (1987), it is early Pliocene. Lothagam 1—at least Member B which yielded both *S. orbus* and *Primelephas gomphotheroides*—is dated about 6 Ma (Maglio, 1973; Behrensmeyer, 1976); the Mpesida Beds are dated 7 Ma (Bishop, 1976). The Kakara and Oluka Formations of Uganda are about 6–9 Ma (Pickford et al., 1993).

Cladistic analyses of elephantids (Tassy, 1990, 1996; Kalb and Mebrate, 1993; Kalb et al., 1996a) give some contradictory results. The only unambiguous apomorphy shared by *S. syrticus* and *S. orbus* is the elongation of lower tusks. As a clade, the genus *Stegotetrabelodon* is considered (Tassy, 1990) as the sister group of (*Primelephas* (*Stegodibelodon* (*Loxodonta* (*Elephas, Mammuthus*)))), while a stem group for this clade (and also of *Anancus* and Stegodontidae) is represented by widespread tetralophodont gomphotheres. Consequently, the differentiation of Elephantidae was considered to have taken place in Africa, its origin being cosmopolitan, at the scale of the Old World. Tetralophodont gomphotheres of middle and late Miocene age are described in Europe (*Tetralophodon longirostris*), in China (*Tetralophodon xiaolongtanensis*), in South East Asia (*Paratetralophodon hasnotensis*), and in Africa (an unamed species described in the late Miocene Namurungule and Ngorora Formations by Nakaya et al., 1984 and Tassy, 1986, respectively).

On the other hand, either *Stegodon* or the clade (*Stegolophodon, Stegodon*) could be included among the Elephantidae according to Kalb and Mebrate (1993) and to Tassy (1996), respectively. Also, according to Tassy (1996), the emergence of *Stegotetrabelodon* outside the tetralophodont gomphotheres is only one among several equally plausible hypotheses. (Since this latter paper was written, however, one unambiguous elephantid synapomorphy—a deep postglenoid fossa—was confirmed to be present in *Stegotetrabelodon syrticus*, so that the elephantid affinities of that species appear to be the better supported hypothesis.) According to Kalb and Mebrate (1993) and Kalb et al. (1996a), the Stegodontidae (genera *Stegolophodon* and *Stegodon*) are not monophyletic and *Stegolophodon* is the sister group of (*Stegotetrabelodon* (*Stegodibelodon* (*Stegodon* (*Primelephas* (*Loxodonta* (*Elephas, Mammuthus*)))))). The inclusion of *Stegodon* among the Elephantidae seems to be based on a questionable hypothesis of homology. Kalb and Mebrate (1993) consider that the plate-like loph(id)s of the molars of stegodonts and elephantids are homologous, and Tassy (1990) recognises for the clade Stegodontidae a distinct "stegolophodont" pattern on the molars. In that case the numerous plate-like loph(id)s of *Stegodon* on the one hand and of Elephantinae on the other would be homoplastic. The palaeobiogeographic conclusions reached by Kalb and Mebrate (1993) and Kalb et al. (1996b) is that the putative common ancestor of elephantids may have immigrated to Africa during the Tortonian and that its descendants—including *Stegotetrabelodon*—are of African origin. The discovery of a stegotetrabelodont in Abu Dhabi indicates African affinities of the Arabian Peninsula fauna during the late Miocene, as well as a plausible extra-African (*sensu stricto*) origin of the Elephantidae, bearing in mind that during the late Miocene the Arabo-African Plate is one unity, palaeogeographically. But the emergence of Elephantinae (that is, the lineage leading to *Primelephas* and other elephantines) seems to have occurred in Africa.

Stegotetrabelodon as a Widespread Old World Genus

In addition to the two unquestioned African species, different authors allocate to *Stegotetrabelodon* various Eurasian species (four or even six species).

European Species

Tobien (1978, 1980) followed by Gaziry (1987) allocate to *Stegotetrabelodon* two European late Miocene species, the original naming being *Mastodon* (*Bunolophodon*) *grandincisivus* Schlesinger, 1917 (also in the Middle East), and *Mastodon longirostris* forma *gigantorostris* Klähn, 1922 from Germany and Austria.

The holotype of *Mastodon* (*Bunolophodon*) *grandincisivus* is found in the late Miocene of Maragheh, Iran, and referred specimens selected by Tobien (1978) come from Hungary, Ukraine, and Bulgaria. An isolated molar from Jebel Barakah (Abu Dhabi) was allocated to that species by Madden et al. (1982). An isolated molar from Zaire was also allocated to that species by Madden (1982). *Mastodon longirostris* forma *gigantorostris* Klähn, 1922—recognised later by Klähn (1931) as a distinct species—is described in the late Miocene of the Dinotheriansande Formation (Vallesian) in Germany, and in Austria.

Asian Species

From the Dhok Pathan Formation of Pakistan (Mahluwal, Jhelum District), Sarwar (1977) recognises the new species *Stegotetrabelodon maluvalensis*. From China, Tobien et al. (1988) allocate to *Stegotetrabelodon* the species *Tetralophodon exoletus* Hopwood, 1935. Also from China three taxa are allocated to "?*Stegotetrabelodon*" by Tobien et al. (1988). They are *Stegotetrabelodon* sp. described by Zhou and Zhang (1983) from the Pliocene of Shensi Province, *Mastodon lydekerri* Schlosser, 1903, and *Stegodon primitium* Liu, Tang, and You, 1973.

The palaeobiogeographic conclusions drawn from these taxonomic statements are as follows: because the late Miocene age of the so-called stegotetrabelodonts from Europe and Pakistan precedes that of the African species, *Stegotetrabelodon syrticus* and *S. orbus*, the origin of *Stegotetrabelodon* was postulated to have taken place in either Europe or Asia (Tobien, 1978, 1980; Gaziry, 1987). According to Gaziry (1987), "*S. grandincisivus*" is the ancestor of the lineage *S. orbus*/*S. syrticus*. Madden et al. (1982) considers that "*S. grandincisivus*" is present in Abu Dhabi and represents the most primitive species of its genus. According to Tobien (1978, 1980) "*S. gigantorostris*" from the Vallesian of Europe is the oldest record of *Stegotetrabelodon* and represents the root of the lineage leading to *S. orbus*/*S. syrticus*. Tobien et al. (1988), who do not identify the genus *Stegotetrabelodon* in the Indo-Pakistanian region, see connections of the Chinese species to the eastern European forms.

On the other hand, Sarwar (1977) postulates the synonymy between the genera *Stegotetrabelodon* Petrocchi, 1941 and *Primelephas* Maglio, 1970, and accepts *Stegotetrabelodon maluvalensis* from the late Miocene of Pakistan as the probable ancestor of the Elephantinae.

Taxonomic Synthesis

Previous computerised cladistic analyses based on parsimony (Tassy, 1985; Tassy and Darlu, 1986, 1987; Kalb and Mebrate, 1993; Kalb et al., 1996a) do not support the idea that the species *Mastodon* (*Bunolophodon*) *grandincisivus* Schlesinger, 1917, *Mastodon gigantorostris* Klähn, 1922, and *Stegotetrabelodon maluvalensis* Sarwar, 1977 belong to the genus *Stegotetrabelodon*. The reasons are summarised in the following discussion.

Status of Mastodon (Bunolophodon) *grandincisivus Schlesinger, 1917*

According to Tobien (1978), supplementary to the holotype (a portion of a lower incisor from Maragheh, Iran described by Schlesinger, 1917), representative specimens of "*S. grandincisivus*" are: (1) a partial mandible with M_3 and lower tusks, but no symphysis, associated to upper tusks and M^3 from Pestszentlörincz, Hungary (Schlesinger, 1922); (2) lower and upper tusks associated with upper and lower M3s from Kertch, Crimea, Ukraine (Pavlow, 1903); and (3) a mandible with lower tusks and M_3 associated with an upper tusk and M^3, from Orjachovo, Bulgaria (Bakalov and Nikolov, 1962).

Distinctive characters are:

1. The lower tusks are among the largest known in the Elephantoidea (width = 195 mm from Pestszentlörincz; see Schlesinger, 1917).
2. The tusks protrude outside the rostrum (a trait checked on isolated tusks because of the very long lateral wear facet).
3. The cross-section of the lower tusks is flattened or even concave dorsally.
4. The dentine in the central part of the lower tusks is tubular.
5. The upper tusks are large.
6. The molars have accessory posttrite cusps, with a pseudo-anancoid pattern (alternate contacts between pretrite and posttrite conules in the interloph(ids)).
7. A thick deposit of cement is present on the upper and lower M3s.

This combination of characters is unique among elephantoids. Only traits 2, 5, and 7 are shared by the African species of *Stegotetrabelodon* (and Abu Dhabi), but these traits are not exclusive of *Stegotetrabelodon*. Characters 2 and 5 are known among trilophodont gomphotheres (for example, *Gomphotherium steinheimensis*) and tetralophodont gomphotheres (for example, *Tetralophodon longirostris*); character 7 is known among various taxa, especially amebelodontids such as *Platybelodon*. Traits 1, 3, 4, and 6 are in contradiction with the definition of *Stegotetrabelodon* based on the type species *S. syrticus*, with other African stegotetrabelodonts, and with the specimen from Abu Dhabi. The proportion and morphology of the lower tusks (figs 18.8 and 18.15) clearly separate "*grandincisivus*" from *S. syrticus* and *S. orbus*. Trait 4 is characteristic of the amebelodontid genera *Platybelodon* and *Torynobelodon* (perhaps synonymous) and traits 2, 3, 6, and 7 (and, of course, 4) are found among amebelodontids. This taxon is also present in the Dhok Pathan Formation of the Middle Siwaliks, Pakistan, described as "cf. ? Amebelodontidae" by Tassy (1983) on the basis of a small—probably juvenile—lower tusk with a cross-section close to that of the holotype and the specimen from Pestszentlörincz.

Finally, this species has also been described in the Sinda Beds of Zaire by Madden (1982) under the binomen *Stegotetrabelodon grandincisivus*. This record is based on an isolated tetralophodont intermediate molar (M_1 or M_2). According to Madden its specific identification is based only on the large size of the molar, assuming it is an M_1 and not an M_2. As this choice cannot be demonstrated, the allocation of this molar to "*grandincisivus*" is more than tentative and is not endorsed here.

No new data refute the former conclusion (Tassy, 1985, 1986), that "*grandincisivus*" is an amebelodontid related to *Platybelodon*. Moreover, I consider that the enigma is somehow resolved by Gaziry's (1987) description for the Sahabi species of "*Amebelodon cyrenaicus* sp. nov". The species is known from molars and tusks. It is a tetralophodont amebelodontid with flat lower tusks with dentinal tubules. The cross-section of the tusks described by Gaziry (1987: 184) is identical (as are the dentinal tubules) to that of the Kertch mandible described by Pavlow (1903) and very close to that of the Orjachovo mandible (personal observation, 1986). As noted earlier, these two specimens were allocated later to "*S. grandincisivus*" by Tobien (1978) followed by Gaziry (1987), though Gaziry himself does not compare these tusks. It is very likely that "*cyrenaicus*" is a junior synonym of "*grandincisivus*". The cross-section of the tusk from Sahabi (as well as every tusk allocated to "*grandincisivus*", including the holotype) is not as flat as that of the amebelodontid genera characterised by dentinal tubules (*Platybelodon* and *Torynobelodon*). The generic allocation of the Sahabi amebelodontid is still open to question. Yet the remarkable discovery of the Sahabi amebelodontid is exactly what was expected to explain the affinities of "*grandincisivus*".

Status of Mastodon mongirostris forma gigantorostris *Klähn, 1922*

According to Tobien (1978, 1980) "*S. gigantorostris* (Klähn, 1922)" is a distinct species and not a morph (variety) of *Tetralophodon longirostris* as Klähn (1922) first hypothesised, and is restricted to the late Miocene of Germany and Austria—Bermersheim (holotype: Klähn, 1922, 1931) and Grossweiffendorf (Steininger, 1965) in Germany, Geiereck (Laarberg), Belvedere Sandpit, Vienna (Schlesinger, 1917), and Kornberg in Austria (Mottl, 1969).

The characters listed by Tobien (1980) do not give a clear distinction between "*S. gigantorostris*" and *T. longirostris*. "*S. gigantorostris*" is merely a larger morph of *T. longirostris*, with upper and lower M3s more complex and with more cement. The lower tusks are more protruding outside the symphysis and the symphysis and lower tusks are less downcurved. Only the strong protrusion of the lower tusks is compatible with the allocation of the "species" to *Stegotetrabelodon*. On the mandible from Bermersheim, Dinotheriensande (holotype of "*gigantorostris*"; see Tobien, 1980) the visible part of the lower tusk is longer than the length of the symphysis but the rostrum is partly restored, so that the respective lengths are conjectural. In a well-preserved mandible of *T. longirostris* from Esselborn, Dinotheriensande (Tobien, 1978: pl. 14) the visible part of the lower tusks is 70% of the length of the symphysis. Moreover, the fact that the symphysis and the lower tusks are less downcurved in "*gigantorostris*" is in contradiction with the definition of *Stegotetrabelodon* (and from this viewpoint, *T. longirostris*—not "*gigantorostris*"—appears to be closer to stegotetrabelodonts from Africa). No recent data refute the early conclusion (Tassy, 1985) that polymorphism of the species *T. longirostris* is an acceptable alternative that explains, in the European tetralophodont sample, the variation of the shape of the mandible and lower tusks, as well as that of the absolute size.

Status of Stegotetrabelodon maluvalensis *Sarwar, 1977*

The species *Stegotetrabelodon maluvalensis* from Mahluwal (Dhok Pathan Formation of Pakistan) is restricted to the holotype, a brevirostrine mandible with M_2–M_3. The species is allocated to *Tetralophodon* by Tobien (1978) and to *Stegolophodon* by Tassy (1983, 1985).

Sarwar (1977) considers that the brevirostrine species *Primelephas gomphotheroides* from the late Miocene of East Africa belongs to *Stegotetrabelodon*. The M_2s from Mahluwal are tetralophodonts, those of African stegotetrabelodonts and of *P. gomphotheroides* are pentalophodont.

The short symphysis clearly distinguishes the mandible from Mahluwal from that of African stegotetrabelodonts and is only compatible with *Stegodibelodon schneideri* from Chad (Coppens, 1972) or with *Primelephas gomphotheroides* though no complete mandible of this latter species has been published so far.

The size and number of plates of the M_3 of the mandible from Mahluwal with eight plates (or 7*x*—*x* expressing a distinct postcingulum, here made of three cusps) recalls African stegotetrabelodonts and *Primelephas gomphotheroides*. The brachydonty, however, contradicts the allocation of the specimen to *Stegotetrabelodon* or *Primelephas* (height of M_3 = 52 mm according to Sarwar, less than the lowest value for *S. syrticus* from Sahabi, which is 57 mm according to Maglio, 1973), while the lowest value for *S. orbus* is 75.1 mm according to Maglio (1970). It must be acknowledged that Maglio (1973: pl. 19, table 3) gives 44.1 mm for the minimum value of *S. orbus*. This measurement is here taken as a mistake as no specimen of *S. orbus* that I studied at the National Museums of Kenya (KNM), Nairobi, reaches this value.

The symphysis has two small alveoli, considered

by Sarwar (1977) to recall *P. gomphotheroides* because this latter species was described as having tusks (Maglio, 1970, 1973; Maglio and Ricca, 1977). This assertion was disputed by Kalb and Mebrate (1993). Personal examination (July 1994 at KNM, Nairobi) of KNM-LT 358—the only partial mandible of *P. gomphotheroides* described by Maglio and Ricca (1977)—does not support the hypothesis that a portion of a tusk is associated with the partial symphysis. There is no alveolus, only a mandibular canal. Because no evidence exists so far in support of the presence of lower tusks in *Primelephas,* an isolated lower tusk from Lukeino allocated to cf. *Primelephas* by Tassy (1986) is called here "Elephantidae indet. *Stegotetrabelodon* seu *Primelephas*".

The mandible from Maluval is considered by Tassy (1983, 1985) to belong to *Stegolophodon* mainly because of the brachydonty of the molars and the even structure of the lophids due to the equal size of the cusps. Yet, according to Sarwar's measurements, the height index of the M_3 is in the range of the lowest molars of *Stegotetrabelodon.*

This mandible is also reminiscent of that of *Stegolophodon hueiheensis* Chow, 1959, a middle or late Pleistocene (?) species (later allocated to a new genus, *Rulengchia,* by Zhou (= Chow) and Zhang (1983), on the basis of both the shape of the symphysis and the presence of small alveoli. Early stegolophodonts from Thailand are known to have lower tusks with oval to piriform cross-sections (Tassy et al., 1992).

Computerised parsimony analyses (Tassy and Darlu, 1986, 1987) do not support a close relationship between *Stegotetrabelodon, Primelephas,* and *S. maluvalensis,* but support a connection of *S. maluvalensis* to (*Stegolophodon, Stegodon*). Yet, posttrite cusps described by Tobien (1978) on three plates of the M_3 of *S. maluvalensis* (not visible on Sarwar's illustration) is a trait in contradiction with the characters of the stegolophodont molars.

Status of Tetralophodon exoletus *Hopwood, 1935*

The holotype of *Tetralophodon exoletus* Hopwood, 1935 is a fragment of mandibular ramus with erupting M_3 from Shaanxi (formerly Shensi) Province, China. The M_3 is huge (length estimated at 295 mm by Tobien et al., 1988). Referred specimens are a portion of a juvenile mandible with P_3, dP_4, P_4, and M_1 described by Hopwood (1935), and isolated dP_2, upper and lower dP3, dP_4, and P_3 described by Tobien et al. (1988). The age of the species is "probably Boadean (= Turolian)", according to Tobien et al. (1988).

Tobien et al. (1988) ascribe *Tetralophodon exoletus* to *Stegotetrabelodon* on the basis of two traits: the large size of M_3 with $7x$ lophids; and intermediate molars tetralophodont (here dP_4 and M_1). The latter characters plus the absence of cement (both plesiomorphic traits) are considered by Tobien et al. (1988) as diagnostic of this somehow primitive species compared with African stegotetrabelodonts.

The lack of cement is in contradiction with the definition of *Stegotetrabelodon* based on the African type species. The tetralophodont grade of dP_4 and M_1 is compatible with *Stegotetrabelodon,* although it is known in all gomphotheres of tetralophodont grade. The large size of the M_3 would be the only trait shared by the Chinese taxon and the large *Stegotetrabelodon syrticus,* but large tetralophodonts are described elsewhere (for example, *Paratetralophodon hasnotensis* (Osborn, 1929) of the Dhok Pathan Formation of the Siwaliks in Pakistan). Moreover, the advanced pattern of the loph(ids) towards the elephantine pattern—a derived trait—seen on the molars of *Stegotetrabelodon syrticus* (as well as of *S. orbus*) is lacking in *Tetralophodon exoletus* Hopwood, 1935. Consequently, as the shape of the mandible and lower tusks of *Tetralophodon exoletus* Hopwood, 1935 is unknown, more material seems to be needed to warrant definitely the allocation of this large Chinese species of tetralophodont grade to the genus *Stegotetrabelodon.*

Other Taxa from China

Following Zhou and Zang (1983), who first described *Stegotetrabelodon* in China, Tobien et al. (1988) ascribe to "*?Stegotetrabelodon*" three taxa: *Stegotetrabelodon* sp. described by Zhou and Zhang (1983) from the Pliocene of Shaanxi (Shensi) Province; *Stegodon primitium* Liu, Tang, and You, 1973; and *Mastodon lydekerri* Schlosser, 1903.

1. *Stegotetrabelodon* sp. from Shaanxi (Zhou and Zhang, 1983) is only a posterior portion of M^3.
2. *Stegodon primitium,* described by Liu et al. (1973) from the Pliocene of Banguo (Yunnan Province), is two M_3s with a stegolophodont/stegodont pattern. Though a stegolophodont—*Stegolophodon banguoensis* Tang, You, Liu, and Pan, 1974—is known from the same area, Zhou and Zhang (1983) allocated the species described by Liu et al. (1973) to *Stegotetrabelodon.* Yet, according to the measurements given by Liu et al. (1973), the height index of the unworn fifth plate of the M_3 from Banguo is low (46.3) and typical of a brachydont molar such as that of stegolophodonts.
3. *Mastodon lydekkeri* Schlosser is restricted to a posterior part of a large M_3 described by Schlosser (1903). According to Tobien et al. (1988), this specimen (destroyed) is only tentatively allocated to *Stegotetrabelodon,* mainly on the basis of its large size.

In short, I follow the conclusion reached by Tobien et al. (1988: 208): "In view of the insufficient, and in parts questionable, documentation of the genus (that is, *Stegotetrabelodon*) in China it seems necessary to avoid further discussion until an improved collection".

Conclusions

The Abu Dhabi discovery brings a remarkable confirmation of the association of stegotetrabelodont characters previously found only in Africa. The molars with a stegotetrabelodont pattern (even if those of the skull from Shuwaihat are rather simple compared with *Stegotetrabelodon syrticus* from Sahabi and *S. orbus* from Lothagam and Mpesida) are associated with a straight upper tusk, with a mandible that has an unabbreviated symphysis, and with long narrow lower tusks, which are oval in cross-section and made of concentric dentinal laminae.

As far as we know, there is no unambiguous datum that supports the hypothesis of the presence

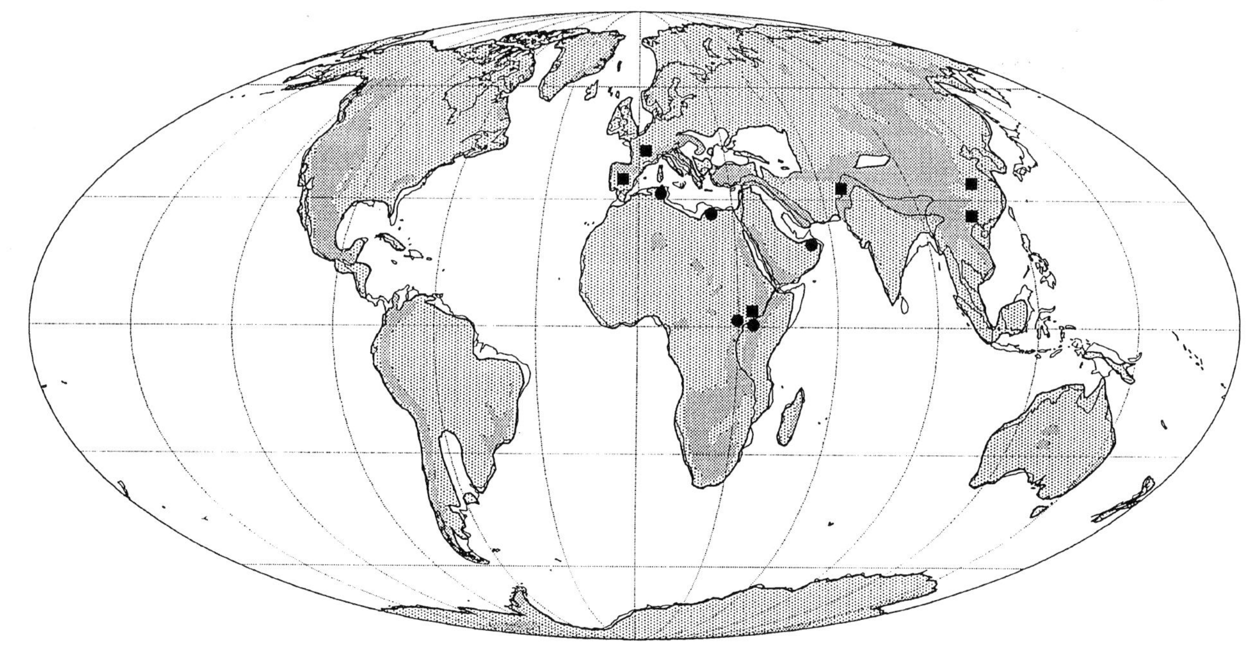

Figure 18.16. African distribution of the elephantid genus *Stegotetrabelodon* (circles) and cosmopolitan distribution of its Eurasian late Miocene stem group: the tetralophodont gomphotheres (squares), based on the Pliocene palaeocoastline at 5 million years ago (map taken from Smith et al., 1994).

of the genus *Stegotetrabelodon* in Eurasia, a conclusion already reached by Tassy (1985) and Kalb and Mebrate (1993). Simply summarised one can conclude that no definitive *Stegotetrabelodon* traits (autapomorphies) are present in those tetralophodont gomphotheres previously held as putative Eurasian stegotetrabelodonts. Therefore the stegotetrabelodont from Abu Dhabi is the first one discovered outside Africa, so far. Importantly, this discovery demonstrates the Arabo-African unity of the elephantid fauna.

The characters described from Abu Dhabi, however, do not permit firm conclusions on the age of the locality and of the geographic origin of the genus. If the evolution of *Stegotetrabelodon* is that of a unique lineage, and if the simple pattern of the molars is considered primitive, the locality can only be of late Miocene age, probably older than the age of Lothagam 1 (Member B)—that is, older than 6 Ma; it could be about 7 Ma or more. The fact that a primitive stage of *Stegotetrabelodon syrticus* lived in Arabia about 7 Ma ago gives further support to the idea of a cosmopolitan origin of the genus, firstly based on the fact that the probable stem group for *Stegotetrabelodon*—tetralophodont gomphotheres—is widespread in Eurasia and Africa (fig. 18.16).

Eurasian affinities of the late Miocene proboscideans from Africa can be hypothesised on the basis of heuristic hypotheses (cladistic patterns) and the geological record. As noted above, the tetralophodont gomphotheres are widespread. Although the search for a primitive Eurasian stegotetrabelodont brings deceptive results, the sister group of the clade (*Stegotetrabelodon,* other Elephantidae) can be located in any place in the Old World, including Eurasia. Yet, to elucidate the phylogeny of late tetralophodonts and early elephantids needs improvements in the description of reported late Miocene tetralophodonts, such as those of the Namurungule Formation, Kenya (Nakaya et al., 1984), and, as usual, the discovery of better specimens (associated crania and mandibles), especially in the middle Miocene (Astaracian) and late Miocene (Vallesian and Turolian) of Eurasia. The mandible of "*S. maluvalensis*" described by Sarwar (1977) also deserves a re-examination.

Association of proboscideans in the late Miocene of Africa also brings some light to the problem. If the isolated molar from Jebel Barakah described by Madden et al. (1982) belongs to the species "*Mastodon*" *grandincisivus,* the association of this species with *Stegotetrabelodon syrticus* recalls a situation known at Sahabi. At Sahabi, *Stegotetrabelodon* is associated with an amebelodontid, "*A. cyrenaicus*", here considered a synonym of "*Mastodon*" *grandincisivus.* As noted above, this latter species is known in Eastern Europe, the Middle East, and the Indian subcontinent. Yet, precise chronological and ecological association of "*Mastodon*" *grandincisivus* and *Stegotetrabelodon syrticus* in the late Miocene of Abu Dhabi is not definitely warranted (the two species, associated at Jebel Barakah, come from two different sites). At Sahabi, Lothagam 1, and Mpesida, *Stegotetrabelodon* is associated with *Anancus,* a widespread genus in Africa and Eurasia. Especially, possible sister-group relationships of the African species *Anancus kenyensis* and the species *A. sivalensis* from the Upper Siwaliks of Pakistan have been hypothesised (Tassy, 1986). At Lothagam 1 (and in the Baynunah Formation of Abu Dhabi), *Stegotetrabelodon* is associated with *Deinotherium,* a widespread genus in the Old World (the phylogeny of which is unknown at the species level).

Consequently, the endemism of proboscideans in the late Miocene of Africa is restricted to the Elephantidae only. Africa then includes the Arabian Peninsula from a palaeobiogeographic viewpoint. It is likely that new discoveries, such as that of the Shuwaihat *Stegotetrabelodon,* may change our image of elephantid diversification. After all, only three important cranial remains of *Stegotetrabelodon* (two associated crania and mandibles, and one mandible) have been discovered to date, including the Shuwaihat skull.

Acknowledgements

I thank Peter Whybrow for having given me the opportunity to study the Abu Dhabi elephantid remains. Peter's help was invaluable during the whole process of the study in The Natural History Museum, London, in 1992 and in 1994. Most of the photographs were made by Harry Taylor (NHM, London). Drawings were made by D. Visset (Université P. and M. Curie, Paris). I especially thank Diana Clements for her careful editing of the manuscript. Observations on KNM-LT 358 were made possible courtesy of Meave Leakey (National Museums of Kenya, Nairobi). In 1977 I was able to observe the specimen 1965 I 112, by courtesy of V. Fahlbusch and K. Heissig (Munich). Many thanks also to Peter Andrews, Alan Gentry, Jon Kalb, and Nancy Todd for fruitful discussions on the Abu Dhabi sites and faunas, and on the evolution of early elephantids.

REFERENCES

Andrews, P. J. 1999. Taphonomy of the Shuwaihat Proboscidean: Emirate of Abu Dhabi, Western Region. Chap. 24 in *Fossil Vertebrates of Arabia*, pp. 338–53 (ed. P. J. Whybrow and A. Hill). Yale University Press, New Haven.

Bakalov, I., and Nikolov, I.V. 1962. *Les Fossiles de Bulgarie. X—Mammifères tertiaries*. Académie Science Bulgarie, Sofia.

Behrensmeyer, A. K. 1976. Lothagam Hill, Kanapoi and Ekora: A general summary of stratigraphy and faunas. In *Earliest Man and Environments in the Lake Rudolf Basin*, pp. 163–70 (ed. Y. Coppens, F. C. Howell, G. Ll. Isaac, and R. E. F. Leakey). University of Chicago Press, Chicago.

Bergounioux, F. M., and Crouzel, F. 1956. Présence de *Tetralophodon longirostris* dans le Vindobonien inférieur de Tunisie. *Bulletin de la Société Géologique de France* 6: 547–58.

Bishop, W. W. 1976. Pliocene problems relating to human evolution. In *Human Origins*, pp. 139–53 (ed. G. Ll. I. Isaac and E. R. McCown). Benjamin, Menlo Park.

Boaz, N. T. 1987. Introduction. In *Neogene Paleontology and Geology of Sahabi*, pp. xi–xv (ed. N. T. Boaz, A. El-Arnauti, A. W. Gaziry, J. de Heinzelin, and D. D. Boaz). Alan R. Liss, New York.

Coppens, Y. 1972. Un nouveau proboscidien du Pliocène du Tchad, *Stegodibelodon schneideri* nov. gen. sp., et le phylum des Stegotetrabelodontinae. *Comptes Rendus de l'Académie des Sciences, Paris* 274-D: 2962–65.

Gaziry, A. 1987. Remains of Proboscidea from the Early Pliocene of Sahabi, Libya. In *Neogene Paleontology and Geology of Sahabi*, pp. 183–203 (ed. N. T. Boaz, A. El-Arnauti, A. W. Gaziry, J. de Heinzelin, and D. D. Boaz). Alan R. Liss, New York.

Geraads, D. 1982. Paléobiogéographie de l'Afrique du Nord depuis le Miocène terminal, d'après les grands mammifères. *Geobios*, mémoire spécial 6: 473–81.

———. 1989. Vertébrés fossiles du Miocène supérieur du Djebel Krechem El Artsouma (Tunisie Centrale). *Geobios* 22: 777–801.

Glennie, K. W., and Evamy, B. D. 1968. Dikaka: Plants and plant-root structures associated with aeolian sand. *Palaeogeography, Palaeoclimatology, Palaeoecology* 4: 77–87.

Hopwood, A. T. 1935. Fossil Proboscidea from China. *Palaeontologia Sinica* 9: 1–108.

Kalb, J. E., and Mebrate, A. 1993. Fossil elephantoids from the hominoid-bearing Awash Group, Middle Awash Valley, Afar depression, Ethiopia. *Transactions of the American Philosophical Society, Philadelphia* 83: 1–114.

Kalb, J. E., Froehlich, D. J., and Bell, G. L. 1996a. Phylogeny of African and Eurasian Elephantoidea of the late Neogene. In *The Proboscidea: Evolution and Palaeoecology of Elephants and their Relatives*, pp. 101–16 (ed. J. Shoshani and P. Tassy). Oxford University Press, Oxford.

———. 1996b. Palaeobiogeography of late Neogene African and Eurasian Elephantoidea. In *The Proboscidea: Evolution and Palaeoecology of Elephants and their Relatives*, pp. 117–23 (ed. J. Shoshani and P. Tassy). Oxford University Press, Oxford.

Klähn, H. 1922. *Die Badischen Mastodonten und ihre süddeutschen Verwandten.* Gebrücher Borntraeger, Berlin.

———. 1931. Rheinhessisches Pliozän besonders Unterpliozän in Rahmen des Mitteleuropäischen Pliozän. *Geologische und Paläontologische Abhandlungen* 18: 279–340.

Liu, H., Tang, Y. J., and You, Y. Z. 1973. A new species of *Stegodon* from Upper Pliocene of Yuanmou, Yunnan. *Vertebrata PalAsiatica* 11: 192–200.

Madden, C. T. 1982. Primitive *Stegotetrabelodon* from latest Miocene of Subsaharan Africa. *Revue de Zoologie Africaine* 96: 782–96.

Madden, C. T., Glennie, K. W., Dehm, R., Whitmore, F. C., Schmidt, R. J., Ferfoglia, R. J., and Whybrow P. J. 1982. *Stegotetrabelodon (Proboscidea, Gomphotheriidae) from the Miocene of Abu Dhabi.* United States Geological Survey, Jiddah.

Maglio, V. J. 1970. Four new species of Elephantidae from the Plio-Pleistocene of Northwestern Kenya. *Breviora* no. 341: 1–43.

———. 1973. Origin and evolution of the Elephantidae. *Transactions of the American Philosophical Society, Philadelphia* 63: 1–149.

Maglio, V. J., and Ricca, A. B. 1977. Dental and skeletal morphology of the earliest elephant. *Verhandelingen der Koninklijke Nederlandse Akademie van Wetenschappen, Afdeeling Natuurkunde, Eerste Reeks* 29: 1–51.

Mottl, M. 1969. Bedeutende Proboscidier-Neufunde aus dem Altpliozän (Pannonien) Südost-Österreichs. *Österreichische Akademie der Wissenschaften, Mathematisch-naturwissenschaffliche Klasse, Denkschriften* 115: 1–50.

Nakaya, H., Pickford, M., Nakano, Y., and Ishida, H. 1984. The late Miocene large mammal fauna from the Namurungule Formation, Samburu Hills, Northern Kenya. *African Study Monograph,* suppl. issue 2: 87–31.

Pavlow, M. 1903. *Mastodon angustidens* Cuv. et *Mastodon* cf. *longirostris* Kaup, de Kertch. *Annuaire Géologique et Minéralogique de la Russie* 4: 129–39.

Petrocchi, C. 1941. Il giacimento fossilifero di Sahabi. *Bollettino della Società Geologia Italiana* 60: 107–14.

———. 1943. Sahabi, eine neue Seite in der Geschichte der Erde. *Neues Jahrbuch für Mineralogie, Geologie und Paläontologie, Stuttgart* 1943-B: 1–9.

———. 1954. I Proboscidati du Sahabi. *Rendiconti dell'Academia Nazionale dei XL,* 4th ser., 4–5: 1–66.

Pickford, M., Senut, B., and Hadoto, D. 1993. *Geology and Palaeobiology of the Albertine Rift Valley, Uganda-Zaire. Volume I: Geology.* CIFEG Publication Occasionnelle no. 24. Centre International pour la Formation et les Echanges Géologiques, Orléans.

Sarwar, M. 1977. Taxonomy and distribution of the Siwalik Proboscidea. *Bulletin of the Department of Zoology, University of the Punjab* 10: 1–172.

Saunders, J. J. 1977. Late Pleistocene vertebrates of the Western Ozark Highlands, Missouri. *Reports of Investigations. Illinois State Museum of Natural History* 33: 1–118.

Schlesinger, G. 1917. Die Mastodonten des K. K. naturhistorischen Hofmuseum. *Denkschriften des Kaiserlich-Koniglichen (Wien) Naturhistorischen Hofmuseum, Geologisch-Paläontologische Reihe* 1: 1–230.

———. 1922. Die Mastodonten der Budapester Sammlungen. *Geologica Hungarica,* editio separata 2: 1–284.

Schlosser, M. 1903. Die fossilen Säugetiere Chinas nebst einer Odontographie der recenten Antilopen.

Abhandlungen der Bayerischen Akademie der Wissenschaften, Mathematisch-Naturwissenschaftliche 2: 1–221.

Smith, A. G., Smith, D. G., and Funnell, B. M. 1994. *Atlas of Mesozoic and Cenozoic Coastlines.* Cambridge University Press, Cambridge.

Steininger, F. 1965. Ein bemerkenwester Fund von *Mastodon* (*Bunolophodon*) *longirostris* Kaup, 1832 (Proboscidea, Mammalia) aus dem Unterpliozän (Pannon) des Hausruck-Kobernausserwald-Gebietes in Österreich. *Jahrbuch der Geologischen Bundesanstalt* 108: 195–212.

Tassy, P. 1983. Les Elephantoidea miocènes du Plateau de Potwar, Groupe de Siwalik, Pakistan. *Annales de Paléontologie* 69: 99–136, 235–97, 317–54.

———. 1985. La place des mastodontes miocènes de l'Ancien Monde dans la phylogénie des proboscidiens (Mammalia) : hypothèses et conjectures. Thèse Dr. ès Sci. *Mémoires des Sciences de la Terre, Université Curie, Paris* 85-34: 1–862.

——— 1986. *Nouveaux Elephantoidea (Mammalia) dans le Miocène du Kenya.* Cahiers de Paléontologie. Editions du Centre National de la Recherche Scientifique, Paris.

———. 1990. Phylogénie et classification des Proboscidea (Mammalia) : historique et actualité. *Annales de Paléontologie* 76: 159–224.

———. 1995. Les proboscidiens (Mammalia) fossiles du Rift Occidental, Ouganda. In *Geology and Palaeobiology of the Albertine Rift Valley Uganda-Zaire. Volume II: Paléobiologie/Palaeobiology,* pp. 215–55 (ed. B. Senut and M. Pickford). CIFEG Publicational Occasionnelle 1994/29. Centre International pour la Formation et les Echanges Géologiques, Orléans.

———. 1996. Who is who among the Proboscidea? In *The Proboscidea: Evolution and Palaeoecology of Elephants and their Relatives,* pp. 39–48 (ed. J. Shoshani and P. Tassy). Oxford University Press, Oxford.

Tassy, P., and Darlu, P. 1986. Analyse cladistique numérique et analyse de parcimonie; l'exemple des Elephantidae. *Geobios* 19: 587–600.

———. 1987. Les Elephantidae : nouveau regard sur les analyses de parcimonie. *Geobios* 20: 487–94.

Tassy, P., Anupandhanat, P., Ginsburg, L., Mein, P., Ratanastien, B., and Sutteethorn, V. 1992. A new *Stegolophodon* (Proboscidea, Mammalia) in the Miocene of northern Thailand. *Geobios* 25: 511–23.

Tobien, H. 1978. On the evolution of mastodonts (Proboscidea, Mammalia), Part 2: The bunodont tetralophodont groups. *Geologisches Jahrbuch Hessen* 196: 159–208.

———. 1980. A note on the mastodont taxa (Proboscidea, Mammalia) of the "Dinotheriensande" (Upper Miocene, Rheinhessen, Federal Republic of Germany). *Mainzer Geowissenschaften Mitteilungen* 9: 187–201.

Tobien, H., Chen, G., and Li, Y. 1988. Mastodonts (Proboscidea, Mammalia) from the Late Neogene and Early Pleistocene of the People's Republic of China. Part 2: The genera *Tetralophodon, Anancus, Stegotetrabelodon, Zygolophodon, Mammut, Stegolophodon;* some generalities on the Chinese Mastodonts. *Mainzer Geowissenschaften Mitteilungen* 17: 95–220.

Whybrow, P. J. 1989. New stratotype; the Baynunah Formation (Late Miocene), United Arab Emirates: Lithology and palaeontology. *Newsletters on Stratigraphy* 21: 1–9.

Whybrow, P. J., Hill, A., Yasin al-Tikriti, W., and Hailwood, E. A. 1990. Late Miocene primate fauna, flora and initial palaeomagnetic data from the Emirate of Abu Dhabi, United Arab Emirates. *Journal of Human Evolution* 19: 583–88.

Zhou, M., and Zhang, Y. 1983. Occurrence of the proboscidean genus *Stegotetrabelodon* in China. *Vertebrata PalAsiatica* 21: 52–58.

Hipparions from the Late Miocene Baynunah Formation, Emirate of Abu Dhabi, United Arab Emirates

19

Véra Eisenmann and Peter J. Whybrow

The names Baynunah Formation (Whybrow, 1989) and the Shuwaihat Formation (Whybrow et al., 1999; Bristow, 1999—Chapters 4 and 6) are given to deposits of mainly clastic, late Miocene sedimentary rocks outcropping in an area of about 16 000 km^2 in the Western Region of the Emirate of Abu Dhabi, United Arab Emirates. It is likely that the fossil-bearing Baynunah Formation is coeval with parts of the Agha Formation of Iraq (Thomas et al., 1980) and Iran (James and Wynd, 1965). From the most western fossil site, Jebel Barakah (fig. 19.1) to the most eastern (Tarif), a distance of about 150 km, and further east to Abu Dhabi city, the regional dip is merely 1° east. All sedimentary units are horizontally-bedded; there is no regional folding and therefore no complicated stratigraphy. The sedimentary associations are, however, complex and tremendously variable (Whybrow et al., 1999; Friend, 1999; Bristow, 1999—Chapters 4–6).

Except for a small collection (1982–83) made by Al Ain Museum and German archaeologists (Vogt et al., 1989) now in Al Ain Museum, Emirate of Abu Dhabi, fossil collections from the Western Region made since 1986 by The Natural History Museum/Yale University team are temporarily housed in the Department of Palaeontology, The Natural History Museum, London: material collected pre-1984 has BMNH numbers. A collection made by Andrew Hill and Walid Yasin in 1984 is stored in Al Ain Museum. All palaeontological material collected by the NHM/Yale team has registration numbers of the Emirate of Abu Dhabi (AUH). Comparative hipparion material is housed in the American Museum of Natural History, New York (AMNH), Muséum National d'Histoire Naturelle, Paris (MNHN), and Ungarische Geologische Reichsanstalt (UGR).

Methods and Material

Bone scraps can be found on the slopes of most of the Miocene exposures. The Natural History Museum/Yale University team has attempted to be nonselective in their collecting to reduce sampling problems (Hill, 1987). All hipparion fossils were collected by close examination of outcrop surfaces. The genus *Hipparion* is represented among the Abu Dhabi fossils by 41 specimens, including 13 upper cheek teeth, 14 lower cheek teeth, and 14 limb bones, nearly all fragmentary, collected at six localities (fig. 19.1). The sizes of bones and teeth show that two species are represented: one small or middle-sized, and one larger. What makes the sample most interesting is the presence of two mandibular fragments belonging to the smaller species. In size and proportions they differ from other *Hipparion* mandibles and warrant the description of a new species.

Mandibles and other bones and teeth were measured according to the recommendations of the New York International Hipparion Conference of 1981 (Eisenmann et al., 1988). Two new mandibular

 ISBN 0-300-07183-3

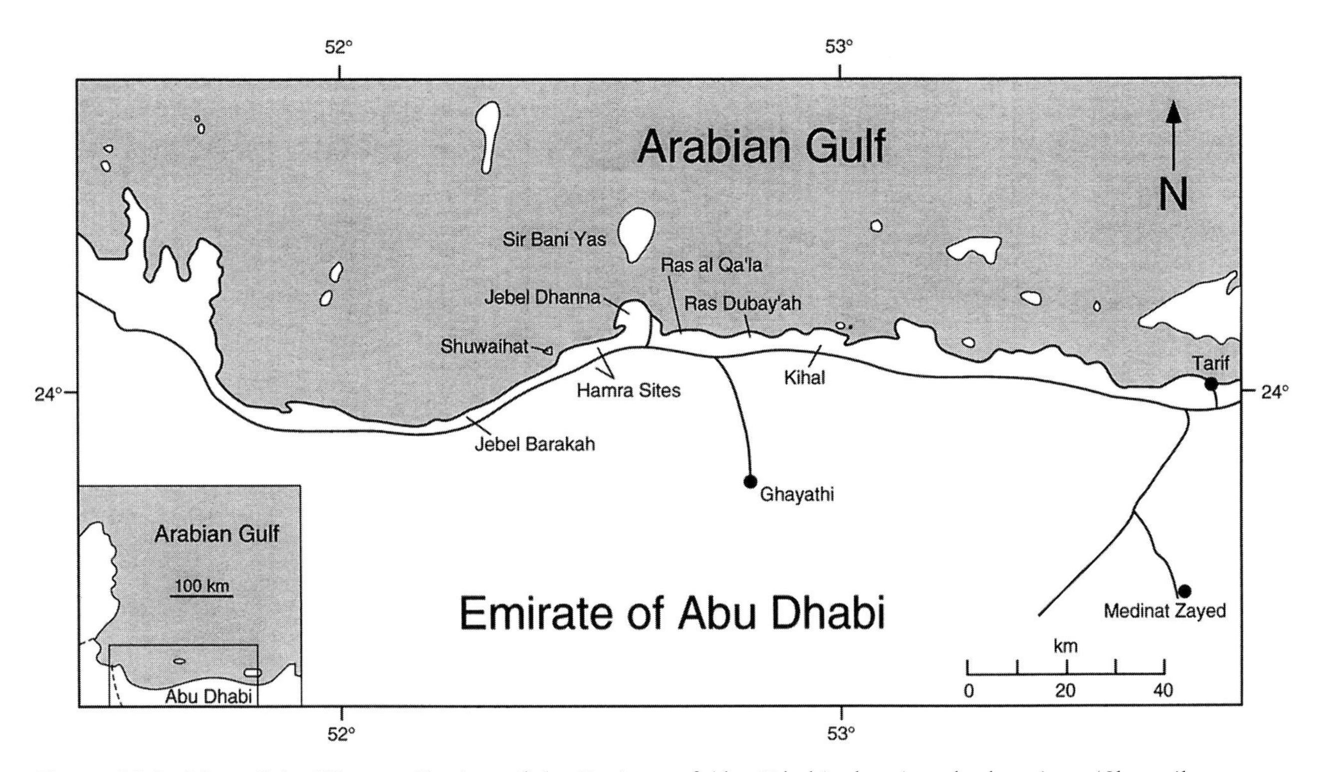

Figure 19.1. Map of the Western Region of the Emirate of Abu Dhabi, showing the locations (Shuwaihat, Hamra, Jebel Barakah, Jebel Dhanna, Ras Dubay'ah, and Kihal) where specimens of *Hipparion* have been found.

measurements were introduced to evaluate the length of a fragmentary muzzle. They are:

A = distance between infradentale (the point between the alveoli of the first incisors) and the symphyseal constriction (where the width of the symphysis is smallest); and

B = the distance between the symphyseal constriction and the second premolar.

The sum of A plus B is the muzzle length. A third measurement comes from the system used by Gromova (1952):

G3 = the distance between the mandibular angle and the front of the second premolar, which gives an idea of size independent of the muzzle length (muzzle length plus G3 is the maximal anteroposterior length of the mandible).

Simpson's ratio diagrams (Simpson, 1941) were used to compare the sizes and proportions of the Abu Dhabi *Hipparion* mandibles to others. Unworn or little worn cheek teeth were sectioned at mid-crown to provide evidence of the enamel pattern.

The ages of *Hipparion* localities given here are according to the updated European Land Mammal Zones (MN) of Mein (1990). Comparisons were made chiefly with hipparions for which the samples are good enough to give points of comparison (mandibles, cheek series), and where the age is at least approximately known. Because of the geographical position of the Emirate of Abu Dhabi in the Arabian Peninsula comparisons were also attempted with some African, Middle Eastern, and Asiatic hipparions.

Systematics

Order Perissodactyla Owen, 1848
Family Equidae Gray, 1821
Genus *Hipparion* de Christol, 1832
Hipparion sp., medium–large size

Referred Material (fig. 19.2.4–7, 9, 10)

Shuwaihat. Site S1, N 24° 06′ 38″, E 52° 26′ 09″: AUH 28, left metatarsal IV, distal; AUH 88, left astragalus; AUH 89, right tibia, proximal.

Hamra. Site H5, N 23° 06′ 04″, E 52° 31′ 38″: AUH 46, left P^2 (fragments); AUH 144, right, lower P (unworn and sectioned). Site H6, N 23° 06′ 43″, E 52° 31′ 28″: AUH 267, left posterior phalanx II.

Jebel Barakah. Site B2, N 24° 00′ 13″, E 52° 19′ 35″ (U.S. Marines Alpha-1 GPS survey point, 1990): BMNH M50664, left P^2 (fragment); BMNH M50667, left pelvis.

Jebel Dhanna. Site JD3, N 24° 10′ 31″, E 52° 34′ 21″: AUH 176, right metacarpal IV, proximal; AUH 177, right metacarpal IV, distal; AUH 246, left tarsal navicular. Site JD5, N 24° 10′ 22″, E 52° 34′ 38″: AUH 178, right P^3 or right P^4. Site JD4, N 24° 10′ 42″, E 52° 34′ 12″: AUH 265, right, lower P (unworn and sectioned).

Ras Dubay'ah. Site R2 (collected 1989, now a Defence Area): AUH 196, right astragalus (fragment); AUH 197, right cuneiform (fragment); AUH 203, right M^1 or right M^2 (sectioned); AUH 205, vestibular half of right P^3 or right P^4; AUH 208, left lower cheek tooth (unworn fragment); AUH 210, vestibular half of right P^3 or right P^4 (unworn); AUH 212, left M^1 or left M^2 (unworn); AUH 216, two lower cheek teeth (other teeth plus mandible fragments).

Description

Size medium to large. Moderately hypsodont teeth. Upper cheek teeth plicated; long protocone, lingually flattened, and with pointed ends (fig. 19.2.7). Left lower cheek teeth with a hipparionine double knot, and shallow ectoflexids on premolars (fig. 19.2.4, 10c).

Comments

The only character that deserves a short comment is the rather elongated, pointed, and flattened protocone. Despite exceptions like specimens from Seu de Urgell and some African forms, it may be said that Vallesian hipparions usually have shorter and more oval (not flattened) protocones. They also tend to have extremely plicated fossettes and multiple plis caballin. In contrast, Pliocene hipparions tend to have elongated, pointed and/or flattened protocones. Both morphologies, however, occur in the Turolian.

The referred limb bones include two fragmentary and badly preserved tali (AUH 88 and 196) measuring about 50 mm (maximum height), a second phalanx (AUH 267) measuring 37.7 mm (maximum height), 28 mm (minimum width), and 33 mm (distal articular width). The distal articular surface of the metatarsal IV from Shuwaihat (AUH 28) is 20 mm long and 10 mm wide. On the metacarpal IV from Jebel Dhanna (AUH 176 and 177) the same measurements are 17.7 mm and 8 mm, respectively.

Hipparion abudhabiense sp. nov.

Holotype

Right mandibular fragment, AUH 270 (fig. 19.3.1a and 1b.

Etymology

From the Emirate of Abu Dhabi, United Arab Emirates.

Type Locality

Jebel Dhanna, site JD3, N 24° 10′ 31″, E 52° 34′ 21″; Western Region of the Emirate of Abu Dhabi, United Arab Emirates.

Age and Geographic Distribution

Late Miocene, probably early to middle Turolian (European Mammal Age), Baynunah Formation of Abu Dhabi (Whybrow, 1989; Whybrow et al., 1990). Possibly related to the early Turolian *H. dietrichi* from Greece—Samos (Q1 and Andrianos) and Lower Axios Valley (Vathylakkos and Ravin des Zouaves); to the small early Turolian hipparion from Iraq (Jebel Hamrin); to the Algerian middle Turolian *H. sitifense;* to the Turolian medium-sized *H.* cf. *dietrichi* from Turkey (Kayadibi, Garkin, and Kinik) and Lebanon; but also resembling an early Pliocene form from La Gloria 4, Spain (Eisenmann and Mein, 1995).

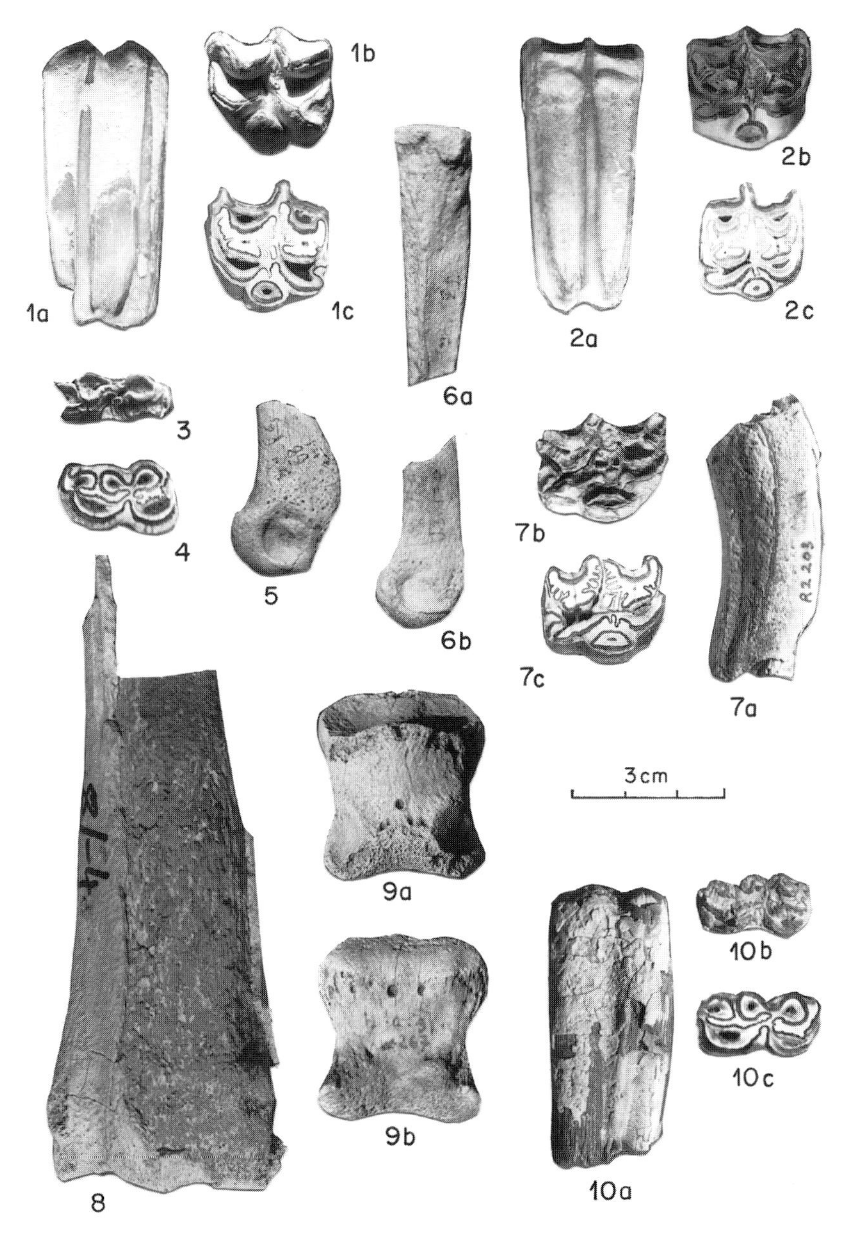

Figure 19.2. *Hipparion abudhabiense* (1, 2, 3, 8) and *Hipparion* sp. (4, 5, 6, 7, 9, 10). Left upper premolar, AUH 264: *1a,* vestibular view; *1b,* occlusal view; *1c,* mid-crown section. Right upper molar, AUH 229: *2a,* vestibular view; *2b,* occlusal view; *2c,* mid-crown section. *3,* Right lower premolar, AUH 174; occlusal view. *4,* Right lower premolar, AUH 265: mid-crown section. *5,* Distal end of a left fourth metatarsal, AUH 28; lateral view. Right fourth metacarpal, medial view: *6a,* AUH 176, proximal end; *6b,* AUH 177, distal end. Right upper molar, AUH 203: *7a,* mesial view; *7b,* occlusal view; *7c,* mid-crown section. *8,* Distal end of left radius, BMNH M50665; dorsal view. Second phalanx of the third digit, AUH 267: *9a,* dorsal view; *9b,* ventral view. Right lower premolar, AUH 144: *10a,* vestibular view; *10b,* occlusal view; *10c,* mid-crown section.

Diagnosis

Skull unknown. Size small or medium; lower cheek teeth series about 140 mm long. Mandibular ramus deep both in front of the second premolars (P_2) and between the premolars and molars (between P_4 and M_1). Broad symphysis at the constriction, probably indicating a broad muzzle. Very short distance between the second premolar and the symphyseal constriction probably indicating a short muzzle. The molar length nearly equals the premolar length. Teeth moderately hypsodont. Double-knot hipparionine. Ectoflexids deep on the molars and at least some premolars. No ectostylids. Frequent protostylids, either plis (P_3, P_4, M_1, and M_2 of the holotype) or isolated columns (M_3 of the holotype). Protocones small (fig. 19.3.2) and rather rounded (fig. 19.2.1c, 2c). Plications and plis caballin are moderately developed.

Referred Material (figs 19.2.1–3, 8 and 19.3.1–5

Shuwaihat. Site S1, N 24° 06′ 38″, E 52° 26′ 09″: AUH 23, left, lower M? (unworn, fragments); AUH 72, half anterior left, lower M; AUH 115, left M_1 or left M_2; AUH 164, phalanx I distal (fragment); AUH 165, phalanx I half proximal; AUH 166, phalanx I half distal. Site S4, N 24° 06′ 44″, E 52° 26′ 12″: AUH 228, vestibular half of left, upper M; AUH 264, left, upper P (unworn and sectioned).

Hamra. Site H1, N 23° 06′ 50″, E 52° 31′ 31″: AUH 174, right, lower P, lingual fragment. Site H5, N 23° 06′ 04″, E 52° 31′ 38″: AUH 231a, left P^3 or left P^4. Site 6, N 23° 06′ 43″, E 52° 31′ 28″: AUH 266a, left M_1 or left M_2 (fragment).

Jebel Barakah. Site B1, N 24° 00′ 24″, E 52° 19′ 48″: AUH 229, right M^1 or right M^2 (sectioned); AUH 230, left M^1 or left M^2. Site B2, N 24° 00′ 13″, E 52° 19′ 35″ (U.S. Marines Alpha-1, 1990); BMNH M50661, left P_3 or left P_4; BMNH M50662, left P_2 (fragment); BMNH M50663, right, lower M (fragment); BMNH M50665, left distal radius.

Jebel Dhanna. Site JD5, N 24° 10′ 22″, E 52° 34′ 38″: AUH 452, right mandible.

Kihal. Site K1, N 24° 07′ 23″, E 53° 00′ 27″: AUH 260, right, upper M? (unworn fragment).

Description and Comparisons of *Hipparion abudhabiense*

Mandibles (tables 19.1 and 19.2)

The two mandibular fragments were collected at Jebel Dhanna. One (AUH 452) is a right ramus lacking most teeth and with very worn P_4 and M_1, but conserving the proximal half of the symphysis. The holotype AUH 270, also a right ramus, lacks the symphysis but has well-preserved teeth. Both mandibles have a relatively deep ramus (fig. 19.3.1a and 5a). The specimen AUH 452 seems to have a short muzzle, or at least a short measurement B (fig. 19.3.5b). Specimen AUH 270 has relatively large molars (fig. 19.3.1b).

Table 19.1. Measurements (in mm) of some African and Eurasian hipparion mandibles (also see table 19.2)

Locality		Abu Dhabi	Abu Dhabi	Samos Q1	Samos Q?
Specimen		AUH 452	AUH 270	AMNH 20655	UGR OK 557
Species		*Hipparion abudhabiense*	*Hipparion abudhabiense*	*Hipparion dietrichi*	*Hipparion matthewi*
I_1–symph. constriction	A	—	—	50.0	41.0
P_2–symph. constriction	B	34.0	—	46.0	31.0
Height in front of P_2	12	49.0	—	56.5	39.5
Alveolar length P_2P_4	3	67.0	72.0	69.0	57.0
Alveolar length M_1M_3	4	—	70.0	63.5	56.0
Height between P_4 and M_1	11	64.0	69.0	67.5	52.0
Width at symph. constriction	14	34.0	—	41.2	27.7
Length P_2 to gonion	G3	—	225.0	215.0	208.0

Abbreviations: A = distance between the infradentale (between the alveoli of the first incisors) and the symphyseal constriction; B = distance between the symphyseal constriction and the second premolars (P_2); 12 = height of the mandibular ramus in front of P_2; 3 = length of the premolars; 4 = length of the molars; 11 = height of the mandibular ramus between P_4 and M_1; 14 = width at the symphyseal constriction; G3 = distance between the mandibular angle and the front of P_2.

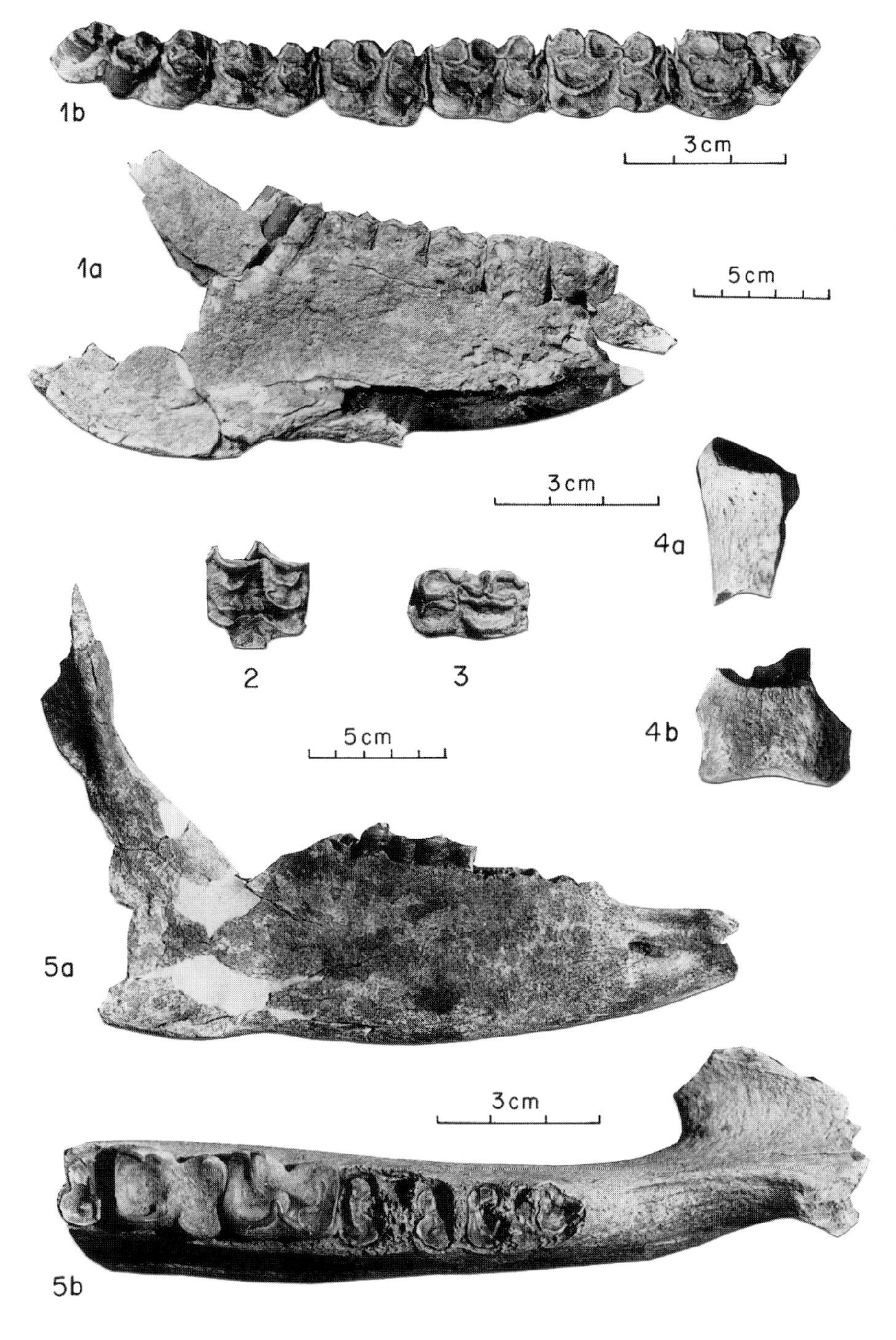

Figure 19.3. *Hipparion abudhabiense*, sp. nov. Holotype. Right mandibular ramus, AUH 270: *1a,* lateral view; *1b,* occlusal view. *2,* Left upper molar, AUH 230; occlusal view. *3,* Left lower premolar, BMNH M50661; occlusal view. Proximal phalanx of the third digit, dorsal view: *4a,* AUH 165, proximal fragment; *4b,* AUH 164, distal fragment. Right mandibular ramus, AUH 452: *5a,* lateral view; *5b,* occlusal view.

A ratio diagram (fig. 19.4) shows the differences in sizes and proportions between the two mandibles of *H. abudhabiense*, the average of two mandibles of *H. africanum* (reference zero line), and the mandibles most resembling *H. abudhabiense:* the type mandible of *H. matthewi* (Samos, quarry unknown), one mandible of *H. dietrichi* AMNH 20665 from Samos, Q1 (Sondaar, 1971), and the average of 8–16 mandibles of *H. dietrichi* from the Thessaloniki area of Greece: Vathylakkos (Koufos, 1988), Ravin des Zouaves (RZO, Koufos, 1987), and "Saloniki" (MNHN 290a and others).

Judging from the mandibular length excluding the muzzle (measurement G3) all mandibles are smaller than in the Vallesian reference specimen of *H. africanum;* all have relatively short muzzles (measurements A and/or B), relatively high rami (12 and 11), relatively broad symphyses (14), and relatively large molar lengths (4) compared with the premolar lengths (3). The Jebel Dhanna mandibles, however, cannot be referred to any of these *Hipparion* species. *Hipparion matthewi* is much smaller but has a relatively longer distance between the P_2 and the symphyseal constriction (B). The last character also distinguishes *H. abudhabiense* from other hipparion species.

Another ratio diagram (fig. 19.5) compares the average of the two mandibles of *H. abudhabiense* with Turkish mandibles referred to *H. mediterraneum* and to *H. matthewi* (Koufos and Kostopoulos, 1994). Only the better preserved and figured material (mandibles D 532 and D 533 from Kemiklitepe D, and mandible A 537 from Kemiklitepe A) has been used so that measurements A, B, and G3 could be estimated on the plates. According to the ratio diagram, neither the mandibles referred to as *H. mediterraneum* (D 532 and 533) nor the mandible referred to as *H. matthewi* (A 537) resemble the mandible of *H. abudhabiense.* The Arabian species has a much higher horizontal ramus (measurements 12 and 11).

Lower Cheek Teeth (tables 19.3–19.5)

The lower cheek teeth have a banal, hipparionine, double knot. This kind of morphology has no biostratigraphical significance outside Africa. It is only in Africa that hipparionines with double knots are no longer found after 6 million years (Ma), being completely replaced at that time (or perhaps even earlier) by more evolved patterns—first the caballoid, then the caballine double knot (Eisenmann, 1995a). In Eurasia, the caballine double knot appears only in MN 15–16 at Villaroya (Spain), Kvabebi (Georgia), Beregovaja (Russia), and Shamar (Mongolia). Even then, it does not replace the hipparionine pattern, which is still found in MN 15 zone at Layna (Spain), Perpignan (France), and Çalta (Turkey), and co-occurs with the caballine pattern at Shamar and Beregovaja. Peculiarly, the lower cheek teeth from Kemiklitepe D (Koufos and Kostopoulos, 1994, pl. III, figs 3 and 4) are rather caballoid, although they belong to an early Turolian hipparion.

Usually protostylids are present in hipparions, especially in moderately worn or very worn teeth. Their occurrence is so commonplace that they hardly have any systematic significance. They are present on the early Pliocene cheek teeth from La Gloria 4 and on the Kemiklitepe lower cheek teeth (Koufos and Kostopoulos, 1994). Protostylids are also said to be present but not strong in *H. dietrichi* from Samos Q1 (Sondaar, 1971), average in *H. dietrichi* from Andrianos Quarry in Samos, and double in very worn teeth (Koufos and Melentis, 1984). They are moderate and rarely isolated in *H. dietrichi* from Ravin des Zouaves and Vathylakkos (Koufos, 1987, 1988).

More important is another character: the deep development of the ectoflexid (vestibular groove) seen not only on the molars but at least on some premolars, such as the P_4 of the holotype and a fragmentary premolar from Hamra (fig. 19.2.3). Deep ectoflexids are the rule on *Hipparion* molars, even if they have a tendency to become shallower as in the evolved African *Hipparion* species (Eisenmann, 1977). But on *Hipparion* premolars, short ectoflexids are the rule, even though they may be deep on very worn teeth, as in some Vallesian species such as *H. depereti* of Montredon (Eisenmann, 1988).

In *H. abudhabiense* the deep ectoflexid on moderately worn premolars is associated with a high (deep) mandibular ramus and a short muzzle. This association of characters was investigated in other hipparions. Unfortunately, the mandible of *H. depereti* (MN 10) is unknown, but Vallesian mandibles seem to have much

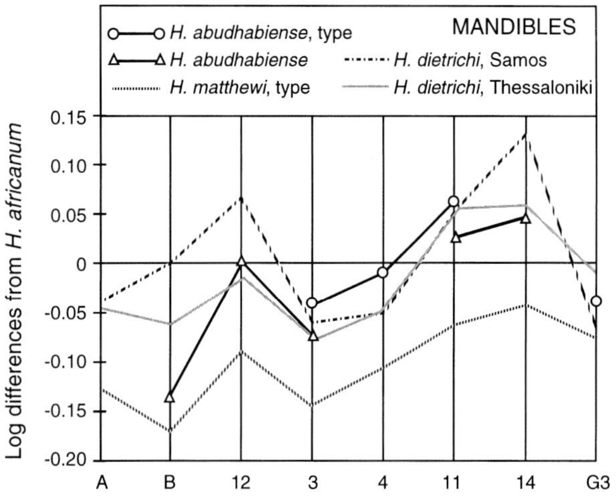

Figure 19.4. Ratio diagrams comparing the mandibles of *Hipparion abudhabiense* of Jebel Dhanna (type specimen AUH 270, and AUH 452), the type mandible of *H. matthewi* from Samos, and the early Turolian *H. dietrichi* from Samos Quarry 1 and the Thessaloniki area (Vathylakkos, Ravin des Zouaves, and specimen MNHN 290a from "Salonique"). The average of two mandibles of the Vallesian *H. africanum* from Bou Hanifia is the reference line. Key: A, distance between the infradentale (between the alveoli of the first incisors) and the symphyseal constriction; B, distance between the symphyseal constriction and the second premolars (P_2); 12, height of the mandibular ramus in front of P_2; 3, length of the premolars; 4, length of the molars; 11, height of the mandibular ramus between P_4 and M_1; 14, width at the symphyseal constriction; G3, distance between the mandibular angle and the front of P_2.

Table 19.2. Measurements (in mm) of some African and Eurasian hipparion mandibles (also see table 19.1)

Locality		Bou Hanifia	Salonique	Ravin Zouaves	Vathylakkos	Thessaloniki
Number		$n = 2$	MNHN 290a	RZO 44 and 76	$n = 2–5$	$n = 3–8$
Species		*H. africanum*	*H. dietrichi*	*H. dietrichi*	*H. dietrichi*	*H. dietrichi*
I_2–symph. constriction	A	55.0	48.0	50.0	—	49.3
P_2–symph. constriction	B	46.0	41.0	39.0	—	40.0
Height in front of P_2	12	48.5	49.0	48.4	43.6	47.0
Alveolar length P_2P_4	3	79.5	66.0	64.0	68.0	66.0
Alveolar length M_1M_3	4	71.5	65.0	61.5	64.0	63.5
Height between P_4 and M_1	11	60.0	66.0	73.0	65.7	68.2
Width at symph. constriction	14	30.5	35.0	35.0	—	35.0
Length P_2 to gonion	G3	248.0	—	—	241.5	241.5

Note: For abbreviations see table 19.1.

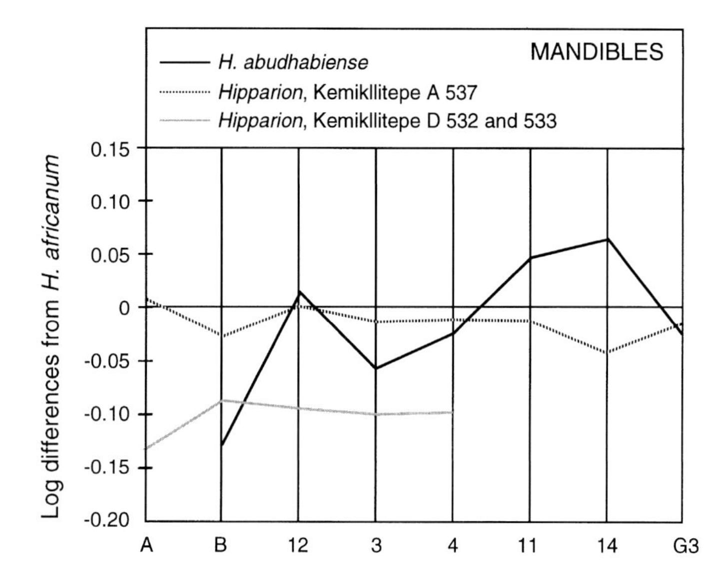

Figure 19.5. Ratio diagrams comparing the average of two mandibles of *Hipparion abudhabiense* of Jebel Dhanna, the average of two mandibles (D 532 and D 533) from Kemiklitepe D, and one mandible (A 537) from Kemiklitepe A. The average of two mandibles of the Vallesian *H. africanum* from Bou Hanifia is the reference line. Data on Kemiklitepe from Koufos and Kostopulos, 1994). Abbreviations as in legend to figure 19.4.

Table 19.3. Measurements (in mm) of the lower cheek teeth of the type mandible of *Hipparion abudhabiense,* with mode of protosylids and occurence of plis caballinid

		P_2	P_3	P_4	M_1	M_2	M_3
Wear stage		IV	III	III	III	III	II
Height		17.0	23.0	—	—	—	31.0
	Length	28.0	23.0	22.0	19.5	20.2	25.0
	L Ant Foss	5.0	7.0	6.2	5.1	6.3	6.5
Occlusal	L Double K	10.0	12.5	12.0	11.0	10.2	9.7
	L Post Foss	11.5	12.7	9.2	7.7	8.6	7.2
	Width	13.0	12.0	12.2	12.0	10.0	9.0
Protostylid		—	Pli	Pli	Pli	Pli	Isolated
Pli caballinid		0	0	0	0	0	0

Abbreviations: L = length; Ant Foss = anterior fossette; Double K = double knot; Post Foss = posterior fossette.

lower rami than *H. abudhabiense*. From an unknown Turolian level of Samos, the type *H. matthewi* has shallow ectoflexids on the premolars, although the teeth are very worn. The situation is unclear for *H. dietrichi*. Mandibles from Samos Q1 (MN 11), which do have short muzzles and high rami, may have deep ectoflexids on very worn premolars (Sondaar, 1971: pl. II, fig. c) and even moderately worn ones (Wehrli, 1941: pl. 23, fig. 6), but at the Andrianos Quarry (also MN 11) the ectoflexid is shallow on premolars (Koufos and Melentis, 1984: pl. VI, fig. 1). The state of this character is unknown for the Lower Axios Valley *H. dietrichi* (MN 11). Only one lower series is figured from the Ravin des Zouaves (Koufos, 1987: fig. 8d); it has deep ectoflexids but is very worn. No lower cheek series is figured from Vathylakkos, and the description states only that ectoflexids are deeper on molars (Koufos, 1988: 17). Among the "Saloniki" material of the MNHN (Paris), only very worn premolars of *H. dietrichi* have deep ectoflexids.

Deep ectoflexids are also seen in the rather worn P_3 and P_4 (but not on the P_2) of a lower cheek series from La Gloria 4, Spain, belonging in the MN 14 zone (Eisenmann and Mein, 1995). Thus, judging from the characters of the mandible, *H. abudhabiense* resembles *H. dietrichi* from Greece (MN 11 zone) while the same kind of deep ectoflexid on premolars is present in *Hipparion* sp. from Spain (MN 14 zone).

Upper Cheek Teeth (tables 19.6 and 19.7)

Three characters are usually considered: the shape of the protocone, the degree of enamel plication, and size. Interpretation of these characters is, however, made difficult by the wide range of variation, intraspecific and interspecific, which does not seem to have any clear biostratigraphical significance. We have nevertheless compared the upper cheek teeth of *H. abudhabiense* with those of hipparions that resemble it most in mandibular and lower cheek teeth characters.

The upper cheek teeth referred to *H. abudhabiense* have a rather small protocone, flattened on the lingual side (fig. 19.3.2) or rounded (fig. 19.2.1c and 2c). The enamel plication is moderate. In the Pliocene hipparion from La Gloria 4, the enamel plication is also moderate but the protocone is long and elliptical. In *H. dietrichi*, the enamel plication is again moderate, and the protocone is said to be elliptical or oval (Wehrli, 1941; Sondaar, 1971; Koufos and Melentis, 1984; Koufos, 1987, 1988). On the upper cheek teeth of Kemiklitepe referred to *H. mediterraneum* and to *H. matthewi* (Koufos and Kostopoulos, 1994: pl. III, fig. 2; pl. IV, fig. 3) protocones are small and rounded but the enamel plication seems stronger than in *H. abudhabiense.*

Figure 19.6 compares the sizes of the upper premolars (P^3 and P^4) and molars (M^1 and M^2). Only measurements of moderately-worn teeth (stages II and III) were considered valid: occlusal or mid-crown lengths in *H. abudhabiense* and *Hipparion* sp. (the other hipparion from Abu Dhabi), and *H. dietrichi* from the MN 11 zone of Greece. The data on the Greek forms were published by Koufos (1987, 1988). *Hipparion abudhabiense* seems to have slightly larger teeth than the average of *H. dietrichi* from Vathylakkos and Ravin des Zouaves. Its teeth are smaller than in *H. mediterraneum* from Kemiklitepe D and larger than in *H. matthewi* from Kemiklitepe A and B (Koufos and Kostopoulos, 1994: tables 6 and 9).

Limb bones

No complete limb bones have been found, so there is no way to calculate gracility. The distal radius fragment (BMNH M50665) has a distal articular width of 46.5 mm and a distal articular anteroposterior diameter of 26 mm. The first phalanx (AUH 164 and 165) fragments seem to have the same size as *H. dietrichi* from Vathylakkos (Koufos, 1988); the distal supra-articular width is 28.5 mm, the distal articular width is 29 mm, and the distal articular anteroposterior diameter is 18 mm.

RELATIONSHIPS WITH AFRICAN, MIDDLE EASTERN, AND ASIATIC HIPPARIONS

Africa

Algeria

There is only one species of hipparion, *H. africanum*, at Bou Hanifia, represented by skulls, mandibles, and limb bones. The mandible (reference line in fig. 19.4) has a longer muzzle and a lower ramus than in *H. abudhabiense*. Moderately worn upper cheek teeth have longer and oval protocones. Ectoflexids are shallow on lower premolars. The age of the hipparion level at Bou Hanifia (MN 9) is estimated at little more than 10.5 Ma (Sen,

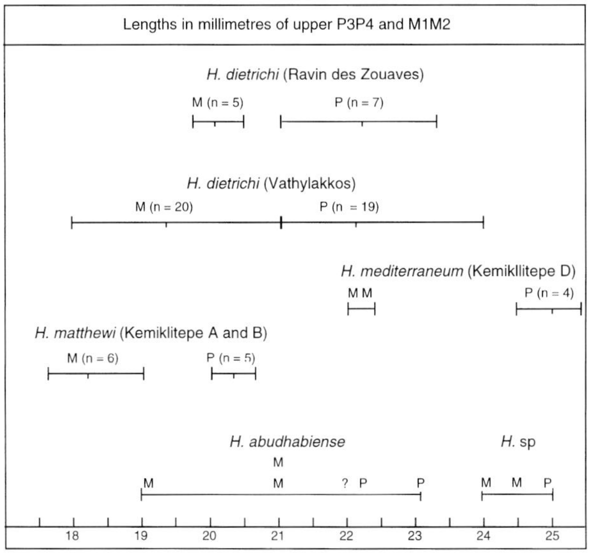

Figure 19.6. Lengths of upper cheek teeth in hipparions from Greece, Turkey, and the United Arab Emirates. The horizontal lines show the observed ranges.

Table 19.4. Measurements (in mm) of the lower cheek teeth of Abu Dhabi hipparions, with mode of protostylids and occurrence of plis caballinid (see also table 19.5)

		Hipparion abudhabiense	*Hipparion abudhabiense*	*Hipparion abudhabiense*	*Hipparion abudhabiense*	*Hipparion abudhabiense*	*Hipparion abudhabiense*	*Hipparion abudhabiense*
Tooth		P_2	p	m	m	p	m	m
Specimen		M 50662	M 50661	AUH 23	M 50663	AUH 174	AUH 115	AUH 72
Wear stage		II	II	I	III	II	II	II
Height		30.0	36.0	50.0	33.0	44.0	39.0	37.5
Occlusal	Length	21.5+	22.0	(23)	—	(25.5)	(21)	—
	L Ante foss	7.2	7.0	9.0	—	7.0	6.8	7.0
	L Double K	12.1	12.0	14.0	—	16.0	12.0	—
	L Post foss	(12)	11.0	10.5	10.0	13.0	9.0	—
	Width	(17)	12.6	—	(12.5)	—	11.0	10.5
Mid-crown	Length	—	21.0	—	—	—	(20)	—
	L Ante foss	—	—	—	—	—	—	—
	L Double K	—	—	—	—	—	—	—
	L Post foss	—	—	—	—	—	—	—
	Width	—	13.7	—	—	—	(12)	—
At 1 cm	Length	—	21.0	—	—	—	—	—
	Width	—	13.7	—	—	—	—	—
Protostylid		—	0	—	—	—	—	—
Pli caballinid		—	0	—	0	0	0	—

Note: For abbreviations see table 19.3. Measurements in parentheses are approximate.

Table 19.5. Measurements (in mm) of the lower cheek teeth of Abu Dhabi hipparions, with mode of protostylids, and occurence of plis caballinid (see also table 19.4)

		Hipparion sp.	*Hipparion* sp.	*Hipparion* sp.	*Hipparion* sp. ?	*Hipparion* sp. ?	*Hipparion abudhabiense*	*Hipparion* sp. ?
Tooth		p	p	p?	p?	m?	m	m1
Specimen		AUH 144	AUH 265	AUH 208	AUH 216	AUH 216	AUH 266a	AUH 677
Wear stage		I	0–I	I	I	0	II–III	0
Height		55+	56.0	48.0	48.0	51.0	23.0	52.0
Occlusal	Length	25.5	27.0	(26)	26.0	25.0	—	—
	L Ante Foss	8.0	11.0	7.5	8.3	—	6.0	—
	L Double K	—	12.3	(13)	13.0	—	11.2	—
	L Post Foss	13.0	11.8	—	11.2	—	8.0	—
	Width	14.0	11.0	—	—	10.0	11.8	—
Mid-crown	Length	26.0	25.0	—	22.5	22.3	—	22.5
	L Ante Foss	7.8	7.0	—	—	—	—	6.7
	L Double K	15.0	13.0	—	—	—	—	14.0
	L Post Foss	13.8	11.0	—	—	—	—	7.5
	Width	13.0	12.5	—	—	11.2	—	12.0
At 1 cm	Length	—	23.5	—	22.0	21.0	—	—
	Width	—	14.0	—	—	11.0	—	—
Protostylid		0	Isolated	—	—	Pli	Isolated	—
Pli caballinid		0	0	—	—	—	0	0

Note: For abbreviations see table 19.3. Measurements in parentheses are approximate.

Table 19.6. Measurements (in mm) of the upper cheek teeth of Abu Dhabi hipparions, with number of plications (see also table 19.7)

		Hipparion sp.	*Hipparion* sp.	*Hipparion* sp.	*Hipparion* sp.	*Hipparion* sp.	*Hipparion* *sp.*	*Hipparion* sp.
Tooth		P^2	P^2	P	M	M	P	P
Specimen		AUH 46	M50664	AUH 210	AUH 212	AUH 203	AUH 205	AUH 178
Wear stage		II–III	II–III	0–I	0	I	?	I?
Height		42.0	40.0	50.0	51.0	54.0	—	44.0
Occlusal	Length	—	—	26.0	26.0	26.0	25.0	—
	Width	(34)	—	—	—	—	—	—
	Protocone L	—	—	—	—	9.0	—	—
	Protocone W	—	—	—	—	4.0	—	—
Mid-crown	Length	—	—	25.0	24.5	24.0	—	—
	Width	—	—	—	—	—	—	—
	Protocone L	—	—	—	—	8.1	—	—
	Protocone W	—	—	—	—	4.1	—	—
At 1 cm	Length	—	—	23.0	—	24.0	—	—
	Width	—	—	—	—	—	—	—
Plis fossette		Few	6–3	—	—	11–10	?–4	9–?
Plis caballin		—	—	—	—	2	—	—

Note: measurements in parentheses are approximate.

Table 19.7. Measurements (in mm) of the upper cheek teeth of Abu Dhabi hipparions, with number of plications (see also table 19.6)

		Hipparion abudhabiense	*Hipparion abudhabiense*	*Hipparion abudhabiense*	*Hipparion abudhabiense*	*Hipparion abudhabiense*	*Hipparion abudhabiense*	*Hipparion abudhabiense?*
Tooth		P	P	?	M	M	M	M2
Specimen		AUH 264	AUH 231a	AUH 260	AUH 229	AUH 230	AUH 228	AUH 676
Wear stage		0	III–IV	0	I	II	I	0
Height		55.0	22.0	48+	50.0	47.0	54.0	55.0
Occlusal	Length	26.0	22.3	25.0	24.2	20.0	23.0	23.0
	Width	23.0	—	—	23.0	20.1	—	19.0
	Protocone L	—	—	—	6.5	6.0	—	8.0
	Protocone W	—	—	—	4.1	3.0	—	2.5
Mid-crown	Length	24.0	—	22.0	21.1	19.0	21.0	22.0
	Width	24.0	—	—	23.0	21.0	—	23.0
	Protocone L	7.0	—	—	7.0	—	—	8.0
	Protocone W	5.0	—	—	4.5	—	—	—
At 1 cm	Length	22.0	22.3	—	20.7	18.0	20.5	—
	Width	22.6	—	—	23.2	20.3	—	—
Plis fossette		14–9	15–3	7–6	9–7	8?–3	—	—
Plis caballinid		1	—	—	1	0	—	—

1990), or about 10.85 Ma by extrapolation based on the sedimentation rate. The morphology of the skull, teeth, and limb bones of *H. africanum* is, however, much more evolved than in the usual Vallesian *H. primigenium*.

Hipparions from the middle Turolian (MN 12) may represent *H. sitifense*. The scanty material possibly belonging to *H. sitifense* was previously figured and discussed in detail by Eisenmann (1980). The type material is lost, and the material collected later at the same locality of Saint-Arnaud is limited to a badly preserved lower cheek tooth and a distal fragment of MT III. The age of the material is unknown. At present, our limited knowledge indicates that *H. sitifense* was a primitive form, with rather short and curved crowns and a rounded protocone (Pomel, 1897: pl. I), smaller than *H. africanum* but not by much. More teeth were collected later in Algeria and referred to *H. sitifense*, although some of them could as well be referred to *H. africanum* if their morphology and not their age was considered. A few teeth come from Amama 2 (collected by J.-J. Jaeger), others probably from Amama 2 level "lower site of Kef el Amama" (collected by L. Ginsburg). According to Ameur-Chehbeur (1991), Amama 2 is Turolian. Some teeth come from Chabet el Maatga (old collections, and new collections by Mr and Mrs. Coiffait, who believe that the age of Chabet el Maatga is close to that of Amama 2). The whole sample is unsatisfactory because of its small size and the poor preservation of the teeth. It does appear, however, that, as at Abu Dhabi, two hipparions may be present, and that one small upper premolar has a small and rounded protocone. Consequently, there is a certain resemblance between *H. sitifense* of the middle Turolian of Algeria and *H. abudhabiense*.

East Africa

The *Hipparion primigenium*, described by Hooijer (1975), from the Ngorora Formation, Kenya, is represented by a few teeth and limb bones. It is larger than *H. africanum*, and does not resemble *H. abudhabiense* at all. There are no hipparion fossils in the type section of the Ngorora Formation, only in younger sites; their age is between 9.0 and 8.5 Ma (A. Hill, personal communication), certainly younger than 10 Ma (Hill et al., 1986), and they probably belong in the MN 11 zone.

From the Namurungule Formation, Samburu Hills, Kenya, Nakaya and Watabe (1990) describe *Hipparion* aff. *africanum*. Even if the skull and some other fossils are related to *H. africanum*, it is clear that the variation within the Samburu sample is too great for a single species. For example, one metacarpal (12272) is much larger than the others, and compares well with the large *H. koenigswaldi* from Nombrevilla, Spain (MN 10). According to the published measurements, the big upper cheek teeth of Samburu are larger than in *H. africanum*, but compare well with teeth from Nombrevilla. It is only the small teeth of Samburu that may be referred to the Bou Hanifia species (Eisenmann, 1995a). The large teeth from Samburu are similar in size to the large species of Abu Dhabi, but the morphology is very different: the figured upper cheek teeth have very plicated enamel, multiple plis caballin, and oval protocones. The small teeth from Samburu have longer protocones than *H. abudhabiense*. The age of the Samburu hipparions is estimated at 9 Ma.

The presence of two species of hipparion at one site was classically considered as an indication of Turolian rather than Vallesian age. Recent studies have shown, however, that there is no monotypy of Vallesian hipparions (Eisenmann, 1995b). In consequence, the presence of two hipparions at Samburu is no longer unique nor surprising.

A primitive, large *Hipparion* was described at Nakali, Kenya (Aguirre and Alberdi, 1974). From its size and morphology, it appears very close to the large *Hipparion* from Samburu. The ages are probably also close, since Nakali is believed to be possibly a little younger than 9 Ma (Hill, 1987).

Another primitive form, *H. macrodon* has been described from the Kakara Formation of Uganda (Eisenmann, 1995a) situated in the base of the Upper Miocene (Pickford et al., 1993). The upper premolars are larger than any of the known species of hipparion. Morphology and size differ completely from those of *H. abudhabiense*.

Between 6.5 and 4 Ma ago there are several East African hipparion localities: Mpesida Beds, Lukeino Formation, Lothagam 1, and lower units of the Chemeron Formation (Hill, 1987; Hill and Ward, 1988). From the literature it is not always clear if the fossils come from these or from younger units (Hill et al., 1992): Lothagam 3 (about 4 Ma) or the upper units of Chemeron (about 4–1.6 Ma). Some of the material was published by Aguirre and Alberdi (1974), Hooijer and Maglio (1974), and Hooijer (1975). In the literature, the hipparion cheek teeth are poorly represented. The material consists of small samples: one tooth at Mpesida, probably no teeth from the lower levels of Chemeron, and very worn teeth in the skull of *H. turkanense* of Lothagam. It is nevertheless clear that, in the East African Miocene—for example, at Lukeino—the first hipparions with a caballoid pattern on the lower cheek teeth appear. This pattern becomes common and more pronounced—"caballine"—later and is made still more original by the development of ectostylids. Whatever their size, these hipparions lack primitive characters and cannot be closely related to *H. abudhabiense.* Since these publications, larger collections have been made from sites in the Mpesida, Lukeino, and Chemeron units by Andrew Hill and colleagues, and from Lothagam by Meave Leakey's team. So far this material is unpublished.

Middle East

Turkey

Staesche and Sondaar (1979) have correlated the Turolian localities of Garkin and Kinik, respectively, with Samos Quarries 1 and 4 (MN 11 zone) and Samos Quarry 5 (MN 13 zone), and found that the medium-sized hipparion remains from Kayadibi, Garkin, and Kinik resembled *H. dietrichi* in various characters. At Kayadibi and Garkin, there is a larger form that makes the comparison with the two hipparions of Abu Dhabi especially interesting. The upper cheek teeth of the Turkish *H.* cf. *dietrichi* are the size of those in *H. abudhabiense* and also have rounded protocones. But the larger Turkish species also has rounded protocones that do not resemble those of *Hipparion* sp. of Abu Dhabi.

Koufos and Kostopoulos (1994) have described several hipparions from the localities of Kemiklitepe. The lower level (Kemiklitepe D) has yielded early Turolian or even Vallesian Carnivora (Bonis,1994) and Proboscidea (Tassy, 1994). According to our ratio diagrams, the hipparion mandibles from this lower level of Kemiklitepe are not very different from those of *H. africanum* from Bou Hanifia (fig. 19.5). The upper levels of Kemiklitepe are believed to belong in the MN 12 or 13 zone. According to our (unpublished) ratio diagrams, the mandibles from the upper levels resemble not the typical *H. matthewi* from Samos, but the small hipparions from Ravin de la Pluie and Dytiko. Neither the upper nor the lower level hipparions resemble *H. abudhabiense.*

Lebanon

Malez and Forsten (1989) have published some material from the Bekaa Valley. The artiodactyls indicate a Turolian age. Two hipparions are present, the dimensions of which fit well with the two forms from Garkin. Morphology and size also appear to fit well with the Abu Dhabi hipparions: protocones tend to be pointed and flattened on the larger teeth, rounded on the smaller. At least one small premolar has a deep ectoflexid. It is possible that both the smaller form (*H.* cf. *dietrichi* ?) and the larger one are related to the Abu Dhabi forms.

Iraq

From the Agha Jari Formation of Jebel Hamrin, Thomas et al. (1980) have described a fauna with two hipparions: a medium-sized *H. mediterraneum* and a larger *H.* cf. *primigenium.* The fauna was believed to be close in age to the Vallesian–Turolian boundary. A well-preserved skull of *Prostreptsiceros zitteli* was found later in the same area, indicating an early Turolian age (Bouvrain and Thomas, 1992).

The smaller upper cheek teeth, with their small and rounded or flattened protocones, fit well with *H. abudhabiense,* but out of five small lower premolars, only one shows a tendency to a deep ectoflexid like the fragmentary premolar of Hamra (fig. 19.2.3). The larger upper cheek teeth have elongated protocones as in *Hipparion* sp. of Abu Dhabi, but with more plicated enamel (Thomas et al., 1980: pl. I, fig. 3). The larger lower cheek teeth (Thomas et al., 1980: pl. I, fig. 1) have a somewhat caballoid pattern, similar to the medium-sized hipparion from Kemiklitepe D. These characters and the relative hypsodonty of both small and large teeth are surprising in Vallesian or early Turolian forms.

Asia

Siwaliks

The upper cheek lengths of hipparions from Abu Dhabi fall inside the ranges of variation given by Hussain (1971) for *H. antilopinum* and *H. theobaldi*. All the upper cheek teeth figured by Hussain, however, have much more plicated enamel than the hipparions from Abu Dhabi and none has the small and rounded protocones of *H. abudhabiense*. Rounded protocones are present in the type of *H. feddeni* from the "middle Siwalik group of Perim Island" (Lydekker, 1884; MacFadden and Woodburne, 1982: fig. 9) but the teeth seem to be larger in *H. feddeni* than in *H. abudhabiense*. Bernor and Hussain (1985) state that one difference between the genera *Cormohipparion* and *Hipparion* is the shape of the protocone: elongate in *Cormohipparion*, rounded (at medium wear) in *Hipparion*. They do not illustrate, however, any Siwalik "Hipparion" with rounded protocones. Moreover, they seem to refer the type of *H. feddeni* (with rounded protocones) to *Cormohipparion perimense*, which is supposed to have elongate protocones.

Iran

There is no mandible—at least in the material of the Muséum National d'Histoire Naturelle, Paris, that has a muzzle as short as *H. abudhabiense*. At least two skulls (MNHN 18 and 359) belong to the short-muzzled *H. dietrichi* type, however, so that it is not unlikely that *H. abudhabiense* has affinities with some Maragheh hipparions (Bernor, 1985).

Discussion

Several observations can be made about the Abu Dhabi hipparion material. Two species of *Hipparion* are represented: they are a small to medium-sized *H. abudhabiense* and a medium-sized to large *Hipparion* sp. They differ not only in size, but in the shape of the protocone—very small and rather rounded in the former and elongate in the latter—and by the depth of the ectoflexid on the lower premolars, which tend to be deep in *H. abudhabiense*. The larger hipparion from the Baynunah Formation is poorly represented so that its affinities are not clear. *Hipparion abudhabiense* compares well with *H. dietrichi* from the MN 11 zone of Greece (Samos and Thessaloniki area) but it may have had an even shorter muzzle. Morphologically similar upper and/or lower cheek teeth may also be present in North Africa (*H. sitifense*), and Turkey, Lebanon, Iraq, and Piram (formerly Perim) Island.

Hipparion abudhabiense probably had a very short and broad muzzle, suggesting grazing rather than browsing habits. This species also shows two apparently contradictory characters: a relatively deep mandibular ramus, suggesting hypsodonty, and deep ectoflexids on some premolars, a primitive character interpreted by Gromova (1952: 92) as a poor adaptation to abrasive food. A Pliocene hipparion from Spain (La Gloria 4) associates deep ectoflexids on some premolars with an extremely slender metatarsal. Here again, there is an apparent contradiction between the dry environment suggested by the metatarsal and the soft food suggested by the primitive teeth.

Because of their geographical location, resemblances between the hipparions from Abu Dhabi and those from Africa and Asia might be expected, and we have already discussed the characters that may indicate relations with North African and Siwalik hipparions. Our best point of comparison, however, remains with *H. dietrichi*, a European species. Three possibilities may explain this. First, in East Africa there is a gap between the Nakali–Samburu faunas (which are probably too old to compare with the fauna from Abu Dhabi) and late Turolian sites (which are probably too young). It is possible that the fauna from Abu Dhabi falls into this gap. Second, in North Africa the Turolian hipparion material is too poor for adequate comparison. Finally, for Siwalik specimens, the problem is a little different as most of the material is unpublished. Skulls have been discussed at length and repeatedly, but there are no data on the mandibles, not enough data on the limb bones, and not enough figures of the cheek teeth. When the Indian and Pakistani hipparions are better known, their relationship to the hipparion fauna from Abu Dhabi may become clear.

A final point should be made. We have already noted that the Vallesian *H. africanum* looks more "evolved" than classical Vallesian hipparions. Evolved characters (caballoid double knots, hypsodonty) may also be observed in Vallesian or early Turolian hipparions from Turkey (Kemiklitepe D) and Iraq (Jebel Hamrin). Do all these discrepancies indicate that age evaluations are wrong, or should biostratigraphic correlations be made only when the palaeoecological conditions are similar?

References

Aguirre, E., and Alberdi, M. T. 1974. Hipparion remains from the Northern part of the Rift Valley (Kenya). *Proceedings Koninklijke Nederlandse Akademie van Wetenschappen* 77: 146–57.

Ameur-Chehbeur, A. 1991. Un nouveau genre de Gerbillidae (Rodentia, Mammalia) du Mio-Pliocène d'El Eulma, Algérie Orientale. *Geobios* 24: 509–12.

Bernor, R. L. 1985. Systematic and evolutionary relationships of the hipparionine horses from Marageh, Iran (Late Miocene, Turolian age). *Palaeovertebrata* 15: 173–269.

Bernor, R. L., and Hussain, S. T. 1985. An assessment of the systematic, phylogenetic and biogeographic relationships of Siwalik hipparionine horses. *Journal of Vertebrate Paleontology* 5: 32–87.

Bonis, L. de. 1994. Les gisements de mammifères du Miocène supérieur de Kemiklitepe, Turquie: 2, Carnivores. *Bulletin du Muséum National d'Histoire Naturelle, Paris,* 4th ser., 16, sect. C, no.1: 19–39.

Bouvrain, G., and Thomas, H. 1992. Une antilope à chevilles spiralées: *Prostrepsiceros zitteli* (Bovidae). Miocène supérieur du Jebel Hamrin en Iraq. *Geobios* 25: 525–33.

Bristow, C. S. 1999. Aeolian and sabkha sediments in the Miocene Shuwaihat Formation, Emirate of Abu Dhabi, United Arab Emirates. Chap. 6 in *Fossil Vertebrates of Arabia,* pp. 50–60 (ed. P. J. Whybrow and A. Hill). Yale University Press, New Haven.

Eisenmann, V. 1977. Les Hipparions africains: valeur et signification de quelques caractères des jugales inférieures. *Bulletin du Muséum National d'Histoire Naturelle, Paris,* 3rd ser. no. 438, Science Terre 60: 69–87.

———. 1980. Caractères spécifiques et problèmes taxonomiques relatifs à certains Hipparions africains. In *Actes du 8e Congrès Panafricain de Préhistoire et des Etudes du Quaternaire,* pp. 77–81 (ed. R. E. Leakey and B. A. Ogot). The International Louis Leakey Memorial Institute for African Prehistory, Nairobi.

———. 1988. Les Périssodactyles Equidae. In *Contribution à l'étude du gisement miocène supérieur de Montredon (Hérault): Les grands Mammifères,* pp. 65–96 (ed. S. Legendre). Palaeovertebrata Mémoire Extraordinaire, Montpellier.

———. 1995a. Equidae of the Albertine Rift Valley, Uganda. In *Geology and Palaeobiology of the Albertine Rift Valley, Uganda-Zaire, Volume II: Paléobiologie/Palaeobiology,* pp. 289–308 (ed. B. Senut and M. Pickford). CIFEG Publicational Occasionnelle 1994/29. Centre International pour la Formation et les Echanges Géologiques, Orléans.

———. 1995b. What metapodial morphometry has to say about some Miocene hipparions. In *Palaeoclimate and Evolution; with Emphasis on Human Origins,* pp. 148–63 (ed. E. S. Vrba, G. H. Denton, T. C. Partridge, and L. H. Burckle). Yale University Press, New Haven.

Eisenmann, V., and Mein, P. 1995. Revision of the faunal list and study of Hipparion of the Pliocene locality of La Gloria 4 (Spain). *Acta Zoologica Cracoviensia* 39 (1): 121–30.

Eisenmann, V., Alberdi, M. T., De Giuli, C., and Staesche, U. 1988. Studying fossil horses. In *Collected Papers after the "New York International Hipparion Conference, 1981"*, p. 71 (ed. M. O. Woodburne and P. Sondaar). Brill, Leiden.

Friend, P. F. 1999. Rivers of the Lower Baynunah Formation, Emirate of Abu Dhabi, United Arab Emirates. Chap. 5 in *Fossil Vertebrates of Arabia*, pp. 38–49 (ed. P. J. Whybrow and A. Hill). Yale University Press, New Haven.

Gromova, V. I. 1952. Gippariony (rod *Hipparion*) po materialam Taraklii, Pavlodara i drugim. *Trudy Paleontologicheskogo Instituta Akademii Nauk SSSR* 36: 1–475.

Hill, A. 1987. Causes of perceived faunal change in the later Neogene of East Africa. *Journal of Human Evolution* 16: 583–96.

Hill, A., and Ward, S. 1988. Origin of the Hominidae: the record of African large hominoid evolution between 14 My and 4 My. *Yearbook of Physical Anthropology* 31: 49–83.

Hill, A., Curtis, G., and Drake, R. 1986. Sedimentary stratigraphy of the Tugen Hills, Baringo District, Kenya. In *Sedimentation in the African Rifts*, pp. 285–95 (ed. L. Frostick, R. W. Renaut, I. Reid, and J.-J. Tiercelin). Geological Society of London Special Publication no. 25. Blackwell, Oxford.

Hill, A., Ward, S., and Brown, B. 1992. Anatomy and age of the Lothagam mandible. *Journal of Human Evolution* 22: 439–51.

Hooijer, D. A. 1975. Miocene to Pleistocene hipparions of Kenya, Tanzania and Ethiopia. *Zoologische Verhandelingen* 142: 1–75.

Hooijer, D. A., and Maglio, V. J. 1974. Hipparions from the late Miocene and Pliocene of Northwestern Kenya. *Zoologische Verhandelingen* 134: 1–34.

Hussain, S. 1971. Revision of *Hipparion* from the Siwalik Hills of Pakistan and India. *Bayerische Akademie der Wissenschaften* 147: 1–68.

James, G. A., and Wynd, J. G. 1965. Stratigraphic nomenclature of Iranian oil consortium agreement area. *Bulletin of the American Association of Petroleum Geologists* 49: 2182–245.

Koufos, G. D. 1987. Study of the Turolian hipparions of the Lower Axios Valley (Macedonia, Greece). 1. Locality "Ravin des Zouaves-5" (RZO). *Geobios* 20: 293–312.

———. 1988. Study of the Turolian hipparions of the Lower Axios Valley (Macedonia, Greece). 3. Localities of Vathylakkos. *Paleontologia y Evolucio* 22: 15–39.

Koufos, G. D., and Kostopoulos, D. S. 1994. The late Miocene mammal localities of Kemiklitepe, Turkey. 3. Equidae. *Bulletin du Muséum National d'Histoire Naturelle, Paris,* 4th ser., sect. C, no.1: 41–80.

Koufos, G. D., and Melentis, J. K. 1984. The Late Miocene (Turolian) mammalian fauna of Samos island (Greece). 2. Equidae. *Science Annals, Faculty of Sciences, University of Thessaloniki* 24 (47): 47–78.

Lydekker, R. 1884. Additional Siwalik Perissodactyla and Proboscidea. *Memoirs of the Geological Survey of India. Palaeontologia Indica* 3 (1): 1–34.

MacFadden, B. J., and Woodburne, M. O. 1982. Systematics of the Neogene Siwalik hipparions (Mammalia; Equidae) based on cranial and dental morphology. *Journal of Vertebrate Paleontology* 2: 185–218.

Malez, M., and Forsten, A. 1989. Hipparion from the Bekaa valley of Lebanon. *Geobios* 22: 665–70.

Mein, P. 1990 (1989). Updating of MN zones. In *European Neogene Mammal Chronology,* pp. 73–90 (ed. E. H. Lindsay, V. Fahlbusch, and P. Mein). Plenum Press, New York.

Nakaya, H., and Watabe, M. 1990. Hipparion from the Upper Miocene Namurungule Formation, Samburu Hills, Kenya: Phylogenetic significance of newly discovered skull. *Geobios* 23: 195–219.

Pickford, M., Senut, B., and Hadoto, D. ed. 1993. *Geology and Palaeobiology of the Albertine Rift Valley, Uganda-Zaire, Volume I: Geology.* CIFEG Publication Occasionnelle no. 24. Centre International pour la Formation et les Echanges Géologiques, Orléans.

Pomel, A. 1897. Homme, singe, carnassiers, equidés, suilliens, ovidés: Les Equidés. *Monographies du Service de la Carte Géologique de Algérie,* n.s. *Paléontologie:* 1–44.

Sen, S. 1990 (1989). Hipparion datum and its chronologic evidence in the Mediterranean area. In *European Neogene Mammal Chronology,* pp. 495–505 (ed. E. H. Lindsay, V. Fahlbusch, and P. Mein). Plenum Press, New York.

Simpson, G. G. 1941. Large Pleistocene felines of North America. *American Museum Novitates* no. 1136: 1–27.

Sondaar, P. Y. 1971. The Samos Hipparion. *Proceedings Koninklijke Nederlandse Akademie van Wetenschappen* 74: 417–41.

Staesche, U., and Sondaar, P. Y. 1979. Hipparion aus dem Vallesium und Turolium (Jüngtertiär) der Türkei. *Geologisches Jahrbuch* 33: 35–79.

Tassy, P. 1994. Les gisements de Mammifères du Miocène supérieur de Kemiklitepe, Turquie: 7. Proboscidea (Mammalia). *Bulletin du Muséum National d'Histoire Naturelle, Paris,* 4th ser., 16, sect. C, no. 1: 143–57.

Thomas, H., Sen, S., and Ligabue, G. 1980. La faune Miocène de la Formation Agha Jari du Jebel Hamrin (Irak). *Proceedings Koninklijke Nederlandse Akademie van Wetenschappen* 83: 269–87.

Vogt, B., Gockel, W., Hofbauer, H., and Al-Haj, A. A. 1989. The coastal Survey in the Western Province of Abu Dhabi. *Archaeology in the United Arab Emirates* V: 49–60.

Wehrli, H. 1941. Beitrag zur Kenntnis der "Hipparionen" von Samos. *Palaeontologische Zeitschrift* 22: 321–86.

Whybrow, P. J. 1989. New stratotype; the Baynunah Formation (Late Miocene), United Arab Emirates: Lithology and palaeontology. *Newsletters on Stratigraphy* 21 (1): 1–9.

Whybrow, P. J., Friend, P. F., Ditchfield, P. W., and Bristow, C. S. 1999. Local stratigraphy of the Neogene outcrops of the coastal area: Western Region, Emirate of Abu Dhabi, United Arab Emirates. Chap. 4 in *Fossil Vertebrates of Arabia,* pp. 28–37 (ed. P. J. Whybrow and A. Hill). Yale University Press, New Haven.

Whybrow, P. J., Hill, A., Yasin al-Tikriti, W., and Hailwood, E. A. 1990. Late Miocene primate fauna, flora and initial palaeomagnetic data from the Emirate of Abu Dhabi, United Arab Emirates. *Journal of Human Evolution* 19: 583–88.

Fossil Suidae from the Baynunah Formation, Emirate of Abu Dhabi, United Arab Emirates

20

Laura Bishop and Andrew Hill

The initial palaeontological investigation of an area provides interesting opportunities and problems. The Baynunah Formation of the Emirate of Abu Dhabi is one such region (Whybrow et al., 1990). The Arabian Peninsula is situated geographically at an intersection of several biogeographic zones, an area with a potentially interesting history during the Messinian Salinity Crisis of the late Miocene. Prior to the research initiated by The Natural History Museum/Yale expedition to Abu Dhabi, the only suid remains of comparable antiquity from Arabia were those from Ad Dabtiyah, in eastern Saudi Arabia (Hamilton et al., 1978; Pickford, 1987; 1978; Whybrow, 1987). Pickford pointed out the similarity of two of the four known specimens to two species of the middle Miocene genus *Listriodon: L. lockharti* common in Europe and *L. akatikubas,* known from eastern Africa and the Congo. These two species are found in middle Miocene sediments, suggesting that the Ad Dabtiyah sequence is middle Miocene in age. These sediments significantly predate the estimated biostratigraphic age for the Baynunah Formation.

Late Miocene Suidae

At the end of the middle Miocene, the archaic suid subfamilies Listriodontinae, Hyotheriinae, and Kubanochoerinae that had been the dominant suids during the earlier part of the Miocene were disappearing (Pickford, 1988, 1993) (table 20.1). The hyotheriines were small animals with generalised dentitions and are thought to be the ancestral suid subfamily (Pickford, 1993). Kubanochoeres ranged from tiny to very large-bodied and had bunodont, brachydont teeth; some lineages developed "horns" (Pickford, 1993). The listriodont pigs were dentally quite specialised with lophodont teeth.

By the later part of the Miocene the dominant suids were the Tetraconodontinae and the Suinae. The former are known from middle Miocene deposits, and the latter are first known after the disappearance of the archaic middle Miocene taxa. Tetraconodonts are characterised by thick enamel, relatively simple postcanine tooth morphology, bunodonty, and enlarged third and fourth premolars. These characteristics are present in part of the Baynunah sample. All modern Eurasian pigs, with the exception of the babirusa (*Babyrousa babyrussa*), belong to Suinae, as does the domestic pig *Sus scrofa*. This subfamily is also represented in the Baynunah Formation.

Africa

Deposits from the late Miocene are relatively rare in Africa (Hill and Ward, 1988; Hill, 1995, 1999—Chapter 29). The suid reported in those that exist, primarily the Mpesida Formation of the Tugen Hills and the Samburu Hills deposits, is *Nyanzachoerus* indet. (Nakaya et al., 1984; Pickford et al., 1984; Pickford, 1986; Hill, 1995). The remains are scarce and fragmentary so it is difficult to draw any conclusions about the morphology of these suids compared with *N. devauxi* and *N. syrticus,* which become common later in time. Metrically, these pigs are quite dissimilar from later ones.

The common eastern African late Miocene suid was originally named *Nyanzachoerus tulotus* on the

Table 20.1. Distribution of fossil suid taxa in the middle and late Miocene

Age (Ma)	Subfamily/tribe	Europe	Siwaliks	Africa
	Suinae	*Korynochoerus*	*Propotamochoerus hysudricus*	
		Microstonyx	*Hippopotamodon*	
		Sus	*? Sus*	
	Hippohyini		*Hippohyus*	
			Sivahyus	
8	Tetraconodontinae		*Sivachoerus*	*Nyanzachoerus*
	Suinae	*Korynochoerus*	*Propotamochoerus hysudricus*	
		Microstonyx	*Hippopotamodon*	
	Tetraconodontinae	*Paraleuchastochoerus*	*Tetraconodon*	*? Nyanzachoerus*
10			*Lophochoerus*	
	Listriodontinae	*Listriodon splendens*	*Listriodon pentapotamiae*	*Listriodon akatikubas*
				Lopholistriodon
	Tetraconodontinae	*Conohyus simmorensis*	*Conohyus sindiensis*	
	Hyotheriinae	*Hippophyus soemmeringi*	*Hippophyus pilgrimi*	
	Kubanochoerinae			*Libycochoerus massai*
15				*Libycochoerus khinzikebirus*

Source: after Pickford (1988).

basis of fossils from Lothagam, Kenya (Cooke and Ewer, 1972). Cooke later determined that the Lothagam suids were indistinguishable from *Sivachoerus syrticus* Leonardi, 1952 described from Sahabi in Libya, but more correctly placed within *Nyanzachoerus* (Cooke, 1987).

Nyanzachoerus syrticus is a large suid of the subfamily Tetraconodontinae. In addition to the Lothagam and Sahabi samples, it has been identified in Kenya from Ekora, Kanam, and the Lukeino Formation of the Tugen Hills, all of which date to the late Miocene (Harris and White, 1979; Hill et al., 1985, 1986, 1992). The species is not currently known south of Kenya, and its last well-dated appearance is at 5.6 million years (Ma) from the Tugen Hills (Hill et al., 1992) though White (1995) points out that the Sahabi occurrence could be slightly younger. A smaller and more primitive taxon, *Nyanzachoerus devauxi,* was described from deposits at Oued al Hamman, Algeria (Arambourg, 1968). This provenance is now in doubt, and the holotype is thought to come from Dublineau (Thomas et al., 1982). *Nyanzachoerus devauxi* is similarly distributed in space and time, and is known from Sahabi, the Tugen Hills, and Lothagam (Cooke, 1987; Hill et al., 1992).

Most late Miocene African localities preserve a maximum of two taxa, both tetraconodonts, *N. devauxi* and *N. syrticus.* The sole exception to this is Sahabi, where the more derived *N. kanamensis,* usually found in later deposits in eastern Africa, is also recovered with the two more primitive forms (Kotsakis and Ingino, 1979; Cooke, 1987). Since this co-occurrence is unique within any single geological stratum in Africa, it may imply that the Sahabi fauna averages a greater period of time or that ecological conditions were unique at the site. Another possibility is that the Sahabi fauna preserves a stage of tetraconodont evolution unsampled in the remainder of the African record.

Asia

Late Miocene deposits of the Indian subcontinent preserve a variety of suids. The large tetraconodont *Sivachoerus,* described by Pilgrim (1926), is common in sediments of the upper part of the Dhok Pathan Formation, and apparently occurs in later sediments of the Siwaliks (Pickford, 1988). It has, however, been hypothesised by Pickford (1988) that its ancestors are not in the tetraconodont and lophodont forms from the earlier part of the Siwalik sequence, such as *Tetraconodon* and *Lophochoerus,* but rather among the earlier tetraconodonts of Africa. *Sivachoerus* has heavily crenellated premolar enamel and encircling cingulation of the P_4. The third and fourth premolars of *Sivachoerus* are even larger, relative to the anterior premolars and molars, than are present in either *Nyanzachoerus* or in the Abu Dhabi pigs. The third molar has an inflated appearance most apparent in the lingual side of the protoconid/metaconid junction. These characters are apparently absent in the Abu Dhabi sample. The mandible is also much more massive in *Sivachoerus* than in *Nyanzachoerus.*

Pigs of the subfamily Suinae are also common in deposits of the Indian subcontinent. *Propotamochoerus hysudricus,* described by Stehlin (1899), is common in the Nagri and Dhok Pathan Formations and is also known from Perim Island (now called Piram Island; Pickford, 1988). This species is not known from Europe (van der Made and Moyà-Solà, 1989). The large suine pig *Hippopotamodon* described by Lydekker (1877) is also known from India in the Nagri and Dhok Pathan Formations, and from sediments that post-date those in the Dhok Pathan type area (Pickford, 1988). Later Siwalik sequence sediments also preserve two genera of relatively hypsodont suine pigs of the tribe Hippohyini Pickford, 1988—*Hippohyus* and *Sivahyus,* described by Falconer and Cautley (in Owen, 1840–45) and Pilgrim (1926), respectively. These pigs, which have complex cusp patterns, are unlike any of the specimens recovered from Abu Dhabi or in any other Old World deposits of similar age.

Europe and the Eastern Mediterranean

In the middle Miocene, European suid diversity is quite high, with numerous taxa of listriodont, hyotheriine, and tetraconodont pigs represented (van der Made and Moyà-Solà, 1989; van der Made, 1989–90), but by the later Miocene there are fewer species representing fewer subfamilies. In recent work, taxonomy has been revised, leading to a picture of less suid diversity in the late Miocene (approximating MN zones 11–13) (van der Made and Moyà-Solà, 1989; van der Made, 1989–90; Fortelius et al., 1996).

Microstonyx, a suine, is the dominant late Miocene genus, present at numerous Eurasian localities. The genus *Korynochoerus* Schmidt-Kittler, 1971, also common in the late Miocene and early Pliocene, has been sunk into *Propotamochoerus,* originally named for Siwalik specimens (Fortelius et al., 1996). One suine pig, *Eumaiochoerus etruscus,* is a highly derived and endemic island form found only in Tuscany (Hürzeler, 1982). By MN 16a (middle Pliocene), European suid diversity is further diminished with most specimens subsumed into *Sus,* a genus of presumably Asian origin (Azzaroli, 1975; van der Made and Moyà-Solà, 1989).

Material from the Baynunah Formation

So far, 17 specimens have been recovered at 10 sites from 6 localities (table 20.2). Both dental and fragmentary postcranial remains are known. There are at least two taxa represented in the sample. The larger and more common of the two is provisionally identified as *Nyanzachoerus syrticus* (Leonardi, 1952 = *N. tulotus* Cooke and Ewer, 1972) on the basis of its dental morphology and size. The smaller taxon is referred to *Propotamochoerus* cf. *hysudricus.*

Table 20.2. Suid specimens from the Baynunah Formation, according to taxonomic assignment; within each taxon, specimens are listed numerically, with their provenance and a brief description of the material

Taxon	Specimen no.	Site no.	Description
Propotamochoerus hysudricus	BMNH M49433	B2	rt mandibular corpus with P_4–M_3
	AUH 55	S1	lt M^3
aff. *P. hysudricus*	AUH 58	S1	Unerupted lt I^2
	AUH 423	H3	rt I^3, labial fragment
	AUH 627	M1	lt I^1
Nyanzachoerus syrticus	AUH 329	H1	lt P_3, partial lt M_3 and associated bone fragments
	AUH 549	S6	rt mandibular corpus with P_3–M_3
aff. *N. syrticus*	AUH 238	TH1	Juvenile lt mandibular corpus with unerupted C, I_3, dP_3, dP_4, M_1 and unerupted M_2 in the crypt
	AUH 328	TH1	rt edentulous mandibular corpus
	AUH 784	R2	Molar crown fragment
	AUH 128	S2	Lower rt canine fragment
	AUH 557	S6	P^4 fragment
	AUH 690	S4	P_3 fragment
Suidae indet.	AUH 3	H1	lt distal humerus
	AUH 626	M1	lt distal humerus
	AUH 172	H1	rt distal humerus
	AUH 331	H4	Rib fragment

Locality abbreviations: B, Jebel Barakah; S, Shuwaihat; H, Hamra; R, Ras Dubay'ah; TH, Thumayriyah; M, Jebel Mimiyah (Al Mirfa).

Fossil collections from the Western Region of the Emirate of Abu Dhabi made since 1986 by The Natural History Museum/Yale University team and others, are temporarily housed in the Department of Palaeontology, Natural History Museum, London. All palaeontological material collected by the NHM/Yale team has registration numbers of the Emirate of Abu Dhabi (AUH). Other material collected has BMNH numbers.

Description

Order Artiodactyla Owen, 1848
Infraorder Suina Gray, 1868
Family Suidae Gray, 1821
Subfamily Tetraconodontinae Simpson, 1945
Genus *Nyanzachoerus* Leakey, 1958
Species *Nyanzachoerus syrticus* (Leonardi, 1952)

Specimen AUH 329: Hamra, Site H1

This specimen consists of a complete left P_3 and partial left M_3 associated with numerous tooth and bone fragments (fig. 20.1.1). At least some of these are not suid, but the specimen has yet to be prepared completely. Both identifiable teeth are preserved as enamel crowns only; the dentine has been removed for the most part during diagenesis or subsequent exposure. The enamel is in fairly good condition, with some exfoliation on its surfaces. P_3 exhibits a typical tetraconodont morphology, being large, massive, and having a relatively high central cusp. The tooth has a well-defined anterior cingulum, and a pronounced posterior cingulum, which extends to the worn occlusal surface of the tooth. The enamel surface is smooth, showing none of the corrugations typical of *Sivachoerus*. There is a raised ridge running down the mesial surface of the main cusp, joining its apex with the anterior cingulum. Planar occlusal wear is marked along the distal face of the tooth. There is slight waisting along the cervix, and a pronounced lingual flare to the P_3 distal to this indentation.

The M_3 is complete except for a portion of the anterior cingulum, the lingual portion of the protoconid, and the central trigonid pillar. It is bunodont and appears slightly puffy. The cusps have some grooves that appear to be deep enough to persist in wear and make indentations in the outlines of the dentine lakes. The anterior cingulum is relatively small, and the cingulum does not encircle the tooth.

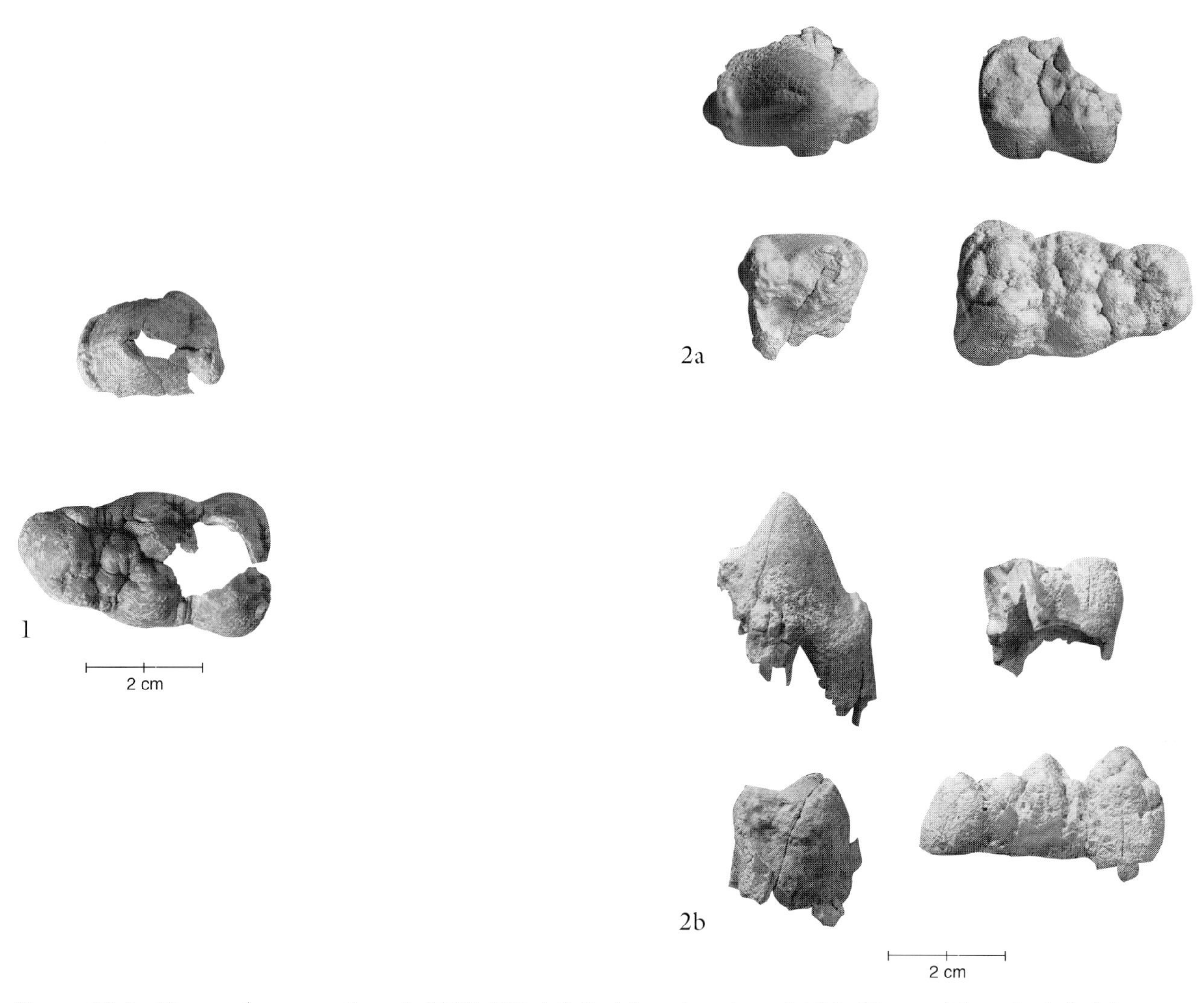

Figure 20.1. *Nyanzachoerus syrticus. 1,* AUH 329, left P_3 (above) and partial M_3 (P_3, mesial to the left; M_3, mesial to the right); occlusal view. *2,* AUH 549, right P_3–M_3 (M_1 not illustrated; order, clockwise from upper left corner, P_3, M_2, M_3, P_4): *2a,* occlusal view (mesial to the left except P_4); *2b,* labial view (mesial to the right except P_4).

M_3 has a small talonid with only one central, well-developed cusp, flanked by other accessory cuspules of which the lingual is the strongest. There is slight waisting between the trigonid and the talonid. Several accessory cuspules are present between the metaconid and the entoconid. Dental measurements for this specimen and others assigned to *Nyanzachoerus syrticus* are found in table 20.3.

Specimen AUH 549: Shuwaihat, Site S6

A partial right hemimandible, with associated P_3, P_4, fragmentary M_1 and M_2, and complete M_3 (figs 20.1.2a, 2b). An incomplete left P_4 was also recovered with the specimen. The hemimandible was found with numerous bone and tooth root fragments, which are in need of preparation. This specimen is the most complete tetraconodont pig recovered from the Baynunah Formation, and is the most diagnostic for the taxon *Nyanzachoerus syrticus.*

The teeth are well preserved and well mineralised, but the enamel surfaces have an unusual, sandblasted appearance so that details of the occlusal morphology are lost. P_3 is typically tetraconodont, and is identical in description and detail to that of the Hamra pig, above. The enamel surface is not as well preserved as in that specimen, so it is impossible to determine whether any enamel crenellations are present. The overall morphology would ally this tooth with *Nyanzachoerus* rather than *Sivachoerus.* Like P_3, P_4 also has a single

Table 20.3. Measurements (in mm) for lower dentition of *Nyanzachoerus syrticus*

	dP_2	dP_3	dP_4	P_2	P_3	P_4	M_1	M_2	M_3
AUH 238 lt (juvenile)	18.65	(15.96)	24.80	—	—	—	20.40	26.05	—
	8.45	9.55	8.45	—	—	—	17.24	20.80	—
AUH 549 rt	—	—	—	—	26.56	22.70	c.19.00	(25.4)	42.59
	—	—	—	—	20.96	(21.5)	(17.3)	(21.6)	25.05
AUH 329 lt	—	—	—	—	25.33	—	—	—	45.60
	—	—	—	—	19.52	—	—	—	(25.4)
AUH 690 lt	—	—	—	—	—	—	—	—	—
	—	—	—	—	(17.4)	—	—	—	—

Note: Measurements were taken according to the scheme described by Harris and White (1979): for each cell, the upper measurement is the basal length and the lower measurement the breadth, taken at the cervix. Estimates are in parentheses.

main conid. The posterior cingulum is more developed in this tooth, however, and a centrally located conid within it approaches the full height of the main conid. This is offset lingually and buccally by pronounced indentations. There is a ridge running up the mesial side of the tooth, connecting the anterior cingulum with the tip of the central conid. The tooth is more broad than P_3, especially distally, where there is a pronounced lingual flare to the tooth.

The molars have simple morphology and are bunodont and brachydont. The enamel appears to be relatively thick. M_1 is heavily worn and broken, missing the anterior portion and most of the lingual side of the tooth except a part of the mesial entoconid. The dentine lakes appear to have a crenellated shape, and there are basins preserved between the protoconid and the hypoconid. M_2 appears similar to M_1, but is larger and better preserved, missing only the distobuccal portion of the hypoconid. The dentine lakes are roughly star-shaped. There appears to be a basin separating the metaconid and entoconid, and another, cuspule-lined one between the protoconid and the hypoconid. The anterior and posterior cingula are well developed, but somewhat obliterated during eruption. M_3 is a triangular tooth with a small, somewhat waisted talonid. The anterior cingulum appears distinct and well developed. The four main trigonid cusps are strong, and there is one major pillar in the talonid, located distolabially. The pillar on the trigonid–talonid junction is also well developed. There are numerous small accessory cuspules in the talonid, and the anterior and posterior portions of the trigonid are separated by labial and lingual indentations and cuspule-lined enamel basins between the major cusps.

aff. *Nyanzachoerus syrticus*

Specimen AUH 238: Thumayriyah, Site TH1

Juvenile left mandibular corpus with I_3, dP_2, dP_3, dP_4, M_1, and unerupted C and M_2 in the crypt (fig. 20.2.1a, 1b). The specimen also has a fragment of the ascending ramus, including a partial mandibular angle. It has been damaged considerably during the taphonomic process, and reconsolidated using hardeners and glue during recovery. The preserved morphology is compatible with an assignment to *Nyanzachoerus syrticus* but the morphology is undiagnostic because there is no comparative juvenile sample assigned to the species. The morphology of the developing permanent dentition is tetraconodont in character. Only the tip of the canine is present, and it exhibits a D-shaped cross-section. The I_3 is robust with a pronounced lingual crest running down the long axis. The dP_1 is missing through apparent breakage. dP_2–dP_4 have long roots, exposed through damage, and relatively thin enamel. dP_2 is long and narrow, with a pronounced central cusp. At the cervix, there is a marked constriction and raised enamel level between the mesial and distal roots. The crown of dP_3 is severely damaged but appears to have had the same overall basal shape as dP_2. dP_4 is well worn and damaged. The enamel appears thick. There are three pillar pairs and the tooth broadens distally. M_1 is in full occlusion and exhibits a tetraconodont pattern, with strong simple cusps indented by grooves and strong anterior and posterior cingula. There is a small basin joining the protoconid and hypoconid, but morphology is difficult to discern due to weathering and breakage. An unerupted M_2 is exposed in the crypt. The crown had not yet fully formed, but visible morphology is similar to M_1 and like *Nyanzachoerus*.

Specimen AUH 328: Thumayriyah, Site TH1

Large right hemimandible in bad condition. Broken tooth crowns and roots are apparently present for P_3–M_3. The mandible is broken, very weathered and cracked, with cortical bone badly damaged on all surfaces. Some tooth is present, although the enamel has been exfoliated from all exposed portions of the tooth crowns. The crowns are still partly covered in matrix and further preparation might reveal more of the occlusal morphology if any enamel remains beneath. The mandible is deep, over 73 mm high in the region of the ascending ramus. The broken teeth were apparently massive and bunodont. The P_3 and P_4 were relatively and absolutely large. The size of the teeth and the robusticity in the premolar region are compatible with those of the tetraconodonts,

Figure 20.2. Aff. *Nyanzachoerus syrticus,* AUH 238, juvenile left mandible with unerupted C, I_3, dP_2, dP_3, dP_4, and M_1 in occlusion and M_2 in the crypt: *1a,* occlusal view; *1b,* labial view.

particularly *Nyanzachoerus.* Further preparation of this specimen might prove useful for a secure taxonomic assessment.

Specimen AUH 784: Ras Dubay'ah, Site R2

This specimen is an unerupted crown fragment, most probably from an M3 according to size, morphology, and enamel thickness. The cusps are relatively low, pyramidal, and have relatively deep grooves. This cusp morphology, as well as the squarish relative position of the cusps, is similar to that of *Nyanzachoerus.*

Specimen AUH 128: Shuwaihat, Site S2

Lower right canine fragment, split longitudinally; this is an inferolateral portion. The canine is large. No wear is apparent on the preserved fragment and the dentine is absent so the tooth might be unerupted. The preserved portion of the canine suggests that it is suboval in cross-section. The enamel is slightly rugose and the tooth has a longitudinal ridge, similar to that seen on some specimens of *Nyanzachoerus.*

Specimen AUH 557: Shuwaihat, Site S6

P^4 fragment. This tooth is very fragmentary, but appears to be the labial cusp with thick enamel and relatively deep grooves on the cusp. Cusp relief is not high, suggesting a bunodont tooth. There are preserved fragments of a relatively high cingulum. The morphology and size of the specimen are compatible with *Nyanzachoerus.* The tooth is not worn, but the enamel is pitted and weathered.

Specimen AUH 690: Shuwaihat, Site S4

Left P_3 fragment consisting of the mesial portion of the tooth crown. The tooth is broken distal to the indented portion of the cervix. The fragment appears to be relatively narrow, and because it is only a mesial fragment the breadth reported in table 20.3 must be considered a minimum value. The anterior cingulum is pronounced with a very jagged appearance on the cranial border. The ridges on the mesial surface of the tooth, which connects the anterior cingulum with the apex of the protocone, is raised and marked, separated off from the buccal face of the tooth by a depressed line in the enamel. Several observations suggest that the tooth might have been unerupted. The central portion of the tooth is missing and there is a large cavity in the dentine. No roots are preserved. There is no apparent wear on the preserved portion, and the anterior cingulum bears no contact facet.

Subfamily Suinae Gray, 1821
Genus *Propotamochoerus* Pilgrim, 1925
Species *Propotamochoerus hysudricus* (Stehlin, 1899)

Specimen BMNH M49433: Jebel Barakah, Site B2

BMNH M49433 is a right mandibular corpus bearing P_4–M_3 (figs 20.3.1a, 1b). The inferior border of the corpus is not preserved. The bone is missing anterior and inferior to the distal roots of the P_4, which are exposed in this specimen. The corpus is also broken posterior to the M_3, preserving only the initial rise of the lateral (buccal) portion of the ascending ramus. The specimen has considerable postmortem damage to the bone at the cranial border of the alveoli; this exposes considerable areas of the tooth roots. Much of the enamel has been lost through wear during life combined with postmortem damage.

The individual was a mature adult, with M_3 in full occlusion, and all the teeth are well worn. The teeth appear to be thick-enamelled, bunodont, and of relatively simple occlusal morphology. There is no indication of lophodonty or other dental specialisation. The mesial portion of P_4 is missing, but the distal portion has a typical suine pattern of morphology and, apparently, of size relative to the molar row. P_4 has a large and pronounced posterior cingulum set off from the central conid by deep, V-shaped lingual and buccal furcula. Unfortunately, its morphology cannot be assessed completely due to breakage, which contributes to some level of uncertainty in attributing the specimen to *Propotamochoerus hysudricus*.

The M_1 appears relatively small, although it is worn and broken to below the cervix. All occlusal morphology is obliterated. M_2 is in a similar, although less worn, state. Breakage has exfoliated much of the enamel. The enamel that remains suggests it was relatively thick, and that there was a well-developed basin joining the protoconid and the hypoconid on the buccal side of the tooth. M_3 is the best preserved of the teeth, although it is missing a large portion of the enamel from the mesiobuccal aspect of the protoconid. It has a small anterior cingulum, and the entoconid is worn to a mushroom-shaped dentine lake. The cusps of the tooth are conical and isolated, and enamel is thick. A cuspule-bordered basin joins the hypoconid and the protoconid, apparently flaring lingually. Cusp morphology seems simple, without deep grooves dividing the surface of the individual cusps. Because the tooth is worn and abraded, these features might have been lost through wear or postmortem abrasion. The talonid is strong, with two very well-developed cusps and a strong cusp on the trigonid–talonid junction. For dental measurements of this and other specimens assigned to *Propotamochoerus hysudricus*, see table 20.4.

Specimen AUH 55: Shuwaihat, Site S1

This specimen is an incompletely erupted left M^3 (figs 20.3.2a, 2b). It is missing the anterior cingulum from just lingual to the midline to the labial border, and the labial aspect of the paracone. The tooth roots are present but fragmentary, and appear to be flaring and extensive. The overall shape of the tooth is like a wedge, with the simple talon comprised of one major pillar in the midline flanked by two rows of labial cusplets. Cusps are conical in shape and simple in morphology; the enamel is thick. The paracone and metacone are well separated, with a small basin between them. On the lingual side of the tooth, the hypoconid is abutted mesially with an accessory cuspule that is equal in height to the central trigone median cusp. The tooth is not worn enough to determine whether this cuspule's apparent differentiation from the hypocone would remain visible in wear. Enamel furrows are present in the tops of the cusps.

aff. *Propotamochoerus hysudricus*

Specimen AUH 58: Shuwaihat, Site S1

Unerupted left I^2. The specimen has a marked groove on the lateral (distal) aspect. The ridges on the occlusal edge of the incisor have not yet been obliterated through wear. The tooth is broken but the crown is not yet filled with dentine, suggesting that the root had not formed. Dimensions at the cervix: anteroposterior, 7.77 mm; mediolateral, 13.37 mm.

Specimen AUH 423: Hamra, Site H3

Labial fragment of a right I^3. Comparable in size and morphology with *Propotamochoerus*, but this specimen is very fragmentary. Approximate (minimum) mediolateral dimension at the cervix is 11.48 mm.

Specimen AUH 627: Jebel Mimiyah, Al Mirfa, Site M1

Left I^1 compatible in size and morphology with *Propotamochoerus hysudricus*. The tooth has pronounced wear, and is broken along the labial surface. No other details of anatomy are preserved.

Postcrania—Suidae indet.

Specimen AUH 3: Hamra, Site H1

Left distal humerus. This specimen appears to be a suid, but the epiphyseal morphology is obliterated because all cortical bone has exfoliated. The flaring and morphology of the epicondyles, however, are suid-like. The mediolateral breadth of the distal humerus is 48.50 mm, about the size of a large bushpig; this must be considered a minimum value because of the damage.

Specimen AUH 626: Jebel Mimiyah, Al Mirfa, Site M1

Left distal humerus with distal third of the humeral shaft. The cortical bone of this specimen is well preserved. It belonged to a mature but relatively small individual, about the size of a warthog. This size is consistent with that of *Propotamochoerus hysudricus*. The trochlea is well defined, although the bone surface is somewhat pitted. It has a short, rounded medial portion. The angle and definition of the median trochlear ridge, the abbreviated, concave lateral

Table 20.4. Dental measurements (in mm) of *Propotamochoerus hysudricus*

Upper teeth	P^2	P^3	P^4	M^1	M^2	M^3
AUH 55 lt	—	—	—	—	—	28.7
	—	—	—	—	—	19.8
Lower teeth	P_2	P_3	P_4	M_1	M_2	M_3
BMNH M49433 rt	—	—	12.7	13.6	17.8	32
	—	—	10.7	11.6	14.5	16

Note: Measurements (all estimates) were taken according to the scheme described by Harris and White (1979): for each cell, the upper measurement is the basal length and the lower measurement the breadth, taken at the cervix.

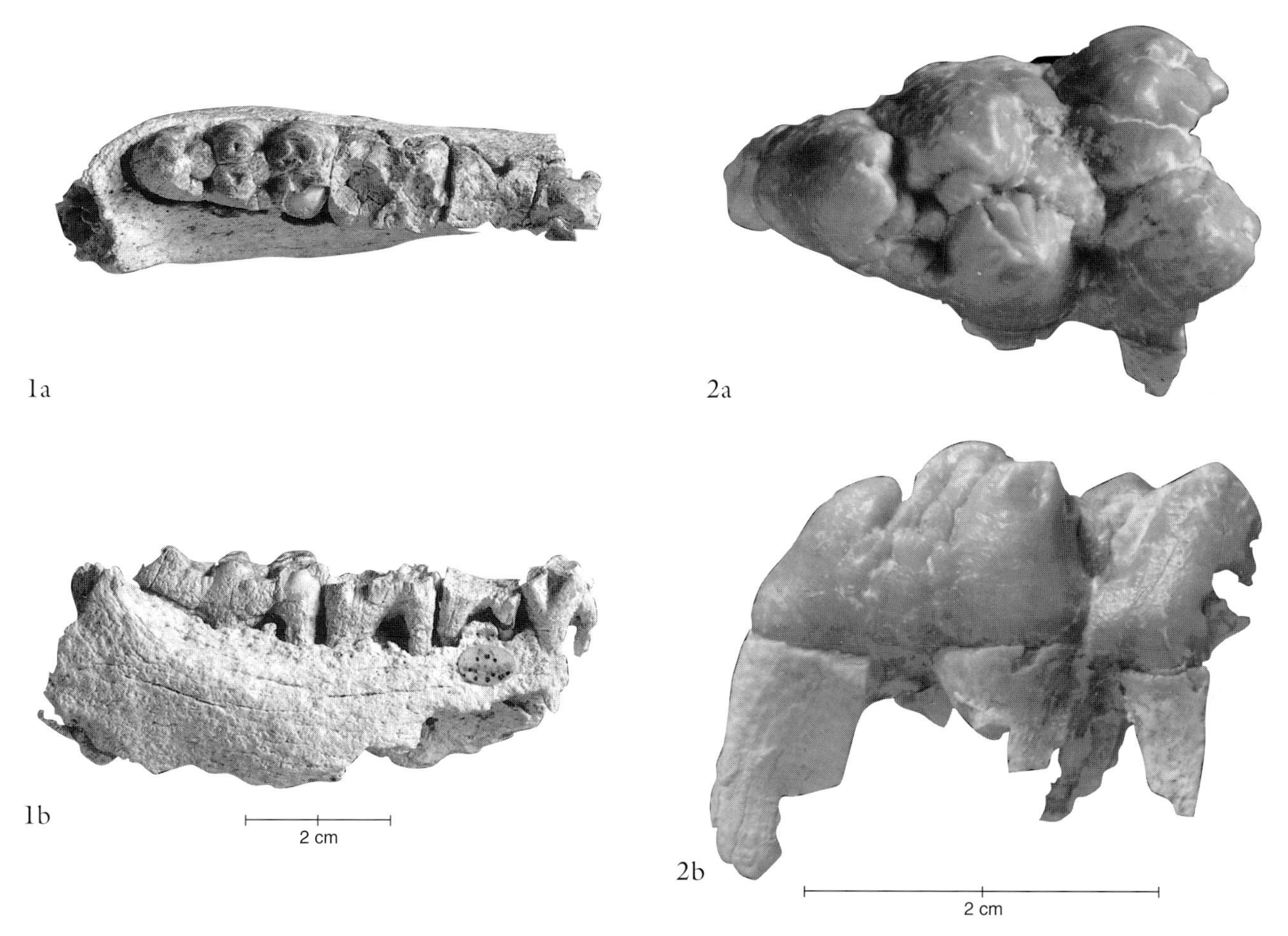

Figure 20.3. *Propotamochoerus hysudricus, 1,* BMNH M49433, right mandibular corpus with P_4–M_3: *1a,* occlusal view; *1b,* labial view. *2,* AUH 55, left M^3: *2a,* occlusal view; *2b,* labial view.

portion of the trochlea, and the epicondylar robusticity and size are typical of Suidae. The measurements are: distal mediolateral = 33.15 mm; distal lateral anterioposterior = 33.06 mm; trochlea maximum superoinferior = 23.42 mm; trochlea mediolateral = 24.89 mm.

Specimen AUH 172: Hamra, Site H1

Incomplete right distal humerus, preserving only the trochlea. This humerus has well-defined trochlear ridging, which conforms with the description of AUH 626, above. It is the largest of the three suid humeri recovered from the Baynunah Formation, a size consistent with *Nyanzachoerus syrticus*. The (estimated) measurements are: trochlea maximum superioinferior 40 mm; trochlea mediolateral 42 mm.

Specimen AUH 331: Hamra, Site H4

Incomplete rib, with very large head and flat shaft. It may not represent a suid.

DISCUSSION

Numerous characters ally the large Abu Dhabi specimens with the African mid-size Tetraconodontinae, and especially with *Nyanzachoerus syrticus*, such as the relatively narrower and more high-crowned P_3, a smooth enamel surface, and a pronounced and sharp ridge on the mesial surface of the P_3. The molars of *Nyanzachoerus syrticus* are triangular, bunodont, and have a waisted talon(id) and compact trigon(id). While there are accessory cuspules on the trigon(id) and trigon(id)–talon(id) junction, these are generally not accompanied by an inflated appearance or by the formation of well-defined basins as they are in later *Nyanzachoerus kanamensis*.

Metrically, the lower teeth of *Nyanzachoerus* cluster together in a bivariate plot (fig. 20.4). Only lower tetraconodont teeth from the Baynunah Formation were complete enough to measure. The measurements are well dispersed for many of the teeth. Individual taxa do not always cluster well, but in each case the Abu Dhabi sample is closest to *Nyanzachoerus syrticus*. Third molars are the tooth most often preserved, collected, and identified. On a bivariate plot of measurements for this tooth the taxa cluster well, and length and breadth measurements of the Abu Dhabi sample are not significantly separable from the *Nyanzachoerus syrticus* metrics (paired *t*-test; $P =$ n.s.) (fig. 20.5). The picture is somewhat complicated by the addition of *Sivachoerus* specimens from the Siwaliks sequence, whose third molar metrics seem to bridge the gap in measurements between *Nyanzachoerus syrticus* and the Abu Dhabi sample and *N. kanamensis* (fig. 20.6). Nonetheless, morphological characters described above link the Abu Dhabi pigs with *N. syrticus*, to the exclusion of the Siwalik sample.

The smaller suids from the Baynunah Formation are represented by a mandible and an isolated M^3 from Shuwaihat. These have been attributed to the Siwalik taxon *Propotamochoerus hysudricus*, which belongs to the subfamily that includes modern wild Eurasian pigs, the Suinae. Metrically, the measurable Baynunah specimen falls within the known *P. hysudricus* sample (fig. 20.7). The characters that link these samples are suine proportioned premolars, with a suine P_4 pattern preserved distally in the mandibular fragment BMNH M49433. Also, the molars possess thick-enamelled, isolated cusps, and the M_3 has relatively few and simple accessory cuspules. Much uncertainty in this attribution is due to nomenclatural problems, discussed below.

Nomenclature

There is great confusion over the status of *Propotamochoerus hysudricus*. The problem is discussed by Colbert (1935) and reviewed by Pickford (1988) and Fortelius et al. (1996). The confusion partly derives from the lack of formality in the original introductions of the generic and the specific names, and partly from great differences in the application of the names by various early authors. In using *Propotamochoerus hysudricus* here we do not claim to have satisfactorily unravelled the taxonomic tangle in favour of this nomen, but hope the following notes will explain the difficulties.

The trivial name *hysudricus* first appeared as a caption to a figure in Falconer and Cautley (1846), where it was used for a species of the genus *Sus*. *Sus hysudricus* was adopted in a subsequent publication by Lydekker (1884). Pickford (1988) indicates that Lydekker's use of the term encompassed specimens that were not in the species envisaged by Falconer and Cautley (1846). Pilgrim (1925) introduced the genus *Propotamochoerus* in a footnote to the text of the Presidential Address to the Geological Section of

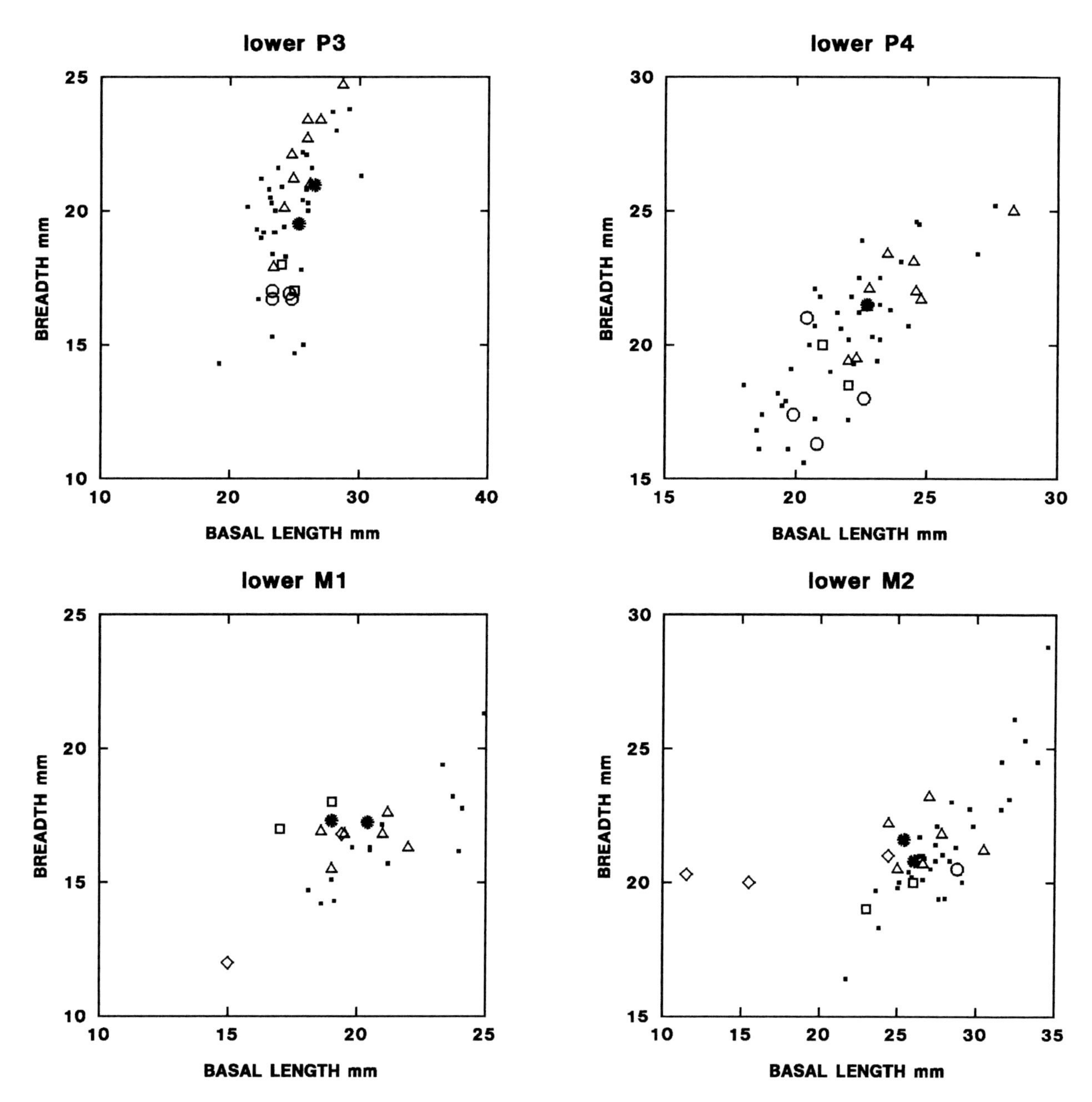

Figure 20.4. Bivariate plots of basal lengths and breadths of the lower dentition of *Nyanzachoerus* from Africa: Emirate of Abu Dhabi suids (asterisks), *Nyanzachoerus devauxi* (large squares), *N. kanamensis* (small squares), and *N. jaegeri* (open circles). The larger diamonds represent a few specimens attributed to *Nyanzachoerus* indet. by Pickford, and which come principally from the Mpesida Formation of the Tugen Hills, Kenya. *Nyanzachoerus devauxi* has been argued to be a more primitive form of suid than *N. syrticus*. *Nyanzachoerus kanamensis* is the type species of the genus and is very well distributed in time and space. *Nyanzachoerus jaegeri* is the most derived of the nyanzachoeres and has very hypsodont and elaborated M3s and reduced premolars. Measurements from Harris and White, 1979; Pickford, 1988; Arambourg, 1968; Leonardi, 1952; and unpublished data using the measurement scheme of Harris and White (1979) (L. Bishop, in preparation).

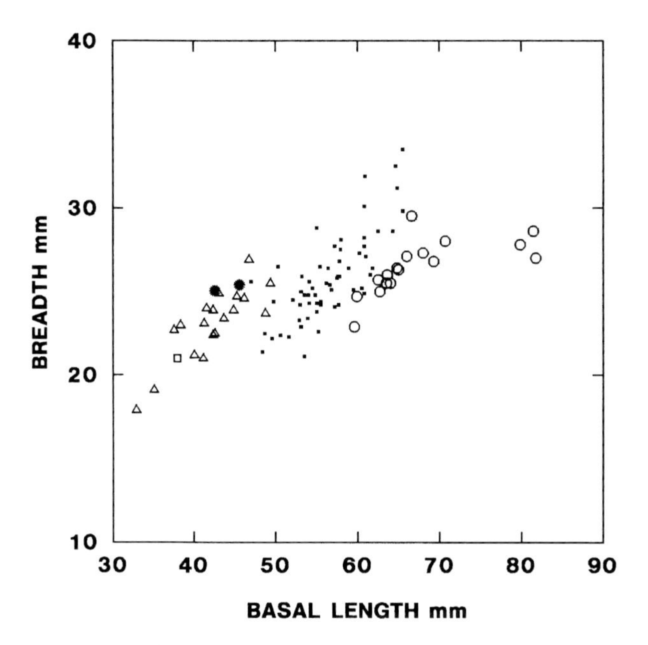

Figure 20.5. Bivariate plots of basal lengths and breadths of lower third molars of African *Nyanzachoerus*. Symbols as described in legend to figure 20.4.

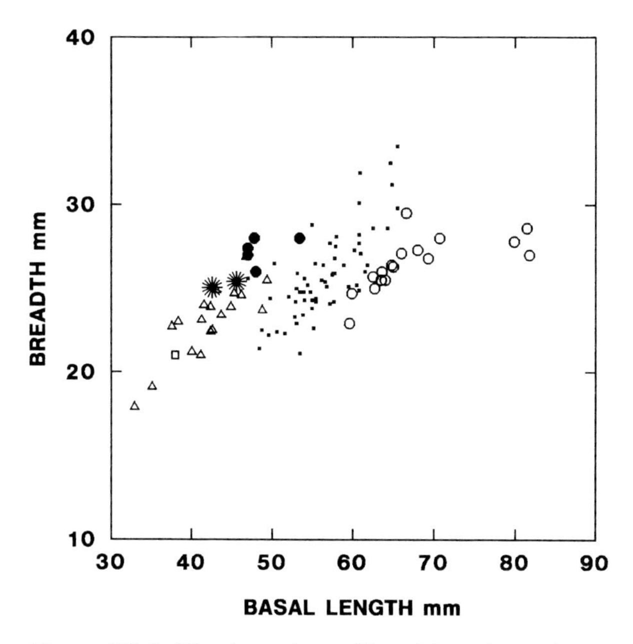

Figure 20.6. Bivariate plots of basal lengths and breadths of lower third molars of tetraconodonts. Symbols as described in legend to figure 20.4, with infilled circles representing *Sivachoerus prior.*

the Twelfth Indian Science Congress. He linked it there to *hysudricus,* designating it as the type species of the new genus. This species included a portion of the material referred to by Falconer and Cautley (1846) as *Sus hysudricus.* A year later, however, in a more detailed monograph, he used the species *salinus* as the type species of *Propotamochoerus* (Pilgrim, 1926). In doing this he introduced more than the obvious confusion because the two different putative type species, *salinus* and *hysudricus,* belong to different subfamilies of Suidae (Pickford, 1988). As shown by Pilgrim's figures, the species *salinus* is a tetraconodont and *hysudricus* is suine. Stehlin (1899) also discusses *hysudricus,* which he uses in combination with *Potamochoerus,* but his treatment is far from clear. First he attributes the species to Falconer, when he can only mean Falconer and Cautley (1846). A little later he attributes it to Lydekker. Further on he refers to *Sus hysudricus* (attributing it to Falconer) and then *Potamochoerus hysudricus* (attributing it to Lydekker). Colbert (1935), however, in a detailed discussion, concludes that Stehlin is technically the author of the species. Stehlin uses part of Lydekker's hypodigm as the type (GSI B30: a mandibular ramus), and Stehlin's description is the type description of the species.

Other developments since Pilgrim's publications have added little clarity to the situation. Arambourg (1968) used the genus *Propotamochoerus* for his species *devauxi.* This taxon is clearly tetraconodont as is properly reflected in its present status as a species of *Nyanzachoerus.* As a partial justification of his allocation, Arambourg figures a *Propotamochoerus hysudricus* mandible after Pilgrim. This mandible is tetraconodont, not suine. In a parallel set of obfuscations Pickford (1986) notes that Golpe-Posse (1979) attributes tetraconodont specimens from Bled ed-Douarah, Tunisia, to *Propotamochoerus.*

In his revision of *Propotamochoerus* Pickford (1988) affirmed the priority of the name *hysudricus* as the type species, and relegated *salinus* to *Conohyus* indet. Pickford bases his decision to uphold the taxon *Propotamochoerus hysudricus* not only on priority but also upon his belief that in setting up the diagnosis of

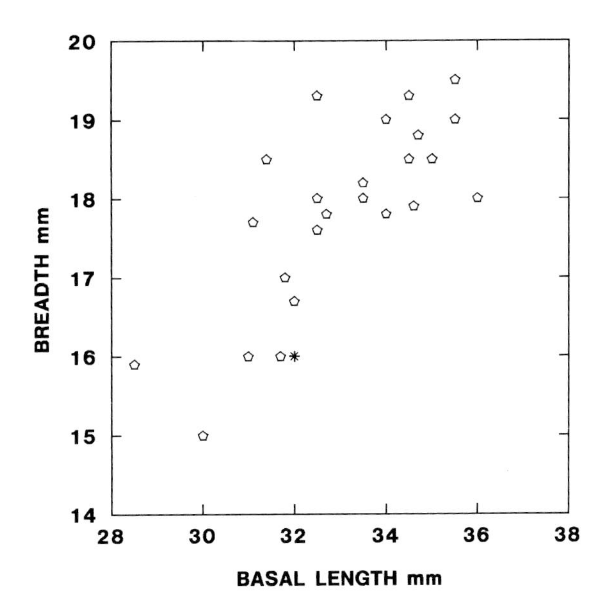

Figure 20.7. Bivariate plots of basal lengths and breadths of lower third molars of *Propotamochoerus hysudricus* from the Emirate of Abu Dhabi (asterisk) and the Indian subcontinent (pentagons).

the genus Pilgrim (1926), despite his claim that the tetraconodont *salinus* was the type species, actually relied upon specimens belonging to *hysudricus*.

Distribution within the Baynunah Formation

The best Baynunah specimen of the large tetraconodont *Nyanzachoerus syrticus*, the fragmentary right mandible AUH 549, is from Shuwaihat, site S6 from whence another tooth fragment attributable to the same species derives. Thumayriyah, site TH1 produced a left juvenile hemimandible, AUH 238, and an edentulous right mandibular corpus, AUH 328, also referable to *Nyanzachoerus*. Hamra provides postcranial remains consistent in size with this identification. *Propotamochoerus* is represented by and identified primarily on the basis of one fragmentary mandible from Jebel Barakah, BMNH M49433. AUH 55, an M^3 from Shuwaihat, may confirm the presence of this taxon. Other postcranial remains from Shuwaihat and perhaps Jebel Mimiyah, Al Mirfa are referable to *Propotamochoerus* on the basis of size.

Both taxa are represented at the localities of Shuwaihat and Hamra, although not at the same sites. Sample sizes at all sites are small, however, so we have no basis for suggesting that both taxa were not present throughout the Baynunah region. Fragmentary remains have been provisionally assigned to one of the two well-represented taxa. Future finds may reveal the presence of other taxa to which the current fragments would better be assigned.

Biostratigraphy

Biostratigraphically, the Abu Dhabi suid fauna does not contain forms with the derived and specialised dentitions of the European, Asian, and African middle Miocene, such as Listriodontinae, Kubanochoerinae, or Hyotheriinae. Small Asian and European tetraconodonts, common in the middle Miocene, are similarly absent. The tetraconodont *Sivachoerus,* only known later than 3.5 Ma ago in the Siwalik sequence (M. Morgan, personal communication), has not been recovered from Abu Dhabi. The Baynunah Formation also does not have the later Miocene–Pliocene hypsodont forms that are common in the later part of the Siwalik sequence. The hippohyine genera *Hippohyus* and *Sivahyus* are unknown in Abu Dhabi, as they are in Africa. In Africa, non-tetraconodont, suine pigs are first known in the middle Pliocene (Harris and White, 1979). No forms similar to these African suines have yet been recovered.

Although the absence of these forms constitutes negative evidence, the suid fauna recovered from the Baynunah Formation indicates a later Miocene age, and a biogeographic affinity with both Siwalik and African provinces. Since East African sequences bearing *Nyanzachoerus syrticus* have been radiometrically dated, this taxon is useful for biostratigraphic correlations. *Nyanzachoerus syrticus* occurs only in sediments older than 5.6 Ma (Hill et al., 1992). The temporal distribution of the taxon appears confined to the end of the late Miocene. This appears consistent with the range of *Propotamochoerus hysudricus* in the Siwalik sequence, from 10.4 to 6.8 Ma ago (M. Morgan, personal communication), so would seem to be a good estimate for the age of the Baynunah Formation.

BIOGEOGRAPHY

The geographic position of the Arabian Peninsula is between the Sub-Paratethyan, Siwalik, and (North) African zoogeographic provinces (Bernor, 1983, 1984). Bernor (1983: fig. 1) included Abu Dhabi in his proposed Sub-Paratethyan province, which includes the late Miocene locality of Maragheh, Iran. At Maragheh, the suid fauna consists of one taxon, *Microstonyx erymanthus*. This species is absent in Africa and India but common in Europe (Bernor, 1986) (table 20.5). It is absent from the Baynunah Formation assemblage. The suid fauna from the Baynunah Formation shows no affinity with suids from Maragheh or other sites in the Sub-Paratethyan province.

For biostratigraphic and biogeographic analysis, Fortelius et al. (1996) have hypothesised that the European suine *Korynochoerus* is allied with *Propotamochoerus* (table 20.5). The possible phylogenetic relationship between the Siwalik *Sivachoerus*, other Eurasian tetraconodonts, and the African *Nyanzachoerus* has been discussed by numerous authors (Leakey, 1958; Cooke and Ewer, 1972; Cooke, 1987; Pickford, 1986, 1988). The Baynunah Formation sample is relevant to any determination of relationship by virtue of geographic position. Tetraconodont suids are common in the Miocene of Europe and Asia, but medium-sized suids of the *Sivachoerus–Nyanzachoerus* clade are restricted to the late Miocene of the Indian subcontinent, northern and northeastern Africa, and now Abu Dhabi. The Siwalik taxon, *Sivachoerus prior*, is known from sediments postdating 3.5 Ma ago (M. Morgan, personal communication).

The Abu Dhabi tetraconodonts, with their affinity to the African form, suggest that the direction of transfer was from south to north. The limited distribution of *P. hysudricus* in the Indian subcontinent and in Abu Dhabi, along with the presence of primitive *Propotamochoerus* species elsewhere in Europe (Fortelius et al., 1996), suggests the hypothesis that this taxon originated in Eurasia and spread southwards to Arabia. The presence of *P. hysudricus* and *N. syrticus* indicate zoogeographic affinities with the Siwalik and North African provinces, respectively (table 20.6). This implies that interchange between these two provinces created the suid fauna of the Baynunah Formation.

CONCLUSIONS

In summary, there are several points to be made about suids recovered so far from the Baynunah Formation. There is a minimum of two genera from two different subfamilies—one tetraconodont taxon and a suine. Fragmentary material suggests the possibility that an additional non-tetraconodont species is represented. This level of suid diversity has not yet been found in the late Miocene of Africa. The suid fauna of the Miocene of Abu Dhabi is more diverse at the generic level than

Table 20.5. Late Miocene distribution of Old World suid taxa

Subfamily/tribe	Europe	Siwaliks	Africa
Suinae	*Korynochoerus*	= *Propotamochoerus*	
	Microstonyx	*Hippopotamodon*	
	Sus	?*Sus*	
Hippohyini		*Hippohyus*	
		Sivahyus	
Tetraconodontinae		*Sivachoerus*	= *Nyanzachoerus*

Sources: distribution is after Pickford (1988); taxonomic equivalencies were suggested by Fortelius et al. (1996) and Cooke (1987).

Table 20.6. Biogeographic relationships of Baynunah Formation Suidae; the arrows between the zoogeographic zones suggest the direction of migration of the taxa

Subfamily/tribe	Siwaliks		Abu Dhabi		Africa
Suinae	*Propotamochoerus*	→→	*Propotamochoerus*		
	Hippopotamodon				
	?Sus				
Hippohyini	*Hippohyus*				
	Sivahyus				
Tetraconodontinae	*Sivachoerus*		*Nyanzachoerus*	←←	*Nyanzachoerus*

known late Miocene African faunas, although less so than in some Asian and European late Miocene contexts where several genera have been recovered from single horizons.

The suid specimens recovered thus far all exhibit relatively primitive morphotypes, with bunodont teeth and generalised dentition. There is no indication of the increase in hypsodonty, enamel crenellation, or molar complexity that characterises the Pliocene suids of the Indian subcontinent or Africa. Neither is there any indication of more archaic species, such as the Listriodontinae or Hyotheriinae, distributed in Africa and/or Asia during the middle Miocene. The suid taxa recovered so far confirm the interchange of some faunal elements between northern and northeastern Africa, the Siwalik zone, and Arabia, and suggest that the age of the formation is late Miocene (8–5.6 Ma).

Acknowledgements

Many individuals and organisations have assisted with the fieldwork that has lead to the retrieval of these specimens. In particular, Andrew Hill thanks the Undersecretary of the Department of Antiquities and Tourism, Al Ain, His Excellency Saif Ali Dhab'a al Darmaki for the help provided by his department, especially from Walid Yasin. Subsequent work has been largely funded by grants to the project from the Abu Dhabi National Oil Company (ADNOC) and the Abu Dhabi Company for Onshore Oil Operations (ADCO). Andrew Hill also received funding from the L. S. B. Leakey Foundation, USA. Laura Bishop thanks the National Science Foundation and the Leverhulme Trust for funding. Bill Sacco (Peabody Museum of Natural History, Yale University) took the photographs.

References

Arambourg, C. 1968. Un Suidé fossile nouveau du Miocène supérieur de l'Afrique du Nord. *Bulletin de la Société Géologique de France* 10: 110–15.

Azzaroli, A. 1975. Remarks on the Pliocene Suidae of Europe. *Sonderdruck aus Zeitschrift fur Säugetierkunde* 40: 355–67.

Bernor, R. L. 1983. Geochronology and zoogeographic relationships of Miocene Hominoidea. In *New Interpretations of Ape and Human Ancestry,* pp. 21–64 (ed. R. L. Ciochon and R. S. Corruccini). Plenum Press, New York.

———. 1984. A zoogeographic theater and biochronologic play: The time/biofaces phenomenon of Eurasian and African Miocene mammal provinces. *Paleobiologie Continentale* 14: 121–42.

———. 1986. Mammalian biostratigraphy, geochronology, and zoogeographic relationships of the Late Miocene Maragheh fauna, Iran. *Journal of Vertebrate Paleontology* 6: 76–95.

Colbert, E. H. 1935. *Palaeochoerus perimensis.* Siwalik mammals in the American Museum of Natural History. *Transactions of the American Philosophical Society, Philadelphia,* n.s.: 1–401.

Cooke, H. B. S. 1987. Fossil Suidae from Sahabi Libya. In *Neogene Paleontology and Geology of Sahabi,* pp. 255–66 (ed. N. T. Boaz, A. El-Arnauti, A. W. Gaziry, J. de Heinzelin, and D. D. Boaz). Alan R. Liss, New York.

Cooke, H. B. S., and Ewer, R. F. 1972. Fossil Suidae from Kanapoi and Lothagam, Northwestern Kenya. *Bulletin of the Museum of Comparative Zoology* 143: 149–296.

Falconer, H., and Cautley, P. T. 1846 (1845–49). *Fauna Antiqua Sivalensis.* Smith, Elder and Co., London.

Fortelius, M., van der Made, J., and Bernor, R. L. 1996. Middle and Late Miocene Suoidea of Central Europe and the Eastern Mediterranean: Evolution, biogeography and paleoecology. In *The Evolution of Western Eurasian Later Neogene Faunas,* pp. 348–73 (ed. R. L. Bernor, V. Fahlbusch, and H.-W. Mittmann). Columbia University Press, New York.

Golpe-Posse, J. M. 1979. Suidés des gisements du Bled Dourah (W de Gafsa, Tunisie). *Notes du Service Géologique, Tunisie* 44: 75–107.

Hamilton, W. R., Whybrow, P. J., and McClure, H. A. 1978. Fauna of fossil mammals from the Miocene of Saudi Arabia. *Nature* 274: 248.

Harris, J. M., and White, T. D. 1979. Evolution of the Plio-Pleistocene African Suidae. *Transactions of the American Philosophical Society* 69: 1–128.

Hill, A. 1995. Faunal and environmental change in the Neogene of East Africa: Evidence from the Tugen Hills Sequence, Baringo District, Kenya. In *Paleoclimate and Evolution,* pp. 178–93 (ed. E. S. Vrba, G. H. Denton, T. C. Partridge, and L. H. Burkle). Yale University Press, New Haven.

——. 1999. Late Miocene sub-Saharan African vertebrates, and their relation to the Baynunah fauna, Emirate of Abu Dhabi, United Arab Emirates. Chap. 29 in *Fossil Vertebrates of Arabia,* pp. 420–29 (ed. P. J. Whybrow and A. Hill). Yale University Press, New Haven.

Hill, A., and Ward, S. 1988. Origin of the Hominidae: The record of African large hominoid evolution between 14 My and 4 My. *Yearbook of Physical Anthropology* 31: 49–83.

Hill, A., Curtis, G., and Drake, R. 1986. Sedimentary stratigraphy of the Tugen Hills, Baringo, Kenya. In *Sedimentation in the African Rifts,* pp. 285–95 (ed. L. E. Frostick, R. Renaut, and J. J. Tiercelin). Geological Society of London Special Publication no. 25, Blackwell, Oxford.

Hill, A., Drake, R., Tauxe, L., Monaghan, M., Barry, J. C., Behrensmeyer, A. K., Curtis, G., Jacobs, B. F., Jacobs, L., Johnson, N., and Pilbeam, D. 1985. Neogene palaeontology and geochronology of the Baringo Basin, Kenya. *Journal of Human Evolution* 14: 759–73.

Hill, A., Ward, S., and Brown, B. 1992. Anatomy and age of the Lothagam mandible. *Journal of Human Evolution* 22: 439–51.

Hürzeler, J. 1982. Sur le suidé du lignite de Montebamboli (prov. Grosseto, Italie). *Comptes Rendus de l'Académie des Sciences, Paris* 295: 697–701.

Kotsakis, T., and Ingino, S. 1979. Osservazioni sui *Nyanzachoerus* (Suidae, Artiodactyla) del Terziario Superiore di Sahabi (Cirenaica, Libia). *Bollettino del Servizio Geologico d'Italia* C: 391–408.

Leakey, L. S. B. 1958. Some east African Pleistocene Suidae. *Fossil Mammals of Africa* no. 14. British Museum (Natural History), London.

Leonardi, P. 1952. Resti fossili di *"Sivachoerus"* del giacimento di Sahabi in Cirenaica (Africa Settentrionale). Notizie preliminari. *Atti della Accademia Nazionale dei Lincei, Rendiconti,* ser. 8vo, 13: 166–69.

Lydekker, R. 1877. Notices of new and rare mammals from the Siwaliks. *Records of the Geological Servey of India* 9: 154.

———. 1884. Indian Tertiary and post-Tertiary Vertebrata. Siwalik and Narbada bunodont Suina. *Memoirs of the Geological Survey of India. Palaeontologia Indica* 3: 35–104.

Nakaya, H., Pickford, M., Nakano, Y., and Ishida, H. 1984. The Late Miocene large mammal fauna from the Namurungule Formation, Samburu Hills, Northern Kenya. *African Study Monographs,* suppl. issue 2: 87–131.

Owen, R. 1840–45. *Odontography.* H. Baillière, London.

Pickford, M. 1986. *A Revision of the Miocene Suidae and Tayassuidae, (Artiodactyla, Mammalia) of Africa.* Tertiary Research Special Paper no. 7. E. J. Brill, Leiden.

———. 1987. Miocene Suidae from Ad Dabtiyah, eastern Saudi Arabia. *Bulletin of the British Museum (Natural History),* Geology 41: 441–46.

———. 1988. Revision of the Miocene Suidae of the Indian subcontinent. *Münchner Geowissenschaftliche Abhandlungen* A12: 1–92.

———. 1993. Old World Suoid systematics, phylogeny, biogeography and biostratigraphy. *Paleontologia i Evolucio* 26–27: 237–69.

Pickford, M., Nakaya, H., Ishida, H., and Nakano, Y. 1984. The biostratigraphic analyses of the faunas of the Nachola area and Samburu Hills, Northern Kenya. *African Study Monographs,* suppl. issue 2: 67–72.

Pilgrim, G. E. 1925. Presidential Address to the Geological Section of the Twelfth Indian Science Congress. *Proceedings of the Twelfth Indian Science Congress,* 206–10.

———. 1926. Fossil Suidae of India. *Memoirs of the Geological Society of India. Palaeontologia Indica* 8: 1–65.

Schmidt-Kittler, N. 1971. Die obermiozäne Fossilagerstatte Sandelzhausen. 3. Suidae (Artiodactyla, Mammalia). *Mitteilungen der Bayerischen Staatssammlung für Paläontologie und historische Geologie* 11: 129–70.

Stehlin, H. G. 1899. Ueber die Geschichte des Suiden-Gebisses. *Abhandlungen der Schweizerischen Paläontologischen Gesellschaft* 26–27: 1–527.

Thomas, H., Bernor, R., and Jaeger, J.-J. 1982. Origines du peuplement mammalien en Afrique du Nord durant le Miocène terminal. *Geobios* 15: 283–97.

van der Made, J. 1989–1990. A range-chart for European Suidae and Tayassuidae. *Paleontologia i Evolucio* 23: 99–104.

van der Made, J., and Moyà-Solà, S. 1989. European Suinae (Artiodactyla) from the Late Miocene onwards. *Bollettino della Società Paleontologica Italiana* 28: 329–39.

White, T. 1995. African omnivores: Global climatic change and Plio-Pleistocene hominids and suids. In *Paleoclimate and Evolution, with Emphasis on Human Origins,* pp. 369–84 (ed. E. S. Vrba, G. H. Denton, T. C. Partridge, and L. H. Burkle). Yale University Press, New Haven.

Whybrow, P. J. 1987. Miocene geology and palaeontology of Ad Dabtiyah, Saudi Arabia. *Bulletin of the British Museum (Natural History),* Geology 41: 367–457.

Whybrow, P. J., Hill, A., Yasin al-Tikriti, W., and Hailwood, E. A. 1990. Late Miocene primate fauna, flora and initial palaeomagnetic data from the Emirate of Abu Dhabi, United Arab Emirates. *Journal of Human Evolution* 19: 583–88.

A Fossil Hippopotamus from the Emirate of Abu Dhabi, United Arab Emirates

21

ALAN W. GENTRY

The Baynunah Formation of the Western Region, Emirate of Abu Dhabi (Whybrow, 1989; Whybrow et al., 1999—Chapter 4), is predominantly composed of poorly consolidated, horizontally bedded sands, and land vertebrates and freshwater molluscs have been recorded and collected from the Baynunah intermittently since 1949. Collecting expeditions of 1984–1996 by a Natural History Museum/ Yale University team have explored numerous localities, particularly Jebel Barakah (stratotype locality), Hamra, Jebel Dhanna, and the islands of Thumariyah and Shuwaihat. The recovery of hipparionine horses has suggested that its age is no earlier than late Miocene.

Among the mammalian fossils are remains of hippopotamuses to be described below. In the following descriptions comparisons are made with fossils in the Palaeontology Department of The Natural History Museum, London (institutional abbreviation BMNH). Localities and deposits mentioned in the text other than those of the Baynunah Formation are:

Baringo deposits, Kenya, middle and late Miocene and including the Ngorora Formation (Hill et al., 1985; Hill, 1995); Beglia Formation, Tunisia, late Miocene (Pickford, 1990); Bône, Algeria, of uncertain age (Howell, 1980); Dhok Pathan, Siwaliks, late Miocene (Barry and Flynn, 1990); Fort Ternan, Kenya, dated to 14.0 million years (Ma) ago (Cerling el al., 1991); Irrawaddy Group, Burma, Pliocene (Bender, 1983); Kaiso Formation, Uganda, Pliocene (Cooke and Coryndon, 1970; Pickford et al., 1988); Kanapoi, Kenya, early Pliocene (Leakey et al., 1996); Kansal (= Pinjor) Formation, Siwaliks, Pliocene (Barry and Flynn, 1990); Koobi Fora Formation, East Turkana, Kenya, Pliocene and early Pleistocene (Harris, 1991); Kuguta, Kenya, Plio-Pleistocene; Lothagam, Kenya, late Miocene (Leakey et al., 1996); Lukeino, Kenya, late Miocene (Coryndon, 1978a); Lusso Beds, Semliki, Zaire, late Pliocene (Boaz et al., 1992); Maboko, Kenya, middle Miocene (Feibel and Brown, 1991); Mpesida, Kenya, late Miocene (Coryndon, 1978a); Nachukui Formation, West Turkana, Kenya, Pliocene and early Pleistocene (Harris et al., 1988); Nakali, Kenya, late Miocene (Aguirre and Leakey, 1974); Ngeringerowa, Kenya, late Miocene (Pickford, 1981; Hill, 1995), c. 9.0 Ma; Nkondo Formation, Uganda, Pliocene (Faure, 1995); Pinjor (or Kansal) Formation, Siwaliks, Pliocene spanning 2.5–1.5 Ma (Barry and Flynn, 1990); Sahabi, Libya, late Miocene (Boaz et al., 1987; Geraads, 1989); Saint Arnaud, Algeria, Pliocene (Arambourg, 1970); Shungura Formation, Omo, Ethiopia, Plio-Pleistocene (Heinzelin, 1983); Siwaliks Group, Pakistan and India, Miocene–Pliocene (Barry and Flynn, 1990); Tatrot Formation, Siwaliks, Pliocene spanning 5.0–2.5 Ma (Barry and Flynn, 1990); Wadi Natrun, Egypt, late Miocene (Howell, 1980); Warwire Formation, Uganda, middle Pliocene (Faure, 1995); Wembere–Manonga Formation, Tanzania, latest Miocene–early Pliocene (Harrison, 1997).

PREVIOUS STUDIES OF FOSSIL HIPPOPOTAMUSES

The extant common hippopotamus, *Hippopotamus amphibius* Linnaeus, 1758, lived during the historic period in sub-Saharan Africa and northwards in the Nile Valley. It also occurred in other parts of northern Africa until the Holocene whenever climatic and environmental conditions allowed. At intervals during the Pleistocene *H. amphibius* or similar species ranged across southern and western Europe and

 ISBN 0-300-07183-3

southwestern Asia. In Africa the Plio-Pleistocene *H. gorgops* Dietrich, 1926, was a specialised large species with elevated orbits. Dwarf hippopotamuses lived on various Mediterranean islands and in Madagascar, and one survived in West Africa until the twentieth century AD. The most completely known remains of earlier hippopotamuses are *Hexaprotodon sivalensis, H. protamphibius,* and *H. harvardi.*

Hexaprotodon sivalensis Falconer and Cautley, 1836

The first adequately preserved pre-Pleistocene hippopotamus to be studied scientifically was *Hexaprotodon sivalensis* from the Pinjor and later part of the Tatrot Formations of the Siwaliks. Hussain et al. (1992: 73, fig. 5) found it to be common throughout and on either side of the Gauss normal palaeomagnetic epoch (3.40–2.48 Ma ago).

The most noticeable feature of *Hexaprotodon sivalensis* is the presence of a total of six upper and six lower incisors instead of the four present in *Hippopotamus amphibius.* On this basis Falconer and Cautley (1836) founded *Hexaprotodon* as a subgenus of *Hippopotamus* for their new fossil. *Hexaprotodon* later became widely used as a generic name and has now come to include species with fewer than six incisors. In the remainder of this chapter *Hippopotamus* will be abbreviated as *Hip.* and *Hexaprotodon* as *Hex.*

The Siwaliks fossil is slightly smaller than *Hip. amphibius.* Other differences are that the incisors are more nearly equally sized; the upper canine has a large and deep posterior groove; the incisors occlude at their tips in a more or less horizontal plane and not along part of their distal courses in a more or less vertical plane; P^1 (or dP^1 if that be its real identity) may have two roots instead of only one and may be more persistent; the lower canines have smooth rather than ridged enamel; the diastemata are shorter; the premolar row is longer relative to the molar row and the molars are less trefoliate (see Coryndon, 1978a: fig. 23.1); P_2s and P_3s are about the same size as P_4s or slightly larger; P^4 has two cusps; the cheek teeth are lower crowned; the tooth rows remain almost parallel instead of diverging anteriorly; the anterior ends of the premaxillae do not diverge from the median line; the mandibular symphysis is less widened; the braincase is less expanded; the sagittal crest is higher and better developed; the lachrymal bone is separated from the nasals by the anterior part of the frontal; the nasals are less widened posteriorly; the preorbital skull length is shorter in relation to the postorbital length; the auditory bullae may be more inflated; the glenoid facets are flatter; the limb bones are more gracile; and the toes less plantigrade. Colbert (1935: 281) itemised these differences in some detail.

Hexaprotodon first appears in the Siwaliks sequence at 5.7 Ma (Barry, 1995). The species is listed as *Hex. sivalensis,* but reasons for this identification have not been given. Later fossil hippopotamuses from the Indian Pleistocene are *Hex. namadicus* Falconer and Cautley, 1847 and *Hex. palaeindicus* (Falconer and Cautley, 1847). They began to evolve tetraprotodonty although reduced I_2s generally remain present. There is no great discrepancy in size between the remaining incisors, unlike *Hip. amphibius.* Hooijer (1950: 39) noted that the paracone on the upper molars of *Hex. palaeindicus* had a posterior lobe projecting outwards instead of backwards, and that the metacone had more of a tetrafoliate (four-leaved) pattern in middle wear—that is, with a slight projection lingually. These facts suggest that the Indian Pleistocene hippopotamus was not conspecific with *Hip. amphibius.*

Hexaprotodon protamphibius (Arambourg, 1944)

This species was originally described from the Shungura Formation, Omo, where it is known from Members A–G. It also occurs in equivalent levels of the Nachukui Formation and in part of the Koobi Fora Formation (Harris, 1991). It was hexaprotodont until the later part of Member B at about 3.0 Ma, and then became tetraprotodont, apparently at least one million years earlier than did the Indian hippopotamus.

There is doubt about whether I_2 or I_3 was the lower incisor lost when *Hex. protamphibius* became tetraprotodont. Arambourg (1947: 321) labels the surviving lower incisors of *Hex. protamphibius* as I_1 and I_3 without further discussion. A Tulu Bor mandible of *Hex. protamphibius* (Harris, 1991: fig. 2.2) shows I_2 as the smallest of the three lower incisors, and with its alveolus displaced above the line of I_1 and I_3. But Coryndon and Coppens (1973: 149) write, apparently in reference to the upper dentition, that the first incisor, the central tooth, was the one lost when *Hex. protamphibius* became tetraprotodont. Gèze (1985: 91) states that *Hex. sivalensis* differs from African hippopotamuses, other than *Hex. coryndoni* (a species from the Hadar Formation), in that I_2 is the incisor to suffer reduction. His fig. 4b of a mandible of *Hex. shungurensis* (probably = *protamphibius;* Harris, 1991: 34) shows I_3 more reduced than I_2. With regard to *Hip. amphibius,* Reynolds (1922: 11) thought that the third incisor had disappeared in both upper and lower jaws, but Harger (1932) recorded an extra lower incisor between the two normal ones (that is, apparently an otherwise lost I_2). Hooijer (1942) recorded instances of anomalous additional lower incisors in *Hip. amphibius* that appeared on one side only and were I_3s. Judging by anomalies in *Hip. amphibius* and the Madagascan *lemerlei,* Hooijer (1943: 289–90) thought that I_3 may be the one that disappeared, but at that stage he had still not seen Harger (1932). Later he was unable to determine from foetuses which incisor is lacking in *Hip. amphibius* (Hooijer, 1950: 10). Kahlke (1985) noted that previous observations of incisor vestiges in *Hip. amphibius* could have been misidentifications of erupting permanent canines.

Apart from the supposed time of the change to tetraprotodonty, there are few other differences of *Hex. protamphibius* from *Hex. sivalensis.* So far as can be deduced, the lachrymal just contacted the nasals in later *Hex. protamphibius,* the sagittal crest may be less raised, the P^1 is single-rooted (Arambourg, 1947: 318) while that of *Hex. sivalensis* looks as if it has two roots, and the P^4 comes to have only one cusp. P_3 may be shorter relative to P_4 than in *Hex. sivalensis.*

Harris (1991: 45) suggested or implied that *protamphibius* evolved into the extinct diprotodont *Hex. karumensis* Coryndon, 1977, known from the upper Koobi Fora and Shungura Formations.

Hexaprotodon harvardi Coryndon, 1977

This is the earliest adequately known fossil hippopotamus, but only an outline account has yet been published. The holotype and other remains come from Lothagam 1 and the species has also been identified at Kanapoi. Harrison (1997) has given a full account of material from the Wembere–Manonga Formation, Tanzania. The skull is about the size of *Hex. protamphibius* or *sivalensis,* but narrower or more gracile. Other likely differences from *Hex. protamphibius* or *sivalensis* are that the nasals are less widened posteriorly, the auditory bullae less inflated, the orbits and sagittal crest less elevated, the face less constricted behind the canines, the upper canines less wide apart, the premolar row longer, the occlusal pattern less trefoliate on the molars, and the mandibular ramus less deep (Coryndon, 1977: table 4). Differences from *Hex. protamphibius* are that the lachrymal fails to contact the nasal, P^1 has two roots, and P^4 is very large and has two subequal cusps. Postcranially it is more gracile than *Hex. sivalensis* or *protamphibius* (Coryndon, 1978b: 491).

Hippopotamuses from Mpesida and Lukeino (Coryndon, 1978a) may be earlier than *Hex. harvardi* from Lothagam, and are more certainly earlier than the first *Hexaprotodon* in the Siwaliks. Teeth from Mpesida are the size of *Hex. harvardi.* Upper canines have a deep posterior groove, lower canines have smooth enamel, premolars are large and pustulate with marked cingula.

At Lukeino there are many remains including a hexaprotodont mandible and a P^4 with two subequal (buccal and lingual) cones surrounded by a strong cingulum. A smaller species with lower-crowned teeth may also be present.

Hippopotamus kaisensis Hopwood, 1926

Hip. kaisensis is a name for hippopotamus teeth from the Kaiso Formation. The molars resemble those of *Hex. sivalensis* and *protamphibius,* but isolated teeth are not easily distinguishable from *Hip. amphibius.* Moreover, some fossils can be very large—for example, mandibular pieces of the O'Brien collection from Kaiso Village (Cooke and Coryndon, 1970: 196, table 36).

Tetraprotodonty can be demonstrated from a symphysis BMNH M25271 (Cooke and Coryndon, 1970: pl. 14B), upper canines can have a shallow groove posteriorly (Cooke and Coryndon, 1970: 192), and a left P_4 is a robust tooth 5–10% wider than in *Hex. sivalensis* or *protamphibius* (Cooke and Coryndon, 1970: 195; my own measurement of BMNH 36722; Harris, 1991: table 2.2).

Hitherto the age of the Kaiso Formation has been established only by faunal correlation; its upper fauna may be a later Pliocene equivalent to Shungura Formation Members F–G, and its lower fauna earlier in the Pliocene. Thus *Hip. kaisensis* could be conspecific with *Hex. protamphibius,* or possibly with the large hippopotamuses that begin to appear higher in the Shungura Formation, such as *Hex. karumensis* and *Hip. gorgops.* Pavlakis (1990), working on hippopotamuses of the Lusso Beds, concluded that *Hip. kaisensis* was not distinguishable from *Hip. amphibius.* Faure (1995) used the name *Hip. kaisensis* for hippopotamus fossils of the Warwire and Nkondo Formations in Uganda, going back perhaps to 5.0 Ma.

Hexaprotodon sahabiensis Gaziry, 1987

This Sahabi species appears to be slightly smaller than *Hex. sivalensis* and the mandible less deep. It is hexaprotodont according to Gaziry (1987) and on the testimony of Petrocchi (1952: 27), who was describing a skull that can no longer be found. The canine teeth have smooth enamel, the cheek teeth are low crowned, P^4 has two cusps, and Gaziry thought the premolar and molar rows were of the same length. The cusps on the molars are only poorly trefoliate, as in *Hex. harvardi.*

The P_4 has two main cusps side by side centrally (metaconid and protoconid) and two subsidiary ones side by side posteriorly. These latter cusps appear not to have survived in later species like *Hex. sivalensis* and *protamphibius,* nor in *Hex. harvardi,* although a syntype P_4 (BMNH M12621) of the dwarf species *Hex. imagunculus* has preserved into late wear a single wide posterior cusp.

The P^3 of *Hex. sahabiensis* has rather clear anterior and posterior lobes with a prominent constriction between them (Gaziry, 1987: fig. 6). A similar morphology appears to be present in one or two *Hex. sivalensis* but is absent in a syntype *Hex. imagunculus* (BMNH M12619) and in other African fossil hippopotamuses. The P^3 of *Hex. sahabiensis* also has a small posterolingual cusp rather than simply a cingular outgrowth as in more advanced hippopotamuses.

Although little is known of *Hex. sahabiensis,* it looks as if the P^3 and P_4 characters do differentiate it from the more advanced African hippopotamuses described above. The anterior and posterior lobes of P^3 may indicate a link to *Hex. sivalensis* in so far as one could postulate a longer surviving primitive characteristic in that persistently hexaprotodont lineage.

Hexaprotodon hipponensis (Gaudry, 1876)

Hex. hipponensis was founded for some hippopotamus teeth (Gaudry, 1876: pl. 18) of uncertain geological age from Pont de Duvivier, south of Bône, Algeria (Joleaud, 1920: 16): they were redescribed by Arambourg (1945). Parts of six more or less equally sized incisors, two canines, two premolars, and half a molar all seemed to have come from the same mandible. Hence the species must have been hexaprotodont, like *Hex. sivalensis,* and it was the first record of an African hexaprotodont. The canines lacked any strong ridging of the enamel.

Pomel (1896) assigned a tetraprotodont fossil from Saint-Arnaud to *Hex. hipponensis,* but Arambourg (1947: 327; 1970: 19) held that it was not conspecific with the Pont de Duvivier fossils.

Hippopotamus remains from Wadi Natrun were attributed by Andrews (1902) to *Hex. hipponensis.* Stromer (1907: 110; 1914: 5) affirmed that the animal was tetraprotodont with two more or less equally sized incisors on each side of its lower jaw, but he did not illustrate this feature. Arambourg

(1947: 328), rejecting a tetraprotodont as conspecific with *Hex. hipponensis,* referred the Wadi Natrun form to a new subspecies, *andrewsi,* of *Hex. protamphibius.* If Wadi Natrun is indeed early Pliocene in age (Geraads, 1987: 22), this would be an exceptionally early tetraprotodont species according to Gèze's (1985: 94) estimate that tetraprotodonty appears in East Africa at around 3.0 Ma ago.

Wadi Natrun hippopotamus teeth and casts thereof in London are small and low crowned. The upper molars have strong cingula, and the cusp shape is triangular rather than trefoliate. According to Stromer's (1914) illustrations the upper canine has a narrow but shallow posterior groove (an advanced character), the P^3 does not have clear-cut front and back halves, and P_4 is without posterior cusps. These characters accentuate the differences between the Wadi Natrun and Sahabi hippopotamuses.

Hexaprotodon crusafonti Aguirre, 1963

Hex. siculus (Hooijer, 1946) and *Hex. pantanellii* (Joleaud, 1920) are possible senior synonyms of this species, which occurs in the latest Miocene and possibly the early Pliocene of southern Europe. No evidence of the number of incisors was found by Aguirre (1963: 218). Pantanelli (1879: pl. 4, fig. 5) illustrated a hexaprotodont premaxilla of *Hex. pantanellii* from Italy, whereas Lacomba et al. (1986: 177, pl. 1) interpreted a Spanish mandible of MN 13 age assigned to *Hex. crusafonti* as tetraprotodont. This claim for tetraprotodonty in another early hippopotamus, in addition to the Wadi Natrun one, is noteworthy. Moreover, this mandible appears from the illustration to have a very narrow symphysis in comparison with *Hex. harvardi, Hex. protamphibius, Hex. sivalensis,* and modern hippopotamuses. We see here, therefore, a combination of a primitive narrow symphysis with precocious, if occasional, tetraprotodonty. The back of the symphysis in Lacomba et al. (1986: pl. 1, fig. 1) is level with the front of P_3. The P_4 (Aguirre et al., 1973: fig. 6) has two posterior cusps like *Hex. sahabiensis.* (The P_4 reported by Aguirre in 1963 was said to be a P^2 by Lacomba et al., 1986.) The lower molars in Lacomba et al. (1986: pl. 1) are low crowned. Lower molars of the specimen figured by Cuvier (1804: pl. 2, fig. 1) as "grand hippopotame fossile" (other specimens of which were Pleistocene hippopotamuses) show narrow elongated cusps in occlusal view with wide transverse valleys in between them. The enamel is very roughened vertically and the hypoconulid of M_3 is narrow and isolated. Faure and Méon (1984) give two M_1–M_3 lengths as 120 mm and 124 mm, seemingly slightly smaller than in *Hex. sivalensis.* The P^3 (Lacomba et al., 1986: fig. 8) does not have very clear front and back halves, nor does it have a posterolingual cusp.

Gaziry (1987: 310) thought that *Hex. crusafonti* differed from *Hex. sahabiensis* in the width of upper canines, larger molars and premolars with less prominent cusps, the M_3 hypoconulid larger and more elongated, and with a shallower horizontal ramus of the mandible. These differences do not appear marked; the shallow ramus of the specimen in Aguirre (1963: pl. 3, fig. 3) could arise from its immaturity, and in any case Gaziry had shown from measurements that the ramus of *Hex. sahabiensis* was not itself very deep.

Coryndon (1978b) thought that *Hex. crusafonti* differed from the hippopotamus at Mpesida and Lukeino (previously mentioned here under *Hex. harvardi*) by being smaller, with higher-crowned molars and proportionately smaller premolars. This is surprising because the premolars of *crusafonti* appear to be relatively large and the molars low crowned. Coryndon took large premolars, low-crowned molars, a deep posterior groove on the upper canines, and smooth lower canines as primitive, so that *Hex. crusafonti* appeared to be more advanced. She thought that the Sahabi and Wadi Natrun hippopotamuses were closer to *crusafonti* than to those of sub-Saharan Africa.

Hexaprotodon iravaticus Falconer and Cautley, 1847

Hex. iravaticus comes from the Irrawaddy Group, Burma (Bender, 1983: 100). The lectotype is a hexaprotodont mandibular symphysis, BMNH 14771 (Falconer and Cautley, 1847: pl. 57, figs 10 and 11), on which the crowns of all teeth have been destroyed. Other pieces in London are a part of a left mandible with broken M_3 and much damaged M_2 (M10525) and part of a left juvenile symphysis (M10526). Colbert (1938: 419, fig. 61) described a partial skull, 20037, in the American Museum of Natural History, New York.

Hex. iravaticus is smaller than *Hex. sivalensis* and its most obvious feature is that the symphysis is narrow, as in the *Hex. crusafonti* mentioned above. The back of the symphysis looks as if it would be level with the back of P_3. I_2 is set slightly above I_1 and I_3. The orbits are little elevated, but perhaps more so than in *Hex. harvardi*. The front of the orbit appears to lie above the back of M^2. The front of the lachrymal looks as if it has only a point contact with the nasal. Upper molars retain wider transverse valleys between the front and back pairs of cusps according to Colbert (1938: 420). An upper canine showed a less-deep posterior groove (less-expanded internal lobe or pillar) and less-heavy longitudinal ridging on its enamel than in *Hex. sivalensis* or *Hip. amphibius*.

Irrawaddy fossils have been collected since 1826–27. The fauna of the lower Irrawaddy beds is equivalent to Dhok Pathan (late Miocene) faunas, but Stamp (1922) and then Colbert (1938, 1943) and Bender (1983) showed the hippopotamus as coming from the upper Irrawaddy beds, taken as equivalent to the Tatrot or even the Pinjor zones of the Siwaliks. Although Colbert (1938: 276, 423) noted that the provenances of specimens from the two faunal levels of the Irrawaddy Group are not always clear, he strongly supported (p.422) a late age for *Hex. iravaticus*. This would make the narrow-muzzled *Hex. iravaticus* a contemporary of more advanced hippopotamuses in Africa and of *Hex. sivalensis* in India. It is also evidently contemporary with *Elephas hysudricus* (see Colbert, 1935: 415), a species that Hussain et al. (1992) date back only to 2.7 Ma. Perhaps *Hex. iravaticus* is a long-surviving relict of a primitive narrow-muzzled grade of hippopotamus evolution, with some advanced characters evolved in parallel with other hippopotamuses. From the illustrations in Colbert (1938) and Coryndon (1978b: fig. 23.5) it looks more advanced than *Hex. harvardi* in more elevated orbits, lachrymal contacting nasals, and tooth row positioned slightly more anteriorly relative to orbits.

Dwarf Hippopotamuses

Coryndon (1977) decided that *Choeropsis* Leidy, 1853, for the extant West African pygmy hippopotamus *C. liberiensis* (Morton, 1849), was a junior synonym of *Hexaprotodon* Falconer and Cautley, 1836. *Hex. liberiensis* differs from *Hip. amphibius* mainly in the following characters:

1. Smaller size.
2. Only two lower incisors (diprotodonty) although the upper incisors remain at four.
3. The upper and lower incisors occlude more or less horizontally at their tips and not interstitially.
4. The upper canines have a deeper posterior groove.
5. The enamel of the lower canines is smooth and lacks ridges anterolabially.
6. Shorter premolar rows.
7. Shorter diastemata.
8. P^4 has two cusps.
9. A simpler occlusal pattern of molar cusps.
10. The anterior ends of the premaxillae remain fused medianly and do not diverge.
11. The orbits are set low.
12. The face is relatively shorter and the orbit is further forwards relative to the tooth row.
13. The lachrymal is not in contact with nasal.
14. The back of the braincase remains level or even curves downwards rather than passing upwards in profile.
15. The limbs are more gracile.

Of these 15 characters 1, 3–5, 7–13, and 15 are probably primitive in *liberiensis* and shared with fossil *Hexaprotodon* species and unlike *Hippopotamus*, while 2, 6, and 14 are advances in *liberiensis*. Harrison (1997: 177) believes that *liberiensis* is a "somewhat specialized derivative of the sister taxon of all other hippopotamuses" and should not be included in *Hexaprotodon*.

Fossil dwarf hippopotamuses are known in addition to those of Madagascar and the Mediterranean islands. *Hex. imagunculus* (Hopwood, 1926) was originally described on some isolated small cheek teeth from the Kaiso Formation, and the name has been used for teeth from other African localities. It was probably hexaprotodont. A hexaprotodont mandibular symphysis in the Kaiso Formation ("*Hippopotamus* sp." of Cooke and Coryndon, 1970: pl. 14A) may be small enough to be part of the *Hex. imagunculus* lineage. Its I_2 is slightly higher than the line from I_1 to I_3, and Cooke and Coryndon (1970: 188) note that I_2 is the smallest incisor. Faure (1995) has recorded it back to 5.0 Ma.

Hex. aethiopicus (Coryndon and Coppens, 1975) is a more completely known species, as small as *Hex. imagunculus* and tetraprotodont. It occurs in the Shungura Formation Members E–L and in the Koobi Fora and Nachukui Formations. It shows additional differences from Kaiso Formation cranial remains of small hippopotamuses, BMNH M14801 and M26336, in more elevated orbits and occiput. It could be a descendant of *Hex. imagunculus*. The P^4 of M14801 has only one main cusp, unlike the two cusps on one of the *imagunculus* syntypes, but the *Hex. aethiopicus* P^4 also may have either one or two cusps (Harris, 1991: 48).

The Earliest Hippopotamuses

Pickford (1983) founded *Kenyapotamus* for early hippopotamids from the Miocene of the Baringo deposits in Kenya. The type species is *K. coryndoni*, and the holotype a right M^2 + M^3 from Ngeringerowa. The Nakali teeth are higher crowned and more robust than those from Ngeringerowa. An earlier species comes from Fort Ternan and possibly Maboko. The *Kenyapotamus* teeth are smaller than in *Hexaprotodon* or *Hippopotamus* and lower crowned than in most or all of them, the P^3 has a posterolingual cusp in addition to its main cusp (similar to *Hex. sahabiensis*), P_4 has no metaconid cusp alongside the protoconid and it does have a posterior subsidiary cusp. Pickford (1983: fig. 18) dated the disappearance of *Kenyapotamus* and appearance of later hippopotamuses in East Africa to around 8–7 Ma. Pickford (1989, 1990) discusses *Kenyapotamus* further and records it at Beglia.

THE ABU DHABI HIPPOPOTAMUS

Order Artiodactyla Owen, 1848
Family Hippopotamidae Gray, 1821
Genus *Hexaprotodon* Falconer and Cautley, 1836
Hexaprotodon aff. *sahabiensis* Gaziry, 1987

The holotype of *Hex. sahabiensis* is a left mandible with P_4–M_3 (Gaziry, 1987: fig. 2), housed in the Department of Geology of the Garyounis University, Benghazi, Libya. The type locality is Sahabi, Libya.

Material from Abu Dhabi

AUH 457, Hamra, H1: lower jaw. Parts of left I_1 and I_2 and right cheek teeth survive.

AUH 481, Shuwaihat, S4: partial immature lower jaw with right and front of left horizontal ramus. Surviving crowns of teeth are right I_1–I_3, erupting P_2 + P_3, worn dP_4; left I_{2+3}, canine, erupting P_2–P_3, much of worn dP_4. Occlusal length right dP_4, 43.4 mm. Figure 21.1.

AUH 235, Shuwaihat, S4: pieces of a mandible with incisors.

BMNH M49464, Jebel Barakah, B2: lower jaw with right and left tooth rows and horizontal rami. Surviving crowns of teeth are right P_3 + P_4, M_3; left I_1, I_3, P_2–M_3. Premolars in early wear, M_3 in early middle wear, M_2 in middle wear, and M_1 in late middle wear. Occlusal width (mesiodistal) I_3, 13.6 mm. Occlusal length P_{2-4}, c. 127.0 mm; occlusal length M_{1-3}, c. 139.0 mm. Occlusal lengths and widths: P_3, 39.2 × 23.0 mm; P_4, 34.0 × 23.8 mm; M_1, c. 35.0 × 25.8 (anterior) × 26.6 mm (posterior); M_2, 45.5 × 31.1 (anterior) × 31.8 mm (posterior); M_3, c. 64.0 × 31.8 mm (anterior). Figures 21.2 and 21.3.

AUH 36, Jebel Dhanna, JD5: unworn right upper molar. Occlusal length and width 42.9 × 44.6 mm (anterior) × 39.3 mm (posterior). Figure 21.4.

AUH 99, Shuwaihat, S1: base of a left M^3 in a fragment of palate.

AUH 110, Shuwaihat, S1: parts of a lower molar.

AUH 262, Kihal, K1: partial left P^4, unworn. Occlusal length c. 27.9 mm.

BMNH M49465, Jebel Barakah, B2: right upper P^3, early wear. Occlusal length and width 38.7 × 26.1 (anterior) × 30.9 mm (posterior).

AUH 446, Jebel Barakah, B2: fragment of a lower canine.

AUH 29, Shuwaihat, S1: canine fragment.

AUH 359, Harmiyah, Y1: canine enamel sheet and a cheek tooth fragment.

AUH 57, Shuwaihat, S1: incisor part.

Figure 21.1. *Hexaprotodon* aff. *sahabiensis*, partial immature, lower jaw, AUH 481; lateral view. Scale bar = 10 cm.

AUH 2, Hamra, H1: incisor piece including the occlusal surface.
AUH 292, Jebel Dhanna, JD3: upper incisor piece.
AUH 37, Jebel Dhanna, JD5: ?incisor piece, possibly hippopotamid.
AUH 312, Kihal, K1: ?incisor piece, possibly hippopotamid.
AUH 60, Shuwaihat, S1: tooth fragments.
AUH 92, Shuwaihat, S1: tooth fragments.
AUH 31, Shuwaihat, S1: tooth fragments.
AUH 421, Jebel Dhanna, JD4: tooth fragment.
AUH 103, Shuwaihat, S1: right ilium with part of the acetabular surface.
AUH 133, Shuwaihat, S2: part of a left ischium.
AUH 59, Shuwaihat, S1: distal left femur, much damaged.
AUH 498, Jebel Dhanna, JD3: damaged pieces of a proximal right tibia.
AUH 499, Jebel Dhanna, JD3: probably part of the same tibia as AUH 498.
AUH 150, Hamra, H3: distal right tibia. Maximum distal width 77.5 mm, maximum anteroposterior width 59.7 mm. Figure 21.5.
AUH 288, Jebel Dhanna, JD3: distal right fibula. Figure 21.5.
AUH 5, Hamra, H1: left astragalus, medial surface missing, lateral and posterior surfaces much damaged. Lateral height c. 84.0 mm.
AUH 368, Jebel Barakah, B1: left astragalus. Lateral height 75.2 mm, medial height 68.7 mm, maximum distal width 54.5 mm.
AUH 44, Hamra, H5: right astragalus. Medial height c. 70.5 mm, maximum distal width 56.4 mm. Figure 21.5.
AUH 637, bin Jawabi, BJ: part of right astragalus.
AUH 243, Thumayriyah, TH1: right metatarsal 3. Maximum length 109.4 mm, width of distal condyles 28.4 mm. Figure 21.6.
AUH 247, Shuwaihat, S4: left metatarsal 3. Maximum length 119.1 mm, width of distal condyles 28.2 mm.
AUH 420, Jebel Dhanna, JD4: proximal left metatarsal 4.
AUH 49, Hamra, H5: stem of a right scapula.
AUH 118, Shuwaihat, S1: stem of a right scapula. Figure 21.7.
AUH 443, Jebel Barakah, B2: part of a right scapula.
AUH 497, Jebel Dhanna, JD3: right humerus in four pieces. Length from the top of lateral tuberosity to the base of lateral groove distally c. 350 mm; minimum transverse width of shaft 37.5 mm. Figure 21.8.
AUH 68, Shuwaihat, S1: damaged medial side of the distal end of a left humerus.
AUH 253, Shuwaihat, S4: most of a proximal right radius. Figure 21.7.
AUH 98, Shuwaihat, S1: distal right radius, damaged posterolaterally. Figure 21.7.
AUH 170, Hamra, H1: right lunate. Figure 21.7.
AUH 252, Shuwaihat, S4: right cuneiform. Figure 21.7.
AUH 248, Shuwaihat, S4: distal left metacarpal 2.
AUH 96 + 97, Shuwaihat, S1: right metacarpal 3 in two pieces. Maximum length 139.7 mm, width of distal condyles 31.0 mm. Figure 21.6.
AUH 53, Shuwaihat, S1: right metacarpal 3, rear of the proximal end broken. Maximum length 142.1 mm, width of distal condyles 31.1 mm.
AUH 154, Hamra, H1: right metacarpal 5 in two pieces. Maximum length 90.5 mm, width of distal condyles 28.0 mm. Figure 21.6.
AUH 84, Shuwaihat, S1: distal metapodial 3 or 4.
AUH 83, Shuwaihat, S1: complete proximal phalanx. Maximum length 59.0 mm, proximal width 34.3 mm, distal width 27.5 mm. Figure 21.6.

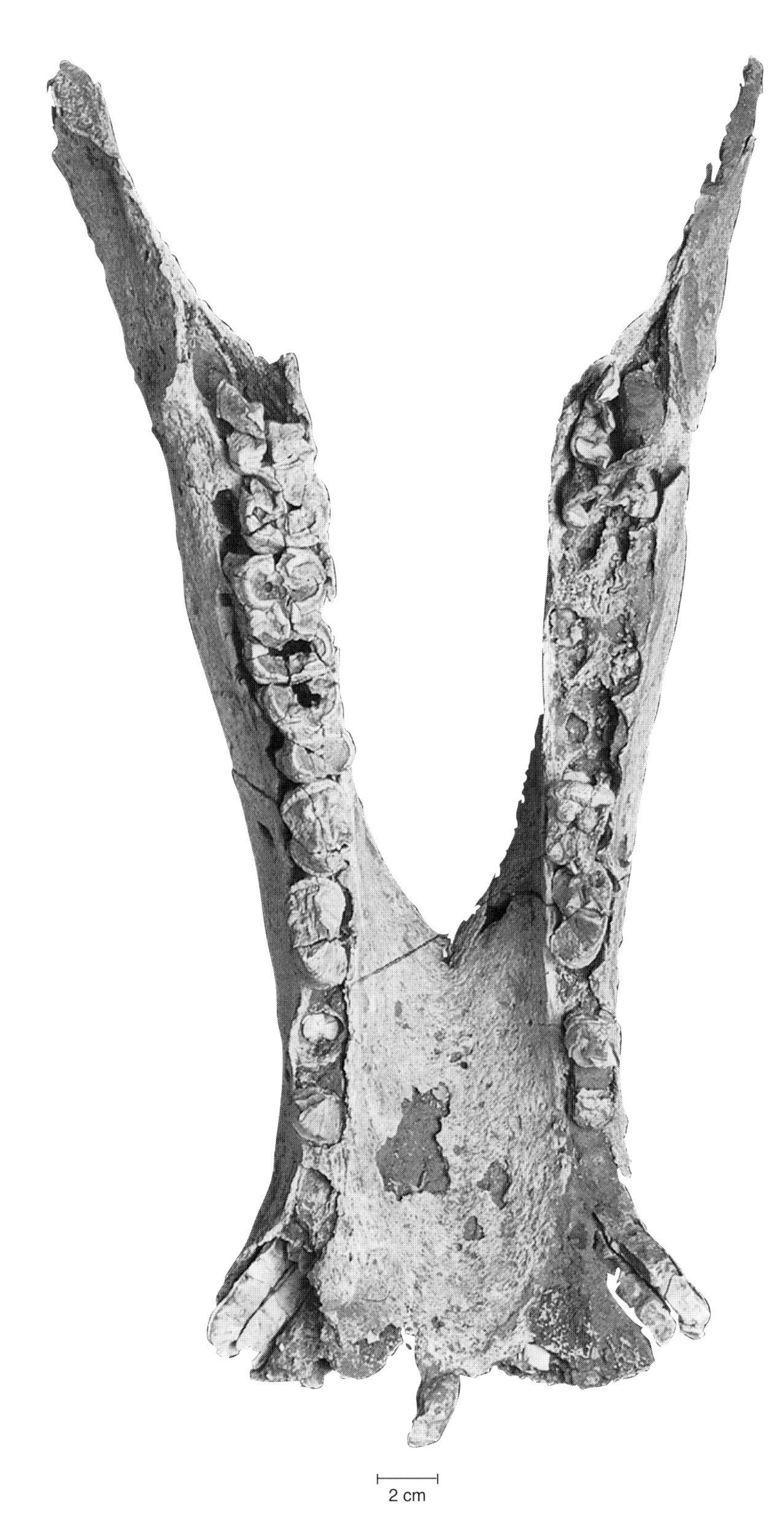

Figure 21.2. *Hexaprotodon* aff. *sahabiensis,* lower jaw, BMNH M49464; occlusal view.

Figure 21.3. *Hexaprotodon* aff. *sahabiensis,* lower jaw, BMNH M49464; right lateral view. Scale bar = 30 cm.

AUH 339, Hamra, H5: distal phalanx. Length 36.0 mm, proximal width 23.8 mm.
AUH 478, Thumayriyah, TH1: distal phalanx. Length 33.5 mm, proximal width 24.2 mm. Figure 21.6.
AUH 431, Ras al Qa'la, Q1: atlas vertebra, lateral part of the wings missing.
AUH 429, Ras al Qa'la, Q1: axis vertebra, damaged posteriorly, presumably the same individual as AUH 431. Width across atlantoid facets 117.4 mm.
AUH 66, Shuwaihat, S1: centrum of a cervical vertebra.
AUH 224, Shuwaihat, S4: centrum of a cervical vertebra.
AUH 105, Shuwaihat, S1: centrum and bases of the neural arch of a thoracic vertebra. Figure 21.5.
AUH 293, Jebel Dhanna, JD3: dorsal part of a left rib.

Description

The Abu Dhabi hippopotamus is a little smaller and less robustly built than *Hex. sivalensis.* The mandible is not deep below the tooth row and the angle does not project sharply downwards. The back of the symphysis is level with the back and front of P_3 in different specimens. It is a hexaprotodont species with more or less equally sized incisors.

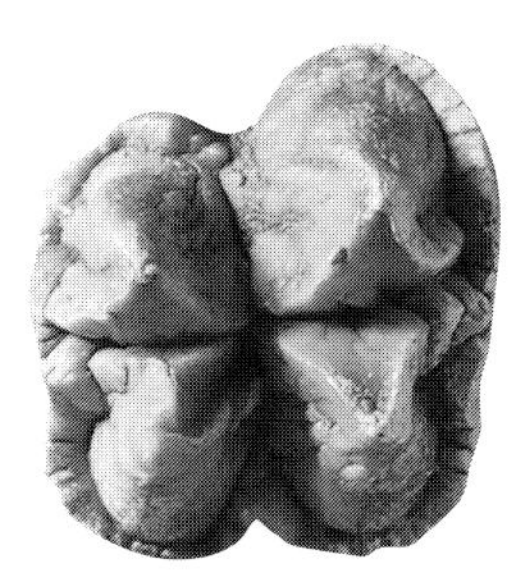

Figure 21.4. *Hexaprotodon* aff. *sahabiensis,* unworn right upper molar, AUH 36; occlusal view, anterior side is to the right.

The I_2s are at a slightly higher level than I_1 or I_3. The lower canines have smooth enamel. A P_1 is present. Diastemata exist on either side of P_1 and between P_2 and P_3. P_4 has a metaconid in addition to its protoconid, and this metaconid is more rounded in occlusal outline than the protoconid. The partial P^4 AUH 262 is low crowned and has two cusps and a strong pustulate cingulum. Most of the labial half of the tooth is missing and only the lingual wall of the labial cusp remains. It looks similar to the Lukeino P^4 shown in Coryndon (1978a: fig. 18.5) and would have been wider than long when complete. Two subequal cusps surrounded by a strong cingulum are characteristic for P^4s of early hippopotamuses (Coryndon, 1978a: 285). P^3 is a wide tooth and is partially divided transversely into anterior and posterior lobes. Its large main cusp spans both lobes and the cingulum is enlarged posterolingually. There is a weak cingulum on the lower molars. The hypoconulid cusp on M_3 rises well above the level of the occlusal surface of the front parts of the tooth.

The unworn upper molar, AUH 36, is a little smaller than the upper molars of *Hex. sivalensis.* It is also lower crowned and relatively shorter and wider. The anterior half of the tooth is markedly wider than the posterior half, a condition seen on geologically earlier rather than later hippopotamuses and on M^3s (Coryndon and Coppens, 1973: pl. IA) rather than M^1s or M^2s. The cingulum is strong and pustulate except posterolabially on the paracone. The cusps are triangular in shape rather than trefoliate.

In general the postcranial bones are a little smaller than those of *Hex. sivalensis* and certainly more gracile. This sometimes gives them a pronounced difference from the corresponding bones of Pleistocene dwarfed hippopotamuses or *Hex. liberiensis,* which are stubby or more massively built. In life the Abu Dhabi hippopotamus would have had longer and more slender legs, perhaps with more of the body proportions of a peccary or wild pig. It would, of course, have been larger than any extant pig or peccary.

The dorsal edge of the ischium AUH 133 is not sharp enough for a pecoran and the bone is quite long anteroposteriorly. These features match a hippopotamus smaller than *Hip. amphibius.* The ilium fragment AUH 103 has little or no sign of the origins of the rectus femoris muscle; it could be of a hippopotamus according to the evident great area and backward-facing rather than lateral-facing plane of the front part of the acetabular articular surface.

The distal femur is smaller than in *Hex. sivalensis.* Little of it survives other than the fossa on the shaft, a part of the medial condyle and a smaller part of the patellar fossa. The fossa is placed too anteriorly on the lateral surface for the bone to be giraffid.

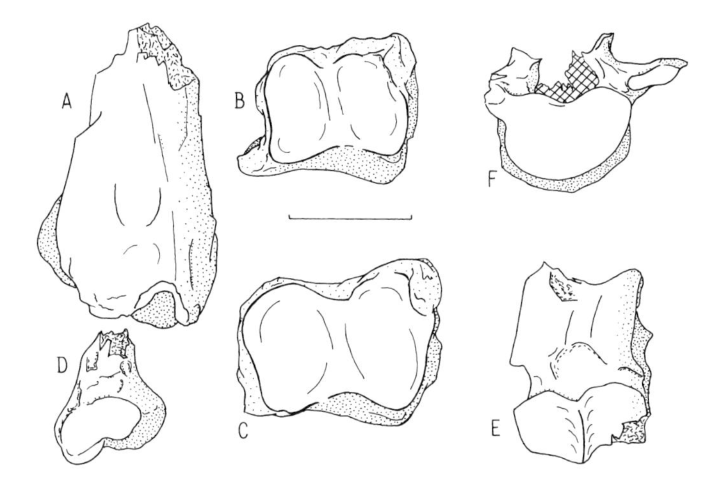

Figure 21.5. Hind limb bones and vertebra. *A, B, D–F, Hexaprotodon* aff. *sahabiensis; C, Hex. sivalensis. A,* Distal right tibia, AUH 150; anterior view. *B, C,* Distal articular surfaces of AUH 150 and of right tibia BMNH 17018a; anterior side towards the top. *D,* Distal right fibula AUH 288; medial view, anterior side towards the left. *E,* Right astragalus, AUH 44; anterior view. *F,* Thoracic vertebra, AUH 105; anterior view. Cross-hatching = matrix. Scale bar = 50 mm.

The articular surfaces of the proximal tibia AUH 498 are more concave than in *Hex. sivalensis.*

The distal tibia is the postcranial bone that least resembles other hippopotamuses. The astragalus facets are narrow and anteroposteriorly long and the flange behind the fibula is much enlarged. The proportions of the astragalus facets are pig-like, but the bone would be large for a pig. Even the large Siwaliks Miocene pig *Hippopotamodon sivalense* has proximal astragalus breadths of only 26.2 mm, 28.2 mm, and 30.5 mm (Pickford, 1988: 76), about half the size needed to fit the Abu Dhabi tibia. Moreover, the large flange behind the fibula is not a pig character. The lateral facet for the astragalus is less expanded anterolaterally than in Madagascan hippopotamuses. The bone is a bit smaller than in *Hex. sivalensis.*

On the distal fibula the facet for the lateral side of the astragalus is curved downwards anteriorly or at any rate it has a ventral edge that is concave downwards.

The astragali are smaller and narrower than in *Hex. sivalensis.* They are not of pigs because the ridge on the ventral condylar articulation is centrally placed and not closer to the lateral edge. Also there is a longitudinally running hollow on the posterior surface medial to the articular facet for the calcaneum.

The metapodials of the Abu Dhabi hippopotamus appear to be about 13% shorter and 28% more slender than in *Hip. amphibius.* They are also longer than metapodials of similar girth from Madagascan hippopotamuses. It seems to be general in hippopotamuses that metacarpals are longer than metatarsals, that distal hollows on anterior surfaces are better developed in metatarsals, and that metatarsals 2 and 5 are more reduced proximally than metacarpals 2 and 5.

The metatarsal 3 has a stronger distal hollow anteriorly than in *Hex. sivalensis,* and the proximal facet for metatarsal 4 is weak, small or not very distinct compared with its condition in *Hex. sivalensis.*

The scapulae are a bit smaller than in *Hex. sivalensis.* The spine is centrally placed between the anterior and posterior edges and distinguishes them from giraffids or bovids. The tuber is narrower in lateral view through being less expanded anterodorsally,

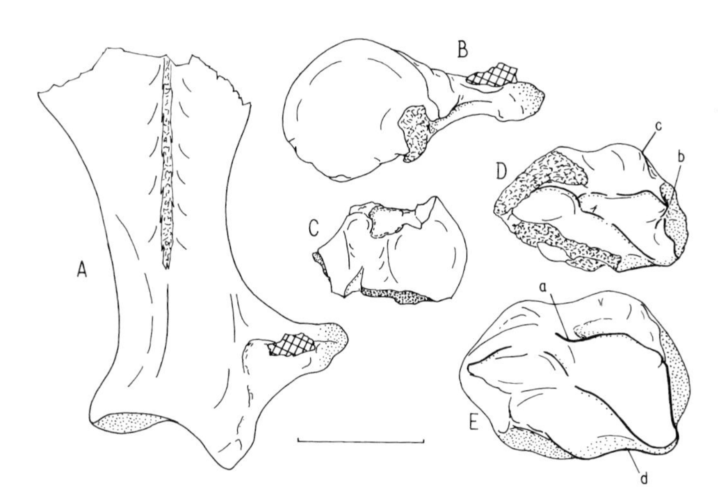

Figure 21.6. Front limb bones. *A–D, Hexaprotodon* aff. *sahabiensis; E, Hex. sivalensis. A,* Right scapula, AUH 118; lateral view. *B,* Right scapula, AUH 118; glenoid view, lateral side towards the top. *C,* Right radius, AUH 253; proximal articular surface, anterior side towards the base. *D, E,* Right radii, AUH 98 and BMNH 16478; distal articular surfaces, anterior sides towards the top. Key: a, indented anterior margin of scaphoid and lunate facets; b, localised tubercle anteromedially to scaphoid facet; c, ridge on medial part of anterior surface; d, concave profile of posterior edge of lunate facet. Scale = 50 mm.

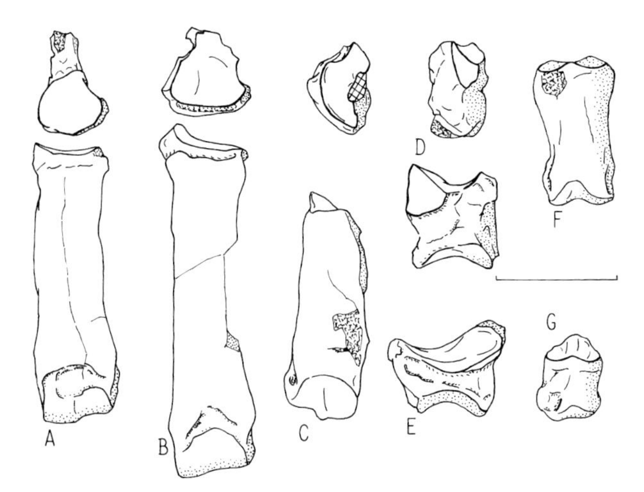

Figure 21.7. *Hexaprotodon* aff. *sahabiensis*. *A,* Right metatarsal III, AUH 243; proximal articular surface; anterior view, anterior side towards the base of drawing. *B,* Right metacarpal III, AUH 96 and 97; same view as A. *C,* Right metacarpal V, AUH 154; same view as A. *D,* Right cuneiform, AUH 252; anterior and medial views. *E,* Right lunate, AUH 170; lateral view. *F,* Proximal phalanx, AUH 83; dorsal view. *G,* Distal phalanx, AUH 478; dorsal view. Scale = 50 mm.

and the articular facet for the humerus is rounded in ventral view rather than being narrower transversely. A hippopotamus scapula from exposure 3 of the Kazinga Channel South in the Kaiso Formation, BMNH M25176, looks intermediate in size and in tuber morphology between the Abu Dhabi examples and *Hex. sivalensis;* its humeral articulation is damaged so that its original shape cannot be determined. Compared with Madagascan fossil hippopotamuses, the Abu Dhabi scapulae have narrower and longer stems below the blade and the blade widens more gradually above the stem.

The humerus AUH 497 has lost the medial tuberosity, the posterior eminence behind the lateral tuberosity, the surface of the bicipital groove, and most of the distal end below the coronoid fossa. It has, however, preserved most of the distal lateral groove. It may be slightly smaller and is certainly more gracile than in *Hex. sivalensis*. Compared with proximal ends BMNH 16467, 16468, and 16670 of *Hex. sivalensis,* the Jebel Dhanna humerus has an infraspinatus insertion set more anteriorly on the lateral tuberosity and standing more proud of the surrounding bone surface. Also the back of the articular head projects further behind the stem or shaft of the humerus in lateral view.

The proximal radius AUH 253 is smaller than *Hex. sivalensis,* having linear dimensions no bigger than 80% of those of the Siwaliks species at best. What remains of the top of the shaft appears considerably more gracile than *Hex. sivalensis*. The back of the lateral facet is slanted from posteromedial to anterolateral, there is a broad and shallow groove between lateral and medial facets, and the medial facet has a considerable anteroposterior dimension. These features are more like hippopotamuses than modern pigs. Modern pigs are too small in size to fit this radius, but a sufficiently large fossil species might have a more similar morphology. For the present this bone is accepted as being of a hippopotamus.

The distal radius AUH 98 is not from a pig because of the indented anterior edge between the scaphoid and lunate facets, and

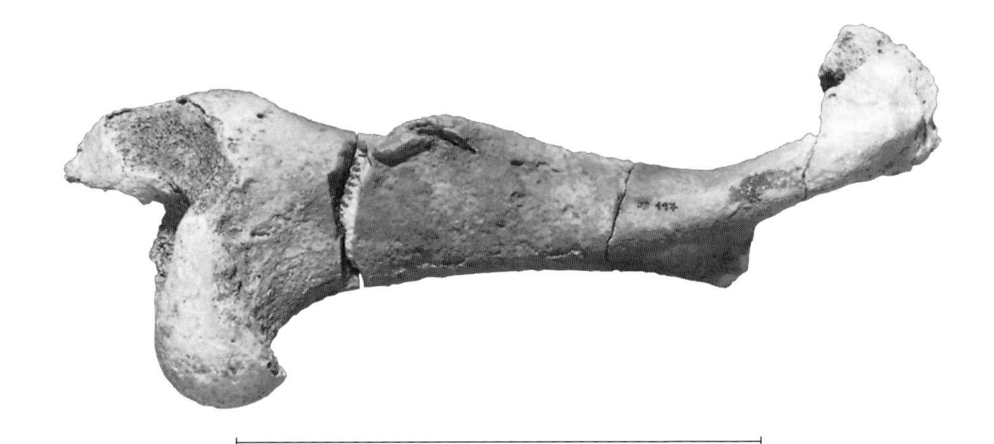

Figure 21.8. *Hexaprotodon* aff. *sahabiensis,* right humerus in four pieces, AUH 497; lateral view. Scale bar = 20 cm.

because this indentation precludes there being room for a prominent concavity in the anterolateral part of the scaphoid facet. It would also be rather large for a pig, and the overall width of the lunate and scaphoid facets together is perhaps too great. It is a little smaller than several examples of *Hex. sivalensis*, the indented anterior margin of the scaphoid and lunate facets is less pronounced, it has a more localised tubercle anteromedially to the scaphoid facet, and a short sharply crested ridge on the medial part of the anterior surface. It also differs from *Hex. sivalensis* in that the scaphoid facet is less expanded posteriorly in ventral view and therefore the posterior edge of the lunate facet (also in ventral view) is less concave in profile. In all these features except size it also differs from the Madagascan Pleistocene hippopotamuses.

The lunate is smaller than in *Hip. amphibius* (no *Hex. sivalensis* lunate was available for comparison) and has less of a ventral articulation on its lateral surface for the cuneiform. It also shows a sharper ridge between the two adjacent facets for articulating on the radius.

The cuneiform is smaller than in *Hex. sivalensis*, less thickened transversely, and with less of a ventral articulation on its medial surface for the lunate.

The distal metacarpal 2, AUH 248, is more slender than an example of *Hex. sivalensis*.

The metacarpals 3 have the anteroposteriorly convex proximal articular surfaces seen in hippopotamuses. There is quite a high-sided facet for the unciform but this facet is little elongated from front to back. The distal hollow on the anterior surface is more marked than in a huge metacarpal 3 of *Hip. gorgops* from Kuguta, BMNH M18504.

The metacarpal 5 has no distal hollow on its anterior surface. Its top articular surface is narrow with the posterior part downturned.

The proximal phalanx is as long as in *Hex. sivalensis* but more slender. The distal phalanges too are very little squashed anteroposteriorly for a hippopotamus. The vertical ridge on the proximal articular surface is well marked.

The atlas vertebra AUH 431 has preserved enough of the posterior part of its left wing to show that it would have been upwardly curved at its lateral edges in the manner of most other hippopotamuses but unlike modern *Hex. liberiensis*.

The axis vertebra AUH 429 is smaller than three examples of *Hex. sivalensis*, BMNH 18018–18020. The anterior articular facets flanking the odontoid process are quite deep in a dorsoventral plane but do not reach up around the sides of the neural arch. The intervertebral canals, anterolaterally on the sides of the neural arch, are preserved on both sides. The top of the neural crest is inclined steeply downwards anteriorly.

The two centra of cervical vertebrae appear to be slightly smaller than *Hex. sivalensis*. The thoracic vertebra AUH 105 has an elongated rib articulation anterolaterally on its transverse process, which is like rear thoracic vertebrae of hippopotamuses and rules out identification as a pecoran. It has no longitudinal ridge ventrally, which is also unlike pecorans. The centrum is dorsoventrally shallow and its dorsal edge at both ends is shallowly indented.

Comparisons

The Abu Dhabi hippopotamus has characters in common with earlier hippopotamuses than *Hex. sivalensis* or *Hex. protamphibius*. It is slightly smaller, the horizontal ramus of the mandible is shallow, the premolar row is longer than in later hippopotamuses, the molars are lower crowned, the molar cusps have less of a trefoil pattern of occlusal wear (because the cusps are less expanded along the median axis of the molars and the central transverse valley is more obvious), the anterior cingulum is smaller, the M_3 hypoconulid is larger and rises well above the occlusal level of the more anterior parts of the tooth, and this hypoconulid has a longer anterior crest in a median position.

From *Hex. harvardi*, so far as it is known from Coryndon's accounts, the Abu Dhabi species differs by its narrower symphysis, a longer premolar row, P_3 longer than P_4, P_4 with two posterior cusps and a stronger metaconid, and the P^3 with a better demarcation between its front and back lobes.

The Abu Dhabi remains are close to *Hex. iravaticus* in both hexaprotodonty and a narrow mandibular symphysis. The back of the symphysis of *Hex. iravaticus* is level with the back of P_3, as in *Hex. harvardi* (Coryndon, 1977: fig. 8) and in one of the three Abu Dhabi specimens. (*Hex. karumensis* varies in this character according to Harris, 1991: figs. 2.8–2.10, 2.15, and so, too, could other hippopotamus species). Little else is known of *Hex. iravaticus* morphology in its mandible, lower dentition, or P^3 to enable comparisons to be made.

In comparison with *Hex. crusafonti* the Abu Dhabi species is definitely hexaprotodont and can be seen to have a P^3 with better demarcation into two lobes, a longer premolar row, and perhaps a less sharply downturned mandibular angle (Aguirre, 1963: pl. 2, fig. 7 and pl. 3, fig. 3).

In comparison with *Hex. sahabiensis* the Abu Dhabi species shows a wider P^3, apparently with less of a posterolingual cusp.

There is no reason to identify the Abu Dhabi remains as *Kenyapotamus*, although a Nakali mandible of the latter (Pickford, 1983: pl. 1, fig. 4) has an angle little extended downwards, and hence like the Abu Dhabi species.

These comparisons of the Abu Dhabi hippopotamus can be summarised in tabular form (table 21.1). The table shows that the Abu Dhabi species has six resemblances and one difference from *Hex. sahabiensis*, six resemblances and three

Table 21.1. Comparison of the Abu Dhabi and other fossil hippopotamuses

	Hex. harvardi	*Hex. iravaticus*	*Hex. crusafonti*	*Hex. sahabiensis*	Abu Dhabi specimens
P^3 width × 100/length	75%	—	87%	72%	80%
P^3 with well-marked front and back lobes	Less so	—	Less so	Yes	Yes
Posterolingual cusp on P^3	No	—	No	Yes	No
Mandible angle turned sharply downwards	—	—	Yes	No	No
Narrow symphysis	No	Yes	Yes	—	Yes
Tetraprotodont incisors	No	No	Can be	No	No
P_{2-4} × 100/M_{1-3}	76%	—	83%	—	91%
Length P_3 × 100/P_4	90%	—	116%	—	115%
P_4 with two posterior cusps	No	—	Yes	Yes	Yes
P_4 with metaconid	Weak	—	—	Yes	Yes

Note: the percentages in the table are quoted from other authors, or calculated from their measurements or illustrations.

differences from *Hex. crusafonti,* and three resemblances and six differences from *Hex. harvardi.* (For the percentages in the table, a separation of 10% or more was counted as a difference.)

Discussion

By sinking *Choeropsis* in *Hexaprotodon,* Coryndon (1977: fig. 6) linked *Hex. liberiensis* with small fossil species, including *Hex. imagunculus, Hex. aethiopicus,* and an early small species cited at Lothagam. But *Hex. liberiensis* also necessarily became linked with *Hex. sivalensis* (type species of the genus) and with other larger species included in *Hexaprotodon.* In line with this Coryndon noted that Anthony (1948) had shown from endocranial casts that *Hex. liberiensis* was more like *Hex. protamphibius* in brain sulci than like *Hip. amphibius.* Coryndon postulated two monophyletic genera sharing an immediate common ancestry: *Hippopotamus* (containing *amphibius, gorgops, kaisensis, aethiopicus,* Madagascan, European, and Mediterranean island species) and *Hexaprotodon* (containing *sivalensis, protamphibius, harvardi, hipponensis, karumensis, imagunculus, primaevus* [= *crusafonti*], and *liberiensis*). It is easily possible, however, that *Hippopotamus* descended from an ancestor like *Hex. protamphibius,* this being a normal or representative large hippopotamus of the middle to late Pliocene of Africa, so that *Hexaprotodon* would become a paraphyletic genus. Other phylogenies too are conceivable—for example, repeated independent origins for successive large or small species. Until these possibilities are analysed more thoroughly, I retain the use of *Hexaprotodon* and *Hippopotamus* largely as in Coryndon and most other recent authors. From figure 21.9 it can be seen that *Hippopotamus* is known back to around 3.0 Ma and *Hexaprotodon* to around 6.0 Ma.

It is easier to arrange fossil hippopotamuses in successive evolutionary grades—namely *Hex. harvardi* to *Hex. protamphibius* to *Hip. amphibius*—than to differentiate species. As noted above, the main distinction between *Hex. sivalensis* and *Hex. protamphibius* is that the one is found in India and the other in Africa. Yet the difference in timing for the acquisition of tetraprotodonty shows that they must have been evolving as separate species. Mazza

(1991) has already pointed to the likelihood of *Hip. gorgops*-like characters developing independently in the Pleistocene hippopotamuses of Europe as another example of parallel evolution in hippopotamuses. Problems of valid species identification is increased when dealing with the main mass of incompletely known fossil species. It is impossible to form any conclusion about the identity or relationships of *Hip. kaisensis*, for example; it could be conspecific with *Hex. protamphibius*, with the tetraprotodont phase of *Hex. karumensis*, with *Hip. gorgops*, or with *Hip. amphibius*. It is a problem with hippopotamus fossils recovered long ago from stratified sequences such as the Kaiso Formation or the Irrawaddy Group to know the correct levels of the few actually diagnostic pieces and what their systematic relationship is to hippopotamuses at other levels.

It is tempting to see a narrow muzzled hippopotamus like that of Abu Dhabi as a more primitive evolutionary grade than *Hex. harvardi*. It appears, however, that *Hex. crusafonti* may sometimes combine precocious tetraprotodonty with a narrow muzzle. Also the Burmese *Hex. iravaticus*,

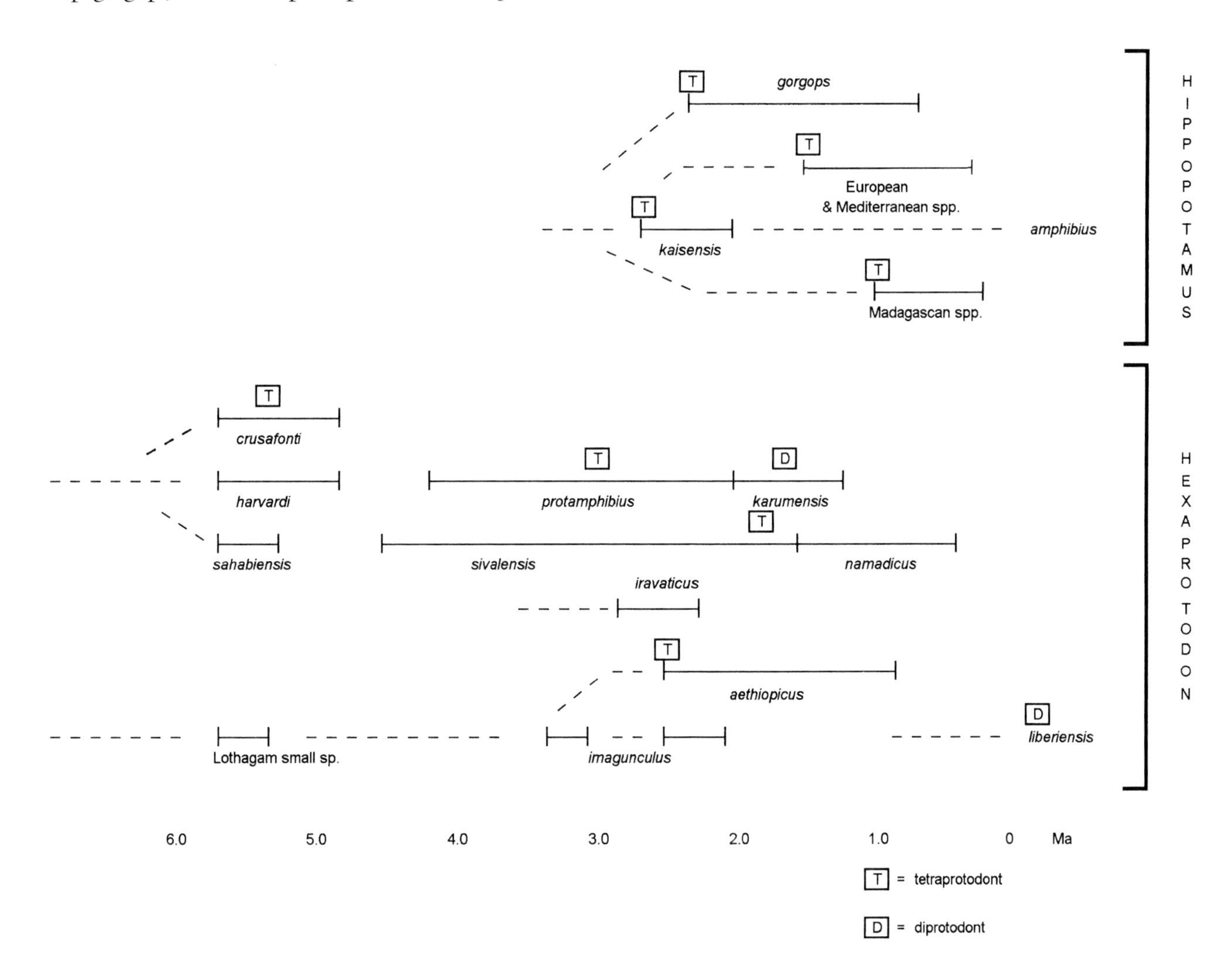

Figure 21.9. Time ranges of hippopotamus species.

another species with a narrow muzzle, is supposed to be of later Pliocene age, thus postdating *Hex. harvardi* in Africa. From the characters of P^3 and P_4 in the Abu Dhabi species, it is evidently as primitive a species as the rather scrappily known *Hex. sahabiensis* from Sahabi. It is not known whether *Hex. sahabiensis* had a narrow muzzle like *Hex. crusafonti* and the Abu Dhabi hippopotamus and unlike its probable contemporary *Hex. harvardi* in East Africa.

ACKNOWLEDGEMENTS

I thank Peter Whybrow for inviting me to work on the Abu Dhabi hippopotamuses, Diana Clements for much assistance including preparation of the diagram, and Ray Bernor for helpful comments. I thank the Keeper of Palaeontology at The Natural History Museum for use of the facilities of his department.

REFERENCES

Aguirre, E. 1963. *Hippopotamus crusafonti* n. sp. del Plioceno inferior de Arenas del Rey (Granada). *Notas y Comunicaciones del Instituto Geologico y Minero de Espana, Madrid* 69: 215–30.

Aguirre, E., and Leakey, P. 1974. Nakali: Nueva fauna de Hipparion del Rift Valley de Kenya. *Estudios Geologicos. Instituto de Investigaciones Geologicas "Lucas Mallada", Madrid* 30: 219–27.

Aguirre, E., Robles, F., Thaler, L., Lopez, N., Alberdi, M. T., and Fuentes, C. 1973. Venta del Moro, nueva fauna finimiocena de Moluscos y Vertebrados. *Estudios Geologicos. Instituto de Investigaciones Geologicas "Lucas Mallada", Madrid* 29: 569–78.

Andrews, C. W. 1902. Note on a Pliocene vertebrate fauna from the Wadi-Natrun, Egypt. *Geological Magazine* 9: 433–39.

Anthony, J. 1948. Etude de moulages endocraniens d'hippopotames disparus. *Mémoires du Museum National d'Histoire Naturelle, Paris,* n.s. 26: 31–56.

Arambourg, C. 1945. Au sujet de l'*Hippopotamus hipponensis* Gaudry. *Bulletin de la Société Géologique de France, Paris* 14: 147–53.

———. 1947. Contribution à l'étude géologique et paléontologique du bassin du lac Rodolphe et de la basse vallée de l'Omo. 2, Paléontologie. In *Mission scientifique de l'Omo, 1932–1933,* vol. 1. *Géologie—Anthropologie,* fasc. 3, pp. 232–562. Muséum National d'Histoire Naturelle, Paris.

———. 1970. Les vertébrés du Pléistocène de l'Afrique du Nord. *Archives du Muséum National d'Histoire Naturelle, Paris* 10: 1–126.

Barry, J. C., 1995. Faunal turnover and diversity in the terrestrial Neogene of Pakistan. In *Paleoclimate and Evolution, with Emphasis on Human Origins,* pp. 115–34 (ed. E. S. Vrba, G. H. Denton, T. C. Partridge, and L. H. Burkle). Yale University Press, New Haven.

Barry, J. C., and Flynn, L. J. 1990 (1989). Key biostratigraphic events in the Siwaliks sequence. In *European Neogene Mammal Chronology,* pp. 557–71 (ed. E. H. Lindsay, V. Fahlbusch, and P. Mein). Plenum Press, New York.

Bender, F. 1983. *Beiträge zur regionalen Geologie der Erde: Geology of Burma.* Gebrüder Borntraeger, Berlin.

Boaz, N. T., Bernor, R. L., Brooks, A. S., Cooke, H. B. S., Heinzelin, J. de, Deschamps, R., Delson, E., Gentry, A. W., Harris, J. W. K., Meylan, P., Pavlakis, P. P., Sanders, W. J., Stewart, K. M., Verniers, J., Williamson, P. G., and Winkler, A. J. 1992. A new evaluation of the significance of the Late Neogene Lusso Beds, Upper Semliki Valley, Zaire. *Journal of Human Evolution* 22: 505–17.

Boaz, N. T., El-Arnauti, A., Gaziry, A., Heinzelin, J. de, and Boaz, D. D. 1987. *Neogene Paleontology and Geology of Sahabi.* Alan R. Liss, New York.

Cerling, T. E., Quade, J., Ambrose, S. H., and Sikes, N. E. 1991. Fossil soils, grasses and carbon isotopes from Fort Ternan, Kenya: Grassland or woodland? *Journal of Human Evolution* 21: 295–306.

Colbert, E. H. 1935. Siwalik mammals in the American Museum of Natural History. *Transactions of the American Philosophical Society, Philadelphia,* n.s. 26: 1–401.

———. 1938. Fossil mammals from Burma in the American Museum of Natural History. *Bulletin of the American Museum of Natural History, New York* 74: 255–436.

———. 1943. Pleistocene vertebrates collected in Burma by the American southeast Asiatic Expedition. *Transactions of the American Philosophical Society, Philadelphia,* n.s. 32: 395–429.

Cooke, H. B. S., and Coryndon, S. C. 1970. Pleistocene mammals from the Kaiso Formation and other related deposits in Uganda. In *Fossil Vertebrates of Africa,* vol. 2, pp. 107–224 (ed. L. S. B. Leakey and R. J. G. Savage). Academic Press, London.

Coryndon, S. C. 1977. The taxonomy and nomenclature of the Hippopotamidae (Mammalia, Artiodactyla) and a description of two new fossil species. *Proceedings Koninklijke Nederlandse Akademie van Wetenschappen* B 80: 61–88.

———. 1978a. Fossil Hippopotamidae from the Baringo Basin and relationships within the Gregory Rift, Kenya. In *Geological Background to Fossil Man,* pp. 279–92 (ed. W. W. Bishop). Geological Society Special Publication no. 6. Scottish Academic Press, Edinburgh.

———. 1978b. Hippopotamidae. In *Evolution of African Mammals,* pp. 483–95 (ed. V. J. Maglio and H. B. S. Cooke).Harvard University Press, Cambridge, Mass.

Coryndon, S. C., and Coppens, Y. 1973. Preliminary report on Hippopotamidae (Mammalia, Artiodactyla) from the Plio/Pleistocene of the lower Omo basin; Ethiopia. In *Fossil Vertebrates of Africa,* vol. 3, pp. 139–57 (ed. L. S. B. Leakey, R. J. G. Savage, and S. C. Coryndon). Academic Press, London.

Cuvier, G. 1804. Sur les ossemens fossiles d'hippopotame. *Annales du Muséum National d'Histoire Naturelle, Paris* 5: 99–122.

Falconer, H., and Cautley, P. T. 1836. Note on the fossil hippopotamus of the Sivalik Hills. *Asiatic Researches, Calcutta* 19: 39–53.

——— 1847 (1845–9). *Fauna Antiqua Sivalensis,* pt 7. Smith, Elder and Co., London.

Faure, M. 1995. Les Hippopotamidae (Mammalia, Artiodactyla) du Rift occidental (Bassin du lac Albert, Ouganda): Etude préliminaire. In *Geology and Palaeontology of the Albertine Rift Valley, Uganda–Zaire, Volume II: Palaeobiology/Paléobiologie,* pp. 321–37 (ed. B. Senut and M. Pickford). CIFEG Publication Occasionnelle 1994/29. *Centre International pour la Formation et les Echanges Géologiques, Orléans.*

Faure, M., and Méon, H. 1984. L'*Hippopotamus crusafonti* de La Mosson (pres Montpellier): Première réconnaissance d'un Hippopotame néogène en France. *Compte Rendu Hebdomadaire des Séances de l'Académie des Sciences, Paris* II, 298: 93–98.

Feibel, C. S., and Brown, F. H. 1991. Age of the primate-bearing deposits on Maboko Island, Kenya. *Journal of Human Evolution* 21: 221–25.

Gaudry, A. 1876. Sur un hippopotame fossile découvert à Bône (Algérie). *Bulletin de la Société Géologique de France, Paris* 4: 501–4.

Gaziry, A. W. 1987. *Hexaprotodon sahabiensis* (Artiodactyla, Mammalia): A new hippopotamus from Libya. In *Neogene Paleontology and Geology of Sahabi,* pp. 303–15 (ed. N. T. Boaz, A. El-Arnauti, A. W. Gaziry, J. de Heinzelin, and D. D. Boaz). Alan R. Liss, New York.

Geraads, D. 1987. Dating the Northern African cercopithecid record. *Human Evolution* 2: 19–27.

———. 1989. Vertébrés fossiles du Miocène supérieur du Djebel Krechem el Artsouma (Tunisie centrale). Comparaisons biostratigraphiques. *Geobios* 22: 777–801.

Gèze, R. 1985. Répartition paléoécologique et relations phylogénétiques des Hippopotamidae (Mammalia, Artiodactyla) du Néogène d'Afrique orientale. In *L'Environnement des hominidés au Plio-Pléistocène,* pp. 81–100 (ed. M. Beden and 18 others). Fondation Singer-Polignac, Paris.

Harger, R. L. 1932. Partial reversion to hexaprotodont dentition in hippopotamus, *H. amphibius* Linn. *Journal of the East Africa and Uganda Natural History Society* 40–41: 129–31.

Harris, J. M. ed. 1991. *Koobi Fora Research Project, Volume 3: The Fossil Ungulates: Geology, Fossil Artiodactyls, and Palaeoenvironments.* Clarendon Press, Oxford.

Harris, J. M., Brown, F. H., and Leakey, M. G. 1988. Stratigraphy and paleontology of Pliocene and Pleistocene localities west of Lake Turkana, Kenya. *Contributions in Science, Natural History Museum of Los Angeles County* no. 399: 1–128.

Harrison, T. J.1997. The anatomy, paleobiology, and phylogenetic relationships of the Hippotamidae (Mammalia, Artiodactyla) from the Manonga Valley, Tanzania. In *Neogene Paleontology of the Manonga Valley, Tanzania: A Window into East African Evolution,* pp. 137–90 (ed. T. J. Harrison). Topics in Geobiology vol. 14. Plenum Press, New York.

Heinzelin, J. de. 1983. The Omo Group. *Annales du Musée Royal de l'Afrique Centrale,* ser. 8vo, *Sciences Géologiques* 85: 1–365.

Hill, A. 1995. Faunal and environmental change in the Neogene of East Africa: Evidence from the Tugen Hills Sequence, Baringo District, Kenya. In *Paleoclimate and Evolution, with Emphasis on Human Origins,* pp. 178–93 (ed. E. S. Vrba, G. H. Denton, T. C. Partridge, and L. H. Burkle). Yale University Press, New Haven.

Hill, A., Drake, R., Tauxe, L., Monaghan, M., Barry, J. C., Behrensmeyer, A. K., Curtis, G., Jacobs, B. F., Jacobs, L., Johnson, N., and Pilbeam, D. 1985. Neogene palaeontology and geochronology of the Baringo Basin, Kenya. *Journal of Human Evolution* 14: 759–73.

Hooijer, D. A. 1942. On the supposed hexaprotodont milk dentition in *Hippopotamus amphibius. Zoologische Mededeelingen* 24: 189–96.

———. 1943. On Recent and fossil hippopotami. *Comptes Rendus de la Société Néerlandaise de Zoologie:* 289–90.

———. 1950. The fossil Hippopotamidae of Asia; with notes on the Recent species. *Zoologische Verhandelingen* 8: 1–124.

Howell, F. C. 1980. Zonation of late Miocene and early Pliocene circum-Mediterranean faunas. *Geobios* 13: 653–57.

Hussain, S. T., van den Bergh, G. D., Steensma, K. J., de Visser, J. A., de Vos, J., Arif, M., van Dam, J., Sondaar, P. Y., and Malik, S. B. 1992. Biostratigraphy of the Plio-Pleistocene continental sediments (Upper Siwaliks) of the Mangla-Samwal Anticline, Azad Kashmir, Pakistan. *Proceedings Koninklijke Nederlandse Akademie van Wetenschappen* 95: 65–80.

Joleaud, L. 1920. Contribution à l'étude des hippopotames fossiles. *Bulletin de la Société Géologique de France, Paris* 20: 13–26.

Kahlke, R.-D. 1985. Untersuchungen zur Incisivenreduktion an altpleistozänen Hippopotamus-Mandibeln von Untermassfeld bei Meiningen (Bezirk Suhl, DDR). *Biologische Rundschau, Jena* 23: 315–21.

Lacomba, J. I., Morales, J., Robles, F., Santisteban, C., and Alberdi, M. T. 1986. Sedimentologia y paleontologia del yacimiento finimioceno de La Portera (Valencia). *Estudios Geologicos. Instituto de Investigaciones Geologicas "Lucas Mallada", Madrid* 42: 167–80.

Leakey, M. G., Feibel, C. S., Bernor, R. L., Harris, J. M., Cerling, T. E., Stewart, K. M., Storrs, G. W., Walker, A., Werdelin, L., and Winkler, A. J. 1966. Lothagam: A record of faunal change in the late Miocene of East Africa. *Journal of Vertebrate Paleontology* 16: 556–70.

Leidy, J. 1853. On the osteology of the head of Hippopotamus and a description of the osteological characters of a new genus of Hippopotamidae. *Journal of the Academy of Natural Sciences of Philadelphia* 2: 207–24.

Mazza, P. 1991. Interrelations between Pleistocene hippopotami of Europe and Africa. *Bollettino della Societá Paleontologica Italiana* 30: 153–86.

Pantanelli, D. 1879. Sugli strati miocenici del Casino (Siena) e considerazioni sul miocene superiore. *Atti della Reale Accademia (Nazionale) dei Lincei. Memorie. Roma* 3: 309–27.

Pavlakis, P. P. 1990. Plio-Pleistocene Hippopotamidae from the Upper Semliki. *Memoirs Virginia Museum of Natural History* 1: 203–23.

Petrocchi, C. 1952. Notizie generali sul giacimento fossilifero di Sahabi: Storia degli Scavi-Risultati. *Rendiconti. Accademia Nazionale dei XL, Roma* 4: 9–34.

Pickford, M. 1981. Preliminary Miocene mammalian biostratigraphy for western Kenya. *Journal of Human Evolution* 10: 73–97.

———. 1983. On the origins of Hippopotamidae together with descriptions of two new species, a new genus and a new subfamily from the Miocene of Kenya. *Geobios* 16: 193–217.

———. 1988. Revision of the Miocene Suidae of the Indian subcontinent. *Münchner Geowissenschaftliche Abhandlungen* A12: 1–92.

———. 1989. Update on hippo origins. *Compte Rendu de l'Académie des Sciences, Paris* 309: 163–68.

———. 1990. Découverte de *Kenyapotamus* en Tunisie. *Annales de Paléontologie, Paris* 76: 277–83.

Pickford, M., Senut, B., Ssemmanda, I., Elepu, D., and Obwona, P. 1988. Premiers résultats de la mission de l'Uganda Palaeontology Expedition à Nkondo (Pliocène du bassin du Lac Albert, Ouganda). *Compte Rendu de l'Académie des Sciences, Paris* 306: 315–20.

Pomel, A. 1896. Les hippopotames. *Carte Géologique de l'Algérie, Paléontologie Monographies, Alger* 8: 1–65.

Reynolds, S. H. 1922. Hippopotamus. In *A Monograph of the British Pleistocene Mammalia,* pp. 1–38 (ed. W. B. Dawkins and S. H. Reynolds). Palaeontographical Society, London.

Stamp, L. D. 1922. An outline of the Tertiary geology of Burma. *Geological Magazine* 59: 481–501.

Stromer, E. 1907. Fossile Wirbeltier-Reste aus dem Uadi Fâregh und Uadi Natrûn in Ägypten. *Abhandlungen hrsg. von der Senckenbergischen Naturforschenden Gesellschaft. Frankfurt a.M.* 29: 99–132.

———. 1914. Mitteilungen über Wirbeltierreste aus dem Mittelpliocän des Natrontales (Ägypten). *Zeitschrift der Deutschen Geologischen Gesellschaft, A. Abhandlungen, Berlin* 66: 1–33.

Whybrow, P. J. 1989. New stratotype; the Baynunah Formation (Late Miocene), United Arab Emirates: Lithology and palaeontology. *Newsletters on Stratigraphy* 21: 1–9.

Whybrow, P. J., Friend, P. F., Ditchfield, P. W., and Bristow, C. S. 1999. Local stratigraphy of the Neogene outcrops of the coastal area: Western Region, Emirate of Abu Dhabi, United Arab Emirates. Chap. 4 in *Fossil Vertebrates of Arabia,* pp. 28–37 (ed. P. J. Whybrow and A. Hill). Yale University Press, New Haven.

Fossil Pecorans from the Baynunah Formation, Emirate of Abu Dhabi, United Arab Emirates

22

Alan W. Gentry

No Tragulidae have been found in the Baynunah Formation, so that this chapter covers only the two pecoran families that are represented in it (for geological details see Whybrow et al., 1999; Friend, 1999—Chapters 4 and 5). In the following descriptions comparisons are made with fossils in the Palaeontology Department of The Natural History Museum, London (institutional abbreviation BMNH). Most of the material is the property of the Emirate of Abu Dhabi (abbreviation AUH). Localities and deposits mentioned in the text other than those of the Baynunah Formation are:

Baringo deposits, Kenya, middle and late Miocene and including the Ngorora Formation (Hill et al., 1985; Hill, 1995); Bugti, Pakistan, early Miocene (Pickford, 1988); Fort Ternan, Kenya, dated to 14.0 million years (Ma) ago (Gentry, 1970); Grays, Essex, United Kingdom, middle Pleistocene (Stuart, 1982); Hasnot, Siwaliks Group, Pakistan, late Miocene (Barry and Flynn, 1990); Hofuf Formation, Saudi Arabia, middle Miocene (Thomas, 1983); Ilford, Essex, United Kingdom, middle Pleistocene (Stuart, 1982); Jebel Hamrin, Iraq, late Miocene (Thomas et al., 1980); Langebaanweg, "E" Quarry, South Africa, earliest Pliocene (Gentry, 1980); Lothagam, Kenya, latest Miocene (Leakey et al., 1996); Maragheh, Iran, late Miocene (Bernor, 1986); Montpellier, France, early Pliocene (Gromolard, 1980); Nakali, Kenya, late Miocene (Aguirre and Leakey, 1974); Pikermi, Greece, late Miocene (Solounias, 1981a); Piram (formerly Perim) Island, India, late Miocene (Prasad, 1974); Sahabi, Libya, late Miocene (Boaz et al., 1987); Samos, Greece, late Miocene (Solounias, 1981b); Sebastopol, Ukraine, late Miocene (Borissiak, 1914); Siwaliks Group, Pakistan and India, Miocene to Pliocene (Barry and Flynn, 1990); Wadi Natrun, Egypt, late Miocene (Andrews, 1902); Wembere–Manonga Formation, Tanzania, latest Miocene–early Pliocene (Harrison, 1997).

Several descriptive phrases exist for the late Miocene (Vallesian and mainly Turolian) faunas that are known from Greece through Turkey to Iran. Knowledge of these faunas originated with the major localities Pikermi, Samos, and Maragheh, known since the last century. Together they form a coherent zoogeographical unit. "Graeco-Iranian" indicates the west to east limits of these faunas but appears to exclude regions immediately north of the Black Sea that have many of the same genera and species. "SE European and SW Asian" is both cumbersome and vague about the geographical regions encompassed. "Sub-Paratethyan Province" (Bernor, 1983) was intended to exclude part of Greece. In this chapter I shall use "Graeco-Iranian".

Systematics

Family Giraffidae Gray, 1821

Remarks

The middle Miocene *Giraffokeryx* Pilgrim, 1911 must have been an immigrant into southeastern Europe and southwestern and eastern Asia, because giraffids or their likely ancestors had been present previously only in Africa, Arabia, and at Bugti in Pakistan. *Injanatherium* Heintz, Brunet, and Sen, 1981 of the Arabian middle Miocene is very similar to *Giraffokeryx* and survived into the late Miocene of Iraq (Brunet and Heintz, 1983: 288).

Late Miocene Graeco-Iranian giraffids include *Palaeotragus* Gaudry, 1861 and *Samotherium* Major, 1888. *Palaeotragus coelophrys* (Rodler and Weithofer, 1890) and its synonyms is a relatively primitive species slightly smaller than the well-known *Samotherium boissieri; Palaeotragus roueni,* type species of its genus, is smaller again, more advanced, and has lengthened legs. A third genus, *Bohlinia* Matthew, 1929, has very long legs and is

 ISBN 0-300-07183-3

ancestral or related to *Giraffa,* despite some differences in its limb bone characters (Geraads, 1979: 380). Its teeth, in so far as they are known, are brachyodont and can be confused with the similarly sized teeth of *P. coelophrys,* but *P. coelophrys* metapodials at Maragheh (Mecquenem, 1924–5: 160) are much shorter than those of *Bohlinia.* Limb bones of *Bohlinia* in fact have articular ends that can be even bigger than in *Samotherium boissieri.*

In the Siwaliks succession the history of giraffes is less clear. *Giraffokeryx punjabiensis* (Pilgrim, 1911) disappears during the span of the Nagri Formation, perhaps around 10 Ma ago. For the rest of the Miocene "large giraffes" are present (Barry et al., 1982: fig. 2) but have not been revised for many years; they were listed as *Bramatherium megacephalum* (Lydekker, 1876b) and *Giraffa punjabiensis* Pilgrim, 1911 by Barry et al. (1991). The latter species is found in the Dhok Pathan from 7.1 until about 5.8 Ma ago (Barry, 1995), and BMNH material comes mostly from Hasnot. The species resembles *Palaeotragus coelophrys* in size and crest morphology on P_4, although the premolars are wider and P_2 relatively longer (compare Colbert, 1935: fig. 193 with Mecquenem, 1924–5: pl. 2, fig. 8). The lower molars also have basal pillars. If *G. punjabiensis* is correctly assigned to *Giraffa,* its metapodials should be much longer than in *P. coelophrys,* but this has not been established in print.

The Baringo deposits of East Africa were once thought to have a sedimentary and time gap between the disappearance of "Palaeotraginae" at about 10 Ma ago and the appearance of Giraffinae at about 6 Ma ago (Hill et al., 1985: fig. 4). Hill (1987: 592), however, later placed Nakali in this gap and Aguirre and Leakey (1974) had discussed giraffid fossils from there. Churcher (1979) identified an upper molar from Lothagam as *Palaeotragus germaini* Arambourg, 1959, a North African Vallesian-equivalent species, but Geraads (1986: 474) preferred to assign the tooth to a form close to *Giraffa.*

Among the large-sized and short-limbed sivatheriine giraffes, the Asian *Bramatherium* Falconer, 1845 and *Hydaspitherium* Lydekker, 1878 are related to (perhaps congeneric with) *Decennatherium* Crusafont Pairo, 1952 of Europe. They have enlarged anterior ossicones and smaller posterior ones, whereas the Spanish *Birgerbohlinia* Crusafont Pairo, 1952 has a large posterior pair and a smaller anterior pair, more reminiscent of later *Sivatherium.* Hence there are two supergeneric groupings of sivatheres. (The state of the cranial appendages is not known in *Helladotherium* Gaudry, 1860.) One or more smaller species of late Miocene sivatheriines exist in the Siwaliks (Matthew, 1929: 541–5; Colbert, 1935: 363). Geraads (1989a: 193) comments that P_3s of the oldest known African sivatheres (Pliocene or late Miocene) are more primitive than those of the late Miocene *Decennatherium* in Europe.

It is evident that the late Miocene giraffes of the Graeco-Iranian region were different from those evolving in Africa and that the taxonomy of Indian fossil giraffids is unclear.

Genus *Palaeotragus* Gaudry, 1861
Type species *Palaeotragus roueni* Gaudry, 1861
?*Palaeotragus* sp.

Material

AUH 405, Ras Dubay'ah, R2: distal metatarsal. Width across condyles 49.5 mm, maximum anteroposterior diameter of condyles 29.4 mm. Figure 22.1.
AUH 289, Jebel Dhanna, JD3: parts of a lumbar vertebra.

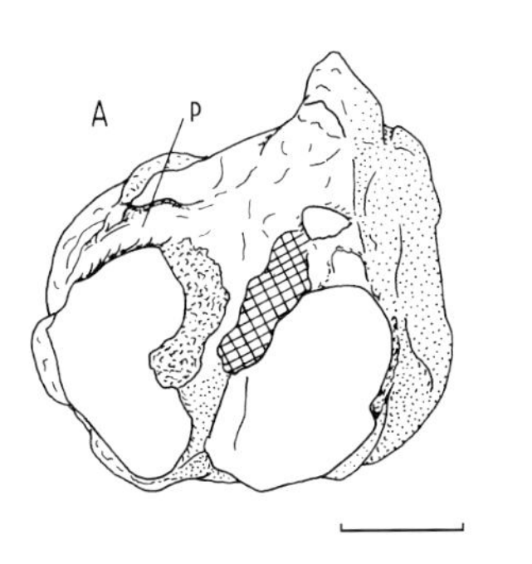

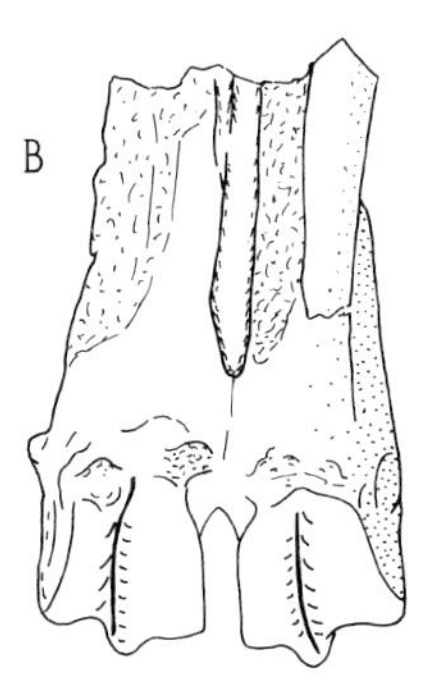

Figure 22.1. *A,* Giraffidae, sp. indet., left naviculocuboid, AUH 209; ventral view, anterior side towards the base of drawing; p = groove for the peronaeus longus tendon. *B,* ?*Palaeotragus* sp., distal metatarsal, AUH 405; anterior view. Scale = 20 mm.

Description

This distal metatarsal is about the same size as in *Palaeotragus roueni*. There is some damage to the anterior surface of the shaft, but it looks as if a longitudinal groove was originally present and was stronger than the narrow and shallow incision evident on some pecoran metacarpals. The ridges flanking this groove would have been too weak for the bone to belong to a bovid metatarsal, and they do not rise as localised flanges at their distal ends as in a bovid. There is very little narrowing of the shaft as it ascends above the condyles, suggesting that, when complete, the metatarsal could have been as long as in *P. roueni*. It is less gracile and slightly larger than the distal end of a cast, BMNH M30210, of a metatarsal belonging to the Fort Ternan giraffid *Giraffokeryx primaevus* (Churcher, 1970) of middle Miocene age.

The partial lumbar vertebra AUH 289 is assigned to this species solely by its appropriate size.

Genus *Bramatherium* Falconer, 1845
Type species *Bramatherium perimense* Falconer, 1845
?*Bramatherium* sp.

Material

AUH 438, Ras Dubay'ah, R2: two pieces of horn cores. Figure 22.2.
AUH 204, Ras Dubay'ah, R2: partial right upper molar, early middle wear. Occlusal length c. 38.0 mm. Figure 22.2.
AUH 206, Ras Dubay'ah, R2: pieces of unworn upper molar(s).
AUH 217, Ras Dubay'ah, R2: left upper molar, late wear.
AUH 372, Ras Dubay'ah, R2: shattered right dP_4, unworn. Occlusal length c. 45.0 mm.
AUH 223, Shuwaihat, S4: thoracic vertebra. Anteriorly placed in the thoracic row, possibly about fifth. The rear epiphysis of the centrum is missing. Figure 22.3.
AUH 225, Shuwaihat, S4: partial centrum of a (?thoracic) vertebra.

Description

The bigger of the two pieces of horns, AUH 438, measures c. 170 mm along its convex edge, c. 85.0 mm maximum width at its base, and c. 30.0 mm thickness or minimum basal width. If the specimen gives an accurate indication of the original shape, the horns would have been short, flattened, curved, and tapering rapidly to the tip. The cortical bone is compact compared with the more cancellous medulla within, as is characteristic of giraffids. The upper molar AUH 204 has lost most of its posterior half. The dP_4 AUH 372 has barely a trace of a posterior basal pillar and no anterior one. The vertebra AUH 223 is definitely not of hippopotamus or rhinoceros. It is about the size of Pleistocene *Bos primigenius* Bojanus, 1827 from Ilford, but more robust. The centra ends, front and back, are possibly less flat than in *Bos* and more curved, as in the Indian late Pliocene *Sivatherium giganteum* Falconer and Cautley, 1836, type species of that genus. Because of this and because of its size in a Miocene context, the vertebra must be a giraffid. The more fragmentary vertebra AUH 225 is probably conspecific.

Discussion

If the bigger of the two pieces of horns, AUH 438, is complete, it would be too small to belong to an adult sivathere. The flattening fails to match known sivathere horns but the irregular knobbiness of the surface suggests a sivatheriine rather than a giraffine giraffid. *Samotherium* horns are longer although also slightly curved backwards. Portions of horns of a Siwaliks *Giraffokeryx*, BMNH M15772, are smoother surfaced and *Injanatherium* has straighter horns. The living *Okapia johnstoni* has much smaller horns but they, too, are short, curved backwards, and somewhat compressed from side to side.

The upper molar AUH 204 is too large to match *Palaeotragus coelophrys* or the Siwaliks *Giraffa punjabiensis*, but is of an appropriate size for *Samotherium boissieri* or M^1s of Miocene sivatheres such as *Birgerbohlinia schaubi* Crusafont Pairo, 1952 or *Helladotherium duvernoyi* (Gaudry and Lartet, 1856). It agrees well with molars of the lectotype left maxilla with P^4–M^3 of *Bramatherium perimense* Falconer, 1845 from Piram Island, BMNH 48933 (Falconer, 1845: pl. 14, figs 4,4a;

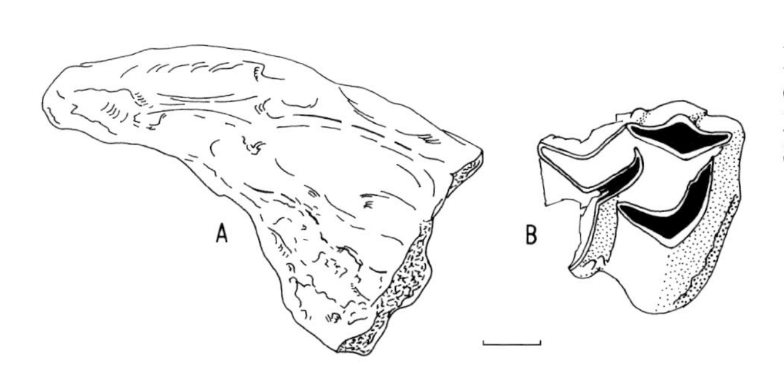

Figure 22.2. ?*Bramatherium* sp. *A*, Piece of horn core, AUH 438. *B*, Partial right upper molar, AUH 204; occlusal view, anterior side to the right. Scale = 20 mm.

Matthew, 1929: fig. 53; Colbert, 1935: fig. 177; also see Matthew, 1929: 550 and Lydekker, 1885: 69 for status of type). The agreement is manifest in size, crown height, and a well-demarcated labial rib on the paracone. Also the M^3 of the *B. perimense* has the posterior crest of the protocone directed posteriorly rather than buccalwards and this agrees with the Abu Dhabi tooth. It is unlikely, however, that these characters would be sufficient to differentiate *B. perimense* from other late Miocene sivatheres and therefore to tie the Abu Dhabi species to India rather than to other zoogeographical regions. *Helladotherium duvernoyi* might be slightly higher crowned (BMNH M4065, cast), but this is uncertain. Teeth of the early African sivathere, *Sivatherium hendeyi* Harris, 1976 from Langebaanweg are larger than AUH 204 (Harris, 1976: table 3), and so is the lower molar of a primitive sivathere from an early Pliocene level of the Wembere–Manonga Formation (Gentry, 1997). The late Pliocene *Sivatherium giganteum* has much larger upper molars with stronger labial ribs between the styles.

The dP_4 AUH 372 could fit a sivatheriine by size—for example, the *Helladotherium duvernoyi,* BMNH M11496. The Spanish *Birgerbohlinia schaubi* of Crusafont Pairo (1952: pls 38 and 41) and a Siwaliks fossil, BMNH M13654, labelled as *Hydaspitherium megacephalum,* both match by size and symmetrically pointed labial lobes but have basal pillars. *Samotherium boissieri* has smaller dP_4s.

The thoracic vertebra AUH 223 is less robust than a *Sivatherium giganteum* thoracic vertebra, BMNH 15289H, from the Siwaliks. Both this and

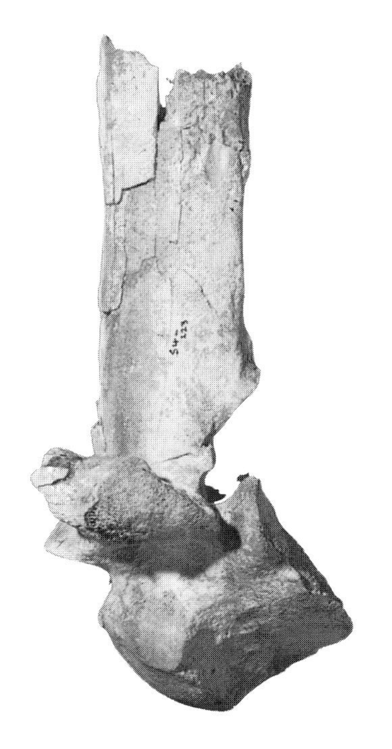

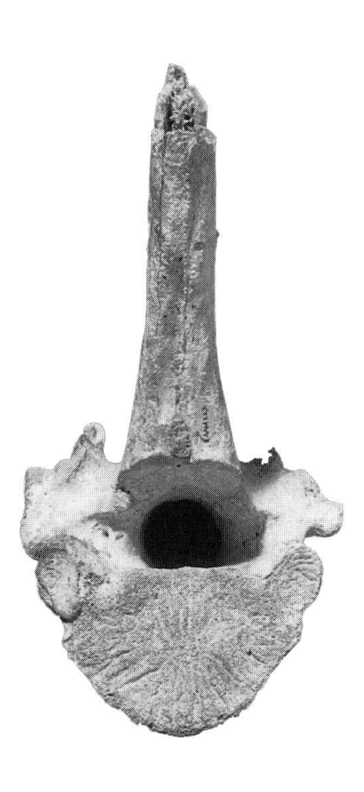

Figure 22.3. ?*Bramatherium* sp., thoracic vertebra, AUH 223; lateral and posterior views. Scale bar = 20 cm.

the other vertebral fragment, AUH 225, from Shuwaihat are bigger than *Samotherium boissieri*. They could belong to a different species from the Ras Dubay'ah horn and dental remains.

Genus indet.
Giraffidae, sp. indet.

Material

AUH 1, Hamra, H1: right occipital condyle.
AUH 211, Ras Dubay'ah, R2: fragments of left upper molars, late wear.
AUH 222, Ras Dubay'ah, R2: part of distal left fibula. Figure 22.4.
AUH 209, Ras Dubay'ah, R2: left naviculocuboid. Maximum width 69.6 mm, maximum anteroposterior diameter 70.4 mm. Figure 22.1.
AUH 358, Ras Dubay'ah, R2: partial right astragalus. Lateral height c. 91.0 mm.
AUH 190, Ras Dubay'ah, R2: much of a wind-damaged right astragalus. Lateral height c. 88.0 mm.
AUH 380, Ras Dubay'ah, R2: right ectocuneiform.
AUH 361, Ras al Aysh, A2: two pieces of a distal left humerus. Width across condyles 90.2 mm.
AUH 198, Ras Dubay'ah, R2: part of a left magnumtrapezoid.
AUH 221, Ras Dubay'ah, R2: part of a left magnumtrapezoid.

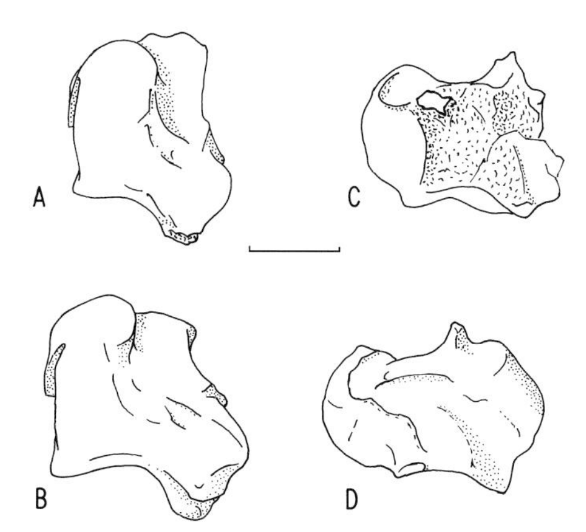

Figure 22.4. Postcranial bones of Giraffidae, sp. indet. *(A, C)* compared with those of *Samotherium boissieri* from Samos *(B, D)*. Scale = 20 mm. *A*, Left cuneiform, AUH 192; lateral view, anterior side to the left. *B*, BMNH M4296; same view as A. *C*, Part of distal left fibula, AUH 222; medial view, anterior side to the right. *D*, Complete distal left fibula BMNH M4286; same view as C.

AUH 192, Ras Dubay'ah, R2: left cuneiform. Figure 22.4.
AUH 387, Ras Dubay'ah, R2: much of the shaft and distal end of a metacarpal.
AUH 432, Ras Dubay'ah, R2: almost complete but fragmented metacarpal with partial distal end. Maximum length c. 350.0 mm, mid-shaft transverse diameter 42.0 mm, width across the distal condyles 78.4 mm.
AUH 61, Shuwaihat, S1: part of the front of an axis vertebra. Width across the atlantoid facet 81.2 mm.
AUH 65 and 78, Shuwaihat, S1: dorsal arches of lumbar vertebrae.
AUH 104, Shuwaihat, S1: centrum and much of the neural arch of a lumbar vertebra.
AUH 189, Ras Dubay'ah, R2: centrum of a lumbar vertebra.
AUH 250, Shuwaihat, S4: most of a centrum of a lumbar vertebra.

Description

A giraffid species, smaller than most sivatheres, is represented in the Baynunah Formation, mainly by postcranial bones.

The rear part of the distal fibula AUH 222 is less bulbous and the top facet is less downturned posteriorly than in three of four Samos examples (BMNH M4286) of *Samotherium boissieri* Major, 1888. *Okapia* is like AUH 222 in that the rear of its top articular facet is little turned down. The distal fibula of modern *Giraffa* is shaped posteriorly more like *Samotherium boissieri* and in side view it looks lower overall than AUH 222. The distal surface of AUH 222 has two components at an angle to one another, and is not straight; in this it is like *Bohlinia attica* (Gaudry and Lartet, 1856), as noted by Geraads (1979: 380), and unlike modern *Giraffa*.

By their size the astragali AUH 190 and AUH 358 agree with *Samotherium boissieri*, but both look slightly more gracile and less squat than in that species.

The naviculocuboid AUH 209 is nearly as big as in *Samotherium boissieri* but slightly less robust. It is considerably bigger (around 20% in linear dimensions) than the middle Miocene *Injanatherium arabicum* Morales, Soria, and Thomas (1987: 453, pl. 4, fig. 6) from the Hofuf Formation, but would probably match the size of the late Miocene *I. hazimi* Heintz, Brunet, and Sen, 1981 of Iraq. It also matches the size of a *Palaeotragus coelophrys* naviculocuboid from Sebastopol (Borissiak, 1914: 121, pl. 3, fig. 7). The facet for the entocuneiform (behind the facet for the ectocuneiform) appears to be smaller than in *Samotherium boissieri* at Samos and much smaller than on Borissiak's example of *Palaeotragus coelophrys*. It is possible, however, that part of its surface has been removed by postmortem erosion. The absence of a rear metatarsal facet is like most *S. boissieri* and unlike bovids such as the large boselaphine *Tragoportax acrae* (Gentry, 1974: 170, fig. 22—metatarsal) of Langebaanweg. Harris (1976: 342, fig. 6B) claims that giraffine naviculocuboids at Langebaanweg have an additional transverse posterior articulation for the metatarsal, but it would be necessary to recheck that these bones could not belong to the early bovine *Simatherium demissum* Gentry, 1980, also present there. (Naviculocuboids of large bovids differ from those of giraffids by being wider across the level of the back of the main metatarsal and ectocuneiform facets.) There is a deep groove for the peronaeus longus tendon behind the front metatarsal facet, as was also noted by Geraads (1979: 381) on a naviculocuboid of *Bohlinia attica*. Geraads (1993: 169) noted the groove as variable in Turkish *Samotherium*.

The extant *Okapia johnstoni* has ectocuneiform and entocuneiform bones fused to the naviculocuboid. Its rear metatarsal facet is variable in size among a sample of three and in whether or not it is joined with the front metatarsal facet. The rear facet is never transversely long and there is no peronaeus groove between the two. A specimen of *Giraffa camelopardalis* has both metatarsal facets joined, as also observed in this species by Geraads (1979: 381). One of six *S. boissieri* specimens from Samos, BMNH M4284B, shows the same condition.

On the cuneiform, AUH 192, the back part of the radius facet is less excavated than in *Okapia* and possibly than in *Samotherium boissieri* from Samos. This facet has a more restricted area than in *Giraffa camelopardalis*. Its pisiform facet is less wide than in *G. camelopardalis*.

The distal metacarpal AUH 387 is about the size of *S. boissieri* and has a strongly hollowed posterior surface as in giraffids generally. Its length was probably no greater than in *S. boissieri*. The much-damaged metacarpal AUH 432 is also about the same size and length as in *S. boissieri* and thus not from a species with metapodials as long as in *Bohlinia* or *Giraffa*.

The partial axis vertebra AUH 61 is the size of this species. It is from a pecoran rather than an equid because the atlantoid facet is complete from one side to the other and not interrupted ventrally, and the odontoid process is too concave dorsally.

The vertebra AUH 104 has no articular facets for ribs and the hollowing for its zygapophyseal articulations is rather deep, so it must be a lumbar rather than a thoracic. It is pecoran and by size must be from a giraffid. Its centrum is larger than in the partial lumbar vertebra AUH 289, assigned above to *Palaeotragus* sp. The strong longitudinal ridge ventrally on the centrum is unlike a hippopotamus, while compared with an equid it is too big and the anterior surface of the centrum insufficiently convex. The more-fragmentary lumbar centrum AUH 250 has a wide and low rear articular surface and would therefore have been not far forward of the sacrum.

The zygapophyses on the two vertebral dorsal arches AUH 65 and 78 are too intricately curved to be equid. Both *Hippopotamus* and *Hexaprotodon* (= *Choeropsis*) can have complexly curved zygapophyses, but AUH 65 and 78 have more the aspect of ruminants than hippopotamus. They are too long to go with the largest *Boselaphus tragocamelus*, 1974.411 in the BMNH Zoology Department or, presumably, with the boselaphine species in the Abu Dhabi fossil fauna. The lumbar centrum AUH 189 is not of an equid, because the front edge of the transverse process (preserved on the right side) is at too high a level relative to the floor of the neural arch. The Baynunah vertebrae are all bigger than modern *Okapia johnstoni*.

Discussion

It is difficult to come to a conclusion about the identity of these giraffid remains. They are probably more gracile than in *Samotherium boissieri*. Their size matches *S. boissieri* or the Iraq *Injanatherium hazimi*, but they could equally belong to some Siwaliks species with metapodials less lengthened than in *Bohlinia* or *Giraffa*. As noted above it has yet to be established that the Siwaliks "*Giraffa*" *punjabiensis* really does have much-lengthened metapodials.

It would be economical with concepts to make the giraffid postcranials at Ras Dubay'ah conspecific with giraffid horn cores and teeth at the same locality. Because most of the latter were assigned to ?*Bramatherium* sp., it is possible either that this ?*Bramatherium* sp. is a modestly sized sivatheriine like some of those in the Siwaliks late Miocene, or that it is not a sivathere after all. In the latter case the larger vertebrae AUH 223 and 225 at Shuwaihat could still represent a sivathere in the Baynunah Formation.

Family Bovidae Gray, 1821
Tribe Boselaphini Knottnerus-Meyer, 1907

Remarks

Boselaphines are known sporadically from the middle Miocene and abundantly from the late Miocene. They can be split into two generic or suprageneric groups (Moyà-Solà, 1983: 198, figs 59, 60):

1. The middle Miocene *Austroportax latifrons* (Sickenberg, 1929) of Europe and later allied forms like *Pachyportax latidens* (Lydekker, 1876a) and *Selenoportax vexillarius* Pilgrim, 1937 that can be associated with modern *Boselaphus* and the bovine *Bubalus*.
2. A group, centring on *Miotragocerus* Stromer, 1928 and *Tragoportax* Pilgrim, 1937, that also appears in the middle Miocene but is best known from the Turolian wherein it is, alongside *Gazella*, the most widespread of all bovids. This second group went extinct around the end of the Miocene.

Genus *Tragoportax* Pilgrim, 1937
Type species *Tragoportax salmontanus* Pilgrim, 1937
Tragoportax cyrenaicus (Thomas, 1979)

Material

AUH 442, Tarif, T2: cranium with horn cores. Figures 22.5–22.8. (The measurements are given later.)

AUH 371, Shuwaihat, S4: two horn cores and pieces of an atlas vertebra. Anteroposterior and mediolateral basal diameters of right horn core = 58.3 × 35.6 mm.
AUH 447, Jebel Barakah, B2: base of a left horn core. Anteroposterior and mediolateral basal diameters = 54.5 × 31.8 mm.
AUH 239, Thumayriyah, TH1: right mandible, P_2–M_3 in late middle wear. (The measurements are given later.) Figure 22.9.
AUH 153, Hamra, H1: part of a vertical ramus, left mandible.
AUH 26, Shuwaihat, S1: part of the labial wall of an upper molar.
AUH 27, Shuwaihat, S1: two pieces of the labial wall of upper molars.
AUH 278, Hamra, H5: fragments of a lower molar.

The remaining fossils in this list are of appropriate size to fit *Tragoportax cyrenaicus* and could well be conspecific.

AUH 183, Jebel Dhanna, JD5: right petrosal. Figure 22.10.
AUH 333, Kihal, K1: part of a left ilium and acetabulum.
AUH 255, Hamra, H6: most of a left acetabulum.
AUH 160, Shuwaihat, S1: right calcaneum. Maximum length 115.1 mm.
AUH 181, Jebel Dhanna, JD5: left calcaneum (with "191" on bone).
AUH 352, Ras al Aysh, A2: left astragalus. Lateral height 60.9 mm, medial height 56.6 mm, distal width 36.2 mm.
AUH 69, Shuwaihat, S1: anterior part of the distal condyles of a left humerus. Width across the condyles 57.6 mm.
AUH 346, Hamra, H1: proximal left radius. Maximum width 62.4 mm, width across the articular surface 59.2 mm. Figure 22.11.
AUH 254, Hamra, H6: left proximal radius. Maximum width c. 64.2 mm, width across the articular surface c. 61.2 mm. Figure 22.11.
AUH 70, Shuwaihat, S1: back part of a left proximal radius. Maximum width 58.6 mm, width across the articular surface 55.8 mm.
AUH 102, Shuwaihat, S1: right proximal radius. Maximum width 59.0 mm, width across the articular surface 58.0 mm. Figure 22.11.
AUH 180, Jebel Dhanna, JD5: lateral part of a left proximal radius.
AUH 232, Hamra, H5: most of a right proximal radius.
AUH 231, Hamra, H5: distal right radius, presumably same bone as AUH 232. Width across the articular surfaces 56.5 mm.
AUH 91, Shuwaihat, S1: right scaphoid.
AUH 169, Hamra, H1: left scaphoid.
AUH 171, Hamra, H1: first phalanx. Maximum length 60.0 mm, proximal width 26.9 mm.
AUH 47, Hamra, H5: partial cervical vertebra.

Description

The fine cranium with practically complete horn cores, AUH 442, is one of the best-preserved specimens of this widespread and common late Miocene genus. It is evidently from a large and advanced species and shows resemblances to *Tragoportax cyrenaicus* from Sahabi.

Its main features are as follows. It is a large *Tragoportax*. The horn cores are long, compressed mediolaterally, the maximum width of cross-section lying rather posteriorly, the lateral surface flat, with an anterior keel, a slightly less-developed posterolateral keel and, from the mid-point upwards, a posteromedial keel. The anterior keel ends about 10.0 cm below the tips, above which the cross-section is much smaller in diameter and no longer mediolaterally compressed. This change in cross-section is often called a "distal demarcation". The horn cores are inserted fairly uprightly in side view and quite close together, strongly diverging at the base, but this divergence diminishing distally, with some backward curvature and a slight torsion, which is anticlockwise from the base up on the right side, There is a shallow postcornual fossa. The frontals rise gradually to a high transverse ridge between the two horn insertions and there is a system of internal sinuses on the broken bone adjacent to the orbits. The temporal ridges are strong and the bone surface between them is rugose. The roof of the braincase is only very slightly inclined and it does not curve downwards posteriorly, the braincase sides are parallel, the mastoid exposure is large and faces posteriorly on the occipital surface, the occipital surface has a horizontal top edge, the basioccipital narrows anteriorly, and the anterior tuberosities are not prominent.

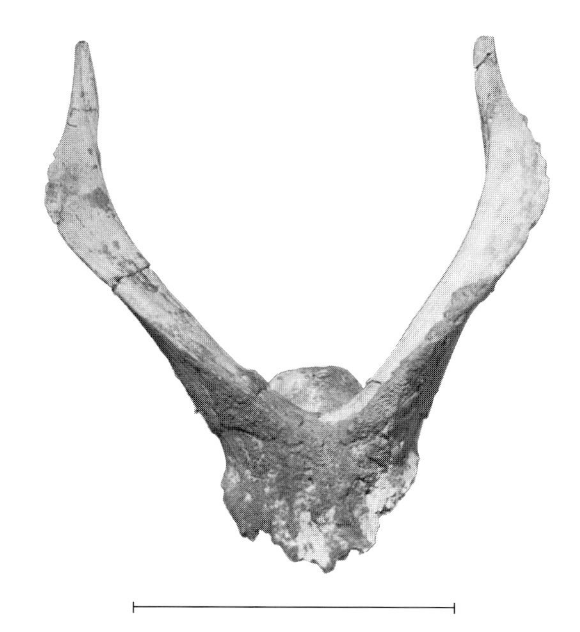

Figure 22.5. *Tragoportax cyrenaicus*, cranium AUH 442; anterodorsal view. Scale bar = 20 cm.

Figure 22.6. *Tragoportax cyrenaicus,* cranium AUH 442; posterior view. Scale bar = 20 cm.

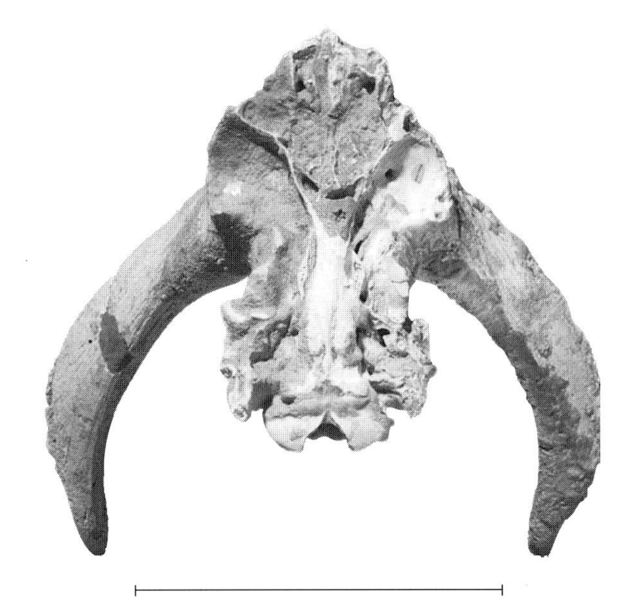

Figure 22.7. *Tragoportax cyrenaicus,* cranium AUH 442; ventral view. Scale bar = 20 cm.

Measurements (in mm) taken on the cranium were:

Length of horn core (right) along anterior edge	382.0
Anteroposterior diameter of horn core at its base	78.0
Mediolateral diameter of horn core at its base	42.0
Minimum width across lateral surfaces of horn pedicels	127.0
Skull width across posterior side of orbits	c. 138.0
Width across lateral edge of supraorbital pits	56.7
Length from back of frontals to top of occipital	c. 83.7
Width across braincase	83.8
Skull width across mastoids immediately behind external auditory meatus	112.7
Occipital height from top of foramen magnum to occipital crest	49.3
Minimum width across temporal lines on cranial roof	31.4
Width across anterior tuberosities of basioccipital	26.8
Width across posterior tuberosities of basioccipital	45.0

The right horn core of AUH 371 is less compressed and slightly smaller than the horn cores on the cranium. The level of maximum transverse thickness is less posterior than in other horn cores. It looks as if the transverse ridge of the frontals between the horn bases would have been high as in the cranium AUH 442. The left horn core is preserved for nearly all its length; it shows lessening divergence distally but almost no backward curvature. There is postmortem surface damage to the base posterolaterally. Its length when complete might have been around 200 mm, and from the base to the demarcation of the cross-section is 140 mm; hence the whole horn core was considerably shorter than on the cranium. It was also not lyrated.

The horn core AUH 447 is again smaller than those on the cranium. It is preserved for about 10.0 cm above the base. The back edge is straight or only slightly curved. Sinuses reach up into the base of the horn core proper.

The mandible AUH 239 has occlusal length and width measurements of the cheek teeth as follows: P_2 14.4 × 8.1 mm; P_3

Figure 22.8. *Tragoportax cyrenaicus,* cranium AUH 442; left lateral view. Scale bar = 20 cm.

17.4 × 10.3 mm; P_4 16.7 × 9.4 mm; P_{2-4} 46.9 mm; M_1 19.9 × 12.0 mm; M_2 22.2 × 13.3 mm; M_3 30.3 × 8.1 mm; M_3 height c. 31.0 mm: M_{1-3} c. 71.2 mm. P_{2-4} as a percentage of M_{1-3} is 66%. The M_1 is in late wear, the M_2 in late middle wear, and the M_3 in middle wear. The P_2 is relatively long; the P_3 metaconid is fairly well directed backwards; the P_4 has a large paraconid flange well separated from the parastylid, and with a metaconid having forwardly and backwardly directed flanges. There is a large basal pillar on M_1 and smaller ones on M_2 and M_3. The M_3 has an anterior transverse (goat) fold that looks broken but was probably once complete nearer the occlusal surface.

Other fossils described below are assigned to *Tragoportax cyrenaicus* mainly on the basis of agreement in size, but some, such as the proximal radii, do match other fossil Boselaphini in certain morphological details.

The petrosal AUH 183 is unlike the petrosal of pigs or hippopotamuses and is probably from a ruminant. It is much smaller than the petrosal of *Samotherium boissieri* visible in the sectioned skull BMNH ?M4216. It is smaller than extant *Boselaphus tragocamelus* and larger than extant *Tetracerus quadricornis.* By size it agrees with a Spanish early late Miocene petrosal attributed to the boselaphine *Protragocerus* aff. *chantrei* by Moyà-Solà (1983: 53, pl. 3, fig. 2), although the morphology looks somewhat different. The promontorium is flatter (less swollen) in tympanic view than in extant *Ammotragus lervia* BMNH Zoology 14.3.22.1. The fossa for the musculus tensor tympani and the fossa anterior to it are shallower and shorter. The same differences can be seen in comparison with *Ovis "musimon" cycloceros* 11.2.24.5 from Persia and *O. m. gmelini* 55.12.26.156, both female skulls. The fossa musculus tensor tympani and the adjacent fossa are deeper than in a *Tetracerus quadricornis* 884a. In endocranial aspect the petrosal AUH 183 is quite like *T. quadricornis.* It is quite similar to an immature *Tragelaphus strepsiceros* 28.11.11.21 in the depth of the fossa musculus tensor tympani and the pattern of ridges in this area of surface. The overall shape is unlike *Damaliscus lunatus* 42.4.11.5 in endocranial view and unlike *D. lunatus* 75.1164 in tympanic view. So few petrosals are visible on extant specimens in museum collections that it is difficult to be sure about individual variation. The Baynunah Formation petrosal appears more like *T. strepsiceros* 28.11.11.21 than *D. lunatus* 24.8.3.14 in shape in tympanic view and in the faint groove passing dorsoposteriorly to anteroventrally over the arched surface anterior to the fenestra rotundum and in other more subtle features.

The pelvis portion AUH 333 is from an animal the size of living *Boselaphus tragocamelus.* The hollow for the medial origin of the rectus femoris just in front of the acetabulum is shallow and more definitely on the medial side of the ilium stem than on a narrow ventral surface.

The calcanea AUH 160 and 181 would fit *Tragoportax cyrenaicus* by size. The astragalus AUH 352 is the size of modern *Boselaphus tragocamelus* but relatively narrower. It is possible that the lateral edge is straighter in anterior view through being less undercut below the lateral condylar rim on the proximal part of the bone.

The partial humerus AUH 69 is the right size for this species.

The radii are too small to fit the giraffid *Palaeotragus* sp. Their morphology agrees with that of *Tragoportax acrae* (Gentry,

Figure 22.9. *Tragoportax cyrenaicus,* right mandible AUH 239; medial view. Scale bar = 2 cm.

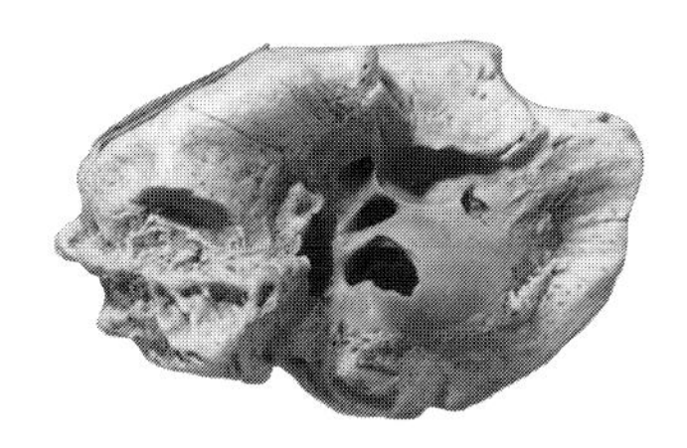

Figure 22.10. *Tragoportax cyrenaicus*, right petrosal AUH 183; tympanic (above) and endocranial (below) views. Scale bar = 5.3 cm.

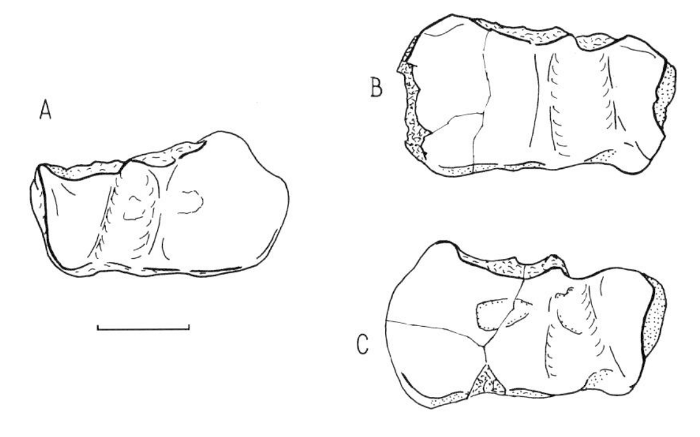

Figure 22.11. *Tragoportax cyrenaicus*, proximal articular surfaces of radii: *A*, right AUH 102; *B*, left AUH 254; *C*, left AUH 346. Anterior sides towards the base of drawing. Scale = 20 mm.

1974: 166, fig. 19), in the lateral facet pointed anteriorly, the back of the lateral facet set well back, and the proximal lateral tubercle of small size and set low. The posteromedial part of the medial facet, however, projects more strongly than in *T. acrae*. The lateral facet of AUH 102 is drawn out posterolaterally, which gives it a diagonal posterior edge, somewhat like *Samotherium boissieri* from Samos. The nonprojection of the posterolateral part of the medial facet in AUH 254 and 102 looks different from *T. acrae* (Gentry, 1974: fig. 19g), but AUH 70 and 346 resemble *T. acrae*. The deep groove between the lateral and medial facets is better marked than in modern *Boselaphus tragocamelus*, so, too, is the anterior point of the lateral facet, the posterior position of the back edge of the lateral facet, and the posterior projection of the posteromedial part of the medial facet. The great width between the ridges on the anterior surface of the distal end AUH 231 and its wide cuneiform facet agree well with *T. acrae*.

The scaphoid AUH 91 is transversely narrow like the African greater kudu *Tragelaphus strepsiceros* but less deep. It is also more indented in dorsal profile than the greater kudu. There is more of a posterior prominence on its medial side than in *Tragoportax acrae* (Gentry, 1974: fig. 20b) but less than in modern *Boselaphus tragocamelus*. Another scaphoid, AUH 169, is transversely wider and much more like modern *B. tragocamelus*. It is also more deeply indented in dorsal profile and has a "normal" posterior prominence medially unlike AUH 91.

The phalanx AUH 171 is more robust and less gracile than in modern *Boselaphus tragocamelus*. It is close in size to a first phalanx of *T. acrae* (Gentry, 1974: fig. 23) but relatively wider.

The cervical vertebra AUH 47 looks as if it would be the fifth in the series. It is rather short front-to-back for this position, but if it were a sixth cervical vertebra then the bridge of bone over the vertebrarterial canal would be too wide. There is no sharp central ridge ventrally behind the front centrum.

Discussion

The more primitive species of *Miotragocerus* and *Tragoportax* occur earlier and overlap with the more advanced ones. They have horn cores that are short, straight, compressed mediolaterally, and without much emphasis on the posteromedial corner or keel. *Miotragocerus monacensis* Stromer, 1928 and *M. pannoniae* (Kretzoi, 1941) from Europe, *M. gradiens* (Pilgrim, 1937) from the Siwaliks, *Tragoportax leskewitschi* (Borissiak, 1914) from Sebastopol, and perhaps *T. pilgrimi* (Kretzoi, 1941) from the Siwaliks are all more primitive than Turolian *T. amalthea* (Roth and Wagner, 1854), *T. gaudryi* (Kretzoi, 1941), and *T. rugosifrons* (Schlosser, 1904) from the Graeco-Iranian Province, and also more primitive than *T. salmontanus* Pilgrim, 1937 and *T. browni* (Pilgrim, 1937) from Turolian-equivalent levels in the Siwaliks. Horn cores of advanced species show backward curvature, enhanced flattening of their lateral surfaces, presence of or approach to a posteromedial corner, and the level of maximum transverse width lying more posterior than centrally. Most, but not all, advanced species also show longer horn cores and increased height for the transverse ridge of the frontals between the horn core bases. This last character is absent in the other group of Boselaphini that survives until the present day. It needs to be remembered that the evolution of these character states is more or less continuous and subject to variation (e.g. Moyà-Solà, 1983: pl. 16, figs1 and 2) and parallelism. Hence the differences between species are less clear-cut than the diagram seems to suggest.

Moyà-Solà's (1983) delimitation and analysis of the two species *T. amalthea* and *T. gaudryi* at Pikermi can be agreed and welcomed. *Tragoportax gaudryi* has shorter horn cores in which divergence may increase distally. In *T. amalthea* the horn cores are longer, with a slight torsion and consequent decrease of divergence distally. The lateral surface may be flatter than in *T. gaudryi* and the frontals are more raised between the horn core bases. According to Bouvrain (1994: 18), *T. gaudryi* occurs at localities wherein Cervidae and sometimes Tragulidae also occur and hence it indicates a more wooded habitat.

Tragoportax rugosifrons has long horn cores, sometimes curving backwards and sometimes diverging increasingly towards the tips. The transverse ridge between the horn cores has a more steeply inclined posterior surface than in *T. amalthea*. The horn cores are usually without a sharp diminution in anteroposterior diameter near their tips, and Bouvrain and Bonis (1984) noted that the anterior keel terminates some distance below the tip. Demarcations in *Tragoportax* are often more pronounced in species where the tips curve upwards. Bouvrain (1988, 1994) contrasted this species' non-survival into the later Turolian with *T. gaudryi*, but Solounias (1981b: fig. 30H) thought it was present in later MN 12. I have been content in common with other authors (e.g. Bouvrain, 1988: 56) to synonymise the more completely known *T. curvicornis* and *T. recticornis* (both of Andrée, 1926) with *T. rugosifrons*. The lectotype skull (Schlosser, 1904: pl. 12, fig. 6) of *T. rugosifrons* is, however, not beyond the range of variation of *Tragoportax amalthea* in the inclination of the posterior surface of its intercornual ridge and in the probable divergence of the horn cores.

The morphology of the Abu Dhabi *Tragoportax* befits a Turolian rather than a Vallesian species, and, moreover, one that is generally advanced on *T. amalthea* and *T. rugosifrons*. Compared with *T. amalthea* its horn cores are definitely more divergent; they are also longer and hence can show more torsion and slightly stronger backward curvature. The skull is wider and the top edge of the occipital surface is flat instead of rounded. Differences from *T. rugosifrons* are horn cores with a larger anteroposterior diameter basally, a distal demarcation, a divergence that diminishes distally, and stronger torsion. The back of the transverse ridge between the horn bases rises less abruptly from the braincase roof. There is a flatter top edge of the occipital and the nuchal crests are not concave upwards (see Bouvrain, 1994: pl. 1, fig. 1b).

No Siwaliks *Tragoportax* is as large as the Abu Dhabi species. From the illustrations of Thomas (1984) it looks as if the divergence of the horn cores of *T. pilgrimi* in the Nagri Formation is less than at Abu Dhabi, the anterior part of their lateral surface is still primitively curved round as it approaches the anterior keel and less flat, and the frontals are less elevated between the horn bases. *Tragoportax salmontanus* is also small and has rather short horns, but it does have a relatively wide cranium, diminishing distal divergence of the horn cores, and a higher level of the frontals between the horn core bases as resemblances to the Abu Dhabi species. *Tragoportax punjabicus* (including *browni*) may be slightly larger than the two preceding Siwaliks species and has longer horns, but compared with the Abu Dhabi *Tragoportax* it has a narrower cranium, less divergence of its horn cores, and no distal demarcation on the horn cores.

The holotype of *Tragoportax perimensis* (Lydekker, 1878) from Piram Island is shown in Lydekker (1878: pl. 28, figs 4 and 5) and there is a right horn core in the London collections,

BMNH 18786. The species is further discussed by Pilgrim (1939: 222), who testifies to the short horn cores and their wide basal separation. This antelope may be more akin to *T. salmontanus* and *T. aiyengari* Pilgrim, 1939 than to other *Tragoportax*. The divergence of the horn cores is much less than at Abu Dhabi, and they are less mediolaterally compressed according to measurements given by Pilgrim. Overall it is smaller in size.

The Abu Dhabi species is most similar to two end-Miocene, extra-Eurasian species described from Langebaanweg and Sahabi, which are large, have wide skulls, a high level of the frontals between the horn bases, and divergent horn cores. The Langebaanweg *T. acrae* is also known to have a horizontal top edge to its occipital (Gentry, 1980: 229, fig. 6). It differs from the Abu Dhabi species in its horn cores being very short, less mediolaterally compressed, and with a longer distal end of small diameter above the top of the anterior keel. They are not curved backwards and the lateral surface is less flattened. The frontals are raised still more strongly between the horn bases. The lower dentitions of *T. acrae* are about 10% longer in linear dimensions (Gentry, 1974: table 1) than those of the mandible AUH 239. The premolar row may be relatively longer, and the cause of this could be disproportionate shortening of the P_4 in the Abu Dhabi species or lengthening of the same tooth in *T. acrae.*

Not a lot is known of the Sahabi *Tragoportax cyrenaicus,* but it is substantially different from *T. acrae* as a species. Measurements of the horn core basal diameters and width apart of the horn cores suggest that it is smaller than *T. acrae* (a view shared by Lehmann and Thomas, 1987). Measurements of mandibular teeth given by Lehmann and Thomas are, however, very close to those on the Abu Dhabi mandible AUH 239. A boselaphine M_3 cast BMNH M8198 from Wadi Natrun (Gentry, 1978: 547), also has a goat fold and is about the size of the M_3 of AUH 239. The Sahabi holotype frontlet of *T. cyrenaicus* has no visible sign of a distal demarcation (Thomas, 1979: pl. 1, fig. 5), and a second Sahabi horn core figured by Harris (1987: figs 1 and 2; also see Geraads, 1989b: 786 for its identity as *T. cyrenaicus*) is too badly preserved for the state of this character to be ascertained. No other difference of the Abu Dhabi specimens from *T. cyrenaicus* can be found, and the Abu Dhabi material will accordingly be referred to this species.

On figure 22.12 *Tragoportax cyrenaicus* and *T. acrae* are shown as advanced on *T. amalthea,* but their actual relationships may lie with Siwaliks species of *Tragoportax,* in which case *T. salmontanus* of the Dhok Pathan with its raised frontals between the horn core bases and diminishing distal divergence could be their more primitive sister species (fig. 22.13). Whatever the phyletic truth may be, it is clear that the Abu Dhabi and Sahabi *Tragoportax* is morphologically advanced, even among Turolian species, and is presumably a late representative of its genus.

Genus *Pachyportax* Pilgrim, 1937
Type species *Pachyportax latidens*
(Lydekker, 1876a)

Remarks

This genus was founded for large boselaphines in the Siwaliks, probably closer to the ancestry of Bovini than is *Tragoportax.* The type species comes from the late Miocene, possibly extending into the Pliocene. Its type specimen is an upper molar, but it is best represented by a cranium of *Pachyportax latidens dhokpathanensis* Pilgrim, 1939 in Calcutta of which there is a cast in London, BMNH M26573. A second species—*P. nagrii* Pilgrim, 1939—occurs earlier in the Siwaliks but has only a hornless female cranium as the holotype and is not adequately differentiated from *Selenoportax vexillarius.* A very similar boselaphine to *Pachyportax* is *Parabos* Arambourg and Piveteau, 1929, of which the type and only species, *P. cordieri* (Christol, 1832), comes from Montpellier, southern France at the start of the Pliocene. Following Gromolard's (1980) revision of *Parabos,* it may be that this generic name will be the senior synonym for *Pachyportax.*

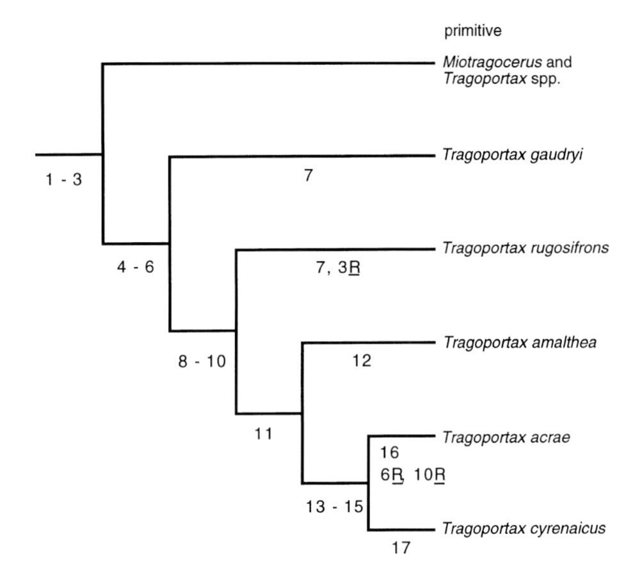

Figure 22.12. Relationships of some of the species of extinct boselaphines related to *Tragoportax cyrenaicus,* mainly according to horn-core characters. The diagram reflects a consensus of recent views rather than the result of a comprehensive analysis and shows how *T. cyrenaicus* and *T. acrae* are advanced on *T. amalthea* and *T. rugosifrons*. *R* = reversals. Numbers indicate the presence of the following advanced character states: 1, anterior and posterolateral keels on horn cores; 2, mediolateral compression; 3, distal demarcation; 4, posteromedial corner or keel; 5, maximum transverse width lying posteriorly rather than centrally; 6, horn cores curved backwards; 7, divergence increases distally; 8, well-raised transverse ridge between horn cores; 9, lateral surfaces flattened with less emphasised curving round anteriorly; 10, horn cores longer; 11, torsion present; 12, stronger mediolateral compression; 13, larger size; 14, increased divergence of horn cores; 15, flat top edge of occipital surface; 16, frontals very strongly raised between horn-core bases; 17, stronger torsion.

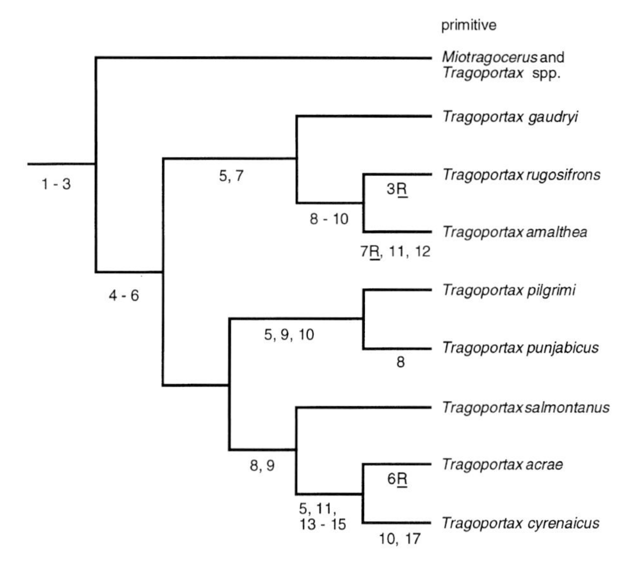

Figure 22.13. An alternative phyletic scheme for extinct boselaphines related to *Tragoportax cyrenaicus,* involving Siwaliks species. This scheme is not the most economical possible and the many parallelisms are indicated. Numbering of character states as on figure 22.12.

Pachyportax latidens (Lydekker, 1876a)

Material

AUH 106, Shuwaihat, S1: partial and much damaged base of a left horn core. Figure 22.14.

AUH 266, Hamra, H1: right mandible with M_{2+3} in middle wear. (The measurements are given later.) Figure 22.15.

AUH 25, Shuwaihat, S1: posterior lingual wall of a left M_3.

AUH 345, Shuwaihat, S4: distal right tibia. Maximum distal width 56.6 mm, maximum anteroposterior width 44.6 mm.

AUH 460, Ras Dubay'ah, R2: part of a right naviculocuboid.

AUH 249, Shuwaihat, S4: distal metatarsal. Width across the condyles 52.7 mm. Maximum anteroposterior diameter of condyles 33.4 mm. Figure 22.16.

Description

On the horn core AUH 106 the frontals between the horn core bases do not attain a high level relative to the dorsal orbital rim. It is possible that there was a temporal ridge running up to the posteromedial side of the horn pedicel.

The two molars on AUH 266 show rugose enamel, basal pillars, outbowed lingual walls, and the M_3 hypoconulid has a small vertical rib on its lingual wall. The M_2 measures 28.1 mm occlusal length × 18.7 mm occlusal width × 13.5 mm height, and the M_3 measures 37.5 × 16.7 × 20.0 mm. The occlusal lengths of P_{2-4} and M_{1-3} would have been c. 54.5 and 84.3 mm, giving a ratio of premolar row to molar row of 65%. The mandibular depth below M_1 = c. 43.5 mm, and below P_2 = c. 31.6 mm. On AUH 25 the entostylid constitutes a more prominent corner immediately in front of the lingual wall of the hypoconulid, and the lingual wall of the hypoconulid lacks a localised vertical rib.

The distal tibia AUH 345 may be too large to fit *Tragoportax cyrenaicus*. It has a definite anterior facet for the distal fibula and is hence unlike giraffids. It is the size of a large *Boselaphus tragocamelus* but the twin facets for the astragalus are wider and more shallow and there is less of a central notch in the posterior edge of the astragalus facets.

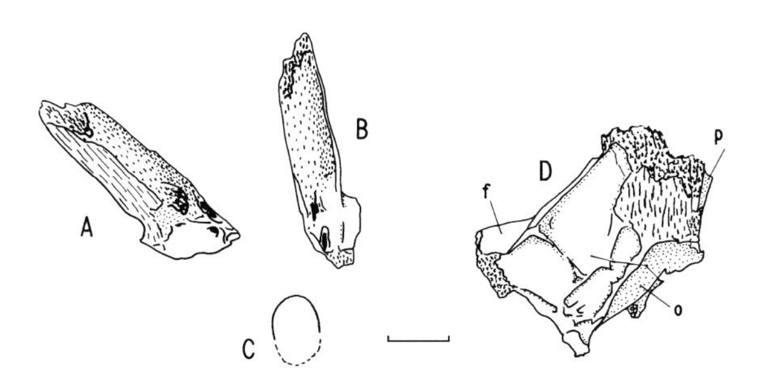

Figure 22.14. *A, B,* Right horn-core base of *Gazella* aff. *lydekkeri,* AUH 389; lateral and anterior views. *C,* Cross-section of AUH 389 just above its base, lateral side to the left, anterior to the foot of the page. *D,* Left horn-core base of *Pachyportax latidens,* AUH 106; anterior view. Key: f, dorsal surface of frontal; p, side surface of pedicel; o, surface of orbit. Scale = 20 mm.

The preservation of the anterior surface on the naviculocuboid AUH 460 leaves open the possibility that it is really a navicular bone and thus not the united naviculocuboid of a ruminant. If it were a ruminant it would fit this species by size.

On the distal metatarsal AUH 249 the flanges on either side of the ventral end of the central vertical gully on the anterior surface have been destroyed. The bone agrees in size with *Palaeotragus roueni,* but the shaft narrows as it rises above the distal end. The distal end is wider than in living *Boselaphus tragocamelus.* Its maximum width across the condyles is about 15% greater than in *Tragoportax acrae* (Gentry, 1974: fig. 21).

Discussion

The low level of the frontals relative to the dorsal orbital rim on AUH 106 is unlike post-Vallesian *Tragoportax* and agrees with the Siwaliks genera *Selenoportax* and *Pachyportax.* The horn core also matches these genera in its probable large size and wide insertion. It agrees with *Pachyportax latidens* rather than *Selenoportax vexillarius* in a very low inclination of the horn core in side view, and less divergence of the horn cores. It could also be

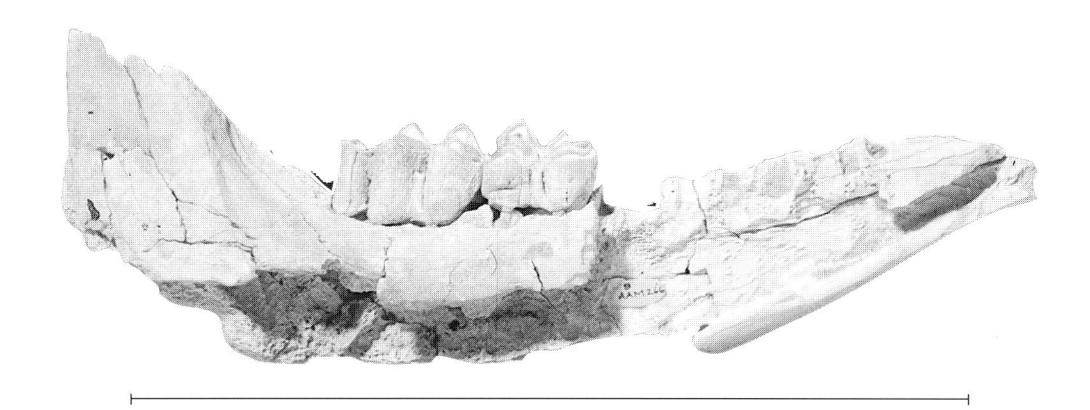

Figure 22.15. *Pachyportax latidens,* right mandible AUH 266; lateral view. Scale bar = 20 cm.

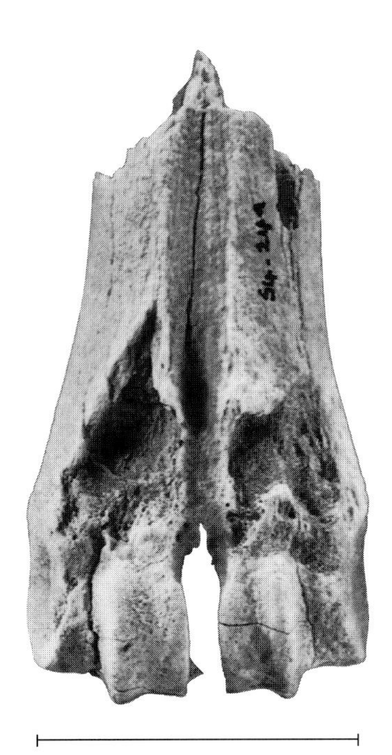

Figure 22.16. *Pachyportax latidens,* distal metatarsal AUH 249; anterior view. Scale bar = 5 cm.

similar to the left horn core of a large boselaphine or bovine from Piram Island, BMNH M2402a, referred by Pilgrim (1937: 746; 1939: 175) to *Selenoportax lydekkeri* (Pilgrim, 1910). The horn cores of *Parabos cordieri* from the start of the French Pliocene (Gromolard, 1980) were similar to those of *Pachyportax.*

The mandible differs from a giraffid mandible like the Pikermi *Palaeotragus roueni* BMNH M8367 in that the teeth are too hypsodont, their basal pillars are too large, and the central fossettes on the occlusal surface are isolated from the exterior even in the middle wear of M_3. It is definitely from a boselaphine bovid but too large to be accommodated in *Tragoportax cyrenaicus.* It is of appropriate size to match the holotype upper molar of *Pachyportax latidens* (BMNH M34567, cast) and is slightly larger than mandibles of *Tragoportax acrae* of Langebaanweg. *Parabos cordieri* appears to have had the lower premolar row with an occlusal length 77% that of the molar row (my own measurements of a specimen in Lyon); it was also slightly smaller than the Arabian fossil.

The mandible is too large to match the upper dentitions on two Pikermi skulls of *Palaeoryx pallasi* (Wagner, 1857) (BMNH M10831, M10832). *Pachytragus ligabue* Thomas (1983) from the Hofuf Formation is about the size of AUH 226 but has the premolar/molar row ratio at c. 55%.

Genus indet.
Bovidae, sp. indet.

Material

AUH 63, Shuwaihat, S1: medial side of right astragalus. Height 48.2 mm.
AUH 407, Harmiyah, Y1: left and right astragali. Much abraded. Lateral height of the left one 50.5 mm, medial height 47.1 mm, distal width 28.9 mm. Lateral height of the right one c. 48.0 mm, medial height c. 45.0 mm, distal width c. 28.0 mm.
AUH 391, Harmiyah, Y1: partial left naviculocuboid.

These tarsal bones are rather small to be conspecific with the other postcranial bones assigned to *Tragoportax cyrenaicus.* They have a size appropriate for a late Miocene bovid such as one of the *Pachytragus* species of Samos or perhaps a species of about the size of the Sahabi ?*Hippotragus* sp. (Lehmann and Thomas, 1987: 328).

Tribe Antilopini Gray, 1821

Remarks

Spiral-horned antilopines of a variety of genera are especially characteristic of the Graeco-Iranian Turolian. Knowledge of them has been much extended in the past 15 years by the work of Bouvrain (see Bouvrain, 1992), and further revision is likely. The type species of *Prostrepsiceros* comes from lower and middle Maragheh, around 9–7 Ma in age. Lehmann and Thomas (1987) assigned to *Prostrepsiceros* a large and late species, *P. libycus,* from Sahabi. The ancestry of *Antilope cervicapra,* the living blackbuck of India, lies within this group.

Genus *Prostrepsiceros* Major, 1891
Type species *Prostrepsiceros houtumschindleri* (Rodler and Weithofer, 1890)
Prostrepsiceros aff. *libycus* Lehmann and Thomas, 1987

Material

AUH 237, Shuwaihat, S4: two pieces of a left horn core. Figure 22.17.
AUH 236, Shuwaihat, S4: pieces of a horn core.
AUH 413, Harmiyah, Y1: horn core fragment, possibly spiralled.
AUH 258, Kihal, K1: left astragalus. Lateral height 43.6 mm, medial height 39.4 mm, distal width 24.6 mm. Figure 22.17.
AUH 191, Ras Dubay'ah, R2: left astragalus, much abraded. Medial height 35.9 mm, distal width 23.0 mm.
AUH 400, Shuwaihat, S2: much of a left astragalus in two pieces. Medial height 37.8 mm, distal width 22.6 mm.
AUH 411, Hamra, H3: much of a left ectocuneiform.
AUH 173, Hamra, H1: part of a proximal left humerus.
AUH 132, Shuwaihat, S2: distal left humerus. Width across articular condyles c. 38.0 mm.
AUH 392, Harmiyah, Y1: proximal left radius. Maximum width 45.8 mm, width across articular surface 42.7 mm. Figure 22.17.
AUH 223a Ras Dubay'ah, R2: right unciform, damaged anteromedially.
AUH 325, Thumayriyah TH1: right lunate.

Description

The pieces of the horn core AUH 237 are much damaged. One piece, about 90 mm long, could be from its base, because there appears to be a boundary between the pedicel and core on one of its surfaces. The anteroposterior and transverse basal diameters would be c. 40.0 × 30.6 mm, giving an index of compression at 76.5%. The presumed lateral surface is less rounded than the medial surface, but most of the medial surface is missing. The level of maximum transverse thickness lies centrally. There is an approach to a posterior keel. There is slight torsion. A distal piece of horn core, about 60 mm long, shows a deep longitudinal groove.

The pieces of horn core AUH 236 are more poorly preserved than AUH 237, but are probably conspecific. AUH 413 is more doubtfully referred to this species.

The postcranial bones listed above are of a suitable size to be conspecific with the horn cores, but otherwise there is no association with the horn cores.

The distal humerus AUH 132 has rolled edges. It is slightly smaller than the Samos partial humerus BMNH M4309, which is of a size to go with the Samos *Pachytragus* radius BMNH M4315.

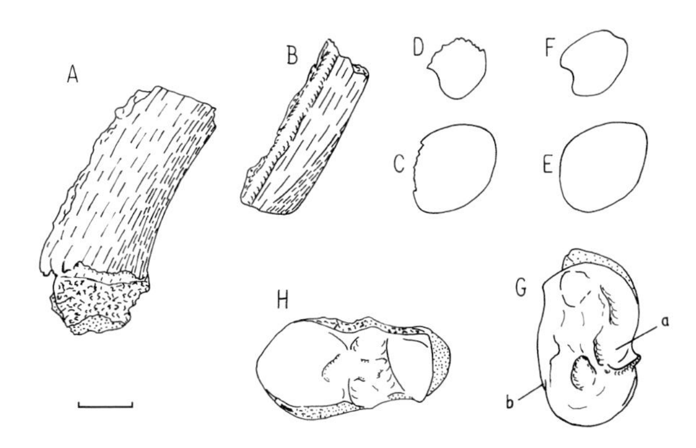

Figure 22.17. *A,* Supposed base of left horn core of *Prostrepsiceros* aff. *libycus,* AUH 237; lateral view. *C,* Cross-section of AUH 237 just above its base, lateral side to the right, anterior to the foot of the page. *B,* A more distal piece of (?the same) horn core, AUH 237. *D,* Cross-section of the more distal horn-core piece. *E,* Basal cross-section of a cast, BMNH M32982, of the paratype left horn core of *Prostrepsiceros libycus* (Lehmann and Thomas, 1987: fig. 7 C, D); lateral side to the right, anterior to the foot of the page. *F,* Distal cross-section of BMNH M32982. *G,* Medial view of left astragalus AUH 258; a = hollowing for medial malleolus of tibia, b = absence of indentation for naviculocuboid. *H,* Proximal articular surface of left radius, AUH 392, assigned by size to *Prostrepsiceros* aff. *libycus,* anterior side to foot of page. Scale = 20 mm for A–F and 13.3 mm for G and H.

The condyles are upright and the hollow for the lateral humeroradial ligament is not very deep as preserved. The lack of steepness of the sides of the ridge between the medial and lateral condyles suggests bovid rather than cervid affinities.

The proximal radius AUH 392 differs from those of *Tragoportax cyrenaicus* by the posteromedial part of the medial facet not projecting posteriorly.

The medial surface of the astragalus AUH 258 shows a large hollowing for the medial malleolus of the tibia and almost no indentation of the back edge for the naviculocuboid (fig. 22.17a, b), both characters being unlike modern *Boselaphus.* AUH 191 looks similar to AUH 258. They are of a suitable size to match a Sahabi metatarsal (BMNH cast M32984) of *P. libycus* (Lehmann and Thomas, 1987: pl. 3, fig. 3).

Discussion

The horn core AUH 237 differs from *Prostrepsiceros houtumschindleri* of lower and middle Maragheh by its greater compression, weaker keels, and less-developed torsion. Among known species of *Prostrepsiceros* it appears closest to *P. libycus* from Sahabi. It is similar to this species by its large size, the level of maximum transverse thickness of the horn core lying centrally, some sign of a posterior keel, weakness of the torsion, and, on the distal piece of horn core, the presence of a deep longitudinal groove. In *P. libycus* this groove lies on the anterior plane or edge. AUH 237 differs, however, in its greater degree of compression; Lehmann and Thomas (1987: table 6) give an average index for Sahabi *P. libycus* of 83.6%. The medial surface of AUH 237 is too poorly preserved for it to be apparent whether there would have been, as in *P. libycus,* well-marked anteromedial and posteromedial surfaces on either side of a narrowly zoned level of maximum transverse thickness of the horn core.

Hitherto *Prostrepsiceros libycus* has been known only from its type site at Sahabi. The closest species among the *Prostrepsiceros* of the Graeco-Iranian Turolian faunas would be *P. fraasi* (Andrée, 1926), which is also large and has more compression in the distal part of its horn cores, but which has the compression oriented anteroposteriorly rather than mediolaterally, no posterior keel, an approach to an anterior keel descending anteromedially rather than anteriorly, no deep longitudinal groove anteriorly, horn cores more uprightly inserted and more divergent, and a higher level of the frontals between the horn cores relative to the orbital rims.

Prostrepsiceros aff. *vinayaki* (Pilgrim, 1939)

Material

AUH 441, Hamra, H6: right horn core. Figure 22.18.
AUH 76, Shuwaihat, S1: lateral side of a left astragalus. Height 33.4 mm.
AUH 279, Jebel Dhanna, JD6: right astragalus. Medial height 30.5 mm.
AUH 415, Shuwaihat, S1: medial side of a right astragalus. Height 29.6 mm.
AUH 322, Thumayriyah, TH1: left naviculocuboid.
AUH 286, Hamra, H5: part of a right calcaneum.
AUH 4, Hamra, H1: part of a left calcaneum.
AUH 364, Jebel Mimiyah, M1: distal metatarsal. Width across condyles 25.0 mm, maximum anteroposterior diameter of condyles 18.7 mm. Figure 22.19.
BMNH M50713, Jebel Barakah, B2: proximal left radius. Maximum width 39.1 mm, width across the articular surface 35.3 mm. Figure 22.19.
AUH 321, Thumayriyah, TH1: epiphysis of a juvenile distal left radius. Width across the articular surfaces 34.8 mm.
AUH 343, Jebel Dhanna, JD3: first phalanx. Length 39.8 mm, proximal width 10.8 mm.

Description

The horn core AUH 441 has anteroposterior and transverse basal diameters of 29.6 × 20.9 mm, giving it an index of compression of 71%. The lateral surface is flatter than the medial surface, there is an anterior keel and an approach to a posterior one, it is inserted above the orbit, slightly inclined in side view, not very divergent, curving backwards in side view, with slight torsion, and there is a shallow postcornual fossa.

The radius BMNH M50713 is smaller than those considered under *Tragoportax cyrenaicus* or *Prostrepsiceros* aff. *libycus.* The lateral facet is not drawn out posteriorly and the lateral tubercle is larger than in the others. The drop in level of the medial relative to the lateral facet in anterior view is much more pronounced than in a Samos radius, BMNH M4315, possibly belonging to *Pachytragus.*

The tarsal bones, being larger than Pikermi *Gazella capricornis* (Wagner, 1848), are thereby of an appropriate size to go into *Prostrepsiceros* aff. *vinayaki.* The naviculocuboid AUH 322 is rather small but not here separated as a smaller species. It, too, is larger than in *G. capricornis.*

The distal metatarsal AUH 364 is slightly smaller than the cast, BMNH M32984, of a Sahabi metatarsal assigned by Lehmann and Thomas (1987: pl. 3, fig. 3) to *P. libycus.* It has well-marked hollows above the condylar keels anteriorly to receive phalanges in

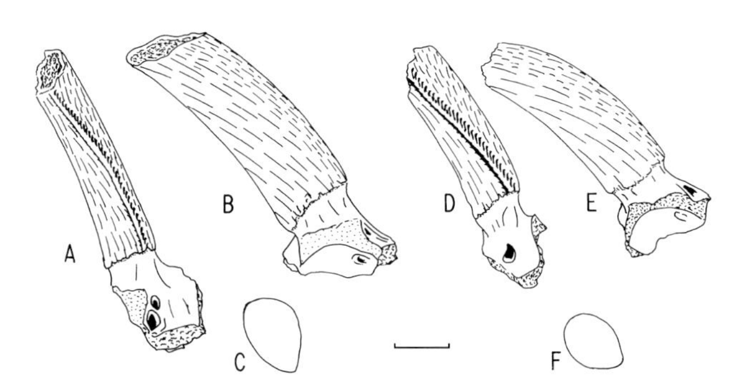

Figure 22.18. *A, B,* Right horn-core base of *Prostrepsiceros* aff. *vinayaki,* AUH 441; anterior and lateral views. *C,* Cross-section of AUH 441 just above its base; lateral side to the left, anterior to the foot of the page. *D, E,* Cast of holotype horn-core base of *Prostrepsiceros vinayaki,* BMNH M42957; anterior and lateral views, drawn as if of the right side. *F,* Cross section of BMNH M42957; lateral side to the left, anterior to the foot of the page. Scale = 20 mm.

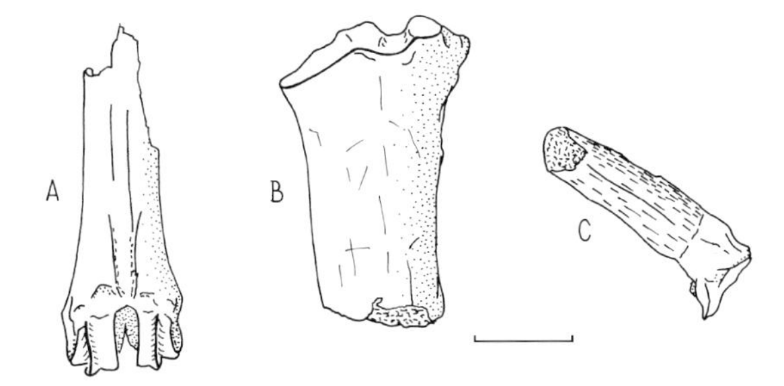

Figure 22.19. *A, B,* Distal metatarsal AUH 364 and proximal left radius BMNH M50713, both of a size to fit *Prostrepsiceros* aff. *vinayaki;* anterior views. *C,* Right horn core AUH 332, a possible female of *Gazella* aff. *lydekkeri;* lateral view. Scale = 20 mm.

extreme extension and also well-marked hollows on the sides of the condyles for ligaments. The aspect of the bone is of an open-country or fleet-footed antelope, not a skulker in thickets.

Discussion

The horn core is from a smaller species than the *Prostrepsiceros* aff. *libycus* just considered, and it also differs in being more compressed and with a better-marked anterior keel. It agrees with the Sahabi *P. libycus* in having a fairly narrow zone of maximum transverse thickness and hence some indication of anteromedial and posteromedial surfaces, in inclination and divergence, and in frontals being at a low level between the horn cores. It is very similar to a cast, BMNH M42957, of the holotype horn core of *P. vinayaki* from the Dhok Pathan of the Siwaliks (see Pilgrim, 1939: 42, pl. 1, fig. 10; also see Thomas, 1984: 41, pl. 3, fig. 7 for additional material). The resemblance is shown in less compression than in *P. libycus,* the level of maximum transverse thickness being fairly narrow and thus separating anteromedial and posteromedial components of the medial surface, the presence of an anterior keel and a slight posterior keel (Thomas maintains that the posterior keel is stronger than the anterior one), insertion above the orbit, in inclination and divergence. It is, however, slightly larger, the anteroposterior and transverse basal diameters of M42957 being 25.5 × 19.5 mm, giving an index of compression of 76%. *Nisidorcas planicornis* (Pilgrim, 1939) of Piram Island and the Graeco-Turkish Turolian (Bouvrain, 1979; Köhler, 1987: 199) is similarly sized, and compressed to about the same extent, but has a posterior keel strengthening distally and no anterior keel.

The drop in level of the medial relative to the lateral facet in anterior view on the radius M50713 could suggest cervid affinities according to the criteria of Heintz (1970). This drop is in fact more pronounced than in British Pleistocene deer radii from Grays, Essex. A drop in level, however, can occur in bovids—for example, two among 16 *Tragelaphus scriptus,* a *Cephalophus sylvicultor* 1961.8.9.80, and a *Capricornis sumatraensis* 1965.3.25.2 in the Department of Zoology, BMNH. I would prefer an antler as evidence for the presence of fossil Cervidae in the Arabian Peninsula. This radius looks like a non-boselaphine bovid smaller than the Samos *Pachytragus* radius BMNH M4315.

Genus *Gazella* Blainville, 1816
Type species *Gazella dorcas* (Linnaeus, 1758)
Gazella aff. *lydekkeri* Pilgrim, 1937

Material

AUH 389, Harmiyah, Y1: right horn core. Anteroposterior and mediolateral basal diameters = c. 22.4 × c. 18.2 mm. Figure 22.14.

AUH 332, Hamra, H2: right horn core. Anteroposterior and mediolateral basal diameters = 12.7 × 12.7 mm. Figure 22.19.

Description and Discussion

The horn core AUH 389 is quite badly preserved with the surface missing over much of its anterior half. Its basal diameters suggest a compression index of 81%. It is smaller and a little more compressed than *Gazella capricornis* of Pikermi. It is also smaller than a cast, BMNH M15760, of *G. lydekkeri* Pilgrim, 1937 (figs 35–39), but resembles the latter species in its weaker backward curvature than in *G. capricornis.* A horn core from Jebel Hamrin named as *Gazella* cf. *deperdita* by

Thomas, Sen, and Ligabue (1980: pl. 1, fig. 7) is more backwardly curved than in AUH 389 or the *G. lydekkeri* cast.

The small horn core AUH 332 is about the same size as a rather large example of the African steinbok, *Raphicerus campestris*. It shows no compression. It has a preserved length of about 35.0 mm and when complete might have been a little over 50.0 mm long. The lateral surface is flat and parts of the other surfaces are probably missing, thereby giving the horn core an irregularly shaped cross-section. It shows a slight backward curvature. A slight indication of the midfrontal plane on the medial side of the horn pedicel suggests that the insertion was probably fairly upright. There was a postcornual fossa.

This horn core could be a female or immature specimen of *Gazella*. *Gazella lydekkeri* in the Dhok Pathan of the Siwaliks had hornless females (Pilgrim, 1937: 800), but possible *Gazella* female horn cores are present in the Samos collection of the BMNH (Pilgrim and Hopwood, 1928: 12, pl. 1, fig. 1). The other alternative for AUH 332 is that it is an early neotragine or a relative of the Nagri (Siwaliks) supposed small boselaphine *Elachistoceras* Thomas (1977). The latter, however, like many neotragines, shows a slightly concave anterior edge of its horn core in lateral profile.

Discussion of the Baynunah Formation Giraffidae and Bovidae

Age and Correlation

The ruminant species identified from the Baynunah Formation are as follows:

Family Giraffidae	?*Palaeotragus* sp.
	?*Bramatherium* sp.
	Giraffidae sp. indet.
Family Bovidae	
Tribe Boselaphini	*Tragoportax cyrenaicus*
	Pachyportax latidens
Tribe indet.	Bovidae sp. indet.
Tribe Antilopini	*Prostrepsiceros* aff. *libycus*
	Prostrepsiceros aff. *vinayaki*
	Gazella aff. *lydekkeri*

This is a respectable ruminant fauna, nine-species strong, from which it is possible to derive an indication of the age of the Baynunah Formation.

Sivatheres are not known from before the late Miocene unless one adopts the doubtful proposition that the four horns of the middle Miocene *Giraffokeryx* indicate sivatheriine affinity. If the ?*Bramatherium* sp. is a sivathere, it agrees better with late Miocene sivatheres than with those of the Pliocene or later.

The postcranial bones assigned to *Giraffidae* sp. indet. reach a larger size than in middle Miocene *Giraffokeryx*, and so are more likely to belong to a late Miocene species. Furthermore, the distal metatarsal of ?*Palaeotragus* sp. is more robustly built than in middle Miocene *Giraffokeryx*, and thus fits better with a late Miocene age.

Tragoportax cyrenaicus at its type locality is a morphologically advanced species of the very late Miocene (Geraads, 1989b: 790). An allied species is found in the presumed early Pliocene of Langebaanweg.

Pachyportax occurs in the Nagri and Dhok Pathan zones of the Siwaliks, and *Proamphibos* replaces it soon after the start of the Tatrot. Hence around 7.0 Ma, or the latest Miocene (see Barry et al., 1991 for the dating of the Siwaliks deposits), would be a likely date for a fauna with *Pachyportax latidens*. The French boselaphine *Parabos cordieri* is also like *Pachyportax* and occurs in European Mammal Zone MN 14 (Mein, 1990: 79), which equals early Pliocene; hence it is younger than most *Pachyportax*.

Prostrepsiceros aff. *libycus* is at least close to the species at Sahabi, while the smaller *Prostrepsiceros* aff. *vinayaki* is close to a late Miocene Siwaliks species, for which Thomas (1984: 42) accepts a date of around 7.0–7.5 Ma. Together, the two Abu Dhabi species suggest a date in the latest Miocene. *Gazella lydekkeri*, which the limited morphological knowledge of the gazelle horn core AUH 389 suggests as a possible near relative, is again a late Miocene species.

The Baynunah ruminants hence support an age for the formation of late in the Miocene, perhaps equivalent to MN 13 in terms of the European Mein zone scale and around 6.0 Ma ago in terms of dating.

Zoogeography and Palaeoecology

The Baynunah ruminants can be compared with those of late Miocene faunas in adjacent continental regions. Making comparisons at regional level rather than with particular localities increases the deficiencies of temporal resolution. The beginning of the late Miocene at 10 Ma ago is twice as old as its end, so if the Baynunah fossils are indeed close to the end of the Miocene in age, we shall be comparing them with some faunas that are twice as old. Hence it becomes more difficult to separate effects of geography from those of time.

Below tribal level the Baynunah pecorans show little resemblance to the rich Turolian assemblage known from the Graeco-Iranian region. The following list of giraffids and bovids for that region is based mainly on Solounias (1981a), Bernor (1986), Köhler (1987), and Köhler et al. (1995).

Family Giraffidae	*Palaeotragus*, 2 spp.
	Samotherium boissieri
	Decennatherium macedoniae
	Helladotherium duvernoyi
	Bohlinia attica
Family Bovidae	
Tribe Boselaphini	*Tragoportax*, 3 spp., not incl. *cyrenaicus*
Tribe ?Bovini	*Samokeros minotaurus*
Tribe Reduncini	?*Redunca* sp.
Tribe Antilopini	*Prostrepsiceros*, 4 spp., not incl. *libycus* or *vinayaki*
	Protragelaphus, 2 spp.
	Palaeoreas, 3 spp.
	Nisidorcas planicornis
	(?*Hispanodorcas*) *rodleri*
	Oioceros, 3 spp.
	Samodorcas kuhlmanni
	Gazella, 2 spp. not incl. *lydekkeri*
Tribe Ovibovini	*Urmiatherium*, 2 spp.
	Plesiaddax inundatus
	Criotherium argalioides
	Palaeoryx pallasi[1]
Tribe Caprini	*Protoryx carolinae*[1]
	Pachytragus, 2 spp.[1]
	Norbertia hellenica
	Pseudotragus, 2 spp.[1]

These faunas contrast with that of Abu Dhabi in the definite presence of the distinctive mixed feeder or grazing giraffe *Samotherium boissieri* (Solounias et al., 1988). The much greater abundance of spiral-horned Antilopini (all the listed Antilopini except *Gazella*) is also noteworthy; they make up 38% of the listed species. Caprini and archaic and specialised Ovibovini are present here but absent at Abu Dhabi.[1] Cervidae occur sparingly in the Graeco-Iranian Turolian and there are even rare records of Tragulidae, but as yet there are no records of these families from the Baynunah Formation. The presence of a reduncine in the Graeco-Iranian list is based only on a P_{2+3} from MN 13 in Turkey (Köhler, 1987: 214), and the tribe was not a normal component of MN 11 and MN 12 faunas in this region. ("*Redunca*" *eremopolitana* Erdbrink, 1982 from the Maragheh area is based on a *Pachytragus* or some later caprine like *Norbertia*.) The differences at species level in *Tragoportax* and *Prostrepsiceros* suggest that the Baynunah Formation is of later date.

At Langebaanweg the following species are present (Harris, 1976; Gentry, 1980):

Family Giraffidae	*Giraffa* sp.
	Sivatherium hendeyi
Family Bovidae	
Tribe Tragelaphini	*Tragelaphus*, ? 2 spp.
Tribe Boselaphini	*Tragoportax acrae*
Tribe Bovini	*Simatherium demissum*
Tribe Reduncini	*Kobus*, 2 spp.
Tribe Alcelaphini	*Damalacra*, 2 spp.
Tribe Neotragini	*Raphicerus paralius*
Tribe Antilopini	*Gazella* sp.
Tribe Ovibovini	Gen. et sp. indet.

This is a fauna with an early bovine and the African tribes Tragelaphini, Reduncini, Alcelaphini, and Neotragini. There are also a boselaphine and ovibovine looking like Miocene immigrants from far to the north. Most of the species show primitive characters and are thereby differentiated from later species of East and South Africa. Only the boselaphine has any resemblance to a species in the Baynunah Formation and this is likely to reflect the fact that both the species concerned are late and advanced representatives of *Tragoportax*. The presence of African tribes of antelopes is more likely to show a zoogeographical than a temporal difference, but the bovine could indicate a slightly later date for Langebaanweg.

In East Africa, at Mpesida, Lukeino, and the Manonga Valley (Thomas, 1980; Gentry, 1997), the combined giraffid and bovid fauna can be given as:

Family Giraffidae	*Giraffa* sp.
	?*Sivatherium* sp.
Family Bovidae	
Tribe Tragelaphini	*Tragelaphus* sp.
Tribe Bovini	*Ugandax* cf. *gautieri*
Tribe Cephalophini	*Cephalophus* sp.
Tribe Reduncini	*Kobus* aff. *subdolus*
	Kobus aff. *porrecticornis*
Tribe Hippotragini	*Praedamalis* sp.
Tribe Alcelaphini	*Damalacra* sp.
	Alcelaphini, 2 more spp.
Tribe Aepycerotini	*Aepyceros* sp.
Tribe Neotragini	*Madoqua* sp.
Tribe Antilopini	?*Gazella* sp.

This assemblage has some resemblance to Langebaanweg, but Cephalophini, Hippotragini, and *Aepyceros* are additional African constituents. The only possible resemblance to the Baynunah Formation pecorans seems to be the presence of a sivatheriine giraffid at a comparable stage of evolutionary development. This ?*Sivatherium* sp. could just as easily belong to *Bramatherium* by what little is known of its primitive characters. There are no Boselaphini or spiral-horned Antilopini, although boselaphines are present earlier in the East African Miocene and *Tragoportax* at Lothagam 1. Also Nakaya et al. (1984: 109, pl. 9, fig. 5)) recorded a possible spiral-horned antilopine (as *Palaeoreas* sp.) in the earlier Namurungule Formation, and Smart (1976) listed *Antilope* sp. for Lothagam 1.

The North African pecorans, as known from Bled ed-Douarah, Sahabi, and Wadi Natrun (Geraads, 1989b; Harris, 1987; Lehmann and Thomas, 1987; also see comments in Gentry, 1980), are:

Family Giraffidae	Giraffidae, sp. indet.
	Sivatherium aff. *hendeyi*
Family Bovidae	
Tribe Boselaphini	*Tragoportax cyrenaicus*
Tribe Bovini	"*Leptobos*" *syrticus*
Tribe Reduncini	*Kobus* sp.
Tribe Hippotragini	*Hippotragus* sp.
Tribe Alcelaphini	*Damalacra* sp.
Tribe Neotragini	*Raphicerus* sp.
Tribe Antilopini	*Prostrepsiceros libycus*
	Gazella sp.

This list is quite close to the Abu Dhabi fauna with a conspecific boselaphine and spiral-horned antilopine, and two giraffids, but it differs in the presence of several African tribes. It needs to be remembered (Geraads, 1989b) that several stratigraphic levels are represented at these localities, probably ranging up into the Pliocene, and some of the African tribes of antelopes may be present only in higher levels.

Towards the end of the Miocene in the Siwaliks, in the Dhok Pathan, one finds the following giraffids and bovids (Pilgrim, 1937, 1939; Barry et al., 1982, 1991; Barry, 1995):

Family Giraffidae	*Giraffa punjabiensis*
	Bramatherium megacephalum
Family Bovidae	
Tribe Boselaphini	*Tragoportax salmontanus*
	Tragoportax browni
	Pachyportax latidens
Tribe Reduncini	*Kobus porrecticornis*
Tribe Antilopini	*Prostrepsiceros vinayaki*
	Gazella lydekkeri

This late Miocene fauna is very like that of Abu Dhabi in the *Pachyportax* and spiral-horned antilopine, but the *Tragoportax* are different at species level. If the giraffine really is a *Giraffa*, the metapodials would be too long to agree with the Baynunah Formation Giraffidae, sp. indet. It is likely that smaller sivatheriines are present in addition to *Bramatherium megacephalum*, and this could be a resemblance to the Baynunah Formation ?*Bramatherium* sp., if the latter should also be a modestly sized species. The presence of late Miocene Reduncini shows both that this tribe had evolved before the end of the Miocene, and that the Siwaliks and, by implication, Arabia and North Africa had a zoological connection with the whole of Africa at this period.

After the end of the Miocene, Cervidae, the bovine *Proamphibos*, and possibly Hippotragini appear in the Siwaliks while boselaphines become rare or temporarily absent. This fauna is much less similar to that of the Baynunah Formation and suggests that the latter cannot be as late as the Pliocene.

One fossil fauna with pecorans that deserves mention in comparison with the Baynunah Formation is that of Piram (formerly Perim) Island, Gulf of Cambay, India (Prasad, 1974). From the collection in the BMNH and from the older literature, especially Pilgrim (1939), the pecorans appear to be:

1. Giraffidae, smaller sp.: a left maxilla with P^3–M^3, BMNH 37259, about the size of *Samotherium boissieri* (see Matthew, 1929: 551).
2. *Bramatherium perimense:* based on skulls and dental material, mentioned above and discussed in Colbert (1935).
3. *Pachyportax* sp.: the main specimens are a left horn core BMNH M2402a (Lydekker, 1886: pl. 3, fig. 5), another left horn core perhaps numbered BMNH 18785, and an immature cranium BMNH 37262, holotype of *Perimia falconeri* (Lydekker, 1886). The lateral surface of the horn cores is more rounded than in *Pachyportax latidens,* thereby giving a cross-section more similar to the bulkier horn cores of the primitive Tatrot bovine *Proamphibos* Pilgrim, 1939. In fact Pilgrim (1939: 174) considered the horn core M2402a to belong to *Selenoportax* and placed it in a species *S. lydekkeri* founded on teeth (Pilgrim, 1910) and of larger size than *S. vexillarius.*
4. *Tragoportax perimensis* (Lydekker, 1878): mentioned above.
5. *Nisidorcas planicornis* (Pilgrim, 1939): a small antilopine with spiralled horn cores, mentioned above. The horn core BMNH 37264 is the holotype, and the immature female cranium M3683, the holotype of *Cambayella watsoni* Pilgrim, 1939, may be the same species.
6. Prasad (1974: 15, pl. 5, fig. 7) referred a small Piram Island horn core to *Ruticeros compressa,* sp. nov. The genus *Ruticeros* is otherwise known by horn cores of *R. pugio* Pilgrim, 1939, supposedly of late Miocene age. They are compressed, with a flat lateral surface, slight backward curvature, diminishing divergence distally, and weak keels anteriorly and posteriorly. They could well be of a small boselaphine, as Pilgrim supposed, or antilopine.

Both the Piram Island and Abu Dhabi faunas contain two or more giraffids, a large and a smaller boselaphine, and one or more antilopines with spiral horns. No African tribes of antelopes, supposed ovibovines of the *Urmiatherium* group, nor caprines of the *Palaeoryx–Protoryx–Pachytragus* group, nor cervids, nor tragulids have been found in either fauna.

The findings of these zoogeographical comparisons can be summarised as follows (see also table 22.1):

- The Baynunah Formation ruminants are very unlike those of the Graeco-Iranian Turolian. Where genera are shared, the species are different, and the Baynunah ones look more advanced.
- They also differ from East and South African faunas in that the giraffid sp. indet. is not a

Table 22.1. Characteristics of regional faunas of pecorans compared with the pecorans of the Baynunah Formation

	Abu Dhabi	Graeco-Iran	South Africa	East Africa	North Africa	Siwaliks	Piram Island
More than two giraffids	Yes	Yes	No	No	No	No	No
Bovini have appeared	No	Yes	Yes	Yes	Yes	No	No
African tribes of bovids	No	Yes	Yes	Yes	Yes	Yes	No
Fewer than three spiral-horned bovids	Yes	No	Yes	Yes	Yes	Yes	Yes
Ovibovines present	No	Yes	Yes	No	No	No	No
Protoryx group present	No	Yes	No	No	No	No	No
Differences (out of six) from Abu Dhabi	—	5	4	3	3	2	1

Giraffa, and in the absence of African tribes of antelopes. The latter may be a genuine zoogeographical difference, but it is also possible that the Baynunah Formation predates the evolutionary appearance of Hippotragini, Alcelaphini, or even Reduncini.

- They show some affinity with North African, Siwaliks, and Piram Island late Miocene faunas and hence look like a latitudinally differentiated fauna lying across the vast tract of land to the south of that inhabited by the Graeco-Iranian faunas. Whereas deer are common in late Miocene faunas north of the present Black Sea, they are rare in Greece, Turkey, and Iran, and as yet absent in the Abu Dhabi, North African, Siwaliks, and Piram Island faunas.
- The presence of the specialised Miocene ovibovines and of caprines of the *Protoryx–Pachytragus* group only in the Graeco-Iranian region could suggest flatter terrain, climatic unsuitability to the south and east, or that Abu Dhabi postdates the demise of those particular ovibovines. The presence of giraffines, boselaphines, spiral-horned antilopines, and a gazelle at Abu Dhabi could indicate rather parkland-like country. There might have been a southern or hot-country climate, but not one with any substantial development of aridity. The gazelle might have had a life-style more like that of the present-day West African *Gazella rufifrons* than like the desert gazelles of North Africa and Arabia. The absence of tragulids in the Abu Dhabi fauna but not in the earlier Miocene of the Arabian Peninsula (Gentry, 1987; Thomas et al., 1982) suggests that the range contraction of that family had proceeded further and that suitable habitats were by now unavailable in the Arabian Peninsula.

Acknowledgements

I thank Peter Whybrow for inviting me to work on the Abu Dhabi pecorans, Diana Clements for much assistance, Peter Forey for preparation of diagrams, and Ray Bernor for helpful comments. I thank the Keeper of Palaeontology at The Natural History Museum for use of the facilities of his department.

Note

1. These species are frequently considered to belong to the African tribe Hippotragini, as, for example, by Solounias (1981b) and Erdbrink (1988).

References

Aguirre, E., and Leakey, P. 1974. Nakali: nueva fauna de *Hipparion* del Rift Valley de Kenya. *Estudios Geologicos. Instituto de Investigaciones Geologicas "Lucas Mallada", Madrid* 30: 219–27.

Andrews, C. W. 1902. Note on a Pliocene vertebrate fauna from the Wadi-Natrun, Egypt. *Geological Magazine* 9: 433–39.

Barry, J. C., 1995. Faunal turnover and diversity in the terrestrial Neogene of Pakistan. In *Paleoclimate and Evolution, with Emphasis on Human Origins,* pp. 115–34 (ed. E. S. Vrba, G. H. Denton, T. C. Partridge, and L. H. Burkle). Yale University Press, New Haven.

Barry, J. C., and Flynn, L. J. 1990 (1989). Key biostratigraphic events in the Siwaliks sequence. In *European Neogene Mammal Chronology,* pp. 557–71 (ed. E. H. Lindsay, V. Fahlbusch, and P. Mein). Plenum Press, New York.

Barry, J. C., Lindsay, E. H., and Jacobs, L. L. 1982. A biostratigraphic zonation of the Middle and Upper Siwaliks of the Potwar plateau of northern Pakistan. *Palaeogeography, Palaeoclimatology, Palaeoecology* 37: 95–130.

Barry, J. C., Morgan, M. E., Winkler, A. J., Flynn, L. J., Lindsay, E. H., Jacobs, L. L., and Pilbeam, D. 1991. Faunal interchange and Miocene terrestrial vertebrates of southern Asia. *Paleobiology* 17: 231–45.

Bernor, R. L. 1983. Geochronology and zoogeographic relationships of Miocene Hominoidea. In *New Interpretations of Ape and Human Ancestry,* pp. 21–64 (ed. R. L. Ciochon and R. S. Corruccini). Plenum Press, New York.

———. 1986. Mammalian biostratigraphy, geochronology, and zoogeographic relationships of the Late Miocene Maragheh fauna, Iran. *Journal of Vertebrate Paleontology* 6: 76–95.

Boaz, N. T., El-Arnauti, A., Gaziry, A., de Heinzelin, J., and Boaz, D. D. 1987. *Neogene Paleontology and Geology of Sahabi.* Alan R. Liss, New York.

Borissiak, A. 1914. Mammifères fossiles de Sebastopol. *Mémoires du Comité Géologique. Saint Pétersbourg,* n.s. 87: 1–154.

Bouvrain, G. 1979. Un nouveau genre de Bovidé de la fin du Miocène. *Bulletin de la Societé Géologique de France, Paris* 21: 507–11.

———. 1988. Les *Tragoportax* (Bovidae, Mammalia) des gisements du Miocène supérieur de Ditiko (Macédoine, Grèce). *Annales de Paléontologie, Paris* 74: 43–63.

———. 1992. Antilopes à chevilles spiralées du Miocène supérieur de la province Gréco-Iranienne: nouvelles diagnoses. *Annales de Paléontologie, Paris* 78: 49–65.

———. 1994. Un Bovidé du Turolien inférieur d'Europe orientale: *Tragoportax rugosifrons. Annales de Paléontologie, Paris* 80: 61–87.

Bouvrain, G., and Bonis, L. 1984. Etude d'un miotragocère du Miocène supérieur de Macédoine (Grèce). In *Actes du symposium paléontologique G. Cuvier, Montbéliard,* pp. 35–49 (ed. E. Buffetaut, J. M. Mazin, and E. Salmon). Montbéliard.

Brunet, M., and Heintz, E. 1983. Interprétation paléoecologique et relations biogéographiques de la faune de vertébrés du Miocène supérieur d'Injana, Irak. *Palaeogeography, Palaeoclimatology, Palaeoecology* 44: 283–93.

Churcher, C. S. 1979. The large palaeotragine giraffid, *Palaeotragus germaini,* from Late Miocene deposits of Lothagam Hill, Kenya. *Breviora* 453: 1–8.

Colbert, E. H. 1935. Siwalik mammals in the American Museum of Natural History. *Transactions of the American Philosophical Society, Philadelphia* 26: 1–401.

Crusafont Pairo, M. 1952. Los jirafidos fosiles de Espana. *Memorias y Comunicaciones del Instituto Geologico, Barcelona* 8: 1–239.

Erdbrink, D. P. B. 1988. Protoryx from three localities east of Maragheh, N. W. Iran. *Proceedings Koninklijke Nederlandse Akademie van Wetenschappen* B 91: 101–59.

Falconer, H. 1845. Description of some fossil remains of *Dinotherium,* giraffe and other Mammalia, from the Gulf of Cambay, western coast of India. *Quarterly Journal of the Geological Society of London* 1: 356–72.

Friend, P. F. 1999. Rivers of the Lower Baynunah Formation, Emirate of Abu Dhabi, United Arab Emirates. Chap. 5 in *Fossil Vertebrates of Arabia* (ed. P. J. Whybrow and A. Hill). Yale University Press, New Haven.

Gentry, A. W. 1970. The Bovidae (Mammalia) of the Fort Ternan fossil fauna. In *Fossil Vertebrates of Africa,* vol.2, pp. 243–324 (ed. L. S. B. Leakey and R. J. G. Savage). Academic Press, London.

———. 1974. A new genus and species of Pliocene boselaphine (Bovidae, Mammalia) from South Africa. *Annals of the South African Museum, Cape Town* 65: 145–88.

———. 1978. Bovidae. In *Evolution of African Mammals,* pp. 540–72 (ed. H. B. S. Cooke and V. J. Maglio). Harvard University Press, Cambridge, Mass.

———. 1980. Fossil Bovidae (Mammalia) from Langebaanweg, South Africa. *Annals of the South African Museum, Cape Town* 79: 213–337.

———. 1987. Ruminants from the Miocene of Saudi Arabia. *Bulletin of the British Museum (Natural History),* Geology 41: 433–39.

———. 1997. Fossil ruminants (Mammalia) from the Manonga Valley, Tanzania. In *Neogene Paleontology of the Manonga Valley, Tanzania*, pp. 107–35 (ed. T. J. Harrison). Plenum Press, New York.

Geraads, D. 1979. Les Giraffinae (Artiodactyla, Mammalia) du Miocène supérieur de la région de Thessalonique (Grèce). *Bulletin du Muséum National d'Histoire Naturelle, Paris* 1: 377–89.

———. 1986. Remarques sur la systématique et la phylogénie des Giraffidae (Artiodactyla, Mammalia). *Geobios* 19: 465–77.

———. 1989a. Un nouveau Giraffidé du Miocène supérieur de Macédoine (Grèce). *Bulletin du Muséum National d'Histoire Naturelle, Paris* 11: 189–99.

———. 1989b. Vertébrés fossiles du Miocène supérieur du Djebel Krechem el Artsouma (Tunisie centrale): Comparaisons biostratigraphiques. *Geobios* 22: 777–801.

———. 1993. Les gisements de mammifères du Miocène supérieur de Kemiklitepe, Turquie: 8 Giraffidae. *Bulletin du Muséum National d'Histoire Naturelle, Paris* 16: 159–73.

Gromolard, C. 1980. Une nouvelle interprétation des grands Bovidae (Artiodactyla,Mammalia) du Pliocène d'Europe occidentale classés jusqu'à présent dans le genre *Parabos: Parabos cordieri* (de Christol) emend., ?*Parabos boodon* (Gervais) et *Alephis lyrix* n. gen., n. sp. *Geobios* 13: 767–75.

Harris, J. M. 1976. Pliocene Giraffoidea (Mammalia, Artiodactyla) from the Cape Province. *Annals of the South African Museum, Cape Town* 69: 325–53.

———. 1987. Fossil Giraffidae from Sahabi, Libya. In *Neogene Paleontology and Geology of Sahabi*, pp. 317–21 (ed. N. T. Boaz, A. El-Arnauti, A. W. Gaziry, J. de Heinzelin, and D. D. Boaz). Alan R. Liss, New York.

Harrison, T. J. ed. 1997. *Neogene Paleontology of the Manonga Valley, Tanzania: A Window into East African Evolution*. Topics in Geobiology vol. 14. Plenum Press, New York.

Heintz, E. 1970. Les cervidés Villafranchiens de France et d'Espagne. *Mémoires du Muséum National d'Histoire Naturelle, Paris* 22: 1–303.

Heintz, E., Brunet, M., and Sen, S. 1981. Un nouveau Giraffidé du Miocène supérieur d'Irak: *Injanatherium hazimi* n. g., n. sp. *Compte Rendu Hebdomadaire des Séances de l'Académie des Sciences, Paris* 292: 423–26.

Hill, A. 1987. Causes of perceived faunal change in the later Neogene of East Africa. *Journal of Human Evolution* 16: 583–96.

Hill, A. 1995. Faunal and environmental change in the Neogene of East Africa: Evidence from the Tugen Hills Sequence, Baringo District, Kenya. In *Paleoclimate and Evolution, with Emphasis on Human Origins*, pp. 178–93 (ed. E. S. Vrba, G. H. Denton, T. C. Partridge, and L. H. Burkle). Yale University Press, New Haven.

Hill, A., Drake, R., Tauxe, L., Monaghan, M., Barry, J. C., Behrensmeyer, A. K., Curtis, G., Jacobs, B. F., Jacobs, L., Johnson, N., and Pilbeam, D. 1985. Neogene palaeontology and geochronology of the Baringo Basin, Kenya. *Journal of Human Evolution* 14: 759–73.

Köhler, M. 1987. Boviden des turkischen Miozäns (Kanozoikum und Braunkohlen der Turkei 28). *Paleontologia y Evolucio* 21: 133–246.

Köhler, M., Moyà-Solà, S., and Morales, J. 1995. The vertebrate locality Maramena (Macedonia, Greece) at the Turolian-Ruscinian boundary (Neogene). 15. Bovidae and Giraffidae (Artiodactyla, Mammalia). *Münchner Geowissenschaftliche Abhandlungen* A 28: 167–80.

Leakey, M. G., Feibel, C. S., Bernor, R. L., Harris, J. M., Cerling, T. E., Stewart, K. M., Storrs, G. W.,

Walker, A., Werdelin, L., and Winkler, A. J. 1996. Lothagam: A record of faunal change in the late Miocene of East Africa. *Journal of Vertebrate Palaeontology* 16: 556–70.

Lehmann, U., and Thomas, H. 1987. Fossil Bovidae from the Mio-Pliocene of Sahabi, (Libya). In *Neogene Paleontology and Geology of Sahabi*, pp. 323–35 (ed. N. T. Boaz, A. El-Arnauti, A. W. Gaziry, J. de Heinzelin, and D. D. Boaz). Alan R. Liss, New York.

Lydekker, R. 1878. Crania of ruminants from the Indian Tertiaries. *Memoirs of the Geological Survey of India. Palaeontologia Indica* 1: 88–171.

———. 1885. *Catalogue of the Fossil Mammalia in the British Museum (Natural History), II: Artiodactyla, Bovidae.* British Museum (Natural History), London.

———. 1886. Siwalik Mammalia—supplement 1. *Memoirs of the Geological Survey of India. Palaeontologia Indica* 4: 1–21.

Matthew, W. D. 1929. Critical observations upon Siwalik mammals. *Bulletin of the American Museum of Natural History, New York* 56: 437–560.

Mecquenem, R. de. 1924–25. Contribution à l'étude du gisement des vertébrés de Maragha et de ses environs. *Annales de Paléontologie, Paris* 13: 135–60; 14:1–36.

Mein, P. 1990 (1989). Updating of MN zones. In *European Neogene Mammal Chronology*, pp. 73–90 (ed. E. H. Lindsay, V. Fahlbusch, and P. Mein). Plenum Press, New York.

Morales, J., Soria, D., and Thomas, H. 1987. Les Giraffidae (Artiodactyla, Mammalia) d'Al Jadidah du Miocène Moyen de la Formation Hofuf (Province du Hasa, Arabie Saoudite). *Geobios* 20: 441–67.

Moyà-Solà, S. 1983. Los Boselaphini (Bovidae Mammalia) del Neogeno de la peninsula Iberica. *Publicaciones de Geología, Universidad Autonoma de Barcelona* 18: 1–236.

Nakaya, H., Pickford, M., Nakano, Y., and Ishida, H. 1984. The late Miocene large mammal fauna from the Namurungule Formation, Samburu Hills, northern Kenya. *African Study Monographs,* suppl. issue 2: 87–131.

Pickford, M. 1988. The age(s) of the Bugti fauna(s), Pakistan. In *The Palaeoenvironment of East Asia from the Mid-Tertiary,* pp. 937–55 (ed. P. Whyte). Centre of Asian Studies; University of Hong Kong, Hong Kong.

Pilgrim, G. E. 1937. Siwalik antelopes and oxen in the American Museum of Natural History. *Bulletin of the American Museum of Natural History, New York* 72: 729–874.

———. 1939. The fossil Bovidae of India. *Memoirs of the Geological Survey of India. Palaeontologia Indica,* n.s. 26: 1–356.

Pilgrim, G. E., and Hopwood, A. T. 1928. *Catalogue of the Pontian Bovidae of Europe.* Trustees of the British Museum (Natural History), London.

Prasad, K. N. 1974. The vertebrate fauna from Piram Island, Gujarat, India. *Memoirs of the Geological Survey of India. Palaeontologia Indica,* n.s. 41: 1–23.

Schlosser, M. 1904. Die fossilen Cavicornier von Samos. *Beiträge zur Paläontologie und Geologie Oesterreich-Ungarns und des Orients* 17: 28–118.

Smart, C. 1976. The Lothagam 1 fauna: Its phylogenetic, ecological, and biogeographic significance. In *Earliest Man and Environments in the Lake Rudolf Basin,* pp. 361–69 (ed. Y. Coppens, F. C. Howell, G. Ll. Isaac, and R. E. F. Leakey). University of Chicago Press, Chicago.

Solounias, N. 1981a. Mammalian fossils of Samos and Pikermi. 2. Resurrection of a classic Turolian fauna. *Annals of Carnegie Museum, Pittsburgh* 50: 231–69.

———. 1981b. The Turolian fauna from the island of Samos, Greece. *Contributions to Vertebrate Evolution* 6: 1–232.

Solounias, N., Teaford, M., and Walker, A. 1988. Interpreting the diet of extinct ruminants: The case of a non-browsing giraffid. *Paleobiology* 14: 287–300.

Stuart, A. J. 1982. *Pleistocene Vertebrates in the British Isles.* Longman, London.

Thomas, H. 1979. *Miotragocerus cyrenaicus* sp. nov. (Bovidae, Artiodactyla, Mammalia) du Miocène supérieur de Sahabi (Libye) et ses rapports avec les autres *Miotragocerus. Geobios* 12: 267–81.

———. 1980. Les bovidés du Miocène supérieur des couches de Mpesida et de la formation de Lukeino (District de Baringo, Kenya). In *Proceedings of the 8th Panafrican Congress of Prehistory and Quaternary Studies, Nairobi, 5 to10 September 1977,* pp. 82–91 (ed. R. E. F. Leakey and B. A. Ogot). The International Louis Leakey Memorial Institute for African Prehistory, Nairobi.

———. 1983. Les Bovidae (Artiodactyla, Mammalia) du Miocène moyen de la formation Hofuf (Province du Hasa, Arabie Saoudite). *Palaeovertebrata* 13: 157–206.

———. 1984. Les bovidés anté-hipparions des Siwaliks inférieurs (plateau du Potwar), Pakistan. *Mémoires de la Société Géologique de France, Paris* 145: 1–68.

Thomas, H., Sen, S., Khan, M., Battail, B., and Ligabue, G. 1982. The Lower Miocene Fauna of Al-Sarrar (Eastern Province, Saudi Arabia). *ATLAL, The Journal of Saudi Arabian Archaeology* 5: 109–36.

Thomas, H., Sen, S., and Ligabue, G. 1980. La faune Miocène de la Formation Agha Jari du Jebel Hamrin (Irak). *Proceedings Koninklijke Nederlandse Akademie van Wetenschappen* B 83: 269–87.

Whybrow, P. J. , Friend, P. F., Ditchfield, P. W., and Bristow, C. S. 1999. Local stratigraphy of the Neogene outcrops of the coastal area: Western Region, Emirate of Abu Dhabi, United Arab Emirates. Chap. 4 in *Fossil Vertebrates of Arabia,* pp. 28–37 (ed. P. J. Whybrow and A. Hill). Yale University Press, New Haven.

Late Miocene Baynunah Formation, Emirate of Abu Dhabi, United Arab Emirates: Fauna, Flora, and Localities

23

Peter J. Whybrow and Diana Clements

The jebels scattered along the coastal plain of Abu Dhabi's Western Region are the only vertebrate-bearing late Miocene rocks known from Arabia. Most of the fossils collected by The Natural History Museum/Yale University team described in this volume come from outcrops in the Baynunah Formation located between the road from As Sila to Abu Dhabi and the coast (Whybrow, 1989; Whybrow et al., 1990; see also Whybrow et al., 1999—Chapter 4). Here, fluvial clastics deposited by a major river system (Friend, 1999—Chapter 5) contain remains of invertebrates, reptiles, birds, and mammals as well as poorly fossilised plant material. Few vertebrate fossils have been found at localities south of the main road. The jebels in this area lack the gravels and coarse sandstone lithologies seen in the lower part of the Baynunah Formation at the coastal exposures; sediments forming these southern jebels might be higher in the Miocene sequence (the "upper" part of the Baynunah in Friend, 1999—Chapter 5). The most eastern fossiliferous locality is south of Tarif. Here a magnificent partial skull of *Tragoportax cyrenaicus* (Gentry, 1999b—Chapter 22) was the only fossil found in homogeneous red-coloured sandstone (a road-cut exposure) by staff of the United States Geological Survey (Hill et al., 1999—Chapter 3).

Excepting the sea cliff outcrops of the Baynunah Formation, "clean" exposures of Miocene sediments are rare. The region is hyperarid and during the 15 years or so of work by the NHM/Yale team torrential rain has occurred just five times. Detritus, a mixture of weathered Miocene sediments with modern wind-blown sand, on the slopes of most jebels can be 20 cm deep before unweathered rock is reached.

Although the Baynunah Formation crops out in an area of about 1800 km^2, the fossils identified so far (more than 900) come from a 560 km^2 area. Except for the disarticulated, scattered bones of a proboscidean found at Shuwaihat (Tassy, 1999; Andrews, 1999—Chapters 18 and 24), almost all vertebrate fossils are found as isolated elements, often fragmented by extremes of temperature and rehydration of microcrystalline gypsum during precipitation of winter fog and dew.

During the period that the NHM/Yale team, and others, have been finding fossils from the Baynunah Formation, we have collected all material from each locality that might be identifiable so that a possible collecting bias is minimised (Hill, 1987). Some bias has been unavoidable, however, because further collecting at some localities has been impossible, especially from 1992. The development of the Western Region of Abu Dhabi proceeds rapidly. Access to some localities that were relatively easy to visit in the early 1980s is now prohibited. We have suggested to the Abu Dhabi authorities that some localities should be conserved for the scientific heritage of the Emirate as well as for international science. This suggestion has been (1995) well received and it is likely that Shuwaihat will be recognised as a locality of scientific heritage.

Here we list and describe the localities and sites (within the localities), from west to east, their fauna and flora, and present a complete list of the

 ISBN 0-300-07183-3

Plate 23.1. Jebel Barakah viewed from the west. The gravel bed, top left, is at the base of the lower part of the Baynunah Formation; the remaining outcrop, below, is the Shuwaihat Formation.

Plate 23.2. Shuwaihat, site S6. Gillian Comerford (The Natural History Museum) begins the excavation of the proboscidean (Tassy, 1999—Chapter 18). The lower jaw is seen at bottom right and the "planed-off" cranium is to the left (Andrews, 1999—Chapter 24).

Plate 23.3. South-facing escarpment of the Hamra locality (site H5). From the rhythmic clay layers, bottom right, palaeomagnetic samples (Hailwood and Whybrow, 1999—Chapter 8) and palaeosol carbonate nodules were collected (Kingston, 1999—Chapter 25).

Plate 23.4. Isolated jebels of the Baynunah Formation, near bin Jawabi.

Plate 23.5. Jebel Mimiyah (Al Mirfa). One of the excavated jebels showing a magnificent exposure of the Baynunah Formation (Friend, 1999—Chapter 5). From left to right, Peter Ditchfield, Anthea and Alan Gentry, Peter Andrews, and Peter Whybrow.

Baynunah Formation biota (fig. 23.1, p. 321; tables 23.1–23.6). The geographical coordinates listed here for sites were obtained using the GPS system. Although the error is said to be in the region of 25 metres (most sites cover a greater area than this) and that the GPS can be subject to degradation by the United States Department of Defense, good coordinates for the main sites (Jebel Barakah, Shuwaihat, and Jebel Dhanna) were obtained during January 1991.

The Localities

Jebel Barakah (JB) (tables 23.1–23.6; fig. 23.1 and plate 23.1)

Jebel Barakah is the most western fossiliferous locality in the Emirate. It is about 60 metres high and its southern flank slopes towards the main east–west road linking the border of the Kingdom of Saudi Arabia at As Sila with the United Arab Emirates. Most fossils have been found on its seaward-facing slopes and cliffs. It is the type section of the Baynunah Formation (Whybrow, 1989; Whybrow et al., 1999—Chapter 4). On the top of the jebel there used to be a UAE trignometrical station (G 111 at 63 metres; U.K. Ministry of Defence Survey Map, 1:100,000, NG-39-159, 1988). This area is site B1 (N 24° 00′ 24.9″,E 52° 19′ 48.6″). Material from the most western part of the jebel is site B2 (N 24° 00′ 13.6″, E 52° 19′ 35.5″) and close to a United States Marines survey point, labelled Alpha-1 1990. Most fossils from Jebel Barakah, such as *Hexaprotodon* aff. *sahabiensis* (see Gentry, 1999a—Chapter 21), were collected from this area. Site B3 (N 24° 06′ 04.7″, E 52° 31′ 38.5″) is on the cliff face from which samples for palaeomagnetic studies were taken (Hailwood and Whybrow, 1999—Chapter 8). Site B4 (N 24° 00′ 23.6″, E 52° 19′ 37.3″) lies between the top of the jebel itself and the top of the sea cliff. Kingston (1999—Chapter 25) collected soil samples at Jebel Barakah.

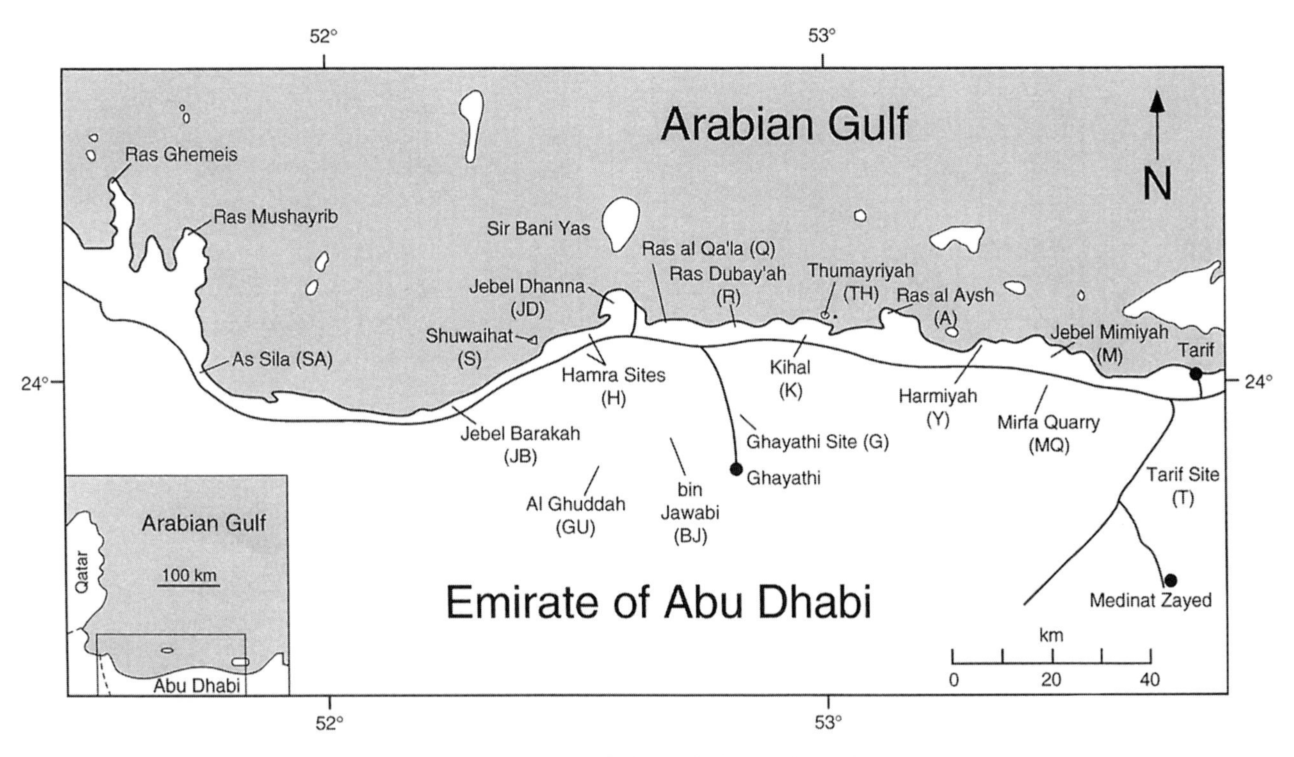

Figure 23.1 Map of the Western Region, Emirate of Abu Dhabi to show fossil localities and their abbreviations.

Shuwaihat (Shouwihat, Shouwehat, or Shuwayhat) (S) (tables 23.1–23.6; fig. 23.1 and plate 23.2)

After leaving the main As Sila to Abu Dhabi road at the signpost to "Jazirat Shuwayhat", the island is approached by a causeway. The island is the type section of the Shuwaihat Formation (Whybrow et al., 1999; Bristow, 1999—Chapters 4 and 6) and most of the Miocene vertebrate fossils from the Emirate of Abu Dhabi have been collected from this locality. A UAE survey point (GHQ 4143, height mapped at 47 metres) exists on the southern top of the westernmost jebel (known locally as Jebel Mershed). Near to this survey point at site S3 (N 24° 06′ 03.5″, E 52° 26′ 19.0″) numerous Miocene bivalves were found (Jeffery, 1999—Chapter 10). Site S1 (N 24° 06′ 38.1″, E 52° 26′ 09.6″) lies on the western face of the island's easternmost jebel. This site was one of the two rodent sampling sites on Shuwaihat (de Bruijn and Whybrow, 1994; de Bruijn, 1999—Chapter 15); it is also where a partial crocodile skull was found (Rauhe et al., 1999—Chapter 14). "Site" S4 (N 24° 06′ 44.7″, E 52° 26′ 12.7″) is a Natural History Museum survey point (marked BM(NH) 1992). It marks the edge of the proboscidean excavation site S6 (Tassy, 1999; Andrews, 1999—Chapters 18 and 24), figure 23.3. Surrounding the excavation is S4, the type locality for *Bagrus shuwaiensis* Forey and Young, 1999. About 20 metres to the northwest (still within site S4) is the second rodent sampling area, the type locality of *Abudhabia baynunensis* de Bruijn and Whybrow, 1999. The remaining fossiliferous site, S2, is the northwestern jebel where much plant material was found (N 24° 06′ 41.7″, E 52° 26′ 04.0″). The last Shuwaihat "site", S5, is the abandoned ADCO oil well SH-4 (N 24° 07′ 06.6″, E 52° 26′ 32.7″).

Al Ghuddah (GU) (table 23.3; fig. 23.1)

Al Ghuddah is an isolated jebel, about 52 metres high, approached by a straight track directly south (14 km) from the main As Sila to Abu Dhabi road. Here a few bone fragments and *Struthio* sp. eggshell were found. This locality is thought to be in the upper part of the Baynunah Formation but there is no hard evidence for this. Micritic limestones contain casts of ?cerithid gastropods (Ditchfield, 1999—Chapter 7).

Hamra Sites (H) (tables 23.1–23.6; fig. 23.1 and plate 23.3)

The Hamra sites can easily be seen from the main road. Site H1 (N 23° 06′ 50.6″, E 52° 31′ 31.0″) is the northern area of an east-facing cliff, part of which is a disused firing range, where *Pachyportax latidens* was found (Gentry, 1999b—Chapter 22). Site H2 (N 23° 06′ 06.9″, E 52° 31′ 42.5″) is the southern part of this cliff. To the west, site H4 from which *Plesiogulo praecocidens* (Barry, 1999—Chapter 17) was discovered in 1989, could not be surveyed as it now a prohibited area. Site H5 (N 23° 06′ 04.7″, E 52° 31′ 38.5″) is a palaeomagnetic sampling site (Hailwood and Whybrow, 1999—Chapter 8). Site H6 (N 23° 06′ 43.7″, E 52° 31′ 28.2″) lies near to the top of the jebel. Remaining sites north of the road are H7 (N 23° 06′ 00.0″, E 52° 31′ 55.9″) a UAE trignometrical station (now demolished) and H9 (N 23° 06′ 08.2″, E 52° 31′ 28.2″). About 2 km south of the road lies Jebel Al Hamra, H8 (N 23° 04′ 24.4″, E 52° 31′ 45.6″, TC 85 survey point; U.K. Ministry of Defence Survey Map, 1:100 000, NG-39-167, 1988). Beneath the survey point lies site H3 (N 23° 04′ 28.7″, E 52° 31′ 37.3″) from where a badly weathered proboscidean skull was found but was not collected as it was badly fractured. Palaeosol carbonate nodules were collected from the Hamra locality (Kingston, 1999—Chapter 25), figure 23.4.

Jebel Dhanna (JD) (tables 23.1–23.6; fig. 23.1)

Most of Jebel Dhanna is now a restricted area. Three (out of four) productive sites exist on the flanks of this salt diapir. Site JD1 (N 24° 08′ 38.4″, E 52° 38′ 05.5″, ADCO GPS) was first found by German archaeologists (Vogt et al., 1989) collaborating on a survey with Al Ain Museum in 1983. A few bone fragments have been collected here. The site is an un-named jebel immediately next to the road about 4 km north of the Jebel Dhanna/As Sila to Abu Dhabi road junction. From site JD3 (N 24° 10′ 31.1″, E 52° 34′ 21.0″)—there is no JD2—comes the second discovery of a primate from the Arabian Peninsula (Hill and Gundling, 1999—Chapter 16), also the holotype of *Hipparion abudhabiense* (Eisenmann and Whybrow, 1999—Chapter 19). Site JD5 (N 24° 10′ 22.9″, E 52° 34′ 38.5″) consists of several hillocks lying to the east of power lines about 1 km east of JD3. Numerous reptile and fish bones have weathered out of soft, red sandstones and the area has now suffered extensively from the use of earth-moving machinery. Site JD4 (N 24° 10′ 42.6″, E 52° 34′ 12.8″) is the north-facing cliff (formerly a sea cliff) within the area now used as the Dalma Island Ferry Terminal. The foreshore has been reclaimed since 1992 and about 2 metres of section are now buried. It is an excellent and readily accessible exposure of the Baynunah Formation river gravels, from which a horse tooth and one proboscidean bone have been recovered. In 1996, new roads, pipelines, and buildings in the northwestern part of Jebel Dhanna had effectively removed most of the fossiliferous sites.

Ras Dubay'ah (R) (tables 23.1–23.6; fig. 23.1)

Ras Dubay'ah has not been surveyed and is now a restricted area. The only evidence of a rhinoceros in the Baynunah fauna, a fragmented molar, comes from this site.

Ghayathi and Bin Jawabi (G and BJ) (tables 23.1–23.3; fig. 23.1 and plate 23.4)

Near the town of Ghayathi lies a small quarry from which unionoid molluscs (Jeffery, 1999—Chapter 10) were collected. It is located (site G1, N 23° 54′ 30.7″, E 52° 48′ 15.9″) about 18 km north, and about 1 km east of the main road from Ghayathi. From bin Jawabi, only a few bone fragments were found and this site might be in the "upper" part of the Baynunah Formation (see Chapter 5).

Kihal (K) (tables 23.2–23.6; fig. 23.1)

The Kihal site, K1 (N 24° 07′ 23.2″, E 53° 00′ 27.9″), can be seen from the main Abu Dhabi road and is approached by a track that leads northwest to Thumayriyah (U.K. Ministry of Defence Survey Map, 1:100 000, NG-39-167, 168, 1988). The main fossil-bearing sediments are those to the east of a large, flat-topped jebel, site K2 (N 24° 06′ 59.5″, E 53° 01′ 07.7″), on top of which is a UAE trignometrical station G 123c. Palaeosol horizons were noted at this locality (Kingston, 1999—Chapter 25) and it is the only site from which soricid rodents have been found (de Bruijn and Whybrow, 1994; de Bruijn, 1999—Chapter 15).

Ras al Qa'la (Q) (tables 23.1–23.6; fig. 23.1)

Ras al Qa'la is the northern part, Q1 (N 24° 09′ 07.1″, E 52° 58′ 32.0″), of the Kihal exposures and can only be visited when the sabkha between it and Kihal is dry. A few fragmented vertebrates were found here but the sea cliff sediments of the Baynunah Formation are well exposed on its northwestern flank.

Thumayriyah (TH) (tables 23.2–23.6; fig. 23.1)

Site TH1 (N 24° 09′ 25.1″,E 53° 00′ 32.6″) where well-preserved suid material was collected (Bishop and Hill, 1999—Chapter 20) lies on private property, entry to which is restricted.

Ras al Aysh (A) (tables 23.5 and 23.6; fig. 23.1)

Ras al Aysh is now difficult to reach owing to fencing and ditching of the intervening sabkha. Sites A1 (N 24° 04′ 59.8″, E 53° 12′ 49.9″) and A2 (N 24° 04′ 18.3″, E 53° 13′ 22.7″).

Table 23.1. Plants and invertebrates from the Baynunah Formation

	B	S	BJ	H	JD	G	Q
Plantae							
Algae gen. et. sp. indet.	■						
Leguminosae							
?*Acacia* sp.		■	■	■			■
Invertebrata							
Mollusca							
Gastropoda							
Buliminidae							
?*Subzebrinus* or ?*Pseudonapaeus*		■					
Bivalvia							
Mutelidae							
Mutela (?subgen. nov. aff. *Chelidonopsis*) sp.	■	■					
Unionidae							
Leguminaia (*Leguminaia*) sp.	■			■	■	■	
Crustacea							
Ostracoda							
Cytherideidae							
Cyprideis sp.	■						

Locality abbreviations: B, Jebel Barakah; S, Shuwaihat; BJ, bin Jawabi; H, Hamra; JD, Jebel Dhanna; G, Ghayathi; Q, Ras al Qa'la.

Table 23.2. Fish, chelonia, and crocodiles from the Baynunah Formation

	B	S	BJ	H	JD	Q	R	K	TH	Y	M	MQ
Pisces												
Siluriformes												
Clariidae												
Clarias sp.	■	■		■	■		■	■			■	
Bagridae												
Bagrus shuwaiensis	■	■		■	■		■			■	■	
Cypriniformes												
Cyprinidae												
Barbus sp.		■										
Reptilia												
Chelonia												
Trionychidae												
Trionyx s.l. sp.	■	■		■	■		■		■			■
Testudinidae												
cf. *Mauremys* sp.	■	■		■	■		■	■			■	
Geochelone (*Centrochelys*) aff. *sulcata*		■		■	■			■				
Crocodilia: gen. et. sp. indet.	■	■	■	■	■	■		■	■	■	■	■
Crocodylidae												
Crocodylus cf. *niloticus*		■										
Crocodylus sp. indet.		■		■								
Gavialidae												
?*Ikanogavialis*		■										
gen. et sp. indet.						■						

Locality abbreviations: B, Jebel Barakah; S, Shuwaihat; BJ, bin Jawabi; H, Hamra; JD, Jebel Dhanna; Q, Ras al Qa'la; R, Ras Dubay'ah; K, Kihal; TH, Thumayriyah; Y, Harmiyah; M, Jebel Mimiyah (Al Mirfa); MQ, Mirfa Quarry.

Table 23.3. Aves from the Baynunah Formation

	B	S	GU	BJ	H	JD	Q	R	K	TH
Aves										
Struthionidae										
Struthio sp.	■	■	■	■	■	■	■	■	■	■
Ardeidae										
Egretta aff. *alba*		■								

Locality abbreviations: B, Jebel Barakah; S, Shuwaihat; GU, Al Ghuddah; BJ, bin Jawabi; H, Hamra; JD, Jebel Dhanna; Q, Ras al Qa'la; R, Ras Dubay'ah; K, Kihal; TH, Thumayriyah.

Harmiyah (Y) (tables 23.2, 23.5, 23.6; fig. 23.1)

Site Y1 (N 24° 04′ 38.4″, E 53° 20′ 23.7″) is a low hill immediately to the north of the main road. Site Y2 (N 24° 04′ 38.4″, E 53° 19′ 29.0″) is a UAE trignometrical station G 130 to the west of Y1.

Jebel Mimiyah (M) (tables 23.2–23.6; fig. 23.1 and plate 23.5)

Site M1 is what remains of Jebel Mimiyah itself (N 24° 04′ 58.2″, E 53° 26′ 07.6″). Known to the NHM/Yale team as "Twin Peaks", this jebel has been bulldozed for roadbuilding material. Parts of it are unsafe but fluvial sediments of the Baynunah river are beautifully exposed. Site MQ (N 24° 02′ 18.6″, E 53° 27′ 45.4″) was a shallow quarry on the south side of the road that could not be found in 1995 and has probably been filled in.

Tarif (T) (table 23.6; fig. 23.1)

Tarif has only one site, T2 (N 23° 56′ 57.4″, E 53° 41′ 32.8″), from which only one fossil, the partial skull of *Tragoportax cyrenaicus* (Gentry, 1999b—Chapter 22) was discovered by staff of the United States Geological Survey. Site T1 (N 23° 57′ 12.3″, E 53° 41′ 21.2″) is a UAE trignometrical station G 138 and site T3 (N 23° 57′ 13.1″, E 53° 41′ 21.4″) is a survey point, TC 72.

The complete flora and fauna recovered from the Baynunah Formation is listed in Table 23.7.

Table 23.4. Small mammals (insectivore and rodents) from the Baynunah Formation

	S	K	M
Insectivora			
Soricidae			
gen. et sp. indet.		■	
Rodentia			
Thryonomyidae			
gen. et sp. indet.	■		
Gerbillidae			
Gerbillinae			
Abudhabia baynunensis	■		■
Myocricetodontinae			
Myocricetodon sp. nov.			■
Muridae			
Murinae			
Parapelomys cf. *charkhensis*	■		■
Dendromurinae			
Dendromus aff. *melanotus*	■		
Dendromus sp.	■		
Dipodidae			
Zapodinae			
gen. et sp. indet.	■		

Locality abbreviations: S, Shuwaihat; K, Kihal; M, Jebel Mimiyah (Al Mirfa).

Table 23.5. Mammals from the Baynunah Formation (see also tables 23.4 and 23.6)

	B	S	H	JD	R	K	TH	A	Y	M
Primates										
Cercopithecidae										
gen. et sp. indet.				■						
Carnivora										
Mustelidae										
Plesiogulo praecocidens			■							
Hyaenidae										
gen. et sp. indet. (very large)				■						
gen. et sp. indet. (medium-sized)										■
Felidae										
Machairodontinae										
gen. et sp. indet.					■	■				
Proboscidea										
Deinotheriidae										
gen. et sp. indet.		■								
Elephantidae										
Stegotetrabelodon syrticus		■	■	■	■	■				
Stegotetrabelodon sp. indet.	■	■	■	■		■	■			■
Family indet.										
gen. et. sp. indet. (possibly "*Mastodon*" *grandincisivus*	■									
Perissodactyla										
Equidae										
Hipparion abudhabiense	■	■	■	■		■				
Hipparion sp.	■	■	■	■	■					
"*Hipparion*" indet.	■	■	■	■	■	■	■	■	■	■
Rhinocerotidae										
gen. et sp. indet.						■				
Artiodactyla										
Suidae										
Propotamochoerus hysudricus	■	■	■							■
Nyanzachoerus syrticus		■	■		■		■			

Locality abbreviations: B, Jebel Barakah; S, Shuwaihat; H, Hamra; JD, Jebel Dhanna; R, Ras Dubay'ah; K, Kihal; TH, Thumayriyah; A, Ras al Aysh; Y, Harmiyah; M, Jebel Mimiyah (Al Mirfa).

Table 23.6. Mammals from the Baynunah Formation (see also tables 23.4 and 23.5)

	B	S	H	JD	Q	R	K	TH	A	Y	M	MQ	T
Artiodactyla													
Hippopotamidae													
Hexaprotodon aff. *sahabiensis*	■	■	■	■	■		■	■		■		■	
Giraffidae													
?*Palaeotragus* sp.				■		■							
?*Bramatherium* sp.		■				■							
gen. et sp. indet.		■	■			■			■				
Bovidae													
gen. et sp. indet.		■								■			
Boselaphini													
Tragoportax cyrenaicus	■	■	■	■			■	■	■				■
Pachyportax latidens		■	■			■							
Antilopini													
Prostrepsiceros aff. *libycus*		■	■			■	■	■		■			
Prostrepsiceros aff. *vinayaki*	■	■	■	■				■			■		
Gazella aff. *lydekkeri*			■							■			

Locality abbreviations: B, Jebel Barakah; S, Shuwaihat; H, Hamra; JD, Jebel Dhanna; Q, Ras al Qa'la; R, Ras Dubay'ah; K, Kihal; TH, Thumayriyah; A, Ras al Aysh; Y, Harmiyah; M, Jebel Mimiyah (Al Mirfa); MQ, Mirfa Quarry; T, Tarif.

Table 23.7. Flora and fauna: composite list for the Baynunah Formation

- Plantae
    - Algae
        - gen. et sp. indet.
    - Leguminosae
        - ?*Acacia* sp.
- Mollusca
    - Gastropoda
        - Buliminidae
            - ?*Sebzebrinus* sp. or ?*Pseudonapaeus* sp.
    - Bivalvia
        - Mutelidae
            - *Mutela* (?subgen. nov. aff. *Chelidonopsis*) sp.
        - Unionidae
            - *Leguminaia* (*Leguminaia*) sp.
- Crustacea[a]
    - Ostracoda
        - Cytherideidae
            - *Cyprideis* sp.
- Pisces
    - Siluriformes
        - Clariidae
            - *Clarias* sp.
        - Bagridae
            - *Bagrus shuwaiensis* Forey and Young, 1999
    - Cypriniformes
        - Cyprinidae
            - *Barbus* sp.
- Reptilia
    - Chelonia
        - Trionychidae
            - *Trionyx* s.l. sp.
        - Testudinidae
            - cf. *Mauremys* sp.
            - *Geochelone* (*Centrochelys*) aff. *sulcata*
    - Crocodilia
        - Crocodylidae
            - *Crocodylus* cf. *niloticus*
            - *Crocodylus* sp.
    - Gavialidae
            - ?*Ikanogavialis*
            - gen. et sp. indet

(*continued*)

Table 23.7. Fauna and flora: composite list for the Baynunah Formation (*continued*)

Aves[a]
- Struthionidae
  - *Struthio* sp.
- Ardeidae
  - *Egretta* aff. *alba*

Mammalia
- Rodentia
  - Thryonomyidae
    - gen. et sp. indet.
  - Gerbillidae
    - Gerbillinae
      - *Abudhabia baynunensis* de Bruijn and Whybrow, 1994
    - Myocricetodontinae
      - *Myocricetodon* sp. nov.
  - Muridae
    - Murinae
      - *Parapelomys* cf. *charkhensis*
    - Dendromurinae
      - *Dendromus* aff. *melanotus*
      - *Dendromus* sp.
  - Dipodidae
    - Zapodinae
      - gen. et sp. indet.
- Insectivora
  - Soricidae
    - gen. et sp. indet.
- Primates
  - Cercopithecidae
    - gen. et sp. indet.
- Carnivora
  - Mustelidae
    - *Plesiogulo praecocidens*
  - Hyaenidae
    - gen. et sp. indet. (very large)
    - gen. et sp. indet. (medium-sized)
  - Felidae
    - Machairodontinae
      - gen. et sp. indet.
- Proboscidea
  - Deinotheridae
    - gen. et sp. indet.
  - Elephantidae
    - *Stegotetrabelodon syrticus*

(*continued*)

Table 23.7. Fauna and flora: composite list for the Baynunah Formation (*continued*)

- Family indet.
  - gen. et sp. indet. (possibly "*Mastodon*" *grandincisivus*)
- Perissodactyla
  - Equidae
    - *Hipparion abudhabiense* Eisenmann and Whybrow, 1999
    - *Hipparion* sp.
  - Rhinocerotidae
    - gen. et sp. indet.[b]
- Artiodactyla
  - Suidae
    - *Propotamochoerus hysudricus*
    - *Nyanzochoerus syrticus*
  - Hippopotamidae
    - *Hexaprotodon* aff. *sahabiensis*
  - Giraffidae
    - ?*Palaeotragus* sp.
    - ?*Bramatherium* sp.
    - gen. et sp. indet.
  - Bovidae
    - gen. et sp. indet.
- Boselaphini
    - *Tragoportax cyrenaicus*
    - *Pachyportax latidens*
  - Antilopini
    - *Prostrepsiceros* aff. *libycus*
    - *Prostrepsiceros* aff. *vinayaki*
    - *Gazella* aff. *lydekkeri*

[a]These groups are not dealt with taxonomically in this volume (Whybrow and Hill, 1999).
[b]One tooth fragment not described in this volume.

Acknowledgements

We thank the Abu Dhabi Company for Onshore Oil Operations (ADCO) for their continuing grant to The Natural History Museum, London, and for their generous support for Miocene studies in the United Arab Emirates since 1991. We also thank ADCO and the Ministry of Higher Education and Scientific Research, United Arab Emirates, for their invitation to participate in the First International Conference on the Fossil Vertebrates of Arabia held in the Emirate of Abu Dhabi, March 1995. This contribution forms part of The Natural History Museum's Global Change and the Biosphere research programme.

References

Andrews, P. 1999. Taphonomy of the Shuwaihat proboscidean, late Miocene, Emirate of Abu Dhabi, United Arab Emirates. Chap. 24 in *Fossil Vertebrates of Arabia,* pp. 338–53 (ed. P. J. Whybrow and A. Hill). Yale University Press, New Haven.

Barry, J. C. 1999. Late Miocene Carnivora from the Emirate of Abu Dhabi, United Arab Emirates. Chap. 17 in *Fossil Vertebrates of Arabia,* pp. 204–208 (ed. P. J. Whybrow and A. Hill). Yale University Press, New Haven.

Bishop, L., and Hill, A. 1999. Fossil Suidae from the Baynunah Formation, Emirate of Abu Dhabi, United Arab Emirates. Chap. 20 in *Fossil Vertebrates of Arabia,* pp. 254–70 (ed. P. J. Whybrow and A. Hill). Yale University Press, New Haven.

Bristow, C. S. 1999. Aeolian and sabkha sediments in the Miocene Shuwaihat Formation, Emirate of Abu Dhabi, United Arab Emirates. Chap. 6 in *Fossil Vertebrates of Arabia* , pp. 50–60 (ed. P. J. Whybrow and A. Hill). Yale University Press, New Haven.

Bruijn, H. de. 1999. A late Miocene insectivore and rodent fauna from the Baynunah Formation, Emirate of Abu Dhabi, United Arab Emirates. Chap. 15 in *Fossil Vertebrates of Arabia,* pp. 186–97 (ed. P. J. Whybrow and A. Hill). Yale University Press, New Haven.

Bruijn, H. de, and Whybrow, P. J. 1994. A Late Miocene rodent fauna from the Baynunah Formation, Emirate of Abu Dhabi; United Arab Emirates. *Proceedings Koninklijke Nederlandse Akademie van Wetenschappen* 97: 407–22.

Ditchfield, P. W. 1999. Diagenesis of the Baynunah, Shuwaihat and Upper Dam Formation sediments exposed in the Western Region, Emirate of Abu Dhabi, United Arab Emirates. Chap. 7 in *Fossil Vertebrates of Arabia,* pp. 61–74 (ed. P. J. Whybrow and A. Hill). Yale University Press, New Haven.

Eisenmann, V., and Whybrow, P. J. 1999. Hipparions from the late Miocene Baynunah Formation, Emirate of Abu Dhabi, United Arab Emirates. Chap. 19 in *Fossil Vertebrates of Arabia,* pp. 234–53 (ed. P. J. Whybrow and A. Hill). Yale University Press, New Haven.

Forey, P. L., and Young, S. V. T. 1999. Late Miocene fishes of the Emirate of Abu Dhabi, United Arab Emirates. Chap. 12 in *Fossil Vertebrates of Arabia,* pp. 120–35 (ed. P. J. Whybrow and A. Hill). Yale University Press, London and New Haven.

Friend, P. F. 1999. Rivers of the Lower Baynunah Formation, Emirate of Abu Dhabi, United Arab Emirates. Chap. 5 in *Fossil Vertebrates of Arabia,* pp.

38–49 (ed. P. J. Whybrow and A. Hill). Yale University Press, New Haven.

Gentry, A. W. 1999a. A fossil hippopotamus from the Emirate of Abu Dhabi, United Arab Emirates. Chap. 21 in *Fossil Vertebrates of Arabia,* pp. 271–89 (ed. P. J. Whybrow and A. Hill). Yale University Press, New Haven.

Gentry, A. W. 1999b. Fossil pecorans from the Baynunah Formation, Emirate of Abu Dhabi, United Arab Emirates. Chap. 22 in *Fossil Vertebrates of Arabia,* pp. 290–316 (ed. P. J. Whybrow and A. Hill). Yale University Press, New Haven.

Hailwood, E. A., and Whybrow, P. J. 1999. Palaeomagnetic correlation and dating of the Baynunah and Shuwaihat Formations: Emirate of Abu Dhabi, United Arab Emirates. Chap. 8 in *Fossil Vertebrates of Arabia,* pp. 75–87 (ed. P. J. Whybrow and A. Hill). Yale University Press, New Haven.

Hill, A. 1987. Causes of perceived faunal change in the later Neogene of East Africa. *Journal of Human Evolution* 16: 583–96.

Hill, A., and Gundling, T. 1999. A monkey (Primates, Cercopithecidae) from the late Miocene of Abu Dhabi, United Arab Emirates. Chap. 16 in *Fossil Vertebrates of Arabia,* pp. 198–203 (ed. P. J. Whybrow and A. Hill). Yale University Press, New Haven.

Hill, A., Whybrow, P. J., and Yasin, W. 1999. History of palaeontological research in the Western Region of the Emirate of Abu Dhabi, United Arab Emirates. Chap. 3 in *Fossil Vertebrates of Arabia,* pp. 15–23 (ed. P. J. Whybrow and A. Hill). Yale University Press, New Haven.

Jeffery, P. A. 1999. Late Miocene swan mussels from the Baynunah Formation, Emirate of Abu Dhabi, United Arab Emirates. Chap. 10 in *Fossil Vertebrates of Arabia,* pp. 111–15 (ed. P. J. Whybrow and A. Hill). Yale University Press, London and New Haven.

Kingston, J. D. 1999. Isotopes and environments of the Baynunah Formation, Emirate of Abu Dhabi, United Arab Emirates. Chap. 25 in *Fossil Vertebrates of Arabia,* pp. 354–72 (ed. P. J. Whybrow and A. Hill). Yale University Press, New Haven.

Rauhe, M., Frey, E., Pemberton, D. S., and Rossman, T. 1999. Fossil crocodilians from the late Miocene Baynunah Formation of the Emirate of Abu Dhabi, United Arab Emirates: Osteology and palaeoecology. Chap. 14 in *Fossil Vertebrates of Arabia,* pp. 163–85 (ed. P. J. Whybrow and A. Hill). Yale University Press, New Haven.

Tassy, P. 1999. Miocene elephantids (Mammalia) from the Emirate of Abu Dhabi, United Arab Emirates: Palaeobiogeographic implications. Chap. 18 in *Fossil Vertebrates of Arabia,* pp. 209–33 (ed. P. J. Whybrow and A. Hill). Yale University Press, New Haven.

Vogt, B., Gockel, W., Hofbauer, H., and Al-Haj, A. A. 1989. The coastal Survey in the Western Province of Abu Dhabi. *Archaeology in the United Arab Emirates* V: 49–60.

Whybrow, P. J. 1989. New stratotype; the Baynunah Formation (Late Miocene), United Arab Emirates: Lithology and palaeontology. *Newsletters on Stratigraphy* 21: 1–9.

Whybrow, P. J., and Hill, A. eds. 1999. *Fossil Vertebrates of Arabia.* Yale University Press, New Haven.

Whybrow, P. J., Friend, P. F., Ditchfield, P. W., and Bristow, C. S. 1999. Local stratigraphy of the Neogene outcrops of the coastal area: Western Region, Emirate of Abu Dhabi, United Arab Emirates. Chap. 4 in *Fossil Vertebrates of Arabia,* pp. 28–37 (ed. P. J. Whybrow and A. Hill). Yale University Press, New Haven.

Whybrow, P. J., Hill, A., Yasin al-Tikriti, W., and Hailwood, E. A. 1990. Late Miocene primate fauna, flora and initial palaeomagnetic data from the Emirate of Abu Dhabi, United Arab Emirates. *Journal of Human Evolution* 19: 583–88.

Dunes near the Liwa Oasis, Emirate of Abu Dhabi.

Proboscidean Taphonomy, Isotopes, and Environments of the Baynunah Formation; Artifacts from the Western Region, Emirate of Abu Dhabi; and Arabian Palaeoenvironments

PART IV

Peter Andrews describes (Chapter 24) the excavation and taphonomy of a single individual of *Stegotetrabelodon syrticus*. This individual was resting on and in a gravel deposit in a dry river channel, having been transported there from a short distance away as a partially articulated carcass. The skull, many ribs, and vertebrae, and most of the major elements of the hind limbs were present, but only two portions of the fore limb were preserved. All of the distal limb elements were missing, but there is no evidence of scavenging. The most likely explanation for the missing bones is through differential transport. Burial was rapid, taking place after disarticulation of all of the bones but before dispersal and before any degree of weathering. Limited weathering of the largest skeletal elements took place, probably because they projected above the early stages of sedimentary burial. Recent erosion has exposed the skull, possibly destroying parts of the skeleton and causing great damage from recent weathering.

John D. Kingston has undertaken carbon isotope studies (Chapter 25) on soil carbonates and herbivore enamel from the Baynunah Formation. Establishing a palaeoecological context for fossil faunal assemblages is critical for understanding the pattern and process of evolution. Inherent in providing such a framework is reconstructing the vegetation. Stable isotopic analyses of palaeosol components and fossil herbivore enamel provides a quantitative means of documenting relative proportions of plants utilising the C_4 and C_3 photosynthetic pathways in the past, reflecting tropical grasslands and wooded/forested habitats, respectively.

Soil carbonates, forming in isotopic equilibrium with CO_2 derived from plant decomposition and respiration, retain a signature of relative amounts of C_4 and C_3 in modern landscapes. This biogenic signal is preserved in the fossil record and can provide estimates of the proportion of C_4 and C_3 plants in the past. Stable carbon isotopic analysis of carbonate nodules from palaeosol horizons in the Baynunah Formation indicate carbonate formation in mixed C_4 and C_3 environments. The lateral variation in the carbon isotopic composition ($\delta^{13}C$) in these nodules most closely resembles that of soil carbonates forming today in grassy woodlands/bushlands.

It has also been well established that the $\delta^{13}C$ of modern tooth apatite is directly related to the isotopic composition of the diet and that this signature remains intact during fossilisation. Palaeodietary reconstructions based on isotopic analyses of the enamel of fossil herbivores are useful in constraining aspects of the vegetation available for feeding in a region. Isotopic analyses of carbonate occluded within the fossil enamel of five herbivore families collected from the Baynunah Formation also suggest a mosaic of C_4 and C_3 vegetation during the late Miocene. The $\delta^{13}C$ of fossil enamel, however, indicates a heavy reliance on C_4 grasses with over a third of the specimens reflecting an exclusive C_4 diet and most of the remaining samples suggesting a significant C_4 component in the diet.

Sally McBrearty in Chapter 26 describes stone artifacts from four localities in the Western Region—Shuwaihat, Hamra, Ras al Aysh, and Jebel Barakah. The artifacts probably occur here because siliceous cap-rocks of the Baynunah Formation provided the raw material from which the tools were made. They occur in surface context, and there is no associated fauna or other cultural debris that would facilitate determining their age or affinities. Technologically the artifacts from Shuwaihat, Hamra, and Ras al Aysh have similarities with a blade industry from the northeastern coast of Qatar. The presence of distinctive crescent-shaped cores links them to the Prepottery Neolithic B of the Levant that dates to about 9300–8300 years before present. The artifacts from Jebel Barakah differ, however, from those at Shuwaihat, Hamra, and Ras al Aysh. There are no blades, but the collection includes radial cores and a single broken biface tip. The techniques represented at Jebel Barakah could date to almost any period, but bifaces and radial cores are known from sites elsewhere on the Arabian Peninsula thought to date to about 7000–4500 b.p. Diagnostic retouched pieces that would be useful in resolving the age of the artifacts are almost entirely lacking at the Abu Dhabi sites. McBrearty notes that they may be absent because the Baynunah Formation outcrops were used as quarries, and the flake or blade blanks made here may have been transported elsewhere for transformation into finished tools.

John D. Kingston and Andrew Hill (Chapter 27) provide a synthesis of the late Miocene palaeoenvironments in Arabia. The palaeoclimate of the Arabian Peninsula during the mid- to late Miocene is poorly known, as are the types of terrestrial palaeohabitats that dominated this region. Lithofacies and associated vertebrate assemblages of the Baynunah Formation (8–6 million years old) provide the only known empirical evidence for terrestrial palaeoenvironments on the Arabian Peninsula during the late Miocene. In their chapter they consider a compilation of palaeoenvironmental data collected from the Baynunah and Shuwaihat Formations exposed in the Emirate of Abu Dhabi and early to middle Miocene deposits in the Kingdom of Saudi Arabia, along with a brief consideration of global and regional scale climatic conditions that may have influenced the movement of faunas across this region.

Taphonomy of the Shuwaihat Proboscidean, Late Miocene, Emirate of Abu Dhabi, United Arab Emirates

24

PETER ANDREWS

During a survey of Shuwaihat Island at the end of the winter field season of 1990, Andrew Hill (Yale University) found fragmented bones lying on the surface; these bones were from the back part of an elephant (proboscidean) mandible. He and Peter Whybrow (The Natural History Museum) partially excavated the jaw but time did not allow further work. Events in the northern Gulf prevented much work in 1991 even though there was a Natural History Museum presence in Abu Dhabi, so in January 1992 excavation of the jaw, by Gillian Comerford (NHM) and Andrew Hill, commenced. During the excavation, the cranium was found and it was realised that much of the skeleton might be present at the site. Consequently, a full excavation had to be postponed until spring 1992.

In April 1992, a NHM team of eight arrived at Shuwaihat. The proboscidean cranium found during the winter was excavated. The mandible and one postcranial bone (tibia) had already been removed. This excavation was extended further in April 1994, to try and recover more of the skeleton and in particular to attempt to find parts of the fore limb and distal limb elements not found earlier. At the end of the April 1994 season a total area of 153 m^2 had been excavated, and it was considered likely that most of the remaining bones had been recovered. During these two seasons' excavation, 35 bones from the proboscidean skeleton were found together with other fauna—mainly crocodile, fish, two suid specimens, a bovid horn core, and a piece of ostrich egg shell. The area of greatest bone concentration has been cut by a recent gully to the south and east of the cranium, and it is probable that the remaining parts of the skeleton were lost by erosion long before our excavation began.

Pascal Tassy has provisionally identified the proboscidean as the genus *Stegotetrabelodon* (Tassy, 1999—Chapter 18). The Abu Dhabi specimen is a primitive species of *Stegotetrabelodon syrticus*.

METHODS

Before beginning the excavation in April 1992, a baseline running southeast at 135° was marked out. The proboscidean cranium was located 1 metre to the north of the baseline. Towards the end of the 1992 excavation a permanent marker was made with concrete and positioned at a measured distance from the baseline, and this marker was used as the site datum for the 1994 excavation. Previous to the excavations, the mandible and several other fossils had been removed from the area to the east of the cranium and within 1 metre of it, and the position and orientation of the mandible could be reconstructed with reference to the cranium by photographs.

The area immediately surrounding the cranium was first excavated by metre squares, laid out with reference to the baseline. In addition to this, a larger area to the east and southeast of the cranium was marked out in metre strips and intensively surface-collected. The excavation around the cranium was later extended westwards, to the north and to

 ISBN 0-300-07183-3

the south, until an area of 153 m^2 was uncovered. The final layout of the excavation and surface collection area are shown in figure 24.1, which is a topographic map of the local gully system with heights tied in to the local sea level (at high tide) of Shuwaihat Island. Also shown on this map is the location of the juvenile hippo jaw (AUH 481; see Gentry, 1999—Chapter 21), which was found on the surface at this locality by Andrew Hill.

Excavation method was by trowel and brush, with bones being measured in two dimensions with reference to the site datum and baseline. No heights were measured because the fossils were all from a single bedding plane, but their positions rel-

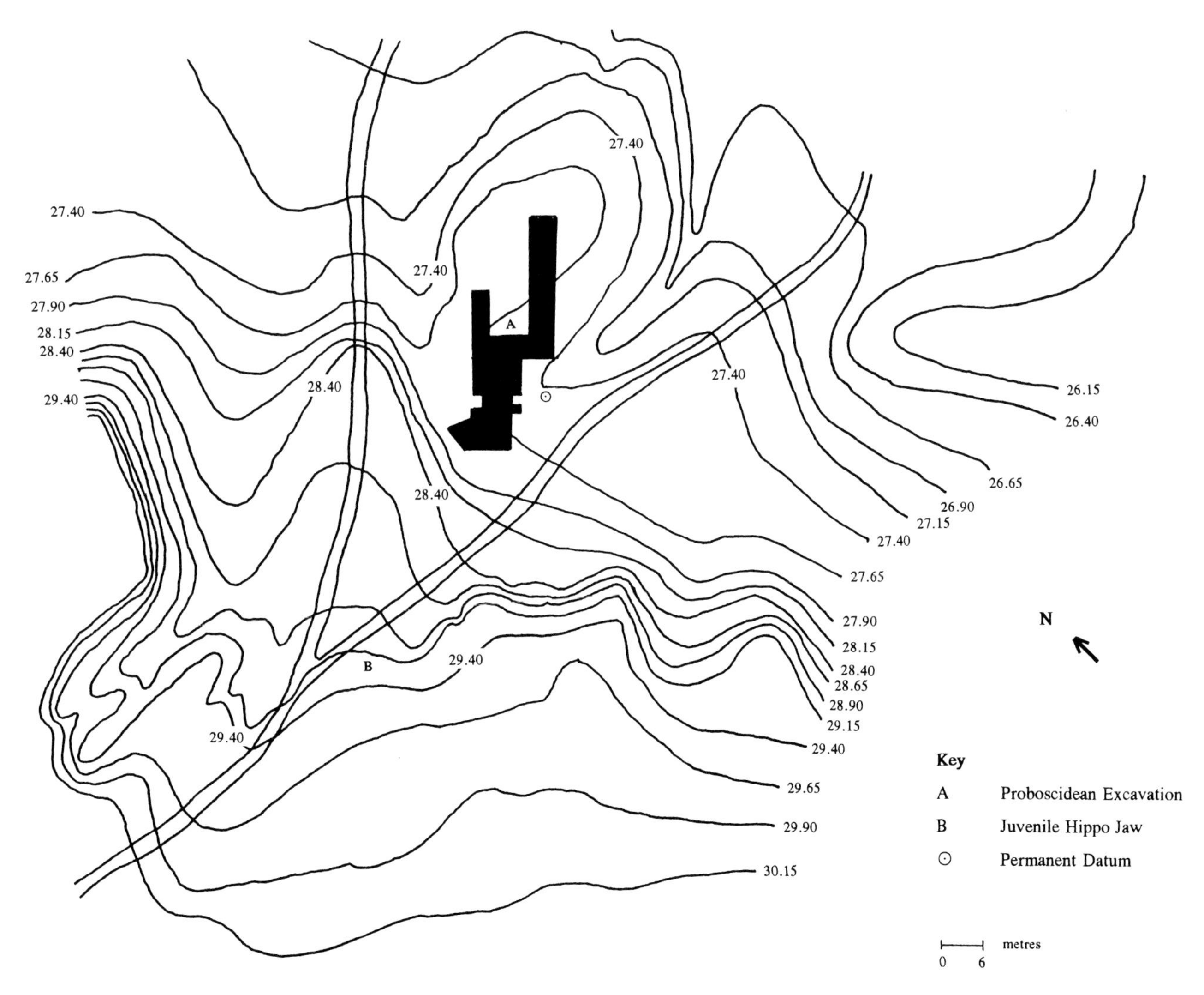

Figure 24.1. Topographic map of the Shuwaihat gully system where the proboscidean (A) and the juvenile hippo jaw (B) were found. Contours are at 25 cm intervals in metres above sea level (high-tide level measured locally). The excavation and surface collection areas are outlined in black, with the lower part being the excavation and the upper extension to the northeast being the surface collection area.

ative to sedimentary features, especially the presence or absence of gravel lenses, was recorded. Fossils were numbered starting from 1, but this has subsequently been changed to 501 to integrate the catalogue with other collections from the area.

Surface collections were made by brush and screen. The 1-metre strips were first picked over by metre square to collect any bone on the surface. The surface debris from each square was subsequently collected by brushing and the residues dry-screened.

Most of the fossils were well mineralised and in good condition and, because they were lying in soft sand, they could be excavated without any problem. The cranium, however, was poorly preserved because the top of it and the tusks were near the present land surface and had been badly affected by recent weathering. The femur crossing the front of the cranium was likewise in poor condition, and both fossils had to be plaster-jacketed before removal. The other fossils were wrapped in acid-free tissue paper on removal and transported back to the laboratory like this.

Surface Collection

Three 1-metre strips were marked out and collected by 1-metre squares over a distance of 13 metres. The strips were located at distances of 15, 16, and 17 metres along the baseline, with the proboscidean cranium located at between 10 and 13 metres along the baseline. The surface collection yielded 55 fossils from an area of 39 m^2, including three fish spines, two crocodile scutes, and three fragments of mammal. None of these could be identified and all show signs of recent weathering, so that they evidently represent the lag deposit from erosion of higher levels of sediment, along with the lag-accumulation of stones that covered the surface of the ground. In the squares closest to the cranium, nine fragments of tusk were recovered, with a further five found in an adjoining square. These tusk fragments represent the modern erosional remnants from the proboscidean cranium, and no further fragments were found below the surface. This area was not excavated further because it produced so few fossils that were taxonomically identifiable. It was found to be stratigraphically below the proboscidean level so that no further remains of the proboscidean could be expected.

Proboscidean Excavation

The proboscidean mandible was the first fossil to be excavated in 1992, and during its removal a tibia, the distal epiphysis of the right radius, and several ribs and vertebrae were found in the same general region. The location of the mandible in relation to the yet-unexcavated cranium has been reconstructed by means of photographs, and it was lying in front of the cranium and parallel to the femur and just to the right of it (fig. 24.2). The tibia, radius, ribs, and vertebrae could not be placed on the excavation plan but are listed on table 24.1 as numbers 500 and 531–536.

The main areas of the excavation were to the west of the proboscidean cranium (fig. 24.2). Cleaning the area exposed two thoracic vertebrae, and further excavation of 12 m^2 produced one further vertebra, two ribs, and a portion of the top of the cranium that had been broken off before the cranium was buried. This fragment of cranium, number 7 on the plan (fig. 24.2),[1] is important because it was excavated in place and must therefore have been displaced from the cranium before burial. The western extension was then extended to the northeast and southwest (fig. 24.2). In the latter direction, a further 81 m^2 were excavated, yielding another five ribs, two vertebrae, and one tibia; and to the northeast another 60 m^2 were excavated, yielding another rib, two vertebrae (including the atlas), and the detached head of the other femur. No sign of the shaft of this bone could be found. Part of a suid mandible (AUH 549; Bishop and Hill, 1999—Chapter 20) was also recovered from this region. As a result of these excavations, there are now 11 complete ribs and 11 vertebrae (9 complete and 2 fragments) of proboscidean, all within 15 metres of each other.

Excavation of the proboscidean cranium itself also produced a femur beneath it, with a scapula

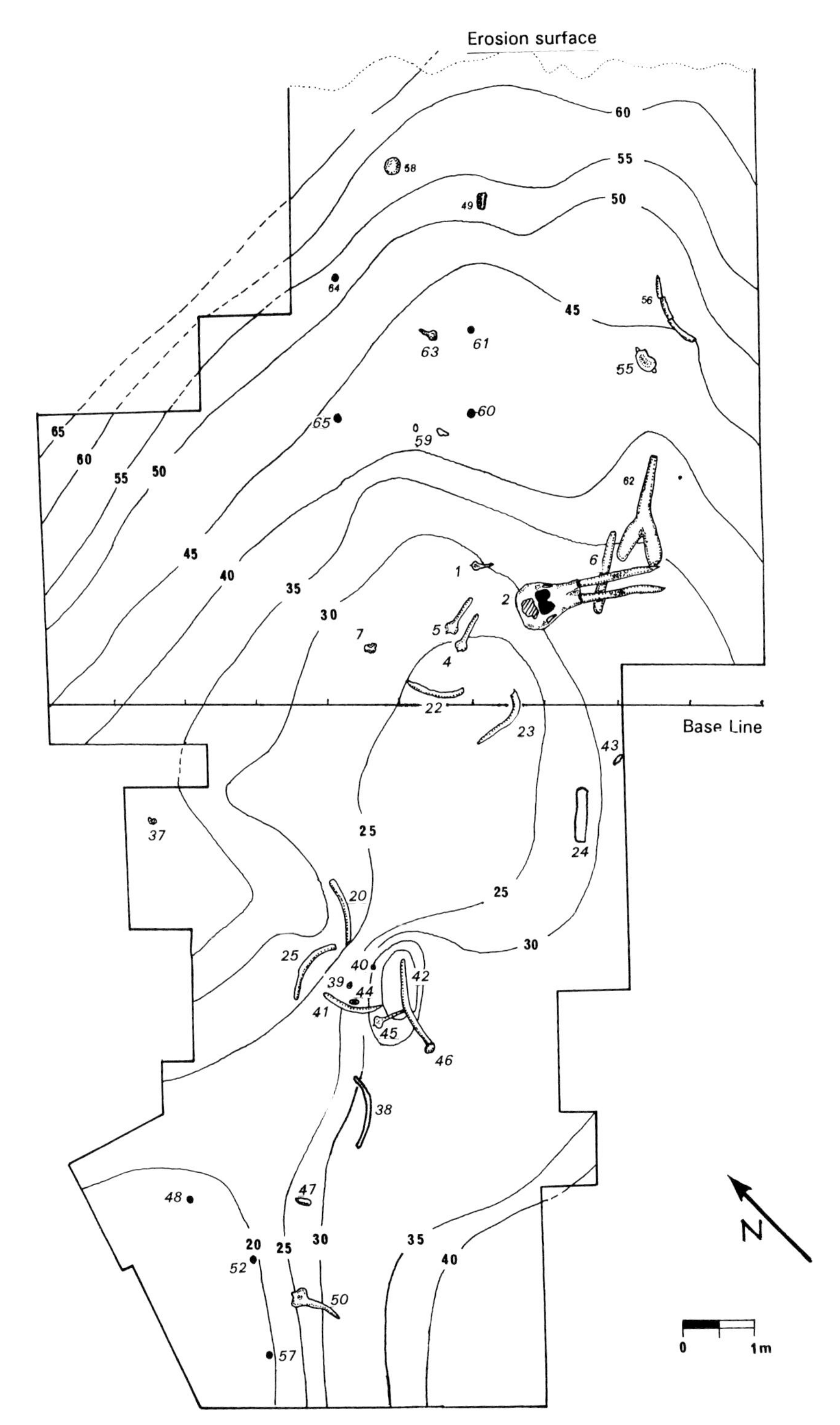

Figure 24.2. Topographic plan of the excavation, with contours at 5 cm intervals expressed as heights in centimetres above the 27-metre contour line from figure 24.1.

squashed between them. One of the teeth had been displaced back on to the vault of the cranium and, 3 metres to the north, one of the teeth detached from the cranium was found. The hind limbs of the proboscidean individual are represented by both femora and two complete tibiae, but the fore limb is represented by only one scapula. Although the atlas vertebra was found close by the cranium, there was no sign of the axis vertebra.

Each of the excavated bones was numbered—shown italicised on figure 24.2. Figure 24.2 also shows the alignment and position of the bones, plotted to scale. The elephant bones are associated with some other animal bones, such as crocodile and catfish, and two other mammal species—two suid specimens and one bovid—were also found at the same level. These fossils and the skeletal elements from the Shuwaihat proboscidean are shown in table 24.1, including the elements from both the excavation and the clearing operations. Contours at 5 cm intervals are shown on figure 24.2, and these are scaled with respect to the 27-metre contour base of figure 24.1: that is, the 30 cm contour is 27.30 metres above sea level.

Position of the Bones within the Sediments

Every one of the fossils recovered during the excavations was found resting on the surface of a pebbly gravel layer (fig. 24.3A,B), and the gravel was also banked up against the sides of some of the larger fossils (fig. 24.3C,D). Above this was a fine sandy silt, apparently deposited as a single unit and covering over all but the largest bones (see below). Some isolated fossils, such as the tibia (number 24) and ribs 22 and 23 (see fig. 24.2), that appeared at first to be enclosed in the silt were found on excavation to be resting on a small patch of the gravel, and all of the other bones were found on the main gravel exposure. Some fossils had one end resting on gravel and the other end dipping down into or resting on the underlying sands (fig. 24.3D). The cranium was also resting on a patch of gravel like the other fossils, but because of its greater size most of it was enclosed by and covered with the overlying sand.

The banking of the gravel against some fossils was consistently on the northeast sides of the bones. This would appear to indicate water flow from that direction and towards the southwest, and this is confirmed by sedimentological measurements of cross-bedding and bedforms, which show water flow towards 231–243° (P. W. Ditchfield, personal communication). This also suggests a direct association between the proboscidean bones and the gravel, for the presence of gravel beneath the bones indicates its prior deposition, while the banking of the gravel against some of the bones indicates its continued movement after the bones had come to rest. The most likely explanation for this combination is that gravel and bone deposition happened simultaneously.

The pebbly gravel layer was discontinuous and had been replaced laterally by a reddish sandy silt. In places it formed a single discrete layer, while in others it appeared to divide, so that two gravel layers were formed with a layer of sand in between. The fossils were associated with either the upper or the lower division. The clasts making up this coarse layer varied from 1 to 28 mm in diameter, with random size distribution. The clasts consist of sand grains fused by a weak carbonate cement. This probably indicates derivation from a soil. There are two possibilities for its accumulation: (1) it may be in place, resulting from a soil formed in situ, with the discontinuities produced by local erosion of the soil; or (2) it may have been transported in from a soil with a carbonate hard pan that formed elsewhere, and when this was eroded the remnants of the hard pan were transported to the proboscidean site. There is some support for the first alternative, in that the pebbles and gravels form a discrete and nearly horizontal layer, but there is no obvious erosional pattern to the discontinuities in the layer. There are discrete islands of gravel and enclosed areas of gravel removal, with no apparent channelling to account for the removal. The horizontal layering of the gravel is also consistent with the second alternative—transport and deposition by water flow—and its patchy distribution is also indicated by water transport.

The association of the fossils with the pebbly gravel layer is consistent in part with both the erosional and the transport interpretations of the sediment. In the former case, the bones may have been laid down on the surface of the carbonate-cemented soil, and their absence from areas where the gravel is missing could be due to the same erosional processes that removed the gravel. In the latter case, the transport of the gravel may be linked with the movement of the bones (see below), and this would explain both their association together and the banking up of gravels against some of the fossils.

Angles of dip of the bones were generally less than 5°, and the actual angle was more a reflection of the uneven surface of the gravel. The most steeply dipping bones were ribs 22 and 25, and these were positioned so that one end was resting on a small gravel bank and the other end dipped down into the silt where the gravel was missing, so that in effect the bones were all in undisturbed positions relative to the sediment. Their alignment appears to be random, with no preferred orientation when all the bones are plotted on a 360° rose diagram, but there is some indication of directionality running from northeast in the region of rib 56 southwards towards vertebra 50 (fig. 24.2).

Condition of the Bones

The excavated fossils were in good condition, with almost no indication of contemporaneous weathering. Some evidence of recent weathering was found; this was not unexpected as the bones were found only just below the surface of the ground. Recent weathering was greatest on the biggest fossils, such as the cranium, the top of which had been planed off level with the present land surface by recent weathering (fig. 24.4). As a result of this, the top of the vault, the premaxilla, and the upper tusks were all in an advanced state of disintegration (fig. 24.5C). Many fragments of bone and dentine littered the surface around the cranium and it was these that were collected in the surface collection (see previous section). In addition, there was some evidence of preburial weathering of the cranium in that some fragments of vault (specimen 7, fig. 24.2) were found in place in the excavation 2.5 metres to the west of the cranium, and these must have been eroded off the top of the vault before burial. Because the cranium is resting on its base, with the top rising well above the highest level of the other bones, it seems likely that after the burial of most of the skeleton, the top of the vault still rose above the blanketing sediments for a period of time before further sediment deposition covered it over. One of the teeth, the upper fourth premolar, was also found detached from the cranium nearly 3 metres away to the north.

Most of the ribs that were recovered were intact, with the only damage being slight abrasion at the ends (fig. 24.5A). They were cracked along their lengths, with transverse and stepped breaks, but there was no abrasion of these broken surfaces, and in most cases there was no displacement of one part relative to another (fig. 24.6A). In the case of specimen 56 (fig. 24.6B), one segment of rib was displaced laterally by about 1 cm across two straight transverse breaks, and both the nature of the breaks and the lack of disturbance of the rest of the specimen indicate that the displacement must have been postdepositional. No evidence could be seen of any carnivore damage to any of the ribs.

Fragile and delicate bone processes, such as vertebral spines and the scapular blade, were slightly abraded, with some loss of surface tissues (fig. 24.5B). This was probably the result of abrasion in situ during burial of the bones. There was no evidence of carnivore activity, either in terms of breakage or of gnawing marks, and so it may be concluded with some degree of certainty that the animal was not killed or scavenged by carnivores.

The major postcranial elements that were found include the femur and two tibiae. The scapula was crushed beneath the cranium and was almost destroyed in the process. The femur shaft is in poor condition, but this is almost certainly the result of recent exposure because it was lying almost at the present surface of the exposure. The epiphyses had become detached (fig. 24.7A), with some loss of the softer tissue of the developing bone, and it seems likely that this can be attributed to abrasion. The tibiae also show evidence of abrasion at their ends, particularly the lower one in figure 24.7B. This was the specimen recovered before the excava-

Table 24.1. Specimens collected from the Shuwaihat elephant site (S6)

AUH no.	Order	Part	Comments
500	Proboscidea	Distal epiphysis of right radius	Exact location not recorded
501	Proboscidea	Thoracic vertebra	
502	Proboscidea	Cranium and tusks	
503	Proboscidea	Mandible	
504	Proboscidea	Thoracic vertebra	
505	Proboscidea	Thoracic vertebra	
506	Proboscidea	Femur	
507	Proboscidea	Cranium fragment	
508–515	Proboscidea	Eight fragments	Clean up around trench 3
516	Proboscidea	Cranium fragment	Clean up
517	Crocodilia	Vertebra	Clean up
518	?	Fragments	Upper levels of trench 1
519	Proboscidea	Cranium fragment	From trench 1
520	Proboscidea	Rib	
521	?	Fragments	Clean up
522	Proboscidea	Rib	
523	Proboscidea	Rib	
524	Proboscidea	Left tibia	
525	Proboscidea	Rib	
526	?	Long bone fragment	Side of trench 3
527	Crocodilia	Tooth	Trench 1 (Cleaning Sect. 2)
528	Proboscidea	Scapula	Found under cranium
529	? Reptilia	Two fragments	Found in matrix of tusks
530	? Mammalia	One fragment	Found in matrix of tusks
531	Proboscidea	Part rib	Exact location not recorded
532	Proboscidea	Rib	Exact location not recorded
533	Proboscidea	Rib	Exact location not recorded
534	Proboscidea	Tibia (two pieces)	Exact location not recorded
535	Proboscidea	Thoracic vertebra	Exact location not recorded
536	Proboscidea	Thoracic vertebra	Exact location not recorded

tion and was lying somewhere near the femur and cranium, close to the present land surface, and the greater damage can be attributed to recent exposure. The lower specimen in figure 24.7B is number 24 on the plan (fig. 24.2), and it was buried more deeply. The proximal end shows some evidence of abrasion, but the distal end is undamaged, and the open epiphysiseal line is sharp and unabraded. This fossil also shows evidence of an early stage of weathering on the midshaft; because it was well below the present land surface this was probably due to preburial weathering.

The lack of weathering on all but the top of the cranium must indicate rapid burial of bones. The degrees of abrasion are also light, with the ends of ribs, some long bones, and delicate vertebral processes showing evidence of abrasion. The fact that most of these bones are still intact, includ-

Table 24.1. (*continued*)

AUH no.	Order	Part	Comments
537	Proboscidea	Vertebral disc	
538	Proboscidea	Rib	
539	?	Bone	
540	?	Bone	
541	Proboscidea	Rib	
542	Proboscidea	Rib	
543	Crocodilia	Lower jaw back with condyle	
544	Crocodilia	Cranial fragment	
545	Proboscidea	Thoracic vertebra	
546	Proboscidea	Vertebral disc	
547	Crocodilia	Lower jaw with teeth	
548	?Crocodilia	Bone fragment	
549	Suid	Mandible, teeth, and frags.	
550	Proboscidea	Thoracic vertebra	
551	?Various	Misc. fragments	Exposed on surface
552	Ostrich	Eggshell	
553	Various	Mostly fish	Sieved sand from baulk
554	Various	Mostly fish (suid, see 549)	Sieved from suid area
555	Proboscidea	Atlas vertebra	
556	Proboscidea	Rib	
557	Suid	Tooth	
558	Proboscidea	Head of femur	
559	Bovid	?Horn core	
560	Proboscidea	Tooth P^4	
561	Proboscidea	Tusk fragments	
562	Coprolite		
563	Proboscidea	Lumbar vertebra	
564	?	Bone-delicate/thin	
565	Proboscidea	Tusk fragments	

ing many ribs and vertebrae, must indicate lack of long-distance transport or any other physically damaging process.

Dispersal of the Skeleton

The bones from the proboscidean excavation are listed in table 24.1, which includes the bones from the excavation as well as those from the cleaning and surface collections. Parts of both hind limbs are present and many thoracic vertebrae and ribs, as well as the cranium and mandible. One cervical vertebra is present, but no lumbar vertebrae or sacrum. Other bones that have been lost include the pelvis, which is one of the bones with a large surface area relative to mass and so is readily moved by water flow, and the distal limb elements, which may be removed by scavenging activities of birds or small carnivores (Haynes, 1991). Both fore limbs are missing except for one scapula that is trapped beneath the cranium and the distal epi-

Figure 24.3. *A*, Head of femur, specimen 58, resting on and at the edge of the pebbly gravel; *B*, thoracic vertebra number 63 resting on thick layer of pebbly gravel; *C*, thoracic vertebra 45 (with a portion of rib 42 bottom left), showing the body of the vertebra resting on a thick layer of pebbly gravel, which thins out to the left and disappears altogether about half way along the vertebral spine; *D*, similar discontinuity of the pebbly gravel for specimen 50. Scales in centimetres.

Figure 24.4. The Shuwaihat proboscidean cranium being excavated by Peter Whybrow and Miranda Armour-Chelu. The upper tusks can be seen in the right foreground and the right molar toothrow centre left. The top of the cranium has been planed off by recent weathering level with the present ground surface.

physis of the radius. The fore limbs are known to disarticulate sooner than the hind limbs (Hill and Behrensmeyer, 1984), and they commonly separate off as a single unit. The scapula is attached to the body only by muscles whereas the humerus, radius, and distal extremities have ligamentous attachments. As the muscles decay or are eaten more rapidly than ligaments, the scapula detaches from the body before the humerus and other limb elements detach from the scapula. The distal elements may be removed by some transport agency, and the small bones of the feet may also disappear relatively quickly, so that it is a commonly observed stage in the dispersal of mammalian skeletons for the cranium, axial skeleton, and proximal hind limbs to be preserved with the fore limbs and feet removed.

The lack of dispersal of the remaining parts of the Shuwaihat proboscidean, in what is clearly a riverine depositional setting, suggests strongly that the bones were buried rapidly. There was clearly no transport of bones from some other place after the complete disarticulation of the skeleton, for it is now in such a compact area. This is consistent with the lack of weathering on the surfaces of the preserved bones, and it could indicate that the animal died at the place where it was found, and the bones that are missing have been moved (or not found in the limited area of the excavation). In this case, a mechanism for the removal of bone must be determined. It is unlikely to have been the result of carnivore activity, because no evidence of carnivores, such as breakage or gnawing of fragile bone processes, has been found. Movement by trampling activity is a possibility, although, again, no direct evidence for this has been found. Removal by water transport is the most likely possibility, although the bones left behind as well as the bones removed do not conform to any hydrodynamic group (Voorhies, 1969).

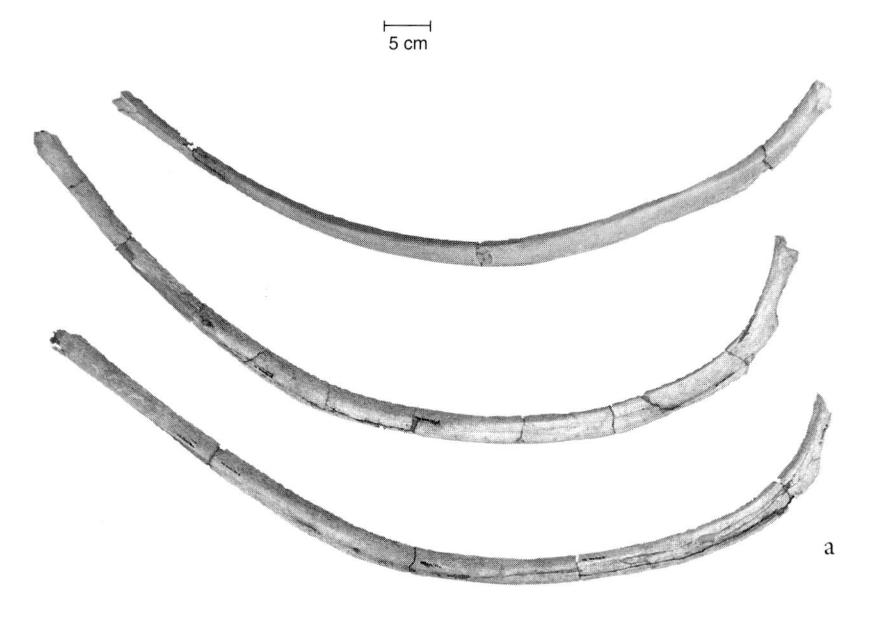

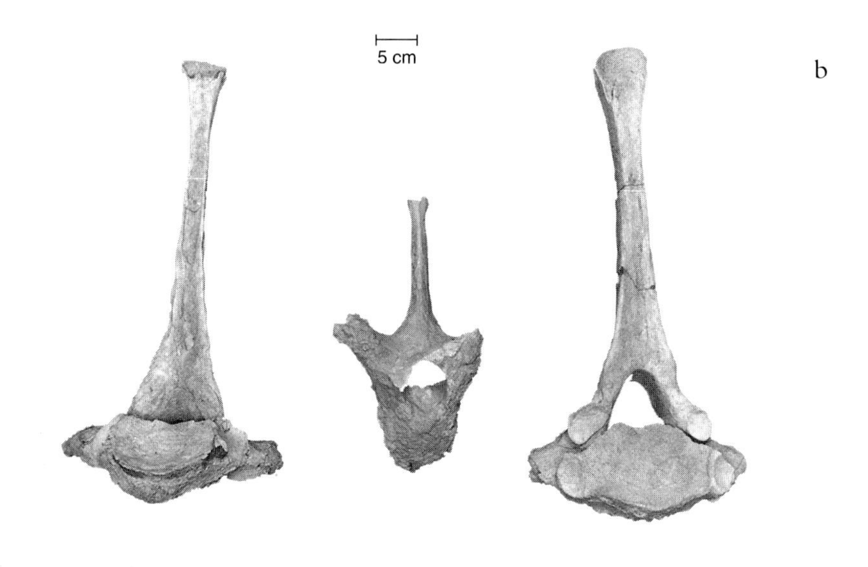

Figure 24.5. *A,* Three restored ribs, from top to bottom numbers 32, 42, and 38; *B,* three restored vertebrae, from left to right numbers 50, 35, and 5; *C,* one of the upper tusks restored but its decayed nature evident.

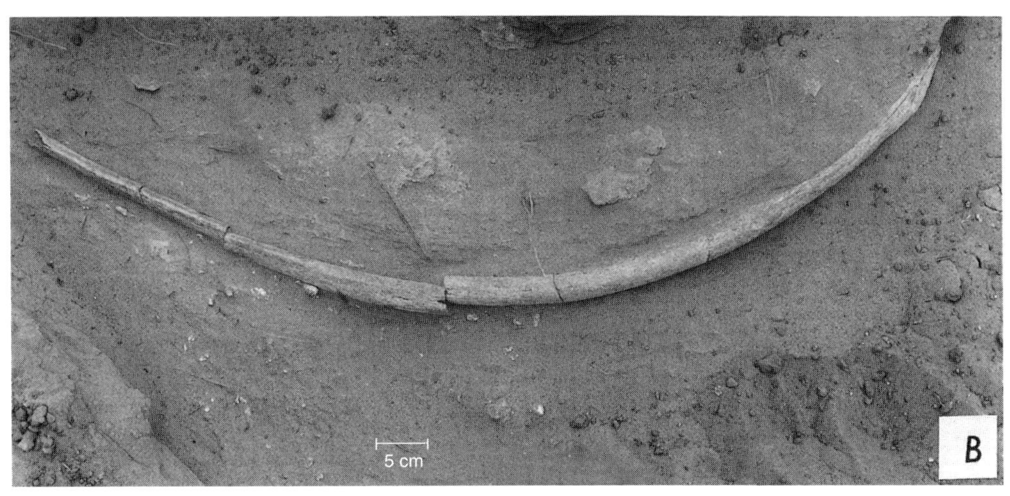

Figure 24.6. *A*, Two ribs and a vertebra in process of excavation; *B*, rib number 56 during excavation, showing the displaced section of rib on the left.

A second possibility for the accumulation of the proboscidean skeleton may be that the animal died elsewhere and the carcass started to break up. The fore limbs may have become detached at this stage, and the distal limb elements removed, and only at this stage was the partially articulated skeleton transported to its present resting place and the missing bones left behind or differentially transported. This is a more reasonable explanation for the loss of these elements, but the anatomical gaps between the skeletal elements preserved in the excavation are hard to explain if the partially articulated carcass was moved as a unit. There are no lumbar vertebrae and no pelvis to connect the thoracic vertebrae with the femur and the two tibiae. The femur represented only by the articular head

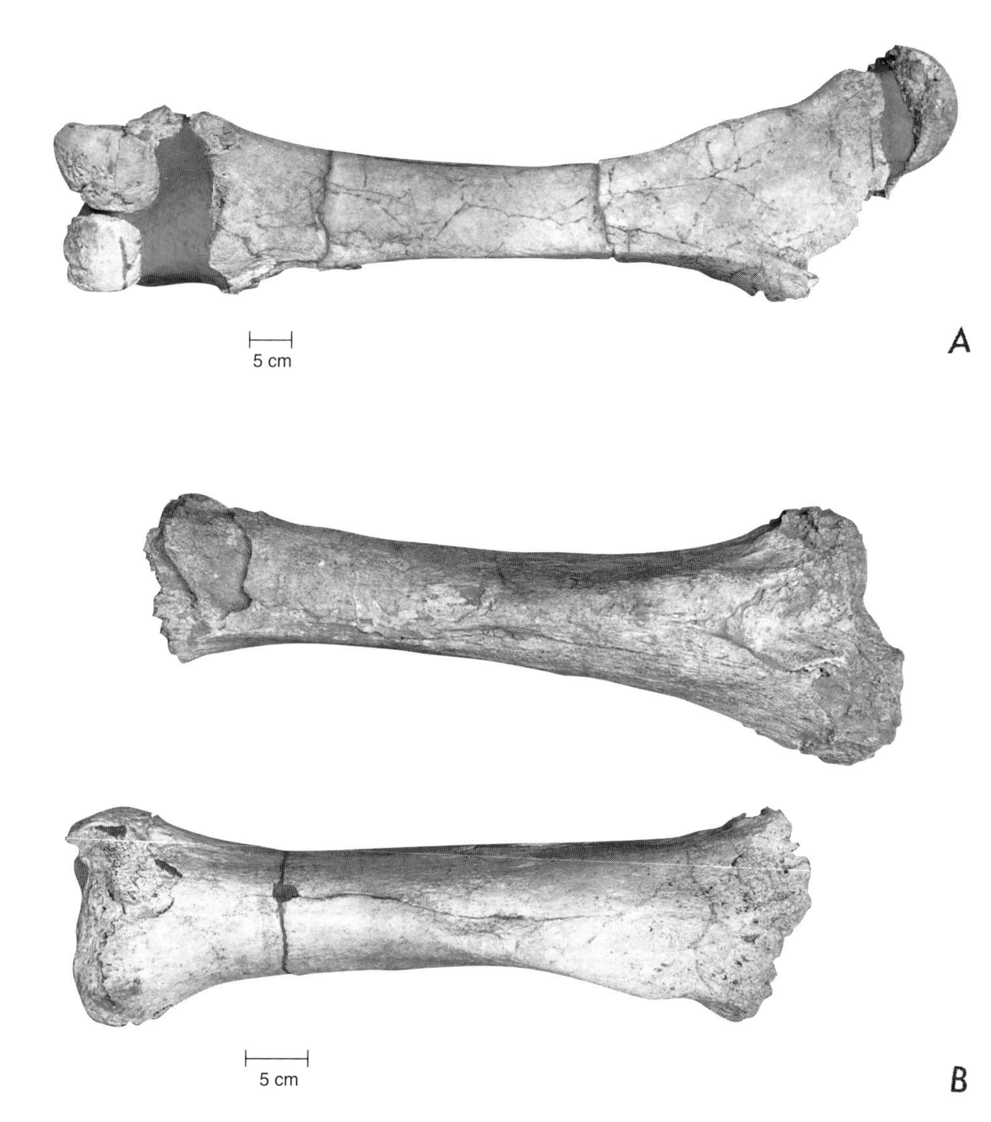

Figure 24.7. *A,* The restored femur, specimen 6, showing damage to the epiphyses and transverse breaks on the shaft. The breaks occurred in place and the separate portions of the bone were still joined together. *B,* The two tibiae, the upper one showing evidence of recent weathering; the lower one is specimen 24, figure 24.2, with a transverse but fused break and a longitudinal weathering crack with incipient exfoliation along it.

must have been moved in one piece and then subsequently destroyed. The fore limbs are represented by a single scapula and the distal epiphysis of one radius, with nothing else remaining. To have the scapula present but nothing else does not fit any predictable disarticulation pattern.

It seems likely that a combination of these two possibilities operated. It has been shown above that the proboscidean bones were deposited at the same time as the gravel on which they are all resting, but the lack of abrasion of the bones, in contrast to the great abrasion of the carbonate gravel clasts, suggests that the bones did not travel far. It is suggested, therefore, that the animal died a short distance away from the fossil site, the soft parts decayed, and the skeleton started to disarticulate before it was then transported to the fossil site. The fore limbs disarticulated first, and were either left behind or transported away, the foot bones were either eaten by scavengers or just dropped off, and many of the vertebrae and ribs were similarly lost. The remainder of the carcass—

the cranium, the major parts of the hind limbs, and an assortment of ribs and vertebrae—was transported in articulation to their present resting place. The presence of one of the scapulae could be due to its position squashed under the cranium, in which case it could be that at least one fore limb was still attached to the carcass, and the scapula was held in place while the rest of the fore limb was removed. The absence of sacrum or pelvis, both of which would have had to be present to link the hind limbs with the vertebral column, is harder to explain. Loss of these elements must indicate subsequent removal after deposition of the main part of the skeleton, as was postulated above for one of the fore limbs.

MODERN ANALOGUES

There is little information on the rates of decay of elephant carcasses at the present time. Rates of decay depend on time of year of death and the numbers of other carcasses available for scavengers. If death is in the dry season, or the body is in a dry situation, the skin dries out and holds the skeleton together longer than when it is wet. Scavenger pressure is also important, for exposed bones may be attacked by scavengers even years after death. Subsequent to large-scale deaths from drought in East Africa—more than 5000 elephants died in Tsavo National Park in 1970–71—it was found that no bodies remained intact by 18 months after death, and by this time most of the skeletons were disarticulated (Coe, 1978). By day 12 after death, the limb bones on the upper side of the carcass were starting to become exposed, and by day 20 all of the limb bones were exposed and the skin gone. Carcasses started to break up after 35 days, and initial signs of weathering were evident after little more than 7 months. After 2 years, the bones were bleached, exfoliation of their surfaces was far advanced, the teeth were shattered, and splits up to 1 mm wide were common. After 5 years, about 15% of bones had been removed from the death sites, although most of the remaining bones were still close to the death sites (Coe, 1978). At this stage, however, many of the bones had been broken, either by trampling by other elephants or by hyaenas scavenging the smaller bones.

The near absence of weathering on the Shuwaihat proboscidean indicates that it was buried before the onset of the first stages of weathering. For most of the bones, weathering had not even reached stage 1 (Behrensmeyer, 1978). This would be well within 6–7 months after death for most of the bones, and the larger bones like the cranium would have been covered in 1–2 years. The close proximity of the bones to each other also suggests that burial took place very soon after disarticulation of the vertebral column and hind limbs or even before if slight post-depositional disturbance is presumed in the sediments. Again, a period of less than 6 months after death is suggested by this evidence. This timetable would apply equally to either depositional possibility described above, whether the partially articulated skeleton was transported to its present resting place or the animal died on the spot and some of its bones were removed. It may be possible to choose between these alternatives if the bone accumulation could be linked with the sedimentological evidence: what is the nature of the pebbly gravel; how was it deposited; how far was it transported, if at all; what sort of soil did it come from; does it indicate a high-energy environment (see below)?

SUMMARY OF POSTMORTEM EVENTS

The distribution of bones from the proboscidean in relation to the carbonate gravel deposits is shown in figure 24.8. The bones are exclusively associated with the gravel deposits, and a provisional interpretation of the taphonomic history of the elephant skeleton is that it accumulated together with the carbonate gravel in an old river channel away from the main course of the river. The gravel bars found during the excavation follow the course of the old river channel, which by this time had become largely silted up and only carried water during exceptional times of high flood.

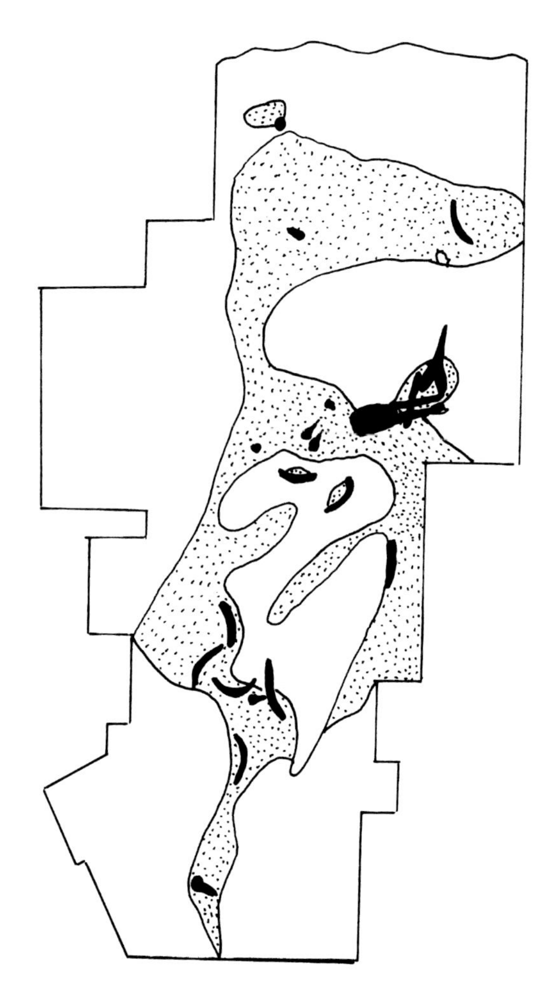

Figure 24.8. Plan of the carbonate gravel showing the relationship of the proboscidean skeletal elements to it. The layout of the excavation is the same as in figure 24.2.

The elephant is thought to have died close to its present resting place when the channel was dry. After decomposition, the skeleton was partially disarticulated, and this was transported as a single unit the short distance to the fossil site, together with gravel clasts derived from the break-up and erosion of a carbonate soil. High-energy conditions in the channel must have persisted for a sufficient time to transport bone and sediment. Subsequent dispersal of ribs, vertebrae, and particularly the pelvis occurred during periods of minor flooding where the water currents were insufficient to move the bones more than a few metres along the surface of the gravel bar on which they were resting but strong enough to transport gravel-sized clasts that banked up around the larger bones. Some bones were moved at this stage, either by water action or by scavengers, although there is little evidence of scavenging (perhaps the carcass was inaccessible, or part of a large die-off so that scavengers were too replete to gnaw the bones). For example, one of the fore limbs must have still been attached to the carcass at this stage, and after the scapula became wedged underneath the cranium, the extremities of the fore limb would subsequently have been removed.

Shortly after disarticulation, certainly within one year, the skeleton was buried during a larger-scale flood, which carried and deposited a thick layer of sandy silt together with additional thin layers of gravel. This buried the skeleton almost completely, but the top of the cranium and tusks were exposed above the sediment for some years, becoming weathered in the process. The weathering began again in recent times as the skeleton again was exposed near the surface of the eroding Miocene deposits.

Acknowledgements

I am very grateful to the many people who helped in the excavation of the proboscidean skeleton. In particular, Diana Clements (The Natural History Museum, London) helped in many ways, both in the field and in the cataloguing and curation of the specimens. She also compiled table 24.1 and helped in the drafting of figures 24.1 and 24.2. David Cameron (NHM) did the final draft of figure 24.2 and Phil Crabb (NHM) took the photographs. I thank both of them. In the field, help is gratefully acknowledged from Libby Andrews, Miranda Armour-Chelu, David Cameron, Phil Crabb, Peter Ditchfield, Alan and Anthea Gentry, and Campbell Smith. I should like to add a special mention for Peter Whybrow, who organised the entire project, directed the fieldwork, excavated all the difficult parts of the proboscidean, and undertook all the liaison with the Abu Dhabi authorities and with the Abu Dhabi Company for Onshore Oil Operations (ADCO), who provided generous and whole-hearted support for the project.

NOTE

1. Field numbers during the excavation commenced at 1, but for registration in the Emirate of Abu Dhabi Collection (AUH), the numbers were required to start from 500. These numbers are shown in table 24.1, so that number 501 in table 24.1 corresponds to number 1 on figure 24.2, 502 to 2, and so on.

REFERENCES

Behrensmeyer, A. K. 1978. Taphonomic and ecological information from bone weathering. *Paleobiology* 4: 150–62.

Bishop, L., and Hill, A. 1999. Fossil Suidae from the Baynunah Formation, Emirate of Abu Dhabi, United Arab Emirates. Chap. 20 in *Fossil Vertebrates of Arabia,* pp. 254–70 (ed. P. J. Whybrow and A. Hill). Yale University Press, New Haven.

Coe, M. 1978. The decomposition of elephant carcases in the Tsavo (East) National Park, Kenya. *Journal of Arid Environments* 1: 71–86.

Gentry, A. W. 1999. A fossil hippopotamus from the Emirate of Abu Dhabi, United Arab Emirates. Chap. 21 in *Fossil Vertebrates of Arabia,* pp. 271–89 (ed. P. J. Whybrow and A. Hill). Yale University Press, New Haven.

Haynes, G. 1991. *Mammoths, Mastodons and Elephants.* Cambridge University Press, Cambridge.

Hill, A., and Behrensmeyer, A. K. 1984. Disarticulation patterns of some modern East African mammals. *Paleobiology* 10: 366–76.

Tassy, P. 1999. Miocene elephantids (Mammalia) from the Emirate of Abu Dhabi, United Arab Emirates: Palaeobiogeographic implications. Chap. 18 in *Fossil Vertebrates of Arabia,* pp. 209–33 (ed. P. J. Whybrow and A. Hill). Yale University Press, New Haven.

Voorhies, M. R. 1969. Taphonomy and population dynamics of an early Pliocene vertebrate fauna, Knox County, Nebraska. *Contributions to Geology, University of Wyoming* 1: 1–69.

Isotopes and Environments of the Baynunah Formation, Emirate of Abu Dhabi, United Arab Emirates

John D. Kingston

While the focus of research on terrestrial fossil vertebrates continues to be morphological and taxonomic, palaeoecological reconstructions have become an established component of site analyses. Embedding faunal evolution within the framework of evolving habitats provides a means of assessing the adaptive significance of morphological and behavioural changes documented in the fossil record. Reconstructing the context of faunal assemblages also provides an opportunity to address one of the central debates in evolutionary theory—the extent to which evolution is driven by perturbations in the physical environment versus biotic interactions. Compilations of site-specific palaeoenvironmental interpretations can ultimately be used to develop detailed regional and global phytogeographic maps of the past and allow us to evaluate how patterns of mammalian evolutionary radiation and intercontinental dispersal are influenced by vegetational changes.

Although it is generally acknowledged that palaeoecological reconstructions are central in understanding evolutionary processes, establishing a detailed and accurate environmental context for fossil assemblages remains problematic. To a large extent, this difficulty is related to the fragmentary nature of the known fossil record, in space and time. Presented with isolated fragments of a modern terrestrial ecosystem such as a random assortment of surface-collected bones, an incomplete assessment of lithofacies, and possibly a leaf or wood specimen, even neoecologists would find it difficult unequivocally to identify the habitat from which the collection was made. Generalisations could be made concerning the temperate or tropical nature of the assemblage and habitat-specific faunal elements might provide further insight, but typically a spectrum of habitats could potentially contribute similar elements. Many animals and plants, and in some cases entire communities, tolerate a wide array of environmental circumstances depending on ecological pressures and can occur in variable settings. Animals with extensive ranging patterns, migratory or daily, typically traverse a number of habitats and it is often difficult to associate them with any single discrete vegetational background. The resolution of the reconstruction improves as the size of the sample set increases, especially if the assumption can be made that the components represent an association of interacting organisms (a community) rather than a fortuitous association of fossil taxa. Uncertainties in interpreting data from modern environments occur in spite of the wealth of detailed data documenting ecological aspects of habitats found today. Interpretation of fossil material typically relies on the assumption that modern habitats can provide a template for palaeohabitat reconstructions. This uniformitarian approach may not in all cases be valid as it is possible that no modern analogues exist for specific ecosystems in the past.

Palaeontologists, working with fragments of palaeoecosystems, face the daunting task of reconstructing the interplay of environment, ecology, and evolution while simultaneously defining each of these parameters. Many different approaches have

 ISBN 0-300-07183-3

been taken in reconstructions of terrestrial palaeoenvironments, including an assessment of lithofacies, ecomorphology (Kappelman, 1991; Plummer and Bishop, 1994), palaeocommunity reconstructions (Andrews et al., 1979; Andrews, 1996), indicator species, palaeobotanical evidence, and inferences based on the global climatic record documented in marine cores. Despite the innovative and circumspect nature of these analyses, taphonomic and interpretive biases are inherent in deciphering the fossil record and it has become increasingly clear that accurate and high-resolution reconstructions need to draw from as many lines of evidence as possible. Within the past decade, stable isotopic analyses of palaeosol components and fossil herbivore enamel have been used to help constrain interpretations of the vegetational physiognomy of palaeohabitats as well as dietary items available for herbivore foraging.

Isotopic Variation and Photosynthetic Pathways

The underlying premise for reconstructing palaeovegetation by isotopic analyses of palaeosol carbonates and tooth enamel is that terrestrial plants using different photosynthetic pathways under varying environmental conditions can be differentiated on the basis of the relative abundance of two naturally occurring stable isotopes of carbon, ^{12}C and ^{13}C (Farquhar et al., 1989). Plants assimilate carbon from the atmospheric CO_2 reservoir by one of three photosynthetic pathways. These pathways, typically referred to as C_3 (Calvin–Benson), C_4 (Hatch–Slack or Kranz), and CAM (crassulacian acid metabolism), represent adaptations to different atmospheric and climatic conditions. In general, carbon incorporated into the organic matrix of vegetation during photosynthesis is significantly depleted in the heavy isotope (^{13}C) relative to atmospheric CO_2, which currently has an isotopic composition ($\delta^{13}C$) of −7.8 per mil (‰) (Keeling et al., 1989). C_3 plants are most depleted whereas plants endowed with the C_4 metabolic pathway are least depleted. Plants that fix CO_2 by CAM display intermediate values overlapping the range of both C_3 and C_4 flora. Essentially all of the isotopic separation or fractionation during plant metabolism is associated with initial phases of carbon fixation involving the uptake of CO_2 into the tissue and subsequent conversion into organic compounds (Craig, 1953; Park and Epstein, 1961; Smith and Epstein, 1971; Farquhar et al., 1982; O'Leary, 1981). Distribution of carbon isotopes among C_3, C_4, and CAM plants is related to a difference in the isotopic fractionation associated with the activity of RuBPC in all plants and PEPC activity in C_4 and CAM plants.

C_3 plants dominate terrestrial environments and account for approximately 85% of all plant species, including almost all trees, shrubs, and high-latitude/altitude grasses preferring wet, cool growing seasons. C_3 flora has a mean $\delta^{13}C$ value of −27.1 ± 2.0‰ (O'Leary, 1988) with a range extending from −22 to −38‰, reflecting genetic and environmental factors (fig. 25.1). Environmental influences affecting the $\delta^{13}C$ of C_3 plants include water stress, nutrient availability, light intensity, CO_2 partial pressure, and temperature (Farquhar et al., 1982; Toft et al., 1989; Tieszen, 1991). Overall, the $\delta^{13}C$ value of C_3 plants tends to be most positive in open, arid, and hot habitats and most negative in cool, moist, and forested environments. In closed-canopy understories, where free exchange with atmospheric CO_2 is restricted, CO_2 can become substantially depleted, resulting in even more negative values (Sternberg et al., 1989; van der Merwe and Medina, 1989). As altitude increases, the partial pressure of CO_2 decreases, resulting in increased CO_2 uptake by plants and more positive $\delta^{13}C$ values for C_3 plants (Tieszen et al., 1979; Korner et al., 1988) with differences of up to 2.6‰ in individual species (Korner et al., 1988). These environmental factors, coupled with genetic differences, result in substantial variations in stable carbon isotopes that need to be considered when attempting to estimate relative proportions of C_3 and C_4 plants in the past either by analysing preserved organic matter or proxies for palaeovegetation.

C_4 physiology is linked almost exclusively to monocots, especially grasses and sedges growing in

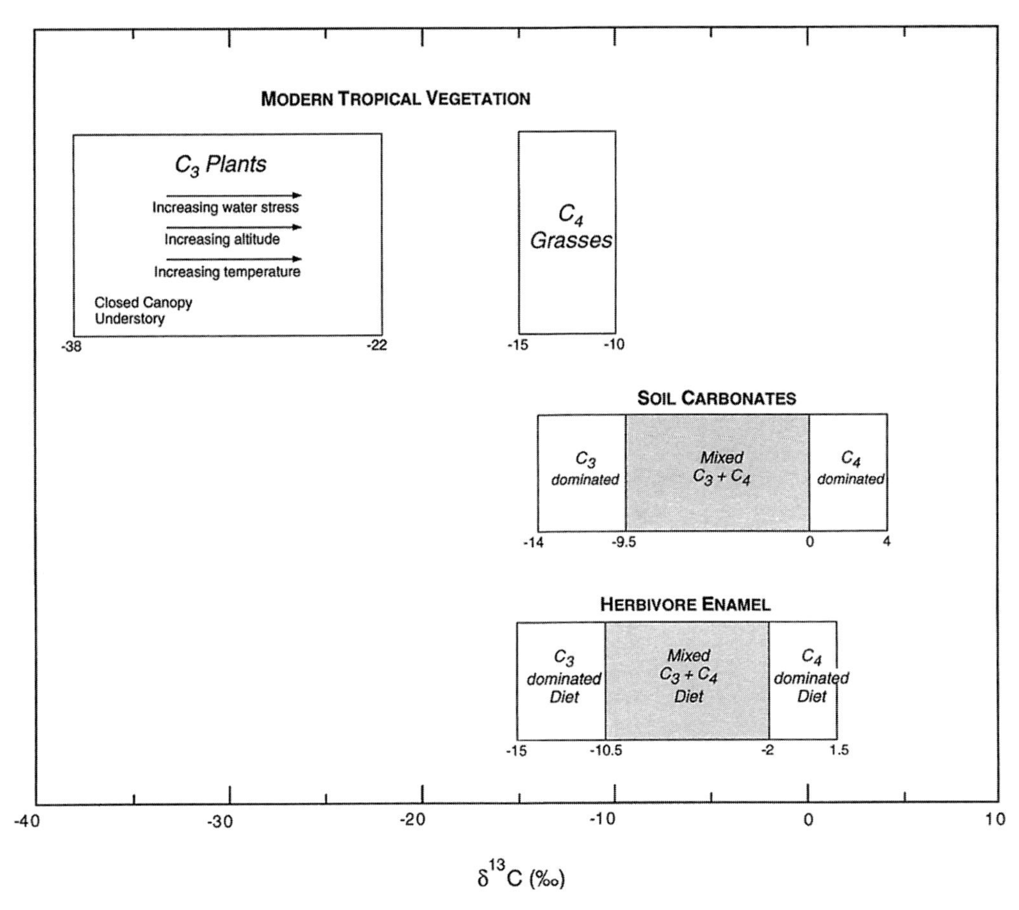

Figure 25.1. Distribution of stable carbon isotopes in terrestrial ecosystems. The depicted ranges of $\delta^{13}C$ values for soil carbonates assumes average $\delta^{13}C$ values of −27 per mil (‰) and −13‰ for C_3 and C_4 plants, respectively, and incorporates variability in the natural range in $\delta^{13}C$ of vegetation as well as ^{13}C-enrichment fractionation effects during gaseous diffusion and carbonate precipitation. Ranges of $\delta^{13}C$ values for herbivore enamel also reflect variability in the isotopic composition of vegetation and deviations from the assumed 13‰ enrichment between diet and enamel. (Adapted from Sikes, 1994.)

hot, arid habitats. A mean $\delta^{13}C$ value of −13.1 ± 1.2‰ has been calculated for C_4 plants (O'Leary, 1988) with a range of −9 to −15‰ (fig. 25.1), or about half that of C_3 plants. The C_4 photosynthetic pathway represents a modification of the C_3 mechanism and is considered to have evolved independently at least 26 times (Peisker, 1986) as a response to either depressed atmospheric CO_2 levels relative to O_2 or to water-stressed environments (Woodward, 1990; Ehleringer, 1991). The CO_2-concentrating mechanism of C_4 plants increases the carbon-fixing efficiency during photosynthesis and C_4 vegetation generally tolerates higher temperatures, drier conditions, and lower atmospheric pCO_2 levels than C_3 species. C_4 photosynthesis, however, is energetically more costly (Salisbury and Ross, 1985) and C_4 vegetation is outcompeted by C_3 plants at temperatures below 25 °C and at higher pCO_2 levels.

Crassulacean acid metabolism (CAM) has evolved independently in many succulent plants including the cacti (Cactaceae) and stonecrops (Crassulaceae). Like C_4 plants, they utilise both the C_3 and C_4 pathways but CAM plants differentially utilise the two- pathways depending on environmental conditions, which results in $\delta^{13}C$ values that span the range of values covered by C_3 and C_4 plants (Deines, 1980; O'Leary, 1981). Under high light intensity or high temperatures, CAM vegetation has C_4-like values whereas under environmental conditions such as low light intensity, cold temperatures, or long days, it has C_3-like values. Although CAM plants can endure extremely xeric conditions their abilility to take in and fix CO_2 is severely limited. In general, therefore, they compete poorly with C_3 and C_4 plants under less-extreme conditions. Generalised habitat reconstructions based on assumed ecological preferences of fossil fauna recovered from the Baynunah Formation suggest that it is highly unlikely that CAM plants comprised a significant component of the biomass during Baynunah times.

As the C_3 and C_4 photosynthetic pathways are associated with different environmental conditions and plant physiognomy, documenting relative proportions of C_3 and C_4 vegetation by isotopic analyses is a useful tool in palaeoenvironmental reconstructions. Specifically, the link between C_4 metabolism and grasses provides a means of differentiating open woodland/grassland landscapes from forested ecosystems in the past. In general, an increase in the proportion of C_4 grasses can be interpreted as representing a decrease in canopy cover. While a C_4 carbon signal implies arid grasslands, a C_3 value can reflect a variety of habitats, ranging from lowland rainforest to arid bushland, which limits the resolving power of a C_3 isotopic signal in reconstructing vegetation. The key to using this relationship is developing a means of retrieving an intact record of the relative proportions of C_3 and C_4 vegetation in the fossil record. As it turns out, several approaches can be used to recover this isotopic record, including isotopic analyses of preserved organic matter, of palaeosol carbonate formed in isotopic equilibrium with palaeovegetation, and of carbon incorporated within fossil bone or enamel—a reflection of available dietary plants.

PALAEOSOLS AND PEDOGENIC CARBONATES IN THE BAYNUNAH FORMATION

Theoretical models and studies of modern soils have established a correlation between the stable carbon and oxygen isotopic composition of soil components and prevailing climatic and ecological conditions (Cerling, 1984; Amundson et al., 1987; Quade et al., 1989; Kelly et al., 1991). In general, the carbon isotopic composition of soil CO_2 and of soil carbonate precipitated in equilibrium with soil CO_2 is controlled by the proportion of surface vegetation utilising the C_3 or C_4 photosynthetic pathway. When $CaCO_3$ precipitates, its stable carbon and oxygen isotope ratios are determined by that of HCO_3^-. The total amount of dissolved carbon in the soil solution, however, is relatively small and it has been demonstrated experimentally that when CO_2 gas is present, the gas phase controls the isotopic composition of the CO_3^{2-} and in turn that of the precipitating carbonate (Bottinga, 1968). Soil CO_2 is a function of mixing from two isotopically distinct sources, biologically respired CO_2 and atmospheric CO_2 (Amundson et al., 1989: Quade et al., 1989). Biologically respired CO_2 refers to CO_2 derived from microbial oxidation of soil organic matter and root respiration. The carbon isotopic composition of biologically derived CO_2 reflects the proportion of C_3 versus C_4 biomass in the local ecosystem and averages about −27‰ when the plant cover is C_3 dominated to about −13‰ when the vegetation is predominantly C_4 grasses (Deines, 1980). At respiration rates typical for temperate and subtropical ecosystems during the growing season (8 mmol/m^2 per hour) (Singh and Gupta, 1977), the carbon isotopic composition of the soil atmosphere is overwhelmingly a function of respired plant CO_2 and closely reflects the C_3/C_4 plant ratio. In arid or semi-arid climates where plant activity is greatly reduced, the soil CO_2 incorporates a larger atmospheric input resulting in more positive $\delta^{13}C$ values for soil CO_2. In addition to the input of atmospheric and biologically respired CO_2, the isotopic composition of soil CO_2 and, ultimately, pedogenic carbonate is a function of several factors and processes, which include soil porosity, soil temperature, mean production depth of soil CO_2, absolute pressure, and the depth within a soil profile (Cerling, 1984). As a result of ^{13}C enrichment due to fractionation during gaseous diffusion and carbonate precipitation, pedogenic carbonates forming at 35 °C and 15 °C are enriched by about 14‰ and 15.5‰, respectively, relative to biologically respired CO_2 (Dorr and Munnich, 1980; Friedman and O'Neil, 1977).

Baynunah Palaeosols

Interbedded within fluvial and floodplain sediments of the Baynunah Formation are a number of palaeosol horizons, which are most conspicuous within the upper portion of the succession. Palaeosols were in general not well developed and in some cases difficult to unequivocally differentiate from fine-grained floodplain facies. Only horizons clearly displaying several pedogenic features were sampled. Criteria for the recognition of palaeosols include destruction of primary bedding resulting in a hackly outcrop weathering pattern, bioturbation, slickensides, root or burrow mottling, carbonate nodule concentrations, and gradational boundaries with underlying lithology. Organic-rich A-horizons were not observed; the palaeosols typically consist of grey or reddish carbonate-leached Bt-horizons ranging from 30 cm to over 2 metres thick. Poorly developed coarse- to medium-grained palaeosols were also evident towards the middle of the formation, suggesting development on aeolian sands. Pedogenic carbonate nodules were associated with a limited number of palaeosol horizons and typically occurred near the base of the palaeosol profile. Nodules were sampled at depths of more than 40 cm in the soil profile to minimise potential mixing of isotopically heavy atmospheric CO_2 with biologically respired CO_2 during carbonate precipitation (Cerling et al., 1989). The carbon isotopic signature of soil organic matter also reflects local plant cover, and analysis of the preserved organic residue in palaeosols can constitute a test of the state of preservation of the original ecological signal in the palaeosol components (Cerling et al., 1989). Preliminary attempts to isolate organic matter from Baynunah palaeosols have been unsuccessful, presumably due to oxidation of the original organic residue.

A prominent feature of some palaeosols are complexes of root casts comprised of celestine (Whybrow and McClure, 1981). Originally described by Glennie and Evamy (1968), these root-like structures were interpreted as having been formed in a wadi environment within a desert. Whybrow and McClure (1981) subsequently suggested that these root structures might represent fossilised mangrove roots, indicating a more tropical rather than arid climatic regime for the region. More recent interpretations of these structures (P. J. Whybrow, personal communication) have cast doubt on a mangrove origin and microscopic examination of root casts has instead revealed morphology with affinities to the family Leguminosae, which includes lianas, laburnums, and acacias (Whybrow et al., 1990).

Methods and Materials

Pedogenic carbonate nodules were rinsed in double-distilled water (ddH_2O) to remove adhering detrital material, soaked in 0.1M HCl for 30 seconds to dissolve potentially contaminating diagenetic surficial $CaCO_3$, rinsed twice more in ddH_2O, and dried in an air convection oven at 75 °C. Nodules were then pulverised to less than 0.5 mm in an agate mortar, which was carefully cleaned with ddH_2O and 0.1M HCl between samples to avoid cross-contamination. Crushed nodules were baked under vacuum at 475 °C for 1 hour (8-hour cool down) to eliminate any organic matter. Carbon within the nodules was then converted to CO_2 by reacting the powder with 100% H_3PO_4 in individual evacuated reaction vessels overnight at 25 °C in a constant temperature waterbath. The CO_2 released was manually collected by cryogenic distillation on a glass vacuum line in 6 mm ampules, which were then analysed on a Finnigan MAT 251 isotope ratio mass spectrometer.

Differences in the carbon isotopic composition of substances are expressed as $\delta^{13}C$ values, which give the per mil deviation of the $\frac{^{13}C}{^{12}C}$ ratio of a sample relative to that of the conventional Pee Dee Belemnite carbonate standard (PDB), which has a $\frac{^{13}C}{^{12}C}$ value of 88.99. Positive values of $\delta^{13}C$ indicate an enrichment of heavy carbon (^{13}C) in the sample relative to the standard whereas negative readings stand for its depletion. The $\delta^{13}C$ ratio is defined by the following equation:

$$\delta^{13}C\ (‰) = \left[\frac{\left(\frac{^{13}C}{^{12}C}\right)_{sample}}{\left(\frac{^{13}C}{^{12}C}\right)_{std}} - 1 \right] \times 1000$$

Analysis of an internal standard in conjunction with carbonate nodule samples yielded a standard deviation of ± 0.03‰ (n = 20). Replicate analysis of four samples resulted in a standard deviation of ± 0.06‰. Overall analytical precision is better than 0.1‰.

Results and Interpretation

Isotopic analyses of 15 palaeosol carbonate nodules collected from four fossil localities—Hamra (site H5), Shuwaihat, Kihal (site K1), and Jebel Barakah—in the Baynunah Formation indicate an average $\delta^{13}C$ of −5.5‰ with a range of −9.0 to −2.2‰ (table 25.1 and fig. 25.2). With the possible exception of three samples from the locality of Hamra (with $\delta^{13}C$ values of −8.6 to −9.0‰), nodular $\delta^{13}C$ values indicate formation in soils in which biologically respired CO_2 was derived from both C_3 and C_4 vegetation. Although it is impossible unequivocally to correlate palaeosol horizons between the different localities, all samples were collected from palaeosol profiles intercalated within fluvial/floodplain facies of the upper Baynunah Formation, less than 15 metres stratigraphically below the gypsiferous cap-rock facies. The extent to which these horizons are correlative depends on whether the cap-rock at the various localities reflects primary

Table 25.1. Stable isotopic composition of pedogenic carbonates from palaeosols within the Baynunah Formation

Sample no.	Locality	Site no.	Sample	$\delta^{13}C$(‰)	$\delta^{18}O$(‰)
AD 636a	Jebel Barakah	JB2	Carbonate nodule	−6.72	−8.53
AD 637a	Jebel Barakah	JB3	Carbonate nodule	−4.89	−8.07
AD 638a	Kihal	K4	Carbonate nodule	−4.87	−6.92
AD 639	Kihal	K5	Carbonate nodule	−2.40	−6.14
AD 640	Shuwaihat	S8	Carbonate nodule	−4.33	−6.37
AD 642	Hamra	H12	Carbonate nodule	−9.04	−9.90
AD 643	Hamra	H12	Carbonate nodule	−8.61	−8.29
AD 644a	Hamra	H12	Carbonate nodule	−8.74	−8.66
AD 645ai	Hamra	H13	Carbonate nodule	−3.14	−8.71
AD 646ai	Hamra	H14	Carbonate nodule	−5.87	−8.13
AD 646aii	Hamra	H14	Carbonate nodule	−6.32	−7.85
AD 647ai	Hamra	H15	Carbonate nodule	−5.53	−9.02
AD 647aii	Hamra	H15	Carbonate nodule	−6.07	−10.63
AD 649a	Hamra	H16	Carbonate nodule	−2.17	−8.70
AD 651a	Jebel Barakah	JB20	Carbonate nodule	−3.31	−7.20

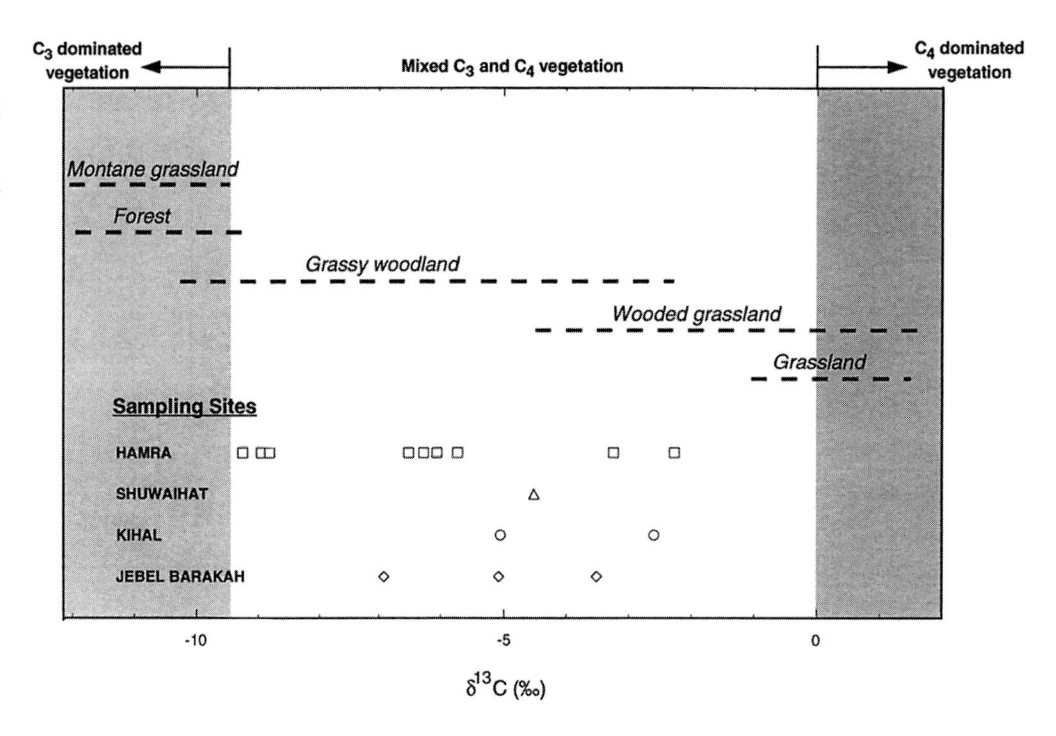

Figure 25.2. The stable isotopic composition of palaeosol carbonate from the Baynunah Formation. Isotopic ranges for the different habitats are based on modern data as well as expected values (Cerling, 1992). (Adapted from Sikes, 1994.)

deposition in a contemporaneous late Miocene sabkha environment or instead reflects a secondary diagenetic feature (Ditchfield, 1999—Chapter 7).

Reconstructions of the Baynunah habitat based on these data depend on assumptions about how the $\delta^{13}C$ variation is partitioned within the sequence. At each locality, except Shuwaihat, carbonates were collected from palaeosol horizons at different stratigraphic levels in the local sections. Although there are distinctive differences in $\delta^{13}C$ values vertically in the section at each locality (table 25.1), there is no consistent trend towards enrichment or depletion of ^{13}C values, reflecting an increase or decrease in C_4 vegetation, respectively, through time, moving upsection at the various localities. At Kihal, for example, the $\delta^{13}C$ value of the youngest measured palaeosol carbonate is 2.5‰ more positive than that of an underlying palaeosol whereas at Hamra, the younger palaeosol has carbonates depleted by over 5‰ relative to an older horizon (table 25.1 and fig. 25.3).

In addition to sampling vertically, at the locality of Hamra, two palaeosol horizons were sampled laterally to assess local heterogeneity in C_3 and C_4 vegetation at each level (fig. 25.3). Carbonates from the upper soil profile (H12) range from −9.0‰ to −5.9‰, indicating a grassy woodland type of environment in which C_3 and C_4 vegetation was unevenly distributed across the landscape. The lower palaeosol sampled at Hamra yielded carbonates with more positive $\delta^{13}C$ values, indicating a significantly greater C_4 component (and more open environment) although only two carbonate nodules were analysed.

Other than closed canopy forest or open grasslands, isotopic analyses of modern vegetation and soil carbonates/organic matter (Tiezen, 1991; Kingston, 1992; Handley et al., 1994; Kingston et al., 1994; Sikes, 1994) indicate that many tropical habitats are isotopically heterogeneous, which reflects localised variation in C_3/C_4 vegetation related to differences in drainage, bedrock, topography, and rainfall. Fluvial/flood plain lithofacies associated with the Baynunah palaeosols indicate that soils formed near a river system that may have supported a mosaic environment ranging from more forested conditions along the river levee to more open woodland/grasslands on adjacent

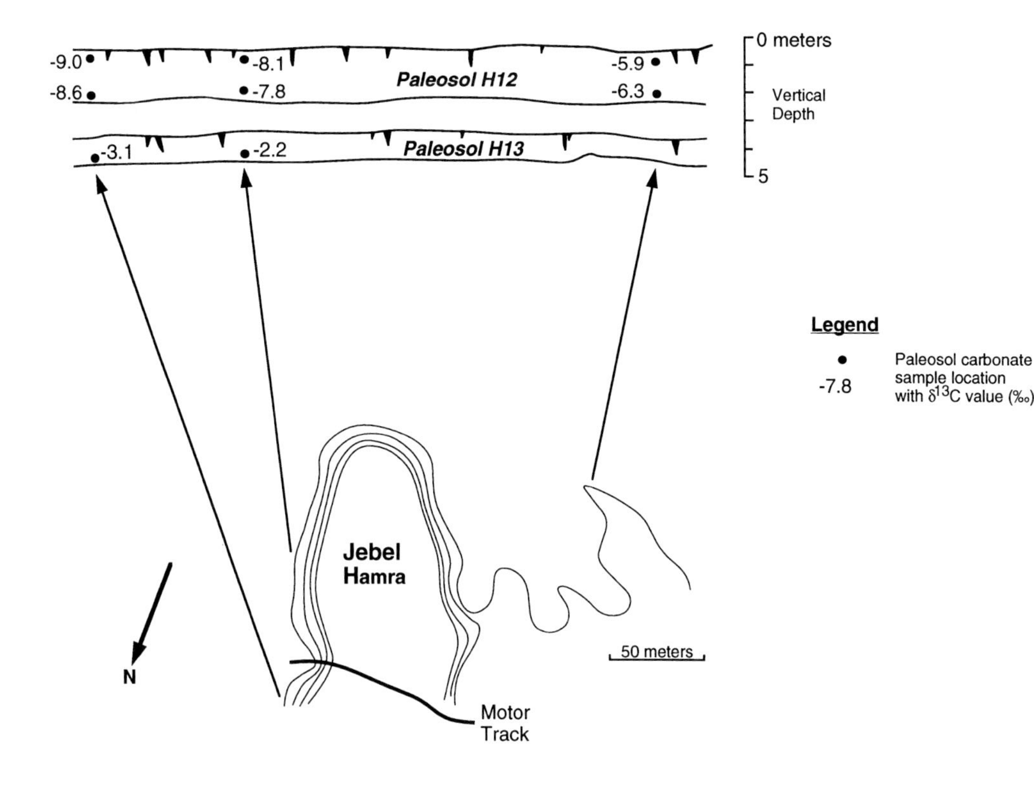

Figure 25.3. Schematic plan map of Jebel Hamra with corresponding partial cross-section indicating the lateral and vertical stratigraphic location of palaeosol carbonate nodules analysed.

floodplains. Evidence for vegetational heterogeneity in the Hamra palaeosol horizon also has potentially important implications for interpreting the vertical variation in the succession. Rather than reflecting vegetational change between the palaeosol levels, these differences may in fact reflect sampling of different microhabitats within a persisting mosaic landscape. A shifting of ecotones associated with lateral migration of the river system within the floodplain could result in a riverine woodland gallery being superimposed over a more open floodplain environment, as implied by the palaeosols at Hamra. Such a scenario does not require invoking a widespread transformation from wooded grassland to more forested vegetation in the relatively short interval of time represented by sediments intercalated between the palaeosols at Hamra as well as the other localities sampled.

Although isotopic values suggest a grassy woodland-type environment, the specifics of the environment remain unknown. Ascribing specific percentages of C_3 and C_4 biomass for the different $\delta^{13}C$ values is problematic because the final isotopic composition of pedogenic carbonate is ultimately a function of the isotopic composition of vegetation, which varies significantly for both C_3 and C_4 vegetation. An attempt to estimate the relative contribution by C_3 and C_4 vegetation for the average isotopic composition of pedogenic carbonate from the Baynunah sequence, –5.5‰, illustrates the difficulties. Assuming an average $\delta^{13}C$ of –23‰ and –13‰ for C_3 and C_4 plants, respectively, and a 15‰ CO_2–$CaCO_3$ phase transformation fractionation (Cerling, 1984), a –5.5‰ value indicates that 75% of the biomass was C_3. This percentage, however, drops to about 50% if C_3 and C_4 vegetation had instead average $\delta^{13}C$ values of –28‰ and –12‰, respectively. Temperature-sensitive fractionation factors during carbonate precipitation (Dienes et al., 1974; Friedman and O'Neil, 1977) and assumptions regarding a correction factor for pre-industrial $\delta^{13}C$ values of atmospheric CO_2 (Marino and McElroy, 1991) can add another 2–3‰ of uncertainty when interpreting the $\delta^{13}C$ of soil carbonates as a proxy for vegetation. Despite these limitations, the link between a C_4 signal and tropical open-habitat grasses provides valuable clues to the physiognomic types of tropical plant palaeocommunities and the extent to which they were dominated by C_3 or C_4 biomass.

Isotopic Analysis of Fossil Enamel and Palaeodietary Reconstructions

A corroborative approach to palaeosol carbonate analysis in documenting relative proportions of C_3 and C_4 vegetation in the past is an analysis of the isotopic signature of carbon incorporated into the inorganic fraction of fossil tooth enamel. It has been well established that the carbon isotopic composition of modern herbivore tissue, including bone and teeth, is directly related to the $\delta^{13}C$ value of the primary photosynthesising plants in the food chain (DeNiro and Epstein, 1978; Tieszen et al., 1983; Ambrose and DeNiro, 1986). Because of physiological fractionation effects, the enamel of mammalian herbivores is consistently enriched by about 12.5‰ relative to diet in terrestrial ecosystems (Krueger and Sullivan, 1984). The relationship between the carbon isotopic composition of body tissue and diet was initially exploited primarily to address archaeological issues such as the introduction of maize, a C_4 domesticate, into previously C_3-dominated New World agricultural economies (Vogel and van der Merwe, 1977; van der Merwe and Vogel, 1978; Schoeninger and Moore, 1992). In almost all of these studies, isotopic analysis focused on collagen, the major constituent of the organic phase of bone. Hydrolysis and dissolution of collagen during fossilisation, however, limits its use to the past few thousand years and in extending these techniques to fossil assemblages, isotopic analysis has concentrated on the mineral portion of bone and teeth. Application of isotopic analyses to fossil specimens has been controversial, primarily due to the difficulty in assessing diagenetic alteration and differential offsets in diet-tissue $\delta^{13}C$ related to trophic-level effects (Sullivan and Krueger, 1981, 1983; Schoeninger and DeNiro, 1982, 1983; Krueger and Sullivan, 1984). Studies have shown that bone apatite is an isotopically unreliable substrate, even for relatively young specimens (5000–10 000 years old), and palaeodietary reconstructions have instead relied on analysis of carbonate occluded within the mineral phase of enamel. Enamel, like the inorganic portion of bone, is a highly substituted, nonstoichiometric hydroxyapatite containing primarily calcium, phosphate, and hydroxide ions ($Ca_{10}(PO_4)_6(OH)_2$) (Eanes, 1979). Carbonate substitutes for both phosphate and hydroxide ions in several positions within the crystal lattice and constitutes between 2 and 4% of the apatite by weight (Brudevold and Söremark, 1967; Rey et al., 1991). Enamel apatite differs significantly from bone apatite in that it is more crystalline, less porous, denser, and has substantially less organic matrix than bone apatite. These features limit potential pathways for infiltration of calcite and minimise the effects of ionic and isotopic exchange during fossilisation (Lee-Thorp, 1989; Glimcher et al., 1991). In addition, the larger crystal size of enamel apatite, relative to bone apatite, provides less surface area per unit weight, which dramatically reduces reactivity. These attributes have led to the recognition of enamel apatite as a much more suitable substrate than bone apatite for dietary reconstructions.

Isotopic analyses of an extensive suite of fossil taxa collected from various South African archaeological and fossil sites (Lee-Thorp, 1989; Lee-Thorp et al., 1989a) provide empirical support for the use of fossil enamel carbonate as a proxy for the relative proportions of C_3 and C_4 components in the diet. Fossil grazers (C_4-dominated diet), browsers (C_3-dominated diet), and mixed or intermediate feeders (combination of C_3 and C_4 diet), distinguished on the basis of microwear analyses, cranial and dental morphology, and taxonomic affinity to extant ungulate species, consistently yielded $\delta^{13}C$ values that reflected the expected dietary signal. Ericson et al. (1981) determined the $\delta^{13}C$ value of enamel apatite from Pliocene herbivore fossils dating to 2 million years ago and reconstructed diets based on isotopic analysis that are in accord with those inferred by analogy to closely related modern taxa. Application of isotopic analyses to fossil enamel strictly for palaeodietary interpretation has been limited, however (Ericson et al., 1981; Lee-Thorp et al., 1989b), and instead its use has been primarily for palaeoecological reconstruction (Thackeray et al., 1990; Kingston, 1992; Quade et al., 1992; Wang et al., 1993; Kingston et al., 1994; MacFadden et al., 1994; Morgan et al., 1994; Quade et al., 1994; Quade et al., 1995).

Methods

Factors controlling the isotopic alteration of biogenic hydroxyapatite are poorly understood. Successful isotopic analysis of enamel apatite for palaeodietary reconstruction depends on the extent to which biogenic structural carbonate is segregated from diagenetic carbonate. In general, the strategy is to digest any organic material associated with the apatite using one of a variety of oxidants such as sodium hypochlorite (NaOHCl), hydrogen peroxide (H_2O_2), or hydrazine (NH_2NH_2) (Koch et al., 1989; Lee-Thorp and van der Merwe, 1991) and to remove any secondary carbonate by dissolution in acetic acid (CH_3COOH), hydrochloric acid (HCl), or triammonium citrate ($(NH_4)_3C_6H_5O_7$) (Hassan et al., 1977; Sillen, 1986; Lee-Thorp and van der Merwe, 1991). Presumably the more exchangeable carbonates (or bicarbonates) associated with the hydration layer, exogenous carbonate such as calcite, and structural apatite close to crystal surfaces or along dislocations that has experienced significant incorporation of secondary carbonate will dissolve during acid treatment, leaving biogenic structural carbonates. Although X-ray diffractrometry and infrared spectroscopy provide a means of monitoring the presence of a diagenetic calcite phase or the degree of apatite recrystallisation, there are no methods for distinguishing between structural biogenic and structural diagenetic carbonate. Different pretreatments can have profound effects on the mineralogy and isotopic composition of the remaining apatite (Lee-Thorp and van der Merwe, 1991). In this study, the basic methodology outlined by Lee-Thorp (1989) has been followed with a few modifications.

Enamel was carefully cleaned of adhering matrix, dentine, and weathering rinds with a high-speed dremel drilling tool and then ground in an agate mortar. Powdered enamel (50–130 mg) was reacted for 24 hours with 2% NaOHCl in 50 ml plastic centrifuge tubes and then rinsed to a pH of 7 by centrifugation with ddH_2O. The residue was treated with 0.1M CH_3COOH for 16 hours under a weak vacuum, rinsed to neutrality by centrifugation with ddH_2O, and freeze-dried. The dried samples (30–100 mg) were reacted with 100% phosphoric acid (H_3PO_4) at 90 °C in sealed borosilicate reaction y-tubes for 48 hours. The liberated carbon dioxide was cleaned and separated cryogenically and then analysed on a MAT-Finnegan 251 mass spectrometer. Precision was ± 0.11‰ for $\delta^{13}C$ ratios of four replicate pairs of fossil enamel. A laboratory standard analysed with the enamel samples yielded a standard deviation of ± 0.07‰ ($n = 5$). Based on these data, overall analytical precision was better than 0.2‰.

Results and Interpretation

Stable carbon isotope data from 33 specimens from at least 10 herbivore species representing 5 families is presented in table 25.2 and in figure 25.4. $\delta^{13}C$ values, ranging from −10.4 to + 0.9‰ suggest that although both C_3 and C_4 plants were available as dietary sources, most of the taxa analysed either relied on a mixed C_3/C_4 diet or were exclusively grazers consuming C_4 grasses. Only several specimens (Giraffidae gen. et sp. indet., one of the *Tragoportax cyrenaicus* samples, and possibly some of the *Hipparion* samples) yielded ^{13}C-depleted values (less than 8‰) consistent with a predominantly browsing (C_3) foraging strategy.

Based on dental and postcranial morphology, the Baynunah hipparion material is attributed to two species (Eisenmann and Whybrow, 1999—Chapter 19), a small to middle-sized hipparion representing a new species (*Hipparion abudhabiense*), and a middle-sized to large *Hipparion* sp. Eisenmann notes that although *H. abudhabiense* had a relatively short and broad muzzle—suggesting grazing rather than browsing habits—it also retains primitive characters, which are interpreted as poor adaptations to abrasive foods such as C_4 grasses.

Table 25.2. Stable isotopic composition of apatite carbonate in fossil enamel collected from the Baynunah Formation

Sample no.	AUH or BMNH* no.	Tooth	Material	Genus and species	Family	Site no.	$\delta^{13}C$(‰)	$\delta^{18}C$(‰)
AD 619a	M49464*	Fragment	Enamel	*Hexaprotodon* aff. *sahabiensis*	Hippopotamidae	B2	–5.55	–8.53
AD 619b	M49464*	Fragment	Dentine	*Hexaprotodon* aff. *sahabiensis*	Hippopotamidae	B2	–5.65	–10.72
AD 620	239	Molar fragment	Enamel	*Tragoportax cyrenaicus*	Bovidae	TH1	–10.36	–2.62
AD 621	609	Fragment	Enamel	"*Hipparion*"	Equidae	JD3	–1.91	–6.10
AD 622a	46	Fragment	Enamel	*Hipparion* sp.	Equidae	H5	–3.61	–6.64
AD 622b	46	Fragment	Dentine	*Hipparion* sp.	Equidae	H5	–3.11	–8.83
AD 623	23	lt lower molar frag.	Enamel	*Hipparion abudhabiense*	Equidae	S1	–6.05	–5.87
AD 624	233	Molar frag.	Enamel	*Stegotetrabelodon* sp.	Elephantidae	H6	0.05	–4.74
AD 625	278	lt molar frag.	Enamel	*Tragoportax cyrenaicus*	Bovidae	H5	0.13	–4.59
AD 626	206	rt upper molar frag.	Enamel	?*Bramatherium* sp.	Giraffidae	R2	–1.62	0.16
AD 627a	211	lt upper molar frag.	Enamel	Gen. et sp. indet.	Giraffidae	R2	–9.16	–4.29
AD 627b	211	lt upper molar frag.	Dentine	Gen. et sp. indet.	Giraffidae	R2	–8.78	–5.64
AD 628	27	Fragment	Enamel	*Tragoportax cyrenaicus*	Bovidae	S1	–6.10	–1.96
AD 629	—	Molar frag.	Enamel	*Stegotetrabelodon* sp.	Elephantidae	B (east)	–4.52	–6.52
AD 630	—	Molar frag.	Enamel	*Hexaprotodon* aff. *sahabiensis*	Hippopotamidae	JD	–0.30	–10.52
AD 631	—	Molar frag.	Enamel	*Hexaprotodon* aff. *sahabiensis*	Hippopotamidae	S1	–1.67	–8.39
AD 632	—	Molar frag.	Enamel	*Stegotetrabelodon* sp.	Elephantidae	H	–1.98	–5.34
AD 633	—	Molar frag.	Enamel	"*Hipparion*"	Equidae	K	0.55	–5.11
AD 634	—	Molar frag.	Enamel	"*Hipparion*"	Equidae	H5	–5.57	–3.97

Table 25.2 (*continued*)

Sample no.	AUH or BMNH* no.	Tooth	Material	Genus and species	Family	Site no.	$\delta^{13}C$(‰)	$\delta^{18}C$(‰)
AD 635	—	Molar frag.	Enamel	Bovid indet.	Bovidae	TH	−6.61	−6.03
AD 750	208	Lower molar frag.	Enamel	*Hipparion* sp.	Equidae	R2	−7.84	−2.74
AD 751	231a	P^3 or P^4	Enamel	*Hipparion abudhabiense*	Equidae	H5	−0.96	−8.25
AD 752	260	Upper ?molar	Enamel	*Hipparion abudhabiense*	Equidae	K1	−2.37	−6.68
AD 753	212	M_1 or M_2	Enamel	*Hipparion* sp.	Equidae	R2	−0.42	−0.50
AD 754	178	rt P^3 or P^4	Enamel	*Hipparion* sp.	Equidae	JD5	−3.32	−3.84
AD 757	205	rt P^3 or P^4	Enamel	*Hipparion* sp.	Equidae	R2	−7.37	−0.65
AD 758	115	M^1 or M^2	Enamel	*Hipparion abudhabiense*	Equidae	S1	−0.28	−6.76
AD 759	M50664*	P^2	Enamel	*Hipparion* sp.	Equidae	B2	0.05	−6.76
AD 760	M60663*	rt lower molar	Enamel	*Hipparion abudhabiense*	Equidae	B2	0.91	−3.11
AD 761	174	rt lower P	Enamel	*Hipparion abudhabiense*	Equidae	H1	−0.73	−6.38
AD 763	72	Lower molar	Enamel	*Hipparion abudhabiense*	Equidae	S1	−3.47	−7.02
AD 764	265	rt lower P	Enamel	*Hipparion* sp.	Equidae	JD4	−1.37	−4.67
AD 765	677	rt lower molar	Enamel	"*Hipparion*"	Equidae	S4	−6.29	−7.26
AD 766	372	rt dP^4	Enamel	?*Bramatherium* sp.	Giraffidae	R2	−5.56	−4.94
AD 767	217	lt upper molar	Enamel	?*Bramatherium* sp.	Giraffidae	R2	−2.69	1.92

Locality abbreviations: B, Jebel Barakah; S, Shuwaihat; H, Hamra; JD, Jebel Dhanna; R, Ras Dubay'ah; K, Kihal; TH, Thumayriyah.

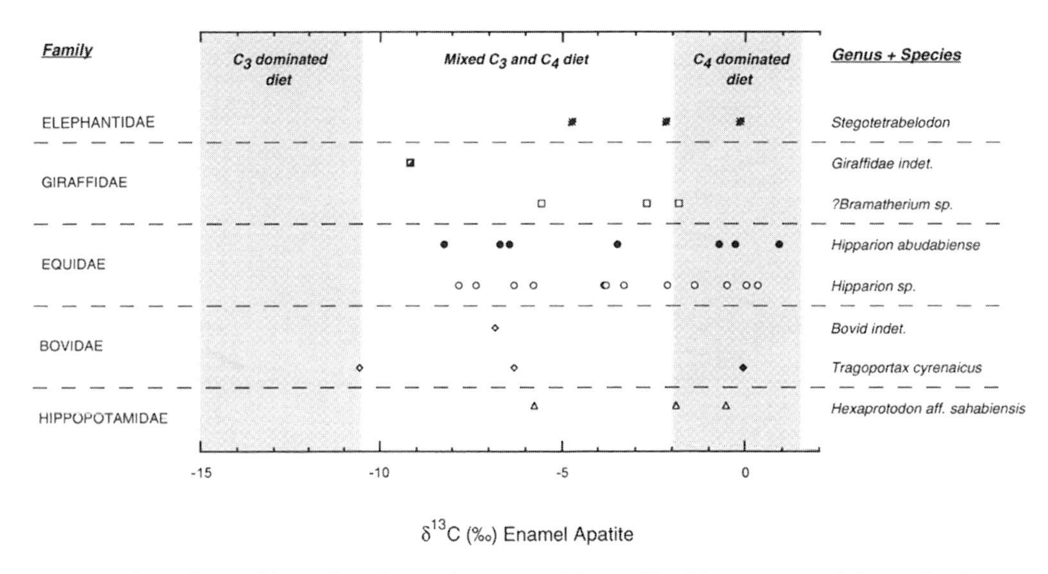

Figure 25.4. The stable carbon isotopic composition of herbivore enamel from the Baynunah Formation.

Isotopic analyses of enamel attributed to *H. abudhabiense* indicates a wide range of food sources in the diet, ranging from exclusively C_4 grasses to predominantly C_3 plants (more than 80% C_3). Three specimens yielded isotopic values consistent with dedicated grazing ($\delta^{13}C$ more than −1‰) while the remaining specimens indicate mixed grazing and browsing strategies with an emphasis on grazing. The range of isotopic values for the larger hipparion species (−7.4 to −0.4‰) is indistinguishable from that of *H. abudhabiense* and also indicates an intermediate grazing strategy.

The isotopic signature of enamel fragments of giraffid species from the Baynunah succession (Gentry, 1999b—Chapter 22)—Giraffidae gen. et sp. indet., and ?*Bramatherium* sp.—appear to indicate ecological partitioning. Two specimens of ?*Bramatherium* sp. from the locality of Ras Dubay'ah yielded $\delta^{13}C_{enamel}$ values of −1.6‰ and −2.6‰, indicating a dietary intake in the range of −16 to −14‰, which is essentially a pure C_4 diet. A third enamel fragment from the same locality had a value of −5.6‰, suggesting a mixed C_3/C_4 diet. These data suggest that, unlike both extant members of the Giraffidae (which are specialised browsers), the foraging strategy of the extinct Baynunah giraffid was dominated by grazing. While these results are in general at odds with conventional assumptions regarding the foraging behaviour of giraffids, a grazing strategy has been suggested for Miocene giraffids. Hamilton (1973) characterised members of the family Sivatheriidae as having short necks and limbs and suggests that they "fed near the ground and grazing forms may have developed". Meladze (1964) hypothesised that the sivatheriids were adapted to life in the savannah. Based on premaxillary shape analysis and quantitative analysis of tooth microwear, Solounias et al. (1988) suggested that the Miocene giraffid *Samotherium boissieri* had dietary adaptations most similar to committed grazers and could have occupied a grazing niche. Of nine late Miocene herbivore species analysed from Samos, Greece (Quade et al., 1994), *Samotherium boissieri* had the most enriched values in ^{13}C (−5.4‰), which was interpreted to indicate that it preferred moisture-stressed C_3 plants, possibly C_3 grasses, or had a C_4 component in its diet. Isotopic analyses of fossil giraffid enamel apatite derived from other sequences, *Samotherium* sp. from the late Miocene of Kemiklitepe, Turkey (Bocherens et al., 1994), *Giraffokeryx punjabiensis* and *Bramatherium megacephalum* from middle–late Miocene strata of the Siwalik sequence in Pakistan (Morgan et al., 1994), Pliocene *Giraffa camelopardalis* from Makapansgat, South Africa (Lee-Thorp and van der Merwe, 1987), and *Giraffa gracilis* from the Pliocene Shungura Formation of Ethiopia (Ericson et al., 1981), yield $\delta^{13}C$ values between –9.8 to –13.2‰, implying that these forms were all obligate browsers. The $\delta^{13}C_{enamel}$ value of the other Baynunah giraffid analysed, Giraffidae gen. et sp. indet., reflects a C_3 diet indicative of a dedicated browser. Isotopic variation in the enamel of the two different giraffids collected from the same locality also provides evidence that diagenetic overprinting is not a factor in the isotopic signal in Baynunah fossil enamel as $\delta^{13}C_{enamel}$ values would be expected to converge rather than yield such contrasting values.

Enamel from an elephantid (*Stegotetrabelodon* sp.) (n = 3; see Tassy, 1999—Chapter 18) and hippopotamus (*Hexaprotodon* aff. *sahabiensis*) (n = 3; see Gentry, 1999a—Chapter 21) analysed from the Baynunah sequence indicates a strong dietary reliance on C_4 grasses. For both of these taxa, two samples plot within or very close to the $\delta^{13}C$ range of an exclusive C_4 grazer, while one specimen for each suggests a more diverse, mixed C_3/C_4 diet. The bovid enamel reflects diverse foraging strategies ranging from committed browsing for *Tragoportax* sp. to a grazing habit for *?Pachyportax latidens.*

CONCLUSIONS

Stable carbon isotopic values of palaeosol carbonate collected from the upper Baynunah Formation record the presence of C_3 and, to a lesser extent, C_4 vegetation at the time during which the soils formed. Lateral variability in the $\delta^{13}C$ of pedogenic carbonate collected from the locality of Hamra implies a heterogeneous environment. None of the palaeosol carbonates analysed yielded $\delta^{13}C$ values indicative of open C_4 grassland. Vertical variability in the isotopic composition of palaeosol carbonates from local sections could reflect shifts in the relative proportions of C_3 and C_4 vegetation through time but this variation most likely represents an artifact of insufficient lateral sampling. If C_3 and C_4 are heterogeneous in the landscape, randomly selected palaeosol carbonates would reflect microhabitats within a mosaic environment rather than provide an overall estimate of vegetation in a region. Without extensive lateral sampling, apparent shifts in the $\delta^{13}C$ of pedogenic carbonate through a local section could simply reflect variable sampling of different microhabitats within similar ecosystems. Palaeosols sampled at the various fossil localities were all less than 15 metres stratigraphically below the cap-rock. If the cap-rock unit is correlative between the localities then the palaeosol levels are roughly penecontemporaneous. As it is unlikely that there were significant changes in habitat during this interval, it is not unreasonable to pool the data from the various levels at the different localities to interpret the data from the upper Baynunah Formation. The range of $\delta^{13}C$ values represented by a compilation of Baynunah palaeosol carbonate analyses is most consistent with a grassy woodland ecosystem.

Unlike isotopic analyses of palaeosol components that directly reflect aspects of palaeovegetation, analyses of fossil herbivore enamel as a proxy of vegetation must be interpreted through a series of filters as there is no definitive correlation between diet and habitat. In general, browsers inhabit forested ecosystems, grazers more open woodland and grassland habitats, and mixed feeders are ecotonal. There are modern grazing ruminants, however, that inhabit forests and browsers occur in open grasslands. In addition to dietary selection by animals, competitive exclusion, migration, and immigration present confounding factors that need to be carefully considered in translating palaeodietary signals into palaeoenvironmental reconstructions. The strategy in reconstructions of the Baynunah environment based on the isotopic composition of herbivore enamel involves sampling a wide range of taxa that would incorporate both grazing and browsing feeding strategies and potentially reflect relative proportions of C_3 and C_4 vegetation in the habitat. While an analysis of 34 specimens representing five herbivore families indicates that both C_3 and C_4 plants were available for consumption, there appears to be a heavy reliance on C_4 grasses with a number of specimens from various taxa falling within the isotopic niche occupied by committed grazers. Only a few specimens plotted within the range of obligate browsers. Of the specimens indicating a mixed grazing and browsing strategy, most of the isotopic values suggest a major C_4 component. These data would appear to imply a more open environment than that reflected by the palaeosol carbonates.

In contrasting environmental reconstructions of the Baynunah habitats based on the palaeosol and enamel isotopic datasets, it should be noted that the two types of data reflect different spatiotemporal aspects of the ecosystem. Palaeosol carbonates typically form over hundreds or even thousands of years and thus preserve a palaeoenvironmental record averaged over an interval spanning many generations of plants. Lateral migration of CO_2 in soil profiles is limited and the $\delta^{13}C$ of pedogenic carbonate is controlled by the local plant cover in an area of less than 10 m^2. Unlike the mineral portion of bone, which is in a constant state of flux during an organism's life span, carbonate is incorporated into the apatite crystal lattice during vertebrate tooth formation only and remains sequestered from subsequent physiological activity during life. As such, the carbon in enamel apatite records the diet for a relatively brief interval of time relative to the formation of palaeosol carbonate. In addition, as noted earlier, herbivores can have extensive ranging patterns and the diet may reflect vegetation sampled from a variety of habitats over a large area. In reconciling the enamel and palaeosol isotopic profiles from the Baynunah Formation two scenarios seem most plausible. Possibly, the herbivore population, dominated by grazers and intermediate feeders, were selectively grazing on C_4 grasses within a regionally extensive grassy woodland environment. Alternatively, and more likely, the environment was heterogeneous and the palaeosol carbonates sampled formed in more wooded environments flanking a river system while many of the herbivores grazed in more open grasslands or wooded grasslands distal to the fluvial environments represented by the Baynunah sediments.

Acknowledgements

I am grateful to Andrew Hill (Yale University, USA) and Peter Whybrow (The Natural History Museum, London) for extending the opportunity to participate in palaeontological investigation in the Emirate of Abu Dhabi, and to the Abu Dhabi Company for Onshore Oil Operations (ADCO) for financial support through their grant to Peter Whybrow. I thank Danny Rye for analytical assistance and access to isotopic facilities in the Kline Geological Laboratory at Yale University. Danny Rye acknowledges NSF grant EAR-9405742.

References

Ambrose, S. A., and DeNiro, M. J. 1986. The isotopic ecology of East African mammals. *Oecologia* 69: 395–406.

Amundson, R. G., and Lund, L. J. 1987. The stable isotope chemistry of a native and irrigated typic natrargid in the San Joaquin Valley of California. *Soil Science Society of America Journal* 51: 761–67.

Amundson, R. G., Sowers, J. M., Chadwick, O. A., and Doner, H. E. 1989. The stable isotope chemistry of pedogenic carbonates at Kyle Canyon, Nevada. *Soil Science Society of America Journal* 53: 201–10.

Andrews, P. 1996. Palaeoecology and hominoid palaeoenvironments. *Biological Reviews* 71: 257–300.

Andrews, P., Lord, J. M., and Nesbit Evans, E. M. 1979. Patterns of ecological diversity in fossil and modern mammalian faunas. *Biological Journal of the Linnean Society* 11: 177–205.

Bocherens, H., Mariotti, A., Fizet, M., Bellon, G., and Borel, J. P. 1994. Les gisements de mammifères du Miocène supérieur de Kemiklitepe, Turqui:10. Biogéochimie isotopique. *Bulletin du Muséum National d'Histoire Naturelle,* Paris 1: 211–23.

Bottinga, Y. 1968. Calculation of fractionation factors for carbon and oxygen exchange in the system calcite–carbon dioxide–water. *Journal of Physical Chemistry* 72: 800–808.

Brudevold, F., and Söremark, R. 1967. Chemistry of the mineral phase of enamel. In *Structural and Chemical Organization of Teeth,* pp. 247–77 (ed. A. E. W. Miles). Academic Press, New York.

Cerling, T. E. 1984. The stable isotopic composition of modern soil carbonates and its relationship to climate. *Earth and Planetary Science Letters* 71: 229–40.

———. 1992. Development of grasslands and savannas in East Africa during the Neogene. *Palaeogeography, Palaeoclimatology, Palaeoecology (Global and Planetary Change Section)* 97: 241–47.

Cerling, T. E., Quade, J., Wang, Y., and Bowman, J. R. 1989. Carbon isotopes in soil and palaeosols as ecology and paleoecology indicators. *Nature* 341: 138–39.

Craig, H. 1953. The geochemistry of the stable carbon isotopes. *Geochimica et Cosmochimica Acta* 3: 53–92.

Deines, P. 1980. The isotopic composition of reduced organic carbon. In *Handbook of Environmental Isotope Geochemistry. V. 1. The Terrestrial Environment,* A, pp. 329–406 (ed. P. Fritz and C. Fontes). Elsevier, New York.

Deines, P., Langmuire, D., and Harmon, R. S. 1974. Stable carbon isotope ratios and the existence of a gas phase in the evolution of carbonate ground-waters. *Geochimica et Cosmochimica Acta* 38: 1147–64.

DeNiro, M. J., and Epstein, S. 1978. Influence of diet on the distribution of carbon isotopes in animals. *Geochimica et Cosmochimica Acta* 42: 341–51.

Ditchfield, P. W. 1999. Diagenesis of the Baynunah, Shuwaihat, and Upper Dam Formation sediments exposed in the Western Region, Emirate of Abu Dhabi, United Arab Emirates. Chap. 7 in *Fossil Vertebrates of Arabia,* pp. 61–74 (ed. P. J. Whybrow and A. Hill). Yale University Press, New Haven.

Dorr, H., and Munnich, K. O. 1980. Carbon-14 and carbon-13 in soil CO_2. *Radiocarbon* 22: 909–18.

Eanes, E. D. 1979. Enamel apatite: Chemistry, structure and properties. *Journal of Dental Research* 58(B): 829–36.

Ehleringer, J. R. 1991. Climate change and the evolution of C_4 photosynthesis. *Trends in Ecology and Evolution* 6: 95–99.

Eisenmann, V., and Whybrow, P. J. 1999. Hipparions from the late Miocene Baynunah Formation, Emirate of Abu Dhabi, United Arab Emirates. Ch. 19 in *Fossil Vertebrates of Arabia,* pp. 234–53 (ed. P. J. Whybrow and A. Hill). Yale University Press, New Haven.

Ericson, J. E., Sullivan, C. H., and Boaz, N. T. 1981. Diets of Pliocene mammals from Omo, Ethiopia, deduced from carbon isotope ratios in tooth apatite. *Palaeogeography, Palaeoclimatology, Palaeoecology* 36: 69–73.

Farquhar, G. D., Ehleringer, J. R., and Hubick, K. T. 1989. Carbon isotope discrimination and photosynthesis. *Annual Review of Plant Physiology and Plant Molecular Biology* 40: 503–37.

Farquhar, G. D., O'Leary, M. H., and Berry, J. A. 1982. On the relationship between carbon isotope discrimination and the intercellular carbon dioxide concentration in leaves. *Australian Journal of Plant Physiology* 9: 121–37.

Friedman, I., and O'Neil, J. R. 1977. Compilation of stable isotope fractionation factors of geochemical interest. *United States Geological Survey Professional Paper* 440-KK.

Gentry, A. W. 1999a. A fossil hippopotamus from the Emirate of Abu Dhabi, United Arab Emirates. Chap. 21 in *Fossil Vertebrates of Arabia,* pp. 271–89 (ed. P. J. Whybrow and A. Hill). Yale University Press, New Haven.

———. 1999b. Fossil pecorans from the Baynunah Formation, Emirate of Abu Dhabi, United Arab Emirates. Chap. 22 in *Fossil Vertebrates of Arabia,* pp. 290–316 (ed. P. J. Whybrow and A. Hill). Yale University Press, New Haven.

Glennie, K. W., and Evamy, B. D. 1968. Dikaka: Plants and plant-root structures associated with aeolian sand. *Palaeogeography, Palaeoclimatology, Palaeoecology* 4: 77–87.

Glimcher, M. J., Cohen-Sokal, L., Kossiva, D., and deRiques, A. 1991. Biochemical analysis of fossil enamel and dentin. *Paleobiology* 16: 219–32.

Hamilton, W. R. 1973. The lower Miocene ruminants of Gebel Zelten, Libya. *Bulletin of the British Museum (Natural History),* Geology 21: 73–150.

Handley, L. L., Odee, D., and Scrimgeour, C. M. 1994. $\delta^{15}N$ and $\delta^{13}C$ patterns in savanna vegetation: Dependence on water availability and disturbance. *Functional Ecology* 8: 306–14.

Hassan, A. A., Termine, J. D., and Haynes, C. V. 1977. Mineralogical studies on bone apatite and their implications for radiocarbon dating. *Radiocarbon* 19: 364–74.

Kappelman, J. 1991. The paleoenvironment of *Kenyapithecus* at Fort Ternan. *Journal of Human Evolution* 20: 95–129.

Keeling, C. D., Bacastow, R. B., Carter, A. F., Piper, S. C., Whorf, T. P., Heimann, M., Mook, W. G., and Roeloffzen, H. 1989. A three-dimensional model of atmospheric CO_2 transport based on observed winds: 1. Analysis of observational data. *Geophysical Monograph* 55: 165–326.

Kelly, E. F., Amundson, R. G., Marino, B. D., and DeNiro, M. J. 1991. Stable carbon isotopic composition of carbonate in Holocene grassland soils. *Soil Science Society of America Journal* 55: 1651–57.

Kingston, J. D. 1992. Stable isotope evidence for hominid paleoenvironments in East Africa. Ph.D. dissertation, Harvard University.

Kingston, J. D., Marino, B. D., and Hill, A. 1994. Isotopic evidence for Neogene hominid paleoenvironments in the Kenya Rift Valley. *Science* 264: 955–59.

Koch, P. L., Fisher, D. C., and Dettman, D. 1989. Oxygen isotope variation in the tusks of extinct proboscideans: A measure of season of death and seasonality. *Geology* 17: 515–19.

Korner, C., Farquhar, G. D., and Roksandic, Z. 1988. A global survey of carbon isotope discrimination in plants from high altitude. *Oecologia* 74: 623–32.

Krueger, H. W., and Sullivan, C. H. 1984. Models for carbon isotope fractionation between diet and bone. In *Stable Isotopes in Nutrition ACS Symposium Series,* no. 258, pp. 205–22 (ed. J. F. Turnland and P. E. Johnson). American Chemical Society, Washington D.C.

Lee-Thorp, J. A. 1989. Stable carbon isotopes in deep time: The diets of fossil fauna and hominids. Ph.D. dissertation, University of Cape Town.

Lee-Thorp, J. A., and van der Merwe, N. J. 1987. Carbon isotope analysis of fossil bone apatite. *South African Journal of Science* 83: 71–74.

———. 1991. Aspects of the chemistry of modern and fossil biological apatite. *Journal of Archaeological Science* 18: 343–54.

Lee-Thorp, J. A., Sealy, J. C., and van der Merwe N. J. 1989a. Stable carbon isotope ratio differences between bone collagen and bone apatite, and their relationship to diet. *Journal of Archaeological Science* 16: 585–99.

Lee-Thorp, J. A., van der Merwe, N. J., and Brain, C. K. 1989b. Isotopic evidence for dietary differences between two extinct baboon species from Swartkrans. *Journal of Human Evolution* 18: 183–90.

MacFadden, B. J., Wang, Y., Cerling, T. E., and Federico, A. 1994. South American fossil mammals and carbon isotopes: A 25 million-year sequence from the Bolivian Andes. *Palaeogeography, Palaeoclimatology, Palaeoecology* 107: 257–68.

Marino, B. D., and McElroy, M. B. 1991. Isotopic composition of atmospheric CO_2 inferred from carbon in C_4 plant cellulose. *Nature* 349: 127–31.

Meladze, G. K. 1964. On the phylogeny of the Sivatheriinae. *International Geological Congress* 22: 47–50.

Morgan, M. E., Kingston, J. D., and Marino, B. D. 1994. Carbon isotopic evidence for the emergence of C_4 plants in the Neogene from Pakistan and Kenya. *Nature* 367: 162–65.

O'Leary, M. H. 1981. Carbon isotope fractionation in plants. *Phytochemistry* 20: 553–67.

———. 1988. Carbon isotopes in photosynthesis. *BioScience* 38: 328–36.

Park, R., and Epstein, S. 1961. Metabolic fractionation of ^{13}C and ^{12}C in plants. *Plant Physiology* 36: 133–38.

Peisker, M. 1986. Models of carbon metabolism in C_3–C_4 intermediate plants as applied to the evolution of C_4 photosynthesis. *Plant, Cell, and Environment* 9: 627–35.

Plummer, T., and Bishop, L. C. 1994. Hominid paleoecology as indicated by artiodactyl remains from sites at Olduvai Gorge, Tanzania. *Journal of Human Evolution* 27: 47–75.

Quade, J., Cerling, T. E., Andrews, P., and Alpagut, B. 1995. Paleodietary reconstruction of Miocene faunas from Pasalar, Turkey using carbon and oxygen isotopes of fossil tooth enamel. *Journal of Human Evolution* 28: 373–84.

Quade, J., Cerling, T. E., Barry, J. C., Morgan, M. E., Pilbeam, D. R., Chivas, A. R., Lee-Thorp, K. A., and van der Merwe, N. J. 1992. A 16 Ma record of paleodiet using carbon and oxygen isotopes in fossil teeth from Pakistan. *Chemical Geology* 94: 183–92.

Quade, J., Cerling, T. E., and Bowman, J. R. 1989. Systematic variations in the carbon and oxygen isotopic composition of pedogenic carbonate along elevation transects in the southern Great Basin, United States. *Bulletin of the Geological Society of America* 101: 464–75.

Quade, J., Solounias, N., and Cerling, T. E. 1994. Stable isotopic evidence from paleosol carbonates and fossil teeth in Greece for forest or woodlands over the past 11 Ma. *Palaeogeography, Palaeoclimatology, Palaeoecology* 108: 41–53.

Rey, C., Renugopalakrishnan, V. M., Shimuzu, M., Collins, B., and Glimcher, M. J. 1991. A resolution-enhanced Fourier transform infrared spectroscopic study of the environment of the CO_3^{-2} ion in the mineral phase of enamel during its formation and maturation. *Calcified Tissue International* 49: 259–68.

Salisbury, F. B., and Ross, C. W. 1985. *Plant Physiology.* Wadsworth Publishing, Belmont.

Schoeninger, M. J., and DeNiro, M. J. 1982. Carbon isotope ratios of apatite from fossil bone cannot be used to reconstruct diets of animals. *Nature* 297: 577–78.

———. 1983. Reply to Sullivan and Krueger. *Nature* 301: 177–78.

Schoeninger, M. J., and Moore, K. 1992. Bone stable isotope studies in archeology. *Journal of World Prehistory* 6: 247–96.

Sikes, N. E. 1994. Early hominid preferences in East Africa: Paleosol carbon isotopic evidence. *Journal of Human Evolution* 27: 25–45.

Sillen, A. 1986. Biogenic and diagenetic Sr/Ca in Plio-Pleistocene fossils of the Omo Shungura formation. *Paleobiology* 12: 311–23.

Singh, J. S., and Gupta, S. R. 1977. Plant decomposition and soil respiration in terrestrial ecosystems. *Botanical Review* 43: 449–528.

Smith, B. N., and Epstein, S. 1971. Two categories of $^{13}C/^{12}C$ ratios for higher plants. *Plant Physiology* 47: 380–84.

Solounias, N., Teaford, M., and Walker, A. 1988. Interpreting the diet of extinct ruminants: The case of a non-browsing giraffid. *Paleobiology* 14: 287–300.

Sternberg, L. S. L., Mulkey, S. S., and Wright, J. S. 1989. Ecological interpretations of leaf carbon isotope ratios: Influence of respired carbon dioxide. *Ecology* 70: 1317–24.

Sullivan, C. H., and Krueger, H. W. 1981. Carbon isotope analysis of separate chemical phases in modern and fossil bone. *Nature* 292: 333–35.

———. 1983. Carbon isotope ratios of bone apatite and animal diet reconstruction. *Nature* 301: 177–78.

Tassy, P. 1999. Miocene elephantids (Mammalia) from the Emirate of Abu Dhabi, United Arab Emirates: Palaeobiogeographic implications. Chap. 18 in *Fossil Vertebrates of Arabia,* pp. 209–33 (ed. P. J. Whybrow and A. Hill). Yale University Press, New Haven.

Thackeray, J. F., van der Merwe, N. J., Lee-Thorp, J. A., Sillen, A., Lanham, J. L., Smith, R. , Keyser, A., and Monteiro, P. M. S. 1990. Changes in carbon isotope ratios in the late Permian recorded in therapsid tooth apatite. *Nature* 347: 751–53.

Tieszen, L. L. 1991. Natural variations in the carbon isotope values of plants: Implications for archaeology, ecology, and paleoecology. *Journal of Archaeological Science* 18: 227–48.

Tieszen, L. L., Senyimba, M. M., Imbamba, S. K., and Troughton, J. H. 1979. The distribution of C_3 and C_4 grasses and carbon isotope discrimination along an altitudinal and moisture gradient in Kenya. *Oecologia* 37: 337–50.

Tiezsen, L. L., Tesdahl, K. G., Boutton, T. W., and Slade, N. A. 1983. Fractionation and turnover of stable carbon isotopes in animal tissues: Implications for the ^{13}C analysis of diet. *Oecologia* 37: 337–50.

Toft, N. L., Anderson, J. E., and Nowak, R. S. 1989. Water use efficiency and carbon isotopic composition of plants in a cold desert environment. *Oecologia* 80: 11–18.

van der Merwe, N. J., and Medina, E. 1989. Photosynthesis and $^{13}C/^{12}C$ ratios in Amazonian rain forests. *Geochimica et Cosmochimica Acta* 53: 1091–94.

van der Merwe, N. J., and Vogel, J. C. 1978. ^{13}C content of human collagen as a measure of prehistoric diet in woodland North America. *Nature* 276: 815–16.

Vogel, J. C., and van der Merwe, N. J. 1977. Isotopic evidence for early maize cultivation in New York State. *American Antiquity* 42: 238–42.

Wang, Y., Cerling, T. E., Quade, J., Bowman, J. R., Smith, G. A., and Lindsay, E. H. 1993. Stable isotopes of paleosols and fossil teeth as paleocology and paleoclimate indicators: An example from the St. David Formation, Arizona. Climate change in continental isotopic records. *Geophysical Monograph* 78: 241–48.

Whybrow, P. J., and McClure, H. A. 1981. Fossil mangrove roots and palaeoenvironments of the Miocene of the eastern Arabian peninsula. *Palaeogeography, Palaeoclimatology, Palaeoecology* 32: 213–25.

Whybrow, P. J., Hill, A., Yasin al-Tikriti, W., and Hailwood, E. A. 1990. Late Miocene primate fauna, flora and initial palaeomagnetic data from the Emirate of Abu Dhabi, United Arab Emirates. *Journal of Human Evolution* 19: 583–88.

Woodward, F. I. 1990. Global Change: translating plant ecophysiological responses to ecosystems. *Trends in Ecology and Evolution* 5: 308–10.

Earliest Stone Tools from the Emirate of Abu Dhabi, United Arab Emirates

26

SALLY MCBREARTY

Artificially fractured fragments of chert signal the first entry of the human genus *Homo* into Abu Dhabi. At most outcrops of the Baynunah Formation, the sequence is capped by a thick layer of resistant tabular chert.[1] The horizontal disposition of this unit lends the characteristic flat-topped appearance to the jebels of Abu Dhabi's Western Region. Weathering of superficial Baynunah Formation sandstones, probably under conditions moister than those at present, resulted in the solution of quartz grains and the redeposition of silica in the form of silcrete (Ditchfield, 1999—Chapter 7). It is unclear whether this process occurred subaerially or at a subterranean bedding plane, but it may have required as much as a million years to complete (C. S. Bristow, personal communication.). This siliceous diagenetic product provided the raw material for implement manufacture by Abu Dhabi's earliest toolmakers.

When humans first occupied the Emirate, Baynunah Formation rocks were already very ancient. The landscape has apparently been attritional for a considerable period of time, as there is no remaining trace of any local sedimentation subsequent to that represented by the late Miocene Baynunah Formation. The location of the shoreline of the Arabian Gulf fluctuated dramatically during the Pleistocene. During the most recent glacial maximum at 18 000 years ago, global sea levels are known to have fallen by about 115 metres, and drops of similar magnitude occurred periodically throughout the Pleistocene at intervals of roughly 100 000 years (Imbrie and Imbrie, 1980; Berger et al, 1984; Ruddiman and Wright, 1987; Johnson and Straight, 1991). This dramatic lowering of base level no doubt vastly accelerated the erosion of Baynunah Formation sediments, which must formerly have covered an area of many tens if not hundreds of thousands of square kilometres (Friend, 1999—Chapter 5).

The Arabian Gulf today is quite shallow, with a mean depth of only about 35 metres, and thus recurring drops in sea level caused its episodic retreat or even disappearance. The Gulf's current maximum depth of about 165 metres lies near the Iranian shore. While relative shifts in plate locations, local uplift, downwarping, and sedimentation may be expected to have affected local topography, a marine remnant or embayment probably persisted here during glacial periods, except when sea levels were at their lowest. During the last glacial maximum, however, the Arabian Gulf is known to have retreated beyond the Straits of Hormuz. At such times western Abu Dhabi would have been separated from Iran by a sandy plain. The combined discharge of the Tigris and Euphrates may have emptied into the Gulf of Oman, though it is not certain that the volume of water would have been sufficient to maintain flow (Kassler, 1973; Rice, 1994).

Occasional periods of increased precipitation in the Late Pleistocene and early Holocene are demonstrated by widespread shallow lacustrine deposits in the interior of the Rub' al Khali, with associated vertebrate fauna and traces of human habitation (Zeuner, 1954; Field, 1958, 1960a,b; Masry, 1974; McClure, 1976, 1988). Similar conditions may have prevailed intermittently at earlier times, and hominid occupation of the Arabian Peninsula in the Early or Middle Pleistocene is

 ISBN 0-300-07183-3

attested by the occurrence of heavily patinated handaxes and other stone tools with a pronounced Middle Palaeolithic aspect at a number of sites in the Kingdom of Saudi Arabia (Masry, 1974; Zarins et al., 1980, 1981). Sea levels in the current interglacial are near their apparent maximum, but even a small rise in sea level, such as the 2-metre rise at about 7000 years ago, resulted in large-scale flooding in this region of low relief. At such times some resistant outcrops of coastal Baynunah Formation rocks may have been isolated as islands. The familiar sabkhas of Abu Dhabi are thought to be fairly recent in origin, the result of a marine transgression perhaps dating to no more than 4000 years ago (McClure, 1976).

ABU DHABI'S STONE ARTIFACT OCCURRENCES

Stone tools made from siliceous cap-rocks are found draped over several Baynunah Formation outcrops (McBrearty, 1993). Isolated artifacts have been observed at other localities in the area, but the largest numbers were encountered at Shuwaihat, Jebel Barakah, Hamra, and Ras al Aysh; discussion here will be confined to these four localities. All artifacts were found on the surface of Baynunah or Shuwaihat Formation outcrops or in very shallow superficial deposits produced by their erosion. Primary artifact collection was uncontrolled, in that objects were selected from the total site area to provide an idea of the range of material present. Frequencies of artifact categories may therefore not be strictly representative, but at all four sites additional small controlled collections were made to determine artifact density and size distribution.

Shuwaihat

Artifacts occur here in the vicinity of the Shuwaihat, site S2, Miocene fossil-collecting area (N 24° 06′ 41.7″, E 52° 26′ 04.0″). They are scattered over an area of about 30 000 m^2 on the sea cliffs and wave-cut platform on the south and west faces of the jebel below an elevation of about 40 metres asl. The raw material at Shuwaihat is a silicified limestone, usually light yellow, but sometimes weathering to a blackish colour. While its quality is poor, the edges of most artifacts are quite sharp.

There are no formal retouched tools in the collection from Shuwaihat. Rather, all artifacts consist of cores and the resulting debitage, though a few elongate high-backed cores on slabs might be classified by some as "pushplanes" or heavy-duty scrapers (fig. 26.1a). Cores are rather large, with a mean size of 116 mm (range 75–163 mm, $n = 16$). Seven of these (43.8%) are radial forms, and include disc, subradial, and high-backed types (fig. 26.1b,c). Raw material at Shuwaihat may seem unpromising for blade manufacture, but two of the Shuwaihat cores (12.5%) are bidirectional blade cores (fig. 26.1d), and one is a crescent-shaped blade core (fig. 26.1f). These distinctive cores are identical to "naviform" cores described by Crowfoot-Payne (1983) from the Pre-Pottery Neolithic B (PPNB) levels at Jericho, Palestine. Technologically they resemble the bifacially prepared *plaquettes* on tile flint described by Inizan (1980b, 1988) for Qatar. The Shuwaihat cores were radially prepared around their circumference and then blades were removed from an axis perpendicular to that of the radial striking platform. The first blade removed from such a prepared core has a dorsal "crest" of intersecting flake scars, the remains of the radial striking (cf. fig. 26.3n). Blades removed subsequently show parallel dorsal scars.

Upon close examination, three additional Shuwaihat cores (fig. 26.1g,i) and two bifacially trimmed slabs (fig. 26.1h) exhibit attempted blade removals and four show breakage that probably resulted from failed attempts at blade removal. Only one true blade was observed. Rather, Shuwaihat debitage consists of flakes with breadth/length (B/L) ratios of approximately 1.0, and a simple pattern of dorsal scars.[2] Although most artifacts at Shuwaihat have a somewhat crude appearance, their lack of elegance reflects primarily the poor quality of the lithic raw material. Indeed, multiple bulbs of percussion show that several blows were sometimes required to detach flakes from the core (fig. 26.1j,k).

Figure 26.1. Lithic artifacts from Shuwaihat: *a*, elongate high-backed core or "pushplane"; *b*, radial core; *c*, high-backed radial core; *d*, bidirectional blade core; *e*, single-platform core; *f*, crescent-shaped or "naviform" blade core; *g*, radial core with attempted blade removal; *h*, bifacially trimmed slab with attempted blade removal; *i*, single-platform core with attempted blade removal; *j*, *k*, flakes with multiple bulbs of percussion.

To determine artifact density at Shuwaihat, a baseline 30 metres long was laid out roughly northwest–southeast (bearing 155°), on the wave-cut platform at the base of the cliff on the south side of the jebel, parallel to the cliff face. All material was collected from five 1-m^2 units, set out at intervals of 5 metres. Most artifacts were found lying directly on the surface of the hard red lithified Shuwaihat Formation sandstone. In two squares, however, they were buried in a thin unconsolidated outwash mantle. Here the sediment was trowel-scraped to a depth of 2 cm and passed through a sieve of 1/8 inch mesh. The deposit consisted of fine pulverised gypsum, sea shell fragments, and medium-fine quartz sand, and was underlain by a thin discontinuous layer of mottled greenish clay. In addition to numerous small artifacts, pieces of naturally fractured stone, modern cormorant bone, and occasional fragments of Miocene fossil bone were encountered in these superficial redeposited sediments. A total of 272 stone objects were found; their mean density is 54/m^2 (range 26–139); of these, 90 (33%) are artifacts. Mean artifact density is 18/m^2 (range 15–29). Most pieces are less than 3 cm in maximum size; one 1-m^2 excavation unit contained five whole flakes less than 2 cm in size. The presence of this microdebitage in addition to large cortical pieces elsewhere at the site demonstrates convincingly that both early and late stages of artifact manufacture took place at Shuwaihat.

Hamra

Artifacts at Hamra are found over an area of about 10 000 m^2, near the Hamra, site H2, Miocene fossil-collecting locality (N 23° 06′ 06.9″, E 52° 31′ 42.5″), on the west side of the jebel, and about 50 metres below the summit. Several further isolated scatters of flaking debris were seen 1.3 km ENE of the summit. The raw material at Hamra is a good-quality yellow to black flint with a fairly deep patina. At the top of the jebel the thick tabular chert outcrop is well exposed, and heat-shattered flint is abundant. Objects whose appearance mimics artificial fracture are numerous, but no true artifacts were observed on the summit itself. About 50 metres downslope, flints are fewer in number but the proportion of artifacts is greater. The slopes are comprised primarily of loose sand derived from Baynunah Formation sandstones; in places they are partially stabilised by ephemeral vegetation.

The methods of artifact production are similar to those seen on Shuwaihat, but because the quality of the lithic raw material at Hamra is excellent, the artifacts have a more elegant appearance. As at Shuwaihat, no incontestable formal tools were observed, though one piece, interpreted here as a radial core, could be identified as a circular scraper (fig. 26.2a), and two elongate multiplatform cores could be interpreted as "limace" forms (fig. 26.2e). Cores include both unidirectional and bidirectional prismatic blade cores (fig. 26.2g–j). Single and multiplatform cores for the production of flakes are present (fig. 26.2f), but flake production was primarily by radial core reduction; 15 (79%) of 19 cores in the collection show a radial flaking pattern. Of these, eight show the removal of flakes or blades from a platform perpendicular to the radial flaking axis, as described above for the Shuwaihat material (fig. 26.2b), and two show breakage probably resulting from failed attempts at blade removal of this kind (fig. 26.2c,d). Whole blades are rather rare (fig. 26.2k); debitage is made up primarily of flakes with simple or radial dorsal scar patterns. Overall artifact dimensions are rather small. The mean size for blade cores is 55 mm (range 46–67 mm; $n = 8$); for all cores, 60 mm (range 46–77 mm; $n = 14$).

Two 1-m^2 units were laid out and all lithic objects within them collected to determine artifact abundance. The first unit was located about 300 metres west of the western edge of the flint-cluttered jebel summit; the second, about a further 120 metres to the NNW. Within the first, upslope unit, 815 flints were collected. Of these, 103 (12.6%) are artifacts. None exceeds 10 cm in size, and, of the artifacts, 70 (68%) are less than 3 cm in size; some of these flakes are less than 1 cm in maximum dimension. Within the second, downslope unit, 62 flints were collected. Of these, 11 (17.7%) are artifacts. Again, none is greater than 10 cm in maximum dimension, and 43 (69.4%) are less than 3 cm in size. The density of artifacts at Hamra is

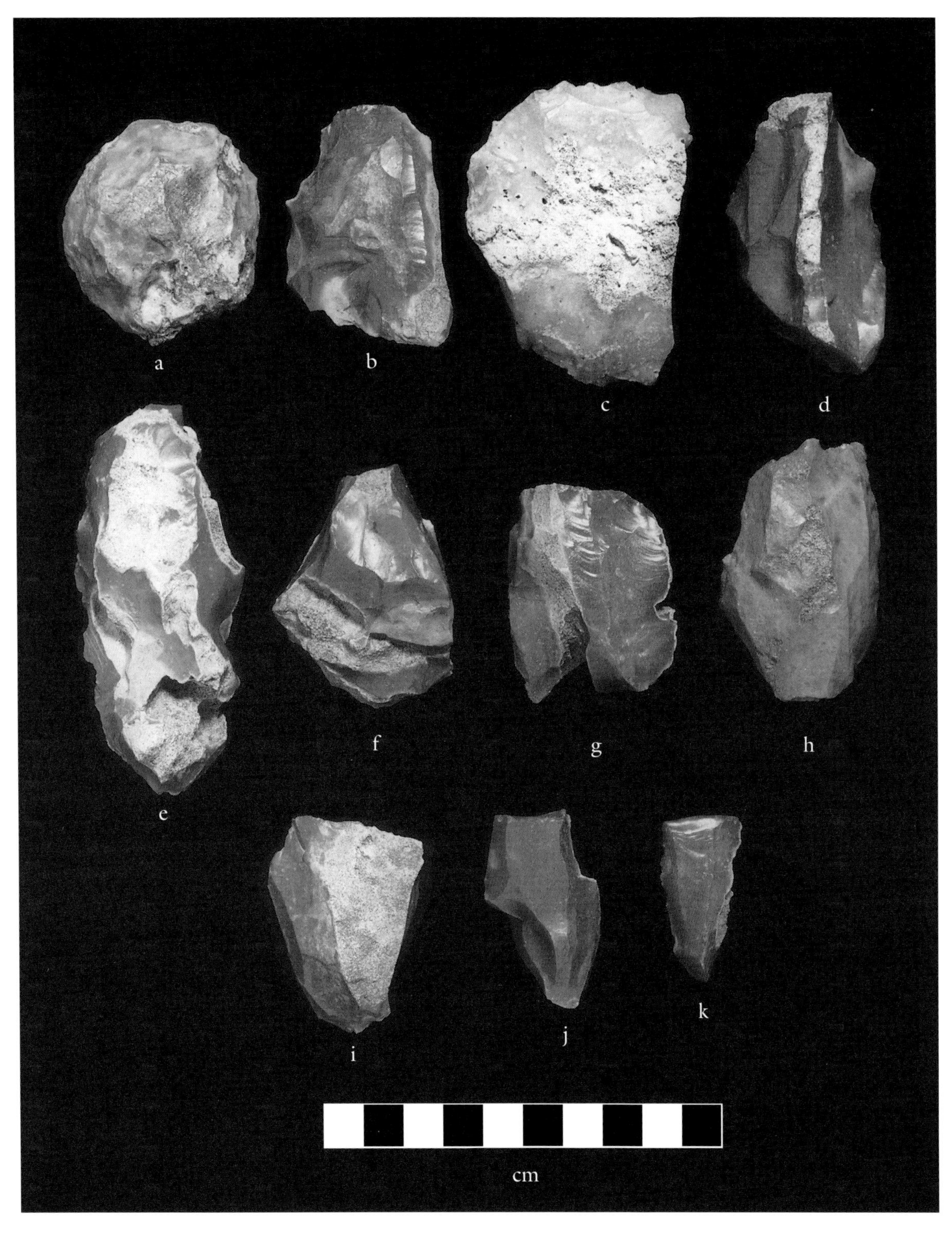

Figure 26.2. Lithic artifacts from Hamra: *a*, high-backed radial core or circular steep scraper; *b*, crescent-shaped or "naviform" blade core; *c*, *d*, radial cores, broken through attempted blade removals; *e*, elongate multiplatform core or high-backed limace; *f*, multiplatform core; *g*, unidirectional blade core; *h*, *i*, *j*, bidirectional blade cores; *k*, plunging blade with parallel bidirectional dorsal scars.

quite staggering (about 100 and 60 per m^2), and the presence of both cortical pieces and microdebitage seems to confirm the impression that all stages of artifact manufacture took place on site.

Ras al Aysh

At Ras al Aysh (N 24° 04′ 59.8″, E 53° 12′ 49.9″) artifacts are found over an area of about 55 000 m^2, about 500 metres northwest of the summit of Ras al Aysh. The scatter has a rather sharply defined southeast boundary, with few worked flints found upslope, near the cap-rock—the source of the lithic raw material. Artifacts occur both on the surface and within a thin superficial mantle of sediments that rests upon Miocene rocks. This mantle is comprised of unconsolidated buff-coloured sands and silts, apparently outwash derived from the Baynunah Formation outcrop that makes up the jebel itself. The Ras al Aysh flint is an ideal raw material for the manufacture of stone tools. All flints, both artifactual and nonartifactual, display a deep green to yellowish-green patina, and many also show pronounced thermal damage in the form of potlid fractures. To determine artifact densities, a north–south baseline 50 metres long was laid out and all lithic material was collected within 11 1-m^2 units, set out at intervals of 5 metres. A collection of 418 lithic objects resulted, of which only 11 (2.6%) are artifacts. The artifacts are not large; only two exceed 10 cm in maximum dimension.

As at Shuwaihat and Hamra, trimmed formal tools are lacking at Ras al Aysh, with the exception of a possible "pushplane" (fig. 26.3a), probably better described as a core. Artifacts include unidirectional and bidirectional cores (fig. 26.3d–f), and the resulting flakes, blades, and flake and blade fragments. As at the sites already described, however, most cores are radial types, including disc and high-backed forms (fig. 26.3b). These make up 11 (65%) of the 17 cores collected. Two of these show blade removals (fig. 26.3c,g). Although neither of these cores has the crescent shape resulting from blade removals perpendicular to the radial flaking axis, as seen at Shuwaihat and Hamra, the presence of "ridge removal" blades (fig. 26.3n) suggests the presence of this technique at Ras al Aysh. Flakes and blades show simple, radial, parallel, and bidirectional dorsal scars. The presence of both cortical flakes and exhausted multiplatform cores indicates that both early and late stages of core reduction are present. The excellent quality of the Ras al Aysh flint, together with the technical skill and methodical regularity of the toolmakers, combine to give the artifacts a somewhat formal or refined appearance.

Jebel Barakah

At Jebel Barakah (N 24° 00′ 23.6″, E 52° 19′ 37.3″) large numbers of stone artifacts occur on the level bluffs on the southeast side of the jebel. A number of Baynunah Formation vertebrate fossils have been found in the gullies leading down to the beach below (Whybrow et al., 1990). The artifact scatter covers a total area of at least 13 000 m^2, and extends from the sea cliffs up to a point about 330 metres ESE of the jebel summit, where there is a pronounced break in slope. On the sea cliffs the lithic artifacts lie directly on Baynunah Formation rocks; upslope they are overlain by a thin superficial layer of soft unconsolidated sediment derived from the exposures of the Baynunah Formation above. The raw material employed at Jebel Barakah is a fairly good quality flint, with a deep black to blue-black patina.[3]

The Barakah artifacts display a very consistent and formalised flaking method, being composed almost entirely of radial cores and the flakes derived from them. There is no trace of any blade element, and this is evident in the character of both the flakes and the cores. All 16 cores collected are radial or high-backed radial forms (fig. 26.4a–i). There are no unidirectional or bidirectional blade cores, and broken radial cores (for example, fig. 26.4j) show no sign of attempted blade removals. Nor are there any blades. Of a sample of 38 measured flakes, mean B/L is near unity (1.012; range 0.55–1.82; s.d. = 0.296). Of the 78 flakes collected, 34 (43.6%) have radial dorsal scars, 32 (29.5%) have simple dorsal scars, and 12 (15.45%) are cortical flakes. None has parallel or bidirectional dorsal scars that would indicate blade production techniques.

Figure 26.3. Lithic artifacts from Ras al Aysh: *a,* trimmed slab or "pushplane"; *b,* radial core; *c,* high-backed radial core with blade removals on ventral (flat) face; *d,* bidirectional flake core; *e,* unidirectional blade core; *f,* bidirectional blade core; *g,* radial core with blade removals; *h,* exhausted high-backed radial core; *i,* exhausted multiplatform core; *j,* unifacial radial core on flake or circular scraper; *k,* blade with unidirectional parallel dorsal scars; *l,* blade with bidirectional parallel dorsal scars; *m,* cortical blade (*lame à crête naturelle*); *n,* "ridge removal" blade from radial core (*lame à crête à deux versants*).

Figure 26.4. Lithic artifacts from Jebel Barakah: *a–h,* radial cores; *i,* high-backed radial core; *j,* broken radial core; *k,* Kombewa flake with ventral flake removal; *l,* biface tip; *m,* flake fragment with marginal unifacial trimming.

At Jebel Barakah are found the only two unambiguous formal tools encountered at any of the Western Region lithic artifact sites. Both are tool fragments, one a biface tip, the other a small flake fragment with marginal unifacial trimming (fig. 26.4l,m). The biface tip has a fairly straight, only slightly sinuous edge when viewed in profile. It has been flaked over the entire surface on each side by direct percussion with soft hammer. It is not particularly small: its breadth at 32 mm from the tip is 47 mm; the projected breadth at 60 mm from the tip is about 70 mm; and the maximum thickness is 14 mm. It is impossible to surmise the absolute original dimensions of the piece, but it seems unlikely that the length of this biface could have been less than 110 mm, and it is quite possible that it is the remains of a much larger object.

Two baselines were laid out at right angles on the sea cliff, one extending 30 metres east–west, the other 20 metres north–south, to determine artifact density. All material was collected within 10 1-m^2 units, set out at intervals of 5 metres. In this area of rapid erosion stone objects are far less dense than at the other sites described here. Within the controlled collection area, 218 objects were encountered, of which only eight (3.6%) are artifacts.

Chronology

The major issues to be considered in interpreting the Abu Dhabi stone artifacts are their age and function, and the identity, affiliations, and way of life of their makers. The age of the material is critical to any reconstruction. The artifacts cannot directly be dated by any conventional method, and the finds lack stratified context or associated material such as ground stone, fossil fauna, or ceramics that could provide an approximate date. Age estimates for stone tools based upon technology alone are by their nature risky and uncertain. Nonetheless, some inferences can be drawn from technological aspects of the artifacts themselves.

The lithic artifact sites described here can be divided into two groups. The first is comprised of Shuwaihat, Hamra, and Ras al Aysh, where blades are present. The second contain only one site, Jebel Barakah, which features the consistent, exclusive use of radial core technology. At Jebel Barakah, there is no blade element, and a single broken biface tip has been found.

The oldest sites previously described for the Gulf region date to the early Holocene. They are characterised by projectile points, usually made on blades, pressure-flaked into elongate lanceolate and foliate forms, often with tangs and barbs. Kapel (1967) first ascribed such sites in Qatar to his "B group". Additional B-group sites were subsequently discovered in Qatar (Inizan, 1978, 1979, 1980a,b, 1988; Tixier, 1980), in Oman (Copeland and Bergne, 1976; Pullar, 1974), and in Saudi Arabia (Masry, 1974). At the stratified site of Aïn Qannas in eastern Saudi Arabia, B-group artifacts lie beneath layers with Ubaid sherds (Masry, 1974), indicating an age in excess of 6000 years. The few reported radiocarbon determinations (Kappel, 1967; Masry, 1974) indicate a date for B-group sites of the order of 7000 years before present (b.p.).

The lack of formal tools at Shuwaihat, Hamra, and Ras al Aysh prevents easy identification with this or any other tradition of lithic technology. The blades found at the Abu Dhabi sites do provide some clues, although blades themselves are of limited use as temporal indicators. In Africa, early blades date to more than 240 000 (Tallon, 1978; McBrearty et al., 1996), and in the Levant blades may be present as early as 200 000 (Grün and Stringer, 1991) or even 300 000 years ago (Mercier et al., 1995). Some European assemblages dating to as much as 115 000 years ago have a blade component (Conard, 1990; Révillion and Tuffreau, 1994; Révillion, 1995), but blades do not become common in Europe until the Upper Palaeolithic period, after 45 000 years ago. Blades formed the basis for subsequent Mesolithic and Neolithic technology, and the technique persisted into historic times; early European visitors to Mesoamerica observed the manufacture of obsidian blades in the sixteenth century; in Europe, flintknappers produced gunflints by means of blade production as late as the nineteenth century.

The pattern of blade production at Shuwaihat, Hamra, and Ras al Aysh is, however, highly distinctive. The crescent-shaped blade cores from these sites show some similarity to those from Açilar, near Khor, on the northeast coast of Qatar, where Inizan (1988) describes a dozen surface occurrences with laminar debitage and pressure-flaked points. Inizan regards the points as the ultimate end product of the manufacturing process at Khor, and she interprets the nonlaminar debitage as the by-product of core preparation for blade production. This may be true for the Abu Dhabi material as well, and the radial flaking may simply function to create a biconvex profile from which blades were detached. In the absence of points or other retouched implements, however, this cannot be asserted with confidence, and it is possible that the large numbers of flakes at Shuwaihat, Hamra, and Ras al Aysh were also intended to be used as implements.

Blade production at Khor, as at the Abu Dhabi sites, consists of bifacial core preparation and the subsequent removal of blades at right angles or near right angles to this prepared surface. Again, as at the Abu Dhabi sites, blade removal at Khor is bidirectional, alternating between platforms at opposite ends of the flaking surface. The tabular flint *plaquettes* at Khor are not radially prepared, however, and thus do not show the crescent shape characteristic of the Abu Dhabi cores. The Abu Dhabi sites also lack the tanged foliate pressure-flaked projectile points found at Khor. In the absence of retouched tools it is impossible positively to identify the Abu Dhabi lithic artifacts with those from Qatar, but the similarity in the general approach to blade production suggests some affinity between the populations at the two groups of sites.

The material from Shuwaihat, Hamra, and Ras al Aysh can be linked more directly to material from the Levant. The crescent-shaped blade cores from these sites show remarkable similarity to the naviform cores from the Pre-pottery Neolithic (PPNB) levels at Jericho in Palestine and Abu Hureyra in Syria (Crowfoot-Payne, 1983). Dates for the PPNB, the "archaic Neolithic" or "stage 2 Neolithic" of Moore (1982) range between c. 9300 and 8300 b.p. (Stager, 1992). Other authors (for example, Pullar, 1974; Copeland and Bergne, 1976; Smith, 1977; Inizan, 1988; Potts, 1992) have noted the similarity of the B-group pressure-flaked points, awls, and backed bladed of the Arabian Peninsula to those of the PPNB of Syria, Jordan, Lebanon, and Palestine, especially those from Beidha (Mortensen, 1970) and Byblos (Cauvin, 1969). In the absence of retouched tools it is impossible positively to identify the Abu Dhabi lithic artifacts with those of the B-group, but the distinctive crescent-shaped or naviform cores indicate a definite tie with the PPNB of the Levant. It is therefore extremely likely that the blades at Shuwaihat, Hamra, and Ras al Aysh were made by people with some connection to the Levant between about 9300 and 8300 years ago.

The technology at Jebel Barakah clearly differs from that at Shuwaihat, Hamra, and Ras al Aysh, and it is more difficult to place temporally. Like the other sites, Jebel Barakah lacks foliates or tanged or barbed arrow heads, and there is no sign of pressure flaking, backed pieces, or ceramics. Neither are there any blade cores or laminar debitage. Rather, Jebel Barakah's technology is characterised by the consistent, exclusive use of radial core technology. The lack of a blade element is not the result of differences in raw material. Although the Jebel Barakah flint is not equal to that at Ras al Aysh or Hamra, it is of fine quality and would present no obstacle to blade manufacture. Indeed, the early inhabitants of Abu Dhabi attempted to make blades from the remarkable poor-quality stone at Shuwaihat, although their efforts met with more limited success.

The Jebel Barakah radial cores are not useful as temporal indicators, as radial cores are present in the world's oldest lithic industries, dating to as much as 2.5 million years ago (Leakey, 1971; Kibunjia, 1994) but persist into the Neolithic. The biface tip from Jebel Barakah, finely worked by direct percussion with a soft hammer, one of only two formal tools found at Abu Dhabi sites, might be expected to provide a more reliable indication of age. But bifaces, too, have a long lifespan. Well-made bifacial handaxes appear first in the Acheulian of Africa as early as 1.8 million years ago (Roche et al., 1994). Small handaxes are common in the Middle Stone Age of Africa and in the Middle Palaeolithic of Europe and the Near East, and larger examples of bifacial points from later Pleistocene or Holocene industries from both the Old and New Worlds may superficially resemble handaxes. Even bifacial rough-cuts for Neolithic polished stone celts may occasionally be mistaken for handaxes by the unwary.

Bifaces, like radial cores, appear to persist into the Holocene on the Arabian Peninsula. Kapel (1967) initially inferred a Palaeolithic age for some sites at Khor, Qatar, from the presence of bifacial implements. Subsequent examination of the sites by the French Mission, however, has shown the presence of Ubaid potsherds (Inizan, 1979, 1980c; Tixier, 1980), and from an examination of the mode of production, Inizan (1980b, 1988) has inferred that the Khor bifaces represent early stages in the manufacture of projectile points.

It is possible that the Jebel Barakah artifacts represent a coastal occurrence of the "western ar-Rub' al Khali Neolithic biface tradition" as defined by Edens (1982, 1988). Sites of this tradition have been described for the interior of Saudi Arabia by Zeuner (1954), Field (1958, 1960a), Gramley (1971), McClure (1976), Gotoh (1981), and others. Those in the western Rub' al Khali are associated with extensive ephemeral Holocene lake deposits, some contain fauna and ground stone vessels. Zeuner (1954) obtained an early radiocarbon date for this material of c. 5235 b.p., but later results suggest an age of between 10 000 and 6000 years (McClure, 1976). Formal tools at these sites include bifacially retouched foliate and lanceolate points, either pressure-flaked or trimmed with a soft hammer, often with stems or barbs. The radial cores and biface tip would fit comfortably into such assemblages, but most of the objects in the collections ascribed by Edens to this tradition are re-touched points, lacking at Jebel Barakah. Lanceolates range in size up to 120 mm in length, consistent with my estimate of greater than or equal to 110 mm for the Jebel Barakah biface, and flake production appears to have been by radial core reduction.

On the basis of similarity in the projectile points, Potts (1992) equates Edens' (1982) biface tradition with the Qatar "D group" of Kapel (1967). Radiocarbon dates for D-group sites range between 7265 and 5515 b.p. (Potts, 1992). At D-group sites in Bahrain, Qatar, and eastern Saudi Arabia, the characteristic lithic artifacts have been found in association with Ubaid sherds (de Cardi, 1974; Masry, 1974; Inizan, 1980c; Potts, 1992), suggesting definite links with Mesopotamia and confirming an age of c. 7500–6000 b.p. (Porada et al., 1992). Potts (1992), however, observes that D-group lithic artifacts persist at sites in the Arabian Gulf region after the disappearance of Ubaid pottery, and Edens (1982) notes that points nearly identical to those of his western Rub' al Khali biface tradition have been found in Sargonid contexts at Ur (4350–4150 b.p.; Porada et al., 1992). The Barakah radial cores and biface tip might fall within the range of artifacts expected at D-group or Neolithic biface tradition sites, but the absence of the diagnostic projectile points, polished stone, fauna, or any distinctive technological feature, precludes identification with this tradition with real confidence. On the other hand, it is quite possible that the lithic artifacts from Jebel Barakah are very ancient, perhaps dating to the Middle Pleistocene. Nothing in their technical execution or state of patination would exclude them from the Acheulian or Middle Stone Age, and sites of both these traditions are known from the interior of Saudi Arabia (for example, Zarins et al., 1980, 1981), and from numerous sites east of the Arabian Gulf.

Discussion

The makers of the artifacts at Shuwaihat, Ras al Aysh, and Hamra probably made their living from products of the sea as well as the hunt, and were part of the Arabian "desert hunter and coastal cruiser" tradition described by Tosi (1986). Their distinctive method of producing blades, however, shows that they were part of a wide early Neolithic tradition stretching all the way to the shores of the Mediterranean. Here, people lived in permanent or semipermanent villages, subsisting on intensively harvested wild grain and legumes, and hunted and trapped gazelle, deer, boar, ass, hare, waterfowl, and marine fish, although experimentation with the domestication of wheat, barley, and wild caprids was well underway (Bar-Yosef, 1981; Bar-Yosef and Cohen, 1992; Bar-Yosef and Meadow, 1995; Moore, 1995). Exotic domesticates may thus have contributed to the Abu Dhabi's early subsistence economy.

The climate in Abu Dhabi when Shuwaihat, Ras al Aysh, and Hamra were occupied was probably more humid than that of the present day. These sites are contemporary with the expansion of ephemeral lakes in the interior of Saudi Arabia (McClure, 1976, 1988; Hötzl et al., 1984). Local proxy evidence comes from the site of Khor 36 in Qatar, where excavations into sabkha sediments revealed subsurface artifacts in clayey deposits interpreted as representing humid conditions (Inizan, 1988). Higher early Holocene sea levels may have rendered the Abu Dhabi sites an offshore archipelago, reachable only by boat.

The very absence of formal tools at any of the sites and their location at sources of raw material suggests a specialised economic function as quarry sites. If they are extractive rather than habitation sites, it is not surprising that potsherds and other domestic items are lacking. The flakes and blades manufactured at the Baynunah Formation outcrops may have been traded or transported elsewhere for transformation into finished tools. Shuwaihat, Ras al Aysh, and Hamra may have provided the raw material for stone tools at the Neolithic occupations on the nearby offshore islands of Sir Bani Yas and Marawah (King, 1992; King et al., 1995).

The picture at Jebel Barakah is more difficult to reconstruct. If the Barakah artifacts date to the Middle Pleistocene, their makers inhabited a world very different from our own. At times, lowered sea levels exposed broad areas of the Arabian Peninsula that are now submerged, and allowed long-range human migration and contact with contemporary populations of Africa and Eurasia. On the other hand, if the Barakah artifacts represent a local manifestation of the Neolithic biface tradition, they could date to the latest mid- to late Holocene, and could perhaps be contemporary with the Ubaid period occupation on the nearby island of Dalma (King, 1992; King et al., 1995). The mid-Holocene landscape was not unlike that of the present day. Such early inhabitants of Abu Dhabi's Western Region may well have been familiar with the peoples and civilisations beyond the Gulf, either by acquaintance with seagoing traders or by travel to the early capitals of the ancient world.

Acknowledgements

I thank Peter Whybrow and Andrew Hill for their kind invitation to participate in the exploration of the Western Region of the Emirate of Abu Dhabi and in the First International Conference on the Fossil Vertebrates of Arabia. I am grateful to the Abu Dhabi Company for Onshore Oil Operations (ADCO) for their support for the project and to His Excellency Sheik Nahayan bin Mubarak al Nahayan for his endorsement of our efforts. Walid Yasin, of the Department of Antiquities and Tourism, Al Ain, was more than generous with his time and attention. I thank also Ofer Bar-Yosef, Chris Edens, Frank Hole, Andrew Moore, and Rita Wright for directing me to essential sources and Geoffrey King for valuable discussion of Abu Dhabi's past.

Notes

1. The terms chert and flint will be used here interchangeably to refer to cryptocrystalline siliceous rocks produced by diagenetic solution.
2. L = length of the flake along its flaking axis; B = maximum breadth of the flake perpendicular to its flaking axis. Blades have B/L ratios of ≤ 0.5. A "simple" pattern of dorsal flake scars implies that all scars originate from the same platform as the flake itself.
3. A superficial deposit of clam shells, less than 5 cm thick, and covering an area of about 10 m^2, is found on the bluffs at Jebel Barakah, about 300 metres east of the summit. A complete shell-tempered brick, measuring about 10 cm × 30 cm × 20 cm, some additional brick fragments, several goat podial bones, and a single Islamic period basal potsherd, indicate a very late date for the debris. No lithic artifacts were

found within or near the shell deposit, and it is judged to be unrelated to those described here.

REFERENCES

Bar-Yosef, O. 1981. The "Pre-Pottery Neolithic" period in the southern Levant. In *Préhistoire du Levant: Chronologie et Organisation de l'Espace depuis les Origines jusqu'au VIe Millénaire*, pp. 555–69 (ed. J. Cauvin and P. Sanlaville). Centre National de la Recherche Scientifique, Paris.

Bar-Yosef, O., and Belfer-Cohen, A. 1992. From foraging to farming in the Mediterranean Levant. In *Transitions to Agriculture in Prehistory*, pp. 21–48 (ed. A. B. Gebauer and T. D.Price). Prehistory Press, Madison.

Bar-Yosef, O., and Meadow, R. H. 1995. The origins of agriculture in the Near East. In *Last Hunters, First Farmers*, pp. 39–94 (ed. T. D. Price and A. B. Gebauer). School of American Research, Santa Fe.

Berger, A., Imbrie, J., Hays, J., Kukla, G., and Saltzmann, B. ed. 1984. *Milankovitch and Climate: Understanding the Response to Astronomical Forcing*. Reidel, Dordrecht.

Cardi, B. de. 1974. The British archaeological expedition to Qatar 1973–1974. *Antiquity* 48: 196–200.

Cauvin, J. 1969. *Les Outillages Néolithic de Byblos et du Littoral Libanais*. Maisonneuve, Paris.

Conard, N. J. 1990. Laminar lithic assemblages from the last interglacial complex in northwestern Europe. *Journal of Anthropological Research* 46: 243–62.

Copeland, L., and Bergne, P. 1976. Flint artifacts from the Buraimi area, eastern Arabia, and their relations with the Near Eastern post-Pleistocene. *Proceedings of the Seminar for Arabian Studies* 3: 40–61.

Crowfoot-Payne, J. 1983. The flint industries of Jericho. In *Excavations at Jericho*, vol. 5, pp. 622–759 (ed. K. M. Kenyon and T. A. Holland). The British School of Archaeology in Jerusalem, London.

Ditchfield, P. W. 1999. Diagenesis of the Baynunah, Shuwaihat, and Upper Dam Formation sediments exposed in the Western Region, Emirate of Abu Dhabi, United Arab Emirates. Chap. 7 in *Fossil Vertebrates of Arabia* , pp. 61–74 (ed. P. J. Whybrow and A. Hill). Yale University Press, New Haven.

Edens, C. 1982. Towards a definition of the western Rub al-Khali "Neolithic". *ATLAL, The Journal of Saudi Arabian Archaeology* 6: 109–24.

———. 1988. The Rub al-Khali 'Neolithic' revisited: the view from Nadqan. In *Araby the Blest: Studies in Arabic Archaeology*, pp. 15–43 (ed. D. T. Potts). Carsten Neibuhr Institute of Ancient Near Eastern Studies, Copenhagen.

Field, H. 1958. Stone implements from the Rub' al Khali, Saudi Arabia. *Man* 58: 93–94.

———. 1960a. Stone implements from the Rub' al Khali, Saudi Arabia. *Man* 60: 25–26.

———. 1960b. Carbon-14 date for a 'neolithic' site in the Rub' al Khali. *Man* 60: 172.

Friend, P. F. 1998. Rivers of the Lower Baynunah Formation, Emirate of Abu Dhabi, United Arab Emirates. Chap. 5 in *Fossil Vertebrates of Arabia* (ed. P. J. Whybrow and A. Hill). Yale University Press, New Haven.

Gotoh, T. 1981. A stone age collection from the Rub' al Khali desert. *Bulletin of the Ancient Orient Museum* 3: 1–15.

Gramley, R. M. 1971. Neolithic flint implement assemblages from Saudi Arabia. *Journal of Near Eastern Studies* 30: 177–85.

Grün, R., and Stringer, C. B. 1991. Electron spin resonance dating and the evolution of modern humans. *Archaeometry* 33: 153–99.

Hötzl, H., Jado, A., Moser, H., Rauert, W., and Zötl, J. 1984. Climatic fluctuations in the Holocene. In *The Quaternary Period in Saudi Arabia*, vol. 2, pp.

301–14 (ed. A. Jado and J. Zötl). Springer-Verlag, Vienna.

Imbrie, J., and Imbrie, J. Z. 1980. Modeling the climatic response to orbital variations. *Science* 207: 943–53.

Inizan, M.-L. 1978. Première mission archéologique française à Qatar. *Paléorient* 4: 347–51.

———. 1979. Troisième mission archéologique française à Qatar. *Paléorient* 5: 277–80.

———. 1980a. Premiers résultats des fouilles préhistoriques de la région de Khor. In *Mission Archéologiques français à Qatar,* vol. 1, pp. 51–98 (ed. J. Tixier). Ministry of Information, Department of Antiquities and Tourism, Doha.

———. 1980b. Sur les industries à lames de Qatar. *Paléorient* 6: 233–36.

Inizan, M.-L. 1980c. Sites à poterie obeidienne à Qatar. *L'archéologie de l'Iraq,* pp. 209–21 (ed. Barrelet, M.). Centre National de la Recherche Scientifique, Paris.

——— 1988. *Mission Archéologiques français à Qatar.* Editions Recherche sur les Civilisations. Centre National de la Recherche Scientifique, Paris.

Johnson, L. L., and Straight, M. 1991. *Paleoshorelines and Prehistory.* CRC Press, Boca Raton.

Kapel, H. 1967. *Atlas of the Stone Age Cultures of Qatar.* Jutland Archaeological Society Publications, Copenhagen.

Kassler, T. 1973. The structural and geomorphic evolution of the Persian Gulf. In *The Persian Gulf: Holocene Carbonate Sedimentation and Diagenesis in a Shallow Epicontinenal Sea,* pp. 11–32 (ed. B. H. Purser). Springer, New York.

Kibunjia, M. 1994. Pliocene archaeological occurrences in the Lake Turkana Basin. *Journal of Human Evolution* 27: 159–72.

King, G. R. D. 1992. Abu Dhabi Islands survey, March–April. Unpublished report.

King, G. R. D., Dunlop, D., Elders, J., Garfi, S., Stephenson, A., and Tonghini, C. 1995. A report of the Abu Dhabi Islands Archaeological Survey (1993–4). *Proceedings of the Seminar for Arabian Studies* 25: 63–74.

Leakey, M. D. 1971. *Olduvai Gorge. Volume 3: Excavations in Beds I and II, 1960–1963.* Cambridge University Press, Cambridge.

Masry, A. H. 1974. *Prehistory in Northeastern Arabia: The Problem of Interregional Interaction.* Field Research Projects, Miami.

McBrearty, S. 1993. Lithic artifacts from Abu Dhabi's Western Region. *Tribulus: Bulletin of the Emirates Natural History Group* 3: 13–14.

McBrearty, S., Bishop, L. C., and Kingston, J. D. 1996. Variability in traces of Middle Pleistocene hominid behavior in the Kapthurin Formation, Baringo; Kenya. *Journal of Human Evolution* 30: 563–80.

McClure, H. 1976. Radiocarbon chronology of Late Quaternary lakes in the Arabian desert. *Nature* 263: 755–56.

———. 1988. Late Quaternary paleogeography and landscape evolution of the Rub' al Khali. In *Araby the Blest: Studies in Arabic Archaeology,* pp. 9–13 (ed. D. T. Potts). Carsten Neibuhr Institute of Ancient Near Eastern Studies, Copenhagen.

Mercier, N., Valladas, H., Valladas, G., and Reyss, J.-L. 1995. TL dates of burnt flints from Jelinek's excavations at Tabun and their implications. *Journal of Archaeological Science* 22: 495–509.

Moore, A. T. M. 1982. A four-stage sequence for the Levantine Neolithic, ca. 8500–3750 B.C. *Bulletin of the American School of Oriental Research* 246: 1–34.

———. 1985. The development in Neolithic societies in the Near East. *Advances in World Archaeology* 1: 1–69.

Mortensen, P. 1970. A preliminary study of the chipped stone industry from Beidha, an early neolithic village in southern Jordan. *Acta Archaeologica* 1: 1–54.

Oates, J. 1982. Archaeological evidence for settlement patterns in Mesopotamia and eastern Arabia in relation to possible environmental conditions. In *Paleoclimate, Paleoenvironments, and Human Communities in the Eastern Mediterranean Region in Later Prehistory*, pp. 359–98 (ed. J. Bintliff and W. van Zeist). BAR International Series 133 (ii). British Archaeological Reports, Oxford.

Porada, E., Hansen, D.P., Dunham, S., and Babcock, S. H. 1992. The chronology of Mesopotamia, ca. 7000–1600 b.c. In *Chronologies in Old World Archaeology*, vol. 1, pp. 122–78; vol. 2, pp. 125–53 (ed. R. W. Erich). University of Chicago Press, Chicago.

Potts, D. T. 1992. The chronology of the archaeological assemblages from the head of the Arabian Gulf to the Arabian Sea, 8000–1750 b.c. In *Chronologies in Old World Archaeology*, vol. 1, pp. 63–76; vol. 2, pp. 77–89 (ed. R. W. Erich). University of Chicago Press, Chicago.

Pullar, J. 1974. Harvard archaeological survey in Oman. *Proceedings of the Seminar for Arabian Studies* 4: 33–48.

Révillion, S. 1995. Technologie du débitage laminaire au paléolithique moyen en Europe septentrionale: État de la question. *Bulletin de la Société Préhistorique Française* 92: 425–41.

Révillion, S., and Tuffreau, A. 1994. *Les industries laminaires au Paléolithique moyen*. Centre National de la Recherche Scientifique, Paris.

Rice, M. 1994. *The Archaeology of the Arabian Gulf*. Routledge, London.

Roche, H., Kibunjia, M., Brugal, J.-P., and Lieberman, D. 1994. New results about the archaeology of West Turkana, Kenya. Paper delivered at the biennial meetings of the Society of Africanist Archaeologists, 29 April, Bloomington, Indiana.

Roche, H., and Tiercelin, J.-J. 1980. Industries lithiques de la formation plio-pléistocène d'Hadar, Ethiopie (Campagne 1976). In *Proceedings of the 8th Panafrican Congress of Prehistory and Quaternary Studies: Nairobi, 5 to 10 September 1977*, pp. 194–99 (ed. B. Ogot and R. E. Leakey). The International Louis Leakey Memorial Institute for African Prehistory, Nairobi.

Ruddiman, W. F., and Wright, H. E. 1987. *North America and Adjacent Oceans during the Last Glaciation*. Geological Society of America, Boulder.

Smith, G. H. 1977. New prehistoric sites in Oman. *Journal of Oman Studies* 3: 71–90.

Stager, L. E. 1992. The periodization of Palestine from Neolithic to Early Bronze times. In *Chronologies in Old World Archaeology*, vol. 1, pp. 22–41; vol. 2, pp. 17–60 (ed. R. W. Erich). University of Chicago Press, Chicago.

Tallon, P. W. J. 1978. Geological setting of the hominid fossils and Acheulian artifacts from the Kapthurin Formation, Baringo District, Kenya. In *Geological Background to Fossil Man*, pp. 361–73 (ed. W. W. Bishop). Scottish Academic Press, Edinburgh.

Tixier, J. 1980. *Mission archéologique française à Qatar, 1976–1977, 1977–1978*, vol. 1. Ministry of Information, Department of Tourism and Antiquities, Doha.

Tosi, M. 1986. The emerging picture of prehistoric Arabia. *Annual Reviews of Anthropology* 15: 461–90.

Uerpmann, M. 1992. Structuring the Late Stone Age of southeastern Arabia. *Arabian Archaeology and Epigraphy* 3: 65–109.

Whybrow, P. J., Hill, A., Yasin al-Tikriti, W., and Hailwood, E. A. 1990. Late Miocene primate fauna, flora and initial palaeomagnetic data from the Emirate of Abu Dhabi, United Arab Emirates. *Journal of Human Evolution* 19: 583–88.

Zarins, J., al-Jawad Murad, A., and Al-Yish, K. S. 1981. The comprehensive archaeological survey program, part I. ATLAL, *The Journal of Saudi Arabian Archaeology* 5: 9–16, 38–42.

Zarins, J., Whalen, N., Ibrahim, M., Mursi, A. al-J., and Khan, M. 1980. Comprehensive archaeological survey program. ATLAL, *The Journal of Saudi Arabian Archaeology* 4: 9–36.

Zeuner, F. E. 1954. Neolithic sites from the Rub Al-Khali, southern Arabia. *Man* 54: 1–4.

Late Miocene Palaeoenvironments in Arabia: A Synthesis

27

JOHN D. KINGSTON AND ANDREW HILL

Faunal interchange, along with biotic interaction in situ, is viewed as a significant process in the evolution and differentiation of terrestrial mammalian communities during the Neogene (Barry et al., 1985, 1990; Flynn et al., 1991; Janis, 1993; Barry, 1995; Opdyke, 1995; Vrba, 1995). The timing and location of these intercontinental migratory events are mediated to an extent by climatic and tectonic events, which can facilitate dispersal by the formation of "corridors" in regions where interchange was previously restricted by geographical or ecological barriers. As barriers are transgressed, speciation and extinction occur as introduced fauna interact with novel ecosystems. Major faunal turnovers in terrestrial successions have been linked with the creation of corridors formed by low sea-level stands or by the tectonic reshuffling of landmasses (Barry et al., 1985; Thomas, 1985). Alternatively, the formation of barriers can separate previously continuous populations, resulting in speciation by vicariance.

Continental plate movement, in addition to controlling the configuration of connections between landmasses (Rögl, 1999—Chapter 35) and development of potential orogenic barriers (Partridge et al., 1995), can contribute to climatic variability, which provides an additional filter to faunal migration. Alteration of the arrangement of marine basins and hence ocean circulation patterns, generation of topographic highs, and formation of volcanic fronts erupting volcanic debris into the atmosphere can all have profound effects on climatic conditions and atmospheric circulation patterns. Shifts in climatic filters that influence dispersal can occur independently of plate tectonic events as a result of orbital perturbations (Milankovitch, 1930; Imbrie and Imbrie, 1980) or meteoric impacts (Hut et al., 1987). Climatic variation influences dispersal by changes in sea level associated with increasing or decreasing glaciation, altering the structure and composition of plant communities and affecting weather conditions or seasonality patterns.

The primary and most obvious means of documenting faunal migrations in the past is by comparing fossil faunal assemblages from sites of various ages and regions (Thomas et al., 1982a; Bernor et al., 1987; Le Loeuff, 1991; Bonis et al., 1992). This approach, however, is complicated by taphonomic effects (Hill, 1987), discrepancies between researchers in assigning taxonomic identification to extinct faunas, and in general by a paucity of information (see Hill and Whybrow, 1999; Hill, 1999—Chapters 2 and 29). Palaeobiogeographical reconstructions benefit greatly from an understanding of the potential dispersal routes and the existence or development of palaeoecological or palaeogeographical obstacles. In this regard, the Arabian Peninsula and the area immediately to the north (the remaining part of the Arabian Plate) is pivotal in assessing faunal interchange in the Old World during the late Miocene. This region forms the intersection of three continents and it is likely to have had a significant role in the movement of fauna between Africa, South Asia, and Europe. Notable palaeontological events in this part of the Old World during the late Miocene include the establishment of the modern East African fauna, which involved the replacement of more archaic middle Miocene forms by taxa more closely related

 ISBN 0-300-07183-3

to extant species (Hill, 1995), a series of significant synchronous first and last appearances in the fossil record of the Siwalik succession in Pakistan (Barry et al., 1990; Barry, 1995), the spread of grassland-adapted fauna (Gabunia and Chochieva, 1982; MacFadden and Cerling, 1994), and the origin of the human lineage (Hill and Ward, 1988; Hill, 1994).

By late Miocene Baynunah times, palinspastic (palaeogeographic) reconstructions of the region place the Arabian Plate roughly in its modern configuration relative to the African and Eurasian Plates (Briggs, 1995). Connections between the Eastern Paratethys and the Indo-Pacific had been severed permanently by the Afro-Arabian Plate impinging onto the Eurasian continent 12–14 million years (Ma) ago (Lyberis et al., 1992; Rögl, 1999—Chapter 35). Development of the modern-day extension of the Red Sea, Arabian Gulf, and Gulf of Aden was incomplete during the mid- to late Miocene and did not form the geographic barriers they do today along the margins of the Arabian Peninsula (Coleman, 1993). No consensus exists regarding the initial stages of rifting in the Red Sea but evidence for extension and widespread volcanism extends back to at least the early Miocene, at which time the structural shape of the Red Sea depositional basin was defined (Coleman, 1993). Although there is evidence that the Red Sea trough occasionally contained deep-water sediments during the Miocene (Crossley et al., 1992), extensive evaporitic sequences throughout this interval suggest episodic batch filling of the basins followed by evaporitic draw-down and shallow-water evaporation associated with sabkhas.

Present Climate

Considering that the regional geographic and tectonic setting of the Arabian Peninsula has remained consistent over the past 15 Ma, the modern climate represents a valuable model for attempts to understand the palaeoclimatic patterns that may have influenced the region in the past. Presently lying between 12° and 38° N, the Arabian Peninsula straddles the subtropical high-pressure belt and much of the area is today climatically arid or semi-arid, dominated by subtropical deserts. Several major climatic regimes interface in this region and the climate and weather of the Arabian Peninsula are influenced by a complex variety of seasonal combinations of high- and low-pressure systems superimposed on annual solar variations. These include the year-round equatorial lows, ridges of the Azores High and seasonal anticyclones, the regional highs lying over the Armenian Plateau in the cool half of the year, summer depressions over Pakistan and the Arabian Gulf, winter lows from the Mediterranean Sea, and depressions from the Sudan in the transition seasons (Hastenrath, 1985; Roberts and Wright, 1993; Schneider, 1996). The Yemen highlands in the extreme southern portion of the Arabian Peninsula manage to intercept some of the moisture borne by the southwesterly South Asian monsoon summer winds. An important factor mediating the interplay between these circulation systems is the relief of the region, characterised by a northeastward inclination with considerable altitude along the western and southern margins. The meridional mountains of Lebanon effectively block the influence of the Mediterranean and the westerly circulation in the winter, most of which travels well north of the Arabian Peninsula anyway, while the Kurdistan and Zagros Mountains check the southward flow of winter air.

The northern portion of the peninsula receives most of its precipitation during the winter half of the year in association with middle- to high-latitude westerly depressions whose tracks are steered by the subtropical jet stream (Wigley and Farmer, 1982). Most of the precipitation falls to the north but occasionally moist air masses penetrate to the interior of the peninsula. Central African depressions lie over the western part of the Arabian Peninsula from October to April but are usually too shallow to bring precipitation. Occasionally, depressions from northern Egypt reach the Arabian Peninsula where they affect the weather along the Red Sea as far as the Kamaran Island and possibly even as far as Bahrain. In the interior of the Arabian Peninsula, rainfall is typically no greater than 50–100 mm and in the Rub' al Khali and along the Gulf of Aden it is even less. Exposed to the minimal influence of the Indian monsoon, the Arabian Sea coast receives little more than 50 mm of rainfall while the Yemen Mountains get 200–400 mm. The interior averages less than 10 rain days a year while the monsoonal part of the Arabian Peninsula coast exceeds 25 days. Mean surface temperatures over the central Arabian Peninsula average 15 °C during the winter with variation associated with elevation (Schneider, 1996).

The thermal low centred over the Arabian Gulf effectively influences the weather and its constancy in summer. The beginnings of this low are felt in April and it is at its deepest in July and August. During the summer, the rather dry northerly and northwesterly etesian winds dominate this region. Thermal stratification in the tropical trade-wind circulation zone does not promote cloud formation between May and November. Very low evaporation also accounts for the inconsiderable cloud cover and rainfall. Over 80–90% of days in Iraq are clear and clouds appear on only 35–40 days in winter. Around the Gulf of Oman and Arabian Gulf, the monsoon increases the relative humidity to over 50–60%. While the mean surface temperature during the summer is 34 °C, the mean maximum surface temperature can reach 45 °C.

Dust storms are characteristic of the arid and semi-arid zones. They are associated both with strong convection along a cold front and with strong, constant winds that transport dust and sand. In the Arabian Peninsula, these storms can affect immense areas, from the Syrian Desert to the Rub' al Khali. They pass over the eastern, less-elevated half of the peninsula, directed from the west by mountains over 1000 metres high. In the south their progress is restricted by the mountains near the Arabian Sea, and in summer by the presence of the intertropical convergence zone and the southwest monsoons associated with it. A prominent physiographic feature of the peninsula is large sandy deserts referred to as sand seas or ergs. Aeolian sands cover 770 000 km^2 or almost 90% of the peninsula's land surface (Whitney et al., 1983).

Past Climates

As the Miocene terrestrial palaeoenvironmental record remains poorly known, interpretations of climatic trends in tropical (low-latitude) terrestrial regions during the past 20 Ma have drawn heavily on global events recorded in the marine record, and to a limited extent from terrestrial sequences elsewhere. This period of time incorporates several major fluctuations in worldwide climate and the onset of Milankovitch mid-latitude northern hemisphere glaciation (deMenocal and Rind, 1993). All of these probably had profound effects on the evolution of the Arabian fauna and flora. In the long-term evolution of global climate, Neogene climatic conditions appear to reflect a continuation of the general trend documented for the past 100 Ma, characterised by a shift from the mid-Cretaceous thermal maximum to a world dominated by bipolar ice sheets. Antarctic and Southern Ocean cryospheric development occurred throughout the Cenozoic while northern hemisphere glaciation developed in the latest Neogene (Miller et al., 1987). This sequential cooling and cryospheric development did not occur uniformly but rather as a series of abrupt shifts representing threshold events (Kennett, 1995). Accompanying this cooling trend was a presumed increase in aridity in low latitudes (Shackleton and Kennett, 1975a). Explanations for the trend are incomplete but research so far suggests that several processes are involved in this long-term evolution of climate. They include shifting orbital parameters, changes in continent–ocean distribution, ocean heat transport, orography, and atmospheric CO_2 levels (Crowley and North, 1991; Prell and Kutzbach, 1992). The following discussion provides an overview of global and continental Miocene climatic trends and changes that may be relevant for interpreting the evolution of landscapes in Arabia during this period.

Early to Middle Miocene (23–12 Ma)

Before the final establishment of the east Antarctic Ice Sheet in the mid- to late Miocene, the early Miocene (23–15.6 Ma) global climate was relatively warmer, and global ice volume was low as is indicated by $\delta^{18}O$ values of planktonic and benthic marine foraminifera (Haq, 1980). $\delta^{18}O$ values between 19.5 and 15 Ma are the lowest in the Neogene, reflecting the climax of Neogene warmth. Antarctica apparently became thermally decoupled from the north, and Antarctic waters continued to cool (Grobe et al., 1990; Kennett and Barker, 1990). Haq (1980) hypothesised that warming of the Atlantic would have been favourable for the existence of widespread lowland forest in Africa. Andrews and Van Couvering (1975) also suggested a homogeneous landscape like the modern Congo Basin across much of eastern and central Africa during the early Miocene. They postulated that as rifting was in its initial stages, orographic barriers associated with crustal doming did not yet exist to prevent moist air masses from the Atlantic reaching the East African plateau, and possibly the Arabian Peninsula. Based on an examination of habitat and ecological diversity spectra, Van Couvering (1980) and Nesbit Evans et al. (1981) also suggested widespread equatorial rainforest communities in eastern Africa at 23–17 Ma. Axelrod and Raven (1978), however, proposed a more complex vegetational history for eastern Africa during this period, in which grassland and woodland communities were established as early as 23 Ma based on microfossil and macrofossil floras from the Ethiopian Highlands that indicate dry-adapted vegetation.

Following this interval of climatic amelioration in the early to middle Miocene was the onset of significant cooling at about 15 Ma. This reflects major expansion of the east Antarctic ice sheet (Kennett and Barker, 1990), renewed cooling at high latitudes and deep oceans, and important changes in deep oceanic circulation (Flower and Kennett, 1994). This dramatic shift to colder climates reflects a critical threshold in climate evolution during the Cenozoic (Kennett, 1995) and represents the onset of climatic and oceanic circulation patterns that characterise and dominate the late Neogene. The expansion of grasslands and of grazing-adapted faunas has been described in South America (Pascual and Juareguizar, 1990), Australia (Stein and Robert, 1985), and North America (MacFadden and Cerling, 1994). The Afro-Arabian Plate converged with the Asian Plate in the middle Miocene bringing to an end the moderating influence of the Tethys Sea on the climate of the Africa/Arabian continent. As the warm Tethys Sea with its associated moist air masses was disrupted, drier conditions and extremes of temperature may have increased over lowland areas of Africa and Arabia (Axelrod and Raven, 1978; Williams, 1994). Although the palaeobotanical record of Africa is poor for the Miocene after about 17 Ma, the current consensus is for a spread of savannah, deciduous forest, thorn forest, and sclerophyllous vegetation at the expense of rainforests (Axelrod and Raven, 1978; Van Couvering, 1980; Bonnefille, 1984; van Zinderen Bakker and Mercer, 1986;). Carbon and oxygen isotopic analyses of palaeosol carbonates and organic matter from various localities in East Africa, however, suggest that while there may have been a gradual increase in C_4 grasses and aridity over the past 20 Ma, Serengeti-type grasslands are a relatively recent phenomena (Cerling, 1992; Kingston et al., 1994).

Late to Terminal Miocene (12–5.5 Ma)

The late Miocene (12–6.5 Ma) represents a prolonged period of cool climate, with average $\delta^{18}O$ values consistently higher than the early Miocene, punctuated by two distinct cooling events recorded in the marine record. The earliest occurred between 12.5 and 11.5 Ma and the second between 11 and 9 Ma. This latter event was manifested by major growth of the Antarctic Ice Sheet (Shackleton and Kennett, 1975a,b), 4–5 °C cooling of deep-ocean bottom water (Miller et al., 1987), and a worldwide temperature drop of 7 °C (Tiwari, 1987). Expansion of polar ice may have caused the dramatic drop in sea level recorded at 11–10 Ma (Moore et al., 1987). Vincent and Berger (1985) suggested that this event was related to a draw-down of atmospheric CO_2 caused by changes in upwelling that increased the removal of carbon from the oceanic sink into sediments. This event was followed by a period of relative warmth during the middle part of the late Miocene, 9–7 Ma (Haq, 1980; Kennett, 1982).

Associated with these Miocene changes was a significant increase in aridity documented by an increase in aeolian deposition throughout the late Cenozoic (Rea et al., 1985), high-latitude shifts in vegetation to more seasonal and arid-adapted flora (Wolfe, 1985), and a hypothesised general transition from forested environments to habitats with abundant grasses (Potts and Behrensmeyer, 1992; Williams, 1994). Such shifts have been documented in both western North America (Axelrod and Raven, 1985) and Australia (Tedford, 1985). At 7.4–7.0 Ma, Quade et al. (1989) detected a dramatic shift from vegetation dominated by C_3 plants (forest/grassland) to one dominated by C_4 (grassland) plants in the Siwalik sediments of Pakistan, possibly correlated with inception or strengthening of monsoonal conditions due to uplift of the Tibetan Plateau or to declining atmospheric pCO_2 (Cerling et al., 1993). In India, humid forest taxa rapidly retreated eastwards to areas of moister climate during this period (Prakash, 1972). Fossil macroflora from the southwestern Cape of South Africa indicate replacement of subtropical rainforest by the present "fynbos" or macchia (Coetzee, 1978) and the transition to a Mediterranean type of climate. Evidence of vegetation in East Africa during this period does not support the widespread replacement of forests by grasslands but rather a persisting heterogeneous landscape (Cerling, 1992; Kingston et al., 1994).

The terminal Miocene to early Pliocene (6.5–5.5 Ma) is characterised by extensive climatic variation, which resulted in significant changes in the size of the polar ice sheet. At the Miocene–Pliocene boundary (5.5 Ma) the Antarctic Ice Sheet may have exceeded its glacial maximum extent by as much as 50% (Shackleton and Kennett, 1975b; Denton, 1985), resulting in an appreciable drop in global sea level of up to 50 metres. The drop in sea level coupled with the closure of the Straits of Gibraltar, due to tectonic impingement of the African Plate against Europe, resulted in the Messinian Salinity Crisis (Hsu et al., 1977; Stein and Sarnthein, 1984; Hodell et al., 1986), which involved the isolation and eventual desiccation of the Mediterranean Sea. The thickness of evaporative sequences in the Mediterranean Basin suggests that the cycle of evaporation must have been repeated about 40 times in the latest Miocene. Climatic deterioration manifested as increasing aridity in low latitudes may have had a significant influence on the African and Arabian flora as the Red Sea Basin was also dry during this interval (van Zinderen Bakker and Mercer, 1986).

Miocene Palaeoenvironments of the Arabian Peninsula

Early to Middle Miocene

Known empirical evidence of terrestrial environments on the Arabian Peninsula during the early to middle Miocene is limited to continental sediments and associated fossil fauna and flora exposed in four areas of eastern Saudi Arabia (Powers et al., 1966; Hamilton et al., 1978; Thomas et al., 1978; Thomas, 1982; Whybrow et al., 1982; Whybrow, 1984; Whybrow, 1987; Whybrow et al., 1990) and the western part of the Emirate of Abu Dhabi (Whybrow et al., 1990; Whybrow et al., 1999—Chapter 4; Bristow, 1999—Chapter 6). These deposits have been divided into the late early Miocene Hadrukh Formation (c. 19–17 Ma), the overlying late early Miocene Dam Formation (c. 17–15 Ma), and the middle Miocene Hofuf Formation in Saudi Arabia and the ?middle Miocene Shuwaihat Formation in Abu Dhabi.

Table 27.1 presents a compilation of palaeoenvironmental data derived from lithofacies studies and analyses of fossil material recovered from these sequences. In general, the data indicate that this portion of the early to middle Miocene Arabian Peninsula was dominated by more open environments than during the Eocene (As-Saruri et al., 1999—Chapter 31) and Oligocene (Thomas et al., 1999—Chapter 30). Interpretations of the fauna

Table 27.1. Palaeoenvironmental reconstructions of pre-Baynunah time in the Arabian Peninsula

	Basis of reconstruction	Formation	Site	Age	Reference
Eocene of Wadi Rayan, and Fayum of Egypt	Fossil fruits and seeds, family ?Nymphaeaceae	Kaninah	Kaninah, southern Yemen	Middle Eocene	As-Saruri et al., 1999 (this vol., Chapter 31)
Tropical lowland evergreen forests	Family Anonaceae	Kaninah	Kaninah, southern Yemen	Middle Eocene	As-Saruri et al., 1999 (this vol., Chapter 31)
Strongly seasonal climate with marked rainy season	Sedimentary lithofacies (lack of evaporites, the presence of siliciclastics, and of broad permanent lakes with carbonate sedimentation)	Ashawq (Shizar Member)	Taqah, Oman	c. 33 Ma	Thomas et al., 1999 (this vol., Chapter 30)
Semi-arid climate	Erycine snake and embrithopod *Arsinoitherium*	Ashawq (Shizar Member)	Taqah, Oman	c. 33 Ma	Thomas et al., 1991
Open savannah	Bovids	Hadrukh	Eastern Province of Saudi Arabia	Late early Miocene	Whybrow et al., 1982
Dry rather than arid (freshwater environment containing dissolved solutes probably concentrated by evaporation)	Fossil fruits (*Midravalva arabica* of the family Potamogetoneae—pondweeds and ditch grasses—and sediments)	Hadrukh	Jabal Midra ash-Shamali, Eastern Province of Saudi Arabia	Late early Miocene	Whybrow et al., 1982; Collinson, 1982
Palm wood	Fossil wood	Hadrukh/Dam	Ad Dabtiyah, Saudi Arabia	19–17 Ma	Thomas et al., 1978 (Whybrow et al., 1987)
Woodland habitat	Rhinoceroses	Hadrukh/Dam	Ad Dabtiyah, Saudi Arabia	19–17 Ma	Gentry, 1987a

Forest	*Dorcatherium*	Hadrukh/Dam	Ad Dabtiyah, Saudi Arabia	19–17 Ma	Gentry, 1987b
Forest if okapi is used as modern analogue, savannah if giraffe is used	*Canthumeryx*	Hadrukh/Dam	Ad Dabtiyah, Saudi Arabia	19–17 Ma	Gentry, 1987b
Tropical to subtropical near-shore environment, tidal flats, and a large estuarine system	Vertebrates (general)	Hadrukh/Dam	Ad Dabtiyah, Saudi Arabia	19–17 Ma	Gentry, 1987b
More open type of woodland and bushland	Microfauna	Dam	As Sarrar, Saudi Arabia	Lower Miocene 17–15 Ma	Thomas et al., 1982b
More forested	Dominance of browsing vertebrate fauna	Dam	As Sarrar, Saudi Arabia	Lower Miocene 17–15 Ma	Thomas et al., 1982b
Mangroves—tropical climate	Fossil roots	Dam	Dawmat al 'Awdah, Saudi Arabia	Lower Miocene 17–15 Ma	Whybrow and McClure, 1981
Palm wood	Fossil wood	Dam	South of Jabal Dawmat al 'Awdah, Saudi Arabia	Lower Miocene 17–15 Ma	Hamilton et al., 1978
Open milieu	Vertebrate fauna	Hofuf	Al Jadidah, Saudi Arabia	c. 14 Ma	Thomas et al., 1978
Open environment with marked aridity	Vertebrate fauna	Hofuf	Al Jadidah, Saudi Arabia	c. 14 Ma	Sen and Thomas, 1979
Alternating interdune and coastal type sabkhas, transverse and barchanoid dunes, and river systems; evidence of marine sediments within the basin	Sedimentary lithofacies	Shuwaithat	Abu Dhabi, UAE	Middle? Miocene (unconformably underlying the Baynunah Formation)	Bristow, 1999 (this vol. Chapter 6)

from the Dam Formation at As Sarrar as browser-dominated (Thomas et al., 1982b), and selected taxa from the Hadrukh/Dam Formation at Ad Dabtiyah as forest dwellers (Gentry, 1987a,b), suggest limited existence of more closed habitats, possibly along river or lake margins. As pointed out by Whybrow and McClure (1981), Whybrow (1984), and Thomas (1982), the suggestion of widespread and persisting aridity during the early to middle Miocene in this region (Kortlandt, 1972; Sen and Thomas, 1979; Thomas, 1979; C. T. Madden, USGS written communication, in Whitney et al., 1983) is unsupported. But aeolian lithofacies reflecting large-scale transverse and barchanoid dunes interbedded with sabkha facies in the ?middle Miocene Shuwaihat Formation (Bristow, 1999—Chapter 6) indicates that arid to hyperarid conditions existed locally if not extensively at times. These dune forms are characteristic of very dry inland desert regions where vegetation is sparse, sand supply is limited, and winds almost unidirectional (Pye and Tsoar, 1990). An evaporitic sequence 2 metres thick at the top of the Dam Formation, Qatar (Whybrow, 1984) also supports intermittent arid conditions.

In contrast to interpretations of widespread equatorial rainforest in Africa (Andrews and Van Couvering, 1975; Nesbit Evans et al., 1981; Van Couvering, 1980) and more forested conditions in the Siwaliks of Pakistan (Quade and Cerling, 1993; Morgan et al., 1994) during the early Miocene, environments of the early Miocene of the Arabian Peninsula appear to have been dominated by more open woodland or bushland ecosystems. Postulated increasing low-latitude aridity during the middle Miocene may be reflected in the dune and sabkha facies of the Shuwaihat Formation.

Late Miocene

The Baynunah succession (8–6 Ma) provides the only late Miocene evidence for palaeoenvironments on the Arabian Peninsula. As the late Miocene terminates with the Messinian Salinity Crisis, there has been speculation that the Arabian Peninsula was characterised by increasingly arid conditions during the time leading up to this event. Noting an increase in aeolian sedimentation in deep-sea cores (Leinen and Heath, 1981; see also Rea et al., 1985), Whitney et al. (1983) suggested that the late Miocene in Arabia may have been more arid and/or experienced greater wind intensities than at present. Also citing evidence for xerophytic vegetation in central Africa in the late Miocene based on pollen studies (Maley, 1980), Whitney et al. (1983) hypothesised that the Arabian Peninsula was also arid as it lies within the same climatic belt. Research on sand seas (ergs) on the Arabian Peninsula led Brown (1960) and Holm (1960) to suggest that the sand bodies formed during a dramatic increase in aridity near the end of the Tertiary or Quaternary. In evaluating the data presented by Whybrow and McClure (1981), C. T. Madden (USGS written communication, in Whitney et al., 1983) interpreted the sediments as having been deposited in a hot arid to semiarid climate. Although these scenarios do not seem unreasonable in the context of apparent global climatic trends during the late Miocene, available evidence from the Baynunah Formation does not corroborate these notions of aridity.

Information bearing on the palaeoenvironment of Baynunah times is summarised in table 27.2 and indicates environments ranging from grassy woodlands to wooded grasslands. As these deposits are primarily fluvial, much of the data provides information about conditions within the river system (Friend, 1999; Jeffery, 1999; Forey and Young, 1999; Lapparent de Broin and Dijk, 1999; Rauhe et al., 1999—Chapters 5, 10, 12, 13, and 14), indicating a major drainage system comprised of diverse microhabitats ranging from deep open bodies of water, to moderately fast-moving currents, to quiet, shallow water. Although the ancient river system undoubtedly had great significance for the regional flora and fauna, these reconstructions do not provide direct information relating to terrestrial environments distal to the ancient river system. The river may have been flowing through an arid region or alternatively more humid, tropical ecosystems. Terrestrial fauna associated with the river system, however, include a wide range of taxa such as large bovids, suids, elephantids, equids, carnivores, and a cercopithecine (Whybrow et al., 1990). These suggest communities relying on extensive resources in a region dominated by wooded grasslands, rather than being confined to a riparian forest isolated within a desert-like region. Isotopic analyses of palaeosol carbonates, herbivore enamel (Kingston, 1999—Chapter 25), and ratite eggshells (Ditchfield, 1999—Chapter 7) support such an interpretation. Interpreted adaptations of pulmonate gastropods (Mordan, 1999—Chapter 11) and species of *Clarias* (Forey and Young, 1999—Chapter 12) hint at the possibility of seasonality in precipitation during Baynunah times.

Discussion

In contrast to the apparently more open environments of the Arabian Peninsula during the Miocene, limited environmental data from the Eocene and Oligocene indicate that the region may have supported more forested habitats. Components of an assemblage of fossil fruits and seeds recovered from the middle Eocene Kaninah Formation, in the southern part of the Republic of Yemen, suggest affinities to modern tropical lowland evergreen forests (As-Saruri et al., 1999—Chapter 31). Aspects of the palaeoflora are similar to those described from the Eocene and Oligocene of Egypt. Terrestrial sediments with a diverse vertebrate record indicate a marginal–littoral depositional environment along the Oligocene coast of Oman (Thomas et al., 1999—Chapter 30) at the site of Taqah. Despite interpretations of a semi-arid palaeoclimate based on calcareous crusts and the presence of an erycine snake and of the embrithopod *Arsinoitherium* (Thomas et al., 1991), overall sedimentological criteria suggest a strongly seasonal climate with a marked rainy season (Thomas et al., 1999—Chapter 30) (see table 27.1). Emphasis is placed on the resemblance of the Taqah faunal assemblage to that of the Oligocene Jebel Quatrani Formation of the Fayum in Egypt (Thomas, 1999—Chapter 30), which has been interpreted as possibly representing a forest assemblage (Fleagle, 1988). Any comprehensive understanding of phytogeographic trends during the Tertiary in this region is compromised by the limited data relating to palaeohabitats, but available information is consistent with the possibility of a transition from forested to more open habitats during the late Oligocene/early Miocene.

Climatic information deciphered from deep-sea cores (Prell et al., 1992) and general circulation models (Ruddiman and Kutzbach, 1989; Prell and Kutzbach, 1992) suggest that the climatic conditions of the Arabian Peninsula have varied considerably in the past. Although most of the peninsula today lies beyond the influence of the South Asian and African monsoon system, the origin and dynamics of the monsoons had a direct if not indirect effect on the evolution of climate there. Appreciable uplift of the Tibetan Plateau about 8 Ma (Molnar et al., 1993) has been linked to an intensification of monsoonal patterns recognised by changes in marine microfossils in the Arabian Sea and western Indian Ocean (Kroon et al., 1991; Prell et al., 1992). This uplift continued intermittently during the Miocene and affected the development of the easterly jet stream that today brings dry subsiding air to the Arabian Peninsula. Major shifts in the fauna and flora from the Siwalik succession of the Potwar Plateau have been linked to climatic change resulting from this uplift (Barry et al., 1990; Quade and Cerling, 1995). Changes in orbitally induced solar radiation associated with this modification of atmospheric circulation patterns no doubt affected environments of the Arabian Peninsula during the late Miocene as did conditions associated with the onset of the Messinian Salinity Crisis.

High-resolution studies of Pleistocene lacustrine deposits in the Arabian Peninsula (McClure, 1978; Kutzbach et al., 1993; Roberts and Wright, 1993) reveal a series of changes in lake levels that reflect shifts in climatic regimes possibly driven by changes in the intensity of the monsoons. Assuming that the complexity of factors controlling the climate and environments of the modern Arabian Peninsula existed in the past, minor perturbations in the circulation patterns across Africa, the Indian Ocean, the Mediterranean, or regions to the north during the Miocene could have had major effects on the habitats present in this region.

Table 27.2. Palaeoenvironmental reconstructions of the Baynunah Formation

Palaeohabitats	Basis of reconstruction	Reference
Diverse river system with large, deep, and open bodies of water (either lakes or slow- to fast-moving rivers) with shallow banks indicated by the gavialis specimens; *Crocodylus* has the ability to negotiate river sections with steep banks	Crocodylidae	Rauhe et al., 1999 (this vol., Chapter 14)
Grassy woodland/bushland vegetational physiognomy associated with the fluvial deposits with possibly more open grasslands distal to the river system	Stable carbon isotopic analysis of palaeosol carbonate and hervibore enamel	Kingston, 1999 (this vol., Chapter 25)
Aeolian sands in arid environment	Plant root moulds	Glennie and Evamy, 1968
Fluvial deposition with no evidence of marine influence; gravels containing transported bones deposited by currents flowing with speeds of tens of centimetres (per second?); rivers appear to have been variable in flow, probably 3–10 metres deep during flood stage, and braided in the sense that they contained many sediment bars, and channels of varied size and form; palaeosols indicate periodic subaerial exposure	Lithology of the lower Baynunah Formation	Friend, 1999 (this vol., Chapter 5)
Dietary reliance on both C_3 and C_4 vegetation (wooded grassland and grassy woodland)	$\delta^{13}C$ of ratite eggshells	Ditchfield, 1999 (this vol., Chapter 7)
Adapted to life in moderately fast-moving streams and rivers rather than lakes or slow-moving rivers	Two species of unionoids (swan mussels)	Jeffery, 1999 (this vol., Chapter 10)
Genera within the Pseudonapaeinai are characteristic of rather xeric habitats, but tend to occur in situations where there is regular seasonal humidity, whether in the form of precipitation at high elevations or lower down close to water at the base of wadis; suggests wetter than present and perhaps seasonal conditions	Pulmonate gastropods	Mordan, 1999 (this vol., Chapter 11)

Table 27.2. Palaeoenvironmental reconstructions of the Baynunah Formation (*continued*)

Palaeohabitats	Basis of reconstruction	Reference
Prefer quiet, shallow waters and/or swamp conditions and can withstand intermittent periods of drought	*Clarias* sp.	Forey and Young, 1999 (this vol., Chapter 12)
Found in a variety of freshwater habitats but most are bottom dwellers in relatively slow-moving waters	Bagridae	Forey and Young, 1999 (this vol., Chapter 12)
Generalised, inhabiting a wide range of habitats	*Barbus* sp.	Forey and Young, 1999 (this vol., Chapter 12)
Terrestrial tortoise is an intertropical form inhabiting savannahs, open forests, and even deserts and can endure aridity; some of the aquatic tortoises prefer flowing wide river systems whereas others are less-good swimmers and prefer muddy waters	Chelonia	Lapparent de Broin and Dijk, 1999 (this vol., Chapter 13)
Similar and possibly ancestral to *Gulo*, which has been considered as indicative of boreal forest or woodland; wolverines are also animals of open tundra and can indicate grassy and open habitats	*Plesiogulo*	Barry, 1999 (this vol., Chapter 17)
Relatively deep mandibular ramus suggest hypsodonty = tough food; short and broad muzzle = grazing; deep ectoflexids on some premolars are interpreted as poor adaptation to abrasive foods	*Hipparion abudhabiensis*	Eisenmann and Whybrow 1999 (this vol., Chapter 19)
Fossil wood	*Acacia*	Whybrow and Clements 1999 (this vol., Chapter 23)

Summary

Although the Baynunah Formation and associated fossil material are limited, temporally and spatially, when viewed in the context of late Miocene palaeoenvironments of the Arabian Peninsula, they provide significant evidence that the region did not present a continuous ecological barrier to intercontinental exchange. Analyses of lithofacies and associated fauna and flora indicate an open woodland environment adjacent to a major river system. Habitats distal to the river may have been more open, perhaps even grasslands. While conceivable, it is unlikely that the Baynunah sediments, representing the only known window into the late Miocene of this area, record a moderate environment set within a region and time interval dominated by arid to semi-arid climatic patterns. Unlike the widespread semi-arid to hyperarid conditions that dominate the Arabian Peninsula today, the environment during the late Miocene may have been more varied and supported a number of different types of habitats. In this case it would be unreasonable to extrapolate the interpreted environments for western Abu Dhabi to a regional scale.

It remains difficult to link global climatic trends and events documented in the marine record with the evolution of terrestrial communities, primarily because of a lack of sites and resolution in continental sediments. Although it is tempting to correlate climatic shifts with faunal interchange, local and regional effects of global shifts on terrestrial ecosystems are intricate and basically unknown. Late Cenozoic cooling at high latitudes clearly had an aridifying effect on low latitudes but the timing and magnitude of these changes on the continents varied greatly depending on buffering by regional and local atmospheric circulation patterns.

References

Andrews, P. J., and Van Couvering, J. H. 1975. Palaeoenvironments in the East African Miocene. In *Approaches to Primate Paleobiology,* pp. 62–103 (ed. F. S. Szalay). Karger, Basel.

As-Saruri, M. L., Whybrow, P. J., and Collinson, M. E. 1999. Geology, fruits, seeds, and vertebrates (?Sirenia) from the Kaninah Formation (middle Eocene), Republic of Yemen. Chap. 31 in *Fossil Vertebrates of Arabia,* pp. 443–53 (ed. P. J. Whybrow and A. Hill). Yale University Press, New Haven.

Axelrod, D. I., and Raven, R. H. 1978. Late Cretaceous and Tertiary vegetation history of Africa. In *Biogeography and Ecology of Southern Africa,* pp. 77–130 (ed. M. J. A. Werger). Junk, The Hague.

Barry, J. C. 1995. Faunal turnover and diversity in the terrestrial Neogene of Pakistan. In *Paleoclimate and Evolution with Emphasis on Human Origins,* pp. 115–34 (ed. E. S. Vrba, G. H. Denton, T. C. Partridge, and L. H. Burckle). Yale University Press, New Haven.

Barry, J. C. 1999. Late Miocene Carnivora from the Emirate of Abu Dhabi, United Arab Emirates. Chap. 17 in *Fossil Vertebrates of Arabia,* pp. 203–08 (ed. P. J. Whybrow and A. Hill). Yale University Press, New Haven.

Barry, J. C., Flynn, L. J., and Pilbeam, D. R. 1990. Faunal diversity and turnover in a Miocene terrestrial sequence. In *Causes of Evolution: A Paleontological Perspective,* pp. 381–421 (ed. R. Ross and W. Allmon). University of Chicago Press, Chicago.

Barry, J. C., Johnson, N. M., Raza, S. M., and Jacobs, L. L. 1985. Neogene faunal change in southern Asia: Correlations with climate, tectonic and eustatic events. *Geology* 13: 637–40.

Bernor, R. L., Brunet, M., Ginsburg, L., Mein, P., Pickford, M., Rögl, F., Sen, S., Steininger, F., and Thomas, H. 1987. A consideration of some major topics concerning Old World Miocene mammalian chronology, migrations and palaeoecology. *Geobios* 20: 431–39.

Bonis, L. de, Bouvrain, G., Geraads, D., and Koufos, G. 1992. Diversity and paleoecology of Greek late Miocene mammalian faunas. *Palaeogeography, Palaeoclimatology, Palaeoecology* 91: 99–121.

Bonnefille, R. 1984. Cenozoic vegetation and environments of early hominids in East Africa. In *The Evolution of the East Asian Environment. II. Palaeobotany, Palaeozoology and Palaeoanthropology*, pp. 579–612 (ed. R. O. Whyte). University of Hong Kong, Centre of Asian Studies, Hong Kong.

Briggs, J. C. 1995. *Global Biogeography.* Elsevier, Amsterdam.

Bristow, C. S. 1999. Aeolian and sabkha sediments in the Miocene Shuwaihat Formation, Emirate of Abu Dhabi, United Arab Emirates. Chap. 6 in *Fossil Vertebrates of Arabia*, pp. 50–60 (ed. P. J. Whybrow and A. Hill). Yale University Press, New Haven.

Brown, G. F. 1960. Geomorphology of Western Central Saudi Arabia. In *Proceedings of the 21st International Geological Congress, Copenhagen*, pp. 150–59.

Cerling, T. E. 1992. Development of grasslands and savannas in East Africa during the Neogene. *Palaeogeography, Palaeoclimatology, Palaeoecology (Global and Planetary Change Section)* 97: 241–47.

Cerling, T. E., Wang, Y., and Quade, J. 1993. Expansion of C_4 ecosystems as an indicator of global ecological change in the late Miocene. *Nature* 361: 344–45.

Coetzee, J. A. 1978. Climate and biological changes in south-western Africa during the late Cainozoic. In *Palaeoecology of Africa*, pp. 13–29 (ed. J. A. Coetzee and Z. van Zinderen Bakker). Balkema, Rotterdam.

Coleman, R. G. 1993. *Geologic Evolution of the Red Sea.* Oxford University Press, New York.

Collinson, M. E. 1982. Reassessment of fossil Potamogetoneae fruits with description of new material from Saudi Arabia. *Tertiary Research* 4: 83–104.

Crossley, T., Watkins, C., Raven, M., Cripps, D., Carnell, A., and Williams, D. 1992. The sedimentary evolution of the Red Sea and the Gulf of Aden. *Journal of Petroleum Geology* 15: 157–72.

Crowley, T. J., and North, G. R. 1991. *Paleoclimatology.* Oxford University Press, New York.

deMenocal, P. B., and Rind, D. 1993. Sensitivity of Asian and African climate to variations in seasonal insolation, glacial ice cover, sea-surface temperature, and Asian orography. *Journal of Geophysical Research* 98: 7265–87.

Denton, G. H. 1985. Did the Antarctic Ice Sheet influence late Cainozoic climate and evolution in the southern hemisphere? *South African Journal of Science* 81: 224–29.

Ditchfield, P. W. 1999. Diagenesis of the Baynunah, Shuwaihat, and Upper Dam Formation sediments exposed in the Western Region, Emirate of Abu Dhabi, United Arab Emirates. Chap. 7 in *Fossil Vertebrates of Arabia*, pp. 61–74 (ed. P. J. Whybrow and A. Hill). Yale University Press, New Haven.

Eisenmann, V., and Whybrow, P. J. 1999. Hipparions from the late Miocene Baynunah Formation, Emirate of Abu Dhabi, United Arab Emirates. Chap. 19 in *Fossil Vertebrates of Arabia*, pp. 234–53 (ed. P. J. Whybrow and A. Hill). Yale University Press, New Haven.

Fleagle, J. G. 1988. *Primate Adaptation and Evolution.* Academic Press, San Diego.

Flower, B. P., and Kennett, J. P. 1994. The Middle Miocene climatic transition: East Antarctica ice sheet development, deep ocean circulation and global carbon cycling. *Palaeogeography, Palaeoclimatology, Palaeoecology* 108: 537–55.

Flynn, L. J., Tedford, R. H., and Zhanxiang, Q. 1991. Enrichment and stability in the Pliocene mammalian fauna of North China. *Paleobiology* 17: 246–65.

Forey, P. L., and Young, S. V. T. 1999. Late Miocene fishes of the Emirate of Abu Dhabi, United Arab Emirates. Chap. 12 in *Fossil Vertebrates of Arabia*, pp. 120–35 (ed. P. J. Whybrow and A. Hill). Yale University Press, London and New Haven.

Friend, P. F. 1999. Rivers of the Lower Baynunah Formation, Emirate of Abu Dhabi, United Arab Emirates. Chap. 5 in *Fossil Vertebrates of Arabia,* pp. 38–49 (ed. P. J. Whybrow and A. Hill). Yale University Press, New Haven.

Gabunia, L. K., and Chochieva, K. I. 1982. Co-evolution of the *Hipparion* fauna and vegetation in the Paratethys region. *Evolutionary Theory* 6: 1–13.

Gentry, A. W. 1987a. Rhinoceroses from the Miocene of Saudi Arabia. *Bulletin of the British Museum (Natural History),* Geology 41: 409–32.

Gentry, A. W. 1987b. Ruminants from the Miocene of Saudi Arabia. *Bulletin of the British Museum (Natural History),* Geology 41: 433–39.

Glennie, K. W., and Evamy, B. D. 1968. Dikaka: Plants and plant-root structures associated with aeolian sand. *Palaeogeography, Palaeoclimatology, Palaeoecology* 4: 77–87.

Grobe, H., Futterer, D. K., and Spiess, V. 1990. Oligocene to Quaternary processes on the Antarctic continental margin. In *Proceedings of the Ocean Drilling Program, Scientific Results,* pp. 121–31 (ed. R. F. Barker and J. P. Kennett). College Station, Tex.

Hamilton, W. R., Whybrow, P. J., and McClure, H. A. 1978. Fauna of fossil mammals from the Miocene of Saudi Arabia. *Nature* 274: 248–49.

Haq, B. U. 1980. Biogeographic history of Miocene calcareous nannoplankton and paleoceanography of the Atlantic Ocean. *Micropaleontology* 26: 414–43.

Hastenrath, S. 1985. *Climate and Circulation of the Tropics.* Reidel, Dordrecht.

Hill, A. 1987. Causes of perceived faunal change in the later Neogene of East Africa. *Journal of Human Evolution* 16: 583–96.

———. 1994. Late Miocene and early Pliocene hominoids from Africa. In *Integrative Paths to the Past: Paleoanthropological Advances in Honor of F. Clark Howell,* pp. 123–45 (ed. R. S. Corruccini and R. L. Ciochon). Prentice-Hall, Englewood Cliffs, N. J.

———. 1995. Faunal and environmental change in the Neogene of East Africa: Evidence from the Tugen Hills Sequence, Baringo District, Kenya. In *Paleoclimate and Evolution, with Emphasis on Human Origins,* pp. 178–93 (ed. E. S. Vrba, G. H. Denton, T. C. Partridge, and L. H. Burkle). Yale University Press, New Haven.

———. 1999. Late Miocene sub-Saharan African vertebrates and their relation to the Baynunah fauna, Emirate of Abu Dhabi, United Arab Emirates. Chap. 29 in *Fossil Vertebrates of Arabia,* pp. 420–29 (ed. P. J. Whybrow and A. Hill). Yale University Press, New Haven.

Hill, A., and Ward, S. 1988. Origin of the Hominidae: The record of African large hominoid evolution between 14 My and 4 My. *Yearbook of Physical Anthropology* 31: 49–83.

Hill, A., and Whybrow, P. J. 1999. Summary and overview of the Baynunah fauna, and its context. Chap. 2 in *Fossil Vertebrates of Arabia,* pp. 7–14 (ed. P. J. Whybrow and A. Hill). Yale University Press, New Haven.

Hodell, D. A., Elmstrom, K. M., and Kennett, J. 1986. Latest Miocene benthic $\delta^{18}O$ changes, global ice volume, sea level and the 'Messinian salinity crisis'. *Nature* 320: 411–14.

Holm, D. A. 1960. Desert morphology in the Arabian Peninsula. *Science* 132: 1369–79.

Hsü, K. J., Montadert, L., Bernoulli, D., Cita, M. B., Erickson, A., Garrison, R. E., Kidd, R. B., Milieres, F., Mueller, C., and Wright, R. 1977. History of the Mediterranean Salinity Crisis. *Nature* 267: 399–403.

Hut, P., Alvarez, W., Elder, W. P., Hansen, T., Kaufman, E. G., Keller, G. R., Showmaker, E. M., and Weissman, P. T. 1987. Comet showers as a cause of mass extinctions. *Nature* 329: 1118–26.

Imbrie, J., and Imbrie, J. Z. 1980. Modeling the climatic response to orbital variations. *Science* 207: 943–53.

Janis, C. M. 1993. Tertiary mammal evolution in the context of changing climates, vegetation, and tectonic events. *Annual Review of Ecology and Systematics* 24: 467–500.

Jeffery, P. A. 1999. Late Miocene swan mussels from the Baynunah Formation, Emirate of Abu Dhabi, United Arab Emirates. Chap. 10 in *Fossil Vertebrates of Arabia*, pp. 111–15 (ed. P. J. Whybrow and A. Hill). Yale University Press, New Haven.

Kennett, J. P. 1982. *Marine Geology*. Prentice-Hall, Englewood Cliffs, N.J.

———. 1995. A review of polar climatic evolution during the Neogene, based on the marine sediment record. In *Paleoclimate and Evolution, with Emphasis on Human Origins*, pp. 49–64 (ed. E. S. Vrba, G. H. Denton, T. C. Partridge, and L. H. Burckle). Yale University Press, New Haven.

Kennett, J. P., and Barker, P. F. 1990. Latest Cretaceous to Cenozoic climate and oceanographic developments in the Weddell Sea, Antarctica. In *Proceedings of the Ocean Drilling Program, Scientific Results*, pp. 937–60 (ed. R. F. Barker and J. P. Kennett). College Station, Tex.

Kingston, J. D. 1999. Isotopes and environments of the Baynunah Formation, Emirate of Abu Dhabi, United Arab Emirates. Chap. 25 in *Fossil Vertebrates of Arabia*, pp. 354–72 (ed. P. J. Whybrow and A. Hill). Yale University Press, New Haven.

Kingston, J. D., Marino, B. D. and Hill A. 1994. Isotopic evidence for Neogene hominid paleoenvironments in the Kenya Rift Valley. *Science* 264: 955–59.

Kortlandt, A. 1972. *New Perspectives on Ape and Human Evolution*. Stichting voor Psychobiologie, Amsterdam.

Kroon, D., Steens, T., and Troelstra, S. R. 1991. Onset of monsoonal related upwelling in the western Arabian Sea as revealed by planktonic foraminifers. *Proceedings of the Ocean Drilling Program, Scientific Results* 117: 257–63.

Kutzbach, J. E., Guetter, P. J., Behling, P. J., and Selin, R. 1993. Simulated climatic changes; results of the COHMAP climate-model experiments. In *Global Climates since the Last Glacial Maximum*, pp. 24–93 (ed. H. E. J. Wright, J. E. Kutzbach, T. I. Webb, T. I. Webb, W. F. Ruddiman, F. A. Street-Perrott, and P. J. Bartlein). University of Minnesota Press, Minneapolis.

Lapparent de Broin, F. de, and Dijk, P. P. van. 1999. Chelonia from the late Miocene Baynunah Formation, Emirate of Abu Dhabi, United Arab Emirates: Palaeogeographic implications. Chap. 13 in *Fossil Vertebrates of Arabia*, pp. 136–62 (ed. P. J. Whybrow and A. Hill). Yale University Press, New Haven.

Leinen, M., and Heath, G. R. 1981. Sedimentary indicators of atmospheric activity in the northern hemisphere during the Cenozoic. *Palaeogeography, Palaeoclimatology, Palaeoecology* 36: 1–22.

Le Loeuff, J. 1991. The Campano-Maastrichtian vertebrate faunas from southern Europe and their relationship with other faunas in the world; palaeobiogeographical implications. *Cretaceous Research* 12: 93–114.

Lyberis, N., Yurur, T., Chorowicz, J., Kaspoglu, E., and Gundogdu, N. 1992. The east Anatolian Fault: An oblique collision belt. *Tectonophysics* 204: 1–15.

MacFadden, B. J., and Cerling, T. E. 1994. Fossil horses, carbon isotopes and global change. *Trends in Ecology and Evolution* 9: 481–86.

Maley, J. 1980. Les changements climatiques de la fin du Tertiare en Afrique. In *The Sahara and the Nile*, pp. 63–86 (ed. M. A. J. Williams and H. Faure). Balkema, Rotterdam.

McClure, H. A. 1978. The Rub' al Khali. In *Quaternary Period in Saudi Arabia,* pp. 252–63 (ed. S. S. Al-Sayyari and J. E. Zotl). Springer-Verlag, New York.

Milankovitch, M. 1930. Mathematische Klimalehre und astronomische Theorie der Klimaschwankungern. In *Handbuch der Kilmatologie,* pp. 1–176 (ed. W. Koppen and R. Geiger). Gebruder Borntraeger, Berlin.

Miller, K. G., Fairbanks, R. G., and Mountain, G. S. 1987. Tertiary oxygen isotope synthesis, sea level history, and continental margin erosion. *Paleoceanography* 2: 1–19.

Molnar, P., England, P., and Marinod, J. 1993. Mantle dynamics, uplift of the Tibetan Plateau, and the Indian Monsoon. *Reviews of Geophysics* 31: 357–96.

Moore, T. C., Loutit, T. S., and Grenlee, S. M. 1987. Estimating short-term changes in eustatic sea level. *Paleoceanography* 2: 625–37.

Mordan, P. B. 1999. A terrestrial pulmonate gastropod from the late Miocene Baynunah Formation, Emirate of Abu Dhabi, United Arab Emirates. Chap. 11 in *Fossil Vertebrates of Arabia,* pp. 116–19 (ed. P. J. Whybrow and A. Hill). Yale University Press, New Haven.

Morgan, M. E., Kingston, J. D., and Marino, B. D. 1994. Carbon isotopic evidence for the emergence of C4 plants in the Neogene from Pakistan and Kenya. *Nature* 367: 162–65.

Nesbit-Evans, E. M., Van Couvering, J. A., and Andrews, P. 1981. Paleoecology of Miocene sites in western Kenya. *Journal of Human Evolution* 10: 98–121.

Opdyke, N. D. 1995. Mammalian migration and climate over the last seven million years. In *Paleoclimate and Evolution, with Emphasis on Human Origins,* pp. 109–14 (ed. E. S. Vrba, G. H. Denton, T. C. Partridge, and L. H. Burckle). Yale University Press, New Haven.

Partridge, T. C., Wood, B. A., and deMenocal, P. B. 1995. The influence of global climatic change and regional uplift on large-mammalian evolution in East and Southern Africa. In *Paleoclimate and Evolution, with Emphasis on Human Origins,* pp. 331–55 (ed. E. S. Vrba, G. H. Denton, T. C. Partridge, and L. H. Burckle). Yale University Press, New Haven.

Pascual, R., and Juareguizar, E. O. 1990. Evolving climates and mammal faunas in Cenozoic South America. *Journal of Human Evolution* 19: 23–60.

Potts, R., and Behrensmeyer, A. K. 1992. Late Cenozoic terrestrial ecosystems. In *Terrestrial Ecosystems through Time: Evolutionary Paleoecology of Terrestrial Plants and Animals,* pp. 419–541 (ed. A. K. Behrensmeyer, J. D. Damuth, W. A. DiMichele, R. Potts, H. D. Sues, and S. L. Wing). University of Chicago Press, Chicago.

Powers, R. W., Ramirez, L. F., Redmond, D., and Berg, E. L. 1966. Sedimentary geology of Saudi Arabia. *United States Geological Survey Professional Paper* 560D: 1–146.

Prakash, U. 1972. Paleoenvironmental analysis of Indian Tertiary floras. *Geophytology* 2: 178–205.

Prell, W. L., and Kutzbach, J. E. 1992. Sensitivity of the Indian monsoon to forcing parameters and implications for its evolution. *Nature* 360: 647–52.

Prell, W. L., Murray, D. W., Clemens, S. C., and Anderson, D. M. 1992. Evolution and variability of the Indian Ocean summer monsoon: Evidence from the western Arabian Sea drilling program. In *Synthesis of Results from Scientific Drilling in the Indian Ocean,* pp. 447–69 (ed. R. A. Duncan). Geophysical Monograph Series. American Geophysical Union, Washington, D. C.

Pye, K., and Tsoar, H. 1990. *Aeolian Sand and Sand Dunes.* Unwin Hyman, London.

Quade, J., and Cerling, T. 1995. Expansion of C_4 grasses in the late Miocene of northern Pakistan: Evidence from stable isotopes in paleosols. *Palaeogeography, Palaeoclimatology, Palaeoecology* 115: 91–116.

Quade, J., Cerling, T. E., and Bowman, J. 1989. Development of Asian monsoon revealed by marked ecological shift during the latest Miocene in northern Pakistan. *Nature* 342: 163–66.

Rauhe, M., Frey, E., Pemberton, D. S., and Rossmann, T. 1999. Fossil crocodilians from the late Miocene Baynunah Formation of the Emirate of Abu Dhabi, United Arab Emirates: Osteology and palaeoecology. Chap. 14 in *Fossil Vertebrates of Arabia*, pp. 163–85 (ed. P. J. Whybrow and A. Hill). Yale University Press, New Haven.

Rea, D. K., Leinen, M., and Janacek, T. R. 1985. Geologic approach to the long-term history of atmospheric circulation. *Science* 227: 721–25.

Roberts, N., and Wright, H. E. J. 1993. Vegetational, lake-level, and climatic history of the Near East and Southwest Asia. In *Global Climates since the Last Glacial Maximum*, pp. 194–220 (ed. H. E. J. Wright, J. E. Kutzbach, T. I. Webb, E. F. Ruddiman, F. A. Street-Perrott, and P. J. Bartlein). University of Minnesota Press, Minneapolis.

Rögl, F. 1999. Oligocene and Miocene palaeogeography and stratigraphy of the circum-Mediterranean region. Chap. 36 in *Fossil Vertebrates of Arabia*, pp. 501–07 (ed. P. J. Whybrow and A. Hill). Yale University Press, New Haven.

Ruddiman, W. F., and Kutzbach, J. E. 1989. Forcing of late Cenozoic northern hemisphere climate by plateau uplift in southern Asia and American west. *Journal of Geophysical Research* 94: 18 409–18 427.

Schneider, S. H. 1996. *Encyclopedia of Climate and Weather*. Oxford University Press, New York.

Sen, S., and Thomas, H. 1979. Découverte de Rongeurs dans le Miocène moyen de la Formation Hofuf (Province du Hasa, Arabie Saoudite). *Comptes Rendus Sommaires de la Société Géologique de France* 1: 34–37.

Shackleton, N. J., and Kennett, J. P. 1975a. Late Cenozoic oxygen and carbon isotopic changes at DSDP Site 284: Implications for glacial history of the Northern Hemisphere and Antarctica. In *Initial Reports of the Deep Sea Drilling Project*, pp. 801–07 (ed. J. P. Kennett and R. E. Houtz). U. S. Government Printing Office, Washington, D. C.

———. 1975b. Paleotemperature history of the Cenozoic and the initiation of Antarctic glaciation: Oxygen and carbon isotope analyses in DSDP Sites 277, 279 and 281. In *Initial Reports of the Deep Sea Drilling Project*, pp. 743–56 (ed. J. P. Kennett and R. E. Houtz). U. S. Government Printing Office, Washington, D. C.

Stein, R., and Robert, C. 1985. Siliciclastic sediments at Sites 588, 590 and 591: Neogene and Paleogene evolution in the southwest Pacific and Australian climate. In *Initial Reports of the Deep Sea Drilling Project*, pp. 1437–54 (ed. J. P. Kennett and C. C. von der Borch). U. S. Government Printing Office, Washington, D. C.

Stein, R., and Sarnthein, M. 1983. Late Neogene events of atmospheric and oceanic circulation offshore northwest Africa: High resolution record from deep-sea sediments. In *Palaeoecology of Africa*, pp. 9–36 (ed. J. A. Coetzee and E. M. van Zinderen Bakker). Balkema, Rotterdam.

Tedford, R. H. 1985. Late Miocene turnover of the Australian mammal fauna. *South African Journal of Science* 81: 262–63.

Thomas, H. 1979. Le rôle de barrière écologique de la ceinture saharo-arabique au Miocène: Arguments paléontologiques. *Bulletin du Muséum National d'Histoire Naturelle, Paris* 1: 127–35.

———. 1982. La peninsule arabique et l'expansion des primates hominoides miocènes. In *International Meeting on Paleontology: Essentials of Historical Geology,* pp. 215–27 (ed. E. M. Gallitielli). STEM Mucchi, Modena Press, Venice.

———. 1985. The early and middle Miocene land connection of the Afro-Arabian plate and Asia: a major event for hominid dispersal? In *Ancestors: The Hard Evidence,* pp. 42–50 (ed. E. Delson). Alan R. Liss, New York.

Thomas, H., Bernor, R., and Jaeger, J.-J. 1982a. Origines du peuplement mammalien en Afrique du Nord durant le Miocène terminal. *Geobios* 15: 283–97.

Thomas, H., Roger, J., Sen, S., Pickford, M., Gheerbrant, E., Al-Sulaimani, Z., and Al-Busaidi, S. 1999. Oligocene and Miocene terrestrial vertebrates in the southern Arabian peninsula (Sultanate of Oman) and their geodynamic and palaeogeographic settings. Chap. 30 in *Fossil Vertebrates of Arabia,* pp. 430–42 (ed. P. J. Whybrow and A. Hill). Yale University Press, New Haven.

Thomas, H., Sen, S., Khan, M., Battail, B., and Ligabue, G. 1982b. The Lower Miocene fauna of Al-Sarrar (Eastern Province, Saudi Arabia). *ATLAL, The Journal of Saudi Arabian Archaeology* 5: 109–36.

Thomas, H., Sen, S., Roger, J., and Al-Sulaimani, Z. 1991. The discovery of *Moeripthecus markgrafi* Schlosser (Propliopithecidae, Anthropoidea, Primates) in the Ashawq Formation (Early Oligocene of Dhofar Province, Sultanate of Oman). *Journal of Human Evolution* 20: 33–49.

Thomas, H., Taquet, P., Ligabue, G., and Del'Agnola, C. 1978. Découverte d'un gisement de vertébrés dans les dépôts continentaux du Miocène moyen du Hasa (Arabie Saoudite). *Comptes Rendus Sommaires de la Société Géologique de France* 1978: 69–72.

Tiwari, R. K. 1987. Higher-order eccentricity cycles of the middle and late Miocene climatic variations. *Nature* 327: 219–21.

Van Couvering, J. A. H. 1980. Community evolution in East Africa during the late Cenozoic. In *Fossils in the Making: Vertebrate Taphonomy, and Paleoecology,* pp. 272–98 (ed. A. K. Behrensmeyer and A. Hill). University of Chicago Press, Chicago.

van Zinderen Bakker, E. M., and Mercer, J. H. 1986. Major Late Cainozoic climatic events and palaeoenvironmental changes in Africa viewed in a world wide context. *Palaeogeography, Palaeoclimatology, Palaeoecology* 56: 217–35.

Vincent, E., and Berger, W. H. 1985. Carbon dioxide and polar cooling in the Miocene: The Monterey hypothesis. In *Geophysical Monograph,* pp. 455–68 (ed. E. T. Sundquist and W. S. Broeker). American Geophysical Union, Washington, D. C.

Vrba, E. S. 1995. On the connections between paleoclimate and evolution. In *Paleoclimate and Evolution, with Emphasis on Human Origins,* pp. 24–45 (ed. E. S. Vrba, G. H. Denton, T. C. Partridge, and L. H. Burckle). Yale University Press, New Haven.

Whitney, J. W., Faulkender, D. J., and Rubin, M. 1983. The environmental history and present conditions of the northern sand seas of Saudi Arabia. Open File Report 83-749. U. S. Geological Survey, Washington, D. C.

Whybrow, P. J. 1984. Geological and faunal evidence from Arabia for mammal "migrations" between Asia and Africa during the Miocene. *Courier Forschungsinstitut Senckenberg* 69: 189–98.

———. 1987. Miocene geology and palaeontology of Ad Dabtiyah, Saudi Arabia. *Bulletin of the British Museum (Natural History),* Geology 41: 367–457.

Whybrow, P. J., and Clements, D. 1999. Arabian Tertiary fauna, flora, and localities. Chap. 33 in *Fossil Ver-*

tebrates of Arabia, pp. 460–73 (ed. P. J. Whybrow and A. Hill). Yale University Press, New Haven.

Whybrow, P. J., and McClure, H. A. 1981. Fossil mangrove roots and palaeoenvironments of the Miocene of the eastern Arabian peninsula. *Palaeogeography, Palaeoclimatology, Palaeoecology* 32: 213–25.

Whybrow, P. J., Collinson, M. E., Daams, R., Gentry, A. W., and McClure, H. A. 1982. Geology, fauna (Bovidae, Rodentia) and flora from the Early Miocene of eastern Saudi Arabia. *Tertiary Research* 4: 105–20.

Whybrow, P. J., Friend, P. F., Ditchfield, P. W., and Bristow, C. S. 1999. Local stratigraphy of the Neogene outcrops of the coastal area: Western Region, Emirate of Abu Dhabi, United Arab Emirates. Chap. 4 in *Fossil Vertebrates of Arabia,* pp. 28–37 (ed. P. J. Whybrow and A. Hill). Yale University Press, New Haven.

Whybrow, P. J., Hill, A., Yasin al-Tikriti, W., and Hailwood, E. A. 1990. Late Miocene primate fauna, flora and initial palaeomagnetic data from the Emirate of Abu Dhabi, United Arab Emirates. *Journal of Human Evolution* 19: 583–88.

Whybrow, P. J., McClure, H. A., and Elliott, G. F. 1987. Miocene stratigraphy, geology, and flora (algae) of eastern Saudi Arabia and the Ad Dabtiyah vertebrate locality. *Bulletin of the British Museum (Natural History),* Geology. 41: 371–82.

Wigley, T. M. L., and Farmer, G. 1982. Climate of the eastern Mediterranean and Near East. In *Palaeoclimates, Palaeoenvironments and Human Communities in the Eastern Mediterranean Region in Later Prehistory,* pp. 3–37 (ed. J. L. Bintliff and W. van Zeist). British Archaeological Reports, Oxford.

Williams, M. A. J. 1994. Cenozoic climatic changes in deserts: A synthesis. In *Geomorphology of Desert Environments,* pp. 644–70 (ed. A. D. Abrahams and A. J. Parsons). Chapman and Hall, London.

Wolfe, J. A. 1985. Distribution of major vegetational types during the Tertiary. In *The Carbon Cycle and Atmospheric* CO_2, pp. 357–75 (ed. E. T. Sundquist and W. S. Broeker). American Geophysical Union, Washington, D. C.

Miocene lake beds containing abundant remains of terrestrial amphibians, Republic of Yemen.

Regional Faunas and Floras from the Sultanate of Oman, the Republic of Yemen, Africa, and Asia

Part V

Lawrence J. Flynn and Louis L. Jacobs, in Chapter 28, comment on our growing knowledge of Miocene small-mammal faunas of the Arabian Peninsula. These indicate that, for microfauna, an African biogeographic affinity predominated until late in that epoch. Late Miocene assemblages include more small mammals that appear to show broader distributions than in previous times. Widespread distribution is clearest among muroid rodents, particularly murines and gerbils. The mouse genera *Parapelomys* and *Saidomys* indicate a late Miocene spread of murines throughout much of southern Asia and Africa. The gerbil *Abudhabia* is a late Miocene element of the Arabian Peninsula fauna, known to have ranged from North Africa into Afghanistan. In this chapter, they describe a species from Pakistan that is closely related to the Arabian form. This new species, from a securely dated locality in the late Miocene Dhok Pathan Formation, coexists with the last cricetids known to occur in the Siwalik sequence, and is an element of general faunal change in that region. For the Siwaliks, *Abudhabia* may be an indicator of increasing seasonality, or possibly less annual moisture; this is supported by isotopic studies of fossil enamel apatite and palaeosols. The diversification of modern gerbils may be correlated with increasing seasonality in climate on a global scale.

The Baynunah fauna has strong African affinities, and probably falls into the time period between 8 and 6 million years ago. Andrew Hill (Chapter 29) briefly reviews late Miocene sites from sub-Saharan Africa that may be comparable in time and faunal composition. They are relatively few in number and are concentrated on the eastern side of the continent. It is difficult to perceive regional faunal differences in Africa during the late Miocene, mainly because of the paucity of information. At a generic level, the Baynunah fauna clearly is allied with that of Africa, but its additional similarities to faunas in Europe and Asia confirm the impression of Arabia as an important area of faunal exchange during the late Miocene.

Herbert Thomas and his colleagues of the Franco-Omani Palaeontology Expedition in the Sultanate of Oman have discovered several early Oligocene and Miocene vertebrate localities. Chapter 30 presents the geodynamic and palaeogeographic context of the localities. The succession of transgressive–regressive cycles along the southern margin of the Arabian Plate determined the distribution of depositional environments from the late Paleocene up to the middle Miocene. Vertebrate accumulations are only observed within littoral marine facies, as at Thaytiniti and Taqah in Dhofar, and in the Huqf area in central Oman. The faunal associations are generally characterised by the juxtaposition of elements indicative of marine, freshwater, and terrestrial environments. The early Oligocene mammal faunas at Thaytiniti and Taqah are dominated by rodents and primates, whereas proboscideans are more abundant in the early Miocene of the Huqf area. The association of marine invertebrates with vertebrate remains and the intercalation of the fossiliferous horizons in marine deposits allow both localities of Dhofar to be assigned to the lower part of the early Oligocene; however, stratigraphic and palaeontological arguments indicate that the Taqah fauna is younger than that of Thaytiniti. The mammalian remains from the Huqf area have close similarities to those from Jebel Zelten in Libya, suggesting an age of 17.5–15.5 million years ago.

From their work in the Republic of Yemen, Mustafa Latif As-Saruri, Peter J. Whybrow, and Margaret E. Collinson (Chapter 31) briefly describe the lithology and the microfacies of the middle Eocene Habshiyah Formation, Kaninah–Jizwal sequence. From this sequence they report the first occurrence in Arabia of Paleogene fruits (?Nymphaeaceae) and seeds (*Anonaspermum*, Anonaceae) associated with vertebrate remains (Mammalia indeterminate, ?Sirenia). The Anonaceae suggest a tropical aspect whilst the distinctive unidentified fruit of ?Nymphaeaceae indicates links with the Eocene and Oligocene palaeofloras of Egypt.

Louis L. Jacobs and colleagues (Chapter 32) describe the discovery in 1990 of a dinosaur in the transitional marine beds of the Mabdi Formation (late Jurassic) of northern Yemen. Most of the specimen remains in situ; however, fragments that have been prepared suggest a sauropod, probably a cervical vertebra, because of the relatively thin bone and complex structure. Although more refined determination must await further excavation and preparation, Yemen presents an opportunity for the study of Mesozoic terrestrial biogeography and endemism, because in the late Jurassic, unlike the Cenozoic, the region of Arabia was at the eastern margin of the African portion of Gondwana and not strategically situated for faunal dispersal with northern continents.

Peter J. Whybrow and Diana Clements (Chapter 33) state that before the mid-1970s no vertebrate faunas had been reported from Arabia; just two vertebrate fossils had been recorded—a hippopotamus skull (?Pleistocene, Saudi Arabia) and a proboscidean tooth (late Miocene, Emirate of Abu Dhabi). Oil-company reports dating from the 1940s had recorded vertebrates but this information was never published. In Chapter 33, all known Tertiary sites in the Arabian Peninsula, and one relevant site in Mesopotamia, are listed with their faunas to provide a comprehensive reference for the Arabian Tertiary terrestrial biota.

Late Miocene Small-Mammal Faunal Dynamics: The Crossroads of the Arabian Peninsula

28

LAWRENCE J. FLYNN AND LOUIS L. JACOBS

As faunal data on the Neogene deposits of the Arabian Peninsula have accumulated, it has become clear that the biogeographic affinities of the region were mainly but not exclusively with the rest of Africa through much of Miocene time. Later in the epoch, physical links were stronger with Asia than the African landmass (Whybrow, 1984), and Eurasiatic elements became more prevalent. Both small and large mammals (Sen and Thomas, 1979; Thomas, 1983) indicate middle Miocene faunal communication with adjacent Asia and, to a lesser extent, with the Indian subcontinent. Until recently, the affinity of fauna of the Arabian Peninsula during the late Miocene could be discussed only indirectly, and by reference to assemblages from elsewhere.

New data indicate a change for small mammals in the later part of the late Miocene. The data come from three sources—the emerging microfauna of Abu Dhabi, and the developing records of East Africa and Pakistan. In the late Miocene, there is increasing evidence of interchange of small mammals between Africa and southern Asia, via the Arabian Peninsula, with the result that the Arabian microfauna became more cosmopolitan. This is most clearly indicated by muroid rodents, especially gerbils.

PREVIOUS WORK

Hard work through the 1970s and 1980s built a good small-mammal record for North Africa, which serves as a standard for comparison with the faunas of the Arabian Peninsula. The monograph of Jaeger (1977) summarised the succession as it was then known of Morocco, Algeria, and Tunisia. Subsequent work has added important new data, but has not negated any of the salient points of that synthesis.

Ameur (1984) added information to the faunal list for the early late Miocene assemblages from Bou Hanifia in Algeria. She recognised both the murine *Progonomys* and the dendromurine *Senoussimys* in North Africa at this time. Slaughter and James (1979) clarified the relationships of late Neogene Egyptian *Saidomys* to modern murines. The late Neogene Sahabi fauna of Libya became available for further comparison. Munthe (1987) presented the Sahabi microfauna, which includes the widespread squirrel *Atlantoxerus* and a ctenodactylid (designated *Sayimys*), the common muroid *Myocricetodon,* a murine identified as *Progonomys* but whose stephanodonty calls this designation into question, and a gerbil. The gerbil was compared with the late Miocene *Protatera* from Algeria, and is discussed below.

Advances in the Arabian Peninsula and nearby Asia included an early Miocene assemblage from the Hadrukh Formation, Saudi Arabia (Whybrow et al., 1982). The most common taxa from there are a dipodoid comparable to Eurasiatic forms, and a cricetid of modern grade, both suggesting immigration to the area, followed by endemic change. The cricetid *Shamalina* was compared favourably to early middle Miocene *Megacricetodon* from Pakistan (Lindsay, 1988).

 ISBN 0-300-07183-3

The early Miocene fauna from the Dam Formation of Saudi Arabia (Thomas et al., 1982) shows clear African affinities. Among the rodents are thryonomyoids, ctenodactylids, a form like *Leakeymys ternani,* and a pedetid. Pedetid rodents spread, surprisingly, into Turkey as well (Sen, 1977). A few rodent specimens were retrieved from the middle Miocene Hofuf Formation of Saudi Arabia, and assigned to the genera *Atlantoxerus* and *Metasayimys* (now *Sayimys*). De Bruijn et al. (1989) identify the same Arabian species, *Sayimys intermedius,* in the early/middle Miocene of Pakistan, but Baskin (1996) notes some morphological differences from specimens of the Indian subcontinent. Indications, therefore, are that the mid-Miocene small-mammal fauna already began to show a more cosmopolitan nature. Winkler (1994) pulls together evidence mainly from Africa and southern Asia to constrain the timing of interchange between those regions for various small mammals.

Because of a lack of later Miocene evidence on Arabian faunas, previous discussions on the palaeogeography of the area brought together evidence from elsewhere. Jacobs (1978), Sen et al. (1979), and Sen (1983) began to recognise muroids of African affinity in the late Neogene of Pakistan and Afghanistan. Jacobs (1978) saw *Parapelomys* as a likely antecedent to early Pliocene African *Pelomys,* and hypothesised a second migration (of *Golunda*) to Africa, in the later Pliocene. The presence of *Golunda* in Africa was brought into question by the analysis of Musser (1987). Brandy et al. (1980) specified an episode of murine immigration from southern Asia to Africa about the Miocene/Pliocene boundary, but did not discuss the two-phased immigration of Jacobs (1978). Recently, *Golunda* was found in the early Pliocene of Ethiopia (WoldeGabriel et al., 1994). For North Africa, Thomas et al. (1982) saw exchange principally with Europe, via the Iberian Peninsula, at the end of the Miocene, but they noted an earlier late Miocene introduction of a form like the Siwalik *Karnimata* from Pakistan.

The lack of late Neogene fossils documenting the faunal history of the region is slowly being remedied by discovery of new late Neogene deposits. Winkler (1997) describes new material from the Miocene/Pliocene Ibole Member of the Manonga–Wembere Formation in Tanzania. The Ibole murine *Saidomys,* together with other species of this genus from elsewhere in Africa and from Afghanistan, suggests derivation from late Miocene south Asian stock, and likely immigration of *Saidomys* to Africa through the Arabian Peninsula before 6 million years (Ma) ago. Winkler (1994) previously noted that rabbits and the porcupine *Hystrix* entered Africa from Asia at about the beginning of Messinian time.

Most important for the Arabian Peninsula is the recent discovery of late Miocene rodents in Abu Dhabi. De Bruijn and Whybrow (1994) presented preliminary findings that are very promising of a new source of data for the biogeography of the Arabian Peninsula. The fauna shows similarity to coeval North African assemblages, but also contains some intriguing elements. The thryonomyid and *Dendromus* are comparable to sub-Saharan fossils, suggesting that the local climate was not as xeric then as at present. The *Myocricetodon* represents a widespread group known from North Africa to Turkey to Pakistan. The presence of *Parapelomys* in the fauna confirms the previous hypothesis of introduction of this group into Africa around the end of Miocene time. De Bruijn and Whybrow (1994) also discuss a new gerbil, *Abudhabia baynunensis,* which they relate to a previously known Afghan form (see Sen, 1983). Here we describe a Siwalik relative.

The Siwalik Gerbil—Systematics

Gerbillidae Alston, 1876
Abudhabia de Bruijn and Whybrow, 1994
Abudhabia pakistanensis sp. nov.

Holotype

GSP 27649, upper right first molar in a maxillary fragment.

Hypodigm

GSP 27649 and GSP 27650, upper left second molar in a maxillary fragment.

Etymology

The species is named after the provenance of the hypodigm.

Type Locality

Y387, a 8.6Ma-old horizon (palaeomagnetic correlation to the time scale of Cande and Kent, 1995) in the Dhok Pathan Formation, Siwalik Group; mottled red-brown clayey siltstone 800 metres west of Mahluwala Village. The site is interpreted as a palaeosol (J. C. Barry, personal communication) and tied to the nearby local measured section, 25 metres above the richer site Y388.

Diagnosis

A large species with a transversely elongated anteroloph on M^1, and a strong anterior cingulum on M^2; no strong posterior cingula on the upper molars.

Description

The upper molars are characterised by low, broad transverse lophs formed by the fusion, through wear, of opposite cusps. Longitudinal connections are undeveloped, although they are in evidence a bit lingual to the midline of the teeth by the proximity of cusps that would unite in late wear. Transverse valleys are not deep; cusps are not greatly elevated or transversely elongated.

The upper first molar (see fig. 28.1; length, width = 2.68, 1.84 mm) includes three transverse lophs. The first (anterior) is composed of a double anterocone, fused at this wear stage, but the dentine still shows an anterior inflection between its two parts. The second loph is composed of protocone and paracone, joined posteriorly and showing a deep anterior infolding, and shallower posterior sculpting. The posterior loph is the broadest loph anteroposteriorly, and shows a deep cleft between hypocone and metacone, but none posteriorly; at this wear stage it is flush with the posterior wall of the molar. The second loph is longest, making the tooth widest at this point. The first loph, although well developed and about three-quarters the length of the second, is not attenuated lingually. Thus, the outline of the tooth is primitive in its anterolingual inflection. There is a weak anterior cingulum and no evidence of a posterior cingulum. The maxillary fragment indicates that the nutrative foramen was posterolingual to this molar, rather than directly opposite the second molar as in *Tatera*. There are three roots visible, the largest being the projecting anterior root. There is a small central lingual root and a small posterolabial root.

Figure 28.1. Holotype of *Abudhabia pakistanensis*, GSP 27649, a right first molar; occlusal view, anterior to the right. Scale bar = 1 mm.

The second molar (1.68, 1.76 mm) shows two lophs, which are composed of protocone–paracone and hypocone–metacone. These opposite cusps are joined more centrally (fig. 28.2) than in the first molar. The first loph shows slight posterior infolding, the second slight anterior sculpting. The shorter second loph approaches the first loph closest lingually (at the hypocone/protocone). A strong anterior cingulum characterises this tooth. It joins the protocone and slopes to the base of the crown in the form of a short lingual crest and a strong labial cingulum that is well separated from the paracone. These features indicate the tooth to be from the left side. A weak posterior cingulum is indicated, but reduced from behind by an appression facet. Three roots are present: a larger, central lingual root, and two labial roots.

The two specimens show moderate wear, although the second molar seems more worn than the first. For this reason and differences in coloration, we do not consider them to represent the same individual. No physical contact is possible with these maxillary fragments.

Comparisons

Tong (1989) added important new observations, particularly on fossil gerbil skulls, that will help ultimately to develop a phylogenetic classification of muroids related to gerbils. Like de Bruijn and Whybrow (1994), however, we do not see that the issues of higher-level systematics within muroids have been settled. We consider gerbils as a family-

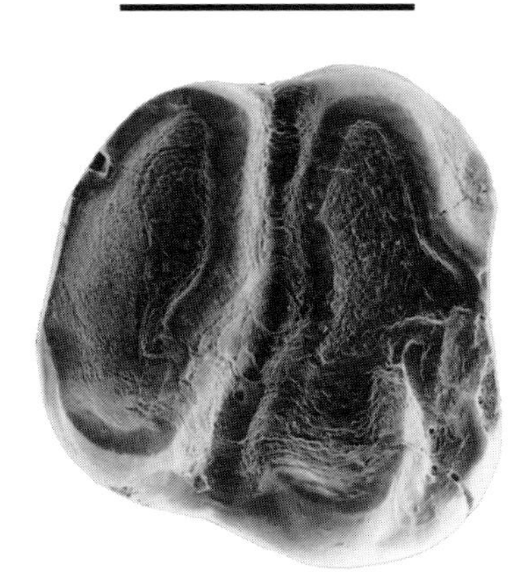

Figure 28.2. Referred upper left second molar of *Abudhabia pakistanensis*, GSP 27650; occlusal view, anterior to the right. Scale bar = 1 mm.

level taxon, consistent with our views in Flynn et al. (1985) wherein we recognise an early Miocene burst in muroid evolution that resulted in distinct radiations of several groups, including gerbils. Whether it will prove more practical to use nomenclature at the subfamily and tribe level for gerbils awaits further analysis, particularly of Miocene fossils. *Abudhabia* and *Protatera* share a lack of longitudinal connections in molars with taterillines; otherwise relationships are unclear.

Known *Abudhabia pakistanensis* teeth are bigger than the molars referred to the type species of the genus. The Siwalik form is further distinguished from *A. baynunensis* in the derived feature of lack of a strong posterior cingulum on the first upper molar. Also in the first molar, the anteroloph is longer; in the type species, the anteroloph is less than two-thirds the second loph in length. Referred second molars in the two species are similar, except that the anterior cingulum in *A. pakistanensis* is heavier. That of *A. baynunensis* is weaker, but does bear a small distinct labial cusp.

Compared with *Abudhabia kabulense* from the late Neogene locality Pul-e Charkhi, Afghanistan, the Siwalik species is smaller and the anterior cingulum on the second molar is stronger. In other features, however, the upper molars of the two species are quite similar. On the first molar, the anteroloph is long, nearly three-quarters the length of the second loph, and the posterior cingulum is reduced. Sen (1983) mentions a weak posterior cingulum, but there is no indication of a median cusp located there. Also, there are no accessory cusps on the anterior cingulum of the second molar in either species.

The importance of these features is difficult to evaluate, given the small sample sizes for all but the Afghan species. The degree of cingulum development, for example, can be weighed better in larger samples. The strong posterior cingulum on the first molar of *Abudhabia baynunensis* seems to distinguish this species from the others, but the referred specimen (specimen AUH 573 of de Bruijn and Whybrow, 1994) may not represent the usual characteristics of the species.

A small sample of gerbils from the late Miocene of Sahabi, Libya, may shed some light on this feature. Munthe (1987) redescribed microfauna that had been collected at Sahabi in 1981. Among the interesting forms there is a gerbil resembling the samples discussed above. The gerbil was attributed to a known genus and named as a new species, *Protatera yardangi*. It is close to the Abu Dhabi form in size and most features, but the upper first molar lacks a posterior cingulum. Despite ambiguity in the report, apparently one lower molar and two upper first molars are figured. The material should be re-examined in comparison with new samples, especially from Abu Dhabi, and renewed screening at Sahabi would clarify variability, for example, in cingulum development. Tentatively, we transfer the species to *Abudhabia*. Future analysis can test whether *A. yardangi* is a senior synonym for the Abu Dhabi sample. Prior to publication of *Abudhabia*, primitive gerbils lacking anteroposterior connections between lophs were generally assigned to the genus *Protatera*. *Protatera algeriensis* had been named for a large sample of isolated teeth from Amama 2, late Miocene of Algeria, (Jaeger, 1977). This species is large, and derived in having elevated, narrow transverse crests on the molars. Illustrated material is not heavily worn, yet the cusps are not distinct. The anteroloph is nearly as long as the second loph on the first molar, and there is no posterior cingulum. The second molar lacks a strong anterior cingulum. All of these derived conditions are shared with *Tatera,* and indeed, this species might be accommodated comfortably in the extant genus. These features serve to distinguish *P. algeriensis* and *Tatera* from species attributed here to *Abudhabia.*

One other late Miocene gerbil named recently, the Spanish *Protatera almenarensis* (Agusti, 1990), shows the advanced features of large size (the first molar is nearly 3 mm long), high cusps joined in transverse crests, broad anterior loph on the first molar, and lack of a posterior cingulum on that tooth. These features argue against transferal of the Spanish species to *Abudhabia.* The lower first molar shows a posterior cingulum and the upper second molar has a weak anterior cingulum.

Discussion

The gerbil *Abudhabia* is one more faunal element that points to homogenisation of small-mammal faunas across northern Africa and through southern Asia in the late Miocene. The similarity of these gerbils is striking. Close, if not conspecific, samples are known from Libya and the Arabian Peninsula, and congeners are known from Afghanistan and Pakistan. These are all about the same age—that is, late Miocene. We have a date for the Siwalik site of 8.6 Ma, and although a bit older than present conjecture, this is a reasonable estimate for Abu Dhabi and Sahabi. The Pul-e Charkhi sample is younger (near the Miocene/Pliocene limit) and distinct morphologically from *Abudhabia pakistanensis.* The differences of larger size and reduced anterior cingulum on M^2 are advanced and consistent with derivation of *Abudhabia kabulense* from *A. pakistanensis.*

Abudhabia pakistanensis, referred to as "cf. *Protatera*" in Flynn et al. (1995), is rare in the Siwaliks as presently known. It is absent from well-sampled older horizons (Y388 and the older "U-level") and from younger sites, for example, around Dhok Pathan Rest House. Whether gerbils were common in the latest Miocene and Pliocene of the Siwaliks will be tested as those horizons are sampled more thoroughly. *Tatera indica* is an element of the modern fauna of the Potwar Plateau.

The geographic distribution (fig. 28.3) of gerbil populations assigned to *Abudhabia* reflects relationship. The samples from Pakistan and Afghanistan appear to be more closely related to each other than to those to the west; the Abu Dhabi and Sahabi forms are likely to be closer to each other than those to the east. The relationships of *Abudhabia* and *Protatera* to other late Neogene and extant gerbils remain to be studied. The late Miocene taxa *Pseudomeriones* and *Epimeriones* are generally considered gerbils, but placed outside extant groups. Whether their presence indicates increasingly xeric conditions is undemonstrated, but possibly the late Miocene spread of derived gerbils across Africa, through southern Asia and into the Iberian Peninsula corresponds to widespread opening of habitat, reflecting increasing seasonality or increasing aridity.

Muroid rodents dominate Neogene microfaunas in the Siwaliks of Pakistan. Through the middle Miocene, cricetids are the most abundant and diverse of the Muroidea. During the late Miocene, however, murids replace them in both numbers of specimens and numbers of species. The last record of cricetids is about 8.6 Ma, at locality Y387, Dhok Pathan Formation, near Khaur Village (Mahluwala Kas), northern Pakistan. That same locality has yielded two teeth of the gerbil *Abudhabia,* which normally would be taken as an indicator of open environment. *Abudhabia* is an element of a major faunal turnover (Barry et al. 1990) and, with other microfauna from later localities, indicates increasing aridity during the late Miocene (Jacobs and Flynn, 1981). Growing databases of both microfauna and megafauna point to sustained change over a period of time perhaps greater than 2 Ma, starting before 9 Ma (Jacobs and Downs, 1994; Morgan et al., 1995).

There is also considerable evidence for habitat change from palaeosol carbonate isotopes (Quade and Cerling, 1995). Whereas initial data suggest that an abrupt change began about 8 Ma ago (present time scale), Morgan et al. (1994) point to a longer trend of change that began before that time.

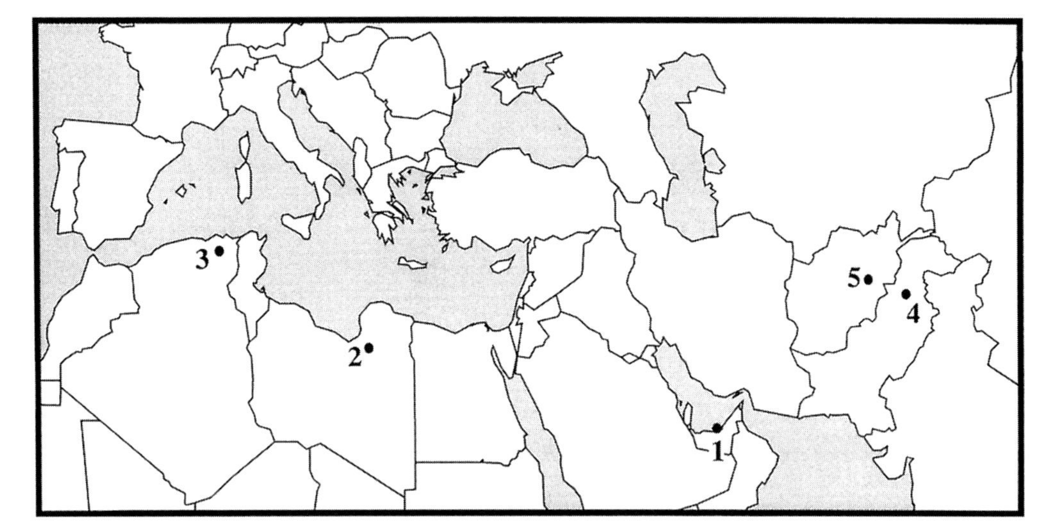

Figure 28.3. Geographic distribution of fossil gerbils discussed in the text. Localities: 1, Emirate of Abu Dhabi; 2, Sahabi, Libya; 3, Amama II, Algeria; 4, Y387, Dhok Pathan Formation, Siwaliks of the Potwar Plateau, Pakistan; 5, Pul-e Charkhi, Afghanistan.

We maintain that change in the Potwar was indeed a long, sustained event. Climatic change, probably involving altered rainfall patterns and increased seasonality, took a toll on the native biota. Plants and animals alike were effected; some adapted, others were eliminated, still others expanded into the area, which meant higher faunal turnover. No two taxa responded exactly alike. Different species have different tolerances and, logically, could be expected to respond at different times in the record. Given long-term climatic change, biotic change would be equally long term. Change among plants to C_4 pathways would be one such biotic change, itself a result of climatic change, but not dating a single climatic event.

ACKNOWLEDGEMENTS

This research was supported in part by NSF grant BNS 8812306. We thank Will Downs, John C. Barry, and S. Mahmood Raza for their assistance, and I. U. Cheema (Pakistan Museum of Natural History), who found the locality Y387. Michèle E. Morgan and Alisa J. Winkler provided welcome advice on the manuscript. Richard H. Tedford (American Muséum of Natural History, New York) provided generous access to unstudied fossils. Sevket Sen (Muséum National d'Histoire Naturelle, Paris) aided in various ways, including access to samples from Pul-e-Charki, providing casts of relevant fossils, and discussion of distinctive features of the Afghan and Pakistan gerbils. Maria Rutzmoser (Museum of Comparative Zoology, Harvard) made comparative samples of modern gerbils accessible to us. Scanning electron photographs were provided by the Shuler Museum, Southern Methodist University; reshot as figures 28.1 and 28.2 by Hillel Burger (Peabody Museum, Harvard). Figure 28.3 was prepared by Will Downs (Bilby Research Center, Northern Arizona University, Flagstaff).

REFERENCES

Agusti, J. 1990 (1989). The Miocene rodent succession in eastern Spain: A zoogeographical appraisal. In *European Neogene Mammal Chronology*, pp. 375–404 (ed. E. H. Lindsay, V. Fahlbusch, and P. Mein). Plenum Press, New York.

Ameur, C. 1984. Découverte de nouveaux rongeurs dans la formation miocène de Bou Hanifia (Algérie Occidentale). *Geobios* 17: 167–75.

Barry, J. C., Flynn, L. J., and Pilbeam, D. R. 1990. Faunal diversity and turnover in a Miocene terrestrial sequence. In *Causes of Evolution: A Paleontological Perspective,* pp. 381–421 (ed. R. Ross and W. Allmon). University of Chicago Press, Chicago.

Baskin, J. A. 1996. Systematic revision of Ctenodactylidae (Mammalia, Rodentia) from the Miocene of Pakistan. *Palaeovertebrata* 25: 1–49.

Brandy, L.-D., Sabatier, M., and Jaeger, J.-J. 1980. Implications phylogénétiques et biogéographiques des derniers découvertes de Muridae en Afghanistan, au Pakistan, et en Ethiopie. *Geobios* 13: 639–43.

Bruijn, H. de, and Whybrow, P. J. 1994. A Late Miocene rodent fauna from the Baynunah Formation, Emirate of Abu Dhabi; United Arab Emirates. *Proceedings Koninklijke Nederlandse Akademie van Wetenschappen* 97: 407–22.

Bruijn, H. de, Boon, E., and Taseer Hussain, S. 1989. Evolutionary trends in *Sayimys* (Ctenodactylidae, Rodentia) from the Lower Manchar Formation (Sind, Pakistan). *Proceedings Koninklijke Nederlandse Akademie van Wetenschappen* B92: 191–214.

Cande, S. C., and Kent, D. V. 1995. Revised calibration of the geomagnetic polarity time scale for the late Cretaceous and Cenozoic. *Journal of Geophysical Research* 100, B4: 6093–95.

Flynn, L. J., Barry, J. C., Morgan, M. E., Pilbeam, D. R., Jacobs, L. L., and Lindsay, E. H. 1995. Neogene Siwalik mammalian lineages: Species longevities, rates of change, and modes of speciation. *Palaeogeography, Palaeoclimatology, Palaeoecology* 115: 249–64.

Flynn, L. J., Jacobs, L. L., and Lindsay, E. H. 1985. Problems in muroid phylogeny. In *Evolutionary Relationships among Rodents,* pp. 589–616 (ed. W. P. Luckett and J. L. Hartenberger). Plenum Press, New York.

Jacobs, L. L. 1978. Fossil rodents (Rhizomyidae and Muridae) from Neogene Siwalik Deposits, Pakistan. *Bulletin. Museum of Northern Arizona, Flagstaff* 52: 1–103.

Jacobs, L. L., and Downs, W. R. 1994. The evolution of murine rodents in Asia. In *Rodent and Lagomorph Families of Asian Origins and Diversification,* pp. 149–56 (ed. Y. Tomida, C. K. Li, and T. Setoguchi). Tokyo National Science Museum Monograph no. 8.

Jacobs, L. L., and Flynn, L. J. 1981. Development of the modern rodent fauna of the Potwar Plateau, northern Pakistan. *Proceedings of the Neogene/Quaternary Boundary Field Conference, India, 1979:* 79–82.

Jaeger, J.-J. 1977. Les Rongeurs du Miocène Moyen et Supérieur du Maghreb. *Palaeovertebrata* 8: 1–166.

Lindsay, E. H. 1988. Cricetid rodents from Siwalik deposits near Chinji Village. Part I. Megacricetodontinae, Myocricetodontinae, and Dendromurinae. *Palaeovertebrata* 18: 95–154.

Morgan, M. E., Badgley, C., Gunnell, G. F., Gingerich, P. D., Kappelman, J. W., and Maas, M. C. 1995. Comparative paleoecology of Paleogene and Neogene mammalian faunas: Body-size structure. *Palaeogeography, Palaeoclimatology, Palaeoecology* 115: 287–317.

Morgan, M. E., Kingston, J. D., and Marino, B. D. 1994. Carbon isotopic evidence for the emergence of C_4 plants in the Neogene from Pakistan and Kenya. *Nature* 367: 162–65.

Munthe, J. 1987. Small-mammal fossils from the Pliocene Sahabi Formation of Libya. In *Neogene Paleontology and Geology of Sahabi,* pp. 135–44 (ed. N. T. Boaz, A. El-Arnauti, A. W. Gaziry, J. de Heinzelin, and D. D. Boaz). Alan R. Liss, New York.

Musser, G. G. 1987. The occurrence of *Hadromys* (Rodentia: Muridae) in early Pleistocene Siwalik Strata in northern Pakistan and its bearing on biogeographic affinities between Indian and northeastern African murine faunas. *American Museum Novitates* no. 2833: 1–36.

Quade, J., and Cerling, T. 1995. Expansion of C_4 grasses in the late Miocene of northern Pakistan: Evidence from stable isotopes in paleosols. *Palaeogeography, Palaeoclimatology, Palaeoecology* 115: 91–116.

Sen, S. 1977. *Megapedetes aegaeus* n. sp. (Pedetidae) et à propos d'autres "Rongeurs Africains" dans le Miocène d'Anatolie. *Geobios* 10: 983–86.

———. 1983. Rongeurs et lagomorphes du gisement pliocène de Pul-e Charkhi, bassin de Kabul, Afghanistan. *Bulletin du Muséum National d'Histoire Naturelle* 5C: 33–74.

Sen, S., and Thomas, H. 1979. Découverte de Ron-geurs dans le Miocène moyen de la Formation Hofuf (Province du Hasa, Arabie Saoudite). *Comptes Rendus Sommaires de la Société Géologique de France* 1: 34–37.

Sen, S., Brunet, M., and Heintz, E. 1979. Découverte de rongeurs "africains" dans le Pliocène d'Af-ghanistan (bassin de Sarobi): Implications paléobiogéographiques et stratigraphiques. *Bulletin du Muséum National d'Histoire Naturelle, Paris,* sect. C, 4(1): 65–75.

Slaughter, B. H., and James, G. T. 1979. *Saidomys natrunensis,* an arvicanthine rodent from the Pliocene of Egypt. *Journal of Mammalogy* 60: 421–25.

Thomas, H. 1983. Les Bovidae (Artiodactyla, Mammalia) du Miocène moyen de la formation Hofuf (Province du Hasa, Arabie Saoudite). *Palaeovertebrata* 13: 157–206.

Thomas, H., Sen, S., Khan, M., Battail, B., and Ligabue, G. 1982. The lower Miocene fauna of Al-Sarrar (Eastern Province, Saudi Arabia). ATLAL, *The Journal of Saudi Arabian Archaeology* 5: 109–36.

Tong, H. 1989. Origine et évolution des Gerbillidae (Mammalia, Rodentia) en Afrique du Nord. *Mémoires de la Société Géologique de France* 155: 1–120.

Whybrow, P. J. 1984. Geological and faunal evidence from Arabia for mammal "migrations" between Asia and Africa during the Miocene. *Courier Forschungsinstitut Senckenberg* 69: 189–98.

Whybrow, P. J., Collinson, M. E., Daams, R., Gentry, A. W., and McClure, H. A. 1982. Geology, fauna (Bovidae, Rodentia) and flora from the Early Miocene of eastern Saudi Arabia. *Tertiary Research* 4: 105–20.

Winkler, A. J. 1994. The Middle/Upper Miocene dispersal of major rodent groups between southern Asia and Africa. In *Rodent and Lagomorph Families of Asian Origins and Diversification,* pp. 173–84 (ed. Y. Tomida, C. K. Li, and T. Setoguchi). Tokyo National Science Museum Monograph no. 8.

———. 1997. Systematics, paleobiogeography, and paleoenvironmental significance of rodents from the Ibole Member, Manonga Valley, Tanzania. In *Neogene Paleontology of the Manonga Valley, Tanzania: A Window into East African Evolution,* pp. 311–32 (ed. T. J. Harrison). Topics in Geobiology vol. 14. Plenum Press, New York.

WoldeGabriel, G., White, T. D., Suwa, G., Renne, P., Heinzelin, J. de, Hart, W. K., and Helken, G. 1994. Ecological and temporal placement of early Pliocene hominids at Aramis, Ethiopia. *Nature* 371: 330–33.

Late Miocene Sub-Saharan African Vertebrates, and Their Relation to the Baynunah Fauna, Emirate of Abu Dhabi, United Arab Emirates

29

Andrew Hill

Initial impressions of the fossil fauna from the Baynunah Formation suggested that its age is between 8 and 6 million years (Ma) (Hill et al., 1999—Chapter 3). This has been confirmed, but not further constrained, by more detailed examinations of the specimens (Whybrow and Hill, 1999—this vol., Part III). This estimate is partly based on faunal comparisons with eastern Africa, a region of the world fairly well calibrated by radiometric means. It is also true, however, that this part of the record in sub-Saharan Africa is relatively poorly known. But enough is known to appreciate that it was an interesting time. Evidence suggests it was during this period that many changes took place in the terrestrial vertebrate fauna of Africa. It signals the beginnings of a shift from the more archaic early Miocene biota to one that incorporates the modern elements characterising the Ethiopian region today (Hill, 1985, 1995). Part of this change was the divergence of the lineages leading to modern chimpanzees, gorillas, and humans. It is also a time when significant climate changes appear to have occurred on a global scale (Kingston and Hill, 1999—Chapter 27). Here I first summarise data concerning the geographical and temporal distribution of relevant fossiliferous horizons and sites in sub-Saharan Africa, to demonstrate the extent to which time and geography are sampled there. Second, I briefly discuss some aspects of faunal change and biogeography in sub-Saharan Africa, and point out some similarities with the Baynunah fauna.

Late Miocene Fossil Sites in Sub-Saharan Africa

In sub-Saharan Africa there are relatively few fossiliferous localities from the late Miocene, even fewer that fall into the same age range as the Baynunah Formation sites in Abu Dhabi, and they are patchily distributed geographically. Most late Miocene sites are concentrated on the eastern side of the continent, predominantly associated with the rift system. There is one in northern Ethiopia, in the Chorora Formation. In Kenya they occur in the Nawata Formation, at Lothagam near Lake Turkana, and in the Ngorora, Mpesida, and Lukeino Formations of the Tugen Hills, west of Lake Baringo. On the eastern side of the rift, there are sites in the Namurungule Formation at Nakali, and then in the west of the country, in the lower parts of the Kanam Formation at Kanam. There are some relevant sites in Uganda and Zaire on opposite sides of the western rift, such as occur in the Nkondo Formation. Possibly the lower portions of the Wembere–Manonga Formation in the Manonga Valley of Tanzania could be as old as 6 Ma. Elsewhere in sub-Saharan Africa, some of the breccias from Berg Aukas, Namibia, may contain late Miocene fossils.

Chorora Formation, Ethiopia

Exposures of the Chorora Formation are found in the Kesem–Kebema basin, south of the Afar depression. One site has produced a small and fragmentary collection of mammals (Sickenberg and Schönfeld, 1975; Tiercelin et al., 1979). Radiometric dates beneath the fossiliferous horizon are given as 9 Ma and 11 Ma (Kunz et al., 1975); Tiercelin et al. (1979) quote bracketing dates of 10.5 Ma and 10.7 Ma. In terms of the Tugen Hills fauna the presence of cf. *Stegotetrabelodon* (Tassy, 1986) and *Hipparion* suggest that these dates may be a little old. Asfaw and colleagues have recently revisited the region (Asfaw et al., 1990).

Tugen Hills, Kenya

The Tugen Hills, west of Lake Baringo in the eastern rift of Kenya, present a succession of fossiliferous rocks ranging in age from about 16 Ma into the Pleistocene, with relatively few significant gaps (Hill et al., 1985, 1986; Hill, 1987, 1995). Since the 1980s they have been investigated by the Baringo Paleontological Research Project (BPRP). The earliest unit relevant here is the Ngorora Formation, bracketed by the Tiim Phonolite Formation at the base, dated at about 13.2 Ma, and the Ewalel Phonolites at the top dated at about 7.6 Ma. Consequently only sites in the very top of the unit may correspond in age to those in Abu Dhabi. Some late Miocene sites occur in the main sequence of the formation, but the succession at Ngeringerowa, to the south, has radiometric dates falling into the 9.5–8 Ma range (Hill, 1995; A. Deino, personal communication). The Mpesida Beds, stratigraphically above the Ngorora Formation, are intercalated within flows of the Kabarnet Trachyte Formation, which in turn overlies the Ewalel Phonolites. The age of the sediments is therefore constrained by K–Ar dates on Kabarnet Trachyte flows below and above sedimentary lenses (Hill et al., 1985, 1986). Trachytic flows near the base of the volcanic succession to the south near Kabarnet have produced ages ranging from 7.5 Ma to 7.3 Ma. In the Yatya area, Chapman and Brook (1978) obtained an age of 7.0 Ma on a flow just beneath Mpesida sediments, while trachytes above the sedimentary sequence have been dated to 6.2 Ma and 6.36 Ma (Hill et al., 1985). More recently single-crystal laser fusion $^{40}Ar/^{39}Ar$ analysis of a lapilli tuff in the upper 5 metres of Mpesida sediments exposed on the southwest flank of Rormuch has produced a date close to these latter estimates (A. Deino, personal communication). The Lukeino Formation occupies a considerable geographic area along the eastern base of the Tugen Hills. The areal extent is probably greater than 150 km^2. The Lukeino Formation overlies the Kabarnet Trachytes (6.36–6.20 Ma) and it is stratigraphically overlain by the Kaparaina Basalts. Early dates on these cover a wide range (Chapman and Brook, 1978), but more recent determinations by BPRP are more focused on 5.6 Ma (Hill et al., 1985, 1986). Ages on tuffs within the formation are 6.06 Ma and 5.62 Ma, consistent with the bracketing dates, and a new set of determinations for the succession is in preparation.

The Lothagam Succession, Kenya

Lothagam is a very rich set of sites to the southwest of Lake Turkana first investigated in any detail by Patterson et al. (1970). Early estimates of the date of the fossils, primarily based on biostratigraphy, centred on 5.5 Ma, and a more recent attempt (Hill et al., 1992) suggested most of the fauna was a little older than this. Work has been renewed in the region by Meave Leakey's team (Leakey et al., 1996), who have established the geological framework in much more detail and greatly enhanced the faunal collections. The fossiliferous part of the succession in the upper Miocene is the Nawata Formation. The Nawata is divided into lower and upper members. In the past reliable radiometric dates have been difficult to obtain, but a series of single-crystal laser fusion $^{40}Ar/^{39}Ar$ determinations are in progress (I. McDougall, in preparation). The lower part of the Nawata Formation probably extends from more than 7.91 Ma to about 6.24 Ma, and the top of the upper unit is about 5.5 Ma old (Leakey et al., 1996). In addition, other areas southwest of Lake Turkana have the promise of fossiliferous sites extending even further back into the later Miocene (C. Feibel, personal communication).

The Namurungule Formation, Kenya

The Namurungule Formation crops out south of Lake Turkana on the east of the rift valley. There are three main areas of outcrop, and the geology has been described by Makinouchi et al. (1984). Dating attempts have not been very satisfactory and are discussed in Matsuda et al. (1984, 1986; see also Hill and Ward, 1988). If correct, radiometric methods constrain the age to between 13 and 6.4 Ma, but palaeontological considerations (Nakaya et al., 1984; Pickford et al., 1984a,b) suggest something near to 9 Ma. Probably 8 ± 2 Ma is the best that can be proposed (Hill, 1994).

Nakali, Kenya

Nakali is a site also on the eastern side of the Kenya Rift. It received some attention from Aguirre and Philip Leakey (Aguirre and Leakey, 1974), and was later re-examined by Richard Leakey and Alan Walker (unpublished). It has only a sparse fauna and is not well dated, but is interesting because the fauna shows some differences from those known at better sites elsewhere. It is possible that it represents some otherwise unsampled time period, perhaps between 8 and 7 Ma.

Kanam, Kenya

Kanam is on the western shores of Lake Victoria, and was originally worked on by Louis Leakey in the 1930s. Tom Plummer, Rick Potts, and colleagues recently renewed investigation. Faunal and initial radiometric results suggest that it extends into the top of the later Miocene, perhaps back to 6 Ma.

Lake Albert Basin, Uganda and Zaire

There have been intermittent investigations of late Miocene fossiliferous rocks in the western rift valley, both on the Uganda and Zaire sides, for many years, but Brigitte Senut and her team recently reinvestigated them more intensely (Pickford et al., 1993; Senut and Pickford, 1994). Based on faunal correlation with the eastern rift, they estimate the age of the sites to range from 8 to 1 Ma. For parts of the section younger than 4 Ma estimates have been reinforced by geochemical correlation of volcanic ash with dated tephra in the eastern rift. In the Nkondo–Kaiso region the upper Miocene part of the succession is represented by the Nkondo Formation. The lower member of this, the Nkondo Member, is assigned an age of about 7–6 Ma. The upper member, the Nyaweiga Member, is probably Pliocene. In the Kisegi Valley, the Kakara Formation is believed by Pickford et al. (1993) to be late Miocene in age, somewhere between 12 and 9 Ma. The presence of *Hipparion,* however, would suggest that if the Kakara fauna is homogeneous, it is nearer the younger end of the range. The lower member of the overlying Oluka Formation provides a fauna that suggests to Pickford et al. (1993) an age younger than the Namurungule Formation (which he gives as 9–8 Ma), and probably near 7–6.7 Ma. The upper member of the formation is suggested to be about 6 Ma, and correlates biostratigraphically with the Nkondo Member of the Nkondo Formation. On the Zaire side of the rift are several units that could extend back into the upper part of the Miocene, judged mainly on the basis of molluscan biostratigraphy. Pickford et al. (1993) suggest that sites near Karugamania are probably older than 7 Ma. And in the Sinda–Mohari area are fossil mammals probably older than 8 Ma. Higher in the succession, at Ongaliba, are sites that may be around 6 Ma.

Manonga Valley, Tanzania

Harrison has reinvestigated a basin of fossiliferous sediments in the Manonga Valley of northern Tanzania (Harrison, 1992, 1997; Harrison and Verniers, 1993). There are about 30 fossil sites in the Ibole Formation and in the overlying Wembere–Manonga Formation, which are exposed over a very extensive lake basin. Unfortunately there are no volcanic horizons to permit radiometric dating, but by comparison with early work on Lothagam faunas Harrison suggests the sites are late Miocene or early Pliocene in age. It seems quite possible, however, that most of the fauna derives from the Pliocene. If it were late Miocene then *Nyanzachoerus syrticus* might be expected to occur as part of the suid sample, for example, whereas it is absent (Bishop, 1997). *Nyanzachoerus syrticus,* a suid that occurs in the Baynunah fauna (Bishop and Hill, 1999—Chapter 20), and which is reasonably common in sub-Saharan Africa, is found earlier than 5.6 Ma in the Tugen Hills succession (Hill et al., 1992). Information from Lothagam and other relevant successions is consistent with this last-appearance datum.

Berg Aukas, Namibia

Several sites in Namibia are breccia cave infillings, dated biostratigraphically (Conroy et al., 1992). They appear to belong to seven different periods of infilling, from early middle Miocene to Holocene. Berg Aukas is one of these sites, and some of the blocks of breccia from the cave appear later Miocene in age. Pickford (1996), for example, alludes to a hyrax, *Heterohyrax,* as coming from blocks estimated to be 10–9 Ma old.

SAMPLING IN TIME AND GEOGRAPHY

Much of late Miocene time is covered by sites in sub-Saharan Africa, though not necessarily well; for example, the Ngorora Formation in the Tugen Hills is the only sequence that has sites in the 12.5–8 Ma range. Some sites around 12.5–11 Ma produce extensive information, but the time between 11 and 9 Ma ago is relatively poorly documented there so far. Chorora may be near 10.5 Ma in age, but its fauna is not extensive. The top of the Tugen Hills Ngorora Formation provides so far exclusive evidence of the 9.5–8 Ma period. Sites in the Namurungule Formation might be 8 Ma old, but they are not well dated. There is a gap in the Tugen Hills succession between about 8 and 6.5 Ma ago, which is not filled by any other sites unless the poorly dated Nakali falls into this category. Sites at Lothagam probably extend below 7.5 Ma, and provide information to the top of the Miocene. The Mpesida Beds and Lukeino Formation in the Tugen Hills provide the same from 6.5 Ma. Some sites in the western rift may reach back as far as 12 Ma. Isolated sites such as Berg Aukas provide small glimpses at about 9 Ma in southern Africa. Parts of the Manonga area may be near the top of the Miocene.

All of these occurrences, except Berg Aukas, are broadly on the eastern edge of the continent, leaving much of its area totally unsampled during this time period.

FAUNAL CHANGE IN THE LATE MIOCENE

On the basis of the Tugen Hills collections I have previously indicated a major change in the sub-Saharan African terrestrial vertebrate fauna between about 8 and 6.5 Ma ago (Hill, 1985, 1987, 1995; Hill et al., 1985, 1986). This shift can now also be seen from the nature of the fauna at Lothagam, where additional taxa contribute to the picture (Leakey et al., 1996). In the Tugen Hills this can be appreciated by comparing faunas from sites in the Ngorora Formation with those from the Mpesida Beds. Intimations of this change already occur when older Ngorora sites are compared with those at the top of the formation. One particularly rich and well-documented site is BPRP#38 in the type section (Hill et al., in press), and this provides a good example of a fauna at the base of the Upper Miocene. By the time of Ngeringerowa, however, around 9.5–8 Ma, we see the first Equidae, in the form of *Hipparion,* and the first cercopithecid monkeys, represented by the colobine *Microcolobus tugenensis.* Among bovids there are possibly the earliest reduncines. But by the time of the overlying Mpesida Beds additional new families have appeared, such as Elephantidae, represented by *Stegotetrabelodon orbus,* and Leporidae. Another genus of proboscidean appears for the first time, *Anancus kenyensis;* also making their appearance are more modern kinds of hippopotamus (*Hexaprotodon*), more advanced kinds of suid, such as *Nyanzachoerus* (the same genus and species that occurs in the Baynunah fauna; Bishop and Hill, 1999—Chapter 20), and a contemporary genus of rhinoceros, represented by *Ceratotherium praecox.*

All these Mpesida creatures are of much more modern aspect than those found in the underlying Ngorora Formation, which have more in common with even earlier Miocene forms. In the more fossiliferous and overlying Lukeino Formation, dated at about 6 Ma, this impression of change is corroborated.

Although a considerable change, it need not be a sudden one. It is probably the greatest change documented in the Tugen Hills sequence between 13 and 1.5 Ma ago, but it also coincides with the greatest gap in the succession. There are faunas at about 8 Ma ago; then nothing is known until about 6.5 Ma ago. Nakali is a site that may fall into this time period, and its fauna suggests that otherwise undocumented events are occurring in this sparsely documented interval. Although the Nakali collection is small, several taxa from there are known from nowhere else in sub-Saharan Africa. Among these unique taxa is a large rhinoceros of a kind otherwise not known in Africa, a rodent with affinities to some in Pakistan, a hyrax, and a hyaenid.

Faunal Shifts and Global Climatic and Environmental Change

Despite these reservations, it is tempting, and even possibly appropriate, to regard this local East African faunal change as part of an apparently global shift in climate and environments. Some of the evidence for this is discussed by Kingston and Hill (1999—Chapter 27). There is some evidence, for example, of a cooling event between 11 and 9 Ma ago, and a dramatic drop in sea level is recorded at 11–10 Ma ago. Such changes as these appear to have been correlated with increasing aridity, reflected in vegetational changes documented in such widely separated regions as western North America and Australia. Between about 6.5 Ma ago and the beginning of the Pliocene we see evidence of extensive climatic variation, the isolation of the Mediterranean Basin, and the onset of the Messinian Salinity Crisis.

One good proxy of local vegetational conditions is provided by palaeosol carbonates, which provide a means of monitoring local change through time. Different kinds of plants fractionate carbon isotopes differently during photosynthesis. In particular, most open-country grasses use the C_4 pathway, whereas trees and shrubs metabolise by the C_3 pathway. This signal is preserved in soil carbonates that formed at the time (see Kingston, 1999—Chapter 25). By analysing soil carbonates Quade et al. (1989) infer a striking shift in vegetation in the Siwalik sequence of Pakistan between about 7.4 and 6 Ma ago. A similar examination of pedogenic carbonates in the Tugen Hills sequence, however, reveals no comparable shift (Cerling, 1992; Kingston, 1992; Kingston et al., 1992, 1994). Certainly there is no sign of the savannah grasslands required by Coppens (1994). But it is also true that Kingston's analyses of carbon isotopes in tooth enamel (Kingston, 1992; Morgan et al., 1994) show that the first herbivores that exclusively consumed C_4 vegetation appear in the Tugen Hills succession at about 7 Ma ago. At a gross level, however, plants show no dramatic shift. Instead, they consistently indicate mosaic or mixed habitats. In the Tugen Hills there is evidence of forest vegetation in the Ngorora Formation at 12.6 Ma. This is in the form of well-preserved entire leaves (Kabuye and Jacobs, 1986; Jacobs and Kabuye, 1987, 1989; Jacobs and Winkler, 1992). Kapturo is a new fossil leaf site (BPRP#133), not yet stratigraphically fitted into the succession, but dated between 7.2 and 6.9 Ma, and hence just before known Mpesida sediments. It contains a rich deciduous woodland palaeoflora (Jacobs, 1994; Jacobs and Deino, 1996). At a little later in time forest is present in the Mpesida Beds (BPRP: work in progress).

Late Miocene African Biogeography

The relative paucity of sites limits the extent to which reliable statements may be made about African late Miocene palaeobiogeography. Biogeography can be considered at various levels. On the continental scale Africa has characteristic faunal elements in the late Miocene when compared to other continents and large regions. It is in this sense that the Baynunah fauna can be said to show strong African affinities. But it is harder to detect regional differences within Africa itself, mainly due to the scarcity of sites and their concentration on particular areas of the continent. Some things are possible. There are differences at a local level in aquatic vertebrates between the Tugen Hills succession, for example, and sites, such as Lothagam, only a few hundred kilometres north. In the late Miocene and at other times in the Tugen Hills there is no *Euthecodon,* a point also mentioned by Leakey et al. (1996). This genus could be missing because of environment and ecology, but a biogeographic explanation is reinforced by the additional absence of the nilotic fish *Lates* in the Tugen Hills, although it is present at sites in the Lake Turkana Basin from Lothagam onwards. This suggests the maintenance of fluvial and lacustrine separation of these neighbouring regions over a long period to the present day. The nilotic affinities of the Lake Turkana Basin appear to have been established early. In terms of terrestrial faunas, however, the two basins are difficult to separate.

There appear to be some differences between the eastern and western rift terrestrial faunas during the late Miocene; Pickford and Senut (1994) comment on the presence of the proboscidean *Stegodon* in the west, for example. But it is hard to judge at present how much this is due to genuine biogeographic separation, or simply to the suggested vegetational and ecological differences that they also discuss.

Thomas (1979; Thomas et al., 1982) has convincingly argued for late Miocene faunal differences between northern Africa and what is now sub-Saharan Africa, leading to the suggestion of a barrier between these areas at that time, and this seems to be substantiated by later work.

Biogeographic Affinities of the Baynunah Fauna

It is possible to see the Baynunah fauna as more part of this North African fauna than the sub-Saharan one, particularly at the level of species. The subtleties of similarity at the species level are discussed by Gentry (1999a,b—Chapters 21 and 22), for example, but the suggestion is still borne out at the level of genera. There are 12 taxa of large mammals in the Baynunah fauna identified to the generic level; 9 are also African; about 8 are shared with North African faunas and about 8 with those from the south. The strong connection with Africa as a whole is shown by the fact that at least three of these genera—*Hexaprotodon, Nyanzachoerus,* and *Stegotetrabelodon* (sensu Tassy, 1999—Chapter 18)—otherwise occur only in Africa at that time (unless the Baynunah fauna postdates the appearance of *Hexaprotodon* in Iberia).

The Baynunah fauna also has 8 of its 12 identified genera in common with Asian faunas, and 7 with those of Europe, but this does not indicate such a strong affinity in either case. None of the genera shared with Europe is endemic to Europe and Arabia. Only one of those shared with Asia otherwise occurs exclusively on that continent (*Propotamochoerus*). These strong shared similarities displayed at the generic level with all adjacent biogeographic regions emphasises forcefully, however, the pivotal role that Arabia played as a junction for faunal migration in the late Miocene.

Acknowledgements

I thank Peter Whybrow for a decade of most enjoyable collaboration and companionship in the United Arab Emirates, and in the Anglesea public house. The work in the Anglesea is financed by ourselves. My initial visits to the Emirate of Abu Dhabi were funded by the Department of Antiquities and Tourism, Emirate of Abu Dhabi, and I thank the Undersecretary, Saif Ali Al Dhab'a al Darmaki for the hospitality of his department, and Walid Yasin for his great kindness and help. Later, funding for the work in the Emirate came largely from the Abu Dhabi Company for Onshore Oil Operations (ADCO), and I also received an air fare from the L. S. B. Leakey Foundation. The investigations of the Tugen Hills succession forms part of the Baringo Paleontological Research Project based at Yale University, USA, working jointly with the National Museums of Kenya. I thank the Office of the President, Republic of Kenya, for research permission. That research has been financed by grants from the National Science Foundation, USA (most recently SBR-9208903), from the Louise Brown Foundation, the L. S. B. Leakey Foundation, and Mr J. Clayton Stephenson. I thank Craig Feibel for advice on the Lothagam succession, and John Kingston, Sally McBrearty, and Tom Gundling for their comments on the manuscript.

REFERENCES

Aguirre, E., and Leakey, P. 1974. Nakali: nueva fauna de Hipparion del Rift Valley de Kenya. *Estudios Geologicos. Instituto de Investigaciones Geologicas "Lucas Mallada", Madrid* 30: 219–27.

Asfaw, B., Ebinger, C., Harding, D., White, T., and WoldeGabriel, G. 1990. Space-based imagery in paleoanthropological research: An Ethiopian example. *National Geographic Research* 6: 418–34.

Bishop, L. 1997. Suidae from the Manonga Valley; Tanzania. In *Neogene Paleontology of the Manonga Valley, Tanzania: A Window into East African Evolution,* pp. 191–217 (ed. T. J. Harrison). Topics in Geobiology vol. 14. Plenum Press, New York.

Bishop, L., and Hill, A. 1999. Fossil Suidae from the Baynunah Formation, Emirate of Abu Dhabi, United Arab Emirates. Chap. 20 in *Fossil Vertebrates of Arabia,* pp. 254–70 (ed. P. J. Whybrow and A. Hill). Yale University Press, New Haven.

Cerling, T. E. 1992. Development of grasslands and savannas in East Africa during the Neogene. *Palaeogeography, Palaeoclimatology, Palaeoecology (Global and Planetary Change Section)* 97: 241–47.

Chapman, G. R., and Brook, M. 1978. Chronostratigraphy of the Baringo Basin, Kenya Rift Valley. In *Geological Background to Fossil Man,* pp. 207–23 (ed. W. W. Bishop). Scottish Academic Press, London.

Conroy, G., Pickford, M., Senut, B., Van Couvering, J. A., and Mein, P. 1992. *Otavipithecus namibiensis,* first Miocene hominoid from Southern Africa (Berg Aukas, Namibia). *Nature* 353: 144–48.

Coppens, Y. 1994. East side story: the origin of mankind. *Scientific American* 270(5): 88–95.

Gentry, A. W. 1999a. A fossil hippopotamus from the Emirate of Abu Dhabi, United Arab Emirates. Chap. 21 in *Fossil Vertebrates of Arabia,* pp. 271–89 (ed. P. J. Whybrow and A. Hill). Yale University Press, New Haven.

———. 1999b. Fossil pecorans from the Baynunah Formation, Emirate of Abu Dhabi, United Arab Emirates. Chap. 22 in *Fossil Vertebrates of Arabia,* pp. 290–316 (ed. P. J. Whybrow and A. Hill). Yale University Press, New Haven.

Harrison, T. J. 1992. Paleoanthropological exploration in the Manonga Valley, northern Tanzania. *Nyame Akuma* 36: 25–31.

——— ed. 1997. *Neogene Paleontology of the Manonga Valley, Tanzania: A Window into East African Evolution.* Topics in Geobiology vol. 14. Plenum Press, New York.

Harrison, T. J., and Verniers, J. 1993. Preliminary study of the stratigraphy and mammalian palaeontology of Neogene sites in the Manonga Valley, northern Tanzania. *Neues Jahrbuch für Geologie und Paläontologie, Abhandlungen* 190: 57–74.

Hill, A. 1985. Les variations de la faune du Miocène récent et du Pliocène d'Afrique de l'est. *L'Anthropologie* (Paris) 89: 275–79.

———. 1987. Causes of perceived faunal change in the later Neogene of East Africa. *Journal of Human Evolution* 16: 583–96.

———. 1994. Late Miocene and early Pliocene hominoids from Africa. In *Integrative Paths to the Past: Pale-*

oanthropological Advances in Honor of F. Clark Howell, pp. 123–45 (ed. R. S. Corruccini and R. L. Ciochon). Prentice-Hall, Englewood Cliffs, N. J.

———. 1995. Faunal and environmental change in the Neogene of East Africa: Evidence from the Tugen Hills Sequence, Baringo District, Kenya. In *Paleoclimate and Evolution, with Emphasis on Human Origins,* pp. 178–93 (ed. E. S. Vrba, G. H. Denton, T. C. Partridge, and L. H. Burkle). Yale University Press, New Haven.

Hill, A., and Ward, S. 1988. Origin of the Hominidae: The record of African large hominoid evolution between 14 My and 4 My. *Yearbook of Physical Anthropology* 31: 49–83.

Hill, A., Curtis, G., and Drake, R. 1986. Sedimentary stratigraphy of the Tugen Hills, Baringo District, Kenya. In *Sedimentation in the African Rifts,* pp. 285–95 (ed. L. Frostick, R. W. Renaut, I. Reid, and J.-J. Tiercelin). Geological Society of London Special Publication, no. 25. Blackwell, Oxford.

Hill, A., Drake, R., Tauxe, L., Monaghan, M., Barry, J. C., Behrensmeyer, A. K., Curtis, G., Jacobs, B. F., Jacobs, L., Johnson, N., and Pilbeam, D. 1985. Neogene palaeontology and geochronology of the Baringo Basin, Kenya. *Journal of Human Evolution* 14: 759–73.

Hill, A., Leakey, M., Kingston, J., and Ward, S. in press. New cercopithecoids and a hominoid from 12.5 Ma in the Tugen Hills succession. *Journal of Human Evolution.*

Hill, A., Ward, S., and Brown, B. 1992. Anatomy and age of the Lothagam mandible. *Journal of Human Evolution* 22: 439–51.

Hill, A., Whybrow, P. J., and Yasin, W. 1999. History of palaeontological research in the Western Region of the Emirate of Abu Dhabi, United Arab Emirates. Chap. 3 in *Fossil Vertebrates of Arabia,* pp. 15–23 (ed. P. J. Whybrow and A. Hill). Yale University Press, New Haven.

Jacobs, B. F. 1994. Paleoclimate reconstructions using middle to late Miocene paleofloras from central Kenya. *American Journal of Botany,* suppl. 81: 94.

Jacobs, B. F., and Deino, A. 1996. Paleoclimatic estimates for the Miocene Kabasero and Kapturo paleobotanical localities: Tugen Hills Kenya, and 40Ar/39Ar dating of the Kapturo site. *Palaeogeography, Palaeoclimatology, Palaeoecology* 123: 259–71.

Jacobs, B. F., and Kabuye, C. H. S. 1987. A middle Miocene (12.2 my old) forest in the East African Rift Valley, Kenya. *Journal of Human Evolution* 16: 147–55.

———. 1989. An extinct species of *Pollia* Thunberg (Commelinaceae) from the Miocene Ngorora Formation, Kenya. *Review of Palaeobotany and Palynology* 59: 67–76.

Jacobs, B. F., and Winkler, D. A. 1992. Taphonomy of a middle Miocene autochthonous forest assemblage, Ngorora Formation, central Kenya. *Palaeogeography, Palaeoclimatology, Palaeoecology* 99: 31–40.

Kabuye, C. H. S., and Jacobs, B. 1986. An interesting record of the genus *Leptaspis,* Bambusoideae from Middle Miocene flora deposits in Kenya, East Africa. *International Symposium on Grass Systematics and Evolution,* pp. 1–32. Smithsonian Institution, Washington, D.C.

Kingston, J. D. 1992. Stable isotope evidence for hominid paleoenvironments in East Africa. Ph.D. thesis, Harvard University.

———. 1999. Isotopes and environments of the Baynunah Formation, Emirate of Abu Dhabi, United Arab Emirates. Chap. 25 in *Fossil Vertebrates of Arabia,* pp. 354–72 (ed. P. J. Whybrow and A. Hill). Yale University Press, New Haven.

Kingston, J. D., and Hill, A. 1999. Late Miocene palaeoenvironments in Arabia: A synthesis. Chap. 27 in *Fossil Vertebrates of Arabia,* pp. 389–407 (ed. P. J. Whybrow and A. Hill). Yale University Press, New Haven.

Kingston, J. D., Hill, A., and Marino, B. 1992. Isotopic evidence of late Miocene/Pliocene vegetation in the east African Rift Valley. *American Journal of Physical Anthropology,* suppl. 14: 100–101.

Kingston, J. D., Marino, B., and Hill, A. 1994. Isotopic evidence for Neogene hominid paleoenvironments in the Kenya Rift Valley. *Science* 264: 955–59.

Kunz, K., Kreuzer, H., and Muller, P. 1975. Potassium-argon age determinations of the trap basalt of the southeastern part of the Afar rift. In *Afar Depression of Ethiopia,* pp. 370–74 (ed. A. Pilger and A. Rosler). Schweizerbart'sche Verlagsbuchhandlung, Stuttgart.

Leakey, M. G., Feibel, C. S., Bernor, R. L., Harris, J. M., Cerling, T. E., Stewart, K. M., Storrs, G. W., Walker, A., Werdelin, L., and Winkler, A. J. 1996. Lothagam, a record of faunal change in the late Miocene of east Africa. *Journal of Vertebrate Paleontology* 16: 556–70.

Makinouchi, T., Koyaguchi, T., Matsuda, T., Mitsushio, H., and Ishida, S. 1984. Geology of the Nachola area and the Samburu Hills, west of Baragoi, northern Kenya. *African Study Monographs,* suppl. 2: 15–44.

Matsuda, T., Torii, M., Koyaguchi, T., Makinouchi, T., Mitsushio, H., and Ishida, S. 1984. Fission-track, K-Ar age determinations and palaeomagnetic measurements of Miocene volcanic rocks in the western area of Baragoi, northern Kenya: Ages of hominoids. *African Study Monographs,* suppl. 2: 57–66.

———. 1986. Geochronology of Miocene hominoids east of the Kenya Rift Valley. In *Primate Evolution,* pp. 35–45 (ed. J. G. Else and P. C. Lee). Cambridge University Press, Cambridge.

Morgan, M. E., Kingston, J. D., and Marino, B. D. 1994. Carbon isotopic evidence for the emergence of C_4 plants in the Neogene from Pakistan and Kenya. *Nature* 367: 162–65.

Nakaya, H., Pickford, M., Nakano, Y., and Ishida, H. 1984. The late Miocene large mammal fauna from the Namurungule Formation, Samburu Hills, northern Kenya. *African Study Monographs,* suppl. 2: 87–131.

Patterson, B., Behrensmeyer, A. K., and Sill, W. D. 1970. Geology and fauna of a new Pliocene locality in north-western Kenya. *Nature* 226: 918–21.

Pickford, M. 1996. Pliohyracids (Mammalia, Hyracoidea) from the upper Middle Miocene at Berg Aukas, Namibia. *Comptes Rendus de l'Académie des Sciences, Paris* 322: 501–5.

Pickford, M., and Senut, B. 1994. Palaeobiology of the Albertine Rift Valley: General conclusions and synthesis. In *Geology and Palaeobiology of the Albertine Rift Valley Uganda-Zaire, Volume II: Palaeobiology/Paléobiologie,* pp. 409–23 (ed. B. Senut and M. Pickford). CIFEG Publication Occasionnelle 1994/29. Centre International pour la Formation et les Echanges Géologiques, Orléans.

Pickford, M., Ishida, H., Nakano, Y., and Nakaya, H. 1984a. Fossiliferous localities of the Nachola-Samburu Hills area, northern Kenya. *African Study Monographs,* suppl. 2: 45–56.

Pickford, M., Nakaya, H., Ishida, H., and Nakano, Y. 1984b. The biostratigraphic analyses of the faunas of the Nachola area and Samburu Hills, Northern Kenya. *African Study Monographs,* suppl. 2: 67–72.

Pickford, M., Senut, B., and Hadoto, D. 1993. *Geology and Palaeobiology of the Albertine Rift Valley, Uganda-Zaire, Volume I: Geology.* CIFEG Publication Occasionnelle no. 24. Centre International pour la Formation et les Echanges Géologiques, Orléans.

Quade, J., Cerling, T. E., and Bowman, J. 1989. Development of Asian monsoon revealed by marked ecological shift during the latest Miocene in northern Pakistan. *Nature* 342: 163–66.

Senut, B., and Pickford, M. eds. 1994. *Geology and Palaeobiology of the Albertine Rift Valley Uganda-Zaire, Volume II: Palaeobiology/Paléobiologie.* CIFEG Publication Occasionnelle 1994/29. Centre International pour la Formation et les Echanges Géologiques, Orléans.

Sickenberg, O., and Schönfeld, M. 1975. The Chorora Formation—lower Pliocene limnical sediments in the southern Afar (Ethiopia). In *Afar Depression of Ethiopia,* pp. 277–84 (ed. A. Pilger and A. Rosler). Schweizerbart'sche Verlagsbuchhandlung, Stuttgart.

Tassy, P. 1986. *Nouveaux Elephantoidea (Mammalia) dans le Miocène du Kenya.* Cahiers de Paléontologie. Centre National de la Recherche Scientifique, Paris.

———. 1999. Miocene elephantids (Mammalia) from the Emirate of Abu Dhabi, United Arab Emirates: Palaeobiogeographic implications. Chap. 18 in *Fossil Vertebrates of Arabia,* pp. 209–33 (ed. P. J. Whybrow and A. Hill). Yale University Press, New Haven.

Thomas, H. 1979. Le rôle de barrière écologique de la ceinture saharo-arabique au Miocène: Arguments paléontologiques. *Bulletin du Muséum National d'Histoire Naturelle, Paris* 1: 127–35.

Thomas, H., Bernor, R., and Jaeger, J.-J. 1982. Origines du peuplement mammalien en Afrique du Nord durant le Miocène terminal. *Geobios* 15: 283–97.

Tiercelin, J.-J., Michaux, J., and Bandet, Y. 1979. Le Miocène supérieur du sud de la depression de l'Afar, Ethiopie: Sediments, faunes, âges isotopiques. *Bulletin de la Société Géologique de France* 7: 255–58.

Whybrow, P. J., and Hill, A. eds. 1998. *Fossil Vertebrates of Arabia.* Yale University Press, New Haven.

Oligocene and Miocene Terrestrial Vertebrates in the Southern Arabian Peninsula (Sultanate of Oman) and Their Geodynamic and Palaeogeographic Settings

Herbert Thomas, Jack Roger, Sevket Sen, Martin Pickford, Emmanuel Gheerbrant, Zaher Al-Sulaimani, and Salim Al-Busaidi

During the last four field seasons of the Franco-Omani Palaeontology Expedition (1986– 92) several Oligocene and Miocene vertebrate fossil sites were discovered in the Sultanate of Oman. Whereas the Oligocene faunas came from localities situated in the southern part of Dhofar, the Miocene faunas were collected in the northern margin of Huqf (Central Oman) near Ghaba (fig. 30.1). These discoveries have filled important gaps in the knowledge of the distribution and the evolution of terrestrial vertebrates during the Tertiary of the Arabian Peninsula. The faunal associations could, among other things, be placed into their geodynamic and palaeogeographic context thanks to the revision of the Tertiary units, carried out by the Bureau de Recherches Géologiques et Minières (BRGM) during geological mapping of Oman at a scale of 1:250 000 (Le Metour et al., 1995).

The initial discovery of Oligocene terrestrial vertebrates in the Sultanate of Oman dates from 1986–87, when fossils were found at two localities—Thaytiniti and Taqah—in the coastal strip of Dhofar, within a littoral marine unit (Shizar Member at the base of the Ashawq Formation) dated to the early Oligocene on the basis of the nummulite *Nummulites fichteli* (Thomas et al., 1989a). The fossiliferous unit, about 100 metres thick, has a siliciclastic and carbonate content and constitutes the transgressive episode of the early Oligocene depositional cycle.

The Miocene localities were found in 1991, also within littoral marine strata that crop out in the northern region of Huqf massif some 50 km southeast of the Ghaba North Oil Field (Roger et al., 1994a). Marine molluscs and foraminifera from the succession are extremely similar to those described by Powers et al. (1966) from the type locality of the Dam Formation in Saudi Arabia and by Cavelier (1975) from the same unit in Qatar, dated as Burdigalian–Langhian. In the Ghaba sector, the Miocene series assigned to the Dam Formation, only 50 metres thick, displays a mixed carbonate, siliciclastic, and evaporitic sedimentation, whose vertical arrangement comprises a complete cycle of transgression and regression.

During the Tertiary, the Arabian Peninsula was affected by two major geodynamic events (fig. 30.2) that played a significant role in sedimentatary processes and in the distribution of marine and terrestrial faunas: the beginning of the opening of the Gulf of Aden in the Rupelian and the Alpine tectonic phase that began in the Burdigalian (Le Metour et al., 1995). The succession of transgressive–regressive cycles that affect the very extensive Arabian Platform determined the distribution of the deposits and, over the course of time, displaced the shoreline of the Arabian Sea by large distances. Because of the large predominance of marine sedimentation in contrast with the poorly developed continental deposition during Paleogene–Miocene

 ISBN 0-300-07183-3

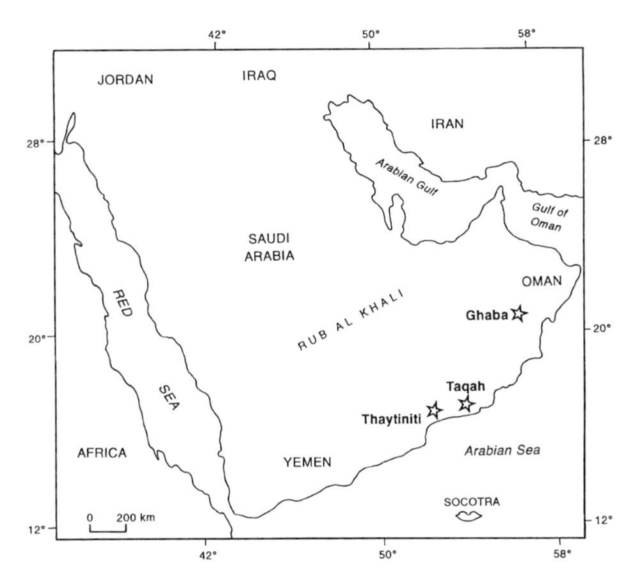

Figure 30.1. Map of the Arabian Peninsula showing fossil vertebrate localities in the Sultanate of Oman.

times, it is unsurprising that terrestrial vertebrates were discovered mainly in littoral facies close to the sea shore where the depositional environment was propitious for concentration and preservation of bones and teeth.

Paleogene Evolution

During the Paleogene, the southeastern belt of the Arabian Peninsula acted as an epeirogenic platform over which were several extensive marine transgressions. Thick accumulations of marine carbonate strata rich in foraminifera, molluscs, and algae were deposited from the end of the Thanetian up to the middle Eocene. Up to now, the oldest transgressive marine facies of coastal plains of Thanetian to Ilerdian age have yielded only marine vertebrates (sirenians), no terrestrial ones. The terminal Bartonian regression initiated a progressive retreat of the platforms. Consequently, during the Priabonian, only the southeastern margin of the plate remained submerged, the rest being subjected to erosion.

Reorganisation of the Afro-Arabian Plate, which began in the Oligocene, profoundly modified the palaeogeographic setting. Along the eastern flanks of Oman this period was characterised by an important lowering of sea level that led to severe erosion of the margins with the formation of thick slope deposits. At the same time in southern Dhofar the opening of the proto-Gulf of Aden began. This event is manifested by the opening of several grabens aligned in a step-like fashion (en echelon), characterised by strong subsidence and the accumulation within them of mixed siliciclastic and carbonate sediments. It is probable that the terminal Bartonian emergence of much of the plate facilitated communication between the eastern parts of Africa (notably Egypt) and the Dhofar region, especially for terrestrial faunas, while the opening of the grabens, partly invaded by the sea, led to the appearance of palaeoenvironments favourable for a diversified terrestrial vertebrate fauna.

Climatic modifications in the Oligocene led to greater rainfall; this favoured the development of vegetation, the extent of which may have been limited to the southern grabens, which were partly inundated by fresh to brackish water. Only the transgressive basal horizons of the graben fillings, consisting of alternating marine and continental sediments, have yielded the remains of terrestrial vertebrates. These fossils appear to be concentrated mainly at the base of transgressive marine sequences that overlie more continental facies according to fluctuations in the littoral zone. Such a model explains the narrow association between the marine and continental faunas. The model also accounts for the concentration of fossils and their fragmentary nature, caused by their immediate but localised reworking by the sea. Rapid subsidence and burial of the deposits ensured the preservation of the fossils.

Dhofar Oligocene Fossiliferous Sites

Thaytiniti occurs in the broad Ashawq Graben, which constitutes the type locality for the Shizar Member (Roger et al., 1994b). The principal fossiliferous level (Thomas et al., 1988, 1992), which is

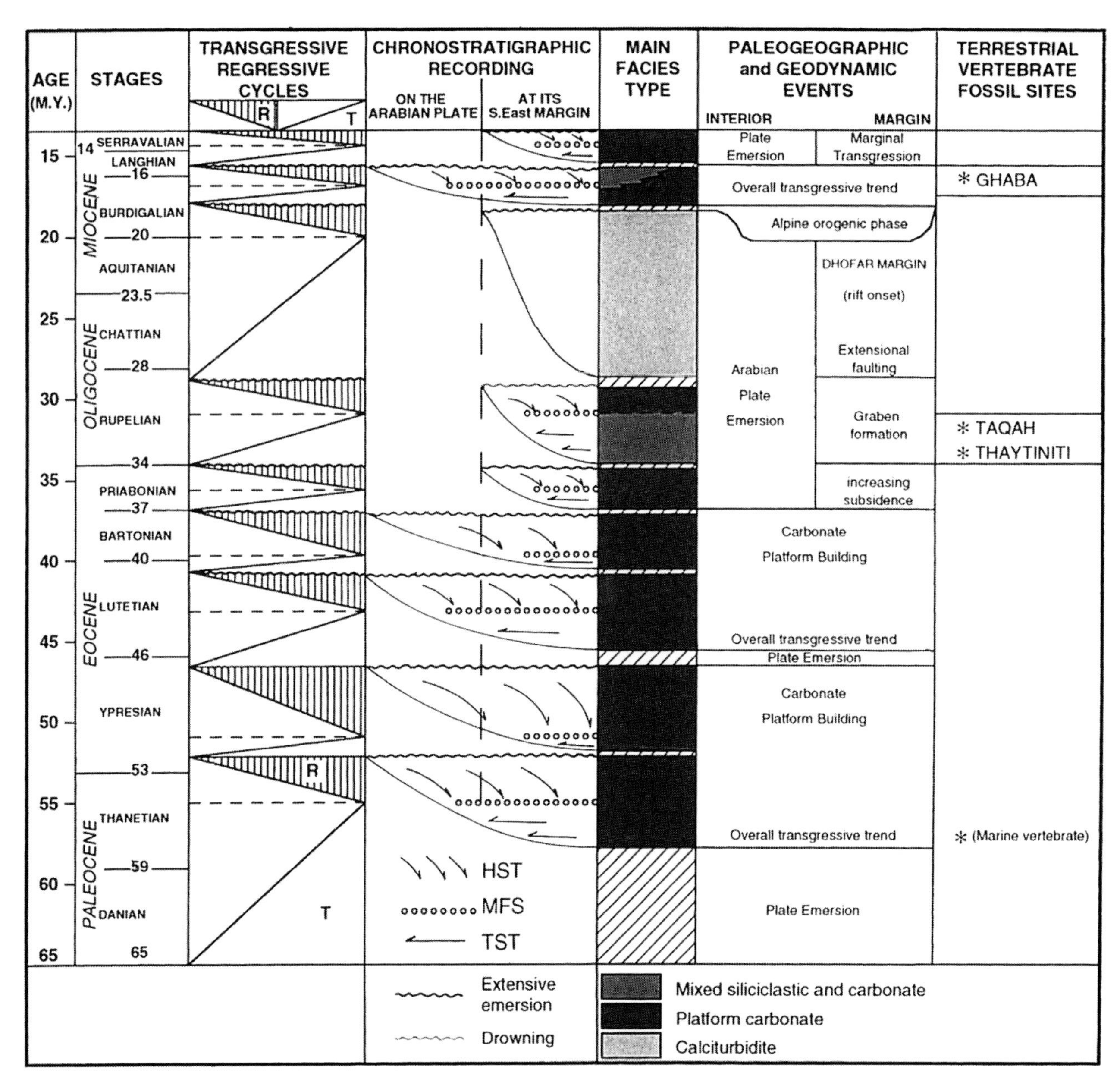

Figure 30.2. Position of fossil sites with respect to principal sedimentary cycles in central and southern Oman. HST = Highstand Systems Tract; MFS = Maximum Flooding Surface; TST = Transgressive Systems Tract.

well defined stratigraphically, could be traced laterally, despite important faults that affect the Thaytiniti area. There is a lateral extension of the fossiliferous level to the north of the site but it has yielded only sporadic concentrations of bone fragments. The locality of Taqah, which lies east of Salalah, is in the coastal plain where it backs onto Jabal Qara. The existence of large faults cutting the area into downthrown and tilted blocks greatly reduces the field investigation of the known fossiliferous level, which is limited, in the present state of our researches, to one site.

Whereas the locality of Thaytiniti occurs right at the base of the first depositional sequence of the Shizar Member (Ashawq Formation), the fossiliferous level of Taqah is intercalated within the second depositional sequence marked by siliciclastic input (fig. 30.3). Taqah is undoubtedly a slightly younger fossil locality—if we take into account the lateral correlations and the stage of evolution of the primates, rodents, and proboscideans. There the sequence that accumulated during a temporary regressive pulse of the sea corresponds to a temporary installation of a more marginal–littoral depositional environment.

The position of both localities close to the sea shore explains the particular faunal and floral associations there (Thomas et al., 1989a, 1991a; Roger et al., 1993). These associations are characterised by the juxtaposition of elements indicative of the different environments: marine (subtidal to intertidal stages)(melobesian algae, benthic foraminifera such as nummulites, corals, bivalves, gastropods, decapods, cirripeds, echinoids, asteroids, and rays, sharks and other fishes), fresh water (charophytes, lungfishes, teleostean fishes, small brevirostran crocodiles, and turtles), and terrestrial (anurans, lacertilians, snakes, tortoises, and numerous mammals) (table 30.1). The mammals, represented by more than a thousand isolated teeth, are marsupials (Crochet et al., 1992), rodents, primates, proboscideans (Thomas et al., 1989b), hyracoids (Pickford et al., 1994), embrithopods, creodonts (Crochet et al., 1990), insectivores, and bats (Sigé et al., 1994). Though poorly known, the proboscidean remains are of interest. None of the four specimens collected in the Dhofar localities matches any of the known Fayum and Algerian Paleogene taxa: *Moeritherium*, *Barytherium*, *Numidotherium*, *Palaeomastodon*, and *Phiomia*. They indicate two or three new proboscidean taxa.

The rodents from Thaytiniti, represented almost exclusively by Phiomyidae, are less diversified than those of Taqah, which comprise at least eight species belonging to four different families. In addition, numerous details of dental morphology clearly indicate the existence of more primitive rodents at Thaytiniti than at Taqah.

The first known Paleogene hyracoids from the Arabian Peninsula are represented by three or four species (Pickford et al., 1994): cf. *Saghatherium bowni*, *Thyrohyrax meyeri*, and two unidentified medium-sized brachyodont species similar to undescribed fossils from Quarry L 41 Fayum, Egypt. As for the bats, eight distinct taxa have been recorded at Taqah, comprising several new genera and/or species. The Taqah bat assemblage displays a high degree of originality in comparison with that of the Fayum, at least at the specific level.

Among the remaining elements, several hundred isolated teeth and a few jaw fragments from the Taqah deposit belong to the oldest anthropoid primates known in the Fayum, including *Oligopithecus savagei*, a new species of *Oligopithecus* (Gheerbrant and Thomas, 1992; Gheerbrant et al., 1995), several new oligopithecines, and the Propliopithecid *Moeripithecus markgrafi* (Thomas et al., 1988; Thomas et al., 1991a; Thomas and Pickford, 1992; Pickford and Thomas, 1994; Senut and Thomas, 1994). Finally, two new genera, *Omanodon minor* and *Shizarodon dhofarensis*, described as adapiform primates, have been recognised in the Taqah material. Because of the fragmentary nature of the material and of possible parallelisms, the systematic position of *Omanodon* and *Shizarodon* within the adapiformes cannot yet be definitively established (Gheerbrant and Thomas, 1992, 1993).

At Taqah, mammals are represented by about 40 species at the current stage of our study (for comparison, about 80 species have been recognised in the Qatrani Formation of the Fayum). Two groups are abundantly represented in terms of both

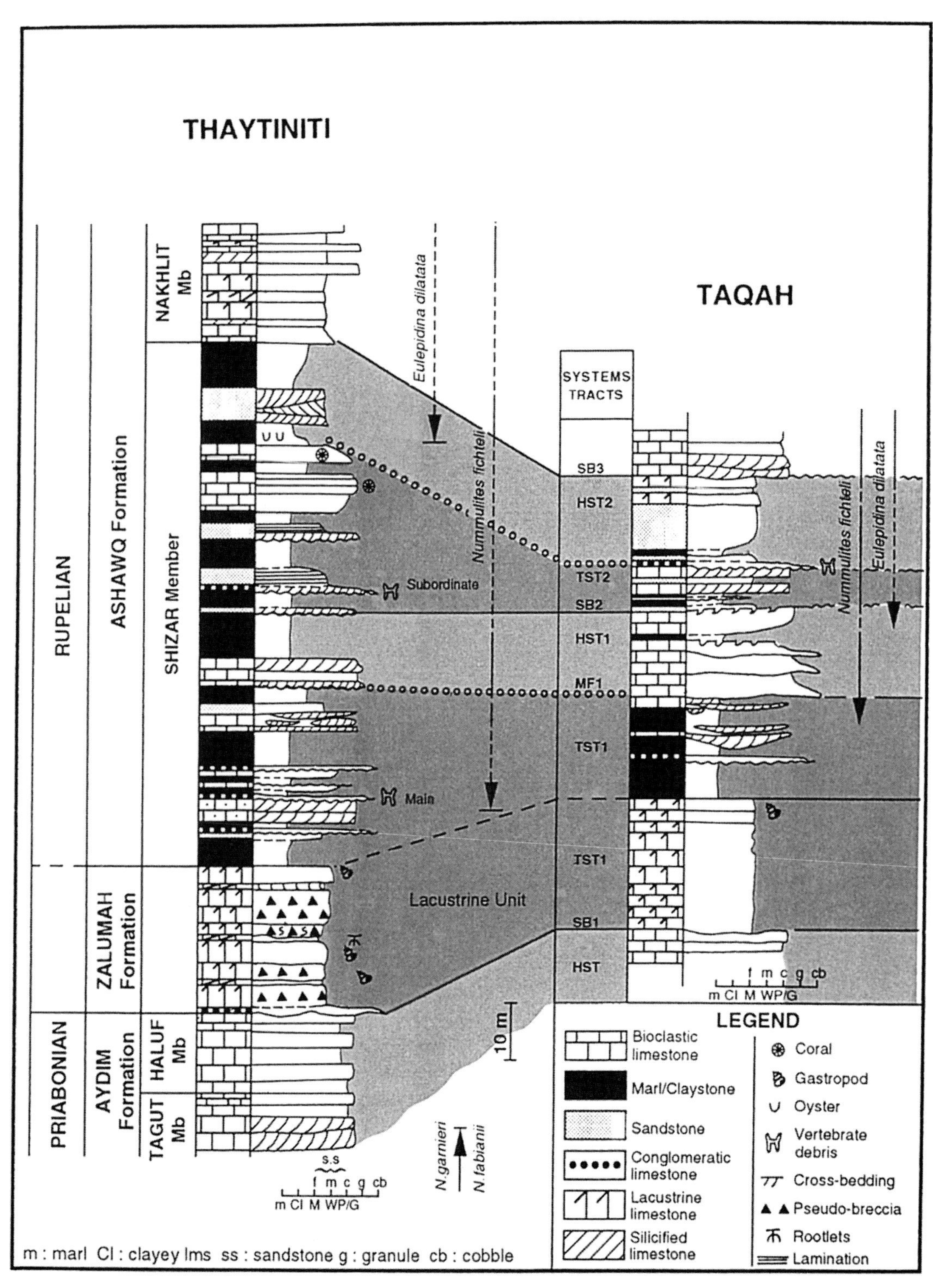

Figure 30.3. Lithostratigraphy and correlations of the Thaytiniti and Taqah sections. HST = Highstand Systems Tract; MF = Maximum Flooding; TST = Transgressive Systems Tract; SB = Sequence Boundary.

Table 30.1. Faunal list from Dhofar (Thaytiniti and Taqah)

INVERTEBRATES	THAYTINITI	TAQAH
Foraminifers		
Nummulites fichteli	●	●
Discorbidae	●	●
Halkyardia minima		●
Vaginulinopsis ?		●
Amphistegina sp.		●
Corals		
Stylophora sp.		●
Diplothelia sp.		●
Bryozoans		
Idmones sp.	●	
Tubucellaria sp.	●	
Calpensia ?	●	
Selenaria ?	●	
Echinoids		
Cidaridae indet.		●
Sismondia cf. *javana*	●	
Clypeaster sp.	●	
Spatangoidea indet.	●	
Stelleroids		
Goniasteridae	●	
Forcipulatida		●
Crinoids indet.	●	
Molluscs		
Bivalvia		
Chlamys sp.	●	
Ostreidae		●
Heterodonta		●
Gastropoda		
Scala sp.		●
Lanistes subcarinatus sp.		●
Hydrobiidae indet.	●	
cf. *Lacuna* sp.		●
Turbonilla sp.		●
Melanoides sp.		●
Pomatias sp.	●	
Potamiidae indet.	●	●
cf. *Murex* (*Chicoreus*) sp.		●
Gastropoda indet.		●
Crustaceans		
Cirripedia *Balanus* spp.	●	●
Ostracoda indet.	●	
Isopoda *Sphaeroma* sp.	●	
Decapoda		
Palaeocarpilius sp.	●	
Atergatis ?	●	
Calappa sp.	●	
Callianassa sp.		●
VERTEBRATES		
Selacians		
Hemipristis cf. *serra*	●	
Hemipristis aff. *curvatus*	●	
Negaprion sp.	●	
VERTEBRATES (continued)	THAYTINITI	TAQAH
Galeocerdo sp.	●	●
Galeocerdo aduncus	●	
Rhizoprionodon/*Sphyrna*	●	
Carcharhinus cf. *amboinensis*	●	●
Nebrius sp.	●	●
Chilosyllium sp.	●	
Rhynchobatus aff. *pristinus*	●	
Rhynobatos sp.	●	
Dasyatis spp. 1 and 2	●	●
Rhinoptera sp.	●	●
Aetobatus aff. *arcuatus*	●	
Myliobatis sp.		●
Dipnoans cf. *Protopterus*	●	●
Teleosts		
Elopiformes : Albulidae	●	
Tetraodontiformes		
Tetraodontoidea	●	
Ostracioidea	●	
Osteoglossiformes : Osteoglossidae	●	●
Characiformes : Alestinae	●	
Siluriformes	●	●
Perciformes		
Sparidae	●	●
Cichlidae	●	
Amphibians		
Anura ?		●
Reptiles		
Lacertilia		
Agamidae spp. 1 and 2		●
Family indet.		●
Serpentes		
Boidae : Erycinae		●
Colubridae indet.		●
Chelonia		
Cryptodira		
Testudinidae indet.	●	
Geochelone sp.		●
Pleurodira		
Podocnemidinae : *Stereogenys* ?	●	
Erymnochelyinae		
Erymnochelys spp. 1 and 2		●
Schweboemys sp.		●
Crocodilia		
Gavialidae : Tomistominae ?	●	
Crocodylidae spp. 1 and 2		●
Mammals		
Marsupialia		
Peradectidae		
Qatranitherium aff. *africanum*		●
Insectivora		
at least four species		●

Table 30.1. Faunal list from Dhofar (Thaytiniti and Taqah) (*continued*)

VERTEBRATES (continued)	THAYTINITI	TAQAH
Chiroptera		
Dhofarella thaleri		●
Hipposideros (*Brachipposideros*) *omani*		●
Hipposideridae gen. et sp. indet.		●
Chibanycteris herberti		●
Philisis sevketi		●
cf. *Philisis* sp.		●
Vespertilionoidea indet.		●
Microchiroptera indet. (var. spp.)		●
Primates		
Adapiformes		
Omanodon minor	cf.	●
Shizarodon dhofarensis		●
Simiiformes		
Oligopithecus rogeri		●
Catopithecus n. sp.		●
Oligopithecinae n. gen., n. sp. 1		●
Oligopithecinae n. gen., n. sp. 2		●
Oligopithecinae indet.	●	
Moeripithecus markgrafi		●
Propliopithecidae gen. et sp. indet.		●
cf. *Apidium*		●
Parapithecidae indet.		●
Primates indet. (var. spp.)		●

VERTEBRATES (continued)	THAYTINITI	TAQAH
Creodonta		
Masrasector ligabuei		●
Proboscidea		
gen. et sp. nov.	●	
cf. *Barytherioidea*	●	
Phiomia sp.		●
Hyracoidea		
cf. *Saghatherium bowni*	●	
Thyrohyrax meyeri		●
cf. *Thyrohyrax meyeri*	●	
gen. et sp. indet. 1		●
gen. et sp. indet. 2	●	
Embrithopoda		
cf. *Arsinoitherium*	●	●
Artiodactyla : Anthracotheriidae		●
Rodentia		
Phiomyidae		
Phiomys cf. *andrewsi*	●	●
Phiomys cf. *lavocati*	●	●
cf. *Metaphiomys* spp. 1 and 2	●	●
Metaphiomys cf. *schaubi*		●
Cricetidae		●
Anomaluridae		●
Family indet. 1	●	
Family indet. 2		●

specific diversity and frequency of remains at Taqah: primates and rodents. Eight rodent species are known from over 1000 isolated teeth; 300–400 specimens were found in each cubic metre of sediment; five species are represented by more than 100 teeth. In contrast, at Thaytiniti where the predominance of rodents and primates in relation to other terrestrial vertebrates is comparable, the diversity of the two groups is much lower. This fact is probably due to the predominantly marine influence at Thaytiniti, itself related to the proximity of the coast, as is especially evident from the abundance of nummulites and shark teeth. At Fayum, the hyracoids are the most diversified group (13 species); from the number of remains, it is much more abundant (Bown et al., 1982). This difference can be attributed to the fact that microvertebrates at Fayum are underrepresented.This underrepresentation is undoubtedly attributable to multiple causes: sedimentary sorting, differential preservation, and differential collecting. By contrast, in the Dhofar localities only small species with a body weight less than 5 kg are well represented. Teeth and bones of the Thaytiniti and Taqah vertebrates are mostly small or very small, a few millimetres in size; large and medium-sized mammals are uncommon, particularly at Taqah.

Considerable sedimentary sorting has thus affected the dental and bone remains of all the vertebrates at both Taqah and Thaytiniti. The original biocoenoses of these deposits are thus imperfectly represented, especially the population structures. In both localities, all elements indicate a warm environment of subtropical to tropical type.

Sedimentological criteria and the presence of a largely derived marine microfauna indicate that the Taqah horizon accumulated in a shallow coastal swamp, sporadically, perhaps even rarely flooded by the sea during tempests or high tides. The formation of calcareous crusts, the presence of an erycine snake, and perhaps of the embrithopod *Arsinoitherium* suggest that the climate was semi-arid (Thomas et al., 1991a). Nevertheless several factors indicate that the climate was strongly seasonal with a marked rainy season; this is shown by the lack of evaporites, the presence of siliciclastics, and the presence in northern Dhofar of broad permanent lakes with carbonate sedimentation (Roger et al., 1994b).

Furthermore, the vertebrate fauna from the two Dhofar localities have a clear endemic Afro-Arabian aspect, indicated in particular by the lungfishes, phiomyids, hyracoids, proboscideans, and catarrhines. The resemblances between the faunas of Taqah and Thaytiniti and those of the Fayum are manifest. The Fayum, it needs to be said, comprises the only important African Oligocene locality if one excludes Malembe (Angola), Zellah (Libya), and Gabal Bou Gobrine (Tunisia), of which the faunas are little known.

Age of the Terrestrial Vertebrates of Dhofar

Among the Dhofari mammals, at the present stage of their study, only a few seem at first inspection to be identical to those of the Fayum, although the relationships between them are manifest. The stage of evolution of the rodents and proboscideans lead us to suggest, however, that the deposits of Thaytiniti, at least, are slightly older than those of the Qatrani Formation or are more or less contemporaneous with Fayum locality L 41. Correlations with marine formations on the basis of characteristically early Oligocene nummulites *(N. fichteli),* which occurs in both of the Dhofari fossiliferous levels, combined with palaeomagnetic data, indicate that the fossil horizon at Thaytiniti is dated around 33 million years (Ma) (chron C 13n).

These nummulites, which are not present in the first 8 metres of the Shizar Member nor in the underlying lacustrine Zalumah Formation, appear for the first time in the vertebrate fossiliferous level at Thaytiniti. Unconformably underlying the Zalumah Formation, the Aydim Formation has in contrast yielded abundant and varied foraminifera, including *N. fabianii, N. garnieri,* and *N. retiatus,* an association characteristic of the Priabonian (Roger et al., 1994b).

In the section exposed at Taqah, 110 km east of Thaytiniti, the regressive sequence enclosing the vertebrate-bearing bed is intercalated between two well-exposed marine units that contain *Nummulites fichteli* and *Lepidocyclina* (*Eulepidina dilatata*): a lower marine unit constituting the basal depositional sequence of the Shizar Member; and an upper marine unit that forms the lower part of the Nakhlit Member. The presence of *N. fichteli,* a terminal form of the *N. fabianii* lineage that appears only at the Eocene–Oligocene boundary, leads us to date the two Dhofar vertebrate localities as early Oligocene (Rupelian). The absence of the foraminifer *Lepidocyclina* from the Thaytiniti fossil bed prompts us to refine the age to basal early Oligocene, since *Lepidocyclina* only appear higher in the series.

At the end of the early Oligocene, the geodynamic conditions that had prevailed until then were interrupted. On the one hand, there was an accentuation of the subsidence of the Dhofar margin. This caused the accumulation of calciturbidites, which continued until the early Miocene (Roger et al., 1994b). On the other hand, the continuous uplift of the interior of the Arabian Plate was responsible for the formation of weathering profiles (silicification) and absence of continental deposition.

Neogene Evolution

The last major marine transgression that flooded the central part of the Arabian Plate covered the Eastern Province of Saudi Arabia, the Rub' al Khali and Central Oman during the Burdigalian–Langhian (Cavelier et al., 1993; Le Metour et al., 1995). An epicontinental sea was formed with lagoonal carbonate facies (Dam Formation) bordered by a vast lacustrine belt in which sedimentation of carbonates also predominated. Deposits exhibit a clear transgressive–regressive cycle marked by the passage from intertidal to subtidal bioclastic marl and carbonate to restricted facies composed of gypsum-interbedded marl (fig. 30.4). The proximity of the emergent

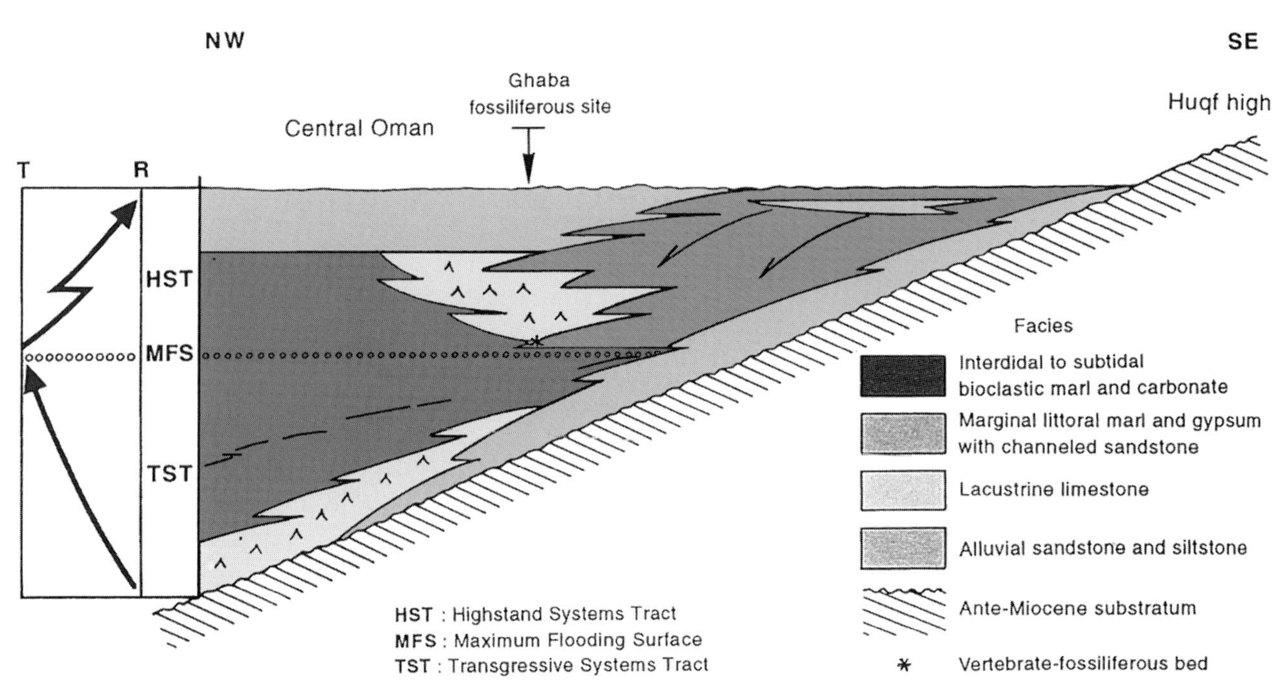

Figure 30.4. Southeastward evolution of the Miocene Dam Formation towards the Huqf high.

Huqf massif explains the intercalation, within the marine succession, of the quartzo-lithic sandstone facies of fluvial origin. In the Ghaba area this has yielded a predominantly terrestrial vertebrate fauna. Whereas numerous fossil sites are known in Saudi Arabia, in Oman only the southeast margin of this intracratonic sea, close to the Huqf high, has yielded terrestrial vertebrates (Roger et al., 1994a).

Ghaba Early–Middle Miocene Fossiliferous Site

Vertebrate remains from Ghaba show limited diversity and are very fragmentary (table 30.2). Besides freshwater fishes such as *Clarias* and *Lates* and certain marine forms (pristids and rays), they include brevirostral and longirostral crocodiles, littoral and terrestrial chelonians, and several mammals—notably two proboscideans (a bunodont mastodont and a deinothere), an anthracotheriid (*Afromeryx zelteni*), and a primitive giraffoid (*Canthumeryx* sp.).

The most striking facet of the Ghaba fauna concerns the minute size of the two proboscideans. Their dwarfing, although relative, could (if not a primitive character) be related to a period of insularity that affected the Ghaba region during the Burdigalian–Langhian for a limited time. This hypothesis is supported by palaeogeographic evidence suggesting the existence of an isolated emergent axis that comprised the Oman Mountains and their extension towards the south (regions of Wahibah and Huqf).

The terrestrial fauna of Ghaba, the African affinities of which are clear, show two taxa (*Canthumeryx* and *Afromeryx zelteni*) very common in the Jebel Zelten fauna of Libya. This enables correlation of the Ghaba sites with faunal zones PIIIa and PIIIb of Pickford (1991), the ages of which are 17.5–15.5 Ma.

Finally, from the middle Miocene until the present day, the region became fully continental, with only rare transgressions along its margins due to fluctuations in sea level. Curiously enough, the

Table 30.2. Faunal list from Ghaba, Sultanate of Oman

Selacians
Pristidae?
Rajiformes

Teleosts
Clarias sp.
Clariidae
Siluroidea indet.
Lates sp.
Perciformes indet.

Crocodilians
?*Crocodylus* sp.
gen. et sp. indet. (*Tomistoma–Euthecodon* group)

Chelonians
Pleurodira
Bothremydidae
gen. et sp. indet.
Podocnemididae
?*Schweboemys* sp.
Cryptodira
Trionychidae
aff. *Cycloderma* sp.
Carettochelyidae
aff. *Carettochelys* sp.
Testudinidae
cf. *Geochelone* sp.

Mammals
Carnivora indet.
Proboscidea
Elephantoidea
gen. et sp. indet.
Prodeinotherium sp.
Rhinocerotidae indet.
Anthracotheriidae
Afromeryx zelteni
Giraffoidea
Canthumeryx sp.
Bovidae indet.
Tragulidae indet.

thick continental Mio-Pliocene deposits that accumulated along the eastern apron of the Oman Mountains have not yet yielded any terrestrial vertebrates, even though the equivalent strata in the Emirate of Abu Dhabi (Whybrow et al., 1990; de Bruijn and Whybrow, 1994; see also Whybrow and Hill, 1999—this volume) are fossiliferous.

CONCLUSIONS

The formation and preservation of Tertiary fossil vertebrate sites in the Arabian Peninsula appear to be closely related to the geodynamic context of the time as well as to geomorphological configurations determined by major regressive/transgressive cycles. Intracratonic zones, far from marine incursions, were emergent for long periods of time. Even though deep weathering occurred in them, they persisted for several million years without showing any fossil accumulations. This is because there were no mechanisms for concentrating fossils and because organic remains would lie exposed at the surface for long periods of time, leading to their destruction. In contrast, the formation of the main continental fossil vertebrate sites in Arabia appears to be directly related to fluctuations in sea level. These sites are preferentially concentrated in belts marking the marine–continent transition, be they transgressive, such as for the Oligocene of Dhofar, or regressive such as for the Miocene of Ghaba.

Not all transgressive/regressive cycles, however, are of the same palaeontological interest. The major cycles that were responsible for flooding the plate and that led to isolation of the continental zones, such as the one that occurred during the Thanetian, Ypresian, and Lutetian/Bartonian, seem to have low potential for the formation of vertebrate fossil sites.

Periods of high sea levels marked by more marginal transgressions that spread over vast interconnected emergent areas appear to be much more favourable for fossilisation. Such a situation occurred during the Priabonian–Rupelian, when the southern margin of Dhofar was progressively downfaulted and flooded due to the tectonic activity that culminated in the opening of the Gulf of Aden and the separation of the Afro-Arabian Plate. It is likely that the onset of this new geodynamic situation, which was extensional, favoured the formation of several continental rift fragments, where the Red Sea would subsequently form (Lorenz et al., 1993). Corridors were created that permitted exchanges or migrations of faunas between Egypt and the southern parts of the Arabian Peninsula. Likewise, the early–middle Miocene transgression, which partly flooded the Arabian Plate, led to the deposition of a vast belt of continental deposits and was responsible for producing conditions suitable for the formation of fossil vertebrate sites, of which several are known in Saudi Arabia and Oman.

References

Bown, T. M., Kraus, M. J., Wing, S. L., Fleagle, J. G., Tiffney, B. H., Simons, E. L., and Vondra, C. F. 1982. The Fayum primate forest revisited. *Journal of Human Evolution* 11: 603–32.

Bruijn, H. de, and Whybrow, P. J. 1994. A Late Miocene rodent fauna from the Baynunah Formation, Emirate of Abu Dhabi; United Arab Emirates. *Proceedings Koninklijke Nederlandse Akademie van Wetenschappen* 97: 407–22.

Cavelier, C. 1975. Le Tertiaire du Qatar en affleurement. In *Lexique stratigraphique internationale,* pp. 89–120 (ed. W. Sugden and A. J. Standring). Centre National de la Recherche Scientifique, Paris.

Cavelier, C., Butterlin, J., Clermonte, J., Colchen, M., Guennoc, P., Guiraud, R. , Lorenz, C., Andreieff, P., Bellion, Y., Poisson, A., Benkhelil, J., Montenat, C., Platel, J. P., and Roger, J. 1993. Late Burdigalian. In *Atlas Tethys Palaeoenvironmental Maps* (ed. J. Dercourt, L. E. Ricou, and B. Vrielynck). BEICIP-FRANLAB, Rueil-Malmaison.

Crochet, J. Y., Thomas, H., Roger, J., Sen, S., and Al-Sulaimani, Z. 1990. Première découverte d'un créodonte dans la péninsule Arabique : *Masrasector ligabuei* nov. sp. (Oligocène inférieur de Taqah, Formation d'Ashawq, Sultanat d'Oman). *Compte Rendu de l'Académie des Sciences, Paris* 311: 1455–460.

Crochet, J. Y., Thomas, H., Sen, S., Roger, J., Gheerbrant, E., and Al-Sulaimani, Z. 1992. Découverte d'un péradectidé (Marsupialia) dans l'Oligocène inférieur du Sultanat d'Oman : Nouvelles données sur la paléobiogéographie des marsupiaux de la plaque arabo-africaine. *Compte Rendu de l'Académie des Sciences, Paris* 314: 539–45.

Gheerbrant, E., and Thomas, H. 1992. The two first possible new adapids from the Arabian Peninsula (Taqah, Early Oligocene of Sultanate of Oman). Paper presented at the 14th Congress of the International Society of Primatology, Strasbourg, 16–21 August 1992. Abstracts, p. 349.

Gheerbrant, E., Thomas, H., Roger, J., Sen, S., and Al-Sulaimani, Z. 1993. Deux nouveaux primates dans l'Oligocène inférieur de Taqah (Sultanat d'Oman): Premiers Adapiformes (?Anchomomyini) de la péninsule Arabique? *Palaeovertebrata* 22: 141–96.

Gheerbrant, E., Thomas, H., Sen, S., and Al-Sulaimani, Z. 1995. Nouveau primate Oligopithecinae (Simiiformes) de l'Oligocène inférieur de Taqah, Sultanat d'Oman. *Compte Rendu de l'Académie des Sciences, Paris* 321: 425–32.

Le Metour, J., Bechennec, F., Berthiaux, A., Chevrel, S., Platel, J. P., Roger, J., and Wyns, R. 1995. *Geology and Mineral Wealth of the Sultanate of Oman.* Oman Ministry of Petroleum and Minerals Geological Document, Muscat.

Lorenz, C., Butterlin, J., Cavelier, C., Clermonte, J., Colchen, M., Dercourt, J., Guiraud, R., Montenat, C., Poisson, A., Ricou, L. E., and Sanulescu, M. 1993. Late Rupelian. In *Atlas Tethys Palaeoenvironmental Maps. Explanatory Notes,* pp. 211–23 (ed. J. Dercourt, L. E. Ricou, and B. Vrielynck). Gauthier-Villars, Paris.

Pickford, M. 1991. Biostratigraphic correlation of the Middle Miocene mammals locality of Jabal Zaltan,

Libya. In *The Geology of Libya,* vol. 4, pp. 1483–90 (ed. M. J. Salem, O. S. Hammuda, and B. A. Eliagoubi). Elsevier, Amsterdam.

Pickford, M., and Thomas, H. 1994. Sexual dimorphism in *Moeripithecus markgrafi* from the early Oligocene of Taqah, Oman. In *Current Primatology,* pp. 261–64 (ed. B. Thierry, J. R. Anderson, J. J. Roeder, and N. Herrenschmidt). Université Louis Pasteur, Strasbourg.

Pickford, M., Thomas, H., Sen, S., Roger, J., Gheerbrant, E., and Al-Sulaimani, Z. 1994. Early Oligocene Hyracoidea (Mammalia) from Thaytiniti and Taqah, Dhofar Province, Sultanate of Oman. *Compte Rendu de l'Académie des Sciences, Paris* 318: 1395–400.

Powers, R. W., Ramirez, L. F., Redmod, D., and Berg, E. L. 1966. Sedimentary geology of Saudi Arabia. *United States Geological Survey Professional Paper* 560D: 1–146.

Roger, J., Pickford, M., Thomas, H., Lapparent de Broin, F. de, Tassy, P., Van Neer, W. , Bourdillon-de-Grissac, C., and Al-Busaidi, S. 1994a. Découverte de vertébrés fossiles dans le Miocène de la région du Huqf au Sultanat d'Oman. *Annales de Paléontologie, Paris* 80: 253–73.

Roger, J., Platel, J. P., Bourdillon de Grissac, C., and Cavelier, C. 1994b. *Geology of Dhofar (Sultanate of Oman).* Oman Ministry of Petroleum and Minerals Geological Document, Muscat.

Roger, J., Sen, S., Thomas, H., Cavelier, C., and Al-Sulaimani, Z. 1993. Stratigraphic, palaeomagnetic and palaeoenvironmental study of the Early Oligocene vertebrate locality of Taqah (Dhofar, Sultanate of Oman). *Newsletters on Stratigraphy* 28: 93–119.

Senut, B., and Thomas, H. 1994. First discoveries of anthropoid postcranial remains from Taqah (Early Oligocene, Sultanate of Oman). In *Current Primatology,* pp. 255–60 (ed. B. Thierry, J. R. Anderson, J. J. Roeder, and N. Herrenschmidt). Université Louis Pasteur, Strasbourg.

Sigé, B., Thomas, H., Sen, S., Gheerbrant, E., Roger, J., and Al-Sulaimani, Z. 1994. Les chiroptères de Taqah (Oligocène inférieur, Sultanat d'Oman): Premier inventaire systématique. *Münchner Geowissenschaftliche Abhandlungen* 26: 35–48.

Thomas, H., and Pickford, M. 1992. New discoveries of *Moeripithecus markgrafi* (Propliopithecidae, Primates) from Taqah (Early Oligocene, Sultanate of Oman). Paper presented at the 14th Congress of the International Society of Primatology, Strasbourg, 16–21 August, 1992. Abstracts, p. 257.

Thomas, H., Roger, J., Sen, S., and Al-Sulaimani, Z. 1988. Découverte des plus anciens "anthropoïdes" du continent arabo-africain et d'un primate tarsiiforme dans l'Oligocène du Sultanat d'Oman. *Compte Rendu de l'Académie des Sciences, Paris* 306: 823–29.

Thomas, H., Roger, J., Sen, S., and Al-Sulaimani, Z. 1992. Early Oligocene vertebrates from Dhofar (Sultanate of Oman). In *Geology of the Arab World,* pp. 283–93 (ed. Ali Sadek). Proceedings of the First International Conference on Geology of the Arab World, Cairo University, January 1992.

Thomas, H., Roger, J., Sen, S., Bourdillon-de-Grissac, C., and Al-Sulaimani, Z. 1989a. Découverte de vertébrés fossiles dans l'Oligocène inférieur du Dhofar (Sultanat d'Oman). *Geobios* 22: 101–20.

Thomas, H., Roger, J., Sen, S., Dejax, J., Schuler, M., Al-Sulaimani, Z., Bourdillon-de-Grissac, C., Breton, G., Broin, F. de, Camion, G., Cappetta, H., Carriol, R. P., Cavelier, C., Chaix, C., Crochet, J. Y., Farjanel, G., Gayet, M., Gheerbrant, E., Lauriat-Rage, A., Noël, D., Pickford, M., Poignant, A. F., Rage, J. C., Roman, J., Rouchy, J. M., Secrétan, S., Sigé, B., Tassy, P., and Wenz, S. 1991a. Essai de reconstruction des milieux de sédimentation et de vie des Primates anthropoïdes de l'Oligocène de Taqah (Dhofar, Sultanat d'Oman). *Bulletin de la Société Géologique de France, Paris* 162: 713–24.

Thomas, H., Sen, S., Roger, J., and Al-Sulaimani, Z. 1991b. The discovery of *Moeripithecus markgrafi*

Schlosser (Propliopithecidae, Anthropoidea, Primates) in the Ashawq Formation (Early Oligocene of Dhofar Province, Sultanate of Oman). *Journal of Human Evolution* 20: 33–49.

Thomas, H., Tassy, P., and Sen, S. 1989b. Paleogene proboscidean remains from the Southern Dhofar (Sultanate of Oman). Paper presented at the Fifth International Theriological Congress, Rome, 22–29 August, 1989, sect. "Evolution and paleoecology of Proboscidea". Abstracts, Papers, and Posters I: 166.

Whybrow, P. J., Hill, A., Yasin al-Tikriti, W., and Hailwood, E. A. 1990. Late Miocene primate fauna, flora and initial palaeomagnetic data from the Emirate of Abu Dhabi, United Arab Emirates. *Journal of Human Evolution* 19: 583–88.

Whybrow, P. J., and Hill, A. eds. 1999. *Fossil Vertebrates of Arabia*. Yale University Press, New Haven.

Geology, Fruits, Seeds, and Vertebrates (?Sirenia) from the Kaninah Formation (Middle Eocene), Republic of Yemen

31

MUSTAFA LATIF AS-SARURI, PETER J. WHYBROW, AND MARGARET E. COLLINSON

A collaborative research project between the Ministry of Oil and Mineral Resources, Mineral Exploration Board, Republic of Yemen, The Natural History Museum, London, and the American Museum of Natural History, New York, was initiated during 1991. The objective of this collaboration was to locate and collect terrestrial vertebrates from the continental Tertiary of the Republic of Yemen. At that time, only amphibian fossils were found in the northern part of the Republic of Yemen, from the Miocene inter-Trap lake deposits. This work resulted, however, in the discovery of terrestrial fruits and seeds of middle Eocene age in the southern part of the Republic. These were collected from the Kaninah Formation, the lateral lithological equivalent of the Habshiya Formation that, in most of Yemen, is usually considered to be marine (Beydoun, 1964). This discovery is the first record of Paleogene plants from the Arabian Peninsula.

The only previous record of fossil fruits from Arabia is represented by a new genus (*Midravalva arabica* Collinson, 1982) of the Potamogetoneae (pondweeds and ditch grasses) described from the early middle Miocene of Saudi Arabia (Collinson, 1982) and found in association with terrestrial vertebrates (Whybrow et al., 1982).

Subsequent work in 1993 resulted in the discovery of fragmented, possibly marine mammal bones also from the Kaninah Formation. Mammalian fossils from the Eocene of the Arabian Peninsula were previously unknown. Other records of terrestrial vertebrates from Arabia are from the Paleocene of the Red Sea coast, Umm Himar Formation, Saudi Arabia (Madden et al., 1978); the Oligocene of Oman (Thomas et al., 1988; Thomas et al., 1999—Chapter 30); the middle Miocene of eastern Saudi Arabia (Thomas et al., 1978; Whybrow et al., 1982); and the late Miocene of the Emirate of Abu Dhabi (Madden et al., 1982; Whybrow 1989; Whybrow et al., 1990; Whybrow and Hill, 1999—this volume). Further work in the Republic of Yemen had to be postponed because of events in the Republic.

Previous Geological Work

Beydoun (1964, 1966) and Beydoun and Greenwood (1968) described the lithology of the type section of the Habshiyah Formation at Jabal Habshiyah located in the Al Jiza' Depression of the Al Mahrah region (formerly part of the "Eastern Aden Protectorate"). Beydoun (1966: H 36) records 224 metres of marine, papery shale, and chalky limestones at the type section. The identified microfauna comprises *Nummulites gizehensis* (Forskel), *Alveolina elliptica* var. *nuttali* Davies; the macrofauna includes *Echinolampas ovalis* B. de St. Vincent, gastropods, and molluscs. The fossils indicate a Lutetian, middle Eocene, age for the Habshiyah

 ISBN 0-300-07183-3

Formation. In the Wadi Hajr Trough, close to the former Western Aden Protectorate border (now Hadramaut District), however, Beydoun (1966) states that the lithology changes to sand and sandy limestone.

Schüppel and Wienholz (1990) studied the Tertiary of the Ataq-Balhaf and Hajr Troughs in the Habban–Al Mukalla region. They did not recognise the regional disconformity between the middle Eocene Habshiyah Formation and the overlying Oligocene–Miocene Shihr Group (Bott et al., 1992). They suggested that the nearshore, continental sediments at the top of the Habshiyah Formation were of late Eocene and Oligocene age and proposed two new units, the Hamarah and Rimah Formations, for that part of the sequence they believed to be late Eocene. They also described part of the younger Shihr Group (marine limestones, conglomerates, gypsum, clay, and sandstones) as forming part of these new formations. They did not, however, offer any palaeontological evidence to justify the age of these new formations, especially that part they considered to be late Eocene.

The Habshiyah Formation

In the northern part of the Republic of Yemen, the Yemen Volcanics (formerly the Aden Trap Series) form most of the Tertiary sequence. The Habshiyah Formation (middle Eocene), however, occurs only in the southern area of the Republic of Yemen, and can be correlated with the Dammam Formation in Saudi Arabia (Powers et al., 1966). The Habshiyah Formation is found in four depressions, which, from the west, are named the Ataq-Balhaf Trough, Hajr Trough, Jiza' Depression, and the Qamar Trough. The maximum thickness of sediments developed in these troughs is about 6 km and the age range is from middle Jurassic to middle Eocene (Jungwirth and As-Saruri, 1990). The Habshiyah Formation becomes progressively more marine to the east owing to the Lutetian transgression from east to west, and conformably overlies evaporites of the early Eocene Rus Formation. The Habshiyah Formation marks the final depositional stage of the Hadramaut group in the Republic of Yemen.

New studies of the Habshiyah Formation by As-Saruri and Langbein (1994, 1995) show that it can be divided into three sequences according to lateral lithological change from east to west. From the comments above and on the basis of the rules of stratigraphic nomenclature (North American Stratigraphic Code, Articles 23, 25, 26: 1983), the Middle Eocene is formally divided into three formations (As-Saruri, in press)—see figure 31.1.

The Habshiyah Formation is mainly developed in the Jiza'–Qamar Basin and in the Mukalla–Sayhut Basin and includes a full marine facies consisting of alternating papery shale and chalky limestone. The thickness of the Habshiya Formation in the area ranges between 50 and 220 metres. At Jabal Al Furt the maximum thickness reached is 310 metres. The Kaninah Formation, developed in the Hajr sector of the Sab'atayn Basin, is well developed at Kaninah itself, where it is 165 metres thick, and at Jizwal, where about 80 metres are exposed (fig. 31.2). This facies is transitional and includes a transitional sand carbonate facies intertonguing between terrestrial and marine carbonates. The Mayfa'ah Formation, developed at Mayfa'ah in Al Qurayn in the Balhalf Basin, where the type section is located, is a continental terrestrial facies about 145 metres thick. The transitional beds are developed as a hypersaline marine facies about 15–20 metres thick and they conformably overlie the gypsum of the Lower Eocene Rus Formation.

Lithology and Microfacies of the Kaninah Formation

The Kaninah Formation is developed in the Hajr sector of the Sab'atayn Basin with the measured stratotype section located at Kaninah itself and at Jizwal. The succession of the Kaninah Formation is characteristic of a transition facies between the fully marine Habshiyah Formation in the east and the continental facies of the Mayfa'ah Formation in the west. The section conformably overlies the gypsum of the Rus Formation (early Eocene) with transitional beds.

The succession begins with carbonates interlayered with papery to marly shales; this reflects a

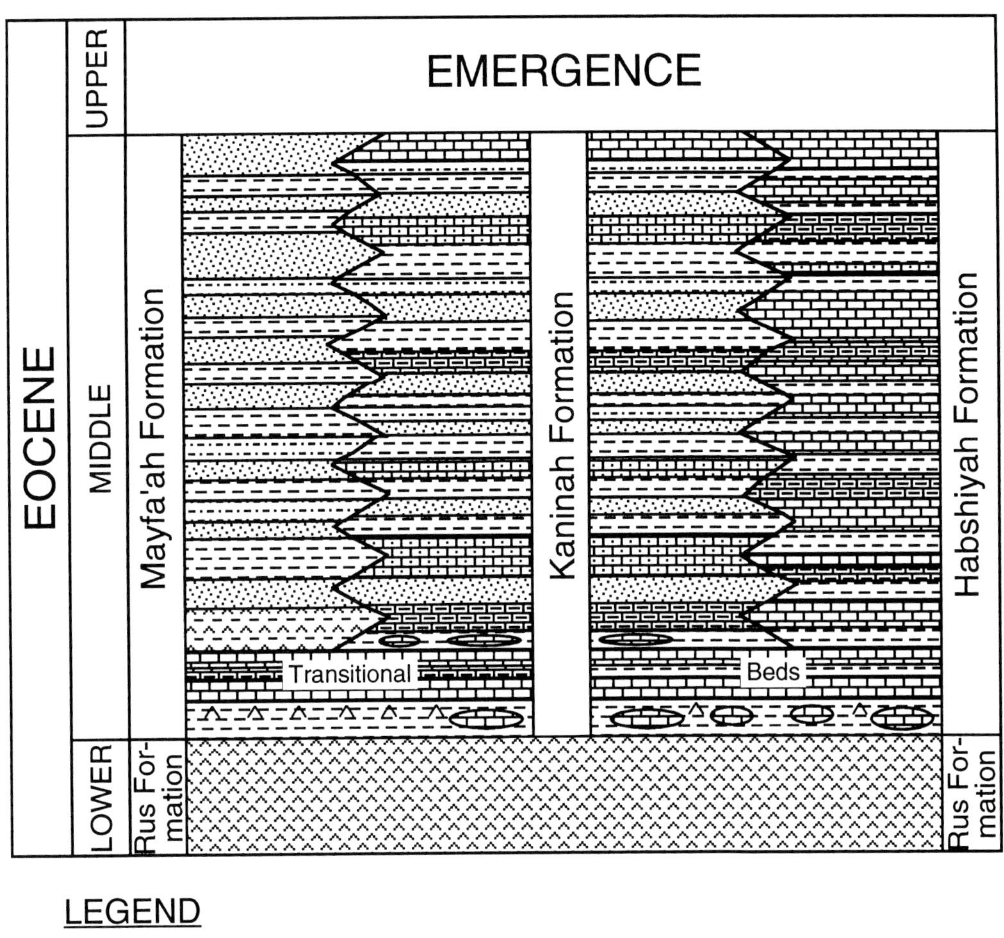

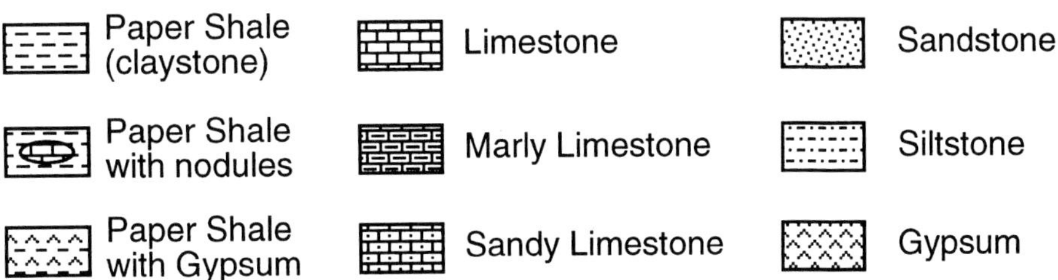

Figure 31.1. Simplified lithostratigraphic subdivision of the Middle Eocene of the Republic of Yemen.

near-shore environment. The middle part of the succession is characterised by sandy carbonates, siliciclastics become more widespread and the papery shale contains more kaolinitic clay. Towards the top of the succession the sandstone, often containing silicified wood, is kaolinitic and has small- to medium-scale cross-bedding. The upper part is also intercalated with clays, limestones, and sandy to marly limestones. Bands of secondary gypsum are often found in the clay horizons. At the top of some sandstone horizons, ferricretes occur with ferruginous casts and moulds of plants and molluscs (fig. 31.3).

Analysis of the Kaninah Formation microfacies displays many differences from its base to its top. At the base, the whole unit is a microfossil packstone

Figure 31.2. Map of the geology around Kaninah–Jizwal and a regional map to show the position of the studied area.

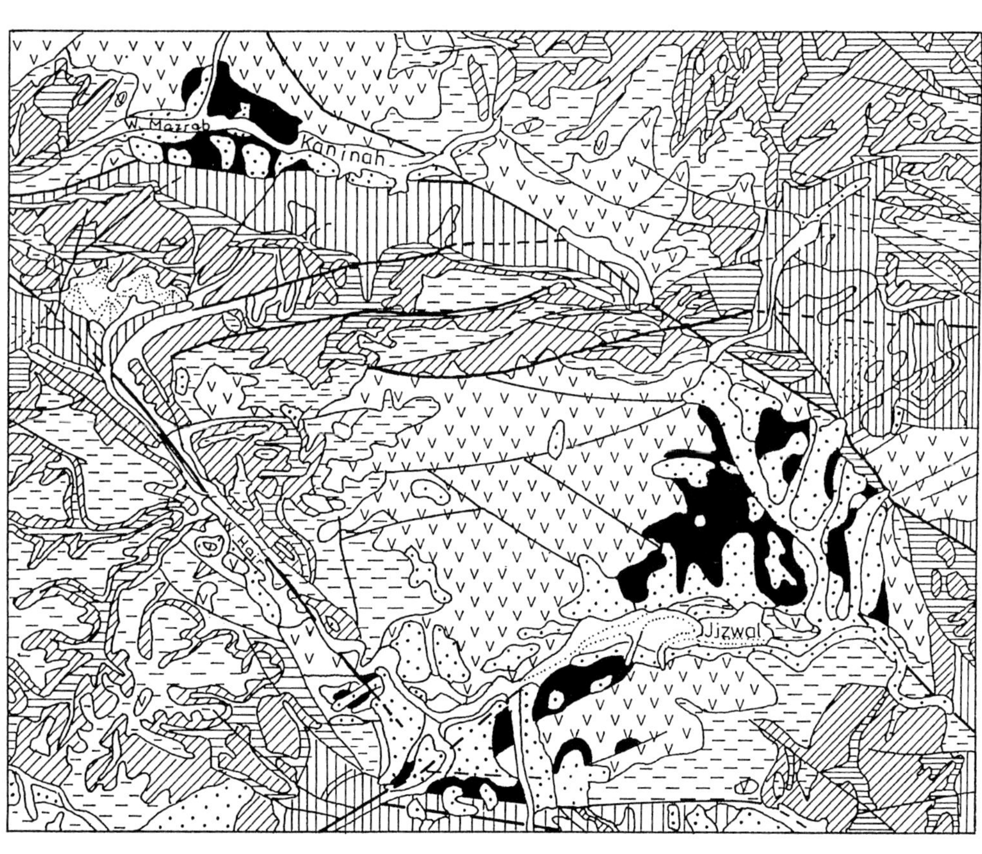

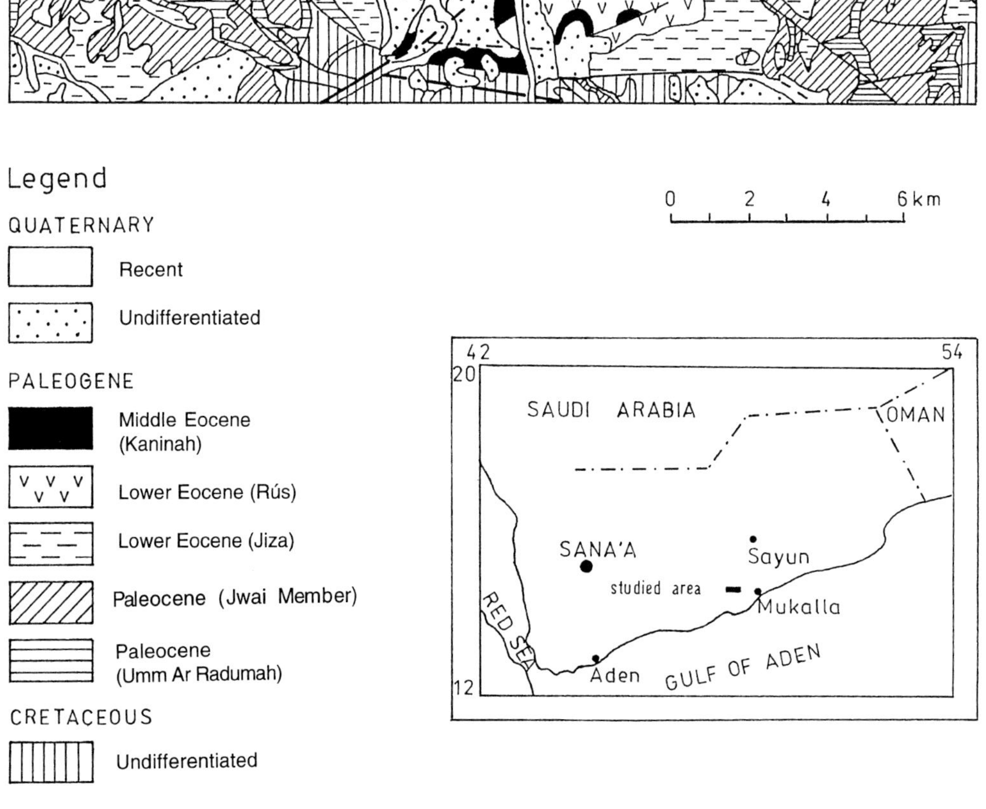

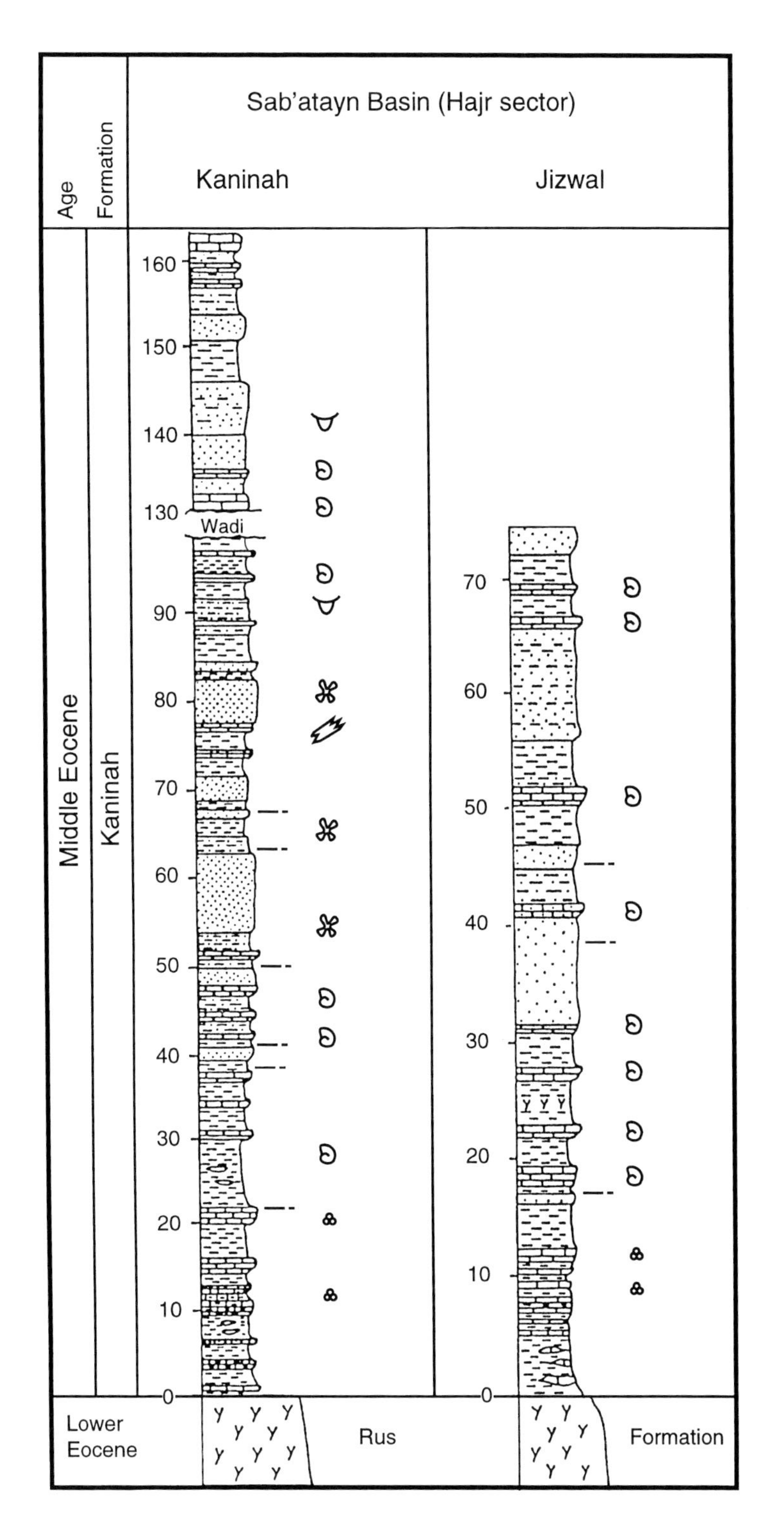

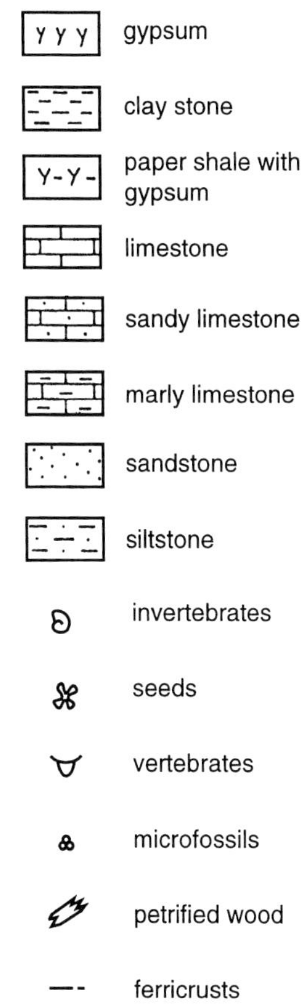

Figure 31.3. Lithological section of the Kaninah Formation. The fruits and seeds figured come from the horizon marked at 80 metres at Kaninah.

containing some foraminifera, gastropods, and algae. This unit is also strongly dolomitised, micritised, and sulphatised. In the middle part of the sequence the bioclastic microfacies become more sandy, usually with brachiopods that are broken through compaction and with stylolites. As well as brachiopods, there are some gastropods, echinoids, and other phosphatised invertebrate fragments. In the upper part of the formation there is an increase in dolomitisation and shell fragments, algae, and quartz sands. The Kaninah Formation is also characterised by the high content of siliciclastics.

Palaeobiology of the Kaninah Formation

Fauna

The invertebrates in the collection have been provisionally identified as bivalve molluscs (*?Caestocorbula, ?Raetomya,* ?lucinoid), a rhynchonellid brachiopod, and a gastropod (*?Tympanotonus*).

Vertebrates consisting of a fragmented vertebra, ribs, and other indeterminate bones of a mammal were collected in a 2 m^2 area, approximately 25 metres stratigraphically above the unit with fruits and seeds (see below). Because of their fragmented condition, contemporary weathering, and the fact that secondary haematite had caused further fragmentation, the bones are difficult to assign to any mammalian order with certainty. The dense bone structure of some rib fragments is suggestive of either Cetacea or Sirenia ribs, probably the latter.

Flora: Wood and Twigs

All of the wood and twig fragments currently available are poorly preserved as iron-rich replacements and lack detailed anatomical structure. It is not even possible to determine if there are growth rings present. Future collecting, however, might locate better preserved material. This would be valuable both for potential palaeoclimate studies and for potential recognition of additional taxa. The reports of silicified wood in the Kaninah Formation (petrified wood, fig. 31.3) offer the best potential for anatomical preservation. As only small axes are present with the fruits and seeds it is impossible to conclude that trees were represented in the palaeoflora but certainly there were woody plants, at least the size of shrubs.

The woody fragments also indicate something of the taphonomy of the flora by showing that it is not merely sorted fruits and seeds. The specimens are not strongly rounded at their broken ends so do not indicate prolonged transport in abrasive conditions. The accumulation may be the result of storm deposition. The presence of twigs of various sizes alongside fruits and seeds, however, is typical of the London Clay Flora of England, which was deposited tens of kilometres offshore, having probably been rafted out to sea (Collinson, 1983). Certainly, the presence of the fossil plants alone cannot be taken to indicate a continental origin for the enclosing sediments.

Fruits and Seeds

One distinctive type of fruit and one determinable type of seed were collected. Preliminary work indicates that the fruit belongs to the ?Nymphaeaceae s.l., the water lilies. Fossil fruits of this family are extremely rare, most of the record being based on seeds (Collinson, 1980) or leaves (Collinson et al., 1993). A fruiting example is known from the middle Eocene of Messel (Collinson, 1988; Schaal and Ziegler, 1992; Collinson et al., work in progress) but this is utterly distinct from the Yemen specimens. The seeds belong to the family Anonaceae whose modern representatives produce edible fruits such as the custard apple. Modern Anonaceae chiefly grow in tropical lowland evergreen forests. The family contains about 120–130 genera, including trees, shrubs, and lianas (Collinson, 1983; Tiffney and McClammer, 1988; Manchester, 1994). Further potential palaeobotanical value of the sequence is indicated by the additional presence of one indeterminable internal seed or endocarp cast and a fragment of an infructescence receptacle very like those of *Epipremnum* (Araceae) from the Fayum, Egypt (Bown et al., 1982).

Anonaceae

The Anonaceae seeds can be assigned to the form genus *Anonaspermum* Ball, 1931 emended by Reid and Chandler, 1933 (see Collinson, 1983). This genus is used for fossil Anonaceae seeds preserved as casts or replacements of the internal storage tissue or endosperm. The endosperm is formed with characteristic ridges, grooves, or punctations (Collinson, 1983) and these have been used to distinguish different fossil species. As reported by Collinson (1983: 73–74), however, "similar endosperm patterns may occur in different living genera and different endosperm patterns may occur within the same living genus". Thus it is impossible to identify the closest living genus or species to which *Anonaspermum* fossils belong.

Anonaspermum fossils were reviewed by Collinson (1983: early Eocene, London Clay flora, England) and Tiffney and McClammer (1988: world wide). Subsequent reports include those of Tiffney et al. (1994: Miocene, Ethiopia) and Manchester (1994: middle Eocene, Clarno Formation, Oregon, USA). Outside Europe and western North America specimens have been described from the Maastrichtian of Nigeria (Chesters, 1955), the Eocene of Egypt (Chandler, 1954), the Miocene of Kenya (Chesters, 1957), the Paleocene of Pakistan (Tiffney and McClammer, 1988), and noted but not described from the Oligocene of the Fayum, Egypt (see Bown et al., 1982; Tiffney et al., 1994).

The Yemen specimens seem to belong to a single overall form although the endosperm casts (10 specimens from Kaninah) vary in size (16–22 mm long and 14–18 mm wide; figs 31.4e–h), quality of preservation, and degree to which the cast is free of adherent matrix or portions of seed coat cast. All are characterised by a strongly bilateral form, ovoid in face view, and lenticular (not four-lobed) in end view, encircled by an equatorial fibre strand (fig. 31.4c) representing the conducting strand or raphe.

Figure 31.4. *c, e–h,* Seeds of *Anonaspermum,* all × 0.9: *c,* lateral view to show encircling raphe and cast of seed coat overlying ridged endosperm; *e, f,* opposite faces of incomplete seed with fragment of seed coat; *g, h,* opposite faces of another complete seed. *a, b, d, i–l,* Fruits of ?Nymphaeaceae all natural size: *a, b,* one fruit; *i–l,* another fruit; *d,* fruit basal fragment; *b, d, j,* basal views; *a, i,* apical views; *k, l,* lateral views. Veins of fruit wall are visible in *a;* equatorial ridge is visible in *k;* small round seeds and depressions where they have been lost (?abraded) are seen clearly in *b* and *l* with carpel walls seen as ridges separating two rows of seeds in *b.*

The endosperm is formed of narrow thin plate-like ridges of varying lengths, some of which originate near (or rarely at) the centre of the face, others part way towards the margin, and others very near the margin. Some also branch once (rarely twice) at various positions across the seed. The ridges are not continuous right across the seed and the central area is formed of separate tubercles, short broken nodular ridges, or rarely a slit-like cavity. These characteristics distinguish the Yemen specimens from those described by Tiffney et al. (1994) from Ethiopia, by Tiffney and McClammer (1988) from Pakistan, and by Chesters (1955) and Manchester (1994); these either have ridges continuous across the seeds or, if discontinuous, have very wide ridges or ridges that are nodular or contorted and not formed of thin smooth plates and/or have a partial reticulum formed from anastomosing ridges in the centre of the seed. These characters also eliminate most species described by Chesters (1957). *Anonaspermum radiatum* Chesters has a somewhat similar form to the Yemen specimens but the ridges are much wider and they radiate from the seed centre, whereas, in the Yemen specimens, the ridges are parallel, at least in the mid-area of the seed. If the key to species of *Anonaspermum* from the London Clay flora is followed in Collinson (1983: 74–75), eliminating successively forms with only ridged endosperm, forms greater than one and a half times longer than broad, forms with partly punctate surface, and forms with ridges that anastomose to produce pits, this leaves *A. complanatum* Reid and Chandler and *A. subcompressum* Reid and Chandler as similar to the Yemen specimens. Both of these species, however, are much smaller (maximum 8 mm long) than the Yemen fossils and the ridges are nodular or irregular. *Anonaspermum aegypticum* Chandler, 1954 (Dano-Montian, Kosseir, Red Sea, Egypt) is based on a single specimen of less than half of a seed. The critical central area is obscured by an adherent seed-coat cast, but the pattern of ridges is very like that on the Yemen specimens.

In summary, the Yemen specimens cannot be included in any described species although they are similar to the inadequately represented *A. aegypticum*. In view of the presence of the distinctive ?Nymphaeaceae fruit in the Fayum flora (B. H. Tiffney, personal communication; Tiffney, Wheeler, and Wing, in preparation), however, the two unpublished Fayum *Anonaspermum* species (Bown et al., 1982; Tiffney et al., in preparation) must also be considered. Comments provided by Tiffney (personal communication, 1996) indicate that one of the Fayum species is very similar to the Yemen specimens and to *A. aegypticum*. In the Republic of Yemen the same *Anonaspermum* species as found at Kaninah has also been found at Al Qurayn (five specimens).

?Nymphaeaceae Fruit

Two large fruits (fig. 31.4a,b,i–l), one central fruit fragment, and one basal portion of the same form of outer fruit wall (fig. 31.4d) have been collected. The complete fruits are about 3.5 cm in diameter and are composed of multiple (about 20–25) fused carpels, each containing two rows of seeds. At present, their family status is uncertain but they appear very similar to *Thiebaudia rayanensis* Chandler described by Chandler (1954) from the Eocene (?Lutetian) of Wadi Rayan, Fayum, Western Desert, Egypt (Chandler, 1954: 149 base). *Thiebaudia* was originally assigned to the Flacourtiaceae but Chesters considered that it belonged in Nymphaeaceae and Chandler later agreed (K. I. M. Chesters, personal communication, 1979; M. E. J. Chandler's written notes on her personal copy of her 1954 paper passed to M.E.C. after she died). The species is represented by a single specimen housed in The Natural History Museum, Department of Palaeontology collections V.31120. Collinson (1980) was unable to determine the genus to which this specimen belonged. Chandler (1954: 183) drew attention to similarities between *Thiebaudia* and *Nymphaeopsis* Kräusel, 1939 based on fruits from the Oligocene of the Mokattam plateau near Cairo, Egypt.

The Yemen specimens are also similar to a single unpublished specimen represented in the flora of the Fayum, Egypt (Bown et al., 1982), which is currently being studied by Tiffney (Tiffney et al., in preparation). Sketches of this specimen were kindly provided to M.E.C. by Bruce Tiffney. It is clear that the geographic proximity (three in Egypt, two in Yemen) of these specimens of such a distinctive fruit form, which is not recorded in other major fruit and seed floras, is of considerable biogeographic and floristic interest. Studies of the seeds (Collinson, 1980) will be critical to confirm determination as Nymphaeaceae and whether or not the fruits all belong to a single species. The limited number of specimens means that studies must be carefully undertaken and it will be necessary to try to bring all the specimens into one place for a final thorough comparison. This it is hoped to achieve in the future (Tiffney and Collinson, work in progress).

Summary of Palaeobiology

Overall, the flora and fauna from the Kaninah Formation (middle Eocene), Lutetian rocks of southern Yemen indicates regional links with the middle Eocene of Egypt (Fraas, 1904: Cetacea, Sirenia, Proboscidea, giant ground birds, and fish) and Somalia (Savage, 1969: Sirenia), and the Paleogene floras of Egypt (see earlier). Thus the Kaninah Formation indicates the proximity of a tropical continental area with woody plants and possible freshwater aquatic waterliles, to a shallow, tropical sea.

ACKNOWLEDGEMENTS

We present our thanks to Ali Gabr Alawi, Othman Nuoman Ahmed, Mohammed Mukred, Mineral Exploration Board, Ministry of Petroleum and Mineral Resources, Sana'a, and Ahmed Shaikh Ba-Abbad, Hassan A. Mokbel, Mineral Exploration Board (Aden Branch), and Mohamed Ba-Sharahil, Mineral Exploration Board (Al Mukalla Branch), for expediting the fieldwork. We also acknowledge the help of Ian Tattersall and James Clark (American Museum of Natural History, New York); Jerry Hooker, Cedric Shute, Tiffany Foster, and Paul Jeffery (The Natural History Museum, Department of Palaeontology, London); and Professor Bruce Tiffney (University of California at Santa Barbara).

REFERENCES

As-Saruri, M. L. in press. Lithostratigraphic subdivision of the Middle Eocene of Southern Arabian Peninsula. *Zeitschrift für Geologischen Gesellschaft, Wuertzburg.*

As-Saruri, L. M., and Langbein, R. 1994. Über einen Marin-terrestrischen Übergangfazies im mittleren Eozän des Jemen. *Sediment 94, Greifwald* 10: 1–12.

———. 1995. The Middle Eocene Habshiyah Formation of the southern Arabian Peninsula. *Zentralblatt für Geologie und Paläontologie,* pt 1, H-1/2: 161–74.

Beydoun, Z. R. 1964. The stratigraphy and structure of the Eastern Aden Protectorate. *Overseas Geological and Mineral Resources Bulletin* 5: 1–107.

———. 1966. Geology of the Arabian Peninsula, Aden Protectorate and part of Dhofar. *United States Geological Survey Professional Paper* 560H: 1–49.

Beydoun, Z. R., and Greenwood, J. G. W. 1968. Aden Protectrate and Dhofar. *Lexique Stratigraphique International* 3: 1–128.

Bott, W. F., Smith, B. A., Oakes, G., Sikander, A. H., and Ibrahim, A. I. 1992. The tectonic framework and regional hydrocarbon prospectivity of the Gulf of Aden. *Journal of Petroleum Geology* 15: 211–43.

Bown, T. M., Kraus, M. J., Wing, S. L., Fleagle, J. G., Tiffney, B. H., Simons, E. L., and Vondra, C. F. 1982. The Fayum primate forest revisited. *Journal of Human Evolution* 11: 603–32.

Chandler, M. E. J. 1954. Some Upper Cretaceous fruits from Egypt. *Bulletin of the British Museum (Natural History),* Geology 2: 147–87.

Chesters, K. I. M. 1955. Some plant remains from the Upper Cretaceous and Tertiary of West Africa. *Annals and Magazine of Natural History* 6: 498–504.

———. 1957. The Miocene flora of Rusinga Island. Lake Victoria, Kenya. *Palaeontographica B* 101: 30–71.

Collinson, M. E. 1980. Recent and Tertiary seeds of the Nymphaeaceae *sensu lato* with a revision of *Brasenia ovula* (Brong.) Reid and Chandler. *Annals of Botany* 46: 603–32.

———. 1982. Reassessment of fossil Potamogetoneae fruits with description of new material from Saudi Arabia. *Tertiary Research* 4: 83–104.

——— 1983. *Fossil Plants of the London Clay.* Field Guides to Fossils no. 1. Palaeontological Association, London.

———. 1988. The special significance of the middle Eocene fruit and seed flora from Messel, West Germany. *Courier Forschunginstitut Senckenberg* 107: 187–97.

Collinson, M. E., Boulter, M. C., and Holmes, P. L. 1993. Magnoliophyta ('Angiospermae'). In *The Fossil Record,* vol. 2, pp. 809–41 (ed. M. J. Benton). Chapman and Hall, London.

Fraas, E. 1904. Neue Zeuglodonten aus dem unteren Mitteleozänen von Mokattam bei Cairo. *Geologische und Palaeontologische Abhandlungen* 10: 1–24.

Jungwirth, J., and As-Saruri, L. M. 1990. Structural evolution of the platform cover on southern Arabian Peninsula. *Zeitschrift für Geologische Wissenschaften* 18: 505–14.

Kräusel, R. 1939. Ergebnisse der Forchungsreisen Prof. E. Stromers in den wusten Ägyptens IV. Die Fossilen Floren Ägyptens 3. Die fossilen Pflanzen Ägyptens E.-L. *Abhandlungen der Bayerischen Akademie der Wissenschaften, Mathematisch-Naturwissenschaftliche* 47: 1–140, pl. 1–23.

Madden, C. T., Glennie, K. W., Dehm, R., Whitmore, F. C., Schmidt, R. J., Ferfoglia, R. J., and Whybrow, P. J. 1982. *Stegotetrabelodon (Proboscidea, Gomphotheriidae) from the Miocene of Abu Dhabi.* United States Geological Survey, Jiddah.

Madden, C. T., Naqvi, I. M., Whitmore, F. C., Schmidt, J. D., Langston, J., and Wood, R. C. 1978. *Paleocene Vertebrates from Coastal Deposits in Harrat Hadan Area, At Taif Region, Kingdom of Saudi Arabia.* United States Geological Survey, Jiddah.

Manchester, S. R. 1994. Fruits and seeds of the Middle Eocene Nut Beds Flora, Clarno Formation, Oregon. *Palaeontographica Americana* 58: 1–205.

North American Stratigraphic Code. 1983. North American Commission on Stratigraphic Nomenclature. *Bulletin of the American Association of Petroleum Geologists* 67: 841–75.

Powers, R. W., Ramirez, L. F., Redmond, D., and Berg, E. L. 1966. Sedimentary geology of Saudi Arabia. *United States Geological Survey Professional Paper* 560D: 1–146.

Savage, R. J. G. 1969. Early Tertiary mammal locality in southern Libya. *Proceedings of the Geological Society of London* 1648: 98–101.

Schaal, S., and Ziegler, W. 1992. *Messel: An Insight into the History of Life and of the Earth.* Clarendon Press, Oxford.

Schüppel, D., and Wienholz, R. 1990. The development of the Tertiary in the Habban-Al Mukalla area, Yemen. *Zeitschrift für Geologische Wissenschaften* 18: 523–28.

Thomas, H., Roger, J. S., and Al Sulaimani, Z. 1988. Découverte des plus anciens "anthropoïdes" du continent arab-africain et d'un primate tarsiiforme dans l'Oligocène du Sultanat d'Oman. *Compte Rendu de l'Académie des Sciences, Paris* 306: 823–29.

Thomas, H., Roger, J., Sen, S., Pickford, M., Gheerbrant, E., Al-Sulaimani, Z., and Al-Busaidi, S. 1999. Oligocene and Miocene terrestrial vertebrates in the southern Arabian Peninsula (Sultanate of Oman) and their geodynamic and palaeogeographic settings. Chap. 30 in *Fossil Vertebrates of Arabia, pp. 430-42* (ed. P.J. Whybrow and A. Hill). Yale University Press, New Haven.

Thomas, H., Taquet, P., Ligabue, G., and Del 'Agnola, C. 1978. Découverte d'un gisement de vertébrés dans les dépôts continentaux du Miocène moyen du Hasa (Arabie Saoudite). *Compte Rendu Sommaire de la Société Géologique de France* 1978: 69–72.

Tiffney, B. H., and McClammer, J. U., Jr. 1988. A seed of the Annonaceae from the Palaeocene of Pakistan. *Tertiary Research* 9: 13–20.

Tiffney, B. H., Fleagle, J. G., and Bown, T. M. 1994. Early to Middle Miocene angiosperm fruits and seeds from Fejej, Ethiopia. *Tertiary Research* 15: 25–42.

Whybrow, P. J. 1989. New stratotype; the Baynunah Formation (Late Miocene), United Arab Emirates: Lithology and palaeontology. *Newsletters on Stratigraphy* 21: 1–9.

Whybrow, P. J., and Hill, A. eds. 1999. *Fossil Vertebrates of Arabia*. Yale University Press, New Haven.

Whybrow, P. J., Collinson, M. E., Daams, R., Gentry, A. W., and McClure, H. A. 1982. Geology, fauna (Bovidae, Rodentia) and flora from the Early Miocene of eastern Saudi Arabia. *Tertiary Research* 4: 105–20.

Whybrow, P. J., Hill, A., Yasin al-Tikriti, W., and Hailwood, E. A. 1990. Late Miocene primate fauna, flora and initial palaeomagnetic data from the Emirate of Abu Dhabi, United Arab Emirates. *Journal of Human Evolution* 19: 583–88.

A Dinosaur from the Republic of Yemen

32

Louis L. Jacobs, Phillip A. Murry, William R. Downs, and Hamed A. El-Nakhal

Dinosaurs are known from all continents. Wherever appropriate rock facies of late Triassic, Jurassic, and Cretaceous ages are found, dinosaur fossils are likely to be discovered. Factors that influence the frequency of dinosaur discoveries, as for most kinds of surface fossils, include the number of people searching a given area and the exposure of the rocks being searched. The Arabian Peninsula is a vast area of often excellent exposures of Mesozoic rock, but seemingly lacking in dinosaurs. That is probably because of the relatively few palaeontologists actually working in these countries, and there is also a possibility that some discoveries go overlooked in unpublished proprietary reports generated in the extensive exploration for oil in Arabia. We note here the discovery of a dinosaur in northern Yemen (fig. 32.1).

The dearth of dinosaur fossils in Arabia derives from the general lack of appropriate depositional environments and the minimal amount of searching in the relatively few formations that might bear dinosaurs. Most of the western part of the Arabian Peninsula is underlain by crystalline rocks of the Arabian Shield, which was continuous with the Precambrian Shield of Africa before the formation of the Red Sea in Tertiary time. Mesozoic sedimentary rocks of Arabia are predominantly shelf carbonates and related facies, well developed over the eastern half of Arabia, varying in position laterally as sea level fluctuated.

Geology

The Triassic Jihl Formation of Saudi Arabia is reported to contain "a few amphibian (?) bone fragments" (Powers et al., 1966), but although the potential exists, no dinosaurs are yet known from it. An ornithischian vertebra and turtle fragments are mentioned without description by Nolan et al. (1990) in their naming of the late Cretaceous (Campanian or Maastrichtian) Al Khawd Formation, Oman.

The Jurassic period was a time of widespread marine transgression, giving rise in Yemen to the Amran Formation, a sequence of carbonates several hundred meters thick. Below the Amran Formation is the terrestrial clastic Kohlan Group, loosely encompassing rocks considered to be Permian through early Jurassic in age. Overlying the Amran Series are the transitional marine beds of the Mabdi Formation and the evaporites and shales of the Sabatain Formation (El-Nakhal, 1987, 1990). The Amran, Mabdi, and Sabatain Formations were included in the Surdud Group by El-Nakhal (1990). Overlying the Surdud Group is the Tawila Sandstone (Geukens, 1966; Mateer et al., 1992), a terrestrial clastic formation reportedly of Cretaceous to Paleocene age.

The Mabdi and Sabatain Formations of the Surdud Group are laterally equivalent units representing the transition from the Amran Limestone to the terrestrial clastics of the Tawila Sandstone (El-Nakhal, 1990). The Mabdi Formation is a variable succession of shallow marine shale, sandstone, and limestone ranging in thickness from about 60 metres to a maximum of 290 metres. The dinosaur reported here is from the Mabdi Formation.

The lateral equivalent of the Mabdi, the Sabatain Formation, consists of evaporites, sandstone, and bituminous shale. A diverse fauna of fish was recovered from the Khulaqah Quarry, a gypsum mine in the Sabatain Formation northeast of Sana'a, including amiiforms, aspidorhynchids, dercetids, enchodontids, ichthyodectids, leptolepids,

 ISBN 0-300-07183-3

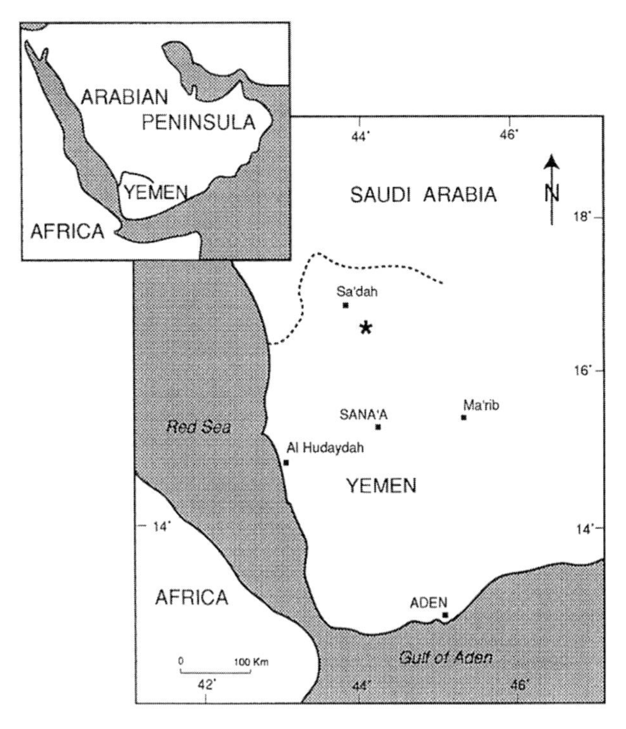

Figure 32.1. Location map. The star shows the position of the dinosaur locality in the Republic of Yemen.

macrosemiids, *Pleuropholis*, pycnodonts, semionotoids, and teleosts (El-Nakhal, 1990). The fish *Pleuropholis bassei* was described from the Sabatain Formation near Al-Harrah village (Basse et al., 1954), and pollen has also been recovered from the Sabatain (El-Nakhal, 1990).

The age of the Mabdi Formation is generally considered Jurassic, Kimmeridgian (Grolier and Overstreet, 1978). The age estimate is based largely on invertebrates collected in former South Yemen (Beydoun, 1964, 1966), although some invertebrate fossils consistent with that age have been identified from correlated beds in the Mabdi further north (Geukens, 1960). More precise age control in northern Yemen is desirable. The Sabatain Formation is of similar age to the Mabdi based on its laterally equivalent position and its contained pollen flora and fish fauna (El-Nakhal, 1990).

Mabdi Formation at the Dinosaur Locality

The Mabdi Formation in the area of the dinosaur locality includes limestone beds with abundant but unstudied marine invertebrates, lime-cemented sandstones, and lesser quantities of green, yellow, or reddish finer-grained clastic beds. All were deposited on or close to the ancient shoreline. The relationship of the Mabdi and Sabatain Formations reflects the conditions present as the sea regressed at the end of the Jurassic. Isolated basins and evaporitic conditions formed as the sea retreated, in which the Sabatain Formation was deposited. Shallow, near-shore, open marine conditions, represented by the Mabdi Formation, prevailed away from the area of the closed basins. A dinosaur preserved in sediments deposited in the Mabdi sea indicates proximity to land. Dinosaurs have been found washed into shallow marine conditions in numerous instances (Horner, 1979; Jacobs et al., 1994). Sauropod footprints are recorded along Cretaceous tidal flats (Winkler et al., 1990).

The area of Mabdi Formation in which the dinosaur was found lies along the Sana'a–Sa'dah highway (fig. 32.2). Exposures are generally flat with low relief. Beds show no consistent structural orientation in this area, although the stratum containing the dinosaur is dipping southeast. Eight field numbers were assigned to fossil localities; locality YMN-3 is that of the dinosaur. Other vertebrates occurring in the Mabdi Formation in this area include hybodont sharks, pycnodont fish, turtle, and crocodilian. Three sections (fig. 32.3) were measured to place the fossil localities in stratigraphic context.

The bones are white and encased in dense, well-indurated, calcite-cemented sandstone. Several blocks of sandstone with bone were removed and subsequently some were sent to the Shuler Museum of Paleontology at Southern Methodist University, Dallas. Bones of the same skeleton remain in the ground for a distance of at least 5 metres from the point at which excavation stopped in 1990. Two subsequent attempts to

Figure 32.2. View looking southeast at YMN-3, the dinosaur locality. The men are standing on blocks containing dinosaur bones.

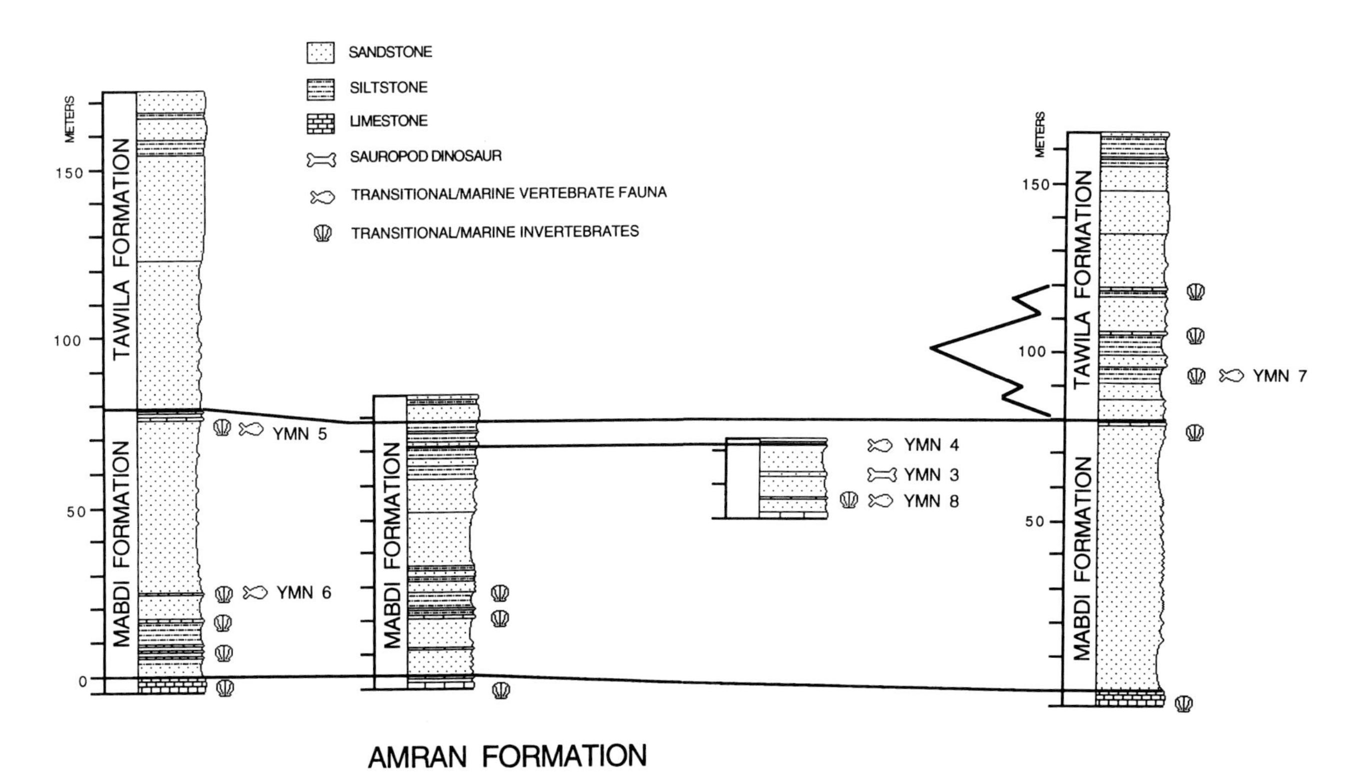

Figure 32.3. Measured sections showing the stratigraphic placement of the dinosaur at locality YMN-3.

return to the field to collect the specimen were cancelled, but a return field season remains on the agenda. The blocks that have been prepared appear to include a sauropod cervical vertebra because the bone is thin and complicated. Precise identification requires further excavation and preparation.

SIGNIFICANCE OF THE DISCOVERY

At this point, given the state of excavation and preparation of the dinosaur from Yemen, its significance lies in its age, its geographic position, and the promise it holds for encouraging further work in Yemen. As seen in the palaeogeographic reconstructions (Scotese and Golonka, 1992), marine regression from the late Jurassic, 152.2 million years (Ma) ago (fig. 32.4) to early Cretaceous, 130.2 Ma (fig. 32.5) exposed the land that is now southwestern Arabia. Gondwana was fragmenting. The Yemen dinosaur is found along the eastern shore of the land mass that was to divide into the South American and African continents during the Cretaceous. Unlike its position throughout the Tertiary, Arabia was not a Mesozoic crossroads between Eurasian and African faunas. Rather, it lay at the eastern extent of the African portion of Gondwana.

ACKNOWLEDGEMENTS

We thank the Yemen Hunt Oil Company, the Institute for the Study of Earth and Man at Southern Methodist University, and the American Institute for Yemeni Studies for supporting our work in Yemen. Preparation of the specimen was by Andrew Konnerth, facilitated by a grant from The Dinosaur Society, for which we are very grateful. We especially thank the government and people of the Republic of Yemen, and in particular the Ministries of Petroleum and Geology and the General Organization for Antiquities and Libraries, for their hospitality, support, encouragement, and friendship.

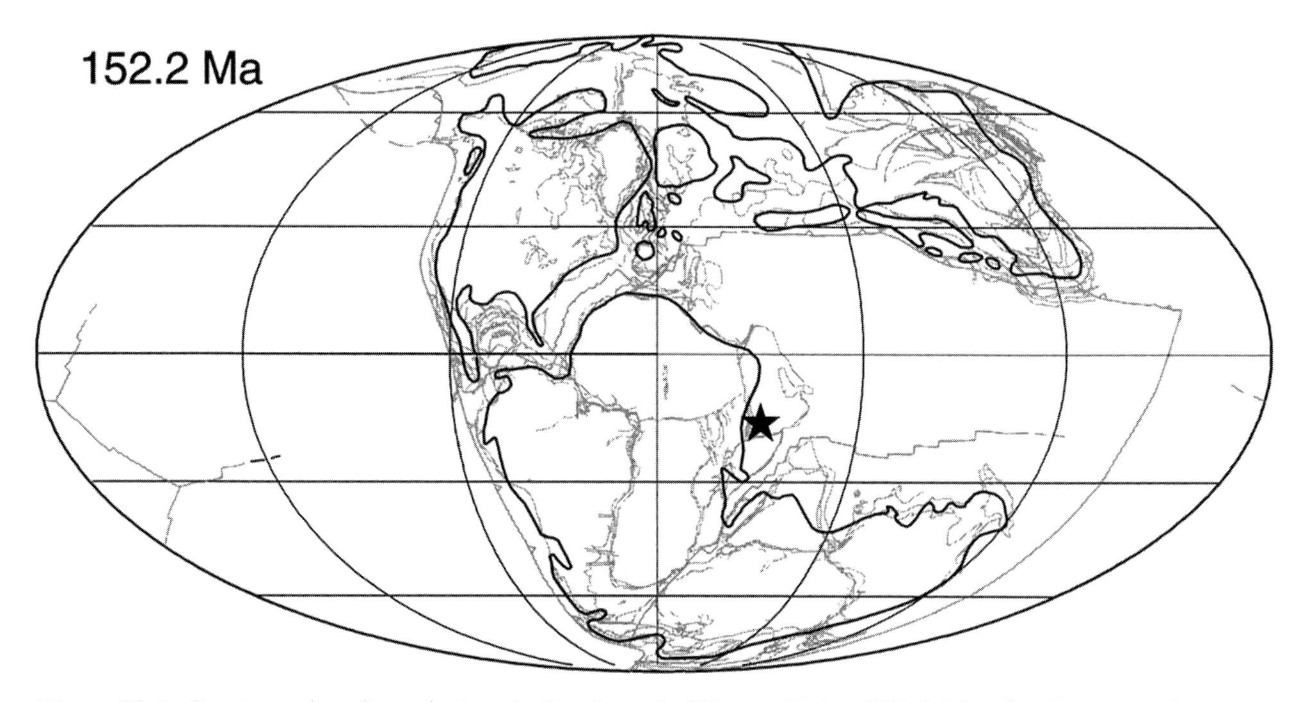

Figure 32.4. Continental outlines during the late Jurassic (Kimmeridgian, 152.2 Ma; after Scotese and Golonka, 1992). The star represents the position of locality YMN-3.

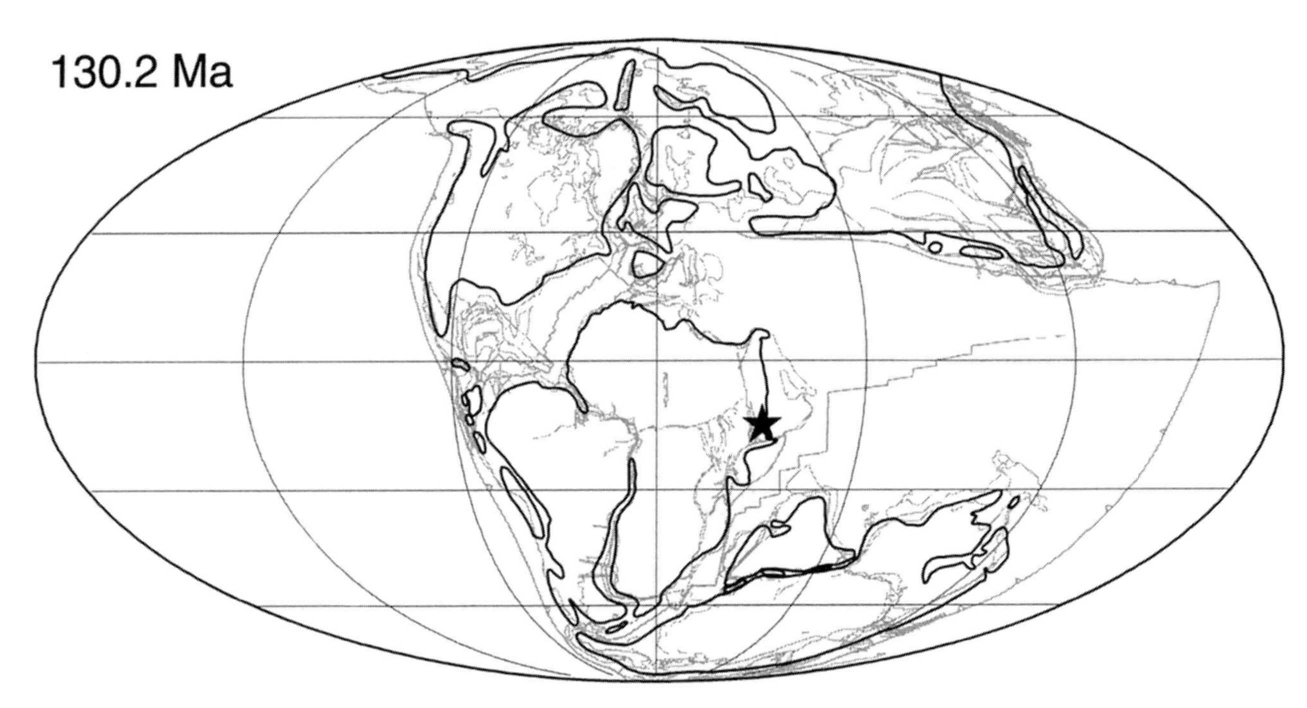

Figure 32.5. Continental outlines during the early Cretaceous (Hauterivian, 130.2 Ma; after Scotese and Golonka, 1992). The star represents the position of locality YMN-3.

REFERENCES

Basse, E., Karrenberg, H., Lehman, J., Alloiteau, J., and Lefranc, L. 1954. Fossiles du Jurassique supérieur et des "Gré de Nube" de la région de Sana (Yemen). *Bulletin de la Société Géologique de France* ser. 6: 655–88.

Beydoun, Z. R. 1964. The stratigraphy and structure of the Eastern Aden Protectorate. *Overseas Geological and Mineral Resources Bulletin* 5: 1–107.

———. 1966. Geology of the Arabian Peninsula, eastern Aden Protectorate and part of Dhufar. *United States Geological Survey Professional Paper* 560H: 1–49.

El-Nakhal, Hamed A. 1987. A lithostratigraphic subdivision of Kholan Group in Yemen Arab Republic. *Iraqi Journal of Science* 28: 149–80.

———. 1990. Surdud Group, a new lithostratigraphic unit of Jurassic age in the Yemen Arab Republic. *Journal of King Saud University, Science* 2: 125–43.

Geukens, F. 1960 Contribution à la géologie du Yémen. *Mémoires de l'Institut Géologiques de l'Université de Louvain* 21: 122–79.

———. 1966. Geology of the Arabian Peninsula; Yemen. *United States Geological Survey Professional Paper* 560B: 1–23.

Grolier, M. J., and Overstreet, W. C. 1978. *Geologic Map of the Yemen Arab Republic (San'a).* United States Geological Survey Miscellaneous Investigations Series. Government Printer, Washington, D.C.

Horner, J. R. 1979. Upper Cretaceous dinosaurs from the Bearpaw Shale (marine) of south-central Montana with a checklist of Upper Cretaceous dinosaur remains from marine sediments in North America. *Journal of Paleontology* 53: 566–77.

Jacobs, L. L., Winkler, D. A., Murry, P. A., and Maurice, J. M. 1994. A nodosaurid scuteling from the Texas shore of the Western Interior Seaway. In *Dinosaur Eggs and Babies,* pp. 337–46 (ed. K. Carpenter, K. F. Hirsch, and J. R. Horner). Cambridge University Press, Cambridge.

Mateer, N. J., Wycisk, P. J., Jacobs, L. L., Brunet, M., Luger, P., Dina, A., Hendriks, F., Weissbrod, T., Gvirtzman, G., Arush, M., Mbede, E., Hell, J., and El-Nakhal, H. A. 1992. Correlation of nonmarine Cretaceous strata of Africa and the Middle East. *Cretaceous Research* 13: 273–318.

Nolan, S. C., Skelton, P. W., Clissold, B. P., and Smewing, J. D. 1990. Maastrichtian to early Tertiary stratigraphy and palaeogeography of the central and northern Oman Mountains. In *The Geology and Tectonics of the Oman Region,* pp. 495–519 (ed. A. H. F. Robertson, M. P. Searle, and A. C. Reis). Geological Society Special Publication no. 49.

Powers, R. W., Ramirez, L. F., Redmond, D., and Berg, E. L. 1966. Sedimentary geology of Saudi Arabia. *United States Geological Survey Professional Paper* 560D: 1–147.

Scotese, C. R., and Golonka, J. 1992. *Paleogeographic Atlas: Paleomap Progress Report no. 20.* Department of Geology, University of Texas at Arlington.

Winkler, D. A., Murry, P. A., and Jacobs, L. L. 1990. Early Cretaceous (Comanchean) vertebrates of Central Texas. *Journal of Vertebrate Paleontology* 10: 95–116.

Arabian Tertiary Fauna, Flora, and Localities

33

PETER J. WHYBROW AND DIANA CLEMENTS

The geological history of Arabia is closely linked to that of Africa and the rifting of Arabia away from Africa and its subsequent anticlockwise rotation closed the Tethys epicontinental seaway by mid-Burdigalian times at the latest (Adams et al., 1983). The existence of a landbridge brought about by the marine disconnection enabled land animals to disperse between Africa and Asia via Arabia. The modern margin of the Arabian Plate includes the peninsula itself, approximately follows the Tigris and Euphrates rivers, parallels the Iraq–Syria–Iraq border to meet the Mediterranean near the Gulf of Alexandretta (Iskenduran). The area to the west of the Levant rift (Palestine and western Jordan) may be part of the Arabian Plate or a separate microplate (Freund et al., 1970). The area of the Arabian Plate approximates the size of the Indian subcontinent. Deposition of mainly marine sediments has been almost constant on the eastern flank of the craton and has resulted in a unique series of sediments being deposited with few major unconformities. Exploration for oil in Arabia has added much knowledge to the geology and palaeontology of the Cenozoic in the Middle East, but as few oil fields are located in Miocene rocks, these deposits have received little study. In fact, subsurface logging does not usually start until the regional (in the Arabian Gulf region and in Mesopotamia) Pre-Neogene Unconformity (PNU) is located—the Neogene is unconformable with middle Eocene or older rocks.

The importance of the Arabian Cenozoic sites with terrestrial vertebrates lies, in part, in the intermediate palaeogeographical position of Arabia between better-known Miocene terrestrial sites found in southwestern Asia and East and North Africa. The lists here of all Arabian Cenozoic sites provide an aid for future studies of African–Arabian–Asian vertebrate links in the Old World (table 33.1).

ARABIAN FOSSIL VERTEBRATE LOCALITIES

Paleocene

Kingdom of Saudi Arabia, Red Sea: locality, Jabal Umm Himar, Harrat Hadan area; Umm Himar Formation (Greenwood, 1983; Madden et al., 1979; Madden, 1982; Whitmore and Madden, 1995).

Fossil vertebrates have been collected from 21 localities at Jabal Umm Himar, near Truabah, Southern Hijaz province, Saudi Arabia, and are the first Paleocene vertebrate fauna known from the Arabian Peninsula. The fossils are from a mudstone 2–3 metres thick in the Umm Himar Formation, which is overall about 20 metres thick. The formation is underlain by Nubian-like quartzose sandstone, which overlies Precambrian rocks. It is overlain by the Harrat Hadan flood basalt of latest Oligocene and early Miocene age. During the early Tertiary the area was a contiguous part of Africa. The palaeoenvironment of Jabal Umm Himar was estuarine and the presence of a coastal environment indicates that at least a shallow seaway must have existed in the region at that time.

 ISBN 0-300-07183-3

Table 33.1. Tertiary vertebrate localities of the Arabian Peninsula

Epoch	Age (Ma)	Formation	Country	Locality	Reference
Late Miocene	?6–8	Baynunah	United Arab Emirates	Western Region, Emirate of Abu Dhabi	Whybrow and Hill, 1999 (this vol.)
Late Miocene	?9–10	Agha Jari	Iraq*	Jebel Hamrin	Thomas et al., 1980
Middle Miocene	?14	Hofuf	Kingdom of Saudi Arabia	Al Jadidah	Thomas et al., 1978
Middle Miocene	15.5–17.5	Dam	Sultanate of Oman	Ghaba	Thomas et al., 1999 (this vol., Chapter 30)
Early middle Miocene	?16–19	Dam	Kingdom of Saudi Arabia	Ad Dabtiyah, Jabal Dawmat al 'Awdah	Whybrow, 1987
Early middle Miocene	?16–19	Dam	Kingdom of Saudi Arabia	As Sarrar	Thomas et al., 1982
Early middle Miocene	?20	Hadrukh	Kingdom of Saudi Arabia	Jabal Midra ash-Shamali	Whybrow et al., 1982
Early Miocene	21–25	Baid	Kingdom of Saudi Arabia	Wadi Sabya	Madden et al., 1983
Early Oligocene	?33	Ashawq (Shizar Member)	Sultanate of Oman	Taqah	Thomas et al., 1999 (this vol., Chapter 30)
Early Oligocene	?33	Ashawq (Shizar Member)	Sultanate of Oman	Thaytiniti	Thomas et al., 1999 (this vol., Chapter 30)
?Middle Eocene	?45	Kaninah	Republic of Yemen	Kaninah, Jizwal	As-Saruri et al., 1999 (this vol., Chapter 31)
Paleocene	?60	Umm Himar	Kingdom of Saudi Arabia	Harrat Hadan	Madden et al., 1979

*The fauna from this locality might relate to the late Miocene fauna in the United Arab Emirates.

- Pisces
  - Selachii
    - *Odontaspis substriata*
    - *Ginglymostoma maghrebianum*
    - *Ginglymostoma sokotoensi*
    - Batoidea
      - *Eotorpedo hilgendorfi*
      - *Myliobatis dixoni*
  - Holostei
    - Pycnodontiformes
      - *Pycnodus* sp. (very large)
      - *Pycnodus* sp. (medium-sized)
      - *Pycnodus* sp. (small)
    - Amiformes
  - Teleostei
    - Amiidae gen. et sp. indet.
    - Siluriformes
    - Perciformes
      - *Parapercichthys arabis*
  - Dipnoi
    - *Ceratodus humei*
- Reptilia
  - Chelonia
    - Pelomedusidae gen. et sp. indet.
  - Crocodilia
    - *Hyposaurus nopcsai*
    - *Rhabdognathus compressus*
    - *Rhabdognathus* sp.
    - *Phosphatosaurus* sp.
    - Dryosauridae gen. et sp. indet.

Middle Eocene

Republic of Yemen: locality, Kaninah, Jizwal; Kaninah Formation (As-Saruri et al., 1999—Chapter 31).

Indeterminate mammal bones (large vertebrae) and a probable sirenian rib. Associated with fruits and seeds that have affinities with middle Eocene floras from Egypt.

Early Oligocene

Sultanate of Oman, Dhofar: locality, Thaytiniti; Shizar Member at the base of the Ashawq Formation (Thomas et al., 1999—Chapter 30; also Crochet et al., 1992; Pickford et al., 1994; Thomas et al., 1988, 1989a,b, 1991b).

Situated on the coastal strip of Dhofar, the locality of Thaytiniti occurs right at the base of the transgressive Ashawq Formation (Shizar Member). Its position explains the juxtaposition of elements indicative of marine, freshwater, and terrestrial environments. The fossil locality is dated at about 33 million years (Ma).

- Pisces
  - Selachii
    - *Hemipristis* cf. *serra*
    - *Hemipristis* aff. *curvatus*
    - *Negaprion* sp.
    - *Galeocerdo* sp.
    - *Galeocerdo aduncus*
    - *Rhizoporionodon/Sphyrna*
    - *Carcharhinus* cf. *amboinensis*
    - *Nebrius* sp.
    - *Chilosyllium* sp.
    - *Rhynchobatus* aff. *pristinus*
    - *Rhynobatos* sp.
    - *Dasyatis* spp. 1 and 2
    - *Rhinoptera* sp.
    - *Aetobatus* aff. *arcuatus*
  - Dipnoi cf. *Protopterus*
  - Teleostei
    - Elopiformes
      - Albulidae
    - Tetradontiformes
      - Tetraodontoidea
      - Ostracioidea
    - Osteoglossiformes
      - Osteoglossidae
    - Characiformes
      - Alestinae
    - Siluriformes
    - Perciformes
      - Sparidae
      - Cichlidae
- Reptilia
  - Chelonia
    - Cryptodira
      - Testudinidae indet.
    - Pleurodira
      - Podocnemidinae
        - ?*Stereogenys*
  - Crocodilia
    - Gavialidae: ?Tomistominae
- Mammalia
  - Rodentia
    - Phiomyidae
      - *Phiomys* cf. *andrewsi*
      - *Phiomys* cf. *lavocati*
      - cf. *Metaphiomys* spp. 1 and 2
    - Family indet. 1
  - Primates
    - Adapiformes
      - ?cf. *Omanodon minor*
    - Simiiformes
      - Oligopithecinae indet.
  - Hyracoidea
    - cf. *Saghatherium bowni*
    - cf. *Thyrohyrax meyeri*
    - gen. et sp. indet. 2
  - Embrithopoda
    - cf. *Arsinoitherium*
  - Proboscidea
    - gen. et sp. nov.
    - cf. Barytherioidea

Early Oligocene

Sultanate of Oman, Dhofar: locality, Taqah; Shizar Member at the base of the Ashawq Formation (Thomas et al., 1999—Chapter 30; also Crochet et al., 1990, 1992; Gheerbrant and Thomas, 1992; Gheerbrant et al., 1993, 1995; Pickford and Thomas, 1994; Pickford et al., 1994; Senut and Thomas, 1994; Sigé et al., 1994; Thomas et al., 1988, 1989a,b, 1991a,b).

Also situated on the coastal strip of Dhofar between sea and continent, the location of Taqah is slightly younger than the Thaytiniti location (around 33 Ma old). The fossil level is intercalated within a regressive clay–grit sequence that caps the deposits of the Shizar Member. The fauna suggest that the horizon accumulated in a shallow coastal swamp, sporadically influenced by the sea. A warm subtropical to tropical climate, strongly seasonal with a marked rainy season, is indicated.

Pisces
- Selachii
  - *Galeocerdo* sp.
  - *Carcharhinus* cf. *amboinensis*
  - *Nebrius* sp.
  - *Dasyatis* spp. 1 and 2
  - *Rhinoptera* sp.
  - *Myliobatis* sp.
- Dipnoi cf. *Protopterus*
- Teleostei
  - Osteoglossiformes
    - Osteoglossidae
  - Siluriformes
  - Perciformes
    - Sparidae

Amphibia
- ?Anura

Reptilia
- Lacertilians
  - Agamidae spp. 1 and 2
  - Family indet.
- Serpentes
  - Boidae: Erycinae
  - Colubridae indet.
- Chelonia
  - Cryptodira
    - *Geochelone* sp.
  - Pleurodira
    - *Ermnochelys* spp. 1 and 2
    - *Schweboemys* sp.
- Crocodilia
  - Crocodylidae spp. 1 and 2

Mammalia

Marsupialia
- Peradectidae
  - *Qatranitherium* aff. *africanum*

Insectivora: at least four species

Chiroptera

Microchiroptera
- *Dhofarella thaleri*
- *Hipposideros* (*Brachipposideros*) *omani*
- Hipposideridae gen. et sp. indet.
- *Chibanycteris herberti*
- *Philisis sevketi*
- cf. *Philisis* sp.
- Vespertilionoidea indet.
- Microchiroptera indet. (various spp.)

Rodentia
- Phiomyidae
  - *Phiomys* cf. *andrewsi*
  - *Phiomys* cf. *lavocati*
  - cf. *Metaphiomys* spp. 1 and 2
  - *Metaphiomys* cf. *schaubi*
- Cricetidae
- Anomaluridae
- Family indet. 2

Primates
- Adapiformes
  - *Omanodon minor*
  - *Shizarodon dhofarensis*
- Simiiformes
  - *Oligopithecus rogeri*
  - *Catopithecus* sp. nov.
  - Oligopithecinae gen. et sp. nov. 1
  - Oligopithecinae gen. et sp. nov. 2
  - *Moeripithecus markgrafi*
  - Propliopithecidae gen. et sp. indet.
  - cf. *Apidium*
  - Parapithecidae indet.
- Primates indet. (various spp.)

Carnivora
- Creodonta
  - *Masrasector ligabuei*

Hyracoidea
- *Thyrohyrax meyeri*
- gen. et sp. indet. 2

Embrithopoda
- cf. *Arsinoitherium*

Proboscidea
- *Phiomia* sp.

Artiodactyla
- Anthracotheriidae

Early Miocene

Kingdom of Saudi Arabia, Red Sea: locality, Wadi Sabya; Baid Formation (Coleman et al., 1979; Madden et al., 1983).

Part of the Jizan group, about 1000 km southeast of Jiddah on the Red Sea coast. The fossil locality is dated at 21–25 Ma. Intruding into the Jizan group are basement rocks K–Ar whole-rock dated at 20–23 Ma.

Pisces
gen. et sp. indet.
Mammalia
Artiodactyla
Masritherium sp.

?Early Middle Miocene

Kingdom of Saudi Arabia, Arabian Gulf: locality, Jabal Midra ash-Shamali; Hadrukh Formation (Collinson, 1982; Hamilton et al., 1978; Lindsay, 1988; Tleel, 1973; Whybrow et al., 1982).

The basal Miocene disconformably overlies middle Eocene rocks. The locality at Jabal Midra ash-Shamali, Damman Dome, Dhahran, is dated as ?early middle Miocene (?20 Ma old) based on stratigraphic position. It may, however, be the basal part of the continental Dam Formation, the marine part of which is dated as Burdigalian, 19–16 Ma old.

Plantae
Potamogetoneae
Midravalva arabica
Mammalia
Rodentia
Zapodidae
Arabosminthus quadratus
Cricetidae
Shamalina tuberculata
Artiodactyla
Bovidae
cf. *Oioceros* sp.

Early Middle Miocene

Kingdom of Saudi Arabia, Arabian Gulf: locality, As Sarrar (Al Sarrar); Dam Formation (continental equivalents) (Thomas et al., 1982).

Shallow marine sediments, with echinoids and molluscs, dated as Burdigalian, middle Miocene interdigitated with vertebrate-bearing continental sediments at three localities: As Sarrar (Al Sarrar), Ad Dabtiyah, and Jabal Dawmat al 'Awdah (unpublished).

Pisces
Selachii
Hemigaleidae
Hemipristis serra
Carcharhinidae
Carcharhinus aff. *priscus*
Carcharhinus aff. *plumbeus*
Galeocerdo cf. *aduncus*
Scoliodon sp.
Negaprion eurybathrodon
Sphyrnidae
Sphyrna sp.
Dasyatidae
Dasyatis sp.
Myliobatidae
Myliobatis sp.
Aetobatus arcuatus
Rhinopteridae
Rhinoptera
Teleostei
Mormyridae
Hyperopisus sp.
Cyprinidae
Barbus sp.
Labeo sp.
Clariidae
Heterobranchus sp.
Clarias sp.
Centropomidae
Lates sp.
Sphyraenidae
Sphyraena sp.
Sparidae indet.
Amphibia
Bufonoidea
Ranoidea
Reptilia
Chelonia
Pelomedusidae
cf. *Schweboemys*
aff. *Stereogenys*
Trionychidae
aff. *Cycloderma*
Testudinidae
Geochelone sp.
Carettochelyidae
Sauria
Lacertidae
Amphisbaenia
Amphisbaenidae
Serpentes
Scolecophidia
Boidae
Python sp.
Eryx/Gongylophis sp.
Colubridae
Elapidae
Naja/Palaeonaja spp.
Viperidae

Crocodilia
Crocodylidae
Crocodylus cf. *pigotti*
Aves
Ciconiidae
Mycteria cinereus
?*Mycteria* sp.
Threskiornithidae
Scolopacidae
Charadriinae indet.
spp. indet.
Mammalia
Insectivora
Erinaceidae
?Primates gen. et sp. indet.
Lagomorpha
Ochotonidae
Rodentia
Ctenodactylidae
Sayimys intermedius de Bruijn et al. 1989 (*Metasayimys* cf. *intermedius*)
Thryonomyidae
Paraphiomys sp.
Pedetidae
Megapedetes cf. *pentadactylus*
cf. *Protalactaga*
Cricetidae
Gerbillidae
Carnivora
Mustelidae
cf. *Martes*
Mionictis sp.
Viverridae
Viverra sp.
Felidae
Pseudaelurus turnauensis
Amphicyonidae
Amphicyon sp.
Hyracoidea
Saghatheriinae
Pachyhyrax aff. *championi*
Proboscidea
Deinotheriidae
cf. *Deinotherium*
Gomphotheriidae
Gomphotherium sp.
?Amebelodontinae
Sirenia indet.
Perissodactyla
Rhinocerotidae
Aceratherium sp.
Dicerorhinus sp.
Artiodactyla
Suidae
Listriodon sp.
gen. et sp. indet.
giant species
Tragulidae
Dorcatherium cf. *libiensis*
Bovidae
gen. et sp. indet.

Early Middle Miocene

Kingdom of Saudi Arabia, Arabian Gulf: localities, Damman Dome (Dhahran), Ad Dabtiyah, and Jabal Dawmat al 'Awdah, Dam Formation (continental equivalents) (Andrews and Martin, 1987; Andrews et al., 1987; Hamilton et al., 1978; Leakey and Leakey, 1986; Thomas, 1982; Tleel, 1973; Whybrow, 1987; Whybrow and McClure, 1981).

Shallow marine sediments, with echinoids and molluscs, dated as Burdigalian, middle Miocene interdigitate with vertebrate-bearing continental sediments at three localities: As Sarrar (Al Sarrar), Ad Dabtiyah, and Jabal Dawmat al 'Awdah (unpublished).

Pisces
Osteichthyes
Cyprinidae
Acanthopterygii
?Centropomidae
Reptilia
Chelonia gen et sp. indet.
Crocodilia
cf. *Crocodylus pigotti*
Mammalia
Hominoidea
Afropithecus turkanensis Leakey and Leakey, 1986 (*Heliopithecus leakeyi* Andrews and Martin, 1987)
Proboscidea
Gomphotheriidae
Gomphotherium cooperi
Perissodactyla
Rhinocerotidae
Dicerorhinus aff. *sansaniensis*
Artiodactyla
Brachypotherium sp.
Suidae
Listriodon cf. *lockharti* or *L.* cf. *akatikubas*
?*Kenyasus* sp.
Tragulidae
Dorcatherium sp.
Dorcatherium (larger sp.)
Giraffoidea
Canthumeryx sp.
Bovidae
Eotragus sp.
Bovid species 2
Bovid species 3

Middle Miocene

Sultanate of Oman, Ghaba: locality, northern margin of Huqf, central Oman; Dam Formation (Thomas et al., 1999—Chapter 30; also Roger et al., 1994).

Miocene series assigned to the Dam Formation, dated as Burdigalian–Langhian (17.5–15.5 Ma), in the State of Qatar, 50 metres thick and consisting of various lithologies, which include limestone, shale, gypsum and anhydrite, siltstones, grits, and conglomerates. The vertical arrangement indicates a complete cycle of transgression and regression.

- Pisces
  - Selachii
    - ?Pristidae
    - Rajiformes
  - Teleostei
    - Siluroidea indet.
      - Clariidae
        - *Clarias* sp.
    - Perciformes indet.
      - *Lates* sp.
- Reptilia
  - Chelonia
    - Pleurodira
      - Bothremydidae gen. et sp. indet.
      - Podocnemididae
        - ?*Schweboemys* sp.
    - Cryptodira
      - Trionychidae
        - aff. *Cycloderma* sp.
      - Carettochelyidea
        - aff. *Carettochelys* sp.
      - Testudinidae
        - cf. *Geochelone* sp.
  - Crocodilia
    - ?*Crocodylus* sp.
    - gen. et sp. indet. (group *Tomistoma–Euthecodon*)
- Mammalia
  - Carnivora indet.
  - Proboscidea
    - gen. et sp. indet.
    - *Prodeinotherium* sp.
  - Perissodactyla
    - Rhinocerotidae indet.
  - Artiodactyla
    - Anthracotheriidae
      - *Afromeryx zelteni*
    - Giraffoidea
      - *Canthumeryx* sp.
      - Bovidae indet.
      - Tragulidae indet.

Middle Miocene

Kingdom of Saudi Arabia, Arabian Gulf: locality, Al Jadidah; Hofuf Formation, continental clastics (de Bruijn et al., 1989; Sen and Thomas, 1979; Thomas, 1983; Thomas et al., 1978).

The locality is in the lower 20 metres of a sequence of homogeneous sandstones 70 metres thick. The vertebrates indicate a "Fort Ternan, Kenya" age of about 14 Ma.

- Indet. Pisces, Chelonia, and Crocodilia
- Mammalia
  - Rodentia
    - Sciuridae
      - *Atlantoxerus* sp.
    - Ctenodactylidae
      - *Sayimys intermedius* de Bruijn et al. 1989 (*Metasayimys intermedius* Sen and Thomas, 1979)
  - Carnivora
    - Hyaenidae
      - *Percrocuta* sp.
  - Proboscidea
    - Gomphotheriidae
      - *Gomphotherium angustidens*
  - Perissodactyla
    - Rhinocerotidae
      - *Dicerorhinus* cf. *primaevus*
  - Artiodactyla
    - Suidae
      - cf. *Lopholistriodon*
    - Giraffidae
      - *Palaeotragus ligabuei*
    - Bovidae
      - *Caprotragoides* aff. *potwaricus*
      - *Protragocerus* sp.
      - cf. ?*Homoiodorcas*

Late Miocene

Mesopotamia, Iraq: locality, Jebel Hamrin; Agha Jari Formation (Thomas et al., 1980).

Vertebrate fossils are found in the region of Injana on the anticline of Jebel Hamrin. The fossiliferous bed that has been researched to date stretches for more than 30 km along the northeast flank. The mammal fauna indicate a late Vallesian age (10–9 Ma).

- Reptilia
  - Chelonia
    - Trionychidae
      - *Trionyx* s.l.
    - Testudinidae
      - *Geochelone* sp.
    - ?Pelomedusidae gen. et sp. indet.
  - Squamata
    - ?Erycinae
  - Crocodilia gen. et sp.
    - *Crocodylus* cf. *niloticus*
    - *Crocodylus* sp.
    - ?*Ikanogavialis*
    - Gavialidae gen. et sp. indet.
    - Crocodylidae gen. et sp. indet.

Aves
Struthionidae
Struthio sp.
Mammalia
Carnivora
Felidae
?*Machairodus* sp.
?Mustelidae gen. et sp. indet.
Proboscidea
Gomphotheriidae
Choerolophodon cf. *pentelici*
Deinotheriidae
Deinotherium sp.
Prodeinotherium sp.
Perissodactyla
Equidae
Hipparion mediterraneum
Hipparion cf. *primigenium*
Rhinocerotidae
Brachypotherium cf. *perimense*
Artiodactyla
Giraffidae
Palaeotragus coelophrys
Samotherium cf. *boissieri*
cf. *Bohlinia attica*
Bovidae
Miotragocerus sp.
Prostrepsiceros aff. *houtumschindleri*
Gazella cf. *deperdita*
?*Gazella capricornis*
Ovibovini gen. et sp. indet.

Late Miocene

United Arab Emirates, Emirate of Abu Dhabi, Western Region: localities, Jebel Barakah, Jazirat Shuwaihat, Al Ghuddah, Hamra, Jebel Dhanna, Ras Dubay'ah, Ghayathi and bin Jawabi, Kihal, Ras al Qa'la, Thumayriyah, Ras al Aysh, Harmiyah, Jebel Mimiyah (Al Mirfa), Tarif; Baynunah Formation (this vol.—Whybrow and Hill, 1999; Barry, 1999; Bishop and Hill, 1999; Lapparent de Broin and Dijk, 1999; de Bruijn, 1999; Eisenmann and Whybrow, 1999; Forey and Young, 1999; Gentry 1999a, b; Hill and Gundling, 1999; Jeffery, 1999; Mordan, 1999; Rauhe et al., 1999; Tassy 1999; Whybrow and Clements, 1999; also de Bruijn and Whybrow 1994; Glennie and Evamy, 1968; Hill et al., 1990; Whybrow, 1989, 1992; Whybrow and McClure, 1981; Whybrow et al., 1990).

Vallesian/Turolian about 8–6 Ma old based on vertebrate faunas.

Plantae
Algae gen. et sp. indet.
Leguminosae
?*Acacia* sp.
Mollusca
Gastropoda
Buliminidae
?*Sebzebrinus* sp. or ?*Pseudonapaeus* sp.
Bivalvia
Mutelidae
Mutela (?subgen. nov. aff. *Chelidonopsis*) sp.
Unionidae
Leguminaia (*Leguminaia*) sp.
Crustacea
Ostracoda
Cytherideidae
Cyprideis sp.
Pisces
Siluriformes
Clariidae
Clarias sp.
Bagridae
Bagrus shuwaiensis Forey and Young, 1999
Cypriniformes
Cyprinidae
Barbus sp.
Reptilia
Chelonia
Trionychidae
Trionyx s.l. sp.
Testudinidae
cf. *Mauremys* sp.
Geochelone (*Centrochelys*) aff. *sulcata*
Crocodilia
Crocodylidae
Crocodylus cf. *niloticus*
Crocodylus sp.
Gavialidae
?*Ikanogavialis*
gen. et sp. indet.
Aves
Struthionidae
Struthio sp.
Ardeidae
Egretta aff. *alba*
Mammalia
Rodentia
Thryonomyidae
gen. et sp. indet.
Gerbillidae
Gerbillinae
Abudhabia baynunensis de Bruijn and Whybrow, 1994
Myocricetodontinae
Myocricetodon sp. nov.

Muridae
Murinae
Parapelomys cf. *charkhensis*
Dendromurinae
Dendromus aff. *melanotus*
Dendromus sp.
Dipodidae
Zapodinae
gen. et sp. indet.
Insectivora
Soricidae
gen. et sp. indet.
Primates
Cercopithecidae
gen. et sp. indet.
Carnivora
Mustelidae
Plesiogulo praecocidens
Hyaenidae
gen. et sp. indet. (very large)
gen. et sp. indet. (medium-sized)
Felidae
Machairodontinae
gen. et sp. indet.
Proboscidea
Deinotheriidae
gen. et sp. indet.
Elephantidae
Stegotetrabelodon syrticus
Family indet.
gen. et sp. indet. (possibly "*Mastodon*" *grandincisivus*)
Perissodactyla
Equidae
Hipparion abudhabiense Eisenmann and Whybrow, 1999
Hipparion sp.
Rhinocerotidae
gen. et sp. indet.
Artiodactyla
Suidae
Propotamochoerus hysudricus
Nyanzochoerus syrticus
Hippopotamidae
Hexaprotodon aff. *sahabiensis*
Giraffidae
?*Palaeotragus* sp.
?*Bramatherium* sp.
gen. et sp. indet.
Bovidae
gen. et sp. indet.
Boselaphini
Tragoportax cyrenaicus
Pachyportax latidens
Antilopini
Prostrepsiceros aff. *libycus*
Prostrepsiceros aff. *vinayaki*
Gazella aff. *lydekkeri*

ACKNOWLEDGEMENTS

We thank the Abu Dhabi Company for Onshore Oil Operations (ADCO) for their continuing grant to The Natural History Museum, London, and for their generous support for Miocene studies in the United Arab Emirates since 1991. This contribution forms part of The Natural History Museum's Global Change and the Biosphere research programme.

REFERENCES

Adams, C. G., Gentry, A. W., and Whybrow, P. J. 1983. Dating the terminal Tethyan events. In *Reconstruction of Marine Environments* (ed. J. Meulenkamp). *Utrecht Micropaleontological Bulletins* 30: 273–98.

Andrews, P. J., and Martin, L. 1987. The phyletic position of the Ad Dabtiyah hominoid. *Bulletin of the British Museum (Natural History)*, Geology 41: 383–93.

Andrews, P., Martin, L., and Whybrow, P. J. 1987. Earliest known member of the great ape and human clade. *American Journal of Physical Anthropology* 72: 174–75.

As-Saruri, M. L., Whybrow, P. J., and Collinson, M. E. 1999. Geology, fruits, seeds, and vertebrates (?Sirenia) from the Kaninah Formation (middle Eocene), Republic of Yemen. Chap. 31 in *Fossil Vertebrates of Arabia*, pp. 443–53 (ed. P. J. Whybrow and A. Hill). Yale University Press, New Haven.

Barry, J. C. 1999. Late Miocene Carnivora from the Emirate of Abu Dhabi, United Arab Emirates. Chap. 17 in *Fossil Vertebrates of Arabia*, pp. 203–08 (ed. P. J. Whybrow and A. Hill). Yale University Press, New Haven.

Bishop, L., and Hill, A. 1999. Fossil Suidae from the Baynunah Formation, Emirate of Abu Dhabi, United Arab Emirates. Chap. 20 in *Fossil Vertebrates of Arabia,* pp. 254–70 (ed. P. J. Whybrow and A. Hill). Yale University Press, New Haven.

Bruijn, H. de. 1999. A late Miocene insectivore and rodent fauna from the Baynunah Formation, Emirate of Abu Dhabi, United Arab Emirates. Chap. 15 in *Fossil Vertebrates of Arabia,* pp. 186–97 (ed. P. J. Whybrow and A. Hill). Yale University Press, New Haven.

Bruijn, H. de, and Whybrow, P. J. 1994. A Late Miocene rodent fauna from the Baynunah Formation, Emirate of Abu Dhabi; United Arab Emirates. *Proceedings Koninklijke Nederlandse Akademie van Wetenschappen* 97: 407–22.

Bruijn, H. de, Boon, E., and Taseer Hussain, S. 1989. Evolutionary trends in *Sayimys* (Ctenodactylidae, Rodentia) from the Lower Manchar Formation (Sind, Pakistan). *Proceedings Koninklijke Nederlandse Akademie van Wetenschappen* B92: 191–214.

Coleman, R. G., Hadley, D. G., Fleck, R. J., Hedge, C. E., and Donato, M. M. 1979. The Miocene Tihama Asir ophiolite and its bearing on the opening of the Red Sea. In *Evolution and Mineralisation of the Arabian–Nubian Shield,* pp. 173–86 (ed. S. A. Tahoun). Pergamon Press, Oxford.

Collinson, M. E. 1982. Reassessment of fossil Potamogetoneae fruits with description of new material from Saudi Arabia. *Tertiary Research* 4: 83–104.

Crochet, J. Y., Thomas, H., Roger, J., Sen, S., and Al-Sulaimani, Z. 1990. Première découverte d'un créodonte dans la péninsule Arabique: *Masrasector ligabuei* nov. sp. (Oligocène inférieur de Taqah, Formation d'Ashawq, Sultanat d'Oman). *Comptes Rendus de l'Académie des Sciences, Paris* 311: 1455–60.

Crochet, J. Y., Thomas, H., Sen, S., Roger, J., Gheerbrant, E., and Al-Sulaimani, Z. 1992. Découverte d'un péradectidé (Marsupialia) dans l'Oligocène inférieur du Sultanat d'Oman: Nouvelles données sur la paléobiogéographie des Marsupiaux de la plaque arabo-africaine. *Comptes Rendus de l'Académie des Sciences, Paris* 314: 539–45.

Eisenmann, V., and Whybrow, P. J. 1999. Hipparions from the late Miocene Baynunah Formation, Emirate of Abu Dhabi, United Arab Emirates. Chap. 19 in *Fossil Vertebrates of Arabia,* pp. 234–53 (ed. P. J. Whybrow and A. Hill). Yale University Press, New Haven.

Forey, P. L., and Young, S. V. T. 1999. Late Miocene fishes of the Emirate of Abu Dhabi, United Arab Emirates. Chap. 12 in *Fossil Vertebrates of Arabia,* pp. 120–35 (ed. P. J. Whybrow and A. Hill). Yale University Press, London and New Haven.

Freund, R., Garfunkel, Z., Zak, I., Goldberg, M., Weisbrod, T., and Derin, B. 1970. The shear along the Dead Sea rift. *Philosophical Transactions of the Royal Society of London* 267: 107–30.

Gentry, A. W. 1999a. A fossil hippopotamus from the Emirate of Abu Dhabi, United Arab Emirates. Chap. 21 in *Fossil Vertebrates of Arabia,* pp. 271–89 (ed. P. J. Whybrow and A. Hill). Yale University Press, New Haven.

———. 1999b. Fossil pecorans from the Baynunah Formation, Emirate of Abu Dhabi, United Arab Emirates. Chap. 22 in *Fossil Vertebrates of Arabia,* pp. 290–316 (ed. P. J. Whybrow and A. Hill). Yale University Press, New Haven.

Gheerbrant, E., and Thomas, H. 1992. The two first possible new adapids from the Arabian Peninsula (Taqah), Early Oligocene of Sultanate of Oman. Paper presented at the 14th Congress of the International Society of Primatology, Strasbourg, 16–21 August, 1992. Abstracts p. 349.

Gheerbrant, E., Thomas, H., Roger, J., Sen, S., and Al-Sulaimani, Z. 1993. Deux nouveaux primates dans l'Oligocène inférieur de Taqah (Sultanat d'Oman): Premiers Adapiformes (?Anchomomyini) de la péninsule Arabique? *Palaeovertebrata* 22: 141–96.

Gheerbrant, E., Thomas, H., Sen, S., and Al-Sulaimani, Z. 1995. Nouveau primate Oligopithecinae (Simiiformes) de l'Oligocène inférieur de Taqah, Sultanat d'Oman. *Comptes Rendus de l'Académie des Sciences, Paris* 321: 425–32.

Glennie, K. W., and Evamy, B. D. 1968. Dikaka: Plants and plant-root structures associated with aeolian sand. *Palaeogeography, Palaeoclimatology, Palaeoecology* 4: 77–87.

Greenwood, P. H. 1983. A new Palaeocene fish from Saudi Arabia. *Journal of Natural History* 17: 405–18.

Hamilton, W. R., Whybrow, P. J., and McClure, H. A. 1978. Fauna of fossil mammals from the Miocene of Saudi Arabia. *Nature* 274: 248–49.

Hill, A., and Gundling, T. 1999. A monkey (Primates, Cercopithecidae) from the late Miocene of Abu Dhabi, United Arab Emirates. Chap. 16 in *Fossil Vertebrates of Arabia*, pp. 198–203 (ed. P. J. Whybrow and A. Hill). Yale University Press, New Haven.

Hill, A., Whybrow, P. J., and Yasin al-Tikriti, W. 1990. Late Miocene primate fauna from the Arabian Peninsula: Abu Dhabi, United Arab Emirates. *American Journal of Physical Anthropology* 81: 240–41.

Jeffery, P. A. 1999. Late Miocene swan mussels from the Baynunah Formation, Emirate of Abu Dhabi, United Arab Emirates. Chap. 10 in *Fossil Vertebrates of Arabia*, pp. 111–15 (ed. P. J. Whybrow and A. Hill). Yale University Press, New Haven.

Lapparent de Broin, F. de, and Dijk, P. P. van. 1999. Chelonia from the late Miocene Baynunah Formation, Emirate of Abu Dhabi, United Arab Emirates: Palaeogeographic implications. Chap. 13 in *Fossil Vertebrates of Arabia*, pp. 136–62 (ed. P. J. Whybrow and A. Hill). Yale University Press, New Haven.

Leakey, R. E., and Leakey, M. G. 1986. A new Miocene hominoid from Kenya. *Nature* 318: 173–75.

Lindsay, E. H. 1988. Cricetid rodents from Siwalik deposits near Chinji Village. Part I. Megacricetodontinae, Myocricetodontinae, and Dendromurinae. *Palaeovertebrata* 18: 95–154.

Madden, C. T. 1982. *Paleocene, Pycnodont Fishes from Jabal Umm Himar, Harrat Hadan Area, At Taif Region, Kingdom of Saudi Arabia*. United States Geological Survey Preliminary Report, Jiddah.

Madden, C. T., Naqvi, I. M., Whitmore, F. C., Schmidt, J. D., Langston, J., and Wood, R. C. 1979. *Paleocene Vertebrates from Coastal Deposits in Harrat Hadan Area, At Taif Region, Kingdom of Saudi Arabia*. United States Geological Survey, Jiddah.

Madden, C. T., Schmidt, D. L., and Whitmore, F. C. 1983. *Masritherium (Artiodactyla, Anthracotheriidae) from Wadi Sabya, Southwestern Saudi Arabia: An Earliest Miocene Age for Continental Rift-valley Volcanic Deposits of Red Sea Margin*. Open-file Report USGS-OF-03-61. Ministry of Petroleum and Mineral Resources, Jiddah, Saudi Arabia.

Mordan, P. B. 1999. A terrestrial pulmonate gastropod from the late Miocene Baynunah Formation,

Emirate of Abu Dhabi, United Arab Emirates. Chap. 11 in *Fossil Vertebrates of Arabia,* pp. 116–19 (ed. P. J. Whybrow and A. Hill). Yale University Press, New Haven.

Pickford, M., and Thomas, H. 1994. Sexual dimorphism in *Moeripithecus markgrafi* from the early Oligocene of Taqah, Oman. In *Current Primatology,* pp. 261–64 (ed. B. Thierry, J. R. Anderson, J. J. Roeder, and N. Herrenschmidt). Université Louis Pasteur, Strasbourg.

Pickford, M., Thomas, H., Sen, S., Roger, J., Gheerbrant, E., and Al-Sulaimani, Z. 1994. Early Oligocene Hyracoidea (Mammalia) from Thaytiniti and Taqah, Dhofar Province, Sultanate of Oman. *Comptes Rendus de l'Académie des Sciences, Paris* 318: 1395–400.

Rauhe, M., Frey, E., Pemberton, D. S., and Rossmann, T. 1999. Fossil crocodilians from the late Miocene Baynunah Formation of the Emirate of Abu Dhabi, United Arab Emirates: Osteology and palaeoecology. Chap. 14 in *Fossil Vertebrates of Arabia,* pp. 163–85 (ed. P. J. Whybrow and A. Hill). Yale University Press, New Haven.

Roger, J., Pickford, M., Thomas, H., Lapparent de Broin, F. de, Tassy, P., Van Neer, W., Bourdillon-de-Grissac, C., and Al-Busaidi, S. 1994. Découverte de vertébrés fossiles dans le Miocène de la région du Huqf au Sultanat d'Oman. *Annales de Paléontologie, Paris* 80: 253–73.

Sen, S., and Thomas, H. 1979. Découverte de Rongeurs dans le Miocène moyen de la Formation Hofuf (Province du Hasa, Arabie Saoudite). *Compte Rendu Sommaire de la Société Géologique de France* 1: 34–37.

Senut, B., and Thomas, H. 1994. First discoveries of anthropoid postcranial remains from Taqah (Early Oligocene, Sultanate of Oman). In *Current Primatology,* pp. 255–60 (ed. B. Thierry, J. R. Anderson, J. J. Roeder, and N. Herrenschmidt). Université Louis Pasteur, Strasbourg.

Sigé, B., Thomas, H., Sen, S., Gheerbrant, E., Roger, J., and Al-Sulaimani, Z. 1994. Les chiroptères de Taqah (Oligocène inférieur, Sultanat d'Oman): Premier inventaire systématique. *Münchner Geowissenschaftliche Abhandlungen* 26: 35–48.

Tassy, P. 1999. Miocene elephantids (Mammalia) from the Emirate of Abu Dhabi, United Arab Emirates: Palaeobiogeographic implications. Chap. 18 in *Fossil Vertebrates of Arabia,* pp. 209–33 (ed. P. J. Whybrow and A. Hill). Yale University Press, New Haven.

Thomas, H. 1982. The Lower and Middle Miocene land connection of the Afro-Arabian plate and Asia: A major event for hominoid dispersal? In *Ancestors: The Hard Evidence,* pp. 42–50 (ed. E. Delson). Alan R Liss, New York.

———. 1983. Les Bovidae (Artiodactyla, Mammalia) du Miocène moyen de la Formation Hofuf (Province du Hasa, Arabie Saoudite). *Palaeovertebrata* 13: 157–206.

Thomas, H., Roger, J. S., and Al-Sulaimani, Z. 1988. Découverte des plus anciens "anthropoïdes" du continent arabo-africain et d'un primate tarsiiforme dans l'Oligocène du Sultanat d'Oman. *Comptes Rendus de l'Académie des Sciences, Paris* 306: 823–29.

Thomas, H., Roger, J., Sen, S., Bourdillon-de-Grissac, C., and Al-Sulaimani, Z. 1989a. Découverte de vertébrés fossiles dans l'Oligocène inférieur du Dhofar (Sultanat d'Oman). *Geobios* 22: 101–20.

Thomas, H., Roger, J., Sen, S., Dejax, J., Schuler, M., Al-Sulaimani, Z., Bourdillon-de-Grissac, C., Breton,

G., Broin, F. de, Camoin, G., Cappetta, H., Carriol, R. P., Cavelier, C., Chaix, C., Crochet, J. Y., Farjanel, G., Gayet, M., Gheerbrant, E., Lauriat-Rage, A., Noël, D., Pickford, M., Poignant, A. F., Rage, J. C., Roman, J., Rouchy, J. M., Secrétan, S., Sigé, B., Tassy, P., and Wenz, S. 1991a. Essai de reconstruction des milieux de sédimentation et de vie des Primates anthropoïdes de l'Oligocène de Taqah (Dhofar, Sultanat d'Oman). *Bulletin de la Société Géologique de France, Paris* 162: 713–24.

Thomas, H., Roger, J., Sen, S., Pickford, M., Gheerbrant, E., Al-Sulaimani, Z., and Al-Busaidi, S. 1999. Oligocene and Miocene terrestrial vertebrates in the southern Arabian peninsula (Sultanate of Oman) and their geodynamic and palaeogeographic settings. Chap. 30 in *Fossil Vertebrates of Arabia,* pp. 430–42 (ed. P. J. Whybrow and A. Hill). Yale University Press, New Haven.

Thomas, H., Sen, S., Khan, M., Battail, B., and Ligabue, G. 1982. The Lower Miocene Fauna of Al-Sarrar (Eastern Province, Saudi Arabia). *ATLAL, The Journal of Saudi Arabian Archaeology* 5: 109–36.

Thomas, H., Sen, S., and Ligabue, G. 1980. La faune Miocène de la Formation Agha Jari du Jebel Hamrin (Irak). *Proceedings Koninklijke Nederlandse Akademie van Wetenschappen* B83: 269–87.

Thomas, H., Sen, S., Roger, J., and Al-Sulaimani, Z. 1991b. The discovery of *Moeripthecus markgrafi* Schlosser (Propliopithecidae, Anthropoidea, Primates) in the Ashawq Formation (Early Oligocene of Dhofar Province, Sultanate of Oman). *Journal of Human Evolution* 20: 33–49.

Thomas, H., Taquet, P., Ligabue, G., and Del'Agnola, C. 1978. Découverte d'un gisement de vertébrés dans les dépôts continentaux du Miocène moyen du Hasa (Arabie Saoudite). *Compte Rendu Sommaire de la Société Géologique de France* 1978: 69–72.

Thomas, H., Tassy, P., and Sen, S. 1989b. Paleogene proboscidean remains from the Southern Dhofar (Sultanate of Oman). Paper presented at the Fifth International Theriological Congress (Rome, 22–29 August, 1989), sect. "Evolution and Paleoecology of Proboscidea". Abstracts, Papers, and Posters I: 166.

Tleel, J. W. 1973. Surface geology Damman Dome, Eastern Province, Saudi Arabia. *Bulletin American Association of Petroleum Geologists* 57: 558–76.

Whitmore, F. C., and Madden, C. T. 1995. *Paleocene Vertebrates from Jabal Umm Himar, Kingdom of Saudi Arabia.* United States Geological Survey Bulletin no. 2093. Government Printing Office, Washington D.C.

Whybrow, P. J. ed. 1987. Miocene geology and palaeontology of Ad Dabtiyah, Saudi Arabia. *Bulletin of the British Museum (Natural History),* Geology 41: 367–457.

Whybrow, P. J. 1989. New stratotype; the Baynunah Formation (Late Miocene), United Arab Emirates: Lithology and palaeontology. *Newsletters on Stratigraphy* 21: 1–9.

———. 1992. Land movements and species dispersal. In *The Cambridge Encyclopedia of Human Evolution,* pp. 169–73 (ed. S. Jones, R. Martin, and D. Pilbeam). Cambridge University Press, Cambridge.

Whybrow, P. J., and Clements, D. 1999. Late Miocene Baynunah Formation, Emirate of Abu Dhabi, United Arab Emirates: Fauna, Flora, and localities. Chap. 23 in *Fossil Vertebrates of Arabia,* pp. 317–33 (ed. P. J. Whybrow and A. Hill). Yale University Press, New Haven.

Whybrow, P. J., and Hill, A. eds. 1999. *Fossil Vertebrates of Arabia.* Yale University Press, New Haven.

Whybrow, P. J., and McClure, H. A. 1981. Fossil mangrove roots and palaeoenvironments of the Miocene of the eastern Arabian peninsula. *Palaeogeography, Palaeoclimatology, Palaeoecology* 32: 213–25.

Whybrow, P. J., Collinson, M. E., Daams, R., Gentry, A. W., and McClure, H. A. 1982. Geology, fauna (Bovidae, Rodentia) and flora from the Early Miocene of eastern Saudi Arabia. *Tertiary Research* 4: 105–20.

Whybrow, P. J., Hill, A., Yasin al-Tikriti, W., and Hailwood, E. A. 1990. Late Miocene primate fauna, flora and initial palaeomagnetic data from the Emirate of Abu Dhabi, United Arab Emirates. *Journal of Human Evolution* 19: 583–88.

Tertiary volcanics, Gulf of Aden, Republic of Yemen.

The Tethyan Arabian Gulf, the Mediterranean, and the World's Tertiary Oceans

PART VI

This part brings together three research projects that focus on the geochronological relationships between the oceans and landmasses. These projects provide an overview of events that relate to the geology of the marine rocks of the Arabian Peninsula during the Tertiary period, 65 to 2 million years ago.

C. Geoffrey Adams, Deryck D. Bayliss, and John E. Whittaker (Chapter 34) describe three modern hypotheses: those of Drooger, of Steininger and Rögl, and of Adams and colleagues. These hypotheses relate to the dating of the event that finally separated the proto-Mediterranean from the Indian Ocean, and they are analysed and dis cussed in the light of data that have become available during the past decade. Several problems of particular importance to the dating and correlation of marine Neogene sediments in the Middle East are also discussed. It is concluded that although the available palaeontological and lithostratigraphical evidence now points most strongly to an early Miocene (Aquitanian) disconnection, and suggests that a later reconnection in middle Miocene times is highly unlikely, further studies of certain sequences in Iran and Iraq will have to be undertaken before a general consensus can emerge.

Fred Rögl (Chapter 35) provides up-to-date information on palaeogeographic reconstructions of the area around the Mediterranean from the Oligocene to the Pliocene. The palaeogeographic reconstruction of the circum-Mediterranean area is dependent on the palinspastic rearrangement of tectonic units and on well-defined stratigraphic correlation between different sedimentary basins. Detailed information for such a rearrangement is, however, limited. For the tectonic requirements, information is available from around the Apennines, Alps, and Carpathians, but is not as accurate for the Eastern Mediterranean region. An updated and refined stratigraphic correlation for the Mediterranean and Paratethys is presented, based on new palaeomagnetic, radiometric, and biostratigraphic data from different research groups. Current palaeogeographic questions that are discussed centre on the intracontinental mammal migrations between Europe, North America, and Africa in the Oligocene and Miocene for specific regions, such as (1) the European–Asian connection, (2) the Bering landbridge, (3) the "*Gomphotherium*" landbridge in the Near East, and (4) the Gibraltar landbridge.

Finally, Norman MacLeod (Chapter 36) describes changes to the world's oceans during the Miocene epoch and earlier. The oceanographic changes in the Oligocene through to the Miocene are among the best understood of any ancient time interval and illustrate the complex interplay of tectonic, circulation, climatic, and watermass factors. Tectonically, the two most important palaeoceanographic events of this period were the progressive isolation of the Antarctic landmass at the South Pole and the sinking of the Iceland–Faeroe ridge between the North Atlantic and Arctic Oceans. The former led to the isolation of the Antarctic continent with its progressive refrigeration as development of a circum-Antarctic current system prevented warm waters from the southern midlatitudes from migrating south. The latter changed the nature of deep-water circulation throughout the Atlantic basin such that relatively warm saline waters originating in the Arctic Ocean surfaced south of the Antarctic Convergence. By the middle Miocene, these two factors combined to produce extensive glaciation of the Antarctic continent, a substantial lowering of sea level, and increase in overall oceanic salinities. These changes in turn affected mid- and low-latitude circulation as well as properties of oceanic water masses.

In the biotic realm several noteworthy changes can be either directly or indirectly tied to these palaeoceanographic events. These include a dramatic increase in plankton productivity in the Southern Ocean, with the accompanying diversification of plankton-feeding clades (for example, mysticete whales); increases in mid- and low-latitude endemism among planktonic biotas; and increases in terrestrial biotas adapted to a mixed woodland–grassland and savannah-like biome as increases in the latitudinal temperature gradient during the Miocene forced these habitats to replace the Oligocene and early Miocene tropical forests.

The Terminal Tethyan Event: A Critical Review of the Conflicting Age Determinations for the Disconnection of the Mediterranean from the Indian Ocean

34

C. Geoffrey Adams,[1] Deryck D. Bayliss, and John E. Whittaker

More than a hundred years ago, Suess (1893) introduced Tethys as a broad seaway separating Europe and North Africa, and stretching eastwards across the area now occupied by Turkey, Syria, Iraq, Iran, Pakistan, northern India, Tibet, Burma, and Thailand, thus uniting the Atlantic with the Indian and Pacific Oceans. Geologists and palaeontologists have since made numerous attempts to date the event that caused the final disruption of Tethys and, with it, the creation of the proto-Mediterranean Sea. It has long been known that the connection with the Pacific was lost at about the end of Cretaceous times and that the relict easterly arm of Tethys finally disappeared during the Eocene when India collided with Asia.

The earliest attempts at dating the terminal disconnection were based on the distribution of plants and terrestrial animals (mainly mammals) in northern Africa, Arabia, and southwest Asia, it being realised that these organisms could not readily cross any seaway of significant breadth. The result was a general consensus of opinion that a landbridge from Arabia to southwest Asia must have come into existence during Miocene times (Termier and Termier, 1960). Dating the relative sedimentary sequences was, however, always a problem and it was not until planktonic foraminifera became widely and successfully used in zonation (from 1957 onwards), and modern chronostratigraphic methods were introduced, that accurate age determinations of the critical marine successions became possible.

In recent years, three attempts have been made to date the disconnection more accurately than hitherto and although all are based on the distribution of marine invertebrates (with or without a consideration of lithofacies), they have produced conflicting results.

Drooger (1979, 1993), on the basis of morphometric data obtained from studies of generally accepted evolutionary lineages in three families of larger foraminifera (the Miogypsinidae, Cycloclypeidae, and Lepidocyclinidae), concluded that the terminal disconnection occurred at about the end of Oligocene times—that is around 23.8 million years (Ma) ago, according to Steininger et al. (1994). Steininger and Rögl (1979) and Rögl and Steininger (1983, 1984), utilising previously published evidence on the distribution of marine and terrestrial sediments plus some scattered palaeontological data, concluded that although a disconnection may have occurred in the early Miocene, the seaway was re-opened later and not finally closed until the middle Miocene. Adams et al. (1983), basing their conclusions on the distributions of all genera of larger foraminifera in the mid-Tertiary of

 ISBN 0-300-07183-3

the Mediterranean, Middle East, and Indian Ocean areas, in the context of the best-known sedimentary sequences, decided that the late early Miocene (Burdigalian) was the most probable time of disconnection.

Clearly, all three conclusions cannot be correct. The purpose of this chapter is, therefore, to review and explain the evidence on which the various conclusions depend and to assess the relative merits of the authors' claims in the light of data obtained since 1983. First, however, it is necessary to consider briefly four important problems that have complicated and confused the literature for the past 50 years.

BIOSTRATIGRAPHIC PROBLEMS

Failure to Identify Critical Species Accurately

When this occurs, sediments are wrongly dated and correlations inaccurate. The foraminiferal genus *Austrotrillina* provides many good examples of inaccurate identifications. Before 1968, only two species of *Austrotrillina* were known from the Middle East, *A. paucialveolata* Grimsdale and *A. howchini* (Schlumberger). But Adams (1968) then showed that the true *A. howchini* from the Indo-West Pacific region was quite different (structurally more advanced), and also younger, than the forms described under this name from the Middle East. He therefore described a new species, *A. asmariensis,* which accommodated most of the wrongly named forms previously described from the Middle East. Despite this, many subsequent authors continued to use *A. howchini* in the wrong sense and to perpetuate the belief that this species is known in the Middle East. There are a few records of *A. howchini* from the western Mediterranean (see Adams, 1968), but these require re-examination in the light of its absence from the Middle East.

Abid and Sayyab (1989), in a valid attempt to determine the age and provenance of the derived blocks in the basal conglomerate of the Euphrates Limestone, claimed to have found four species of *Austrotrillina* including *A. howchini* in the conglomerates. The most cursory examination of the specimens they illustrate as *A. howchini* show that they in no way resemble the true *howchini*. Compare, for example, Adams (1968: pl. 2, figs 3, 4, and 7) with Abid and Sayyab (1989: pl. 1, fig. 2), which, in fact, is a perfectly good *A. asmariensis.* All records of *A. howchini* from the Middle East should therefore be treated with suspicion, unless they can be verified.

Assumptions that the Ranges of Certain Marker Taxa Are Well Established

Most palaeontologists appear to believe that the first and last appearances of certain marker species are well established, but this is not true. For example, they believe that *Orbulina suturalis* Brönnimann and *Borelis melo curdica* Reichel in the Middle East appeared simultaneously and that they, by definition, mark the base of the middle Miocene (N 9). While this is true for *O. suturalis,* it is not true for *B. melo curdica,* which is present in the upper part of N 8 and thus straddles the boundary (Adams, 1983, 1992; Adams et al., 1983). *Borelis melo curdica* can thus be regarded as a reliable middle Miocene marker only when it is accompanied by other marker species. Other so-called markers are far less well known than *B. melo curdica.*

Absolute Ranges of Larger Foraminiferal Species

For some strange reason, micropalaeontologists tend to accept statements about the ranges of important taxa when they are made by well-known specialists, even when they are unaccompanied by proof of any kind. Adams (1992) has discussed this problem with reference to *Miogypsina* and has questioned the upward extension of its range in the Indo-Pacific region from N8 to N14. Drooger (1963), in a well-known review article, stated that the evolution of the *Miogypsina* lineage appeared to be over in the Mediterranean area just before the appearance of *Orbulina,* and no subsequent evidence has appeared to prove him wrong. Yet Prazak (1978) and other authors, while accepting that the Chattian/Aquitanian faunas of northern Iraq are clearly Mediterranean in type, then go on to record *Miogypsina* (species usually unnamed) from strata of middle Miocene or younger age (see Prazak, 1978: tables 2 and 5). They explain this anomaly by citing the already questionable extended range for the genus in the Indo-Pacific, thus implying a northward transgression of the Indian Ocean during the middle or later Miocene that allowed one Indo-Pacific genus to migrate northwards, but not any of the others. Such a scenario is improbable to say the least.

THE OLIGOCENE/MIOCENE BOUNDARY

This has always been a difficult boundary to define in shallow-water marine sediments because the larger foraminiferal faunas in the two stages are very similar, and, for our purposes here, the recently published proposals of Steininger et al. (1994) are accepted. These state (inter alia) that the boundary coincides approximately with the FAD (first appearance datum) of the planktonic foraminifer *Paragloborotalia kugleri* (Bolli). This requires some modification to the same boundary as defined in the East Indies Letter Classification (for example, Adams, 1970, 1983) for shallow-shelf, carbonate environments, but the necessary changes can be easily accommodated. Basically, the boundary would be marked by the first appearances of *Miogypsina gunteri* Cole and/or *Miogypsinoides bantamensis* Tan Sin Hok.

THE THREE DISCONNECTION HYPOTHESES

Disconnection at the End of Oligocene Time (Drooger's Hypothesis)

Drooger (1979) regarded the disconnection as having occurred at about the end of the Chattian. He used data from morphometric studies of the larger foraminiferal genera *Miogypsina* s.s. and *Lepidocyclina* (*Nephrolepidina*), together with the extinction of *Cycloclypeus* in the Mediterranean region, as the basis for his conclusion. He repeated his claim in 1993, but without mentioning that there were two alternative hypotheses.

The three taxa on which Drooger relied all appeared in the Mediterranean region in Oligocene times and all three are known to have survived far longer in the Indo-Pacific region. All show gradual evolutionary changes (trends) in the juvenile parts of the shell.

The morphometric approach to taxonomy involves the study of the evolutionary trends shown in various shell features and their quantification based on counts or measurements of different parameters. Given sufficient specimens, a statistically significant mean can be calculated for each parameter and the results plotted in the form of graphs for assemblages from different stratigraphic horizons. Using this method, results have so far been obtained from lineages within genera belonging to several different families, the best coming from the main *Miogypsina* lineage (several other lineages occur within this genus but for various reasons are not applicable to the solution of the disconnection problem; see Drooger, 1993).

Drooger has been able to show that the Mediterranean lineage of *Miogypsina* differs from the its equivalent in the Indian Ocean. Similarly, the lineages within *Lepidocyclina* (*Eulepidina*) and *L.* (*Nephrolepidina*) seem to differ in the two regions after the end of Oligocene times. These differences do not, however, become significant until the Burdigalian.

Cycloclypeus was represented in both regions during the Oligocene, but became extinct in the western Tethys during Chattian time; it lives on in the Indo-Pacific today. The reason for the extinction of this genus in the western Tethys is unknown. It could, as Adams (1967) suggested, and as Drooger (1993) now appears to believe, be linked to the disconnection of the two areas. But it also disappeared from the western side of the Indian Ocean at about the same time and this can hardly be attributed to the closure of the Tethys seaway. Our knowledge of *Cycloclypeus* in the Mediterranean region is probably now fairly complete thanks to the intensive research undertaken in this area during the past 60 years. The same cannot be said of the Indian Ocean where much work remains to be done. Comparison of *Cycloclypeus* lineages in the two areas is therefore impossible at present.

Species based on morphometric studies have the advantage of being sharply defined. Thus, *Miogypsinoides bantamensis* Tan is described by Drooger (1993) as possessing from 10 to 13 chambers in the initial whorl whereas *Miogypsinoides deharti* van der Vlerk has 8 to more than 10. The element of subjectivity in identification is therefore removed, but morphometric species are not Linnean species and no biological reality is implied by these names. Whether Linnean names necessarily imply a greater degree of reality is an open question.

Lepidocyclina is the other genus in which the evolution of the embryonic apparatus has been shown to be susceptible to morphometric investigation, and work in the East Indies and in the Mediterranean regions appears to indicate that the lineages were different in the two regions (see Drooger, 1993, for a summary). Morphometric studies of the subgenera *Eulepidina* and *Nephrolepidina* have not, however, yet yielded the fine morphometric distinctions claimed for *Miogypsina* and correlation is therefore less precise.

Drooger's hypothesis (1979, 1993) for the dating of the terminal disconnection may be said to rely on two unproven assumptions:

1. That the main *Miogypsina* lineage developed differently in the Mediterranean and Indo-Pacific regions after the end of Oligocene times. This is true, but the main changes were in post-Aquitanian times and not, therefore, necessarily relevant to this discussion, although, as Drooger says, a time-lag might be expected to occur.
2. That *Lepidocyclina* (*Nephrolepidina*) and *L.* (*Eulepidina*) also developed differently in the two areas after Oligocene times. Unfortunately, no comparative study of the lepidocyclines of the Middle East and the western part of the Indian subcontinent yet exists. Beretti and Ambroise's (1980) work on the Madagascar faunas fails to elucidate the situation.

Disconnection in Middle Miocene Times (Steininger and Rögl's Hypothesis)

Steininger and Rögl approached the disconnection problem in a quite different way. Following a preliminary paper (Steininger and Rögl, 1979), they (Rögl and Steininger, 1983, 1984) subsequently constructed a series of impressive palaeogeographic maps for the Mediterranean area and Middle East on the basis of previously published lithological data, but apparently without checking whether the sediments concerned were accurately dated. They concluded that a marine connection between the Indian Ocean and the Mediterranean area existed throughout the Oligocene, that it was disrupted in the Middle East at some time during the early Miocene, when evaporites were widely deposited, and briefly restored again in early middle Miocene times. No lithostratigraphic evidence was presented for the middle Miocene reconnection of the two areas and indeed none exists, reliance being placed on the supposed occurrence of two or three Indo-Pacific species of planktonic foraminifera in Paratethys. This hypothesis, as Adams et al. (1983) pointed out, requires a deep-water seaway of mid-Miocene age along the line of the present Zagros Range. While being theoretically possible, this is highly unlikely, especially as simpler explanations are available.

Another objection made to the Steininger and Rögl hypothesis is that the deep-water trough necessary for the passage of planktonic foraminifera could not have existed without the presence of shallow-water facies on either side. Larger foraminifera would certainly have "migrated" along such shallow-water shelves, but there is little evidence that this occurred. Rögl and Brandstätter (1993) reported *Amphistegina mammilla* (Fichtel and Moll) from central Poland and Austria, but of all possible larger foraminifera, *Amphistegina* is the least impressive as a potential migrant from the Indian Ocean because the genus was present in the Mediterranean region at a much earlier date and could therefore have evolved in situ. In fact, these authors admit that there is at least one questionable record of this species from Israel.

Convincing evidence of reconnection can be provided only by the discovery of Indo-Pacific genera such as *Alveolinella* or *Flosculinella* or of species of *Lepidocyclina* or *Cycloclypeus*, the immediate ancestors of which are known not to have been present in the Mediterranean region previously. Such evidence has not yet been forthcoming.

Disconnection During the Early Miocene (Adams, Gentry, and Whybrow's Hypothesis)

Adams et al. (1983) investigated and correlated the distribution of critical marine carbonates and terrestrial sediments from the northern limits of the Indian Ocean to the Mediterranean and tabulated their associated faunas. They concluded that the continuity of marine sedimentation across the Middle East appears to have been interrupted during Aquitanian times. By the mid-Burdigalian at the latest, a definite barrier to the dispersal of marine organisms existed between the Indian Ocean and the Mediterranean, and that this could have been the landbridge needed for the Orleanian dispersal of mammals. They also firmly refuted Steininger and Rögl's contention that a middle Miocene reconnection occurred.

In the past 12 years new information has been published that has a critical bearing on the arguments and this is added below.

In 1989 Abawi firmly dated the Euphrates Formation of northwest Iraq showing that it falls within the planktonic zones N 7–N 8 (late Burdigalian–early Langhian), but the depositional environment indicated by the smaller benthic foraminifera was shallow-water, open marine. This is further evidence that there was no deep-water connection between either the Indian Ocean or the Mediterranean in Burdigalian times. The overlying Jeribe Limestone is dated not by planktonic foraminifera but by larger foraminifera, the assumption being that *Borelis melo curdica* Reichel is a marker for the middle Miocene. This is not true, however, as this taxon occurs on both sides of the early/middle Miocene boundary (Adams, 1983, 1992; Adams et al., 1983). It is clear, therefore, that further evidence is needed before a middle Miocene age can be ascribed to this limestone unit.

In this context, it should also be remembered that the uppermost part of the Gachsaran Formation (Lower Fars of many authors) in Iran (Lurestan and Khuzestan) also contains *B. melo curdica* (the furthest south this Mediterranean subspecies occurs) and is thus virtually coeval with the Jeribe Limestone.

McCall et al. (1994) described the early Miocene (Aquitanian) coral limestones from the Makran Mountains of southern Iran. Situated within an arm of the "proto-Arabian Gulf", the corals (and associated foraminifera examined and identified in part by Adams) have especial palaeobiogeographical and palaeogeographical interest because they were collected from localities close to areas of Miocene uplift associated with the severance of the Tethyan seaway. Both corals and foraminifera show an almost entirely Indo-Pacific affinity, indicating significant biogeographical separation of the Mediterranean from the Indo-Pacific region even before the final closure occurred.

In addition to this coral and foraminiferal evidence, a study of hemiasterid echinoids by Néraudeau (1994) concluded that the Indian Ocean became disconnected from the Mediterranean in early Miocene (Aquitanian) times.

Robba (1987) and Piccoli et al. (1991) presented results of detailed studies of molluscan faunas in relation to the final occlusion of Tethys, and these should be noted here. Robba's findings are based on a faunal analysis at supraspecific level from Italy and Indonesia (Java) and Piccoli et al.'s mathematical modelling is constructed from identifications from many scattered parts of Tethys and beyond (20 areas), from the Paris Basin in the west to Fiji in the east. In his abstract, Robba states "the final Tethyan event was already concluded in the Burdigalian, between 18 and 19 Ma ago"; further on (p. 409), he writes that the data provided by the benthic molluscs are in agreement with the larger foraminiferal evidence of Adams et al. (1983), indicating faunal divergence took place during the Aquitanian and that the Tethyan seaway (quoting the latter) "ceased to be an effective dispersal route for most marine taxa, and that a land-bridge between Arabia and Southwest Asia existed by mid-Burdigalian times". In their text, Piccoli et al. (1991: 187) agree with Robba (1987) about the final occlusion of Tethys occurring about 18.5 Ma ago, but their palaeogeographical sketch maps, prepared on the basis of recent literature, for the late Oligocene (25 Ma), early Miocene (20 Ma), and the middle Miocene (15 Ma) all show an interrupted, if restricted seaway through the Middle East connecting eastern and western Tethys. Occlusion is indicated on the map only for the Messinian (5.5 Ma). Their three land/sea distribution maps (late Oligocene to middle Miocene) have an "unreal" appearance and are unconvincing, especially the long, narrow seaways and the parallel, elongate peninsulas.

Finally, it must be said, in fairness to others, that Adams et al. (1983) based their conclusions on the distribution of all published records of larger foraminifera across the Middle East known up to that time. The weakness of that approach is that some records are not verifiable, because material is unavailable, so certain identifications are questionable.

Adams et al. (1983) accepted that there was never a good marine connection between the Mediterranean and Indian Ocean after the Oligocene, Aquitanian time being represented mainly by evaporitic deposits. Nevertheless, they thought that the Burdigalian faunas and sediments across the region provided possible evidence for the continuance of a marine link. As a consequence of the absence of the true *Austrotrillina howchini* (Schlumberger) in the region, the failure of *Flosculinella,* and associated genera such as *Cycloclypeus, Alveolinella,* and *Lepidocyclina* (*Nephrolepidina*), to penetrate further north than the southern part of the Persian/Arabian Gulf (Guri Limestone) and of *Borelis melo curdica* Reichel (a Mediterranean subspecies) to occur in the Indian Ocean, it is now thought that there was no continuous Burdigalian seaway across the Middle East.

On present evidence, effective disconnection of the Mediterranean from the Indian Ocean occurred, if not precisely at the end of the Chattian (as Drooger, 1979, postulated on other evidence), soon after in early Miocene (Aquitanian) times.

Such evidence seems to fit well with Whybrow's notion (1984: 195) on the age of the mammal-bearing Hadrukh Formation of Saudi Arabia (Whybrow et al., 1982). Thus an early Miocene connection between Arabia and Iran/Iraq might have been available for "inter-migrations" of Afro-Arabian and Southwest Asian continental faunas.

SUMMARY AND CONCLUSIONS

Drooger's hypothesis (1979, 1993) has the merit of internal consistency. It relies only on morphometric studies of a few taxa, disregards the rest and takes no account of the sedimentary record.

Steininger and Rögl (1979; Rögl and Steininger, 1983, 1984) accept the sedimentary record at its face value and do not question the age determinations of previous authors. They postulate a mid-Miocene reconnection of the Mediterranean and Indian Oceans, arguing that the marine sediments that would provide the necessary evidence are now buried below the Zagros Range.

NOTE

1. C. Geoffrey Adams died on 6 February 1995. His unfinished manuscript (to have been presented by C.G.A. in a complete form at the First International Conference on the Fossil Vertebrates of Arabia, held in the Emirate of Abu Dhabi, March 1995) has been completed and updated by two of his colleagues as a mark of respect for his considerable knowledge and his unswerving pursuit of accuracy and precision in Cenozoic stratigraphy. The opinions expressed in this chapter, however, are those of C. Geoffrey Adams.

REFERENCES

Abawi, T. S. 1989. Foraminifera, stratigraphy and sedimentary environment of the Euphrates Formation, Lower Miocene, Sinjar area, northwestern Iraq. *Newsletters on Stratigraphy* 21: 15–24.

Abid, A. A., and Sayyab, A. S. 1989. *Austrotrillina* species of "the basal conglomerates" at Khan Al-Baghdadi area, West Iraq. *Journal of the Geological Society of Iraq* 22: 18–34.

Adams, C. G. 1967. Tertiary foraminifera in the Tethyan, American and Indo-Pacific provinces. In *Aspects of Tethyan Biogeography,* pp. 195–217 (ed. C. G. Adams and D. V. Ager). Systematics Association Publication no. 7. London.

———. 1968. A revision of the foraminiferal genus *Austrotrillina* Parr. *Bulletin of the British Museum (Natural History),* Geology 16: 73–97.

———. 1970. A reconsideration of the East Indian Letter Classification of the Tertiary. *Bulletin of the British Museum (Natural History),* Geology 19: 87–137.

———. 1983. Speciation, phylogenesis, tectonism, climate and eustasy: Factors in the evolution of Cenozoic larger foraminiferal bioprovinces. In *Evolution, Time and Space: The Emergence of the Biosphere,* pp. 255–89 (ed. R. W. Sims, J. H. Price, and P. E. S. Whalley). Academic Press, New York.

———. 1992. Larger foraminifera; the dating of Neogene events. In *Pacific Neogene: Environments, Evolution and Events,* pp. 221–35 (ed. R. Tsuchi and J. C. Ingle). University of Tokyo Press.

Adams, C. G., Gentry, A. W., and Whybrow, P. J. 1983. Dating the terminal Tethyan events. In *Reconstruction of Marine Environments* (ed. J. Meulenkamp). *Utrecht Micropaleontological Bulletins* 30: 273–98.

Beretti, A., and Ambroise, D. 1980. Mise en évidence d'une lignée phyletique dans le genre *Lepidocyclina* Gümbel à Madagascar, mesure du degré d'évolution, application stratigraphique. Actes du VI(e) Colloque Africaine de Micropaléontologie; Tunis, March 1974. *Annales des Mines et de la Géologie* 28: 61–83.

Drooger, C. W. 1963. Evolutionary trends in the Miogypsinidae. In *Evolutionary Trends in Foraminifera,* pp. 315–49 (ed. G. H. R. von Koeningswald et al.). Elsevier, London.

———. 1979. Marine connections of the Neogene Mediterranean, deduced from the evolution and distribution of larger foraminifera. *Annales Géologiques des Pays Helléniques* hors sér. 1979 (1): 361–69.

———. 1993. Radial foraminifera; morphometrics and evolution. *Verhandelingen der Koninklijke Nederlandse Akademie van Wetenschappen, Afdeeling Natuurkunde, Eerste Reeks* 41: 1–242.

McCall, G. J. H., Rosen. B. R, and Darrell, J. G. 1994. Carbonate deposition in accretionary prism settings: Early Miocene coral limestones and corals of the Makran Mountain Range in southern Iran. *Facies* 31: 141–78.

Néraudeau, D. 1994. Hemiasterid echinoids (Echinodermata: Spatangoida) from the Cretaceous Tethys to the present-day Mediterranean. *Palaeogeography, Palaeoclimatology, Palaeoecology* 110: 319–44.

Piccoli, G., Sartori, S., Franchino, A., Pedron, R., Claudio, L., and Natale, A. R. 1991. Mathematical model of faunal spreading in benthic palaeobiogeography (applied to Cenozoic Tethyan molluscs). *Palaeogeography, Palaeoclimatology, Palaeoecology* 86: 139–96.

Prazak, J. 1978. The development of the Mesopotamian Basin during the Miocene. *Journal of the Geological Society of Iraq* 9: 170–89.

Robba, E. 1987. The final occlusion of Tethys: Its bearing on Mediterranean benthic molluscs. In *Shallow Tethys 2* (Proceedings of the International Symposium on Shallow Tethys 2: Wagga Wagga, September

1986), pp. 405–26 (ed. K. G. McKenzie). Balkema, Rotterdam.

Rögl, F., and Brandstätter, F. 1993. The foraminifera genus *Amphistegina* in the Korytnica Clays (Holy Cross Mts, Central Poland) and its significance in the Miocene of the Paratethys. *Acta Geologica Polonica* 43: 121–46.

Rögl, F., and Steininger, F. F. 1983. Vom Zerfall der Tethys zu Mediterran und Paratethys: Die neogene Paläogeographie und Palinspastik des zirkum-mediterranen Raumes. *Annalen des Naturhistorischen Museums, Wien* 85/A: 135–63.

———. 1984. Neogene Paratethys, Mediterranean and Indo-Pacific seaways: Implications for the paleobiogeography of marine and terrestrial biotas. In *Fossils and Climate,* pp. 171–200 (ed. P. Brenchly). Wiley, London.

Steininger, F. F., and Rögl, F. 1979. The Paratethys history—a contribution towards the Neogene geodynamics of the Alpine Orogene (An Abstract). *Annales Géologiques des Pays Helléniques* hors sér. 1979 (3): 1153–65.

Steininger, F. F., Aubry, M. P., Biolzi, M., Borsetti, A. M., Cati, F., Corfield, R., Gelati, R., Iaccarino, S., Napoleone, C., Rögl, F., Roetzel, R., Spezzaferri, S., Tateo, F., Villa, G., and Zevenboom, D. 1994. *Proposal for the Global Stratotype Section and Point (GSSP) for the base of the Neogene (the Paleogene/Neogene Boundary).* Institute of Paleontology, University of Vienna.

Suess, E. 1893. Are great ocean depths permanent? *Natural Science, London* 2: 180–87.

Termier, H., and Termier, G. 1960. *Atlas de Paléogeographie.* Masson, Paris.

Whybrow, P. J. 1984. Geological and faunal evidence from Arabia for mammal "migrations" between Asia and Africa during the Miocene. *Courier Forschungsinstitut Senckenberg* 69: 189–98.

Whybrow, P. J., Collinson, M. E., Daams, R., Gentry, A. W., and McClure, H. A. 1982. Geology, fauna (Bovidae, Rodentia) and flora from the Early Miocene of eastern Saudi Arabia. *Tertiary Research* 4: 105–20.

Oligocene and Miocene Palaeogeography and Stratigraphy of the Circum-Mediterranean Region

Fred Rögl

The interaction of an exact biostratigraphic timing of geological events, the interpretation of regional tectonics and plate tectonics, and the development of marine and continental faunas through time form the basis for attempts at palaeogeographic reconstructions. In the past many such attempts were restricted by static reconstructions and other factors such as the time slice of reconstruction having too long a timespan, the palaeogeography of the Mediterranean and Paratethys corresponding to sediment distribution maps (for example, Hamor, 1988; Cahuzac et al., 1992), or else the reconstruction for late Neogene still showing a circum-Alpine sea (for example, Dercourt et al., 1985). Tectonically based palinspastic reconstructions of different Mediterranean regions are an important step forward for a general circum-Mediterranean palaeogeography. One of these regions is the Italian peninsula and the reconstruction of the Apennines and the Tyrrhenian Sea (Boccaletti et al., 1986, 1990). Another example is the rearrangement of tectonic units in the central Mediterranean–Alpine area in different time slices by Balla (1987). An attempt at considering plate tectonic movements and the shortening between the Afro-Arabian and Eurasian Plates is necessary. Such sketches of the marine development of the Mediterranean and Paratethys seas during the Neogene, taking into account the mammal migration waves, were presented by Rögl and Steininger (1983), and Steininger et al. (1985). The late Eocene to early Miocene palaeogeography of the Paratethys, also based on tectonic processes, is given by Baldi (1980, 1986). Another method, using a mathematical model, takes into account plate tectonic reconstructions and associations of marine molluscs for reconstructing the development of the Paleogene/Neogene Tethys (Piccoli et al., 1991). A new general palaeogeographic reconstruction will be possible only after fresh attempts are made to disentangle the arrangement of the tectonic units and microplates around the Mediterranean.

At present we can only try to give some new ideas about the key features of landbridges and connecting seaways for the Oligocene to Miocene of the circum-Mediterranean realm. The areas of interest are:

—the Europe–Asia migration pathways;
—the Eurasian–North American Bering landbridge;
—the landbridges in the Near and Middle East for the African–Eurasian connection, the "*Gomphotherium*" landbridge; and
—the Gibraltar landbridge for African migrations in the west.

Important examples of mammal migrations over those continental pathways are given in figure 35.1a,b.

 ISBN 0-300-07183-3

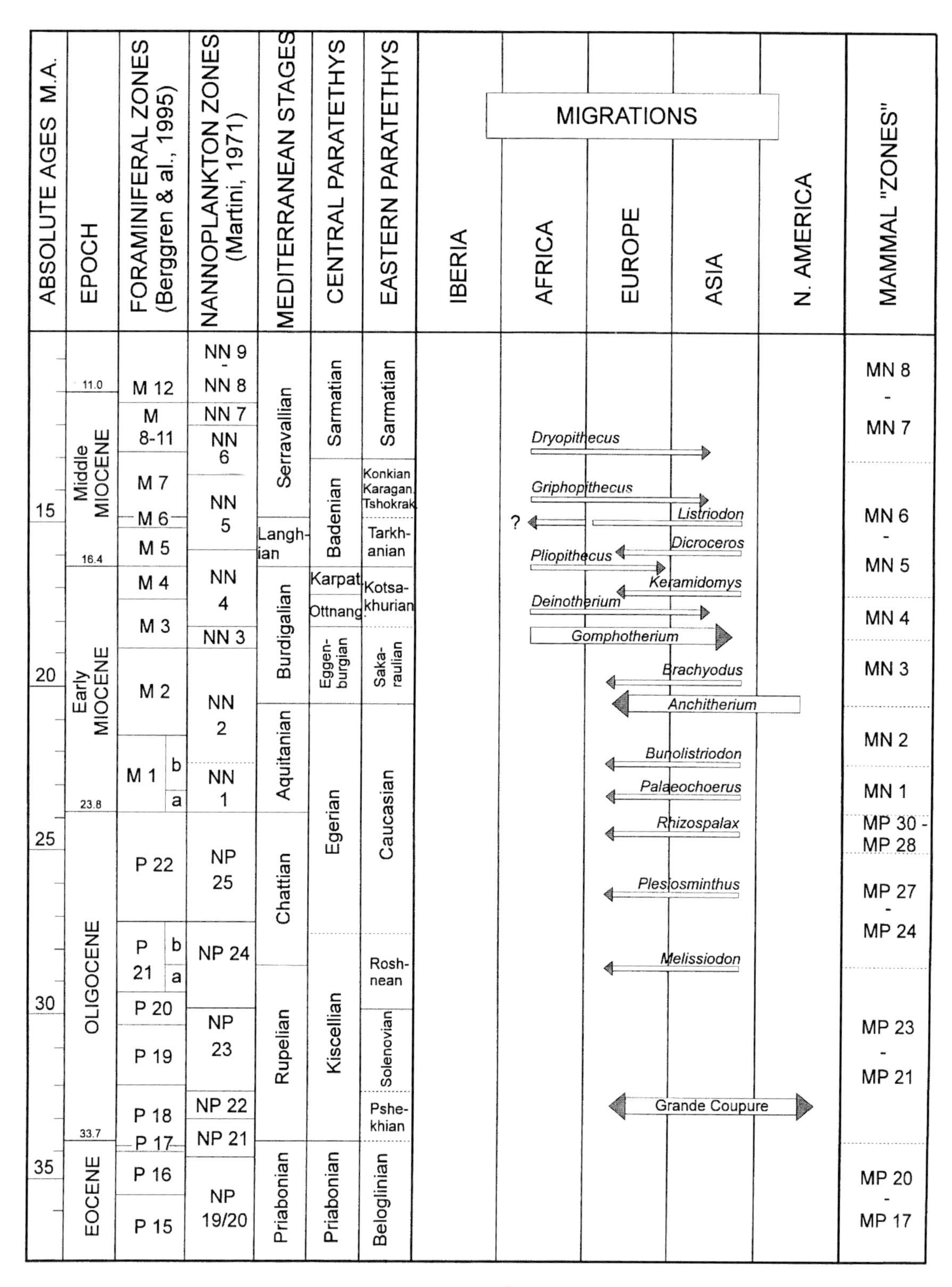
ABSOLUTE AGES M.A.
EPOCH
FORAMINIFERAL ZONES (Berggren & al., 1995)
NANNOPLANKTON ZONES (Martini, 1971)
MEDITERRANEAN STAGES
CENTRAL PARATETHYS
EASTERN PARATETHYS
MIGRATIONS
IBERIA
AFRICA
EUROPE
ASIA
N. AMERICA
MAMMAL "ZONES"
15
20
25
30
35
11.0
16.4
23.8
33.7
Middle MIOCENE
Early MIOCENE
OLIGOCENE
EOCENE
M 12
M 8-11
M 7
M 6
M 5
M 4
M 3
M 2
M 1
b
a
P 22
P 21
P 20
P 19
P 18
P 17
P 16
P 15
NN 9 - NN 8
NN 7
NN 6
NN 5
NN 4
NN 3
NN 2
NN 1
NP 25
NP 24
NP 23
NP 22
NP 21
NP 19/20
Serravallian
Langhian
Burdigalian
Aquitanian
Chattian
Rupelian
Priabonian
Sarmatian
Badenian
Karpat
Ottnang
Eggenburgian
Egerian
Kiscellian
Priabonian
Sarmatian
Konkian
Karagan
Tshokrak
Tarkhanian
Kotsakhurian
Sakaraulian
Caucasian
Roshnean
Solenovian
Pshekhian
Beloglinian
Dryopithecus
Griphopithecus
?
Listriodon
Dicroceros
Pliopithecus
Keramidomys
Deinotherium
Gomphotherium
Brachyodus
Anchitherium
Bunolistriodon
Palaeochoerus
Rhizospalax
Plesiosminthus
Melissiodon
Grande Coupure
MN 8 - MN 7
MN 6 - MN 5
MN 4
MN 3
MN 2
MN 1
MP 30 - MP 28
MP 27 - MP 24
MP 23 - MP 21
MP 20 - MP 17

A

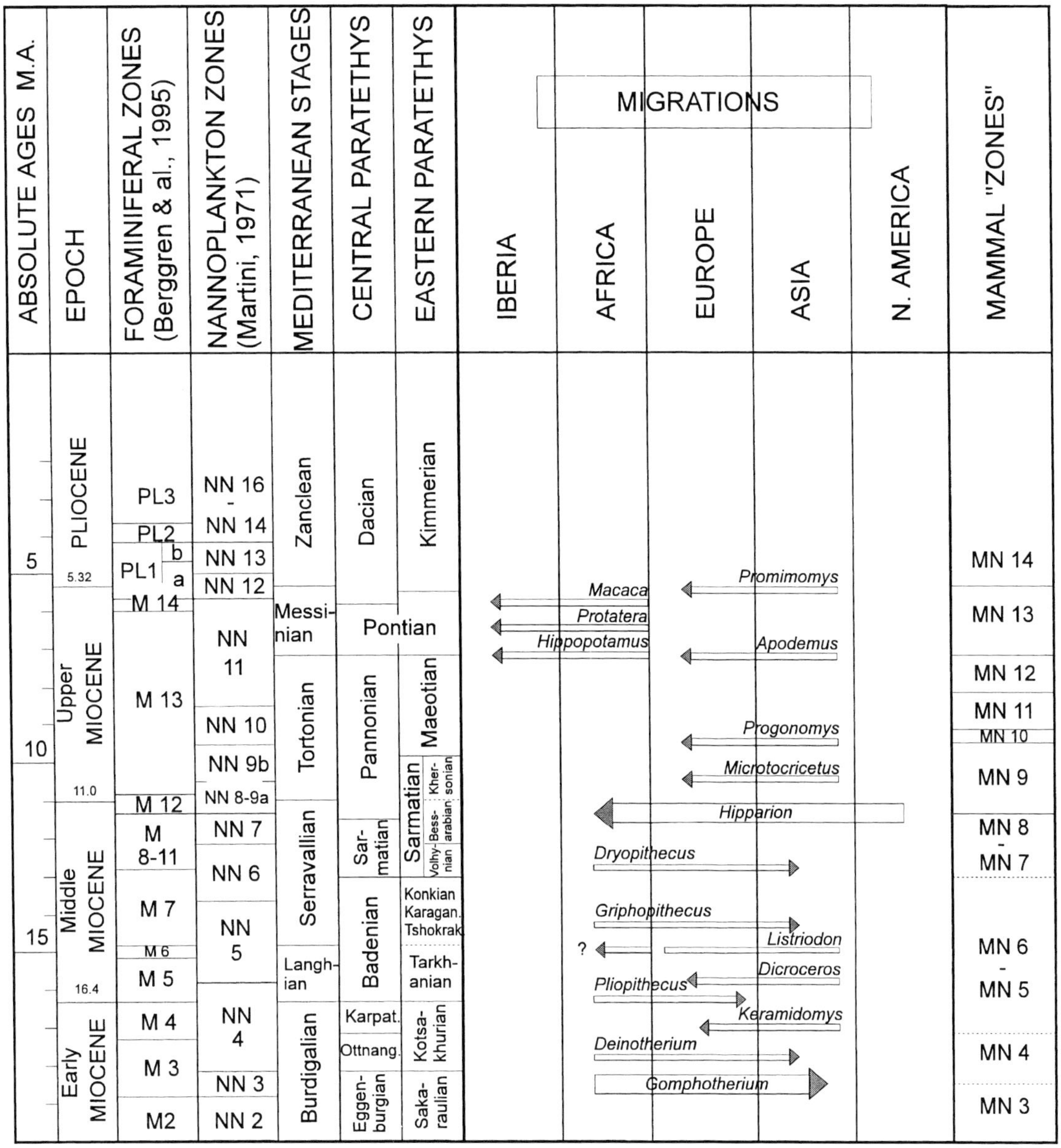

ABSOLUTE AGES M.A.	EPOCH	FORAMINIFERAL ZONES (Berggren & al., 1995)	NANNOPLANKTON ZONES (Martini, 1971)	MEDITERRANEAN STAGES	CENTRAL PARATETHYS	EASTERN PARATETHYS	MAMMAL "ZONES"
5	PLIOCENE	PL3	NN 16 - NN 14	Zanclean	Dacian	Kimmerian	MN 14
5.32		PL2	NN 14	Messi-nian	Pontian		MN 13
		PL1 b	NN 13	Tortonian	Pannonian	Maeotian	MN 12
		PL1 a	NN 12	Serravallian	Sar-matian	Sarmatian: Kher-sonian	MN 11
10	Upper MIOCENE	M 14	NN 11	Langh-ian	Badenian	Sarmatian: Bess-arabian	MN 10
11.0		M 13	NN 10	Burdigalian	Karpat.	Sarmatian: Volhy-nian	MN 9
	Middle MIOCENE	M 12	NN 9b		Ottnang.	Konkian	MN 8 - MN 7
15		M 8-11	NN 8-9a		Eggen-burgian	Karagan.	MN 6 - MN 5
16.4	Early MIOCENE	M 7	NN 7			Tshokrak	MN 4
		M 6	NN 6			Tarkh-anian	MN 3
		M 5	NN 5			Kotsa-khurian	
		M 4	NN 4			Saka-raulian	
		M 3	NN 3				
		M2	NN 2				

MIGRATIONS: IBERIA, AFRICA, EUROPE, ASIA, N. AMERICA

Promimomys; Macaca; Protatera; Hippopotamus; Apodemus; Progonomys; Microtocricetus; Hipparion; Dryopithecus; Griphopithecus; Listriodon; ?; Dicroceros; Pliopithecus; Keramidomys; Deinotherium; Gomphotherium

B

Figure 35.1. Intercontinental migration waves of mammals and biostratigraphic correlation of the Oligocene and Miocene between the Mediterranean and Paratethys. Correlations and time scale according to Berggren et al. (1995), Rio and Fornaciari (1994), Rögl (1996), Rögl and Daxner-Höck (1996), Schmidt-Kittler (1987), and Steininger et al. (1990).

An important part of this chapter is the newly standardised and correlated time scale of the Mediterranean and Paratethys. It is based primarily on the latest palaeomagnetically calculated absolute ages and correlations of Berggren et al. (1995). It follows in part Steininger et al. (1990, 1995), Rögl and Daxner-Höck (1996), and Schmidt-Kittler (1987) for Oligocene mammal ages and migrations, and corresponds to the correlation of Rögl (1996).

Late Eocene—The European Archipelago

The final stage of the Tethys is characterised by the Eocene collision of the Indian and Iranian Plates with Asia, and the Alpine system, including the Adriatic promontory, with Europe. The former Tethys Ocean was split at its western end into a Mediterranean Sea and a European–West Asian intracontinental sea, the Paratethys. In western and middle Europe the Eocene sea surrounded a system of peninsulas and islands (Krutsch and Lotsch, 1958; Ziegler, 1982). Northern Europe was separated from the French–German archipelago by the Danish–Polish trough. To the northnorthwest, Europe remained connected with the North American continent until the Eocene (Königswald, 1981). The North Atlantic opened continuously, shifting North America to the NNW–NW. The Labrador Basin opened to its present size in late Eocene/early Oligocene. The opening of the Norwegian–Greenland Sea gave way to a surface-water exchange between the Arctic and Atlantic Oceans. The Iceland–Faroe Ridge prevented the cold deep-water outflow from the Polar region throughout most of the Cenozoic (Thiede, 1979). The northern De Geer route over Svalbard–Greenland probably acted as a landbridge for later Eocene faunal exchanges. The Thule landbridge via Iceland was submarine from middle Eocene times.

To the east, the northern Tethys covered southern Russia and was connected by the Turgai Strait with the Arctic Ocean on the eastern side of the Ural mountain chain. The Turgai Strait connection between the Tethys and Arctic Ocean, working as a kind of warm-water gyre around Asia, was probably the cause for the temperate Eocene climate in the north, which provided the habitat for the Elsmere Island vertebrate fauna. Before the NNW–NW drift of North America, a broad seaway between the Polar Sea and the North Pacific must have existed. No continental migrations of vertebrate faunas between Europe and Asia and further into America was possible until the Oligocene. A distinct change in benthic foraminiferal fauna (from calcareous to agglutinated assemblages) is recorded from Kamchatka and the North Pacific at the Eocene/Oligocene boundary by Serova (1986). These assemblages correspond to a change to cooler water masses. The worldwide oceanic faunal turnover was more gradual from the late middle Eocene to the earliest Oligocene, corresponding to a gradual change in bottom-water conditions (Miller et al., 1992).

Early Oligocene—The Birth of the Paratethys and the "Grand Coupure"

A distinct change of development from open ocean conditions to restricted marine circulation is observed at the Eocene/Oligocene boundary along the Alpine–Carpathian mountain range and further to the east to Crimea, the Caucasus, and on to Mangyshlak. In Central Asia the marine phase ended at the Eocene/Oligocene boundary. The restricted circulation and cold water from the North Sea produced, in the earlier Oligocene, anoxic bottom water conditions and various phenomena in the surface-water bioproduction in the early Paratethys (Krhovsky et al., 1991). The connection to northern seas is demonstrated by mollusc faunas as far east as the Ustyurt Basin in the easternmost Paratethys, with about 80% of the assemblages corresponding to northern European Rupelian faunas (Nevesskaya et al., 1984).

In the open oceans a worldwide late Eocene extinction of warm-water planktonic foraminifera and a turnover of deep-water benthics is observed, combined with changes of stable isotopes and a decrease in siliceous productivity; and in the early Oligocene there was development of small-sized, low-diversity planktonic faunas (Boersma et al., 1987; Miller et al., 1992).

The Alpine orogenic development is characterised by west to east stretching basins. Sedimentation had already ended in the Alpine Flysch and Gosau Basins during the middle to late Eocene. In the western Alps the *Globigerina* marl sedimentation transgresses the Eocene/Oligocene boundary in the Prealps (Charollais et al., 1980) and is attributed to the North Helvetic Flysch Basin. The first intercalations of anoxic sedimentation occur in the early Oligocene. In the Bavarian Helvetic Basin a transition to the Molasse sedimentation is observed in the Katzenloch beds with flyschoid facies (Hagn, 1978). A very similar development is observed from Austria to northern Hungary and along the Carpathian arch (Baldi, 1980), as in the Prealps, with a change from yellowish *Globigerina* marls to a dark anoxic environment of dark clayey layers with pyritised microfaunas. In the east (Crimea, Caucasus, Mangyshlak) a distinct change from the light-coloured calcareous sediments of the Beloglinskian horizon to the black clays of the Khadum horizon occurs (Veselov, 1979; Voronina and Popov, 1985).

During the early to middle Oligocene the facies of the entire Paratethys is influenced by anoxic conditions, caused by stratification of the water masses. Beginning with dark clays and marls (commonly laminated fish clays), there follow widespread monospecific nanno oozes with a distinct pteropod horizon, and a horizon with small endemic molluscs (for example, *Janschinella* and *Cardium lipoldi*) from Austria to the Caucasus (Baldi, 1984). In the Carpathians the nanno marls laterally interfinger with cherts and diatomites (Kotlarczyk and Lesniak, 1990; Krhovsky et al., 1991). Clay and marl sedimentation followed upwards in the upper NP 23–24 zones, followed by a return to normal marine conditions. After the isolation and restricted connections, new seaways open to Atlantic and Mediterranean bioprovinces in the later Oligocene.

The connection of the newly formed Paratethys with the Arctic Ocean via the Turgai Strait ceased during the early Oligocene. The Danish–Polish Strait remained open for some time and connected to the North Sea. A further restricted connection to the north followed a seaway along the Rhine Graben (Martini, 1990). The first continental migration pathway between Asia and Europe opened after the closure of the Turgai Strait and gave rise to the "Grande Coupure" in the European mammal faunas. By discussing extinctions and migrations Russell and Tobien (1986) mention the Eocene extinctions to be already more prominent between the middle and late Eocene. The Eocene/Oligocene boundary is better characterised by mammal newcomers of Asian and American origin (16 new families according to Savage, 1990: for example, *Anthracotherium* from Asia, Eomyidae and Cricetidae from North America; compare with Schmidt-Kittler, 1987). These American mammal immigrants demonstrate the first closure between Asia and Alaska by the formation of the Bering landbridge. An important point is the timing of the "Grande Coupure", as within nannoplankton zone NP 22, between the middle and upper Pechlbronn Formation in the Rhine Graben and Mainz Basin (Tobien, 1987), corresponding to an age of 33–32 million years (Ma).

LATE OLIGOCENE TO EARLY MIOCENE INDO-PACIFIC SEAWAY

The next prominent horizon in the Paratethys is found in the late Egerian during a time of general warming. This worldwide climatic event is shown in Australia at the base of N 4 (*Globorotalia kugleri* zone) by a distinct excursion of larger foraminifera (McGowran, 1979). Common occurrences of larger foraminifera are reported from northern Hungary, southern Slovakia, Austria, and Slovenia. Besides occurrences of *Miogypsinoides* and *Miogypsina*, the *Lepidocyclina* horizon at Novaj in Hungary is remarkable (Baldi et al., 1961). This horizon probably corresponds to the *Lepidocyclina* (*Eulepidina*) *elephantina* horizon in the Mediterranean and in the Iranian Qum Basin. Such an Indo-Persian connection is also noted in the Eastern Paratethys Sakaraulian mollusc faunas (Nevesskaya et al., 1975, 1984). A reconstruction of a Qum Basin as a Paratethys embayment and the closure of a Mediterranean gateway (Whybrow, 1984: fig. 3) do not bring the necessary Indo-Pacific marine faunal exchange. This landbridge in the Near East corresponds with mid-Burdigalian times.

Connections of the Paratethys to the Mediterranean existed by the "Trans-Tethyan Trench Corridor" in Slovenia, south of the Alps, and along transform faults in the Alps from the Po Basin to the Molasse. During the late Oligocene/earliest Miocene (Egerian) the marine connection along the North Alpine foredeep closed in the west and reopened during the Eggenburgian. A continuing Indo-Pacific connection in the early Miocene is demonstrated by the Indian Ocean crocodile *Gavialosuchus* in Eggenburg (Austria), and the remarkable mollusc fauna with the worldwide horizon of giant pectinids, large bivalves of the same species composition from California to the Eastern and Central Paratethys (Addicott, 1974; Baldi, 1979; Voronina and Popov, 1985). Open seaways between the Indo-Pacific and the Mediterranean via the Arabian Plate existed until the early Burdigalian (Adams, 1976; Adams et al., 1983).

Throughout the Oligocene there was a continuous, but not very impressive, appearance of new Asian mammal immigrants in Europe (fig. 35.1a). An African–Eurasian migration was impossible because of the Tethys seaway. A *Gomphotherium angustidens* molar is reported from Sicily (Burgio, Agrigento province) from biomicritic limestones with Aquitanian larger foraminifers such as *Eulepidina dilatata, Nephrolepidina morgani,* and *Miogypsina gunteri* (Checchia-Rispoli, 1914; Torre et al., 1995). This proves the African origin of this Sicilian microplate (Boccaletti et al., 1990), and the early existence of this African proboscidean species. Finally, the Bering landbridge was active again. The strongest marine regression is shown in the late Oligocene (Haq et al., 1988), and is somewhat less in the Aquitanian. Around the Aquitanian/Burdigalian boundary the North American horse *Anchitherium* appeared in Europe together with the anthracotheriid *Brachyodus* at the base of mammal zone MN 3 (the African origin of *Brachyodus* considered by Pickford, 1990 is questioned because of the Asian origin of anthracotheriids).

First Neogene Africa–Eurasia Mammal Migration—The "*Gomphotherium*" Landbridge

The northward movement and rotation of Africa and the Arabian Plate closed the seaway between the Mediterranean and Arabian Gulf by the middle Burdigalian (Adams et al., 1983). An interesting addition to this landbridge is represented by the Rotem mammal fauna in the Negev directly south of the collision zone (Goldsmith et al., 1988), which was dated as MN 3b. Absolute ages date the base of the Hazeva Formation containing the fauna as 20.7 Ma. The description of the fauna by Tchernov et al. (1987) does not allow a direct correlation with European MN 3a or 3b zones. In any case, in this fauna the first *Gomphotherium, Prodeinotherium,* and *Dorcatherium* appear, which are components of the African Rusinga fauna (dated at 18 Ma and thereby comparable with an MN 4 correlation); the rhinocerotid *Dicerorhinus* demonstrates Eurasian connections. Regional barriers such as the high trans-Jordanian mountains in the east and a southward extended Mediterranean gulf in the west prevented the migration between the Negev and Arabia as well as with Gebel Zelten.

The new landbridge between Africa and Eurasia was used immediately by an intensive mammal exchange (Thenius, 1979). The first impressive arrival in the Old World was the proboscidean *Gomphotherium* at the base of MN 4. Shortly afterwards *Deinotherium* followed, and in the lower MN 5 the first African primate, *Pliopithecus*. Immigrations of Asian elements still continued (fig. 35.1a).

The period from the top of MN 3 to the base of MN 5 is one of the most tectonic active phases in the Mediterranean and Central Paratethys, where mammal zones are now fairly well correlated to the marine scale (Mein, 1989; Steininger et al., 1990; Fejfar, 1990). The basins that stretched from west to east in the Paratethys were cut off from marine connections. New intramountain basins formed as a result of the extension of the Pannonian Basin. The northward drift and rotation of the Apulian Plate ended the marine phases in the Hellenides and Thracian Basins during the late Burdigalian *Globigerinoides bisphericus* zone (Fermeli and Ioakim, 1993). During the Oligocene and early Miocene in Greece the Ionian–Lycian arc shows a strong compression with overthrusts of the Pindos and Ionian zones, and the Lycian nappes in the east, and a northward movement of the Aegean region of about 10° latitude (Kissel et al., 1989). The Aegean region became dry land.

Middle Miocene Marine Highstand

The beginning of the middle Miocene Langhian is correlated with a marine highstand and worldwide transgressive phase (Haq et al., 1987, 1988). All over the Mediterranean, the Central, and Eastern Paratethys the sea is transgressive, especially in the Paratethys where new basins and old foredeeps are flooded. The Ponto-Caspian region was a separate marine basin with a different faunal development, influenced by anoxic bottom conditions during the Tarkhanian, later changing to reduced salinity and endemic faunas. For the Tarkhanian stage a Mediterranean and Indo-Pacific connection is required by Nevesskaya et al. (1984, 1987); an Aegean seaway was not present during the Langhian as it was dry land. In the Central Paratethys, the lower Badenian sea was well oxygenated and had subtropical faunas with rich mollusc assemblages, corals, and a bloom of shallow warm-water foraminifera such as *Amphistegina* and *Heterostegina*.

The Serravallian regression may have caused the closure of the Mediterranean–Central Paratethys connection, and thereby the middle Miocene Paratethys "salinity crisis" (Rögl et al., 1978). This regression is directly related to the distinct stable isotope shift in benthic foraminifera, the mid-Miocene event near the Langhian–Serravallian boundary (Miller et al., 1991; Rio and Fornaciari, 1994). The marine connections of the Paratethys in the middle Miocene are still unresolved. One small passage from the Mediterranean was the "Trans-Tethyan Trench Corridor" in Slovenia, but another passage for an open circulation, and the supply of the Carpathian foredeep from Moravia to Poland, Ukraine, and Bulgaria, is missing. We have proposed, in earlier publications, a connection via the Black Sea–eastern Anatolia–Indian Ocean, an area now subducted under the Zagros range (Rögl and Steininger, 1983; Steininger and Rögl, 1984). The differences between Central and Eastern Paratethys are too strong for one common seaway for both regions, according to investigations of their microfaunas and mollusc assemblages. The southern border of the Black Sea Plate still remained as a barrier in the middle Miocene, and may be the dividing sill between the two regimes. It is a suture zone where part of the Neotethys is already subducted (Erkal, 1987; Görür, 1989). Nevesskaya et al. (1984) have proposed an Indo-Pacific connection of the Eastern Paratethys in the southeast between the Anatolian and Iranian Plates. Such short connections opened and closed repeatedly during the middle Miocene, with the best-developed marine faunas in eastern Georgia and in the Trans-Caspian region closed off from any Mediterranean or Central Paratethys seaway.

A middle Miocene reopening of the Mediterranean–Indo-Pacific connection in southern Anatolia proposed by us was strongly opposed by Adams et al. (1990) because of the differences in larger foraminiferal distribution. In our opinion the climatic optimum at the beginning of the middle Miocene can be triggered only if a circumequatorial current system of the world oceans is re-established. The Caribbean passage is the only oceanic equatorial gateway that has not changed since the mid-Burdigalian cooling period, and would not explain the warming in Eurasia. A new field proof of marine mid-Miocene deeper-water connections in eastern Anatolia, as reported by Gelati (1975), is impossible. In contrast to Adams we are still considering a Mediterranean–Indo-Pacific connection in the middle Miocene, but it must be located around the northern border of the Arabian Plate. Seaways in southeast Anatolia are indicated according to sediment distribution maps of Erentöz

(1956), and by the occurrence of *Flosculinella* in the lower Langhian of the Budakli section near Maras, Adana Basin (Dizer, 1991).

To explain the Indo-Pacific connections, more than one group of fossils must be considered. Adams et al. (1983, 1990) had to admit that the present-day distribution of larger foraminifera in the Indian Ocean, unhindered by physical barriers, is incongruous and distinctly poorer along the African coast. Some climatic differences and influencing currents may also be possible for the Mediterranean and Paratethys in the Langhian/ Badenian. The Central Paratethys immigration of *Amphistegina mammilla* (= *A. hauerina*), a relative of the Indo-Pacific *A. radiata* may be such an example of selective migration (Rögl and Brandstätter, 1993). This species is restricted to the Mediterranean–Indo-Pacific region, does not occur in the Atlantic–Caribbean bioprovince (Larsen, 1978), and is present in large numbers as a rock-forming component.

The first proposals for a Central Paratethys–Indo-Pacific connection were given by Dumitrica et al. (1975), Dumitrica (1978), Rögl et al. (1978), and Popescu (1979) because of differences in the different microfossil groups between the Paratethys and the Mediterranean. Foraminifera and radiolarians of the early/middle Badenian (Moravian, Wielician) correspond fairly well to Mediterranean assemblages. The first strong differences appear in the late Badenian (Kosovian), the time equivalent of the Eastern Paratethys Konkian stage. For calcareous nannoplankton, common *Rhabdosphaera poculi* and *Nannocorbis challengeri* (Rögl and Müller, 1976), frequent syracosphaeroids (Dumitrica et al., 1975) as well as nearly the entire absence of discoasterids are different to the Mediterranean, but the mentioned species are common in the Indo-Pacific deep-sea cores. Siliceous microfossils are rare in the mid-Miocene Mediterranean. Radiolarian Shales of the lower Kosovian (upper Badenian), containing microfossils that are frequently siliceous, are widespread in the Carpathian foredeep of Poland, the Ukraine, and Romania, and in the eastern part of the intra-Carpathian region (Transylvanian, Maramures, and Ukrainian Transcarpathian basins). Diatoms with *Coscinodiscus lewisianus* are from the Pacific zones XX–XXI of Schrader (1973). The silicoflagellates correspond to those of the Pacific *Distephanus stauracanthus* zone, and the radiolarians, with *Lithoptera renzae* and *L. neotera,* correlate with the Pacific *Dorcadospyris alata* zone (Dumitrica et al., 1975). Apart from *Amphistegina,* correlations using other foraminiferal species are more difficult because of missing comparative studies. One such interesting form is *Ammomassilina alveoliniformis* (Millett) in the Kosovian, described from the Malaysian archipelago (Popescu, 1979); another is the common occurrence of *Borelis melo melo* and *B. melo curdica* in the Badenian and lower Sarmatian (Gagic, 1983). The interesting planktonic foraminiferal genus *Velapertina* is reported from the Kosovian (Popescu, 1976), and may occur in the Indo-Pacific obscured under *Praeorbulina/Candorbulina* (for example, Lipps, 1964 figured a comparable trilobate specimen on pl. 2, fig. 5a,b as *Candorbulina universa* Jedlitschka). Sediments with mass occurrences of pteropods (planktonic gastropods), as in the Kosovian and Konkian *Spirialis* Marls, are lacking in the Mediterranean. For other molluscs a reliable comparison is missing. Otherwise an Indo-Pacific connection is documented (Bellwood and Schultz, 1991) by the occurrence of parrotfishes (*Calotomus and Asima*) from upper Badenian sediments of the Leitha Mountains (Austria) and Neudorf sandhill (Slovakia).

An interesting factor concerning this problem is the separate mammal immigration waves from Africa in a succeeding order from MN 4 to MN 7/8: for example, *Gomphotherium–Deinotherium–Pliopithecus–Griphopithecus–Dryopithecus* (Thenius, 1979; Pickford, 1990; de Bruijn et al., 1992). The suids *Bunolistriodon* and *Listriodon* appear in MN 4 and MN 6, respectively, in Africa (Pickford, 1990), where the occurrence of true *Listriodon* in Africa is questioned by Made (1992). This gives the impression of opening and closing barriers.

Another barrier for mammal migration has been the seaways of the early Konkian and early Sarmatian (later middle Miocene) times connecting the southeastern Paratethys. These barriers at least limited mammal migrations between Greece–Anatolia and Pakistan in MN 6–8, and could be the proposed barrier of de Bonis et al. (1992). As an important Asian element, the cervid *Dicroceros* appears in Europe in the middle Miocene.

Late Miocene Paratethys Isolation, the Messinian Event, and the Continental Faunal Turnover

During the Serravallian a general regression occurred in the Mediterranean. Through neotectonic activities the Aegean land was faulted and flooded during the Tortonian transgression, which spread north into the Eastern Paratethys and gave rise to the Maeotian sea, extending into the Dacian and Caspian Basin around 10 Ma ago. The tectonic phases of the Aegean development were presented by Böger and Dermitzakis (1987). The extension of the Aegean land to Crete in the south is shown by the occurrence of the Vallesian "*Hipparion*" and MN 9 micromammals in Kastellios (de Bruijn and Zachariasse, 1979). In the Central Paratethys only the extensional Pannonian Basin existed as an aquatic realm; it was an oligohaline lake with endemic faunas. Changes in the Paratethys paralleled Mediterranean events, and it became a giant oligohaline to freshwater lake over the entire area from the Pannonian to the Caspian Basin during the Pontian. This Pontian lake had an outflow southwards into the Aegean Basin at the time of the Messinian regression, bordered by the present-day islands of the Aegean arc. The sediments of the Pontian are sporadically preserved and their mollusc and ostracod faunas correlate directly to the early Pontian (Rögl et al., 1991).

The mammal faunal turnover in the late Miocene started with the ingression of "*Hipparion*" from North America. This event was caused by a major fall in sea level near the middle/late Miocene boundary that exposed the Bering landbridge once more (Haq et al., 1988). A continuous climatic change throughout the late Miocene produced intra-Eurasian migrations. The migration pathways in Asia, mainly for micromammals, had been restricted as long as the Eastern Paratethys sea extended in the Caspian Basin to the foothills of the Ural mountains. The murids are a good example of these successive migration waves with *Progonomys, Parapodemus,* and *Apodemus* from MN 10 to MN 13, and the invasion of arvicolids with *Promimomys* at the beginning of the Pliocene.

In the Neogene Mediterranean the most impressive event was the Messinian Salinity Crisis and desiccation of the marine basins. For mammal migrations the opportunities were nevertheless restricted. Navigating the deep evaporite basins must have been perilous, and only the old exposed sills could have acted as landbridges: the Gibraltar landbridge was used, for example by different forms such as *Macaca* and *Hippopotamus,* as African immigrants. An interesting Asian immigrant to Iberia is *Protatera* coming by way of northern Africa (Agusti, 1989). A slight possibility of stepping-stone bridges from northern Africa is provided, in Serravallian/Tortonian times, by the movement of Africa along transform faults and its counterclockwise rotation, so that some blocks of the Rif and Betic cordillera may have been sheared off (compare with Cahuzac et al., 1992, where no palinspastic model is used). This seems to be the first opportunity for crossing since the time of a hypothetical Eocene bridge. Other landbridges were present in the middle part of the Mediterranean as a connection between Tunisia, Malta, and Sicily (Torre et al., 1995), as well as the trans-Adriatic Apulian–Dinarides connection (De Giuli et al., 1987).

The modern Mediterranean Sea came into existence with the beginning of the Pliocene transgression, where the deep basins were flooded and covered by pelagic sediments.

Summary

1. During the Eocene the European continent was an archipelago and northwestern Europe was connected with North America. The Tethys Ocean separated Europe from Africa. To the east the Turgai Strait connected the Arctic and

Tethys Oceans, preventing continental faunal exchanges between Asia and Europe. Continental collisions of Asia with India and of Europe with Afro-Arabia occurred, and these collisions initiated orogenic phases that eliminated the Tethys Ocean at the Eocene/Oligocene boundary. The Tethys was split at the western end into a northern inner-continental Paratethys and, in the south, the Mediterranean Sea. The Turgai Strait closed, and the northwestern movement of North America closed the gap between Alaska and East Asia. The Bering landbridge came into existence. A first mammal migration wave between North America, Asia, and Europe characterises the "Grande Coupure" near the Eocene/Oligocene boundary in nannoplankton zone NP 22, around 33–32 Ma ago.

2. The Paratethys developed as an enclosed basin during the Oligocene, during the time that the Mediterranean was connected with the Indo-Pacific and Atlantic. Eurasian vertebrate faunas show no migration waves during this long time. At the end of Oligocene to the beginning of Miocene a broad seaway opened from the Indian Ocean to the Paratethys. At the base of mammal zone MN 3 in the early Miocene the North American horse *Anchitherium* appeared in Europe.
3. The northward movement and rotation of Africa and Arabia closed the Mediterranean–Indo-Pacific seaway by mid-Burdigalian times. The "*Gomphotherium*" landbridge in the Near East opened the African–Eurasian mammal migration pathway. These migrations are best exemplified by proboscideans and primates. In succession, *Gomphotherium, Deinotherium, Pliopithecus, Griphopithecus,* and *Dryopithecus* appear during zones MN 4 to MN 7 in Europe.
4. The beginning of middle Miocene shows a worldwide warming and marine transgression. In the Paratethys the Badenian and Tarkhanian seas cover dry and old foredeeps and new intramountain basins from Vienna to the Transcaspian Basin. The Indo-Pacific connection reopened. A possible reopening from the Indian Ocean to the Mediterranean cannot be proved at present. The Paratethys–Indian Ocean seaway was actively changing during the middle Miocene with repeated closing and reopening and this activity may have prompted immigration waves of continental vertebrates.
5. Near the base of the late Miocene there appears the new North American immigrant "*Hipparion*". The Bering landbridge was again active as indicated by a large fall in sea level near the Serravallian/Tortonian boundary. During the late Miocene the Paratethys marine phases ended. The closure of open seaways and strong regressions in the western part caused endemic aquatic faunas to be divided into restricted basins. New migrations from the east are observed by a succession of micromammals such as *Microtocricetus, Progonomys,* and *Apodemus,* and, at the Miocene/Pliocene boundary, by *Promimomys.* In the Mediterranean, the Tortonian transgression entered the Aegean land mass along new fault zones.
6. By the time of the Messinian Salinity Crisis and the Mediterranean desiccation new landbridges opened between Africa and Europe. For example, the Gibraltar bridge to Iberia enabled new migrations of continental vertebrates. This event also influenced the Paratethys seas, and caused a strong refreshening during the Pontian. The Pontian lake spilled over to the Aegean Basin, and into the Mediterranean Basin.
7. The Pliocene transgression flooded the deep basins of the Mediterranean, and its modern outline was then formed. In the Paratethys the regression continued in the Pannonian Basin in the west and oligohaline conditions prevailed in the Eastern Paratethys.

Acknowledgements

To present this compilation I thank Peter Whybrow (The Natural History Museum, London) and Ray L. Bernor (Howard University, Washington, D.C.), who pushed me into it and also helped in the discussion of unresolved problems. The article would have not been completed without the valuable help of G. Daxner-Höck (Vienna), advising me on mammalian systematics and relationships. For discussions and suggestions on palaeogeographic problems, migrations, and stratigraphy I thank my colleagues, the late C. G. Adams (NHM), J. Agusti (Sabadell), M. Aubry-Berggren (Marseilles), H. de Brujin (Utrecht), G. Clauzon (Marseilles), R. Daams (Madrid), V. Fahlbusch (Munich), O. Fejfar (Prague), L. F. Flynn (Cambridge, Mass.), A. Gentry (NHM), E. P. J. Heizmann (Stuttgart), J. van der Made (Madrid), I. Mouzylev (Moscow), C. Müller (Paris), L. Rook (Florence), O. Schultz (Vienna), F. F. Steininger (Vienna/Frankfurt), and P. Tassy (Paris). My special thanks go to W. A. Berggren (Woods Hole) for the constant use of his unpublished manuscripts on the Paleogene and Neogene stratigraphy.

This study is part of IGCP Project no. 326 "Oligocene/Miocene Transition in the Northern Hemisphere", and was supported by the facilities of the Naturhistorisches Museum Wien.

References

Adams, C. G. 1976. Larger foraminifera and the late Cenozoic history of the Mediterranean region. *Palaeogeography, Palaeoclimatology, Palaeoecology* 20: 47–66.

Adams, C. G., Gentry, A. W., and Whybrow, P. J. 1983. Dating the terminal Tethyan events. In *Reconstruction of Marine Environments* (ed. J. Meulenkamp). *Utrecht Micropaleontological Bulletins* 30: 273–98.

Adams, C. G., Lee, D. E., and Rosen, B. R. 1990. Conflicting isotopic and biotic evidence for tropical sea-surface temperatures during the Tertiary. *Palaeogeography, Palaeoclimatology, Palaeoecology* 77: 289–313.

Addicott, W. 1974. Giant pectinids of the eastern North Pacific margin: Significance in Neogene zoogeography and chronostratigraphy. *Journal of Paleontology* 48: 180–94.

Agusti, J. 1989. On the peculiar distribution of some muroid taxa in the Western Mediterranean. *Bollettino della Società Paleontologica Italiana* 28: 147–54.

Baldi, T. 1979. Changes of Mediterranean/?Indopacific/and boreal influences on Hungarian marine mollusc faunas since Kiscellian until Eggenburgian times; the stage Kiscellian. *Annales Géologiques des Pays Helléniques* 1: 39–49.

———. 1980. The early history of the Paratethys. *Földtani Közlöny. A Magyar Földtani Tarsulat Folyóirata* 110: 456–72.

———. 1984. The terminal Eocene and early Oligocene events in Hungary and the separation of an anoxic, cold Paratethys. *Eclogae Geologicae Helvetiae* 77: 1–27.

——— 1986. *Mid-Tertiary Stratigraphy and Paleogeographic Evolution of Hungary.* Akadémiai Kiadó", Budapest.

Baldi, T., Kecskemeti, T., Nyirö, M. R., and Drooger, C. W. 1961. Neue Angaben zur Grenzziehung zwischen Chatt und Aquitan in der Umgebung von Eger (Nordungarn). *Annales Historico-Naturales Musei Nationalis Hungarici, Mineralogica et Palaeontologica* 53: 67–132.

Balla, Z. 1987. The middle section of the Alpine–Mediterranean belt in the Neogene. *Annales Instituti Geologici Publici Hungarici, A Magyar Allami Földtani Intezet Evkönyve* 70: 301–6.

Bellwood, D. R., and Schultz, O. 1991. A review of the fossil record of the parrot fishes (Labroidei: Scaridae) with a description of a new *Calotomus* species from the middle Miocene (Badenian) of Austria. *Annalen Naturhistorisches Museum Wien* 92A: 55–71.

Berggren, W. A., Kent, D. V., Swisher, C. C., III, and Aubry, M.-P. 1995. A revised Cenozoic geochronology and chronostratigraphy. In *Geochronology, Time Scales and Global Stratigraphic Correlations: A Unified Temporal Framework for an Historical Geology*, pp. 129–212 (ed. W. A. Berggren, D. V. Kent, and J. Hardenbol). SEPM (Society for Sedimentary Geology) Special Publication no. 54. Tulsa.

Boccaletti, M., Calamita, F., Centamore, E., Chiocchini, U., Deiana, G., Micarelli, A., Moratti, G., and Potetti, M. 1986. Evoluzione dell'Appennino toscoumbro-marchigiano durante il Neogene. *Giornale di Geologia* 48: 227–33.

Boccaletti, M., Ciaranfi, N., Cosentino, D., Deiana, G., Gelati, R., Lentini, F., Massari, F., Moratti, G., Pescatore, T., Ricci Lucchi, F., and Tortorici, L. 1990. Palinspastic restoration and paleogeographic reconstruction of the peri-Tyrrhenian area during the Neogene. *Palaeogeography, Palaeoclimatology, Palaeoecology* 77: 41–50.

Böger, H., and Dermitzakis, M. 1987. Neogene palaeogeography in the central Aegean region. *Annales Instituti Geologici Publici Hungarici, A Magyar Allami Földtani Intezet Evkönyve* 70: 217–20.

Boersma, A., Premoli Silva, I., and Shackleton, N. J. 1987. Atlantic Eocene planktonic foraminiferal paleohydrographic indicators and stable isotope paleoceanography. *Paleoceanography* 2: 287–331.

Bonis, L. de, Brunet, M., Heintz, E., and Sen, S. 1992. La province greco-irano-afghane et la répartition des faunes mammalienes au Miocène supérieur. *Paleontologia i Evolucio, Institut de Paleontologia "Miguel Crusafont"* 24–25: 103–12.

Bruijn, H. de, and Zachariasse, W. J. 1979. The correlation of marine and continental biozones of Kastellios Hill reconsidered. *Annales Géologiques des Pays Helléniques* 1: 219–26.

Bruijn, H. de, Daams, R., Daxner-Höck, G., Fahlbusch, V., Ginsburg, L., Mein, P., Morales, J., Heinzmann, E., Mayhew, D. F., Meulen, A. J. van der, Schmidt-Kittler, N., and Telles Antunes, M. 1992. Report of the RCMNS working group on fossil mammals, Reisenburg 1990. *Newsletters on Stratigraphy* 26: 65–118.

Cahuzac, B., Alvinerie, J., Lauriat-Rage, A., Montenat, C., and Pujol, C. 1992. Palaeogeographic maps of the northeastern Atlantic Neogene and relation with the Mediterranean Sea. *Paleontologia i Evolucio, Institut de Paleontologia "Miguel Crusafont"* 24–25: 279–93.

Charollais, J., Hochuli, P., Oertli, H. J., Perch-Nielsen, K., Toumarkine, M., Rögl, F., and Paris, J.-L. 1980. Les marnes à foraminifères et les schistes à Meletta des Chaines subalpines septentrionales (Haute-Savoie, France). *Eclogae Geologicae Helvetiae* 73: 9–69.

Checchia-Rispoli, G. 1914. Sul "mastodon angustidens" Cuvier dei dintorni di Bugio in provincia di Girgenti. *Giornale di Scienze Naturali ed Economiche di Palermo* 30: 285–96.

De Giuli, C., Masini, F., and Valleri, G. 1987. Paleogeographic evolution of the Adriatic area since Oligocene to Pleistocene. *Rivista Italiana di Paleontologia, Milano* 93: 109–26.

Dercourt, J., Zonenshain, L. P., Ricou, L.-E., Kazmin, V. G., Le Pichon, X., Knipper, A. L. , Granjacquet, C., Sborshchikov, I. M., Boulin, J., Sorokhtin, O., Geyssant, J., Lepvrier, C., Biju-Duval, B., Sibuet, J.-C., Savostin, L.A., Westphal, M., and Lauer, J.-P. 1985. Présentation de 9 cartes paléogéographiques au 1/20.000.000 s'étendant de l'Atlantique au Pamir pour la période du Lias à l'Actuel. *Bulletin de la Société Géologique de France, Paris* 1: 637–52.

Dizer, A. 1991. The biostratigraphy of the Langhian and Serravallian in North Kahramanmaras. *Proceedings, Ahmet Acar Geology Symposium,* pp. 71–81.

Dumitrica, P. 1978. Badenian radiolaria from Central Paratethys. In *M4—Badenian (Moravien, Wielicien, Kosovien,* pp. 231–61 (ed. A. Papp, I. Cicha, J. Senes, and F. Steininger). Chronostratigraphie und Neostratotypen no. 6. VEDA, Bratislava.

Dumitrica, P., Gheta, N., and Popescu, Gh. 1975. New data on the biostratigraphy and correlation of the Middle Miocene in the Carpathian area. *Dari de Seama ale Sedintelor, Institutl de Geologie si Geofizica* 61 (1973–1974): 65–84.

Erentöz, L. 1956. Stratigraphie des bassins néogènes de Turquie, plus spécialement d'Anatolie Méridionale et comparaisons avec le Domaine Méditerranéen dans son ensemble. *Publications de l'Institut d'Etudes et de Recherches Minières de Turquie* no. 3: 1–54.

Erkal, T. 1987. Sedimentation in the strike-slip North Anatolian fault zone, Thrace, Turkey. *Annales Instituti Geologica Publici Hungarici, A Magyar Allami Földtani Intezet Evkönyve* 70: 235–44.

Fejfar, O. 1990 (1989). The Neogene VP sites of Czechoslovakia: A contribution based in rodents. In *European Neogene Mammal Chronology,* pp. 211–36 (ed. E. H. Lindsay, V. Fahlbusch, and P. Mein). Plenum Press, New York.

Fermeli, G., and Ioakim, Ch. 1993. Biostratigraphy and palaeoecological interpretation of Miocene successions in the molassic deposits of Tsotylion, Mesohellenic Trench (Grevena area, northern Greece). *Paleontologia i Evolucio, Institut de Paleontologia "Miguel Crusafont"* 24–25 (1992): 199–208.

Gagic, N. 1983. Representatives of the genus *Borelis* in the Badenian and lower Sarmation of Yugoslavia. *Anuarul Institutului de Geologie si Geofizica, Bucaresti* 59: 169–81.

Gelati, R. 1975. Miocene marine sequence from the Lake Van area, eastern Turkey. *Rivista Italiana di Paleontologia, Milano* 81: 427–52.

Goldsmith, N. F., Hirsch, F., Friedman, G. M., Tchernov, E., Derin, B., Gerry, E. 1988. Rotem mammals and Yeroham crassostreids: Stratigraphy of the Hazeva Formation (Israel) and the paleogeography of Miocene Africa. *Newsletters on Stratigraphy* 20: 25.

Görür, N. 1989. Timing the opening of the Black Sea: Sedimentological evidence from the Rhodope–Pontide fragment. In *Tectonic Evolution of the Tethyan Region,* pp. 131–36 (ed. A. M. C. Sengör). NATO Advanced Study Institutes, ser. C, no. 259. Kluwer Academic, Dordrecht.

Hagn, H. 1978. Die älteste Molasse im Chiemgau/östliches Oberbayern (Katzenloch-Schichten, Priabon). *Mitteilunge der Bayerischen Staatssammlung für Paläontologie und Historische Geologie* 18: 167–235.

Hamor, G. 1988. *Neogene Palaeogeographic Atlas of Central and Eastern Europe.* Hungarian Geological Institute, Budapest.

Haq, B. U., Hardenbol, J., and Vail, P. R. 1987. Chronology of fluctuating sea levels since the Triassic. *Science* 235: 1156–67.

———. 1988. Mesozoic and Cenozoic chronostratigraphy and cycles of sea-level change. In *Sea-level Changes: An Integrated Approach,* pp. 71–108. SEPM Special Publication no. 42. Society of Economic Paleontologists and Mineralogists, Tulsa.

Kissel, C., Laj, C., Mazaud, A., Poisson, A., Savascin, Y., Simeakis, K., Fraissinet, C., and Mercier, J. L. 1989. Paleomagnetic study of the Neogene formations of the Aegean Sea. In *Tectonic Evolution of the Tethyan Region,* pp. 137–57 (ed. A. M. C. Sengör). NATO Advanced Study Institutes, ser. C, no. 259. Kluwer Academic, Dordrecht.

Königswald, W. 1981. Paläogeographische Beziehungen der Wirbeltierfauna aus der alttertiären Fossillagerstätte Messel bei Darmstadt. *Geologisches Jahrbuch Hessen* 109: 85–102.

Kotlarczyk, J., and Lesniak, T. 1990. Lower part of the Menilite Formation and related Futoma diatomite member in the Skole unit of the Polish Carpathians. *Akademia Gorniczo-Huntincza, Instytut Geologii i Surowcow Mineralnych,* pp. 1–74.

Krhovsky, J., Adamova, J., Hladikova, J., and Maslowska, H. 1991. Paleoenvironmental changes across the Eocene/Oligocene boundary in the Zdanice and Pouzdrany units (Western Carpathians, Czechoslovakia): The long-term trend and orbitally forced changes in calcareous nannofossil assemblages. In *Proceedings 4th INA Conference, Prague 1991—Knihovnicka Zemni Plyn Nafta.-Hodonin,* pp. 105–87 (ed. B. Hamrsmid and J. Young).

Krutsch, W., and Lotsch, D. 1958. Übersicht über die paläogeographische Entwicklung des zentraleuropäischen Alttertiärs (ohne Tethys-Raum). *Bericht der Geologischen Gesellschaft in der Deutschen Demokratischen Republik, Berlin* 3: 99–110.

Larsen, A. R. 1978. Phylogenetic and paleobiogeographical trends in the foraminiferal genus *Amphistegina. Revista Española Micropaleontología* 10: 217–43.

Lipps, J. H. 1964. Miocene planktonic foraminifera from Newport Bay, California. *Tulane Studies in Geology and Paleontology* 2: 109–33.

Made, J. van der. 1992. African lower and middle Miocene Suoidea (pigs & peccaries). Paper presented at VIII Jornadas de Paleontología Barcelona, 8–10 de Octobre de 1992, Barcelona, sect. 2: Taxonomy of the Listriodontinae (Suidae, pigs). Summaries, pp. 91–92, Barcelona.

Martini, E. 1971. Standard Tertiary and Quaternary calcareous nannoplankton zonation. In *Proceedings IInd Planktonic Conference, Roma,* 1970, pt 2, pp. 738–85 (ed. A. Farinacci). Technoscienza, Rome.

———. 1990. The Rhinegraben system, a connection between northern and southern seas in the European Tertiary. *Veröffentlichungen aus dem Übersee-Museum Bremen* A 10: 83–98.

McGowran, B. 1979. Some Miocene configurations from an Australian standpoint. *Annales Géologiques des Pays Helléniques* 3: 767–79.

Mein, P. 1989. Die Kleinsäugerfauna des Untermiozän (Eggenburgien) von Maigen, Niederösterreich. *Annalen des Naturhistorischen Museums, Wien.* 90A: 49–58.

Miller, K. G., Feigenson, M. D., Wright, J. D., and Clement, B. M. 1991. Miocene isotope reference section, Deep Sea Drilling Project Site 608: An evaluation of isotope and biostratigraphic resolution. *Paleoceanography* 6: 33–52.

Miller, K. G., Katz, M. E., and Berggren, W. A. 1992. Cenozoic deep-sea benthic foraminifera: A tale of three turnovers. In *Studies in Benthic Foraminifera,* BENTHOS *'90, Sendai,* pp. 67–75. Tokai University Press.

Nevesskaya, L. A., Bagdasarjan, K. G., Nosovosky, M. F., and Paramonova, N. P. 1975. Stratigraphic distribution of Bivalvia in the Eastern Paratethys. In *Report on Activity of the R.C.M.N.S Working Groups (1971–1975),* pp. 48–74 (ed. J. Senes). Bratislava.

Nevesskaya, L. A., Goncharova, I. A., Iljina, L. B., Paramonova, N. P., Popov, S. V., Voronina, A. A., Chepalyga, L., and Babak, E. V. 1987. History of the Paratethys. *Annales Instituti Geologici Publici Hungarici, A Magyar Allami Földtani Intezet Evkönyve* 70: 337–42.

Nevesskaya, L. A., Voronina, A. A., Goncharova, I. A., Iljina, L. B., Paramonova, N. P., Popov, S. V., Chepalyga, L., and Babak, E. V. 1984. Istoriya Paratetisa (History of the Paratethys). *Doklady, 27th International Geological Congress Moscow, Paleo-okeanologiya, Koll. 03* 3: 91–101.

Piccoli, G., Sartori, S., Franchino, A., Pedron, R., Claudio, L., and Natale, A. R. 1991. Mathematical model of faunal spreading in benthic palaeobiogeography (applied to Cenozoic Tethyan molluscs).

Palaeogeography, Palaeoclimatology, Palaeoecology 86: 139–96.

Pickford, M. 1990 (1989). Dynamics of Old World biogeographic realms during the Neogene: Implications for biostratigraphy. In *European Neogene Mammal Chronology*, pp. 413–42 (ed. E. H. Lindsay, V. Fahlbusch, and P. Mein). Plenum Press, New York.

Popescu, Gh. 1976. Phylogenetic remarks on the genera *Candorbulina, Velapertina* and *Orbulina. Dari de Seama ale Sedintelor, Institutl de Geologie si Geofizica* 62 (1974–75), 3, Paleont.: 61–167.

———. 1979. Kossovian foraminifera in Romania. *Mémoires, Institut de Géologie et Géophysique, Bucarest* 29: 5–64.

Rio, D., and Fornaciari, E. 1994. Remarks on middle to late Miocene chronostratigraphy. *Neogene Newsletter* 1: 26–34.

Rögl, F. 1996. Stratigraphic correlation of the Paratethys Oligocene and Miocene. *Mitteilungen der Gesellschaft der Geologie und Bergbaustudenten in Österreich, Wein* 41: 65–73.

Rögl, F., and Brandstätter, F. 1993. The foraminifera genus *Amphistegina* in the Korytnica Clays (Holy Cross Mts, Central Poland) and its significance in the Miocene of the Paratethys. *Acta Geologica Polonica* 43: 121–46.

Rögl, F., and Daxner-Höck, G. 1996. Late Miocene Paratethys correlations. In *The Evolution of Western Eurasian Neogene Mammal Faunas*, pp. 47–55 (ed. R. L. Bernor, V. Fahlbusch, and H.-W. Mittmann). Columbia University Press, New York.

Rögl, F., and Müller, C. 1976. Das Mittelmiozän und die Baden-Sarmat Grenze in Walbersdorf (Burgenland). *Annalen des Naturhistorischen Museums, Wien* 80: 221–32.

Rögl, F., and Steininger, F. F. 1983. Vom Zerfall der Tethys zu Mediterran und Paratethys: Die neogene Paläogeographie und Palinspastik des zirkum-mediterranen Raumes. *Annalen des Naturhistorischen Museums, Wien* 85A: 135–63.

Rögl, F., Bernor, R. L., Dermitzakis, M. D., Müller, C., and Stancheva, M. 1991. On the Pontian correlation in the Aegean (Aegina Island). *Newsletters on Stratigraphy* 24: 137–58.

Rögl, F., Steininger, F. F., and Müller, C. 1978. Middle Miocene salinity crisis and paleogeography of the Paratethys (Middle and Eastern Europe). *Initial Reports Deep Sea Drilling Project* 42: 985–90.

Russell, D. E., and Tobien, H. 1986. Mammalian evidence concerning the Eocene-Oligocene transition in Europe, North America and Asia. In *Terminal Eocene Events*, 299–307 (ed. Ch. Pomerol and I. Premoli Silva). Elsevier, Amsterdam.

Savage, R. J. G. 1990 (1989). The African dimension in European early Miocene mammal faunas. In *European Neogene Mammal Chronology*, pp. 587–99 (ed. E. H. Lindsay, V. Fahlbusch, and P. Mein). Plenum Press, New York.

Schmidt-Kittler, N. ed. 1987. *International Symposium on Mammalian Biostratigraphy and Paleoecology of the European Paleogene—Mainz, February 18–21 1987. Münchner Geowissenschaften Abhandlungen*, ser. A, 10: 1–312.

Schrader, H.J. 1973. Cenozoic diatoms from the northeast Pacific, Leg 18. *Initial Reports Deep Sea Drilling Project* 18: 673–797.

Serova, M.Y. 1986. Karagansky section (USSR Karagansky Island, East Kamchatka). In *Terminal Eocene Events*, pp. 147–51 (ed. Ch. Pomerol and I. Premoli Silva). Elsevier, Amsterdam.

Steininger, F. F., and Rögl, F. 1984. Paleogeography and palinspastic reconstruction of the Neogene of the Mediterranean and Paratethys. In *The Geological Evolution of the Eastern Mediterranean,* pp. 659–68 (ed. J. E. Dixon and A. H. F. Robertson). Special Publication of the Geological Society no. 17. Blackwell, Oxford.

Steininger, F. F., Berggren, W. A., Kent, D. V., Bernor, R. L., Sen, S., and Agusti, J. 1996. Circum-Mediterranean Neogene (Miocene and Pliocene) marine-continental chronologic correlations of European mammal units. In *The Evolution of Western Eurasian Neogene Mammal Faunas,* pp. 7–46 (ed. R. L. Bernor, V. Fahlbusch, and H.-W. Mittmann). Columbia University Press, New York.

Steininger, F. F., Bernor, R. L., and Fahlbusch, V. 1990 (1989). European Neogene marine/continental chronologic correlations. In *European Neogene Mammal Chronology,* pp. 15–46 (ed. E. H. Lindsay, V. Fahlbusch, and P. Mein). Plenum Press, New York.

Steininger, F. F., Rabeder, G., and Rögl, F. 1985. Land mammal distribution in the Mediterranean Neogene: A consequence of geokinematic and climatic events. In *Geological Evolution of the Mediterranean Basin,* pp. 559–71 (ed. D. J. Stanley and F.-C. Wezel). Springer Verlag, New York.

Tchernov, E., Ginsburg, L., Tassy, P., and Goldsmith, N. F. 1987. Miocene mammals of the Negev (Israel). *Journal of Vertebrate Paleontology* 7: 284–310.

Thenius, E. 1979. Afrikanische Elemente in der miozaenen Saegetierfauna Europas (African elements in the Miocene mammalian fauna of Europe). *Annales Géologiques des Pays Helléniques* 3:1201–8.

Thiede, J. 1979. History of the North Atlantic Ocean: Evolution of an asymmetric zonal paleoenvironment in a latitudinal ocean basin. Deep drilling results in the Atlantic Ocean: Continental margins and paleoenvironment. *Maurice Ewing Series (AGU)* 3: 275–96.

Tobien, H. 1987. The position of the "Grande Coupure" in the Paleogene of the Upper Rhine Graben and the Mainz Basin. In *International Symposium on Mammalian Biostratigraphy and Paleoecology of the European Paleogene—Mainz, February 18–21 1987* (ed. N. Schmidt-Kittler). *Münchner Geowissenschaften Abhandlungen,* ser. A, 10: 197–202.

Torre, D., Ficcarelli, G., Rook, L., Kotsakis, T., Masini, F., Mazza, P., and Sirotti, A. 1995. Preliminary observations on the paleobiogeography of the Central Mediterranean during the late Miocene. Paper presented at the International Conference on the Biotic and Climatic Effects of the Messinian Event on the Circum Mediterranean, Benghazi. Technical programme and abstracts, 63–64.

Veselov, A. A. 1979. To the accurate definition of the stratigrafical correlation of the Oligocene—lower Miocene border-marking horizons of the Eastern and Central Paratethys. *Annales Géologiques des Pays Helléniques,* hors sér., fasc. 3: 1243–52.

Voronina, A. A., and Popov, S. V. 1985. Main features of the evolution of the Eastern Paratethys in the Oligocene and lower Miocene. *Annales Universitatis Scientiarum Budapestensis de Rolando Eotvos Nominatae,* sect. Geologica 25 (1983): 87–95.

Whybrow, P. J. 1984. Geological and faunal evidence from Arabia for mammal "migrations" between Asia and Africa during the Miocene. *Courier Forschungsinstitut Senckenberg* 69: 189–98.

Ziegler, P. A. 1982. *Geological Atlas of Western and Central Europe.* Shell International Petroleum Maatschappij B.V., The Hague.

Oligocene and Miocene Palaeoceanography—A Review

36

NORMAN MACLEOD

Much of our present understanding of Oligocene and Miocene climates and palaeoceanography stems from the highly successful Cenozoic Palaeoceanography Project (CENOP). CENOP was a multidisciplinary, multi-institutional effort charged with the generation of palaeoceanographic reconstructions of the Miocene ocean utilising information gained from the study of microfossils collected as a result of the Deep Sea Drilling Project (DSDP). The Oligocene–Miocene was chosen as the focus of CENOP investigations because it represented the interval during which modern ice sheets first became established in the polar regions. As such, a detailed appreciation of overall Miocene palaeoceanographic and climatic history was thought to be crucial to the formulation and testing of hypotheses seeking to explain the development and climatic importance of regional events—for example, the closure of the Tethyan gateway and Paratethyan physiographic evolution.

TECTONIC SETTING

Oceanic circulation patterns are primarily controlled by the distribution of continental landmasses over the Earth's surface, eustatic sea level, and various climatic factors (for example, evaporation) that might determine the density of large water masses. In turn, patterns of oceanic circulation strongly influence the Earth's climate by determining the manner in which heat is distributed within the coupled oceanic–atmospheric system. The single most important factor thought to be responsible for late Oligocene–Miocene oceanographic and climatic changes is the progressive isolation of Antarctica at the southern rotational pole and the consequent initiation of circum-Antarctic circulation. Shallow-water circulation around Antarctica developed in two stages: (1) the opening of a shallow proto-Tasman Sea in the late Cretaceous about 80 million years (Ma) ago and (2) the opening of a shallow proto-Scotia Sea in the middle–late Eocene (40 Ma). Before 70 Ma southern high-latitude deep water flowed to the north of Australia.

A deep-water connection within the Tasman Sea dates from the early Oligocene (30 Ma) when a space developed between South Tasman Rise and East Antarctica. Meanwhile, the Drake Passage within the Scotia Sea opened to shallow-water flow in the late Oligocene (22 Ma) when the Antarctic peninsula moved past the tip of southern South America. Lawver et al. (1992) estimate that by middle Miocene (20 Ma) a vigorous circum-Antarctic deep-water circulation had undoubtedly developed.

At the same time as the Scotia and Tasman Seas were opening to circum-Antarctic circulation,

 ISBN 0-300-07183-3

the Tethyan Sea was closing to circum-equatorial circulation. The initial phase of this event was the elimination of circum-equatorial deep-water circulation due to deep closure of the Gibraltar gateway in the late Cretaceous. Before this closure circum-equatorial shallow and deep-water circulation was unimpeded, resulting in the establishment of a "Supertethyan" climatic belt. Progressive deep-water closure of the Gibraltar gateway during the Maastrichtian resulted in degradation of the Supertethys and initiation of the Cretaceous–Tertiary faunal turnover event (Johnson and Kauffman, 1996). Closure of a deep-water connection between the Tethys and the Indian Oceans followed in the latter part of the early Miocene (18 Ma) leading to development of the Paratethys and progressive isolation the Mediterranean Basin.

Oceanographically important events were also occurring in the North Atlantic during the earliest Neogene. The Iceland–Faeroe Ridge, which had previously isolated the Arctic Ocean from the Atlantic, began to subside below sea level during the early Miocene. By the middle Miocene no barrier existed to communication between the Atlantic, Arctic, and North Sea waters at either shallow or intermediate depths. Finally, deep-water circulation through the Indonesian gateway stopped during the middle Miocene as a result of tectonic activity associated with subduction of the Pacific plate interacting with Australia's northward drift.

Changes in Oceanic Circulation Patterns

While cyclonic gyres probably existed in the southern high latitudes during the Cretaceous and Paleocene, these would have been eliminated by the advent of circum-Antarctic circulation in the Oligocene or early Miocene. Before the opening of the Tasman Sea the Ross Sea would have been fed by relatively warm waters from the East Australian Current. With the Oligocene opening of the Tasman Sea, though, cool surface waters from the southern Indian Ocean would have moved over the Ross Sea embayment triggering the formation of sea ice as well as enhancing the production of cold bottom waters. Subsequent refrigeration of the entire Antarctic region would steepen the global latitudinal thermal gradient and intensify oceanic circulation patterns world wide.

Progressive glaciation of Antarctica, together with intensification of the circum-Antarctic Current, resulted in the formation of the Antarctic convergence during the early Miocene. At that time the Cretaceous–Paleogene pattern of low-latitude interocean circulation was reorganised into the modern geometry of surface-water mass circulation belts. This circulation pattern extends to the present day although the exact latitudinal position of these belts has varied during the Neogene.

In the far north, subsidence of the Iceland–Faeroe Ridge was also drastically changing the nature of deep-water oceanic circulation. By the late Oligocene the early stages of ridge subsidence allowed North Atlantic surface waters to flow into the Norwegian Sea. This water (which was derived from the warm, salty, shallow-water outflow of the Tethys at Gibraltar) became much denser as it cooled, sank, and made its way back into the Atlantic as new bottom water. The increased density of this new Norwegian overflow resulted in enhanced circulation velocities and fundamentally changed the character of deep-sea sedimentation patterns in the North Atlantic.

During the middle Miocene the Iceland–Faeroe Ridge subsided to the point where circulation between the Arctic and Atlantic Oceans was more or less unimpeded. This event changed the nature of Atlantic deep-water circulation via the creation of North Atlantic Deep Water (NADW). Before this event water drawn north from the topics was cooled in the North Atlantic, mixed with Antarctic Bottom Water (AABW) and flowed back south as North Atlantic Intermediate Water (NAIW) to the equatorial regions. Once a connection between the North Atlantic and the Arctic Ocean was established, warm Atlantic water flowing north to higher latitudes was cooled more intensively, becoming NADW on re-entry into the Atlantic Ocean Basin. By the middle Miocene the volume of NADW was such that it flowed the entire length of the Atlantic, where it surfaced as warm, saline water south of the Antarctic Convergence.

Climatic Changes

Stable isotopic data suggest that the oceanographic isolation of Antarctica did not result in the immediate formation of a glacial ice cap. The most convincing direct evidence for only partial Antarctic glaciation during the Oligocene comes from Antarctic palynological data. These indicate that substantial areas of Antarctica were covered with a low-diversity *Nothofagus* and *Podocarpus* flora (Kemp, 1975) during the late Oligocene and early Miocene. Taxonomically similar floras are found in modern New Zealand, which is not extensively glaciated. A variety of additional isotopic, faunal, and lithological evidence supports this interpretation.

Although there is an abundance of evidence against the development of extensive Oligocene Antarctic glaciation, establishment of the East Antarctic ice sheet does seem to have taken place by the middle Miocene (14 Ma ago). Evidence for this event includes a sharp increase in the $\delta^{18}O$ values of planktic and benthic foraminifera, the common occurrence of ice-rafted sediments in the regions around Antarctica, and an increase in the frequency of deep-sea hiatuses. Presence of these hiatuses indicates a shoaling of the calcite compensation depth (CCD) that may be connected to decreased temperature of AABW. Moreover, Brewster (1980) suggested that atmospheric circulation intensified during this time as reflected by an increase in Antarctic biogenic sedimentation (see below).

The apparent delay of 20 Ma between the advent of circum-Antarctic circulation and the formation of extensive Antarctic ice sheets may seem puzzling at first, but can be accounted for by examining the effect of oceanographic events taking place in the northern hemisphere. Submergence of the Iceland–Faeroe Ridge in the early–middle Miocene resulted in the creation of NADW and the ultimate appearance of this relatively warm saline bottom water in a zone of upwelling south of the Antarctic Convergence. Heat contained in this water upwelling in a cold area would have been converted to latent heat on its evaporation, resulting in even higher evaporation rates. This created a positive feedback mechanism that provided a new source of atmospheric moisture in the high southern latitudes from the middle Miocene onwards. Schnitker (1980) argued that upwelling NADW provided the moisture necessary to construct the Antarctic ice sheet. Before the introduction of NADW, prevailing climatic conditions (which had been in place since the late Oligocene) were unable to produce extensive glaciation in Antarctica because there was no source for precipitation. Once NADW began to upwell around Antarctica, however, conditions favouring the development of extensive ice sheets quickly established themselves and forced a progressive intensification of the entire process.

As glacial conditions became established in Antarctica during the middle Miocene the global climate cooled. Sea level also began to fall as large amounts of water were removed from the oceanic circulation system. These factors, in turn, intensified marine circulation patterns. Cooling was particularly pronounced in the waters off Antarctica, resulting in a northward migration of the Antarctic Convergence (see below).

The late Miocene glacio-eustatic fall in sea level has been variously estimated at 40–50 metres. Although a eustatic sea-level drop of this magnitude would be expected to have major sedimentological, biotic, and climatic repercussions, fortuitous combination of this oceanographic event with middle European tectonism may have contributed to desertification of the Mediterranean Ocean basin (Ryan, 1973; Hsü et al., 1977; Ryan and Cita, 1978). Aside from the profound local and regional climatic implications surrounding the recognition of extensive late Miocene deposition of marine evaporites in the Mediterranean Basin, removal of these salts from the global marine system would have lowered salinity of the world ocean by approximately 6%, to 2 per mil. This, in turn would have raised the freezing temperature of sea water, thereby increasing production of sea ice, raising the Earth's albedo and, through these factors, further lowering the Earth's mean surface temperature. Thus, glacio-eustatic lowering of sea level during the middle Miocene precipitated the onset of another positive feedback mechanism that helped move the Earth's climate towards a late Miocene glacial maximum.

Water Temperatures and Vertical Structure

General aspects of the physical structure of Cenozoic oceans can be inferred via comparative stable isotopic analysis of calcareous microfossil tests from different localities. The CENOP project subdivided the late Oligocene–early Miocene interval into three time slices, faunas from which were analysed from multiple deep-sea cores, and 19 time series. Time slices provide detailed information on spatial variation among water masses while the time series provide an overview of temporal changes and the character of global ocean throughout the sampled time interval. Although there is some ambiguity in the interpretation of these data (for example, $\delta^{18}O$ values may change as a function of temperature variation and/or removal of ^{16}O due to glaciation), they remain the standard palaeoceanographic reference for the early Neogene.

If changes in the stable isotopic ratio due to glaciation are ignored, Pacific surface waters appear to have warmed during the early middle Miocene (18–14.5 Ma ago) throughout the middle latitudes while retaining a differentiation between eastern (cool) and western (warm) regions. This warming occurs just before a cooling of deep waters in these same ocean basins. Increasing latitudinal temperature gradients in surface waters are evident in the middle Miocene where western tropical sites show a warming and high southern latitude sites show a cooling that continues into the late Miocene. At the same time, the east–west temperature differentiation largely disappears. These patterns are thought to reflect the onset of major continental accumulation of ice sheets, beginning in the middle Miocene (see above).

The vertical structure of water masses can also be inferred via comparative stable isotopic data if the comparisons are made between different planktic and benthic foraminiferal species within the same fauna. These data indicate that, as latitudinal temperature gradients increased throughout the Miocene, surface waters also become more stratified into surface, intermediate, and deep (= near thermocline) zones. Early Miocene surface waters were uniformly warm with little temperature differentiation between surface and deep regions. During the middle Miocene, however, an intermediate zone developed. The size of this intermediate zone apparently fluctuated throughout the middle and late Miocene, expanding during warm climatic intervals and contracting during cool climatic intervals. An east–west biogeographic structure is also present in the planktic foraminiferal faunas of the early–middle Miocene that Keller (1985) has interpreted as reflecting intensified equatorial circulation during cool climatic events. Disappearance of east–west faunal differentiation within the Pacific during the late Miocene is thought to stem from circulation changes brought about by Antarctic glaciation and tectonic closing of the Indonesian gateway.

Biogeography and Biotic Events

Throughout the late Oligocene–Miocene interval the spatial organisation of planktic faunas underwent a dramatic change. Oligocene and early Miocene tropical Pacific planktic foraminiferal faunas were well differentiated along an east–west gradient. During this interval western localities had a warm-water fauna dominated by shallow-dwelling species (for example, *Globigerinoides*) while eastern faunas were dominated by cooler-water species (for example, *Globorotalia*). Kennett et al. (1985) infer that these longitudinal distinctions resulted from sluggish gyral circulation within the Pacific Ocean basin coupled with either warmer western surface waters or a distinctly deeper thermocline along this basin's western margin.

Beginning in the middle Miocene this biogeographic structure began to be replaced by progressively stronger distinctions between equatorial and tropical planktic foraminiferal faunas. By the late Miocene little trace of the previous east–west spatial differentiation existed among these populations while a distinct, warm-water, shallow-dwelling equatorial assemblage (for example, *Globorotalia menardii, G. limbata,* and *Globigerinoides*) had developed. These changes are thought to result from a strengthening of gyral circulation caused by steeper contrasts in latitudinal temperatures and intensification of the equatorial countercurrent after the closure of the Indonesian gateway.

The distribution of siliceous plankton also underwent major changes during the Miocene. During the late Oligocene siliceous oozes were only deposited close to the margins of Antarctica. But throughout the Miocene these sediments came progressively to dominate more northerly areas. In modern oceans the zone of enhanced siliceous productivity marks the approximate position of the Antarctic convergence. North of this zone carbonate productivity predominates whereas south of this zone sediments are mostly siliceous in composition. This zone of siliceous productivity is closely tied to the upwelling of nutrient-rich intermediate water (= NADW) just south of the convergence. As the rate of this upwelling increases so does the level of siliceous productivity. Moreover, as the convergence shifts to more northerly or southerly latitudes, the zone of siliceous sedimentation undergoes a parallel migration. This allows the approximate position of the convergence to be mapped through time.

Rates of upwelling intensity (= siliceous productivity) reflect climatic cycles that alter the strength of westerly winds that move surface waters away from Antarctica. The space left by these surface waters is filled by the upwelling NADW. Throughout the Miocene the rate of siliceous productivity is also inversely correlated with the rate of equatorial planktic productivity, once again reflecting the intensified nature of oceanic circulation that resulted from the increase in latitudinal thermal gradients.

An interesting evolutionary by-product of enhanced Miocene high-latitude productivity can be seen in the cetacean fauna of this interval. Baleen whales (suborder Mysticeti) first appear in the middle Oligocene in the south Atlantic, which suggests that at least some localised areas of high planktic productivity existed in this area. Nevertheless, the first major radiation of this group does not take place until the early Miocene. This correlates well with initial phases of the northward expansion of siliceous oozes following the submergence of the Iceland–Faeroe Ridge and increased glaciation of Antarctica. In this case, it would seem as though abiotic tectonic, oceanographic, and climatic events can be closely tied to major evolutionary diversification in a convincing empirical manner.

Finally, it is worth noting that Miocene climatic changes driven by a combination of tectonic and palaeoceanographic events also had substantial implication for the evolutionary history of our own lineage. The Miocene increase in latitudinal temperature gradients signalled the entry of the Earth's present climatic regime. This global cooling trend strongly affected the biogeography of most plant groups, one of the most obvious examples of which was the replacement of the East African tropical forests with mixed woodland and grassland habitats (Andrews and Van Couvering, 1975). Correlated with this shift in the character of the East African plant biota is the first appearance of bipedal primates.

Conclusions

The reconstruction and integration of Oligocene and Miocene palaeoceanographic, palaeoclimatic, and biotic events constitute one of the most thoroughly documented and compelling examples of the power of global change research programmes to gain insight into the processes that have shaped (and will continue to shape) Earth history. While development of Oligocene–Miocene oceans, climates, and biotas could have been studied in isolation from one another, the union of these traditionally separate research programmes has provided a level of understanding that is demonstrably greater than the sum of its parts.

Early Neogene tectonic events in the northern and southern high latitudes altered the structure of marine ocean basins in such a way as to force the reorganisation of marine circulation patterns. These changes should be seen as the natural consequences of tectonic trends that had been in operation throughout the Paleogene, late Mesozoic, and beyond. Nevertheless the fact that these trends culminated in the simultaneous modification of Miocene ocean basins at either pole, coupled with the presumably fortuitous isolation of a large continental landmass at the southern pole, provided sufficient synergistic momentum to "flip" the Earth's oceanic and climate system between two mutually exclusive stable states in a remarkably short period of time. The fact that there was a long-term drift of the Earth's climate toward a progressively cooler state throughout the Paleogene, including intervals when this process underwent substantial periods of intensification (for example, the Paleocene/Eocene and terminal Eocene events), should in no way detract from the significance of the Miocene catharsis.

Since the world ocean functions as the principal mediator of our planet's heat budget, fundamental changes in the ocean's circulation pattern inevitably precipitate correspondingly intense climatic changes. Effects of these climatic changes in the Oligocene–Miocene were magnified via the operation of at least two major positive feedback mechanisms: upwelling and evaporation of NADW along the Antarctic coast; and isolation of the Tethys basin with consequent changes in salinity values of the world ocean. Both of these operated in such a way as to intensify further the latitudinally organised temperature gradients.

Interestingly, overall biotic response to these climatic changes was not as profound as might have been expected. To some extent this may be linked to the fact that an extended faunal turnover had just occurred in the late Eocene–early Oligocene (see MacLeod, 1990 and references therein). Regardless, the late Oligocene–Miocene represents an interval of sustained evolutionary diversification to which the concurrent expansion of climatic heterogeneity could only have contributed. Indeed, several prominent evolutionary events—for example, diversification of filter-feeding whales and the evolution of bipedal primates—can be linked to specific and predicable environmental changes that resulted (at least in part) from specific contemporaneous tectonic and oceanographic changes.

Acknowledgements

This contribution is part of The Natural History Museum/University College London Global Change and the Biosphere Project.

References

Andrews, P. J., and Van Couvering, J. H. 1975. Palaeoenvironments in the East African Miocene. In *Approaches to Primate Paleobiology*, pp. 62–103 (ed. F. S. Szalay). Karger, Basel.

Brewster, N. A. 1980. Cenozoic biogenic silica sedimentation in the Antarctic Ocean, based on two Deep Sea Drilling Project sites. *Bulletin of the Geological Society of America* 91: 337–47.

Hsü, K. J., Montadere, L., Bernoulli, D., Cita, M. B., Erickson, A., Garrison, R. E., Kidd, R. B., Milieres, F., Mueller, C., and Wright, R. 1977. History of the Mediterranean salinity crisis. *Nature* 267: 399–403.

Johnson, C., and Kaufmann, E. G. 1996. Maastrichtian extinction patterns of Caribbean province rudistids. In *The Cretaceous–Tertiary Mass Extinction: Biotic and Environmental Events*, pp. 231–73 (ed. N. MacLeod and G. Keller). Norton, New York.

Keller, G. 1985. Depth stratification of planktonic foraminifers in the Miocene ocean. In *The Miocene Ocean: Paleoceanography and Biogeography* (ed. J. P. Kennett). *Memoirs of the Geological Society of America, Washington* 163: 177–95.

Kemp, E. M. 1975. Palynology of Leg 28 drill sites, Deep Sea Drilling Project. In *Initial Reports of the Deep Sea Drilling Project*, pp. 599–623 (ed. D. E. Hayes, L. A. Frakes, L. A. Barrettt, D. A. Burns, C. Pei-Hsin, A. B. Ford, A. G. Kaneps, E. M. Kemp, D. W. McCollum, D. J. W. Piper, R. E. Wall, and

P. N. Welds). United States Government Printing Office, Washington, D.C.

Kennett, J. P., Keller, G., and Srinivasan, M. S. 1985. Miocene planktonic foraminiferal biogeography and paleoceanographic development in the Indo-Pacific region. In *The Miocene Ocean: Paleoceanography and Biogeography* (ed. J. P Kennett). *Memoirs of the Geological Society of America, Washington* 163: 197–236.

Lawver, L. A., Gahagan, L. M., and Coffin, M. F. 1992. The development of paleoseaways around Antarctica. In *The Antarctic Paleoenvironment: A Perspective on Global Change,* part 1, pp. 7–30 (ed. J. Kennett and D. A. Warnake). American Geophysical Union, Washington, D.C.

MacLeod, N. 1990. Effects of late Eocene impacts on planktic formanifera. In *Global Catastrophes in Earth History: An Interdisciplinary Conference on Impacts, Volcanism and Mass Mortality* (ed. V. L. Sharpton and P. D. Ward). *Geological Society of America Special Paper* no. 247: 595–606.

Ryan, W. B. F. 1973. Geodynamic implications of the Messinian crisis of salinity. In *Messinian Events in the Mediterranean,* pp. 26–38 (ed. C. W. Drooger), North Holland, Amsterdam.

Ryan, W. B. F., and Cita, M. B. 1978. The nature and distributions of Messinian erosional surfaces—indicators of a several-kilometre-deep Mediterranean in the Miocene. *Marine Geology* 27: 193–230.

Schnitker, D. 1980. Quaternary deep-sea benthic foraminifers and bottom water masses. *Annual Review of Earth and Planetary Science* 8: 343–70.

INDEX

Numbers in *italics* indicate pages on which information is given only in an illustration or table; footnotes are indicated by "n" following the page number.

الفقاريات الأحفورية
في
الجزيرة العربية

مع التركيز على جيولوجية حيوانات العصر الميوسيني المتأخر
والبيئات القديمة في
إمارة أبوظبي بدولة الإمارات العربية المتحدة

إعداد بيتر جيه وايبرو وأندرو هيل
بالتعاون مع شركة أبوظبي للعمليات البترولية البرية
ووزارة التعليم العالي والبحث العلمي في دولة الإمارات العربية المتحدة

مطبعة جامعة ييل

نيو هيفن ولندن

المحتويات

كلمة شكر

حظي المحررون وغيرهم من الاختصاصيين الذين قاموا بالعمل الميداني في إمارة أبوظبي بدعمٍ كبير ومشكور من العديد من المؤسسات في الإمارة. ونادراً ما كان مثل هذا الدعم والاهتمام متوافرين خلال تاريخ اكتشاف الحيوانات والنباتات البرية التي تعود إلى العصر الميوسيني في العالم القديم.

ونتقدم بخالص شكرنا إلى رئيس دولة الإمارات العربية المتحدة صاحب السمو الشيخ زايد بن سلطان آل نهيان على مساندته القيمة لنا واهتمامه المتواصل بعملنا الذي يمكن الآن إضافته إلى المعلومات المحلية المتعلقة بأنظمة الأنهار القديمة في شرق الجزيرة العربية.

كما نشعر بامتنان عميق نحو وزير التعليم العالي والبحث العلمي في دولة الإمارات العربية المتحدة، معالي الشيخ نهيان بن مبارك آل نهيان، لموافقته على رعاية المؤتمر العالمي الأول حول الفقاريات الأحفورية في الجزيرة العربية الذي عُقد في إمارة أبوظبي خلال شهر مارس ١٩٩٥ ونحو وزارته، وخاصة سعادة سيف راشد السويدي، على تنظيم المؤتمر بالتعاون مع شركة أبوظبي للعمليات البترولية البرية (أدكو).

وفي أدكو نفسها، ما كان بإمكاننا القيام بعملنا لولا التشجيع الذي لقيناه من المديرين العامين المتعاقبين تيري آدامز وديفيد وودوارد وكيفن دان وموافقتهم على المنحة التي قدمتها أدكو إلى متحف التاريخ الطبيعي لمساندة المشروع. وكذلك، تلقينا عوناً كبيراً من نبيل زخور في إدارة الشؤون العامة في أدكو، ومن حسن الصيقل في العلاقات العامة، ومن البدري خلف الله في دائرة علم المساحة التطبيقية، ومن ناصر الشامسي في إدارة العلاقات الحكومية ونصر السلامين الذي تولى القيام بأغلبية أعمال الترجمة والتدقيق اللازمة لهذا المشروع.

لقد حظي العمل المبكر المتعلق بهذا المشروع بمساعدة كبيرة من إدارة الآثار والسياحة في العين، ونتقدم بالشكر إلى أمين السر، سعادة سيف علي ضبع الدرمكي على حسن الوفادة واللطف اللذين استقبلتنا بهما إدارته في ذلك الحين، لاسيما من جانب الدكتور وليد ياسين. كما نشعر بالامتنان نحو العاملين في فندق شاطىء الضافرة في جبل الظنة ومديره العام السيد ساشي بنيكار، على المساعدة اللوجستية التي قدموها لنا طيلة سنوات عديدة وعلى الجهود التي بذلوها حتى تكلل المؤتمر العالمي الأول حول الفقاريات الأحفورية في الجزيرة العربية المنعقد في مارس ١٩٩٥ بمثل هذا النجاح العظيم.

ولم يكن بالإمكان إنتاج مثل هذا الكتاب دون المساعدة التي تلقيناها من أناس عديدين. ونتوجهِ بالشكر إلى فاليري وايبرو، التي كانت تعمل سابقاً في متحف التاريخ الطبيعي في لندن، على العمل الأولي الذي قامت به فيما يتعلق بالاعداد الإلكتروني لشكل المخطوطات. كما إننا مدينون بشكل خاص إلى ديانا كلمنت من متحف التاريخ الطبيعي على عملها الدؤوبِ والمتواصل في مجال النصوص والرسومات وخصوصاً المراجع. ومن بين الزملاء الآخرين العاملين في متحف التاريخ الطبيعي الذين قدموا لنا يد العون نخص بالذكر نورمان ماكلويد وجيرمي يونغ وألن جنتري ومايك هوارث وبيتر فوري وفيل كراب وهاري تايلر وبول لاند، ويعمل الثلاثة الأخيرون في وحدة التصوير في متحف التاريخ الطبيعي.

وأخيراً، نشكر فريق مطبعة جامعة ييل المؤلف من جين تومسون – بلاك المحرر العلمي وماري باستي المحررة الأولى للمخطوطات وبول رويستر مدير التصميم والإنتاج وجويس إيبوليتو، محررة الإنتاج، على نصائحهم المتعلقة بإنتاج هذا الكتاب في الوقت المحدد. كما نتقدم بشكر خاص إلى محررة المواد سارة باني. فقد ساعدت بخبرتها الطويلة في الكتب التي تتناول بحوث الحفريات القديمة وبمعرفتها الميدانية بالجزيرة العربية على تحسين مضمون هذا الكتاب بشكل كبير.

مقدمة

بقلم كن دبليو. غلني

عادة لا يجد المرء ترابطاً بين الأرض المغطاة بالكثبان الرملية والسباخ المغطاة بالملح وبين تشكيلة واسعة من الفقاريات الأحفورية، لاسيما عندما يكون الكثير منها مما يرد ذكره في متن هذا الكتاب – كالتماسيح وأفراس البحر، مثلاً – يحتاج بوضوح إلى الماء على نطاق لا يتوافر في الجزيرة العربية اليوم. ويمثل كتاب الفقاريات الأحفورية في الجزيرة العربية مجموعة مهمة من البيانات المتعلقة بالحفريات القديمة التي تغطي تجمعاً من الفقاريات، الكبيرة منها والصغيرة، المائية والبرية. وقد عثر على معظم هذه المستحجرات في كتلة صخرية واحدة من العصر الميوسيني، وهي تشكيلات بينونة في المنطقة الغربية من أمارة أبوظبي في دولة الإمارات العربية المتحدة.

مع أن عدة مؤلفين شاركوا في إعداد الكتاب، إلا أنه لا يشكل الخلاصة المعتادة لمقالات ليس بينها قاسم مشترك، بل هو ثمرة تعاون خُطط له بعناية بين علماء من الجزيرة العربية وأوروبا (معظمهم من المملكة المتحدة) والولايات المتحدة الأمريكية. والقوة الدافعة في تأليف الكتاب وكذلك في التأكد من تنفيذ العمل بصورة فعالة كانت اجتماع كلا من بيتر جيه. وايبرو من متحف التاريخ الطبيعي في لندن وأندرو هيل من جامعة ييل في الولايات المتحدة الأمريكية؛ وقد كتب هذان الرجلان فيما بينهما بصورة منفصلة أو مشتركة حوالي ثلث الفصول البالغ عددها ٣٦ فصلاً.

لقد زار كل من وايبرو وهيل بصورة منفصلة موقع الاكتشاف الأصلي في جبل بركة دون أن يدري أحدهما بعمل الآخر في المنطقة. وما أن أدركا وجود اهتمام مشترك بينهما، حتى قام بينهما تعاون تلقائي. وجرت بعثات أخرى للعثور على مزيد من المستحجرات ولدراستها دراسة وافية. كما تأكدا من وصف وتقييم ترسب غالبية صخور العصر الميوسيني المكشوفة على الوجه الصحيح لتوفير إطار جغرافي سليم للمستحجرات التي عثرا عليها. وبرغم التركيز على المستحجرات الفقارية، فقد استعانا بعلماء الحفريات الذين كانوا يقومون بعمل يتعلق بالأحفوريات غير الفقارية المصاحبة في تشكيلات بينونة ولعلماء الحفريات النباتية الذين كانوا يعملون في أجزاء أخرى من الجزيرة العربية، حيث قدم عملهم إسهامات قيمة في علم البيئة القديمة بصورة عامة.

لا حاجة لنا لتأكيد أهمية فقاريات أبوظبي في إماطة اللثام بشكل أكبر عن نمط هجرة أنواع مختلفة من الحيوانات بين أفريقيا وأوروبا وآسيا خلال الحقب الثالث. وقد لعبت الجزيرة العربية في هذه العملية دوراً محورياً خلال العصر الميوسيني. أولاً، أقيم حاجز بحري جزئي في وجه الهجرة بين أفريقيا والجزيرة العربية بفتح البحر الأحمر في مطلع العصر الميوسيني. وثانياً كذلك أقيم خلال مطلع العصر الميوسيني، حاجز بري في وجه هجرة حيوانات بحر تيثس الجيولوجي بين ما يعرف الآن بالمحيط الهندي والبحر الأبيض المتوسط نتيجة اصطدام الجزيرة العربية وآسيا، وبذلك أتيح تبادل الحيوانات الفقارية البرية بين هاتين المنطقتين لأول مرة. وما من شك في أن طريق الهجرة الذي تشكل حديثاً إلى آسيا قد اتسع نتيجة لانخفاض كبير في مستوى مياه البحار في العالم خلال العصر الميوسيني المتأخر، ربما بسبب الازدياد السريع في الغطاء الجليدي في الدائرة المتجمدة الجنوبية. وربما لن يتسنى لنا أن ندرك ادراكاً تاما الدور الأساسي الذي أدته الحيوانات الفقارية في العصر الميوسيني في المنطقة الغربية من إمارة أبوظبي في فهم الهجرات بين أفريقيا وآسيا بوجه خاص، ما لم نستوعب بصورة تامة محتويات هذا الكتاب ونقارنها مقارنة شاملة بالدراسات الأخرى في الشرق الأوسط.

لعله أمر نادر أن يُدعى عالم في غير الحفريات لكتابة مقدمة كتاب مهم حول الفقاريات الأحفورية (١). غير أنه يبدو أن إشارتي أنا وزميلي برايان إيفامي في مطبوعة صادرة في عام ١٩٦٨ إلى اكتشاف سن حيوان من فصيلة الفيلة في بقعة مغطاة بالحصى في جبل بركة، دفعت بيتر وايبرو إلى زيارة الموقع في عام ١٩٧٩ والعثور على أدلة على وجود فقاريات أحفورية أخرى. وكما يذكر هيل ووايبرو في الفصل الثالث عام ١٩٨٢، فقد كنت أنا ووايبرو مؤلفين ثانويين في كتاب يعيد تقييم تلك السن من إعداد مادن وآخرين (١٩٨٢). أما البقية فقد أصبحت جزءاً من التاريخ (انظر الفصل الثالث، هيل وآخرون، ١٩٩٧).

عندما كتبت أنا وإيفامي دراستنا القصيرة لعام ١٩٦٨ بعنوان الدكاكة، لم نكن ندري أنها تضمنت أول تحقيق ينشر حول مستحجرة فقارية من العصر الميوسيني في الجزيرة العربية، وأن هذا البحث العلمي سيعقبها في النهاية. وفي فترة سابقة، وبالتحديد في

عام ١٩٦٥، خلال أول رحلة ميدانية لي إلى جنوب شرقي الجزيرة العربية، لم أكن أدري مطلقاً بأنني سأشارك بعدها بصورة متقطعة على مدى أكثر من ثلاثين سنة في محاولة اكتشاف بعض من أسرارها الجيولوجية.

كباحث جيولوجي لدى شركة شل في عام ١٩٦٥، كان علي ان أدرس الصحاري الحديثة كي أستطيع أن اكتسب فهماً أفضل لمكامن الغاز (روتليجند) في حقل غاز داتش غرونينغن والتي تعود إلى العصر البرمي والتي لم يتم إدراك ضخامة حجمها إلا في الآونة الأخيرة. كما استخدمت هذه المعرفة أيضاً في الاكتشافات التي جرت التوصل اليها آنذاك في جنوبي بحر الشمال. وسبق أن أخذتنا الرحلة الميدانية عام ١٩٦٥ أنا وإيفامي، عبر جزء كبير من المناطق الداخلية في عمان والإمارات المتصالحة (التي تعرف الآن بدولة الإمارات العربية المتحدة).

كان هدفنا المباشر في الجزء الغربي من إمارة أبوظبي اكتساب فهم أفضل لسبخة مطي، وهي منطقة تنتشر فيها المسطحات الرملية على نطاق واسع، لإجراء مقارنة بينها وبين السباخ الساحلية التي كان يدرسها علماء جيولوجيا آخرون من الكلية الإمبراطورية في لندن ومن شركة شل (انظر مثلاً بيرسر، ١٩٧٣). وكان جبل بركة بمثابة منارة، اجتذبتنا كمحطة جيولوجية قصيرة بعد الاستحمام، وهو ما كنا بأمس الحاجة إليه، والتزود بكميات من الماء العذب والوقود من مقر شركة نفط العراق (التي سبقت شركة أبوظبي للعمليات البترولية البرية - أدكو وهي شركة حديثة) في جبل الظنة. وما لفت نظرنا على الفور هو الكثبان الرملية الرخوة ذات اللون المائل إلى الحمرة التي تتخللها كتل جذورية متحجرة (دكاكا، الموضوع الرئيسي للدراسة التي أصدرناها في عام ١٩٦٨) في الجرف الساحلي، مما يدل على القرب الشديد لمستوى سطح المياه الجوفية في بيئة جافة بخلاف ذلك. وأدى التسلق إلى الحصباء الكائنة في أعلى الجرف إلى اكتشاف سن الحيوان الشبيه بالفيل المشار إليه أعلاه (من جانب إيفامي كما أذكر إن لم أكن مخطئاً).

غير أن الحدث الفردي الذي دفعني إلى العودة النشطة إلى جيولوجية الشرق الأوسط، كان دعوة تلقيتها عام ١٩٩٠ من الدكتور تيري أدامز (المعروف جيداً من قبل وايبرو وهيل) الذي كان في حينه مديراً عاماً لأدكو، لإلقاء كلمة حول جيولوجية جبال عمان أمام جمعية المستكشفين في الإمارات. وأدى ذلك إلى القيام بدراسات ميدانية في كل من الإمارات العربية المتحدة وسلطنة عمان (مع أنني لم استطع العودة إلى جبل بركة إلى أن تم إخلاء بطارية مدفعية كانت منصوبة عليه بعد حرب الخليج)، وإلى الإشراف على طلبة الدكتوراة الذين كانوا يدرسون الترسبات الصحراوية في الإمارات العربية المتحدة والصخور الجليدية العائدة إلى العصر البرمي في سلطنة عمان وإلى الدعوة المشتركة (وقيادة رحلتين ميدانيتين) لعقد مؤتمر عالمي حول صحاري الحقب الرابع وتغيّراته المناخية في العين، بإمارة أبوظبي في شهر ديسمبر عام ١٩٩٥.

هنا يبدو أنني عدت إلى حيث بدأت، فالصخور في تشكيلات بينونة في جبل بركة لاتشير فقط إلى مناخ أكثر رطوبة بكثير في العصر الميوسيني المتأخر مما تشهده المنطقة هذه الأيام، بل أيضاً مع وجود فجوة زمنية محتملة تبلغ زهاء ٩ ملايين سنة، تكمن تحتها رمال الكثبان وترسبات السبخة في تشكيلات الشويهات التي تعود إلى العصر الميوسيني المبكر، والتي تشبه أكثر نتاج مناخ اليوم. ولعله في صميم موضوعنا أن نذكر أن مناخ الحقب الرابع المتأخر في الإمارات قد تقلب بين الجفاف الشديد في ذروات تكون الأنهار الجليدية في المناطق البعيدة من خط الاستواء وبين مناخ أكثر رطوبة منه بين الأدوار الجليدية في عصرنا الحالي (مثلا، ما يُعرف بالوضع المناخي الأمثل منذ حوالي ١٠,٠٠٠ إلى ٥٠٠٠ سنة). وفي هذا الصدد، يبدو أن رمال كثبان الشويهات التي اتجهت جنوباً بفعل رياح شمالية في العصر الميوسيني، تشبه كثيراً الرياح السائدة اليوم التي تحمل معها الرمال.

فضلاً عن أهميتها من حيث نشوء الفقاريات وهجرتها في المنطقة، فإن الإسهامات الواردة في هذا الكتاب، تقدم معلومات عظيمة القيمة إلى علماء الجيولوجيا مثلي ممن لديهم اهتمام بتاريخ العصر النيوجيني في الجزيرة العربية عموماً، وفي إمارة أبوظبي تحديداً. وبتضمين الكتاب دراسة للأشغال اليدوية العائدة إلى المنطقة المرتبطة بأكثر الفقاريات تخريباً، وهو الإنسان، يتطرق الكتاب مباشرة إلى اهتمامي الخاص بتاريخ المنطقة في الحقب الرابع المتأخر.

إنه لمن دواعي سرور المرء أن يرى دولة

الإمارات العربية المتحدة تحتل موقع الصدارة في نواح عديدة من البحوث الجيولوجية في الجزيرة العربية. وهذا ليس بالشيء القليل، نظرا للدعم الذي تقدمه حكومة دولة الإمارات العربية المتحدة (انظر المقدمة بقلم معالي الشيخ نهيان بن مبارك آل نهيان وزير التعليم العالي والبحث العلمي) والقطاع الصناعي المحلي. وما كان للعمل الذي قمت به مؤخراً ولعمل وايبرو وهيل أن يبصرا النور لولا الدعم الذي تلقيناه من إدارة شركة أبوظبي للعمليات البترولية البرية (أدكو).

وتشكل الإسهامات المقدمة في هذا الكتاب دليلاً ناصعاً على التعاون القوي القائم اليوم بين علماء ينتمون إلى العديد من المعارف والعلوم. وهذا البحث المبني على عدة علوم أصبح الآن شرطاً مسبقاً لإماطة اللثام عن تاريخ الغلاف الحيوي والغلاف الصخري الناشئين في بقاع عديدة من العالم – وعلى وجه الخصوص في الجزيرة العربية التي أصبحت الآن منطقة مهمة للدراسات التي تتناول التغير المناخي في الماضي والحاضر.

الهوامش

١– تلقى كن غلني علومه في جامعة أدنبره (دكتوراة علوم عام ١٩٨٤) وأمضى أكثر من ٣٢ سنة في العمل كعالم جيولوجي مختص بالاكتشافات في شركة شل في نيوزيلندا وكندا ونيبال والهند والشرق الأوسط ولندن ولاهاي. وتتمثل اهتماماته البحثية الرئيسية في جيولوجيا الصحراء (في الحاضر والماضي) وجيولوجيا جبال عمان وجيولوجيا بحر الشمال. ومنذ تقاعده في عام ١٩٨٧، واصل أنشطته في هذه المناطق. وهو محاضر فخري في دائرة الجيولوجيا والبترول في جامعة أبردين.

٢– المراجع: انظر الجزء الانكليزي من الكتاب.

الرسم ١ - جزيرة الشويهات. يكشف الجزء الأدنى من الجرف البحري تشكيل الشويهات الذي تقع فوقه رواسب تشكيل بينونة التي تحتوي على مستحجرات.

الرسم ٢ - أجزاء من سن فيل بدائي عثر عليها في رأس دُبيعة وقد عُرضت على رئيس الدولة صاحب السمو الشيخ زايد بن سلطان آل نهيان في جبل الظنة.

الرسم ٣ – فك سفلي لحصان بدائي عثر عليه بجبل الظنة في العام ١٩٨٤.

الرسم ٤ – الدكتور وليد ياسين، من متحف العين، يحمل إحدى عظام ساق فيل بدائي عثر عليها بالحمرا في العام ١٩٨٤.

الجزء الأول، الفصل الأول: مدخل إلى الفقاريات الأحفورية في الجزيرة العربية

بقلم بيتر جيه. وايبرو وأندرو هيل

يبدو أن المناخ يحد من أنواع العديد من الحيوانات، مع أن هناك ما يدعو للاعتقاد بأن ما يحقق هذا الأثر في حالات كثيرة، ليس المناخ بحد ذاته بقدر ما هو التغيير في النباتات، الناتج من المناخ، ... وحيث قامت الحواجز من عصر بعيد، فإنها كانت ستحول دون احتكاك بعض الحيوانات ببعضها البعض، لكن عندما أصبح تجمع الكائنات الحية على جانبي الحاجز، بعد عصور عديدة، يشكل كلا عضوياً متوازناً، فإن تدمير الحاجز قد يؤدي إلى امتزاج جزئي جداً للأشكال المعينة من المنطقتين.

–والاس (١٨٧٦)

في عام ١٨٧٦، عندما نشر ألفرد راسل والاس أفكاره حول العلاقات المتبادلة العالمية للغلاف الجيولوجي المتغير مع الغلاف الحيوي، كانت جغرافية المناطق الداخلية من الجزيرة العربية وحيواناتها وجيولوجيتها لا تزال مجهولة. وقد أشارت المطبوعات القديمة (توماس ١٨٩٤، ١٩٠٠؛ يربري وتوماس، ١٨٩٥) بصورة عرضية إلى انتشار بعض الحيوانات اللبونة الأفريقية والآسيوية الموجودة حالياً في الجزيرة العربية، لكن لم يكن بالإمكان قول شيء في ذلك الوقت حول انتشار الحيوانات القارية التي تعود إلى الحقب الثالث في الجزيرة العربية أو خروجها منها، لأنه لم يتم اكتشاف أية حيوانات متحجرة حتى عام ١٩٧٤ (غير مسمى، ١٩٧٥). وفي الثلاثينيات، قدم البحث عن النفط العربي معطيات وفيرة إلى كل من الخرائط الطوبوغرافية والجيولوجية، وأعطت الاكتشافات التي توصل إليها علماء التاريخ الطبيعي فكرة واضحة عن الحيوانات والنباتات التي تعيش في منطقة ذات جغرافيا جافة متنوعة. ويعود العمل المتعلق بالحفريات القديمة في جبال الهملايا وشرق أفريقيا إلى الاكتشافات التي ترتبط بالثدييات العائدة إلى الحقب الثالث في تلك المناطق والتي جرت في القرن التاسع عشر وعشرينيات القرن العشرين على التوالي. لكن حتى الآونة الأخيرة، كانت الجزيرة العربية تشكل فجوة على صعيد الجغرافيا الحيوية القديمة فيما يتعلق بمعلوماتنا حول الحيوانات الفقارية القارية التي تعود إلى الحقب الثالث في المناطق القريبة من خط الاستواء في العالم القديم.

يجمع هذا الكتاب الذي يتضمن عدداً كبيراً من الصور والرسومات، لأول مرة بين بحوث حول الفقاريات الأحفورية للقارة العربية التي تم أكتشافها في الإمارات العربية المتحدة وسلطنة عمان والجمهورية اليمنية (١). ويقدم الكتاب معلومات حديثة ليس حول الحيوانات والنباتات في الجزيرة العربية فحسب، بل أيضاً حول البيئات القديمة في الجزيرة العربية والترابطات المغنطيسية القديمة الميوسينية في الجزيرة العربية وتحليلات النظائر المستقرة وبعض الأدوات البحرية الأقدم عهداً في الجزيرة العربية. وعلى المستوى الإقليمي، يتم بحث الحيوانات الفقارية للحقب الثالث في الجزيرة العربية في إطار حزام للحيوانات يوصف بأنه شمال أفريقي – عربي – جنوب غرب آسيوي، وتقدم هنا السجلات الأحفورية الجديدة من شرق الجزيرة العربية لأول مرة. وإضافة إلى ذلك، يبحث الكتاب بشكل شامل توقيت إغلاق ممر بحر تيثس الجيولوجي (البحر الأبيض المتوسط القديم في الشرق الأوسط). وقد أدى هذا الفصل إلى تكوين أول اتصال بري لأفريقيا – الجزيرة العربية مع آسيا. ومع أن الكتاب يركز بصورة رئيسية على حيوانات ونباتات العصر الميوسيني المتأخر في إمارة أبوظبي، حيث قام علماء حفريات بارزون منذ عام ١٩٧٩ بعمل حول التعاقب القاري للعصر الميوسيني، فقد ربطنا هذا البحث بالدراسات الجيولوجية المتعلقة به، (الرسومات ١–١ إلى ١–٣) في العصر الميوسيني في أبوظبي والمناطق المجاورة لها التي أجراها علماء الكيمياء الجيولوجية وعلماء طبقات الأرض وعلماء تحديد تواريخ الأحداث في دراسة المجال المغناطيسي للأرض.

سيثير هذا الكتاب اهتمام الاختصاصيين الذين كانت دراساتهم حتى الآن حول انتشار الحيوانات الأفريقية والآسيوية في الحقب الثالث غير مكتملة بسبب

الافتقار إلى التفاصيل من شبه الجزيرة العربية، المتوافرة هنا في مطبوعة واحدة. وسيجد دارسو جيولوجية الشرق الأوسط وجيولوجيو التنقيب عن النفط في هذا الكتاب خلاصة مفيدة، لأننا حاولنا إقامة العلاقة المتبادلة الصعبة بين الأحداث البرية والبحرية في الحقب الثالث.

ينقسم الكتاب بصورة طبيعية إلى ستة أجزاء، جرى تقديم كل منها مع ملخص للفصول. كما تظهر مقدمات كل جزء باللغة العربية أيضاً. ويتألف الجزء الأول من خلاصة ولمحة عامة، مع سرد تاريخ مشروع أبوظبي، ويقدم الجزء الثاني الإطار الجيولوجي الذي جُمعت منه الأحفوريات الفقارية ووصفا لطبقات الأرض المحلية. وتصف فصول الجزء الثالث الدراسات المنهجية حول لافقاريات وزواحف وثدييات إمارة أبوظبي (يشكل عدد منهم فصائل جديدة) في العصر الميوسيني المتأخر وتبحث العلاقات الجغرافية القديمة للمستحجرات. وتتناول الفصول الأربعة في الجزء الرابع مواضيع مرتبطة بالحيوانات والنباتات الأحفورية – علم عمليات التحجر ونظائر الكربون والبيئات القديمة في الجزيرة العربية – وقد أضفنا دراسة حول الأشغال اليدوية في العصر الحجري وأقدم دليل على وجود الجنس البشري في المنطقة. ثم يربط الجزء الخامس مشروع أبوظبي مع البحوث الأخرى في آسيا وأفريقيا ويتضمن دراسات للحيوانات الأقدم عهداً في سلطنة عمان والجمهورية اليمنية. وأخيراً، تعرض في الجزء السادس صورة أشمل للجزيرة العربية في إطار العالم القديم. ونتناول توقيت انفصال بحر تثيس، وكذلك الأحداث التي وقعت في البحر المتوسط وفي براتيثس وتتعلق بانتشار الثدييات. ويقدم الفصل الأخير عرضاً حديثاً لعوامل جغرافية المحيطات القديمة الضرورية لفهم التغير المناخي خلال العصر الميوسيني.

لا تقدم الجزيرة العربية بسهولة إلى علماء الحفريات شواهد بينة على الحيوانات والنباتات البرية التي كانت تعيش فيها في الماضي. وإذا نحينا جانباً قسوة عملية الاكتشاف، حيث يمكن بالصدفة العثور على استثناء، فإن جميع صخورها تقريباً هي صخور بحرية (٢). وحيث أصبحت الترسبات القارية النادرة مكشوفة الآن، لا توجد أنهار وليس هناك إلا قليل من المطر لتفتيت الصخور وإزالة المستحجرات عنها. وعندما تكشف عملية التفتيت فقارية أحفورية، يمكن أن تتحطم عظامها وتتفتت بفعل درجات الحرارة الشديدة وبالتالي تتبعثر وتندثر بفعل الرياح التي تحمل الرمال. ونتبين مدى ندرة الفقاريات الأحفورية في الجزيرة العربية من حقيقة أنه برغم حجم اللوح الجيولوجي في الجزيرة العربية – الذي هو أكبر قليلاً من لوح شبه القارة الهندية – فلا يوجد إلا ١١ موقعاً فقط تم حتى الآن العثور فيها على حيوانات فقارية ذات قيمة وتعود إلى أي عصر جيولوجي.

كما ذكرنا سابقاً، تنبثق دراسات هذه الحيوانات والنباتات من الاكتشاف الذي توصل إليه في عام ١٩٧٤ علماء الحفريات التابعون لمتحف التاريخ الطبيعي في لندن الذي كان في حينه يسمى المتحف البريطاني (التاريخ الطبيعي – لأول حيوانات فقارية برية من العصر الميوسيني في شبه الجزيرة وذلك في الجزء الشرقي من المملكة العربية السعودية. ومنذ السبعينيات، تولى العمل بشكل رئيسي فريقان أوروبيان بالتعاون مع مؤسسات عربية متمثلة في شركات النفط في الدول التي جرت فيها الأبحاث. وقد ركز الفريق الأول بقيادة بيتر وايبرو من متحف التاريخ الطبيعي في لندن(٣) وأندرو هيل من جامعة ييل (بالتعاون مع شركة أبوظبي للعمليات البترولية البرية) على العصر الميوسيني المتأخر في إمارة أبوظبي؛ أما الفريق الثاني بقيادة هربرت توماس من لابوراتوار دوباليونترو بولوجي إيه بر إيستوار، كوليج دو فرانس، باريس (بالتعاون مع وزارة البترول والمعادن والمديرية العامة للمعادن في سلطنة عمان)، فقد ركز على الصخور الأوليغوسينية والميوسينية في سلطنة عمان. كذلك فإن العمل الذي قامت به مجموعات بحثية مختلفة في الجمهورية اليمنية (بالتعاون مع وزارة النفط والثروة المعدنية وجامعة صنعاء) أعطى البوادر الأولى على وجود حيوانات ونباتات برية تعود إلى الحقب الثالث، وأعطى تسلسل العصر الجوراسي المتأخر ما يعتقد بأنه الحدث المدون الثاني لوجود نوع من الديناصورات في الجزيرة العربية.

لقد تغيّر علم الحفريات بشكل هائل منذ اكتشاف الفقاريات الأولى في الجزيرة العربية. وكما يدرك قراء هذا الكتاب، لم يعد علم الحفريات الفقارية علماً مستقلاً. فاليوم لكي نفهم – قدر المستطاع – مواطن الحيوانات والنباتات المنقرضة، يتعاون علماء الحفريات مع غيرهم من الاختصاصيين الجيولوجيين؛ فمشروع مثل البحث

الذي نجريه في إمارة أبوظبي يصبح عمل فريق يضم اختصاصيين من علوم ومعارف متعددة. وفي حين يظل دائماً التعرف المنهجي على الأحفوريات عمود علم الحفريات الفقارية، إلا أن الدراسات التي يجريها أختصاصيون آخرون تقدم الآن عدداً كبيراً من المواضيع العلمية التي تتراوح من علم الأحياء الارتقائي مروراً بعلم الحفريات الجغرافية الحيوية وحركة الألواح القارية وانتهاء بالتصخر والنظائر داخل الصخور والمستحجرات نفسها.

تتواصل عمليات اكتشاف تراث الحفريات القديمة في الجزيرة العربية، بفضل الدعم الواعي الذي تقدمه المؤسسات الحكومية داخل دول شبه الجزيرة العربية. وهذا الكتاب ليس الخطوة الأولى في إعلان نتائج بحوث الحفريات القديمة والبحوث الجيولوجية فحسب، لمصلحة شعوب الجزيرة العربية، لكنه يقدم أيضاً شهادة متعمقة للدور الأكاديمي الناشىء الذي تؤديه الجزيرة العربية الآن للربط بين الدراسات المتعلقة بحيوانات الحقب الثالث وبيئاته في العالم القديم.

الهوامش

١- ينبع زخم هذا الكتاب من المؤتمر العالمي الأول حول الفقاريات الأحفورية في الجزيرة العربية الذي عُقد في إمارة أبوظبي بدولة الإمارات العربية المتحدة في شهر مارس ١٩٩٥.

٢- التعبير الإنجليزي (سارينديتي) قد استحدثه هوريس والبول، الإيرل الرابع لأوكسفورد (١٧١٧-١٧٩٧)، من الحكاية الخرافية الفارسية الأمراء الثلاثة لسرنديب، حيث يتمتع أبطال الحكاية بهذه الموهبة.

٣- مشروع أبوظبي، الذي يستمر حتى العام ٢٠٠٠، يشكل جزءاً من برنامج بحث التغيّر العالمي والغلاف الحيوي الذي ينفذه متحف التاريخ الطبيعي.

٤- المراجع: انظر الجزء الانكليزي من الكتاب.

الجزء الأول، الفصل الثاني: موجز ولمحة عامة حول حيوانات بينونة، في إمارة ابوظبي، وإطارها

أندرو هيل وبيتر جيه. وايبرو

كما يستحيل علينا أن نخمن، في أغلب الحالات، الطبيعة الدقيقة للقوى التي تحد من مجموعة بعض الأنواع وتتسبب بندرة أنواع أخرى أو بانقراضها! وكل ما نأمل عموماً بفعله هو أن نتتبع، نظرياً إلى حد ما، بعض التغيرات الكبيرة في الجغرافية الطبيعية التي حدثت خلال العصور التي سبقت مباشرة عصرنا وأن نعطي تقديراً للأثر الذي يحتمل أن تكون قد أحدثته في توزيع الحيوانات. وقد نستطيع حينها، بمساعدة تلك المعرفة المتعلقة بالتحويل العضوي الماضي والتي يزودنا بها السجل الجيولوجي، أن نحدد مسقط الرأس المحتمل والهجرات اللاحقة لأهم الأجناس والفصائل.

– ولاس (١٨٧٦)

إن إحدى أهم الخصائص البيولوجية الأكثر مثاراً للاهتمام في شبه الجزيرة العربية هي الدور الذي لعبته هذه الجزيرة في الجغرافيا الحيوية للعالم القديم. وتقع الجزيرة العربية عند تقاطع الانقسامات الجغرافية الحيوية الكلاسيكية للعالم القديم. المناطق الإثيوبية والقطبية الشمالية والمعتدلة والشرقية (والاس، ١٨٧٦). كما تشكل شبه الجزيرة العربية منطقة عالمية كبيرة (١) تقع بين ١٢ درجة و٣٠ درجة شمالاً وبين ٣٥ درجة و٦٠ درجة شرقا، وتحتل مساحة تزيد قليلاً على ٣ ملايين كيلومتر مربع تقريباً – كامل اللوح القاري للجزيرة العربية. لذلك فإن مساحة الجزيرة العربية توازي تقريبا مساحة شبه القارة الهندية. وتتوافر اليوم في شبه الجزيرة مجموعة متنوعة من المواطن التي تتراوح من المناطق الجبلية في الجنوب الغربي ذات الهضاب العالية حيث يصل ارتفاع بعض القمم فيها إلى حوالي ٣٨٠٠ متر إلى الصحاري الرملية المنخفضة التي تحتل معظم المنطقة الشرقية. وفي الشمال، على حدود الخليج العربي، نجد المسطحات الملحية التي ينخفض مستوى بعضها عن مستوى سطح البحر.

كما تتميز المنطقة بمناخ متغير، ضمن حدود المناخ الجاف إلى الجاف للغاية (تكاهاش وأراكاوا، ١٩٨١)، فمثلاً، في الجنوب الغربي المرتفع، هناك ٥٠٠ ملم من المطر سنوياً، مع حدود منخفضة لدرجات الحرارة وهطول الثلج على بعض الجبال المرتفعة في فصل البرد. وفي المنطقتين الوسطى والشرقية التي تضم بعضاً من أكثر الصحاري حرارة وجفافاً على سطح الكرة الأرضية، يصل متوسط درجات الحرارة اليومية إلى ما يقرب من ٤٠ درجة مئوية ويقل هطول الأمطار.

إن وصف الرحلات الشاقة التي قام بها المستكشفون الغربيون الأوائل للجزيرة العربية مثل تشارلز داوتي وسانت جون فيلبي ووليام بلغريف وويلفرد ثسيغر يعطي تأكيداً واضحاً لشيوع الاعتقاد بندرة الحيوانات البرية في الجزيرة العربية. بيد أن البحوث اللاحقة أثبتت أن الجزيرة العربية تتمتع بحيوانات ثديية ونباتات متنوعة (هاريسون وبيتس، ١٩٩١). ولا يكتفي هاريسون وبيتس بدراسة شبه الجزيرة العربية، بل ايضا الدول الواقعة إلى الغرب في البحر الأبيض المتوسط وإلى الشمال عبر العراق. وبالتالي فإن الأرقام التالية مبالغ فيها قليلاً بالنسبة لشبه الجزيرة العربية. لكن على مدى هذه المنطقة بأكملها، نجد تمثيلاً لثماني رتب للثدييات، تضم ٢٩ فصيلة وتشتمل على ٨٢ جنساً. وحتى دون الخفافيش وغيرها من الحيوانات الثديية الصغيرة (آكلات الحشرات واللاجومرفا [رتبة من القوارض] والقوارض)، تظل هناك ١٤ فصيلة و٢٩ جنساً. غير أنه يصح القول إن الجزيرة العربية تفتقر إلى تشعب الأنواع، إذ لا يوجد فيها إلا زهاء ٤٠٪ من عدد الأنواع المتوافرة مثلا في منطقة شرق أفريقيا الأصغر مساحة؛ لكن هناك تمثيلاً جيداً لمستويات تصنيفية أعلى، حتى في منطقة شبه الجزيرة. ومن بين الفقاريات الأخرى غير البحرية، تتردد الأسماك البرمائية والنهرية على الجداول والبرك المستديمة في المناطق الجبلية.

الجغرافية الحيوية للجزيرة العربية

إن المناقشات الجادة حول الجغرافية الحيوية للجزيرة العربية – والارتباطات الممكنة لشبه الجزيرة مع آسيا –

قد بدأت في مطلع الثلاثينيات، وتصادفت مع اكتشاف النفط. وجاءت المعطيات حول الصخور البحرية المغمورة في منطقة الخليج من آبار النفط المتزايدة. وعلى أساس الأحفوريات الصغيرة، رأى ديفيز (١٩٣٤) أنه قد يصعب التعرف على جسر بري بين الجزيرة العربية وجنوبي غرب آسيا. واستنتج سافيدج (١٩٦٧) أن هجرات الثدييات من العصر النيوجيني كانت الأقوى بين أوروبا وآسيا والأضعف بين أفريقيا وآسيا. وكان هذا تطويراً لاعتقاد والاس (١٨٧٦) بأن أسلاف الحيوانات الثديية الحديثة في العالم القديم انتشرت من الشمال إلى الجنوب. ولم يُشر والاس قط في هذا العمل إلى الجزيرة العربية، وهو يقسم المنطقة، على نحو يبعث على الغرابة، بين منطقتيه الإثيوبية والمتوسطية (المنطقتين الباردة والمعتدلة). وقد طور برنور (١٩٨٣) هذه الفروقات الإقليمية بدرجة أكبر بالنسبة لحيوانات العصر الميوسيني.

حتى السبعينيات، كانت التعليقات حول الجغرافيا الحيوية للجزيرة العربية تتضمن بمعظمها استنتاجات من التسلسلات البحرية ومن تغير الحيوانات البرية الموثقة في نجاحات الحفريات القديمة المعروفة بشكل أفضل والتي تحققت في الكتل البرية المجاورة. ولم تكن الحيوانات الفقارية البرية من العصر النيوجيني معروفة من الجزيرة العربية نفسها. لكن في عام ١٩٨٤، عُثر على ثدييات من العصر الميوسيني المتوسط في المملكة العربية السعودية (غير مسمى ١٩٧٥؛ أندرو وآخرون، ١٩٧٨؛ هاملتون وآخرون، ١٩٧٨؛ وايبرو، ١٩٨٧). واستند وايبرو (١٩٨٢) بصورة جزئية إلى هذه المعلومات الجديدة لكي يطوروا لاحقاً مفاهيم ديفيز وسافيدج، وأبدوا ملاحظات على احتمال وجود ارتباط بري في العصر الميوسيني المبكر بين شرق الجزيرة العربية وجنوبي غرب آسيا. وقد تمعن آدامز وآخرون (١٩٨٣) بشكل أوفى في هذه الملاحظات من المنظور البحري، وتوسع بها وايبرو (١٩٨٤) في إطار بري. كما استخدم برنور (١٩٨٣، الرسم ٣) هذه المعطيات الأولية. وتشير الدلائل البحرية إلى أن هذا الارتباط البري حدث على أبعد تقدير منذ ما يترواح بين ١٩ – ١٦ مليون سنة العصر الميوسيني.

أعطت هذه التكهنات، مقرونة ببعض الدراسات الميدانية الأولية قبل مطلع الثمانينات، الزخم لجمع مزيد من المعلومات الجيولوجية والحفرية العائدة إلى الحقب الثالث من الجزيرة العربية. لكن برغم الاهتمام الشديد بالمنطقة وموقعها الجغرافي المهم بالنسبة للقارات الأخرى، لا يُعرف إلا القليل عن الحيوانات الأحفورية والتاريخ البيئي القديم في شبه الجزيرة. وفي الفصل الثالث والثلاثين، يشير وايبرو وكليمنتس (١٩٩٨أ) إلى ندرة حدوث اكتشافات للفقاريات الأحفورية المعروفة في الجزيرة العربية. وفي ظل هذه الخلفية يمكن أن نقدّر على أفضل وجه أهمية الاكتشافات الحديثة في إمارة ابوظبي، لاسيما حيوانات بينونة. فهي تقدم معلومات مهمة حول البيئات والحيوانات والنباتات في الحقبة الماضية.

الفقاريات الأحفورية في الجزيرة العربية

يقدم هذا الكتاب معطيات قيمة حول الحيوانات في العصرين الميوسيني والأوليغوسيني في سلطنة عمان (توماس وآخرون، ١٩٩٨– الفصل ٣٠)، وأول تسجيل لحيوانات ونباتات العصر الإيوسيني من الجمهورية اليمنية (السروري وآخرون، ١٩٩٨ – الفصل ٣١) وثاني تسجيل لديناصور من الجزيرة العربية (جيكوبس وآخرون، ١٩٩٨– الفصل ٣٢). وتتضمن الفصول ٣٤ (آدامز وآخرون، ١٩٩٨) و٣٥ (ماكلويد، ١٩٩٨) و٣٦ (روغل ، ١٩٩٨) المعلومات الكتابية الإقليمية ويتناول الفصل ٢٦ أقدم المعلومات حول الإنسان في الإمارات (ماكبريرتي، ١٩٩٨). لكن الموضوع الرئيسي في الكتاب هو جيولوجية وحفريات العصر الميوسيني القاري المتأخر اللتان لم يعثر عليهما في الجزيرة العربية حتى الآن إلا في إمارة أبوظبي.

تشتمل الاكتشافات المصاحبة على أول تعرّف على تشكيل السدود في الإمارات (وايبرو وآخرون، ١٩٩٨– الفصل ٤)، الذي لم يكن معروفاً سابقاً إلا في المملكة العربية السعودية وقطر. وهو وحدة بحرية تعود إلى العصر الميوسيني الأدنى أو الأوسط (بورديغاليان)، عمره ١٩–١٦ مليون سنة العصر الميوسيني.

لقد جرى تنقيح وصف طبقات الأرض الذي أعطاه أساساً وايبرو (١٩٨٩). وتنقسم تشكيلات بينونة لديه الآن إلى تشكيلات الشويهات والتي يغلب عليها العصر الأكولي في الأسفل وتشكيلات بينونة التي هي بمعظمها نهرية وأحفورية في الأعلى. وتتميز تشكيلات الشويهات التي وُصفت حديثا (وايبرو وآخرون، ١٩٩٨،

بريستو ١٩٩٨-الفصلان ٤ و٦) بتكوُّن عرضي للطبقات من العصر الأكولي، يمثل كثبان الرمل التي تفصل بينها الحجارة الطينية والرمال الناعمة التي تفسر بأنها بيئة من السبخة. ومن المحتمل أن تعود إلى العصر الميوسيني الأدنى إلى المتوسط ويُشار إلى أن عمرها يبلغ حوالى ١٥ ± ٣ مليون سنة على أساس المغناطيسية القديمة (هيلوود ووايبرو، ١٩٩٨ - الفصل الثامن). وكما يشير كينغستون وهيل (١٩٩٨) في الفصل ٢٧، يتسم هذا التشكيل بالأهمية لإعطائه أدلة ملموسة على الأحوال الجافة في المناطق البعيدة عن خط الاستواء خلال منتصف العصر الميوسيني. كذلك إذا امتدت أحوال مشابهة عرضياً عبر أفريقيا، تعطي تفسيراً لطبيعة الحاجز المفترض المفضي إلى التمييز بين حيوانات العصر الميوسيني في شمال أفريقيا وجنوب الصحراء الأفريقية (توماس، ١٩٧٩، توماس وآخرون، ١٩٨٢، هيل ١٩٩٨ - الفصل ٢٩). ومن حيث تناظر الطبقات والعمر يمكن ربط تشكيلات الشويهات بشكل عام بتشكيلات الهفوف ذات الصخور الفتاتية في المملكة العربية السعودية.

وتتألف تشكيلات بينونة، كما أعيد تعريفها (وايبرو وآخرون، ١٩٩٨ - الفصل ٤) من تسلسل من الترسبات النهرية في معظمها تشير إلى تشكل نهر ذي درجة ميل منخفضة يتألف من عدد كبير من الأقنية الصغيرة التي تفصل بينها الضفاف الرملية المنخفضة. وربما لم تكن تلك الأقنية أعمق من ٣ أمتار، لكن عرض كامل شبكة النهر المجدولة كان من عشرات إلى مئات الأمتار (فريند ، ١٩٩٨، ديتشفيلد، ١٩٩٨ - الفصلان ٥ و٧). وكان هذا النظام النهري يُصرِّف الماء في منطقة تقع في داخل شبه الجزيرة العربية إلى الشمال الغربي من إمارة أبوظبي الحديثة، وربما كان جزءاً من نظام أكبر يضم نهري دجلة والفرات الحديثين. وفي ذلك الوقت كان مستوى سطح البحر أدنى كثيراً مما هو عليه اليوم، ويعتقد أن الخط الساحلي البحري كان عند مسافة ٣٠٠ كلم تقريباً إلى الشرق من موقعه الحالي.

ومن الواضح أن المناخ قد تغير بشكل ملموس بين زمني الشويهات وبينونة. فقد خفت الأحوال السابقة الجافة للغاية، وتطور النظام النهري لبينونة وأتاح موئلاً للرخويات النهرية (جفري ١٩٩٨ - الفصل ١٠) وبطنيات الأقدام البرية (موردن ، ١٩٩٨ - الفصل ١١) والأسماك والزواحف المائية والطيور والثدييات. ويبدو واضحاً أنه كان هناك جريان دائم للمياه في هذا النهر من وجود سلاحف نهرية كبيرة (لاباران دوبروان وديجك، ١٩٩٨ - الفصل ١٣) والتماسيح بما في ذلك تمساح غريال (روش وآخرون ، ١٩٩٨ - الفصل ١٤)، لكن وجود سمك السلور يوحي بأن التدفق كان بطيئاً أو متقطعاً في بعض الأقنية (فوري ويونغ ، ١٩٩٨ - الفصل ١٢). ويُستدل على وجود تدفق عرضي ذي سرعة أعلى من الكتل المختلفة الأكثر خشونة في بعض الأقنية ومن كون بعض عظام المستحجرات مفصولة ومتجزئة (فريند، ١٩٩٨ - الفصل ٥) . وكانت درجات الحرارة دافئة خلال أزمنة بينونة، وتشير العينة المحفوظة في الترسبات إلى أن المناخ كان شبه جاف، مع معدل سنوي لهطول المطر لا يزيد على ٧٥ ملم (ديتشفيلد، ١٩٩٨ - الفصل السابع). وتألفت النباتات من خليط من العشب (النجيل) والشجيرات والأشجار، ومن بينها أشجار الأقاقيا. وربما كانت الأشجار والشجيرات متجمعة بالقرب من ضفاف النهر، في حين نمت نباتات عشبية أكثر رحابة على مسافة أبعد من النهر نفسه (كينغستون، ١٩٩٨ - الفصل ٢٥).

وقد ساد هذا الموطن مجموعة غنية ومتنوعة من الحيوانات، بما فيها الأشكال القديمة من الفيلة وأفراس البحر والخيول والظباء وحيوانات شبيهة بالذئاب والضباع والقطط ذات الأنياب الطويلة. وفي الفصل ٢٣، يُعدِّد وايبرو وكليمنتس (١٩٩٨ب) عناصر حيوانات بينونة. وقد جمع مشروع متحف التاريخ الطبيعي في لندن/ جامعة ييل اكثر من ٩٠٠ عينة. وبالإجمال هناك ٤٣ نوعاً من الفقاريات المنتمية إلى ٢٦ فصيلة. وتضم ثلاثة أنواع جديدة وجنساً واحداً جديداً : سمك برغيد (فوري ويونغ، ١٩٩٨ - الفصل ١٢) وجربوع أبو ظبي - بينونة (دوبرويجن ووايبرو ، ١٩٩٤) ونوع من الحصان البدائي (أيزنمان ووايبرو، ١٩٩٨ - الفصل ١٩). والحيوانات غير الثديية الأخرى التي تم التعرف عليها هي ثلاثة أنواع من السمك وثلاثة أنواع من السلاحف (تمثل كلا من الأشكال البرية والمائية)، وثلاثة أنواع من التماسيح (بما فيها غريال) ونوعان من الطيور. ومن بين الثدييات هناك ما مجموعه ٣١ نوعاً موثقاً مؤلفاً من ١٧ أو ١٨ فصيلة.

ويشير النظام النهري نفسه والحيوانات الأحفورية التي يحتويها إلى بيئة ومناخ في أزمنة بينونة مختلفين بشكل ملحوظ عنهما في الوقت الحاضر، أو

خلال أزمنة الشويهات التي سبقتها. ويمكن أن نعزو الانتقال من بيئات الشويهات إلى بيئات بينونة إلى عدة عوامل عالمية. وقد بحث بعضها كينغستون وهيل (١٩٩٨) في الفصل ٢٧.

الجغرافية الحيوية القديمة وطبيعة التغير الحيواني

الحيوانات الثديية هي أساساً أفريقية بطابعها (هيل، ١٩٩٨، الفصل ٢٩) ولاسيما من شمال أفريقيا (دجنتري، ١٩٩٨أ،ب – الفصلان ٢١–٢٢)، لكن، كما يمكن أن نتوقع، فإنها تتضمن عناصر آسيوية، كما يتبين من بعض القوارض (دو برويجن ووايبرو، ١٩٩٤) والخنازير (بيشوب وهيل، ١٩٩٨–الفصل ٢٠) والبقريات (دجنتري، ١٩٩٨ب–الفصل ٢٢). كما توجد بعض الأجناس في أوروبا، ولكن ليست هناك أية صلات مؤكدة على مستوى الأنواع مع الحيوانات الأوروبية من العصر الميوسيني المتأخر، مثل تلك التي تعود إلى اليونان وتلك المعروفة شرقاً عبر تركيا إلى شمالي غرب إيران. فليس هناك غزلان، مثلاً، في مجموعة بينونة، ولا يُعرف في الواقع أي أيليات من أية مواقع ميوسينية وبليستوسينية في شبه الجزيرة العربية، مع أنه جرى تسجيل وجود الغزال البني المرقط في بلاد ما بين النهرين واليحمور حتى أزمنة حديثة في شمالي – غرب إيران وجنوبي تركيا وفي فلسطين.

وعوضاً عن ذلك تبدو حيوانات بينونة جزءاً من حزام الحيوانات في العصر الميوسيني المتأخر المتجه غرباً وشرقاً بين ١٥ درجة شمالاً و٣١ درجة شمالاً على وجه التقريب ويشتمل على مواقع في شمال أفريقيا والجزيرة العربية والباكستان والهند وربما أفغانستان. وهذا شكل جغرافي حيواني مختلف عن الأقاليم في شمال أفريقيا وجنوب براتيش التي أشار إليها برنور (١٩٨٣). ويوحي بأنه خلال أزمنة بينونة كان بمقدور الحيوانات الهجرة بسهولة أكبر في اتجاه شرق–غرب، لكن الحركة في اتجاه شمال – جنوب ربما كانت مقيدة بحواجز تمثلها الصحاري أو الجبال أو أنظمة الأنهار القديمة.

وتتيح المراجعة والمقارنة المنهجيتان لأنواع من المناطق المعنية، لاسيما شمال أفريقيا، إزالة الشوائب من الأفكار الحديثة حول تأثير المناخ على التنوع والانتشار أو بدحض تلك الأفكار، كتلك التي افترضتها فربا (١٩٩٥). وهناك مشاكل عديدة مرتبطة بمثل هذا العمل المقارن وبالمفاهيم المتعلقة بطبيعة ودرجة تأثير العوامل الدخيلة الطبيعية على الحيوانات، وتنوعها وهجرتها (هيل، ١٩٨٧، ١٩٩٥، وايت، ١٩٩٥). والمشكلة الرئيسية هي النقص الذي يشوب سجل المستحجرات في ± كلا الزمان والمكان. وبالنسبة للجزيرة العربية، يوضح وايبرو وكليمنتس (١٩٩٨أ – الفصل ٣٣) مدى ضآلة سجل الفقاريات الأحفورية، وهو سجل لم يُكتشف إلا خلال العشرين سنة الماضية تقريباً. وفي جنوب الصحراء الأفريقية أيضاً، هناك عدد قليل من المواقع التي تعود إلى العصر الميوسيني المتأخر (هيل، ١٩٩٨– الفصل ٢٩). ولم تؤخذ عينات عادية أو عينات جيدة من كل الأزمان، وتتجمع المواقع بمعظمها في مساحة صغيرة على الجانب الشرقي من أفريقيا وليس لدينا معلومات مطلقاً تتعلق بالأحداث التي وقعت في العصر الميوسيني المتأخر في حوالى ٩٩٪ من القارة الأفريقية. ويمكن إجراء حسابات مشابهة للمناطق الأخرى؛ فمثلا، التغطية المساحية في جنوبي غرب آسيا ليست أكبر بكثير.

تحد قيود السجل الأحفوري من قدرتنا على إعطاء تحديد دقيق للظهور الأول والأخير للحيوانات وهذه معطيات أساسية لتقييم نمط التغير الحيواني وترابطه بالعوامل الدخيلة. وتترجم هذه التقديرات إلى خطوط إسناد زمنية لظهور المستحجرات في منطقة ما، وإضافة إلى تلك، فإن مكان الظهور الأول للحيوان غالباً ما يفترض بأنه قريب من مركز منشئه.

ومن هذه النماذج قليلة الأهمية نوعاما، جرى تركيب روايات رئيسية حول أقاليم توزيع بقايا الحيوانات الأحفورية في العالم القديم وأحداث التوزيع، وحول نمط التغير الحيواني وطابعه على مر الزمن. فمثلاً، تنبأت فربا (١٩٩٣) أن نبضات التحول يجب أن تحدث ضمن سجل المستحجرات. وافترضت أن معظم تحولات السلالات حدثت في نبضات، بشكل متزامن تقريباً عبر المجموعات المتنوعة للكائنات الحية، وبتزامن يمكن التنبؤ به مع التغيرات في البيئة الطبيعية. ومع أن فربا قد عدلت نظريتها فيما بعد (١٩٩٥)، إلا أنها نادت بإكراه بيئي مباشر وقوي لتغير الأنواع. وهناك مشاكل في بيان نبضات التحول، بالنظر إلى طبيعة سجل الأحفوريات (هيل، ١٩٨٧، ١٩٩٥). وهذا النوع من التغير ليس من السهل رؤيته في أفريقيا أو الجزيرة العربية أو آسيا،

وبالتأكيد في أفريقيا فإن بعض السلالات مثل الخنازير، لا تبدو أنها تتقيد بأفكار فربا (بيشوب، ١٩٩٣، ١٩٩٤، وايت ١٩٩٥). وإن الأحداث المتعلقة بالجبال والتي حدثت على مدى فترة طويلة من الزمن، مثل ارتفاع هضبة الهملايا – التيبت، الذي ربما يتوسط المناخ المحلي، يحتمل أيضاً أن يكون عاملاً مؤثراً (كينغستون وهيل، ١٩٩٨–الفصل ٢٧).

لكن المعلومات المتعلقة بالحيوانات التي تم الحصول عليها حتى الآن من الجزيرة العربية، لها انعكاسات عدة. أحدها هو الإيحاء القوي والمستجد بأنه في العصر الميوسيني المتأخر، امتد حزام من النسب الحيواني عبر شمال أفريقيا والجزيرة العربية وإلى أجزاء من آسيا. وهذه الحيوانات تختلف عن تلك المجاورة، باستثناء شمال زغروس في إيران، وعن جنوب الصحراء الأفريقية. وتتطلب إزالة الشوائب من هذه الإيحاءات قدراً أكبر من المواد الأحفورية، لكن عملية استكشاف الجزيرة العربية على صعيد الحفريات القديمة هي في بدايتها. وإذا أجريت المزيد من الأبحاث في إمارة أبوظبي وشبه الجزيرة العربية ككل فإنها تبشر بآمال عريضة.

الهوامش

١– تعتبر شبه الجزيرة العربية هنا بأنها تضم الدول التالية : دولة الكويت ودولة قطر والمملكة العربية السعودية والإمارات العربية المتحدة وسلطنة عمان والجمهورية اليمنية.

٢– لم يكن ممكناً لعلماء الحفريات الأوروبيين والأمريكيين إجراء مزيد من الاكتشافات في مواقع الفقارياتالميوسينية في المنطقة الشرقية من المملكة العربية السعودية بسبب سياسة القيود على مثل هذا العمل التي فرضتها السلطات السعودية منذ عام ١٩٧٤. وحسب علمنا، لم ينفذ أي عمل تفصيلي حول هذه المواقع المهمة من أي جانب منذ منتصف الثمانينيات. وتظل الدراسات الوحيدة التي أصدرتها الفرق البريطانية والأمريكية (انظر الفصل ٣٣) ناقصة.

٣– المراجع: انظر الجزء الانكليزي من الكتاب.

ملخص الجزء الثاني: جيولوجية العصر الميوسيني في المنطقة الغربية، في إمارة أبوظبي، بدولة الإمارات العربية المتحدة

عندما بدأ وايبرو أول دراساته الجيولوجية للطبقات الصخرية الميوسينية المكشوفة في المنطقة الغربية بين عامي ١٩٧٩-١٩٨٤، كانت المطبوعات الوحيدة المتوافرة عبارة عن خرائط جيولوجية ذات مقياس كبير للمنطقة وتقرير موجز أعده خبراء الجيولوجيا البترولية. وكان جزء كبير من عمله ينحصر بالطبقات الساحلية المكشوفة، حيث كان بالإمكان عموماً رؤية الغلافات المعقدة للصخور بدرجة أكبر من السهولة. والطبقات الصخرية الميوسينية في الداخل بعيداً عن الساحل، في الماضي والحاضر، مغطاة بمعظمها برمال تذروها الرياح وبترسبات ميوسينية متفتتة. ومنذ الثمانينيات، أدت دراسات أكثر تفصيلاً قام بها فريق متحف التاريخ الطبيعي في لندن وجامعة ييل إلى الحصول على البيانات الأساسية اللازمة لاية بحوث ميوسينية مستقبلية في المنطقة. ولا يفوتنا ذكر حقيقة أنه، برغم أن الطبقات الصخرية المكشوفة لا تكشف إلا حوالي ٦٠ متراً من الترسبات الميوسينية، إلا أن التعاقب في المنطقة الغربية يندرج في إطار فترة من العصر الميوسيني تعود إلى زهاء ١٩-٦ ملايين سنة ماضية. ويعتقد أن عمر الترسبات التي تحمل الفقاريات في الجزء الأدنى من تشكيلات بينونة يتراوح من ٨ إلى ٦ ملايين سنة.

الخطوة الأولى في أية دراسة جيولوجية لمنطقة غير معروفة سابقاً هي تحديد التعاقب وتقديم الأسماء للوحدات الاستراتيجرافية لطبقات الأرض) إذا تعذر ربطها بسهولة بالوحدات الأخرى في المنطقة. ويقدم بيتر جيه. وايبرو وبيتر أف. فرند وبيتر دبليو. ديتشفيلد وتشارلي أس. بريستو (الفصل الرابع) معلومات جيولوجية تؤيد أسماً استراتيجرافياً جديداً، هو تشكيلات الشويهات، للترسبات التي هي بغالبيتها ذات أساس ريحي والتي تكمن فوق تشكيلات السدود البحرية، التي سبق وصفها من المملكة العربية السعودية. واسم تشكيلات بينونة الذي جرى تعريفه سابقاً يقتصر الآن على الوحدة العليا التي هي بمعظمها نهرية والتي تحتوي على الحيوانات الأحفورية غير البحرية المهمة التي ورد وصفها في الجزء الثالث ويتم عرض التسجيلات الترسبية التي قيست في ١٤ موقعاً.

يصف بيتر أف. فرند (الفصل الخامس) الترسبات التي شكلها النظام النهري الذي كان يجري فيما مضى عبر المنطقة المعروفة اليوم باسم بينونة.

وتتألف تشكيلات بينونة السفلى أساساً من ذرات رمال ناعمة، لكنها تحتوي أيضاً على مجار كثيرة مغطاة بالحصباء وبعض الوحدات الوحلية المميزة. كما أن حيواناتها الفقارية الأحفورية المهمة، فضلاً عن بعض رخوياتها وعلامات الجذور والأتربة الوفيرة تؤكد أن الوحدة تشكلت (تكونت) في معظمها بفعل الترسب النهري. وليس هناك دليل إيجابي على التأثير البحري.

يقدم الفحص التفصيلي للطبقات الصخرية المكشوفة والنظيفة بشكل غير اعتيادي في فرضة دلما للمعديات في جبل الظنة وفي جبل ميمية (المرفا) معلومات حول الأنهار التي كونت بعض الترسبات. ويبدو أن هذه الأنهار كانت متفاوتة في تدفقها وربما يتراوح عمقها بين ٣ و١٠ أمتار في وقت الفيضان. وكانت مجدولة من حيث إنها احتوت على العديد من القضبان الترسبية والأقنية المتفاوتة الحجم والشكل.ويبدو أن اتجاهات التدفق كانت على العموم في اتجاه شرقي و جنوب شرقي. وربما يكون ذلك النظام النهري سابقا لنظام دجلة والفرات الحالي، لكن هذا مجرد إيحاء قائم على التكهن.

لقد قام تشارلي أس. بريستو (الفصل السادس) بدراسة الترسبات في تشكيلات الشويهات التي تقع تحت ترسبات النهر في تشكيل بينونة. وتهيمن الترسبات الهوائية والسبخا على تشكيلات الشويهات التي كانت سابقاً جزءاً من تشكيلات بينونة. وإن تكوينات القاع الأكولية مكشوفة جيداً وتشير إلى أن تشكل الكثبان كان مستعرضاً وبرخانيا. وغالباً ما تتخلل رمال الكثبان وتقطعها ترسبات السبخة التي يمكن ربطها على مسافة عدة كيلومترات على طول خط الاتجاه. وتم التعرف على نوعين من السبخة : السباخ الصغيرة بين الكثبان (عرضها ١٠ - ١٠٠ متر) والسباخ الساحلية الأكبر مساحة (عرضها أكثر من ١٠٠٠ متر). وأحياناً تتخلل هذه الترسبات ترسبات نهرية وتعلوها بفعل التفتت. وتشير قياسات التيارات القديمة في الحجارة الرملية الهوائية إلى أن الريح القديمة السائدة كانت من الشمال إلى الجنوب. وتشير اتجاهات التيارات القديمة المستمدة من الترسبات النهرية إلى انتشار أوسع بكثير للتيارات مع جريان الأنهار من الغرب إلى الشرق ومن الجنوب إلى الشمال، عبر اتجاه النقل الهوائي أو بعكسه. وتشير عمليات إعادة تشكيل الجغرافية القديمة إلى أن تشكيلات

الشويهات قد ترسبت في حوض يضم كتلاً صخرية قديمة مزود بنظام صرف شبه مغلق حيث كانت الأنهار تصب في سبخة داخلية أو بحرية شاطئية. وتعزى التغيرات في مستوى البحيرة التي أثرت بشدة في أنظمة الكثبان إلى التقلبات المناخية خلال العصر الميوسيني المتأخر. ويوحي وجود منخربات مشكلة من جديد في بعض الرمال الهوائية إلى أن الأحوال البحرية كانت قائمة داخل الحوض، بحيث أن الإطار العام قد يكون سهلاً ساحلياً ذا تضاريس منخفضة مع أنه لم يتم العثور على أدلة على وجود ترسبات بحرية في تشكيلات الشويهات في وجه الطبقة الصخرية.

لفهم البيئة القديمة التي حدثت فيها الترسبات، من الضروري إجراء تحاليل تفصيلية للصخور التي تشكلت من تلك الترسبات. وقد تم تحليل مجموعة عينات بيتر دبليو. ديتشفيلد (الفصل ٧) المأخوذة من التشكيلات الميوسينية المكشوفة من المنطقة الغربية تحليلاً بتروجرافيا وكيميائياً – جيولوجياً لتحديد درجة تغير التصخر ولمحاولة الحصول على معلومات حول البيئات الترسبية لهذه الترسبات. وتبين النتائج البتروجرافية أن تشكيلات بينونة قد خضعت لطمر جزئي فقط ولتصخر موضعي فقط. والملاط الكربوناتي الموجود عُثر عليه في بيئة نيزكية، وفي بعض الحالات فوق أي سطح ماء باطني دائم. كما لا تعطي تشكيلات الشويهات الواقعة تحتها دليلاً يذكر على وجود طمر أو تراص ملموس، مع أنها تحتوي على كميات ملموسة من ملاط الجبس داخل بعض الواجهات. ويبين تشكيل السد تاريخاً تصخرياً أكثر تعقيداً ، حيث أصبحت الترسبات الكربوناتية الآن ذات أشكال دولوميتية على نطاق واسع. وإن الدولوميتات من تشكيل السد متميزة من الناحية الكيميائية – الجيولوجية عن الدولوميتات النادرة التي عُثر عليها في الوحدات الواقعة في الأعلى وتشير إلى أصل تبخري للموائع الدولوميتية. وتشير التحاليل النظائرية لعينات قشر البيض الكالسيتية الراتيتية التي هي بحالة جيدة المأخوذة من تشكيلات بينونة إلى وجود مجموعات نباتية تحتوي على عنصري الكربون – ٣ والكربون – ٤ بصورة مختلطة (منطقة أعشاب ذات غابات وغابات ذات أعشاب) خلال ترسب تشكيلات بينونة.

لقد قام إرني إيه. هيلوود وبيتر جيه. وايبرو (الفصل ٨) بإجراء تحريات للمغناطيسية القديمة والبنية المغناطيسية على خمسة أجزاء تمتد معاً على طول معظم الجزء المكشوف من التعاقب الترسبي الميوسيني في المنطقة الساحلية المطلة على الخليج العربي في إمارة أبوظبي. وكانت أهداف الدراسة مزدوجة : (١) لتحديد التغيرات في قطبية الحقل المغناطيسي الجيولوجي خلال عملية الترسب وللإسهام في الترابطات الاستراتيجرافية بين الأجزاء و(٢) لاشتقاق موضع قطبي مغناطيسي قديم لهذه التشكيلات بما يساعد على حصر وتحديد عصورها الجيولوجية. لقد جرى تقسيم التعاقب الترسبي الميوسيني في منطقة بينونة في أبوظبي إلى تشكيلين منفصلين، تشكيل شويهات الأكولي في الأسفل وتشكيل بينونة النهري بمعظمه في الأعلى.

ويفصل بين هذين التشكيلين عدم توافق يبدو واضحاً في جزء جبل بركة لكن موقعه يصبح مشكوكاً فيه أكثر في الجزء الكائن على جزيرة الشويهات وفي غيرها من الأماكن في المنطقة.

تشير التحريات الاستراتيجرافية – المغناطيسية المذكورة في الفصل الثامن إلى أن عدم التوافق بين تشكيلي الشويهات وبينونة يكمن ضمن غلاف مغناطيسي طويل ذي قطبية عكسية في جزء جبل بركة. والغلاف المغناطيسي ليس جيد التحديد في جزء جزيرة الشويهات، لكن موقعه الذي تم استنتاجه محاط بفواصل قطبية عادية فوقه وتحته. ويعني ضمناً هذا الترابط الاستراتيجرافي – المغناطيسي بأن الحدود بين تشكيلي الشويهات وبينونة يجب أن تقع في مكان ما في الفاصيل ٦.٧٥ – ١٦ متراً فوق مستوى سطح البحر على جزيرة الشويهات.

إن إجراء مقارنة لمواقع القطبية المغناطيسية القديمة المستمدة من تشكيلات الشويهات وبينونة مع منحنى التحرك القطبي الظاهر من العصر الحديث في الجزيرة العربية يشير إلى عمر محتمل قدره ١٥ ± ٣ ملايون سنة (أي العصر الميوسيني المتوسط) بالنسبة لتشكيل الشويهات وحوالي ٦ ± ٣ ملايين سنة (أي العصر الميوسيني المتأخر) بالنسبة لتشكيل بينونة. وهذا الفرق في العصور المحددة بالمغناطيسية القديمة بالنسبة للوحدتين يؤيد انفصالهما المفترض إلى تشكيلين جيولوجيين مختلفين.

وأخيراً، في الجزء الثاني (الفصل ٩)، يذكر روس بيبلز أنه يمكن العثور على العصر الميوسيني في أبوظبي

على شكل منكشفات صخرية وهضبات قائمة الجوانب وبوتات متبعثرة وفي الطبقة تحت السطحية قليلة العمق. وفي الغرب، فإن الصخور السليكاوية الرضيخية والفتاتية من العصرين الميوسيني الأعلى والمتوسط في تشكيلات بينونة والشويهات تشكل ميزات وجبالاً نراها على طول الساحل، بينما في كربونات وتبخيرات العصر الميوسيني الأدنى إلى الأوسط في الشرق في تشكيلات غاخساران تشكل بوتات داخلية ومنكشفات صخرية منعزلة صغيرة وتقع على عمق بضعة أمتار فقط تحت سطح السبخة. كما أن الصخر الميوسيني مكشوف في العديد من جزر أبوظبي، حيث ظهر إلى السطح بواسطة اختراق الطبقة المحدبة الملحية. وقد جرى تحليل عينات من مواقع مختارة عبر الإمارة لتحديد نسب النظائر Sr^{87}/Sr^{86} وC^{13} وO^{18} وS^{34} لتحديد خصائص مميزة نظائرية ترسبية متكاملة للتشكيلات الميوسينية في أبوظبي. وقد جرت مقارنة هذه الخصائص المميزة بالمنحنيات النظائرية الشاملة لتحديد الموقع الاستراتيجرافي المتعلق بعمر الطبقات الصخرية لهذه التشكيلات. واستناداً إلى خصائصها النظائرية ، يمكن نسبة تشكيل غاخساران إلى العصر الميوسيني المبكر– المتوسط (بورديغليان – لانغيان). وإن الخصائص النظائرية لتشكيلات بينونة والشويهات التي تبين أنها نتيجة التعديل التصخري، لا تعكس الأحوال الترسبية، وبالتالي لا يمكن استخدامها لتحديد الموقع الاستراتيجرافي المتعلق بعمر الطبقات الصخرية.

ملخص الجزء الثالث : المستحاثات الحيوانية الميوسينية من تكوينة بينونة الجيولوجية في إمارة أبوظبي بدولة الامارات العربية المتحدة.

يصف بول أ. جيفري (الفصل – ١٠) قسما من مستحاثات الحيوانات اللافقارية المجموعة من تكوينة بينونة الجيولوجية. وتعزى الحيوانات ثنائية الأصداف الى نوعين بفصيلتين عليتين منفصلتين. ويعزز وجود مثل هذه الرخويات الأحفورية وطبيعة الترسبات المغلفة لها الاستنتاج العلمي بوجود نظام نهري سريع التغير وربما كان ذلك بصورة موسمية في منطقة بينونة المذكورة خلال العصر الحديث الوسيط المعروف بالعصر الميوسيني.

ويستدل من حقيقة عدم معرفة هذه الأصناف في المنطقة من قبل وعدم وجودها بين الحيوانات العربية الحديثة ما مفاده بأن مناخ وجغرافية المنطقة خلال العصر الميوسيني الأخير قد أتاحا الاتصال مع مناطق جغرافية أصبحت شبه الجزيرة العربية الآن معزولة عنها بصورة فعلية.

وبعدها يصف بيتر ب. موردان (الفصل ـ١١) هيكلا داخليا لصدفة حلزون أرضي متحجرة. وقد جرت نسبة هذه بصورة مؤقتة الى فصيلة الرئويات من رتبة الرخويات بطنية الأقدام. ولا يتعارض وجود مثل هذا الكائن مع أنماط التوزع الجغرافي في الأوقات الحالية على مستوى العائلة أو الفصيلة لهذا النوع من المخلوقات، كما يستدل من ذلك على احتمال وجود أوضاع موسمية أكثر رطوبة خلال العصر الميوسيني المتأخر في المنطقة المذكورة.

ومن بين الحيوانات المائية التي كانت سائدة في نظام بينونة النهري، ثمة ثلاث مجموعات مصنفة من الفصيلة العليا للأسماك المسماة أوستاوريوفيسان (Ostariophysan) وقد تم وصفها من قبل بيتر ل. فوري وسالي في.ت. يونغ (الفصل ـ١٢). وترتكز هذه المجموعات التصنيفية جميعها على مجموعات كبيرة من بقايا الشظايا الأحفورية، ومن بينها نوع جديد من سمك البياض أو الفتيل باغروس من فصيلة السلوريات المسماة علميا سيلوريفورميس: باغريدا (Siluriformes: Bagridae) ونوع آخر من سمك السلور أو القرقوط المنسوب الى العائلة المعروفة بالاسم العلمي سيلوريفورميس كلاريدا (S. Clariidae) وتتمثل الدلالة على وجودها بصورة رئيسية من خلال البقايا القحفية والنطاق الصدري في حين أن السمك الشبوطي البني من عائلة الشبوطيات المسماة علميا سيبرينيفورميس : سيبرينيدا (Cypriniformes: Cyprinidae) لم يستدل عليه الا من خلال الأسنان البلعومية.

ويستفاد من هذه المكتشفات الأحفورية أن مياه النهر الذي كانت تعيش فيه هذه الأسماك قد كانت بطيئة الجريان مع حصول فيضانات سريعة ومباغتة فيها بعض الأحيان. ومن شأن وجود صنف الأسماك السلورية في مواضع عديدة باالترسبات الأفريقية ـ العربية خلال العصر النيوجيني الذي يشمل العصرين الميوسيني والباليوسيني أن يلقي بظلال كثيفة من الشك على النظريات التقليدية الزاعمة بهجرة هذه الأنواع من أسماك السلور الى أفريقيا من آسيا خلال العصر الحديث القريب الباليوسيني في حين أن تاريخ نشأة وتطور أسماك السلور البياضية يشير بقوة الى حصول تشتت وانتشار لها من آسيا الى أفريقيا خلال الفترة السابقة للحقبة الميوسينية.

ومن بين مستحاثات الزواحف التي استوطنت نظام نهر بينونة والمناطق المحيطة، ثمة ثلاث مجموعات من السلاحف المصنفة علميا موصوفة باسهاب وتفصيل من قبل فرانس دي. لابارينت دي. بروين و بيتر بول فان ديجك (الفصل ـ١٣). وتبدو هذه المجوعات المصنفة أقدم من أصناف العصر البليوسيني الموجودة حاليا ولكنها متأخرة عن الأشكال التي كانت سائدة في العصر الميوسيني الوسيط في منطقة الجزيرة العربية.

وينتمي النوع المسمى سينتروشيليس (*Centrochelys*) من السلاحف الأرضية المخططة الى مجموعة متوطدة منذ العصر الميوسيني القديم في المنطقة الأفريقية ـ العربية وتتطابق في أوصافها مع نوع آخر موجود في منطقة الصحابي في ليبيا. ويبدو أن كلا النوعين الآخرين من السلاحف التي تعيش في المياه العذبة يعتبران في عداد الحيوانات التي هاجرت مؤخرا خلال العصر الميوسيني المتأخر الى منطقة الخليج العربي أو أنها قد وصلت أخيرا الى أفريقيا وذلك حسب العمر الزمني للموضع الذي اكتشفت فيه.

والشظايا المحفوظة للسلحفاة النهرية المسماة الترسة (*Trionyx*) لا تتيح المجال لتقرير الجنس أو النوع الفرعي الذي تنتمي له (مثل رافيتس (*Rafetus*) أو أي جنس آخر غير مصنف بعد) كما لا يمكن اجراء مقارنة دقيقة مع مواقع أخرى على الرغم من أن الأنواع (أو النوعين) تبدو مستجدة. وينتمي الصنف المعروف باسم

موريميس (Mauremys) الى مجموعة أنواع غريبة وهي غير موضحة بصورة تفصيلية ولكن حالة نشوئها وتطورها تعزى الى نمط العصر الميوسيني المتأخر.

ومن المحتمل أن لهذا الصنف صلاتا قوية مع الأصناف السائدة في شمالي أفريقيا (تونس، ليبيا ومصر) والحوض الشرقي للبحر الأبيض المتوسط (العراق)، ولكن لا تظهر هناك أية صلات أو وشائج قريبة مع الأصناف المعروفة في شبه القارة الهندية وأفغانستان. وتشير سلاحف المياه العذبة الى حدوث تغير كامل متصل بحيوانات العصر الميوسيني الوسيط في منطقة الجزيرة العربية بسبب انعدام وجود أي شكل من الأشكال الضخمة أو المفلطحة أو ذوات الأعناق المائلة الى الجانب في الأصداف. سيكلانورباين، كاريتوكيليد وبليورودايـر (cyclanorbine, carettochelyid, pleurodire) التي تعتبر الآن من الأنواع الاستوائية.

ويصف مايكل روهي، ايبرهارد (دينو) فري، دانيال س. بيمبرتون وتورستين روسمان (الفصل - ١٤) البنية العظمية للمستحاثات التمساحية المأخوذة من التكوينة المتقدمة الذكر. وقد تمت مقارنة الأوضاع النظامية والتصنيفية لهذه المستحاثات مع التماسيح السائدة في أفريقيا وآسيا وأمريكا الجنوبية. وقد تم تحديد مواد الشظايا العظمية على أنها تخص نوعا من التماسيح المسماة كروكودايلس (Crocodylus) مقارنة مع تماسيح نهر النيل وبحر ايجه والهند.

ويستفاد من هذه النتائج أنه يمكن استخلاص استنتاجات بيئية قديمة لتوضيح الارتباط بين نوعي التمساح الهندي غافياليد (Gavialid) والتمساح النهري كروكودايليد (Crocodylid) في بيئة نظام بينونة النهري.

وبعد القيام بتنخيل وفرز كيلوغرامات عديدة من الرواسب المأخوذة من تكوينة بينونة، عثر هانس دي.برويجن (الفصل - ١٥) على ثلاثة أسنان قاطعة منفصلة عن بعضها لاحدى آكلات الحشرات و٤٥ سنا من الأسنان الوجنية المنفصلة عن بعضها لبعض أنواع القوارض. وتمثل هذه القوارض خمس فصائل فرعية وسبعة أنواع. وتشير هذه العينة الصغيرة الى وجود لفيف متنوع من هذه الحيوانات يسودها جربوع [أبوظبي - بينونة].

ويستفاد من وجود جربوع أبوظبي الذي يعد من جهة أخرى ضربا من الأنواع المعروفة بحوض كابول في أفغانستان ووجود صنف آخر من النوع المسمى زابودينا (Zapondinae) الى وجود مؤثرات آسيوية في حين أن وجود نوعين من الصنف النباتي المسمى ديندرومس (Dendromus) يشير الى وجود تأثير أفريقي أيضا في حيوانات هذه المنطقة.

ولا تتيح مرحلة النشوء والتطور للقوارض الأحفورية المجموعة من تكوينة بينونة الجيولوجية اقامة علاقة ترابط حيوية طبقية دقيقة نظرا لندرة وضآلة المعلومات المتوفرة بشأن تعاقب تطور وتبدل الحيوانات العربية في العصر النيوجيني الذي يشمل العصرين الميوسيني والباليوسيني. ويمكن الاستنتاج من وجود الفأر الصغير المسمى بارابيلوميس (Parapelomys) ويربوع أبوظبي البدائي أو الأولي بأن الجزء النهري من التكوينة يعود الى العصر الطوروني وأوائل العصر الرسكاني. كما يستفاد من وجود البيئة الترسبية ومقارنة المتطلبات البيئية للأصناف التي لا تزال موجودة على قيد الحياة حاليا والتي تعد ذات صلة وثيقة بالقوارض الأحفورية بأن ثمة بيئة قديمة ذات نظام نهري كبير يشتمل على حياة نباتية كثيفة من القصب والشجيرات النامية على ضفاف القنوات القائمة في سهل مفتوح شبه جاف وقاحل لولا وجود ذلك النظام.

ومن المعروف بصورة جلية ندرة وجود الرئيسات من الحيوانات الثديية التي تشمل الهباريات والقرديات والبشريات في أية مجموعة من حيوانات العصر الميوسيني. وفي الفصل - ١٦، يصف أندرو هيل وتوم غوندلينغ الحيوان الثديي الرئيسي الوحيد من بين مختلف المستحاثات التي عثر عليها في الامارات العربية المتحدة - والذي يمثل بالفعل القرد الأحفوري الوحيد بين سائر المستحاثات التي تم العثور عليها في سائر أنحاء شبه الجزيرة العربية. وتتمثل هذه العينة الأحفورية من خلال سن لأحد السعادين المذنبة تم العثور عليها في تكوينة بينونة. وقد تم ادراجها في سياق تطور القرود في العصر الميوسيني المتأخر كما هو معروف حاليا.

ويورد جون سي. باري (الفصل - ١٧) بأن آكلات اللحوم من العصر الميوسيني في امارة أبوظبي تشتمل على نوعين من فصيلة الضباع المسماة بليسيوغولو برايكوسايدنس (Plesiogula praecocidens) والسنور الضخم المسيّف الأسنان. وتعتبر هذه المجموعة متطابقة

بصورة نموذجية مع مستحاثات العصر الميوسيني المتأخر والباليوسيني التي أمكن العثور عليها في مواقع بأوراسيا وأفريقيا كما أنها قد تدل على احتمال وجود بيئات برية مكشوفة نوعا ما في ذلك الوقت في شبه الجزيرة العربية.

وتشتمل آكلات الأعشاب الكبيرة على الفيلة والدينوثير والاسطغور وربما المستودن (*Mastodon*) الشبيه بالفيل ذي القواطع الكبيرة. وربما لا يوجد سوى نوع واحد من الفيل البائد ستيغوتترابيلودون (*Stegotetrabelodon*) وتتألف هذه العنية الموصوفة من قبل باسكال تاسي في الفصل ـ ١٨ من بقايا متناثرة مأخوذة من الموقع (س ـ ٦) في شويهات وهي متعلقة بحيوان واحد فقط.

ويعطي هذه الاكتشاف مثالا فريدا وقيما للصفات والخواص المترابطة لمختلف الأنظمة التشريحية الوصفية سواء في المجالات: السنية أو القحفية أو ما بعد القحف اذ أن الخواص الفكية والأنياب السفلية والأضراس تتطابق مع الخواص الموجودة في الأنواع الأفريقية لجنس الفيل البائد ستيغوتترابيلودون ومن أنواع ضروبه سيرتيكس وأوربس في حين لا توجد تطابقات في السمات مع الفيلة البائدة ستيغوتترابيلودونتس التي يظن أنها غير أفريقية والموصوفة من قبل مؤلفين عديدين.

وتجدر الاشارة بصورة خاصة الى أن السمات الملحوظة على بقايا الحيوان البائد في موقع شويهات (س ـ ٦) تدحض النظرية القائلة بنسبة حيوان المستودن الشبيه بالفيل ذي القواطع الضخمة (شليزنجر ـ ١٩١٧) الى جنس الحيوان البائد ستيغوتترابيلودون وقد تم توصيف المستودن الشبيه بالفيل ذي القواطع الضخمة بادىء الأمر في ايران وبعدها في أوروبا الشرقية وامارة أبوظبي وجمهورية الكونغو الديموقراطية (زائير سابقا) ويجري اعزاؤه في بعض الأحيان الى أنه بمثابة السلف أو الجد الأعلى لجنس الفيل الأفريقي البائد ستيغوتترابيلودون وهي نظرية لا ندعمها هنا.

ومع أن أصل منشأ الفيلة الأساسية وفصائلها قد يكون من خارج أفريقيا، الا أن أفريقيا تعتبر بمثابة المهد الذي تمت فيه عملية تمايزها وتفرقها. كما أن فهم تفاصيل نشأة وتطور جنس الفيلة القديمة ستيغوتترابيلودون في حد ذاته بحاجة الى المزيد من الاكتشافات.

وفي الفصل ـ ١٩، يفيد كل من فيرا ايزنمان و بيتر جي. وايبرو بأن جنس الحصان المسمى هيباريون (*Hipparion*) قد تمثل في تكوينة بينونة من خلال عظام الفكين والأسنان وشظايا عظام الأطراف. ويبدو من حجم العظام والأسنان أنها تمثل نوعين أحدهما صغير أو متوسط الحجم والآخر أكبر. وقد جرت مقارنة شظيتين فكيتين مميزتين بصورة جلية من النوع الصغير مع مواد أحفورية مشابهة مأخوذة من أوراسيا وأفريقيا. وتبين من خلال حجمهما وتناسقهما أنهما مختلفتان عن عظام أفكاك الحصان المندثر هيباريون وقد أشير اليهما على أنهما يمثلان نوعا جديدا لفصيلة أو جنس حصان الهيباريون البائد.

وتقدم فصيلة الخنزيريات الموصوفة من قبل لورا بيشوب وأندرو هيل (الفصل ـ ٢٠) فرصة متميزة لاستقصاء ودراسة الجوانب المتعلقة بتاريخ تطور الأنسال والسلالات وتنوع التصنيفات الأحيائية والجغرافية الحيوية بشأن الخنزيريات السائدة في العصر الميوسيني المتأخر. ويتناول الفصل المذكور بايجاز نشأة وتوزيع أصناف الخنازير الأحفورية في العصر النيوجيني من العالم القديم كما يصف العينات والنماذج التي أمكن الحصول عليها من تكوينة بينونة.

وتعزى هذه العينات الى مجموعتين تصنيفيتين اثنتين على الأقل حيث تنتسب الخنزيريات الكبيرة الى نوع رباعي القواطع الأمامية يدعى تيتراكوندونت من فصيلة نيانزاكورس سيرتيكس (*Nyanzachoerus syrticus*) (ليوناردو ـ ١٩٥٢). وهذا النوع معروف في العديد من المواقع الأحفورية في أفريقيا.

وأما النوع الأصغر سوايني، فانه مصنف بالاسم العلمي بروبوتاموكرس هيسندريكس (*Propotamochoerus hysudricus*) (ستيهلن ـ ١٨٩٩) وهو معروف من خلال الحفريات في توالي طبقات سيواليك في شبه القارة الهندية. ولا توجد عناصر أوروبية متطابقة أو متشابهة بين المستحاثات أو الأحافير الخنزيرية في أبوظبي. ويستفاد من وجود هذين النوعين أن هناك تبادلا حيوانيا بين آسيا وأفريقيا والمنطقة العربية خلال العصر الميوسيني المتأخر. ويعتبر كلا هذين النوعين متسقين مع العصر الميوسيني النهائي للتكوين.

ويصف ألان و. جينتري (الفصل ـ ٢١) بقايا

أفراس النهر التي يعود تاريخها الى العصر الميوسيني والتي تنتمي الى نوع أكثر بدائية من النوع سداسي القواطع المسمى علميا هيكسا بروتودون هارفاردي *(Hexaprotodon harvardi)* المكتشف بمنطقة لوثاغام في كينيا. وبالمقارنة مع مستحاثات أفراس النهر الأخرى غير المعروفة جيدا، فانها تبدو أقرب ما تكون من النوع سداسي القواطع المسمى هيكسا بروتودون ساهابيانسيس *(H. sahabiensis)* وهيكسا بروتودون ايرافاتيكس *(H. iravaticus)*.

وبعد ذلك يصف ألان جينتري (الفصل ـ ٢٢) الأنواع المجترة. وثمة ثلاثة أنواع من فصيلة الزرافيات (تشتمل على النوع القديم المسمى بالاتراغس *(Palaeotragus)* ونوع آخر يدعى براماثيريوم *(Bramatherium)* ونوع ثالث غير مصنف بعد) بالاضافة الى ستة أنواع من فصيلة البقريات تشتمل على النوع المسمى تراغوبورتاكس سيرانيكس *(Tragoportax cyrenaicus)* ونوع آخر يدعى باشيبورتاكس لاتيدينس *(Pachyportax latidens)* وكلاهما من القبيلة المسماة بوسيلافيني (Boselaphini) ونوع بقري ثالث من قبيلة غير مصنفة في حين أن النوع الرابع الذي يدعى بروستريبسيكيروس ليبيكس *(Prostrepsiceros libycus)* والخامس المسمى بروستريبسيكروس فيناياكي *(P. vinayaki)* والنوع السادس المسمى الغزال ـ ليديكيري *(Gazella lydekkeri)* تنتمي بأنواعها الثلاثة الى قبيلة الظباء).

وعند أخذ هذه الأنواع مع بعضها البعض، فانها تدعم النظرية التي تعزو تشكل تكوينة بينونة في العصر الميوسيني الأخير وربما كان ذلك قبل ١٣ مليون سنة بمقياس مين الأوروبي المستخدم لقياس نطق أو طبقات الأحافير وحوالي ٦ ملايين سنة خالية.

وتظهر هذه الأنواع تشابها وتقاربا في الصفات مع مجموعات حيوانات العصر الميوسيني المتأخر المكتشفة مستحاثاتها في مناطق شمالي أفريقيا وتوالي طبقات سيواليك في شبه القارة في الهندية وجزيرة بيرام أكثر من تشابهها مع مجموعات الحيوانات الاغريقية ـ الايرانية الطورونية أو المناطق الواقعة تحت نطاق الصحراء الكبرى في أواسط أفريقيا.

وقد تعتبر هذه الأنواع جزءا من الحيوانات المميزة بخطوط العرض والتي تقع عبر امتداد الأراضي الشاسعة الى الجنوب من المناطق المأهولة بمجموعة الحيوانات الاغريقية ـ الايرانية. كما أن وجود الزراف والأبقار والظباء ذات القرون اللولبية يدل على أن هذه البلاد قد كانت عبارة عن مناطق رعوية. كما أن من المحتمل شيوع مناخ مداري جنوبي أو حار ولكن بدون أية عوامل أو مؤثرات كبيرة من العوامل المسببة للجفاف.

وفي نهاية المطاف، يورد بيتر جي. وايبرو وديانا كليمينتس في الفصل ـ ٢٣ قائمة مستحاثات الحيوانات المكتشفة في موقع تكوينة بينونة. وتعتبر هذه القائمة ذات فائدة قصوى في الدراسات المستقبلية وبمثابة سجل تاريخي هام وقيم أيضا. وتجدر الاشارة الى أن وتيرة التطور في امارة أبوظبي تعتبر سريعة للغاية وهناك بعض المواقع المتوفرة حاليا لجمع المزيد من المستحاثات منها. وقد تم تبيان الأوصاف الجغرافية لهذه المواقع مع ادراج الاحداثيات السمتية لتحديد المواقع لبعض المواضع بداخلها.

ملخص الجزء الرابع : طريقة استحاثة الخرطوميات، النظائر الكربونية وبيئات المستحاثات في تكوينة بينونة والمشغولات اليدوية القديمة المكتشفة بالمنطقة الغربية في امارة أبوظبي والبيئات الأحفورية العربية.

يصف بيتر اندروز في الفصل - ٢٤ عملية الحفريات وعملية الاحاثة لإحدى أنواع الخرطوميات المسماة ستيغوتيترابيليدون سيرتيكس (*Stegotretabelodon syrticus*) وقد عثر على هذا الحيوان المستحاث البائد مستقرا داخل ترسبات رملية في قناة نهر جاف حيث جرى نقله الى هناك من مسافة قريبة كجثة متصلة جزئيا بالمفاصل. وقد كانت الجمجمة وكثير من الأضلاع والعمود الفقري وأغلب أجزاء الأطراف الخلفية موجودة ولكنه لم يتم حفظ وبقاء سوى جزئين من الطرف الأمامي. ومع أن كافة أجزاء الأطراف القصوى كانت مفقودة، الا أنه لم يظهر على الجثة أية آثار تشير الى تعرضها للنهش من قبل حيوانات أخرى. ولعل أرجح التفسيرات للعظام المفقودة يتمثل في تمايز وتفاوت عملية النقل لها. وقد تمت عملية الطمر والدفن بسرعة بعد تفكك كافة العظام مباشرة ولكن قبل تبعثرها وقبل أن تعتريها أية درجة من درجات التلف بفعل عوامل الطقس. وقد لحقت بعض أعراض التلف المحدودة بالأجزاء القحفية الكبيرة ومن المحتمل أن يكون ذلك بسبب بروزها فوق المراحل الأولى لعملية الطمر والدفن الترسبية. كما أن عوامل التعرية الحاصلة مؤخرا قد عملت على كشف وتعريض الجمجمة واحتمال اتلاف أجزاء منها والتسبب في الحاق أضرار كبيرة بها بفعل تعريضها للعوامل الجوية.

وقد تولى جون د. كينغستون مهمة القيام بالدراسات المتعلقة بالنظائر الكربونية (الفصل - ٢٥) بشأن كربونات الأتربة ومينا آكلات الأعشاب المأخوذة من تكوينة بينونة. وتعتبر عملية تحديد البيئة الأحفورية الأثرية لمجموعات المستحاثات الحيوانية مسألة حيوية في سياق محاولة فهم نمط وعملية النشوء والارتقاء للأحياء. وتعمل تحاليل النظائر المستقرة للمكونات الأحفورية ومينا أسنان مستحاثات آكلات الأعشاب على توفير وسائل تحليل وتحديد قيم كمية لتوثيق المقادير النسبية لاستخدام النباتات لعنصري الكربون -٤ والكربون -٣ اللذين يمثلان مسارات التمثيل الضوئي في الماضي مما يعكس ويبين وجود بيئات مناطق أعشاب وغابات استوائية على التوالي.

وتعمل كربونات الأتربة المتشكلة جراء التعادل النظائري أو المتكافىء مع ثاني أكسيد الكربون المستخلص من خلال تحلل النباتات وتنفسها، على استبقاء بصمة أو شارة دلالية على الكميات النسبية لعنصري الكربون -٤ والكربون -٣ في التضاريس الأرضية الحديثة. ويتم الاحتفاظ بشارة النشوء الأحيائي في السجل الأحفوري كما تعمل على اعطاء تقديرات لنسب وكميات النباتات المشتملة على عنصري الكربون -٤ والكربون -٣ في الأزمنة القديمة. وتشير تحاليل النظائر الكربونية المستقرة للعقد الكربونية المأخوذة من النطق الأحفورية لتكوينة بينونة الى تشكل الكربون في بيئات تشتمل على عنصري الكربون -٤ والكربون -٣ بصورة مختلطة. ويشبه الاختلاف الجانبي في تركيب النظائر الكربونية (لعنصر الكربون -١٣) في هذه العقد الى حد قريب جدا من ذلك الاختلاف في كربونات الأتربة المتشكلة خلال الأوقات الحالية في أراضي المناطق العشبية والغابات والأجمات.

ومن المقرر علميا بصورة جلية أن نظير الكربون -١٣ في أباتيت الأسنان الحديثة يعزى مباشرة الى التركيب الآيسوتوبي أو التناظري لنوع الغذاء وأن هذه السمة تظل سليمة كما هي خلال عملية التحجر والاحاثة. وتعتبر عمليات اعادة تشكيل أنواع وأنماط التغذية للحيوانات القديمة البائدة والمرتكزة على أساس التحاليل النظائرية لمينا أسنان آكلات الأعشاب المتحجرة ذات فائدة قصوى في حصر ملامح أو جوانب الحياة النباتية المتوفرة للتغذية في منطقة ما. كما تفيد التحاليل النظائرية للكربون المختزن في مينا أسنان المستحاثات لخمس عائلات آكلات أعشاب مجموعة من تكوينة بينونة بوجود حياة نباتية فسيفسائية من عنصري الكربون -٤ والكربون -٣ خلال العصر الميوسيني الأخير. كما يشير نظير الكربون -١٣ في مينا أسنان الأحافير الى اعتماد كبير على الأعشاب المشتملة على نظير الكربون -٤ اذ أن أكثر من ثلث العينات تعكس نمطا غذائيا يشتمل على عنصر الكربون -٤ بصورة تامة في حين أن أغلب العينات المتبقية تشير الى وجود مركب الكربون -٤ بصورة ملحوظة في نمط الغذاء الخاص بحيواناتها.

وتقوم سالي ماكبريرتي في الفصل - ٢٦ بوصف المشغولات اليدوية المكتشفة في أربعة مواقع بالمنطقة الغربية في كل من شويهات، الحمراء، رأس العش وجبل بركة. ومن المحتمل اعزاء وجود هذه المشغولات اليدوية الى أن الصخور السطحية السليكونية قد عملت على توفير المواد الأولية التي صنعت منها تلك الأدوات. وقد تم العثور عليها في البيئات السطحية ولم يرتبط وجودها

بأية بقايا حيوانية أو وسائل مستحدثة مما قد يسهل تحديد عمرها أو صلاتها. ومن الناحية التقنية، فان المشغولات اليدوية المكتشفة في مناطق شويهات والحمراء ورأس العش تتسم بأوجه شبيهة بصناعة الأنصال والخناجر التي راجت في الساحل الشمالي الشرقي من قطر.

كما أن وجود قضبان مشكلة بصورة مميزة على شكل هلال يربطها بالعصر الحجري الحديث (ب) قبل العصر الفخاري في المشرق والذي يعود تاريخه الى حوالي ٩,٣٠٠ ـ ٨,٣٠٠ سنة قبل الميلاد. ومع ذلك فان الأدوات المكتشفة في جبل بركة تختلف عن تلك المكتشفة في مناطق شويهات والحمراء ورأس العش حيث لا توجد في مكتشفات جبل بركة أية أنصال ولكن المجموعة تشتمل على قضبان شبه قطرية وطرفا وحيدا مستدقا ثنائي الوجه ومكسورا. ويمكن اعزاء التقنيات المتمثلة في مكتشفات جبل بركة الى أي عصر ولكن الأدوات ثنائية الأوجه والقضبان نصف القطرية تعتبر معروفة في مواقع أحفورية أخرى في شبه الجزيرة العربية ويعتقد بأن تاريخها يعود الى ما بين ٧,٠٠٠ ـ ٤,٥٠٠ سنة قبل الميلاد. ولكن القطع المشذبة التشخيصية والتي قد تفيد في تحديد عمر المشغولات المكتشفة تعتبر غير متوفرة بصورة تامة في المواقع الأحفورية في أبوظبي. ويعزو ماك بريرتي السبب في عدم وجود مثل هذه القطع الى احتمال استخدام النتوءات الصخرية في تكوينة بينونة كمحاجر ولكنه جرى نقل الرقائق والأنصال غير المنجزة هنا الى أماكن أخرى لتشكيلها في أدوات ناجزة وجاهزة للاستخدام.

ويقدم جون د. كينغسون و أندرو هيل في (الفصل ـ ٢٧) تصورا للبيئات القديمة خلال العصر الميوسيني الأخير في شبه الجزيرة العربية. وتعتبر المعلومات المتوفرة عن المناخ القديم في شبه الجزيرة العربية خلال العصر الميوسيني الوسيط والأخير ضئيلة للغاية كما هو الحال مع المعلومات المتوفرة عن البيئات الحيوية القديمة التي كانت سائدة بالمنطقة في تلك الحقبة.

كما تعتبر السحن الصخرية ومجموعات المستحاثات الفقارية المرتبطة بها في تكوينة بينونة (والتي يبلغ عمرها نحو ٨ ـ ٦ ملايين سنة) بمثابة الدلالة الوضعية الوحدية المعتمدة على التجربة أو الملاحظة والاختبار بشأن البيئات القديمة في شبه الجزيرة العربية خلال العصر الميوسيني الأخير. ويتناول العالمان المذكوران أعلاه في الفصل المفرد لهما جملة من البيانات المتعلقة بالبيئات القديمة التي تم جمعها من تكوينتي بينونة وشويهات المكشوفتين في امارة أبوظبي مع الرواسب المتوضعة في المملكة العربية السعودية بالاضافة الى استعراض وجيز للأوضاع المناخية العاليمة والاقليمية والتي قد تكون احدى العوامل التي أثرت في حركة وانتشار الحيوانات عبر المنطقة.

ملخص الجزء الخامس : الحيوانات والنباتات الاقليمية بمناطق سلطنة عمان والجمهورية اليمنية وأفريقيا وآسيا

يعلق لورنس جي. فلين. و لويس ل. جاكوبس في الفصل ـ ٢٨ على المعلومات والمعارف المتنامية لدينا عن الحيوانات الثديية الصغيرة خلال العصر الميوسيني في شبه الجزيرة العربية. ويستفاد من ذلك أنه بالنسبة للأحياء المجهرية أو الصغيرة فان ثمة تشابها أفريقيا في التوزع الجغرافي الحيوي قد كان سائدا حتى نهاية تلك الحقبة. وتشتمل المجموعات الأحيائية الميوسينية على عدد أكبر من الحيوانات الثديية الصغيرة التي تظهر توزعات أوسع من الأوقات السابقة. ويبدو التنوع الواسع الانتشار أكثر وضوحا بين القوارض الفئرانية وخاصة الجراذين والجرابيع. ويدل وجود الجنسين المسميين علميا بارابيلوميس *(Parapelomys)* وسايدوميس *(Saidomys)* على انتشار الفأريات خلال العصر الميوسيني المتأخر عبر مناطق شاسعة من جنوب آسيا وأفريقيا.

ويعتبر جربوع أبوظبي أحد عناصر العصر الميوسيني المتأخر بين حيوانات شبه الجزيرة العربية كما أن من المعروف أنه قد تحدّر من شمال أفريقيا الى أفغانستان. ويصف العالمان المذكوران أعلاه في هذا الفصل نوعا من الفأريات من باكستان ينسب عن كثب الى النوع العربي. وهذا النوع الجديد المكتشف في موقع مؤرخ بدقة في تكوينة دوك باثان يتزامن في وجوده مع آخر الأنواع المعروف وجودها في سلسلة تلال سيواليك الواقعة عند سفوح جبال الهملايا شمال الهند، كما أنه يعد عنصرا من عناصر الدلالة على تغير حيواني عام في تلك المنطقة. وبالنسبة للأنواع المكتشفة في تكوينات السيواليك، فان الجربوع الظبياني قد يعتبر مؤشرا على وجود ظاهرة متزايدة وربما رطوبة سنوية أقل. وتدعم هذه النظرية دراسات النظائر الآيسوتوبية للأباتيت الفسفوري لمينا أسنان المستحاثات والأتربة القديمة. وقد يعزى التنوع في الجرابيع الحديثة الى زيادة مضطردة في المواسم المناخية على مستوى الكرة الأرضية.

وتشتمل مجموعة حيوانات بينونة على تشابهات قوية مع الحيوانات الأفريقية وقد تعزى الى الفترة الزمنية الممتدة بين ٨ ـ ٦ ملايين سنة سابقة. ويستعرض آندرو هيل، في الفصل ـ ٢٩ المواقع الميوسينية المكتشفة في المناطق الأفريقية الواقعة دون نطاق الصحراء الكبرى والتي قد تتطابق في تكوينها الزمني والحيواني مع حيوانات بينونة. وهذه المواقع قليلة العدد نسبيا وتتركز في الطرف الشرقي من القارة الأفريقية.

ومن الصعب أو المتعذر على المرء ملاحظة فوارق حيوانية محلية في أفريقيا خلال الحقبة الميوسينية المتأخرة، ويرجع السبب الرئيسي في ذلك الى ندرة المعلومات المتوفرة. وعلى مستوى الأجناس الحيوانية، فان حيوانات بينونة ترتبط مع الحيوانات الأفريقية، ولكن التشابهات الاضافية لها مع المجموعات الحيوانية الأوروبية والأفريبية تدعم وتؤكد الانطباع السائد بأن الجزيرة العربية تعتبر منطقة هامة في مجال التبادل الحيواني خلال الحقبة الميوسينية المتأخرة.

وقد عثر هيربرت توماس وزملاؤه في البعثة الأثرية الفرنسية ـ العمانية في سلطنة عمان على العديد من المواقع الأثرية للحيوانات الفقارية التي يعود تاريخها الى العصر الباليوسيني اللاحق وحتى العصر الميوسيني.

ويتناول الفصل ـ ٣٠ الديناميكية الأرضية أو القوى الباطنية الكائنة تحت القشرة الأرضية والجغرافية القديمة لهذه المواقع. وقد أدى تعاقب دورات تجاوز طغيان البحر بالانتشار فوق الأرض أو الانحسار والتراجع الى المواضع الأولية على امتداد الخط الجنوبي لهضبة الجزيرة العربية الى تحديد توزع البيئات الترسبية بدءا من العصر الباليوسيني المتأخر وحتى العصر الميوسيني المتوسط. ولا تلاحظ تراكمات الفقاريات الا في السحن البيولوجية القائمة في المناطق البحرية الساحلية مثل منطقتي ثيطينيطي وطاقة في ظفار ومنطقة الحقف (ولعلها منطقة الحجر ـ المترجم) في وسط عمان. وتمتاز الروابط الحيوانية بصورة عامة بتقارب العناصر التي تدل على وجود بيئات بحرية ومياه عذبة وأرضية. ومن بين مجموعة الحيوانات الثديية في العصر الصخري أو الأوليغوسيني الباكر بمنطقتي ثيطينيطي وطاقة، شاع وجود القوارض والرئيسات في حين أن الخرطوميات تعد أكثر تواجدا في العصر الميوسيني الباكر بمنطقة الحقف. كما أن ترابط الحيوانات اللافقارية مع بقايا الحيوانات الفقارية وتداخل الآفاق الأحفورية في الترسبات البحرية يتيحان المجال لاعزاء كلا الموقعين في ظفار الى الجزء السفلي من العصر الأوليغوسيني الباكر. ومع ذلك، فان المناقشات المتعلقة بالطبقات

الأرضية والمستحاثات تشير الى أن مجموعة حيوانات منطقة طاقة أصغر عمرا من مثيلاتها في منطقة ثيطينيطي.

وتشتمل بقايا الحيوانات الثديية المكتشفة بمنطقة حقف على أوجه قريبة الشبه من تلك الحيوانات المكتشفة بمنطقة جبل زيلتون في ليبيا مما يشير الى أن تاريخها يعود الى ما بين ٧,٥ - ٥,٥ مليون سنة سابقة.

ومن خلال مهمتهم في الجمهورية اليمنية، يصف مصطفى لطيف السروري و بيتر جي. وايبرو ومارغريت ي. كولينسون في الفصل - ٣١ بصورة مقتضبة الخصائص الصخرية والسحن المجهرية لتكوينة الحبشية في تتابع طبقات كنينة - جيزوال التي يعود تاريخها الى العصر الأيوسيني الوسيط. ومن خلال هذا التسلسل، يشيرون الى أول وجود فواكه نيمفاسيا (Nymphaeaceae) والبذور آنوفاسبيرومام - آنوفاسيا (Anonaspermum - Anonaceae) التي يعود تاريخها الى حقب الحياة الحديثة والتي ترتبط ببقايا الحيوانات الفقارية ماماليا (غير محددة) - سيرينيا (Sirenia) ويستدل من وجود البذور آنونيسيا على وجود ملامح استوائية في حين أن الوجود المميز للفواكه نيمفاسيا يشير الى روابط مع النباتات القديمة التي يعود تاريخها الى العصرين الأيوسيني والأوليغوسيني في مصر.

ويصف لويس ل. جاكوبس وزملاؤه بالفصل - ٣٢ الاكتشاف الذي توصلوا اليه في عام ١٩٩٠ لديناصور في الطبقات البحرية الترسبية لتكوينة مابدي التي يعود تاريخ تشكلها الى العصر الجوراسي الأخير في شمال اليمن. وقد ظلت معظم أجزاء العينة المكتشفة في موضعها الأصلي الا أن الشظايا التي تم اعدادها للدراسة تشير الى وجود حيوان فقاري عظائي الأرجل وربما كان ذلك نظرا لنحافة العظام النسبية والتركيب المعقد لها.

ومع أن التحديد الأدق يظل رهنا باجراء المزيد من الحفريات والاستكشافات والاعداد للدراسة، فان اليمن توفر فرصة لدراسة الجغرافيا الحيوية والاستيطانات الأرضية في حقب الحياة بالدهر الوسيط وذلك لأنه خلال العصر الجوراسي الأخير وعلى النقيض من عصور حقب الحياة الحديثة، فان المنطقة العربية قد كانت واقعة على الخط الشرقي للجزء الأفريقي من الكتلة الأرضية القديمة المسماة غوندوانا التي يعتقد العلماء باستمرار وجودها في النصف الجنوبي من الكرة الأرضية حتى فترة حقب الحياة القديمة وكانت تلك الكتلة تشمل أفريقيا والهند واستراليا وأمريكا الجنوبية، كما أن المنطقة العربية لم تكن قائمة في موقع استراتيجي للتمدد والانتشار الحيواني مع القارات الشمالية.

ويشير بيتر جي. وايبرو و ديانا كليمنتس في الفصل - ٣٣ الى أنه لم يتم الابلاغ عن اكتشاف مستحاثات حيوانات فقارية في المنطقة العربية قبل أواسط السبعينات من القرن الحالي كما لم يسجل هناك سوى اكتشافين لمستحاثات فقارية وهما عبارة عن جمجمة لفرس النهر القديم العائد الى العصر البليستوسين في المملكة العربية السعودية وناب فيلي يعود تاريخه الى العصر الميوسني القديم في امارة أبوظبي. وتشير تقارير الشركات البترولية التي يعود تاريخها الى الأربعينات من هذا القرن الى العثور على مستحاثات فقاريات ولكن هذه المعلومات لم يتم نشرها. ويشتمل الفصل - ٣٣ على سرد لمواقع المستحاثات الأثرية في شبه الجزيرة العربية التي يعود تاريخها الى عصور الحقب الثالث بالاضافة الى موقع أحفوري واحد ذي صلة في بلاد ما بين النهرين مع ثبت المستحاثات الحيوانية المكتشفة فيها وذلك من أجل توفير مرجع شامل لأحياء البيئة الأرضية العربية خلال الحقب الثالث.

ملخص الجزء السادس : المناطق البحرية القديمة (التيطسية) في الخليج العربي، البحر الأبيض المتوسط ومحيطات الحقب الثالث العالمية

يجمع هذا الجزء في ثناياه ثلاثة مشاريع أبحاث تسلط الضوء على علاقات التسلسل الزمني الجغرافي بين المحيطات والكتل الأرضية. وتقدم هذه المشاريع فكرة شاملة عن الأحداث المتعلقة بجيولوجيا الصخور البحرية لشبه الجزيرة العربية خلال الحقب الثالث الذي يعود تاريخه الى ما بين ٦٥ ـ ٢ مليون سنة خالية.

ويصف سي. جيوفري آدامز، ديريك د. بيليس و جون ي. ويتكر في الفصل ـ٣٤ الفرضيات الحديثة المرجحة من قبل دروغر، ستينينغر و روغل ومن جانب آدامز وزملائه. وتتعلق هذه الفرضيات بتأريخ الأحداث التي أدت في نهاية المطاف الى فصل البحر الأبيض المتوسط البدائي أي الأولي عن المحيط الهندي، وقد تم تحليل ومناقشة هذه الفرضيات في ضوء البيانات التي أصبحت متوفرة خلال العقد المنصرم. كما جرت مناقشة العديد من القضايا ذات الأهمية الخاصة بشأن التأريخ والعلاقة المتبادلة للترسبات البحرية خلال العصر النيوجيني في الشرق الأوسط. ويستنتج من هذه الدراسات أنه على الرغم من أن الدلائل الأحفورية وأوصاف الطبقات الأرضية الصخرية المتوفرة تشير بقوة الآن الى ترجيح احتمال حصول الانفصال خلال العصر الميوسيني الباكر مع استبعاد احتمال عودة الاتصال خلال فترات العصر الميوسيني الوسيط، فانه يتعين اجراء المزيد من الدراسات لبعض التسلسلات في كل من العراق وايران قبل التوصل الى اجماع في الرأي حول هذه القضية.

ويقدم فريد روغل في الفصل ـ٣٥ أوفى وأحدث المعلومات المتعلقة باعادة التشكيلات الجغرافية القديمة للمنطقة المحيطة بالبحر الأبيض المتوسط من العصر الأليغوسيني وحتى العصر البليوسيني. وتتوقف عملية اعادة تشكيل الجغرافية القديمة للمنطقة المحيطة بالبحر الأبيض المتوسط على اعادة ترتيب أوضاع الصخور قبل حركات التصدع والطي للوحدات التكتونية التشكيلية للصخور وعلى العلاقة الطبقية المحددة بوضوح بين مختلف الأحواض الترسبية.

ولكن المعلومات التفصيلية المتوفرة بشأن اعادة الترتيب هذه تعتبر ضئيلة للغاية. وبالنسبة للمتطلبات التشكيلية التكتونية، فان المعلومات متوفرة عن المناطق المحيطة بسلاسل جبال الأبنين والآلب والكرباثيان (الواقعة بين بولندا وتشيكوسلوفاكيا) ولكنها غير دقيقة بالنسبة لمنطقة البحر الأبيض المتوسط. وقد تم تقديم دراسة بشأن العلاقة الصخرية المتبادلة للبحر الأبيض المتوسط والمناطق المحيطة وذلك بناء على البيانات الحديثة بشأن المغناطيسية القديمة والقياسات الاشعاعية ودراسة الطبقات الحيوية التي تبين تكوين الصخور الرسوبية من بقايا الأحياء فيها من بين الدراسات التي أجرتها فرق أبحاث مختلفة. وتتركز الأسئلة الحالية المتعلقة بالجغرافية القديمة على هجرات الثدييات العابرة للقارات بين أوروبا وأمريكا الشمالية وأفريقيا خلال العصرين الأوليغوسيني والميوسيني في بعض المناطق المحددة مثل: ١) الاتصال الأوروبي ـ الآسيوي، ٢) جسر بيرينغ الأرضي بين آسيا وأمريكا والذي يربط بين البحر البيرنغي والمحيط القطبي، ٣) جسر غومفوثيريوم الأرضي في الشرق الأدنى، و ٤) جسر جبل طارق الأرضي.

وأخيرا يصف نورمان ماكليود التغيرات في محيطات العالم خلال الحقبة الميوسينية وما قبلها. وتعتبر التغيرات في جغرافية المحيطات خلال العصر الأوليغوسيني وحتى العصر الميوسيني احدى أفضل الفترات المعروفة من بين سائر الفترات الزمنية القديمة كما أنها توضح الأدوار المتداخلة المعقدة التي تلعبها العوامل التشكيلية التكتونية والدورانية والمناخية والكتل المائية. ومن الناحية التكتونية التشكيلية، فان أهم حدثين في جغرافية المحيطات القديمة خلال تلك الفترة قد كانا متمثلين في العزلة المتواصلة للكتلة الأرضية بالقطب الجنوبي وغرق الحيد الواصل بين آيسلندا وجزر فأيروس بين المحيط الأطلسي الشمالي والمحيط المتجمد الشمالي. وقد أدى الحدث الأول الى انفصال وعزلة القارة القطبية الجنوبية مع استمرار تجمدها بصورة متوالية عندما عمل تطور نظام تيارات في محيط القطب الجنوبي على منع المياه الدافئة لدى خطوط العرض الوسطى من التحرك في الاتجاه الجنوبي في حين عمل الحدث الثاني على تغيير طبيعة سريان المياه العميقة عبر حوض المحيط الأطلسي بحيث تظهر المياه المالحة الدفيئة نسبيا والمتولدة بالأصل في المحيط المتجمد الشمالي على السطح جنوب نقطة التقاء القطب الجنوبي.

وبحلول العصر الميوسيني الوسيط، اتحد هذان العاملان معا لتشكيل التجلد الواسع في القارة القطبية الجنوبية وحدوث انخفاض كبير في مستوى البحر، وزيادة في مستويات الملوحة الشاملة في المحيطات. وقد

أثرت هذه التغييرات بدورها في توزيع الخطوط العرضية الوسطى والسفلى وفي خواص كتل المياه المحيطية.

وفي المجال الحيوي، فان ثمة تغييرات عديدة جديرة بالملاحظة ويمكن اعزاؤها وربطها سواء بصورة مباشرة أم غير مباشرة مع هذه الأحداث المتعلقة بجغرافية المحيطات القديمة. ومن بين هذه التغييرات الزيادة المثيرة في انتاج العوالق المائية أو الكائنات الحية المعلقة بالماء في المحيط الجنوبي مع مصاحبة ذلك بالتنوع في الحيوانات التي تتغذى على العوالق المائية (مثل الحيتان). كما أن الزيادات في معدلات الاستيطان بمناطق الخطوط العرضية الوسطى والسفلى بين الأحياء التي تتغذى على العوالق المائية والزيادات في الأحياء النباتية والحيوانية الأرضية المتكيفة للعيش في بيئات مختلطة تشتمل على غابات وأعشاب أو مجتمعات حيوانية ونباتية شبيهة بالأقاليم المدارية التي تشتمل على أعشاب خشنة في نطاق التدرج الحراري العرضي خلال العصر الميوسيني قد اضطرت تلك الكائنات من حيوانات ونباتات الى استبدال الغابات المدارية التي يعود تاريخها الى العصرين الأوليغوسيني والميوسيني الباكر.